29TH EUROPEAN SYMPOSIUM ON COMPUTER AIDED PROCESS ENGINEERING

PART B

29TH EUROPEAN SYMPOSIUM ON COMPUTER AIDED PROCESS ENGINEERING

PART B

Edited by

Anton A. Kiss
The University of Manchester, United Kingdom
Manchester, M13 9PL

Edwin Zondervan
University of Bremen, Germany
D. Bremen, 28359

Richard Lakerveld
The Hong Kong University of Science and Technology
Clear Water Bay, Kowloon, Hong Kong

Leyla Özkan
Eindhoven University of Technology, The Netherlands
Postbus 513, 5600 MB

ELSEVIER

Amsterdam – Boston – Heidelberg – London – New York – Oxford
Paris – San Diego – San Francisco – Singapore – Sydney – Tokyo

Elsevier
Radarweg 29, PO Box 211, 1000 AE Amsterdam, Netherlands
The Boulevard, Langford Lane, Kidlington, Oxford OX5 1GB, UK
50 Hampshire Street, 5th Floor, Cambridge, MA 02139, USA

Notices
Knowledge and best practice in this field are constantly changing. As new research and experience
broaden our understanding, changes in research methods, professional practices, or medical treatment
may become necessary.

Practitioners and researchers must always rely on their own experience and knowledge in evaluating
and using any information, methods, compounds, or experiments described herein. In using such
information or methods they should be mindful of their own safety and the safety of others, including
parties for whom they have a professional responsibility.

To the fullest extent of the law, neither the Publisher nor the authors, contributors, or editors, assume
any liability for any injury and/or damage to persons or property as a matter of products liability,
negligence or otherwise, or from any use or operation of any methods, products, instructions, or ideas
contained in the material herein.

British Library Cataloguing in Publication Data
A catalogue record for this book is available from the British Library

Library of Congress Cataloging-in-Publication Data
A catalog record for this book is available from the Library of Congress

ISBN (Part B): 978-0-12-819940-4
ISBN (Set) : 978-0-12-818634-3
ISSN: 1570-7946

For information on all Elsevier publications visit our
website at https://www.elsevier.com/

Working together
to grow libraries in
developing countries

www.elsevier.com • www.bookaid.org

Publisher: Joe Hayton
Acquisition Editor: Kostas Marinakis
Editorial Project Manager: Kelsey Connors
Production Project Manager: Paul Prasad Chandramohan
Designer: Greg Harris

Typeset by SPi Global, India

Contents

Process control and operations

CAPE-PSE in hi-tech micro, nano-devices and processes

Education in CAPE-PSE and knowledge transfer

Anton A. Kiss, Edwin Zondervan, Richard Lakerveld, Leyla Özkan (Eds.)
Proceedings of the 29th European Symposium on Computer Aided Process Engineering
June 16th to 19th, 2019, Eindhoven, The Netherlands. © 2019 Elsevier B.V. All rights reserved.
http://dx.doi.org/10.1016/B978-0-128-18634-3.50190-9

Fuzzy Control Applied to Combustion in Sugarcane Bagasse Boilers

Fernando M. de Mello[a,b], Antonio J. G. da Cruz[b,c], Ruy de Sousa Jr.[b,c,*]

[a]*Tereos, Rod. Assis Chateaubriand km 155, Olimpia 15400-000, Brazil*

[b]*Graduate Program in Chemical Engineering, Rod. Washington Luís km 235, São Carlos 13565-905, Brazil*

[c]*Department of Chemical Engineering, Federal University of São Carlos, Rod. Washington Luís km 235, São Carlos 13565-905, Brazil*

ruy@ufscar.br

Abstract

In the sugarcane industry, boilers are widely used for cogeneration. This industry uses mainly sugarcane bagasse as the primary fuel source, followed by other types of biomass. Fuel price has a major impact on the industry results, due to the scenario of the energy market. Therefore, optimizing the combustion is an important lever to reduce this cost. Real-time measurement of oxygen content in the boilers flue gas is one of the requirements for this optimization. Due to the capital cost and the requirements for personnel qualification for maintenance and calibration, sugarcane mills rarely apply such analyzers in closed loop. In this work, a fuzzy control was implemented in the combustion air flowrate of a large boiler in the Tereos group, to maintain the oxygen content in the combustion products within the optimum range. A tool was used to identify the process dynamics and to tune the control loop in real time. After downloading the new strategy in the factory control system and tuning the controller, the oxygen control loop started operation in automatic mode. After twenty-four hours of continuous operation, regulatory and servo controls were able to keep the desired setpoint with low variability (± 1.2% variation in excess air coefficient) and respond to setpoint changes with good response speed (1.5-20 minutes) and overshoot (8-56%). No previous references were found in the literature on the use of fuzzy logic for the combustion control of sugarcane bagasse-fired boilers, especially high pressure and high steam throughput boilers. Results reached so far demonstrate the applicability of fuzzy control for combustion in boilers on sugarcane bagasse, with 40% less variability than a PID controller, and 69% less variability compared to operation in manual mode.

Keywords: sugarcane bagasse boiler, oxygen control, fuzzy logic.

1. Introduction

In the sugarcane industry, boilers are widely used for cogeneration (i.e., simultaneous generation of thermal and electrical energy from the same fuel source). The main fuel used is sugarcane bagasse, followed by other types of biomass, such as cane straw, woodchips and corn stover. According to the Brazilian Sugarcane Industry Union (UNICA), and the National Electric Energy Agency (ANEEL), the granted capacity of electricity supply from biomass represents 9% of the Brazilian energy matrix. The cane harvest period, which occurs between the months of April and November in the southeast of Brazil, is an advantage for electricity generation from bagasse, since it matches the drier seasons of the year. During this period, the generation demand from thermoelectric power plants is increased. In this scenario, the price of sugarcane biomass (bagasse and straw) has a significant impact on the financial results of sugarcane mills. Since the process of manufacturing sugar and ethanol consumes large amounts of steam, the production of electricity from the very steam consumed at the mills is an inexpensive way of generating electricity. Therefore, it is essential that plants with a focus on bioelectricity generation invest in energy efficiency, to maximize the biomass surplus and its conversion into electricity, which reduces the risk exposure to the biomass market. Excess air is an impact factor on combustion efficiency: boilers use an air/fuel curve to maintain excess air as predicted by the manufacturer. However, variations in fuel quality, mainly humidity, require corrections on this curve. The main mechanism to correct the air/fuel curve is the oxygen control in the outlet duct of the effluent gases. This type of control is still little used in sugarcane bagasse boilers. Advanced control may have advantages over PID in this type of application. Thus, this work presents a fuzzy logic approach for the combustion control.

2. Literature Review

Fuel composition defines the combustion's stoichiometry, which in turn defines the theoretical air for the fuel's complete combustion. However, due to furnace mixing deficiencies, an excess amount of air is required to maintain a sufficient amount of oxygen until the end of the burning. One way of expressing the air/fuel ratio is the excess air coefficient, which is the ratio of the total amount of air used in the combustion to the amount of stoichiometric air. The excess air controls the volume and the enthalpy of the flue gases, which are determinant for the boiler's efficiency. A large excess of air is undesirable because it lowers the flame temperature and increases heat losses by enthalpy of the dry gas. On the other hand, a low excess of air can result in incomplete combustion and the formation of CO, soot and smoke, in addition to an accumulation of unburnt fuel, causing risk of explosion. Therefore, an optimum excess air coefficient is required to minimize both losses. In a boiler's design, manufacturers define a curve known as the air/fuel curve. This curve relates the air inlet as a function of the fuel throughput, and aims to maintain the excess air coefficient within the targeted range for each type of boiler. However, variations in fuel composition cause deviations from this curve, and the boiler no longer operates with the best combustion efficiency. Humidity is the main factor of variation in fuel composition throughout the crop. This is mainly due to limitations in the control of imbibition water in the sugar extraction process, cane throughput variability, wear of the drying/dewatering mills. An oxygen level control in the flue gas can correct the air/fuel curve in real time.

Zadeh (1965) introduced the concept of fuzzy sets, which, as the name suggests, is a theory of sets (or classes) whose boundaries are indefinite. He established that a fuzzy set would be a class with varying membership degrees. That is, each element would be allowed a certain degree of membership to the set. Fuzzy logic provides a means of quantifying states and overlapping between them (eg, High and Low), assigning degrees of membership (truth) to each of them (Akisue et al., 2018). Bhandari et al. (2016) simulated the combustion control of a low pressure (20 kgf/cm^2) and low capacity (30 t/h) coal boiler using fuzzy logic, and applied this strategy to the equipment's PLC for testing during load variations. In another work, Li and Chang (1999) combined fuzzy logic control with neural networks in the combustion in coal-fired boilers. This approach aimed to adapt a base of membership functions and rules through a neural network, from previous system responses, until the desired control performance was achieved. As mentioned, the works referenced in this document are based on the operation of coal boilers. No references were found in the literature on the use of fuzzy logic for the combustion control of sugarcane bagasse-fired boilers, especially high pressure (above 65 bar) and high steam throughput (from 200 t/h) boilers. The present work aims to fill this gap, with the implementation of a fuzzy logic controller in the combustion control loop of a bagasse-fired boiler with 67 bar pressure and 200 t/h of steam throughput.

3. Methodology

The implementation of the closed loop oxygen control, controller tuning and data collection was performed in a high-pressure boiler with an oxygen analyzer at the flue gas outlet. The boiler used in this work, model ZS-2T-200/67-525, is a ZANINI-SERMATEC steam generator with a pin hole grate. It has water wall panels, single pass convection beam, two drums supported by the top of the metal structure, with natural circulation and a bagasse burning system in suspension.

Among the boiler's main control loops is the combustion control: variable frequency drives on primary and secondary air fans adjust the airflow based on the air/bagasse curve, in order to guarantee the required excess air under different operating regimes. The air/bagasse curve operates according to a curve determined by the manufacturer. This curve is corrected by an oxygen control loop by fuzzy logic, which was implanted as object of the present work.

The gas analyzer installed in the boiler is the EL3040 model, manufactured by ABB. It consists of two probes, one infrared probe Uras26 for CO analysis, and the other an electrochemical probe Endura AZ20 for O_2 analysis. The O_2 sensor, object of this work, is based on a zirconium oxide cell. This cell is mounted on the tip of the probe that is inserted into the gas duct. The oxygen concentration in the effluent gases is measured using an in situ wet analysis method. The output signal generated by the zirconium cell is processed in the transmitter, providing a local O_2 reading and a 4-20 mA retransmission signal covering the 0-100% O_2 range.

The plant's main control system is Emerson's DeltaV®. It has development tools that allowed the addition of fuzzy logic blocks to the oxygen control loop, and a tuning application (DeltaV Insight®) adjust the scaling factors. All data was stored using OSIsoft's PI System®.

The controller uses, as input signals, the error and change in error of the O_2 concentration, and the change of the control action for correcting the air/bagasse curve as the output signal (input and output linguistic variables). The relationships between these three variables represent a nonlinear controller. The two linguistic values for error and change in error are Negative and Positive. For the output change, a third linguistic value, Zero, is added. The values of error and change in error, along with the membership scaling, respectively, Se and SΔe, determine the degree of membership (memberships to Negative linguistic values are maximum for –Se and –SΔe; memberships to Positive linguistic values are maximum for +Se and +SΔe). The membership functions of the output change are singletons. A singleton represents a fuzzy set whose support is a single point with a membership function equal to 1. Singletons are usually employed in the industry for output changer in order to simplify the computational demand for defuzzification, since they eliminate the need for numerical integration. The membership scaling (SΔu) determines the magnitude of the output change, positive or negative, for a given error and change in error. There are four fuzzy logic rules that the controller uses for a reverse control action. These rules are described in Table 1, with N, P and ZO being Negative, Positive and Zero, respectively.

Table 1. Fuzzy logic rules in DeltaV.

Rule 1	If error is N and change in error is N, make change in output P.
Rule 2	If error is N and change in error is P, make change in output ZO.
Rule 3	If error is P and change in error is N, make change in output ZO.
Rule 4	If error is P and change in error is P, make change in output N.

Tuning can be used to establish the scaling factors (Se, S∆e and S∆u). For small control errors and setpoint changes smaller than a nominal value (∆Ysp), the scaling factors of the fuzzy controller are related to the proportional gain (Kp) and the integral time (Ti) that would be used in a PI block being executed with a scan time (∆t) of 1 second to control the same process. Those relationships are expressed in Eq. (1), (2) and (3).

$$S\Delta e = \beta \Delta Y_{SP} \tag{1}$$

$$S\Delta u = 2S\Delta e K_p \tag{2}$$

$$Se = Se_0 = TiS\Delta e \tag{3}$$

Where:

S∆e	=	change in error scaling factor
Se	=	error scaling factor
S∆u	=	change in output scaling factor
Se_0	=	error scaling factor for a 1 second scan time

Beta is a function of the process dead time and time constant, and ranges between 0.2 and 0.5. The nominal setpoint change value for ∆Ysp is 1%. When the change in setpoint is higher than the nominal setpoint change value (∆Ysp), scaling factors are internally augmented by the FLC block.

Once a control loop is set up to govern a process, it needs to be tuned. The tuning application used in this work applies an on-demand process test to provide tuning recommendations automatically. This on-demand tuning is based on Aström-Hägglund's patented algorithm to calculate the tuning parameters of a closed loop (Aström and Hägglund, 1988). Emerson enhanced this algorithm with the incorporation of a patented technique to identify the process' dead time. During tuning, the output of the controller is determined by a known function that acts as a relay with hysteresis. This relay provides a two-state control and causes a process oscillation with a small, controlled amplitude. From the amplitude and frequency of this oscillation, the tuning software calculates the process' critical gain and critical time. The controller parameters are then computed based on the classical Ziegler-Nichols' rules.

To follow the process variable, manipulated variable and setpoint of the boiler's combustion control loop, OSIsoft's PI System® was used. It is a Plant Information Management System (PIMS), used for plant information management, which collects process data residing in distinct sources and stores it in a single database.

4. Results

After the construction of the oxygen control loop with the FLC block (fuzzy logic controller) and the implementation in the DCS (distributed control system), the combustion control loop was switched to automatic. After starting the operation with this new strategy, the tuner was used to obtain the controller's scaling factors. This step resulted in the following recommendation of scaling factors of the membership functions for the linguistic variables error, change in error and change in output:

• Se (error) = 28.74%; SΔe (change in error) = 0.4%; SΔu (change in output) = 0.68

The controller's parameters recommended by the tuning were updated in the DCS, and the controller returned its operating mode to automatic. A specialized company carried out a combustion regulation of the ZS-2T-200 boiler, measuring gas concentration (O_2, CO and CO_2) at various points in the circuit with portable analyzers, as well as an evaluation of the inlet and exhaust fans. A 2.5% setpoint for the O_2 concentration in the boiler effluent was recommended due to an offset of 2 percentage points caused by the probe's position. This corresponds to a 4.5% O_2 content in the center of the gas duct.

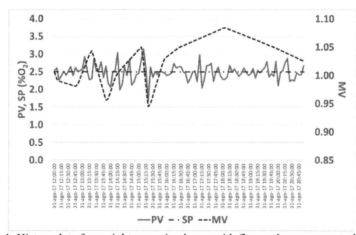

Figure 1. History data from eight operating hours with flue gas' oxygen control loop in automatic mode with the fuzzy controller. PV, SP data on left axis, and MV on the right axis. Source: PI System.

Figure 1 shows the historical data of setpoint (SP), process variable (PV) and manipulated variable (MV) during eight hours of operation of the boiler with the control loop in automatic. During this period, the setpoint was set at 2.5% and the variability of the PV (regulatory control) was evaluated. The mean value of PV in this period was 2.49%, with a standard deviation of 0.22. This deviation represents a variation of ± 1.2% in the excess air coefficient. One can also observe, in Figure 1, that the MV acted without saturation during the entire period.

After eight hours of automatic operation with fixed setpoint, tests were performed to evaluate the controller's performance to setpoint changes (servo control), and to maintain the new setpoint values. Figure 2 shows the total interval of 24 hours of testing between servo and regulatory controls. After twenty-four hours of continuous operation, regulatory and servo controls were able to keep the desired setpoint with low variability (± 1.2% variation in excess air coefficient) and respond to setpoint changes with good response speed (1.5-20 minutes) and overshoot (8-56%).

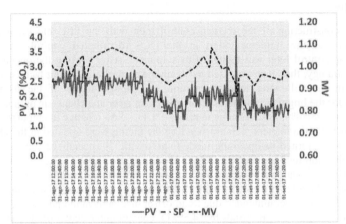

Figure 2. Controller's performance test in servo and regulatory modes over 24 hours.
PV, SP data on left axis, and MV on the right axis. Source: PI System.

Compared to previous PID and manual control results (de Mello, 2018) from Tereos, it was observed 40% less variability than the PID controller, and 69% less variability than operation in manual mode.

5. Conclusions

The implementation of a closed loop oxygen control, controller tuning and data collection was performed in a high-pressure boiler with an oxygen analyzer at the flue gas outlet. Results presented here demonstrate the applicability of fuzzy control for combustion in boilers on sugarcane bagasse.

References

R. A. Akisue, A. C. L. Horta and R. Sousa Jr., 2018, Development of a fuzzy system for dissolved oxygen control in a recombinant *Escherichia coli* cultivation for heterologous protein expression, Computer Aided Chemical Engineering, v. 43, p. 1129-1134.

K. J. Aström and T. Hägglund, 1988, Automatic Tuning of PID Controllers, ISA, RTP.

S. S. Bhandari, P. Makwana and D. K. Maghade, 2016, Solid fuel fired boiler combustion control using fuzzy logic algorithm, ICACDOT, Pune, India, p. 473-476.

W. Li and X. Chang, 1999, A neuro-fuzzy controller for a stoker-fired boiler, based on behavior modeling, Control Engineering Practice, v. 7, p. 469-481.

F. M. de Mello, 2018. Controle nebuloso ("fuzzy") aplicado à combustão em caldeiras a bagaço de cana. Master's Thesis. Federal Universiy of São Carlos - Brazil.

L. A. Zadeh, 1965, Fuzzy Sets, Information and Control, v. 8, n. 3, p. 338–353.

Acknowledgements

To Tereos and São Paulo Research Foundation, FAPESP (grant number 2011/51902-9).

Anton A. Kiss, Edwin Zondervan, Richard Lakerveld, Leyla Özkan (Eds.)
Proceedings of the 29th European Symposium on Computer Aided Process Engineering
June 16th to 19th, 2019, Eindhoven, The Netherlands. © 2019 Elsevier B.V. All rights reserved.
http://dx.doi.org/10.1016/B978-0-128-18634-3.50191-0

Dynamics and Control of a Fully Heat-Integrated Complex Distillation Column

Manuel Rodríguez[a*], Ignacio P. Fernández Arranz[a]

[a] *Universidad Politécnica de Madrid, José Gutiérrez Abascal 2, 28006 Madrid, Spain*

Abstract

Distillation is the most widely used in the separation section of a process. It consumes an important amount of energy (close to 40 % of the energy of chemical process). Many different configurations to increase its thermodynamic efficiency have been proposed. In this paper a new distillation column configuration developed by Toyo Engineering Corporation called SuperHIDiC is analysed. This new column is simulated and compared with a conventional distillation column. Its dynamic performance and control structure performance are also analysed. Results show that SuperHIDiC achieves an important reduction in energy consumption and that a conventional control configuration can be used in order to have a stable operation.

Keywords: Heat integration, complex distillation, process control

1. Introduction

Energy consumption is one of the problems that our world faces in this century. In 2017 energy consumption increased 2.3 % versus 1.1 % increase in 2016 according to the Global Energy Statistical Yearbook 2018, and it is expected to expand by 30 % between today and 2040 as indicated by the International Energy Agency. United Stated energy consumption by use is depicted in Fig 1. industrial consumption amounts to more than 30 % of the total being almost half of it consumed in separation process being distillation the consumer of half of the energy of the separation processes (i.e. around 8 % of the total energy consumption in the U.S.).

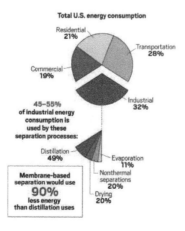

Figure 1. United States energy consumption (Ritter, 2017)

Conventional distillation is a very inefficient process in terms of energy, being its efficiency below 10 %. The idea of more efficient distillation columns is not new, in 1977 (Mah et al. 1977) proposed the first method to save energy through heat integration. In the following years based on Mah's idea, different researchers proposed and patented heat integrated distillation configurations like Seader (1980), Aso (1996), Nakaiwa (2000), etc. These types of columns called HIDiC (*Heat Integrated Distillation Column*) have been studied and its feasibility and profitability demonstrated for low relative volatility separations and tight product specifications (Leon Pulido et al., 2010). Although HIDiCs can achieve (theoretically) almost 60 % energy reduction, their use in industry is far from being high, mainly due to complex (and costly) construction and high maintenance costs. In this paper we explore a new HIDiC developed by Toyo Corp. denominated SuperHIDiC. Next section this new type of column is described, then a simulation of its performance and energy savings is presented, section four shows a dynamic model and a valid control structure and finally in the last section comments on the obtained conclusions.

2. SuperHIDiC column concept

SUPERHIDiC® is HIDiC configuration developed by Toyo Inc. In this setup the stripping and rectifying columns switch positions, being the stripping section at the top and the rectifying section at the bottom, side heat exchange is accomplished by thermo-siphon and/or gravity without pumping. The operating pressure in rectifying section is higher, this makes a transfer of the excess heat duty in the rectifying section to the stripping one possible. Fig 2. shows the configuration of the SuperHIDiC column.

Feed goes into the stripping section. This section operates only with the reboiler, and it has (at least) one side heat exchanger to take advantage of heat integration. The stream at the top of the column goes to a compressor and then exchanges heat before going to the rectifying column. This, high pressure, column operates only with the condenser and it has (at least) one side heat exchanger. The number of side heat exchangers, which can be up to five, depends on the type of separation and operating conditions. Reboiler and condenser duty are significantly reduced due to the heat integration achieved with the side heat exchangers. There is no need of special equipment, so maintenance can be conducted

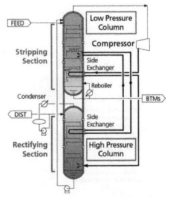

Figure 2. SuperHIDiC distillation column concept (Toyo Engineering, 2017)

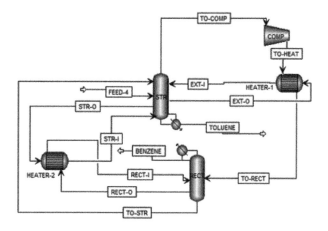

Figure 3. SuperHIDiC simulation configuration using Aspen Plus

by conventional means. Toyo engineers (Toyo Engineering Corporation, 2018) claim to achieve up to a 50 % energy reduction. Main applications are in installations where:

- Temperature difference between the overhead and the bottom is lower than 80 °C.
- Utilities are expensive.
- Condenser and/or reboiler duties are large.

3. SuperHIDiC Simulation

The system to separate using this technology is a stream composed of benzene and toluene. Feed composition is equimolar and its temperature and pressure are 94 °C and 100 kPa respectively. Total flow is 100 kmol/h. Molar purity of both products is 99.5 % molar. The fully heat integrated distillation column has been simulated using Aspen Plus as shown in Fig 3. and the physical properties packages used is RK-SOAVE. The configuration is equivalent to the SuperHIDiC column, it has two columns (stripping and rectifying) with side extractions to the heat exchangers.

The stripping column has 15 stages with side extractions in stages 4 and 13. The rectifying column has also 15 stages with a side stream in stage 4, it operates with a reflux ratio of 2.2. The compressor discharge pressure is 350 kPa.

Table 1. Energy consumption comparison between conventional distillation and SuperHIDiC

	CONVENTIONAL	SUPERHIDIC	REDUCTION (%)
CONDENSER	1.395 MW	0.951 MW	31.87
REBOILER	1.240 MW	0.619 MW	50.10
COMPRESSOR	-	0.275 MW	-
TOTAL	2.635 MW	1.844 MW	30.01

The separation has also been implemented using a conventional distillation column. The conventional column has 30 stages and it operates at a reflux ratio of 2. The obtained results show a significant reduction in energy consumption, see Table 1.
Around 32 % reduction in the condenser and a 50 % reduction in the reboiler. Adding the power needed for the compressor the overall reduction is 30 %, it appears that even considering the extra capital expenses of this new type of column, it can be a better option than the conventional approach. Besides other aspects should be considered as the environmental impact which is much lesser.

The energy savings are 0.791 MW, taking 8760 h/y the resulting annual energy savings are 6929.2 MWh. Considering an emission factor of 0.39 kg CO_2/kWh (Ministry of agriculture, 2018) the greenhouse gas emission reduction is 2702.4 t CO_2/y. The OPEX savings obtained considering an energy price of 52.24 €/MW (OMIE, 2018) are 3.6M €/y.

4. SuperHIDiC Control

According to Toyo Inc. a conventional control structure allows for a stable operation of the distillation column. Pressure control in the stripping column is obtained manipulating compressor speed, and composition control is achieved through inference with temperature. In the rectifying column pressure is control with the condenser duty and distillate flow is also controlled. Inventories in both columns are controlled. This conventional control scheme shown in Fig 4.

It has been implemented in a dynamic simulation in order to check its validity. The simulation has been done with Aspen Dynamics (Fig 5.) using the steady state model developed previously and adding the dynamic information needed. Feed composition and feed flow disturbances have been applied to the process.

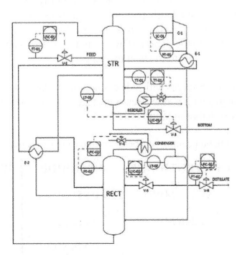

Figure 4. Control scheme to operate the heat integrated column

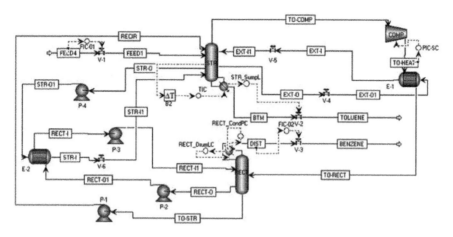

Figure 5. Dynamic simulation of the SuperHIDiC and its control structure

Fig 6. and 7 show the behaviour of the control system flow disturbance (changed from 100 kmol/h to 150 kmol/h which is higher than expected in a real plant as these columns are design for a turndown between 70-100%) and composition disturbance (change of 10% in benzene composition). Both are rejected and the control system exhibits good performance.

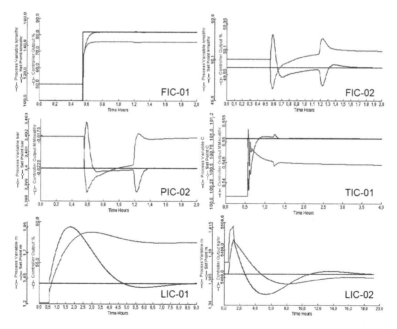

Figure 6. SuperHIDiC control performance under feed flow disturbance

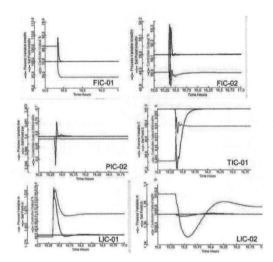

Figure 7. SuperHIDiC control performance under composition disturbance

5. Conclusions and further work

In this work a new distillation configuration to save energy consumption called SuperHIDiC has been tested. A comparison with a conventional distillation column results in significant energy savings (although not so high as claimed by the designers Toyo Inc.) and an important reduction in greenhouse gas emission. A dynamic simulation has also been implemented with its associated control structure. The conclusion is that a conventional control configuration allows good operation of the column. Further work will compare the capital expenses of this configuration with the conventional approach to have a complete economic comparison and will implement advanced control to the column to compare it with the conventional control scheme applied in this work.

References

K. Aso, H. Matuso,T. Noda, N. Takada, N. Kobayashi, 1996. U.S. Patent n° 5.788.047.

J. L. Pulido, M. W. Maciel, R. M. Filho, 2010. Nuevas perspectivas en procesos de separación: Simulación Columna de Destilación con Integración Interna de Calor. Revista ION, 7-12.

R. Mah, J. Nicholas, R. Wodnik, 1977. Distillation with Secondary Reflux and Vaporization, a Comparative Evaluation. AIChE J, 651-658.

Ministry of Agriculture, Fisheries and Food of Spain, 2018. Factores de Emisión: Registro de Huella de Carbono, Compensación y Proyectos de Absorción de Dióxido de Carbono.

M. Nakaiwa, K. Huang, K. Naito, A. Endo, M. Owa, T. Akiya, T. Takamatsu, 2000. A New Configuration of Ideal Heat Integrated Distillation Columns. Computers and Chemical Engineering, 239 - 245.

OMIE, 2017. Informe de Precios (Electricity price report, Spain).

S.K. Ritter, 2017, Putting distillation out of business in the chemical industry, Chemical and Engineering News, vol. 95, 25, 18-21

J. Seader, 1978, U.S. Patent n° 4.284.891.

Toyo Engineering, 2017, Toyo Communications. TOYO TIMES, 10.

Toyo Engineering Corporation, 2018,
http://www.toyoeng.com/jp/en/products/environment/superhidic/

Anton A. Kiss, Edwin Zondervan, Richard Lakerveld, Leyla Özkan (Eds.)
Proceedings of the 29th European Symposium on Computer Aided Process Engineering
June 16th to 19th, 2019, Eindhoven, The Netherlands. © 2019 Elsevier B.V. All rights reserved.
http://dx.doi.org/10.1016/B978-0-128-18634-3.50192-2

The impact of sustainable supply chain on waste-to-energy operations

Maryam Mohammadi,[a] Iiro Harjunkoski[a,b,*]

[a]*Department of Chemical and Metallurgical Engineering, School of Chemical Engineering, Aalto University, FI-00076 Espoo, Finland*

[b]*ABB Corporate Research Center Germany, Wallstadter Str. 59, 68526 Ladenburg, Germany*

iiro.harjunkoski@aalto.fi; iiro.harjunkoski@de.abb.com

Abstract

This paper addresses the optimal planning of an integrated supply chain (SC) network for waste-to-energy (WtE) systems. A mixed integer linear programming (MILP) model is presented, which simultaneously considers the tactical and operational decisions related to transportation, inventory, production, and distribution. The objective of the proposed model is to find a balance between SC costs, waste reduction, and using the generated waste efficiently. It also ensures environmental sustainability in the operations of WtE systems and continuous feedstock supply. The considered waste type is combustible non-biodegradable municipal solid waste (MSW) with low moisture content, and the employed technologies are pyrolysis, gasification, and combined heat and power (CHP) to produce electricity and heat. The proposed MILP model is formulated and solved using GAMS/CPLEX.

Keywords: municipal solid waste, waste-to-energy, supply chain, sustainability, mixed integer linear programming.

1. Introduction

WtE is an energy recovery process that utilizes waste conversion technologies to generate heat and electricity from non-reusable and non-recyclable waste materials. WtE excels the other waste management (WM) strategies (e.g., reusing and recycling) as it can be used to meet the increasing energy demands and to reduce the dependency on energy imports and natural resources, in addition to reducing the waste amount for disposal. WtE also provides superior economic and environmental benefits because of the produced products through the WtE processes (e.g., transportation fuels, synthetic natural gas, chemicals, ferrous and non-ferrous metals, energy), emitting less carbon dioxide (CO_2) compared to power plants using non-renewable resources, and producing less greenhouse gas emissions (such as methane) compared to landfilling. It can also supply base-load power throughout the year to the community. Hence, WtE has become a promising alternative for many countries as an effective WM solution in a sustainable way and as the superior method of waste disposal. The three main thermochemical technologies for extracting energy from waste include conventional incineration, pyrolysis, and gasification. Moreover, CHP, used for simultaneous electricity and heat generation, can significantly increase operational efficiency of the plant, more than the facilities only generating electricity, and thereby decrease the energy costs.

WM can be considered as an SC problem since it involves determining the number, location, and size of production facilities, stocking and sourcing points, capacity allocation, inventory control, distribution and transportation planning, and order fulfilment. Therefore, to optimize the waste flows and decrease the negative impacts of involved processes, as well as ensuring the sustainability of the system, it is crucial to integrate the SC strategies into energy system planning and management. The literature study shows that many models exist, where the production and logistics problems in WtE practices are separately considered (Ahn et al., 2015; Balaman et al., 2018; Mohammadi et al., 2018; Quddus et al., 2018), but not the integrated WtE SC problem for sustainable heat and electricity production. A well-designed WtE SC results in a smooth flow of waste to treatment plants, balancing the waste supply lots and vehicle loads, reducing inventory levels and the stock-out risks, and better re-use of waste for cost-effective renewable energy production. It also contributes to a significant reduction in the total cost of the SC network and environmental emissions. Accordingly, this paper presents an integrated SC model that aims to maximize the total system profit by producing sustainable energy from waste materials while satisfying existing capacity and environmental constraints.

2. Mathematical Model Formulation

The proposed model covers waste collection at cities, separation and pre-treatment of waste in separation centers, processing for energy recovery in WtE plants, and selling the energy products to the markets, as shown in Figure 1.

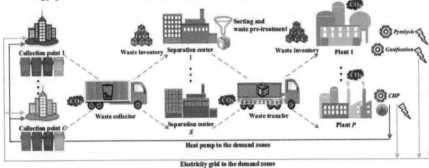

Figure 1. The addressed WtE SC network

The integrated WtE SC problem is formulated as an MILP model. The indices used in the formulation are: w (waste), $\tilde{w}$ (treated waste), c (city), s (separation center), p (plant), j^s, j^{int}, j^e (technologies for waste pre-treatment, intermediate production, and energy production), n^{int} (intermediate product), n^{el} (electricity), n^h (heat), and t (time period). Equation (1) shows the objective function of the proposed model, where the revenue from the selling of heat and electricity is maximized, from which the total cost of the entire SC network is deducted. The selling price of electricity and heat is shown by P_{nct}. The total cost includes waste collection (C_{ct}^{col}), waste separation (C_{wst}^{sep}), waste pre-treatment (C_{wst}^{treat}), waste storage in separation centers ($C_{wst}^{stor}, C_{\tilde{w}st}^{stor}$), waste storage in plants ($C_{\tilde{w}pt}^{stor}$), waste processing for intermediates production in plants ($C_{\tilde{w}pj^{int}{}_t}^{oper}$), intermediates processing for energy production in plants ($C_{n^{int}pj^e{}_t}^{oper}$), fixed and variable transportation (C^{fix}, C^{var}), and heat and electricity distribution costs (C_{npct}^{dist}).

$$
\begin{aligned}
Max\, f = &\sum_{n \in \{n^{el} \cup n^h\}} \sum_c \sum_t P_{nct} \cdot q_{nct} - \Big(\sum_w \sum_c \sum_t C_{ct}^{col} \cdot A_{wct} + \sum_w \sum_c \sum_s \sum_t C_{wst}^{sep} \cdot q_{wcst} \\
&+ \sum_w \sum_s \sum_t C_{wst}^{treat} \cdot q_{wst}^{sep} + \sum_w \sum_s \sum_t C_{wst}^{stor} \cdot i_{wst} + \sum_{\tilde{w}} \sum_s \sum_t C_{\tilde{w}st}^{stor} \cdot i_{\tilde{w}st} + \sum_{\tilde{w}} \sum_p \sum_t C_{\tilde{w}pt}^{stor} \cdot i_{\tilde{w}pt} \\
&+ \sum_{\tilde{w}} \sum_p \sum_{j^{int}} \sum_t C_{\tilde{w}pj^{int}{}_t}^{oper} \cdot q_{\tilde{w}pj^{int}{}_t} + \sum_{n^{int}} \sum_p \sum_{j^e} \sum_t C_{n^{int}pj^e{}_t}^{oper} \cdot q_{n^{int}pj^e{}_t} + \sum_c \sum_s \sum_t (C^{fix} + C^{var} \cdot d_{cs}) \cdot v_{cst} \\
&+ \sum_s \sum_p \sum_t (C^{fix} + C^{var} \cdot d_{sp}) \cdot v_{spt} + \sum_{n \in \{n^{el} \cup n^h\}} \sum_p \sum_c \sum_t C_{npct}^{dist} \cdot q_{npct} \Big)
\end{aligned} \tag{1}
$$

The produced MSW in a city (A_{wct}) is distributed to different separation centers (q_{wcst}) as given in Eq. (2). Equation (3) shows that the total amount of waste inlet to a separation center should not exceed its maximum input capacity (SC_{wst}^{max}). Equation (4) shows the amount of separated waste (q_{wst}^{sep}) using a separation rate for each waste type (R_{ws}^{sep}). Equation (6) shows the amount of waste shipped to the pre-treatment technologies in separation center ($q_{wsj^s{}_t}^{sep}$), which is limited by the technology capacity ($SC_{wsj^s}^{low}, SC_{wsj^s}^{up}$) as given in Eq. (7). The binary variable $z_{wsj^s{}_t}^{treat}$ indicates whether the waste is sent for pre-treatment to a technology or not. Equation (8) shows the amount of treated waste ($q_{\tilde{w}st}^{treat}$) using a pre-treatment rate for each waste type ($R_{w\tilde{w}sj^s}^{treat}$). Equations (5) and (9) define the inventory levels of initial waste and treated waste in a separation center ($i_{wst}, i_{\tilde{w}st}$), which cannot exceed its storage capacity ($S_{ws}, S_{\tilde{w}s}$).

$$
A_{wct} = \sum_s q_{wcst} \qquad\qquad \forall w \in W, c \in C, t \in T \tag{2}
$$

$$
\sum_c q_{wcst} \le SC_{wst}^{max} \qquad\qquad \forall w \in W, s \in S, t \in T \tag{3}
$$

$$
q_{wst}^{sep} = \sum_c q_{wcst} \cdot R_{ws}^{sep} \qquad\qquad \forall w \in W, s \in S, t \in T \tag{4}
$$

$$
i_{wst} = i_{ws(t-1)} + q_{wst}^{sep} - \sum_{j^s} q_{wsj^s{}_t}^{sep} \text{ and } i_{wst} \le S_{ws} \qquad \forall w \in W, s \in S, t \in T \tag{5}
$$

$$
\sum_{j^s} q_{wsj^s{}_t}^{sep} \le i_{ws(t-1)} + q_{wst}^{sep} \qquad\qquad \forall w \in W, s \in S, t \in T \tag{6}
$$

$$
SC_{wsj^s}^{low} \cdot z_{wsj^s{}_t}^{treat} \le q_{wsj^s{}_t}^{sep} \le SC_{wsj^s}^{up} \cdot z_{wsj^s{}_t}^{treat} \qquad \forall w \in W, s \in S, j^s \in J^s, t \in T \tag{7}
$$

$$
q_{\tilde{w}st}^{treat} = \sum_w \sum_{j^s} q_{wsj^s{}_t}^{sep} \cdot R_{w\tilde{w}sj^s}^{treat} \qquad\qquad \forall \tilde{w} \in \tilde{W}, s \in S, t \in T \tag{8}
$$

$$
i_{\tilde{w}st} = i_{\tilde{w}s(t-1)} + q_{\tilde{w}st}^{treat} - \sum_p q_{\tilde{w}spt} \text{ and } i_{\tilde{w}st} \le S_{\tilde{w}s} \qquad \forall \tilde{w} \in \tilde{W}, s \in S, t \in T \tag{9}
$$

Equation (10) shows the amount of treated waste that can be shipped from separation centers to plants ($q_{\tilde{w}spt}$). Equation (11) calculates the number of vehicles used for the transportation of waste from cities to separation centers (v_{cst}) and from separation centers to plants (v_{spt}) considering the vehicle capacity respecting the waste volume ($VC_w, VC_{\tilde{w}}$). Equation (12) limits the CO_2 emissions emitted from transportation by $TE_{st}^{CO_2}$ considering the distance (D_{cs}, D_{sp}), fuel consumption (F), and emission emitted from per liter of fuel (E^{CO_2}). Equation (13) shows that the total waste inlet to a plant is limited by the input capacity of the plant ($PC_{\tilde{w}pt}^{max}$). Equation (14) presents the waste inventory level in a plant ($i_{\tilde{w}pt}$), which cannot exceed the storage capacity ($S_{\tilde{w}p}$) of the plant. Equation (15) limits the amount of waste that can be transferred to processing technologies ($q_{\tilde{w}pj^{int}_t}$), which cannot exceed the capacity limits of the technology ($PC_{\tilde{w}pj^{int}}^{low}, PC_{\tilde{w}pj^{int}}^{up}$) as shown in Eq. (16). $z_{\tilde{w}pj^{int}_t}^{conv}$ is a binary variable that equals 1 if the waste is sent to conversion technologies for the production of intermediate products.

$$\sum_p q_{\tilde{w}spt} \leq i_{\tilde{w}s(t-1)} + q_{\tilde{w}st}^{treat} \qquad \forall \tilde{w} \in \tilde{W}, s \in S, t \in T \qquad (10)$$

$$v_{cst} - 1 \leq \sum_w \frac{q_{wcst}}{VC_w} \leq v_{cst} \text{ and } v_{spt} - 1 \leq \sum_{\tilde{w}} \frac{q_{\tilde{w}spt}}{VC_{\tilde{w}}} \leq v_{spt} \qquad \forall c \in C, s \in S, p \in P, t \in T \qquad (11)$$

$$\left(\sum_c D_{cs} \cdot v_{cst} + \sum_p D_{sp} \cdot v_{spt}\right) \cdot F \cdot E^{CO_2} \leq TE_{st}^{CO_2} \qquad \forall s \in S, t \in T \qquad (12)$$

$$\sum_s q_{\tilde{w}spt} \leq PC_{\tilde{w}pt}^{max} \qquad \forall \tilde{w} \in \tilde{W}, p \in P, t \in T \qquad (13)$$

$$i_{\tilde{w}pt} = i_{\tilde{w}p(t-1)} + \sum_s q_{\tilde{w}spt} - \sum_{j^{int}} q_{\tilde{w}pj^{int}_t} \text{ and } i_{\tilde{w}pt} \leq S_{\tilde{w}p} \qquad \forall \tilde{w} \in \tilde{W}, p \in P, t \in T \qquad (14)$$

$$\sum_{j^{int}} q_{\tilde{w}pj^{int}_t} \leq i_{\tilde{w}p(t-1)} + \sum_s q_{\tilde{w}spt} \qquad \forall \tilde{w} \in \tilde{W}, p \in P, t \in T \qquad (15)$$

$$PC_{\tilde{w}pj^{int}}^{low} \cdot z_{\tilde{w}pj^{int}_t}^{conv} \leq q_{\tilde{w}pj^{int}_t} \leq PC_{\tilde{w}pj^{int}}^{up} \cdot z_{\tilde{w}pj^{int}_t}^{conv} \qquad \forall \tilde{w} \in \tilde{W}, p \in P, j^{int} \in J^{int}, t \in T \qquad (16)$$

Equation (17) calculates the amount of produced intermediates ($q_{n^{int}pt}$) using the conversion rate of $R_{\tilde{w}n^{int}pj^{int}}^{conv}$. Equation (18) shows that the total amount of intermediate product is transferred to energy conversion technologies ($q_{n^{int}pj^e_t}$), which is restricted to the capacity limits of the technology ($PC_{n^{int}pj^e}^{low}, PC_{n^{int}pj^e}^{up}$) as shown in Eq. (19). The binary variable $z_{n^{int}pj^e_t}^{energy}$ indicates whether the intermediate product is sent to energy production technologies or not.

$$q_{n^{int}pt} = \sum_{\tilde{w}} \sum_{j^{int}} q_{\tilde{w}pj^{int}_t} \cdot R_{\tilde{w}n^{int}pj^{int}}^{conv} \qquad \forall n^{int} \in N^{int}, p \in P, t \in T \qquad (17)$$

$$\sum_{j^e} q_{n^{int}pj^e_t} = q_{n^{int}pt} \qquad \forall n^{int} \in N^{int}, p \in P, t \in T \qquad (18)$$

$$PC_{n^{int}pj^e}^{low} \cdot z_{n^{int}pj^e_t}^{energy} \leq q_{n^{int}pj^e_t} \leq PC_{n^{int}pj^e}^{up} \cdot z_{n^{int}pj^e_t}^{energy} \qquad \forall n^{int} \in N^{int}, p \in P, j^e \in J^e, t \in T \qquad (19)$$

Equation (20) shows the conversion of intermediate products to electricity ($q_{n^{el}pt}$) considering the conversion rate of $\tau_{n^{int}n^{el}pj^e}$ and efficiency of $\varphi_{n^{int}n^{el}pj^e}$. Since an intermediate can be converted to both heat and electricity, the coefficient $\lambda^{el}_{n^{int}pt}$ is used, which is the percentage of intermediate product conversion to electricity, and the rest can be converted to heat. Equation (21) shows the heat production considering the conversion rate of $\tau_{n^{int}n^hpj^e}$ and efficiency of $\varphi_{n^{int}n^hpj^e}$. Equation (22) indicates the amount of CO_2 emissions from operating waste ($E^{CO_2}_{\tilde{w}pj^{int}}$) and intermediates ($E^{CO_2}_{n^{int}pj^e}$), which is limited to the pre-determined emission level ($OE^{CO_2}_{pt}$). Equation (23) limits the electricity and heat generation to the production capacity of plant (PC^{max}_{npt}), and the binary variable z^{prod}_{npt} is to check if energy products are produced by the plant or not. Equation (24) shows the total energy transmitted from a plant to cities (q_{npct}), and Eq. (25) shows the energy products received by cities (q_{nct}), which cannot exceed the product demand (D_{nct}).

$$q_{n^{el}pt} = \sum_{n^{int}}\sum_{j^e} \lambda^{el}_{n^{int}pt} \cdot q_{n^{int}pj^e_t} \cdot \tau_{n^{int}n^{el}pj^e} \cdot \varphi_{n^{int}n^{el}pj^e} \qquad \forall n^{el}, p \in P, t \in T \tag{20}$$

$$q_{n^hpt} = \sum_{n^{int}}\sum_{j^e} (1-\lambda^{el}_{n^{int}pt}) \cdot q_{n^{int}pj^e_t} \cdot \tau_{n^{int}n^hpj^e} \cdot \varphi_{n^{int}n^hpj^e} \qquad \forall n^h, p \in P, t \in T \tag{21}$$

$$\sum_{\tilde{w}}\sum_{j^{int}} q_{\tilde{w}pj^{int}_t} \cdot E^{CO_2}_{\tilde{w}pj^{int}} + \sum_{n^{int}}\sum_{j^e} q_{n^{int}pj^e_t} \cdot E^{CO_2}_{n^{int}pj^e} \leq OE^{CO_2}_{pt} \quad \forall p \in P, t \in T \tag{22}$$

$$q_{npt} \leq PC^{max}_{npt} \cdot z^{prod}_{npt} \qquad \forall n \in \{n^{el} \cup n^h\}, p \in P, t \in T \tag{23}$$

$$q_{npt} = \sum_{c} q_{npct} \qquad \forall n \in \{n^{el} \cup n^h\}, p \in P, t \in T \tag{24}$$

$$q_{nct} = \sum_{p} q_{npct} \text{ and } q_{nct} \leq D_{nct} \qquad \forall n \in \{n^{el} \cup n^h\}, c \in C, t \in T \tag{25}$$

3. Results and Discussions

The considered multi-echelon SC problem consists of two cities as waste sources and consumer locations, two separation centers for waste separation and pre-treatment operations with two pre-treatment technologies, and three WtE plants to generate electricity and heat. The total MSW generation is 1,164 kt during a year. The presented MILP model is solved efficiently in a reasonable computational time (0.22 CPU-s) using GAMS/CPLEX. Considering the pyrolysis and gasification with the energy potential of 660 kWh/ton waste and the CHP plant with the capacity of 585 kWh/ton waste, 383 GWh of electricity and 171 GWh of heat are produced in a year. Figure 2 (a) presents the generated electricity and heat by the considered WtE plants. As demonstrated, pyrolysis produced the smallest amount of electricity, meeting 21 % of electricity demand, and the gasification plant met 49 % of electricity demand. Operation of CHP plant depends on the heat demand, and afterward, the range of electricity generation is determined; i.e., the more heat is produced the less energy is available for electricity production. The CHP generated 113.85 GWh of electricity during a year and satisfied 100 % of the heat demand. Therefore, CHP is a more profitable conversion technology than the other two WtE alternatives. However, it is most cost-effective when there is a constant heat demand, e.g., from adjacent industrial plants or district heating systems. The feedstock in pyrolysis was mostly plastic waste with lower calorific value compared to the considered mixed waste, which justifies the lower electricity generation by pyrolysis compared to the other two technologies. As shown in Figure 2 (b), the SC performance is greatly affected by the changes in electricity demand. Regarding the environmental aspects, pyrolysis and gasification are promising technologies in terms of clean energy production and having lower negative environmental impacts, whereas CHP led to the production of higher CO_2 emissions.

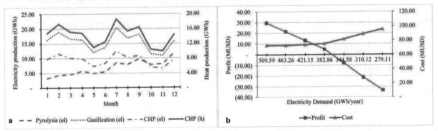

Figure 2 (a): Energy generation by WtE plants; 2 (b): Effect of fluctuations in electricity demand on the SC performance

4. Conclusions

The purpose of this study is to measure the environmental impacts and economic costs of the WtE SC network in order to check the economic and ecological feasibility of the entire system. In such systems, the production process, waste availability, energy demand, plant capacity, and level of energy efficiency affect the performance of WtE plants. Moreover, the coordination between SC entities has a substantial impact on the system performance by minimizing the overall costs, satisfying capacity restrictions at each level of the SC network, and fulfilling the demand requirements. The proposed model provides quantifiable information to make effective and optimum decisions regarding the treatment of waste while ensuring the most feasible (sustainable and cost-efficient) operations.

References

Y.C. Ahn, I.B. Lee, K.H. Lee, J.H. Han, 2015, Strategic Planning Design of Microalgae Biomass-To-Biodiesel Supply Chain Network: Multi-period Deterministic Model, Applied Energy, 154, 528-542.

Ş.Y. Balaman, D.G. Wright, J. Scott, A. Matopoulos, 2018, Network Design and Technology Management for Waste to Energy Production: An Integrated Optimization Framework under the Principles of Circular Economy, Energy, 143, 911-933.

M. Mohammadi, I. Harjunkoski, S. Mikkola, S-L, Jämsä-Jounela, 2018, Optimal Planning of a Waste Management Supply Chain, 13th International Symposium on Process Systems Engineering - PSE 2018, San Diego, California, USA, 1609-1614.

M.A. Quddus, S. Chowdhury, M. Marufuzzaman, F. Yu, L. Bian, 2018, A Two-stage Chance-constrained Stochastic Programming Model for a Bio-fuel Supply Chain Network, International Journal of Production Economics, 195, 27-44.

Anton A. Kiss, Edwin Zondervan, Richard Lakerveld, Leyla Özkan (Eds.)
Proceedings of the 29th European Symposium on Computer Aided Process Engineering
June 16th to 19th, 2019, Eindhoven, The Netherlands. © 2019 Elsevier B.V. All rights reserved.
http://dx.doi.org/10.1016/B978-0-128-18634-3.50193-4

Automating a shuttle-conveyor for multi-stockpile level control

Jeffrey D. Kelly,[a,*] Brenno C. Menezes[b]

[a]*Industrial Algorithms Ltd., 15 St. Andrews Road, Toronto M1P 4C3, Canada*

[b]*Center for Information, Automation and Mobility, Technological Research Institute, Av. Prof. Almeida Prado 532, São Paulo 05508-901, Brazil*

jdkelly@industrialgorithms.ca

Abstract

We describe the application of an online dynamic and discrete scheduling optimization, also known as real-time hybrid model predictive control, applied to a shuttle-conveyor / tripper car intermittently delivering crushed-ore containing copper, iron, etc., to several stockpiles. Each stockpile continuously feeds an apron feeder located in a tunnel underneath the stockpile where multiple of these are combined to a belt feeder to charge a grinding mill which ultimately produces concentrates of minerals. Crushed-ore from a mine several kilometres upstream is transported by both conveyor belt and automatic trains. The shuttle-conveyor is the discrete actuation or manipulated variable depositing the solids onto each stockpile where we vary the time over each stockpile to automatically control to a setpoint or target their levels sensed by industrial radar. In the field, the motion of the shuttle-conveyor is known as sweeping as it moves from one stockpile (or tunnel) to the next in a fixed sequence. The purpose of our application is to improve the stockpile level control performance by automatically adjusting the up-time, fill-time or run-length of the shuttle-conveyor tripper car over each stockpile based on real-time level measurement feedback and the application which is referred to as *smart sweeping*.

Keywords: Mixed-integer linear programming (MILP), Hybrid dynamic control, Stockpile live inventory, Smart sweeping.

1. Introduction

Advances in manufacturing toward the Industry 4.0 (I4) age push the re-examination of industrial systems to identify opportunities to (re) execute them into an improved production state. For such, the information and communication technologies (ICT), high-performance computing (HPC) and mechatronics (MEC) are evolving together with the advances in modelling and solving algorithms (MSA). Considering this I4 era in manufacturing, we explore the MSA aspects in an application found in the mining field (see the Future of Mining in the 21st Century in Lottermoser, 2017). In the system, shuttle-conveyors are modeled in terms of their sequencing and timing as they travel from one position or location to the next on the shuttle-carriage, whereby stockpile levels are used as setpoints to influence the movements of the shuttle-conveyors. Automated belt conveyor systems integrated with intelligent decision-making core is a successful application of I4 in the mining value chain (Braun et al., 2017).

In the proposed work, the design basis of the advanced control system for managing simultaneously multiple stockpile levels for each feeder in the ore land field, warehouse, bunker or barn uses the mixed-integer linear programming (MILP) optimization for the positioning, placement, movement or manipulation of the shuttle-conveyors / tripper cars, the discrete (integer), whereas the stockpile levels are continuous. In addition, the supply (filling) and demand (drawing) of crushed-ore in the ore barn is semi-continuous or intermittent especially considering the nature of how crushed-ore travels from the mine to the barn and how crushed-ore is feed to autogenous mills via the feeders.

The proposed problem is known in the control literature as hybrid model predictive control (HMPC) as there is a mix of discrete and continuous variables with a receding or moving prediction time-horizon. The MILP solution defines the shuttle positions which will collectively manage the stockpile levels over each feeder given the supply and demand of crushed-ore. MILP is a search-based algorithm known as branch-and-bound (B&B) which will evaluate quickly many possible scenarios, samples or situations to find the best solution that optimizes the objective function also taking into consideration both the continuous and discrete nature of the controlled and manipulated variables subject to hard constraints (i.e., lower and upper bounds) and soft constraints (target).

Supported by the I4 fundaments, a high-end radar sensing system (for virtual reality) measures the mineral stockpile level (opening inventory) in real-time for the determination of discrete positions and time-slots of the reversing shuttle-conveyor that creates the stockpiles (ST1 to ST8) for feeding multiple fixed discharge points. A reversing shuttle-conveyor in Figure 1 is a belt conveyor sweeping onto a rail system in both directions which manipulates, belt direction and shuttle movement, and can be automated or controlled manually by an operator.

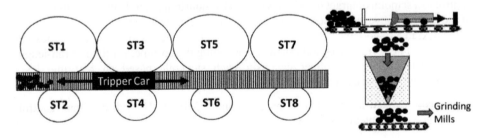

Figure 1. Tripper car shuttle-conveyor schematics.

2. Problem statement

The objective function will minimize the absolute deviations (1-norm performance) of the stockpile estimated *live* inventories (controlled variables) as well as minimizing the absolute excursions or penalties from its lower and upper holdup bounds should any infeasibilities be detected both in the present and in the future. The decision or manipulated variables are the shuttle-conveyor positions (in time-steps) subject to their sequence and timing constraints. In the problem, the shuttle-conveyor (SC) supplies crushed-ore to the ore barn which contains the wet-mills Mill1 and Mill2 as seen in Figure 2. To these mills are fed crushed-ore by eight (8) feeders with corresponding stockpiles (ST1, ST2, ..., ST8) connected to them by tunnels or zones below the stockpiles where apron-feeders gravity load crushed-ore onto belt-conveyors on an incline to be sent to the wet-mills for grinding.

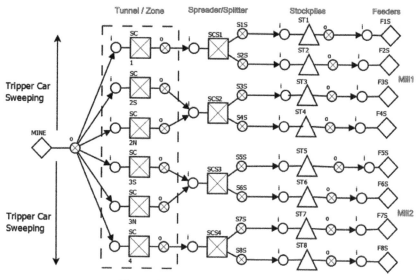

Figure 2. Stockpile control system.

The shuttle-conveyor position sequencing policies are based on the notion of whether the placement sequence from one tunnel to the next and the amount of time the shuttle-conveyor spends over each tunnel are fixed or variable. Sweeping is the name given to a fixed-sequence / variable-time policy with SC traveling forward and backward between tunnels 1 to 4. Figure 2 above depicts a flowsheet for how crushed-ore material from the mine can be moved from SC (tunnels) to the stockpiles (ST1, …, ST8) and ultimately to the wet-mill feeders via the spreader / splitter objects shown. These separation structures or shapes are required to model the non-ideal flow of how crushed-ore flow distributes or spreads over to both stockpiles at each tunnel. It should be emphasized that since SC is a unary or single-use resource only one of the six operations configured (1, 2S, 2N, 4, 3S, 3N) can be setup, active, on, open, etc., within any given time-period. Sweeping is similar to the notion of *phasing* found in project scheduling (Menezes et al., 2015) as there is a fixed-sequence of how each shuttle-conveyor traverses its area of the ore warehouse or barn, i.e., from one phase, task or operation to the next and so on.

3. Mathematical formulation

The network in Figure 2 uses the unit-operation-port-state superstructure (UOPSS) formulation (Kelly, 2005). The objects are: a) unit-operations m for sources and sinks ($\diamond$), stockpiles ($\triangle$) and continuous-processes ($\boxtimes$) and b) the connectivity involving arrows ($\rightarrow$), in-port-states i ($\bigcirc$) and out-port-states j ($\otimes$). Unit-operations and arrows have binary and continuous variables (y and x, respectively). In the mixed-integer linear (MILP) problem (P), the objective function (1) minimizes the absolute deviation of the stockpile inventories, levels or holdups $xh_{m,t}$. The performance constraint (2) defines the lower and upper deviation variables ($xh_{m,t}^{LD}$ and $xh_{m,t}^{UD}$, respectively) around the stockpile holdup target ($\overline{xh}_{m,t}$) with w^D as weight. The semi-continuous constraint (3) limits the stockpile holdup by its lower and upper bounds ($\overline{xh}_{m,t}^L$ and $\overline{xh}_{m,t}^U$), whereby the excursion variables or penalties $xh_{m,t}^{LE}$ and $xh_{m,t}^{UE}$ are included to avoid infeasibilities (with w^E as weight). Similarly, there are semi-continuous constraints for flows of process-units $x_{m,t}$ and arrows $x_{j,i,t}$, although without excursion variables as for the stockpile holdups. Unit-operations m for the mine, shuttle-conveyor, spreader, stockpile, and feeder belong, respectively, to the sets M_{MN}, M_{SC}, M_{SD}, M_{ST}, and M_{FD}. For $x \in \mathbb{R}^+$, $y, yg \in \{0,1\}$ and $zsu_{m,t}, zsd_{m,t}, zsw_{m,t} \in [0,1]$:

$$(P) \ Max \ Z = \sum_t \sum_{m \in M_{ST}} \left[w^D \left(xh_{m,t}^{LD} + xh_{m,t}^{UD} \right) + w^E \left(xh_{m,t}^{LE} + xh_{m,t}^{UE} \right) \right] \quad s.t. \tag{1}$$

$$xh_{m,t} - \overline{xh}_{m,t} + xh_{m,t}^{LD} - xh_{m,t}^{UD} = 0 \ \forall \ m \in M_{ST}, t \tag{2}$$

$$\overline{xh}_{m,t}^L \, y_{m,t} - xh_{m,t}^{LE} \leq xh_{m,t} \leq \overline{xh}_{m,t}^U \, y_{m,t} + xh_{m,t}^{UE} \ \forall \ m \in M_{ST} \ t \tag{3}$$

$$\frac{1}{\overline{x}_{m,t}^U} \sum_{j \in M_{MN}} x_{j,i,t} \leq y_{m,t} \leq \frac{1}{\overline{x}_{m,t}^L} \sum_{j \in M_{MN}} x_{j,i,t} \quad \forall \ (i,m) \in M_{SC}, t \tag{4}$$

$$\frac{1}{\overline{x}_{m,t}^U} \sum_i x_{j,i,t} \leq y_{m,t} \leq \frac{1}{\overline{x}_{m,t}^L} \sum_i x_{j,i,t} \quad \forall \ (m,j) \in (M_{MN}, M_{SC}, M_{SD}), t \tag{5}$$

$$\frac{1}{r_{i,t}^U} \sum_j x_{j,i,t} \leq x_{m,t} \leq \frac{1}{r_{i,t}^L} \sum_j x_{j,i,t} \quad \forall \ (i,m) \in (M_{SC}, M_{SD}), t \tag{6}$$

$$\frac{1}{r_{j,t}^U} \sum_i x_{j,i,t} \leq x_{m,t} \leq \frac{1}{r_{j,t}^L} \sum_i x_{j,i,t} \quad \forall \ (m,j) \in (M_{SC}, M_{SD}), t \tag{7}$$

$$xh_{m,t} = xh_{m,t-1} + \sum_{j_{up}} x_{j_{up},i,t} - \sum_{i_{do}} x_{j,i_{do},t} \ \forall \ (i,m,j) \in M_{ST}, t \tag{8}$$

$$\sum_{m \in M_{SC}} y_{m,t} \leq 1 \ \forall \ t \tag{9}$$

$$\sum_{m \in M_{SC}} y_{m,t} \geq 1 \ \forall \ t \tag{10}$$

$$y_{m_{up},t} + y_{m,t} \geq 2 y_{j_{up},i,t} \ \forall \ (m_{up}, j_{up}, i, m), t \tag{11}$$

$$y_{m,t} - y_{m,t-1} - zsu_{m,t} + zsd_{m,t} = 0 \ \forall \ m \in M_{SC}, t \tag{12}$$

$$y_{m,t} + y_{m,t-1} - zsu_{m,t} - zsd_{m,t} - 2zsw_{m,t} = 0 \quad \forall \ m \in M_{SC}, t \tag{13}$$

$$zsu_{m,t} + zsd_{m,t} + zsw_{m,t} \leq 1 \ \forall \ m \in M_{SC}, t \tag{14}$$

$$\sum_{g \in G_{SC}} yg_{u,g,t} = 1 \ \forall \ u \in U_{SC}, t \tag{15}$$

$$\sum_{(m \in M_{SC}) \subset G_{SC}} y_{m,t} \leq yg_{u,g,t} \ \forall \ u \in U_{SC}, g \in G_{SC}, t \tag{16}$$

$$yg_{u,g,t} - yg_{u,g,t-1} \leq zsu_{m \subset g,t} \ \forall \ u \in U_{SC}, g \in G_{SC}, t \tag{17}$$

$$yg_{u,g,t-1} - yg_{u,g,t} \leq zsu_{m \subset g',t} \ \forall \ u \in U_{SC}, g \in G_{SC}, g' \in G_{SC}, t \tag{18}$$

Equations (4) and (5) represent, respectively, the sum of the arrows arriving in the in-port-states i (or mixers) or leaving from the out-port-states j (or splitters) and their summation must be between the bounds of the unit-operation m connected to them. Equations (6) and (7) consider bounds on yields, both inverse ($r_{i,t}^L$ and $r_{i,t}^U$) and direct ($r_{j,t}^L$ and $r_{j,t}^U$), since the unit-operations m can have more than one stream arriving in or leaving from their connected ports. The quantity balance of inventory or holdup for unit-operations of tanks ($m \in M_{ST}$) in Eq. (8) considering initial inventories $xh_{m,t-1}$ and inlet and outlet streams of the tanks, whereby the port-states j_{up} and i_{do} represent upstream and downstream ports connected, respectively, to the in-port-states i and out-port-states j of the stockpile unit-operations m as represented in Figure 2. In Eq. (9), for all physical units, at most one unit-operation m (as $y_{m,t}$ for procedures, modes or tasks) is permitted in U_{SC} at a time t. Equation (10) is the zero-downtime constraint to activate SC in all time t. Equation (11) is the structural transition logic valid cut of 4 objects (m', j, i, m) to reduce the tree search in branch-and-bound methods.

The sequencing operation of the multiple SC is controlled by the temporal transition constraints (12) to (14) from Kelly and Zyngier (2007). The setup or binary variable $y_{m,t}$ manages the dependent start-up, switch-over-to-itself and shut-down variables ($zsu_{m,t}$, $zsw_{m,t}$ and $zsd_{m,t}$, respectively) are relaxed in the interval [0,1] instead of considering them as logic variables. Equation 12 is necessary to guarantee the integrality of the relaxed variables. By fixing the sequence (neighbor to neighbor) of the SC modes of operation (time-slots of the positions), the grouping constraints (15) to (18) coordinate the grouping setup variable $yg_{u,g,t}$ to update the shuttle-conveyor operation-group to operation-group sequence-dependent *phasing* (or fixed transition between neighbor modes of the SC, e.g., 1→2S→3S→4→3N→2N→1). Other constraints related to uptime of the SC and the shutting-when-full relationship are found in Zyngier and Kelly (2009).

4. Example

The example in Figure 2 uses the structural-based UOPSS framework found in the semantic-oriented modelling and solving platform IMPL (Industrial Modeling & Programming Language) from Industrial Algorithms Limited using Intel Core i7 machine at 3.4 GHz (in 8 threads) with 64 GB of RAM. The optimization for the proposed MILP for a 2 hours time-horizon with 2 minute time-steps (30 time-periods) is solved in 4.07 seconds (with GUROBI 8.1.0). There are 5,472 constraints (1,593 equality) for 2,823 continuous variables and 900 binary variables with 2,130 degrees-of-freedom. Figure 3 shows the Gantt chart for the shuttle-conveyor and spreader time-slot positions (in min).

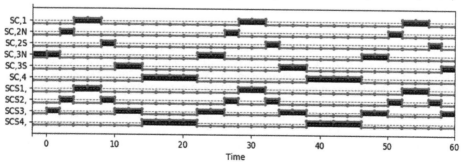

Figure 3. Gantt chart for shuttle-conveyor (SC) and spreader (SCS) time-slot positions.

5. Conclusions

By automating a shuttle-conveyor position in variable time-slots of minutes, a hybrid dynamic control problem coordinates the stockpile *live* inventory of crushed-ore of minerals in the mining field. With advanced automation of mining sub-systems, it may be possible to replace human operators in the field in certain areas and functions. One place to start can be the integration of the synchronization of the sensing, calculating and actuating cycle (within minutes) as in the mining system proposed in this paper. The position of the tripper car reached the level of sensing and calculating technologies to be integrated to automatically program the time-slot of positions of the shuttle conveyor system.

Integrated to the decision-making core proposed in this paper (the hybrid dynamic control), for the advances in the infrastructure of cyber-physical systems (CPS) for a digital twin of the reality in the mining system there are needs of other three smart manufacturing foundations. They are: a) information and communication technologies (ICT), with the ecosystem match of the internet of the things and computing edges as the cloud and fog (Bittencourt, 2018), to capture the *live* inventory using the radar apparatus; b) the high-performance computing (HPC), to solve extremely fast the discrete and continuous problems in distributed machine cores; and c) the mechatronics (MEC) to actuate automatically considering an integrated envelop of the main I4 attributes around the sensing, calculation and actuation cycle of an I4-age manufacturing system.

Acknowledgements

The first author acknowledges the financial support from the São Paulo Research Foundation (FAPESP) under grant #2018/26252-0 for the project over the application of the Industry 4.0 concepts in the mining field.

References

L. Bittencourt, R. Immich, R. Sakellariou, N. Fonseca, E. Madeira, M. Curado, L. Villas, L. Da Silva, C. Lee, O. Rana, 2018, The Internet Of Things, Fog and Cloud Continuum: Integration And Challenges. Internet of Things, 3-4, 134-155.

T. Braun, A. Henning, B.G. Lottermoser, 2017, The Need for Sustainable Technology Diffusion in Mining: Achieving the Use of Belt Conveyor Systems in the German Hard-Rock Quarrying Industry, Journal of Sustainable Mining, 16, 24-30.

J.D. Kelly, 2005, The Unit-Operation-Stock Superstructure (UOSS) and the Quantity-Logic-Quality Paradigm (QLQP) for Production Scheduling in The Process Industries, In Multidisciplinary International Scheduling Conference Proceedings: New York, United States, 327-333.

J.D. Kelly, D. Zyngier, 2007, An Improved MILP Modeling of Sequence-Dependent Switchovers for Discrete-Time Scheduling Problems, Industrial Engineering Chemistry Research, 46, 4964.

B.G. Lottermoser, 2017, The Future of Mining in the 21st Century. Georesources Journal, 2, 5-6.

B.C. Menezes, J.D. Kelly, I. E. Grossmann, A. Vazacopoulos, 2015, Generalized Capital Investment Planning of Oil-Refineries using MILP and Sequence-Dependent Setups, Computer and Chemical Engineering, 80, 140-154.

D. Zyngier, J.D. Kelly, 2009, Multi-product Inventory Logistics Modeling in The Process Industries. In: Wanpracha Chaovalitwongse, Kevin C. Furman, Panos M. Pardalos (Eds.) Optimization and Logistics Challenges in the Enterprise. Springer optimization and its Applications, 30, Part 1, 61.

Anton A. Kiss, Edwin Zondervan, Richard Lakerveld, Leyla Özkan (Eds.)
Proceedings of the 29th European Symposium on Computer Aided Process Engineering
June 16th to 19th, 2019, Eindhoven, The Netherlands.
http://dx.doi.org/10.1016/B978-0-128-18634-3.50194-6

A simple modeling approach to control emulsion layers in gravity separators

Christoph Josef Backi[a,b,*], Samuel Emebu[b], Sigurd Skogestad[b] and Brian Arthur Grimes[b]

[a]BASF SE, 63056 Ludwigshafen am Rhein, Germany
[b]Department of Chemical Engineering, Norwegian University of Science and Technology (NTNU), 7491 Trondheim, Norway
christoph.backi@ntnu.no

Abstract

This paper presents an extension to a gravity separator model that has been previously introduced by some of the authors. It expands the existing model by adding differential states for emulsion layer thickness between the oil- and the water-continuous phases in addition to the gas pressure, the overall liquid level and the water-continuous layers. These differential states can ultimately be used for controller design in order to regulate the layer thickness by adding demulsifiers to the inflow. Due to the fact that demulsifiers are a primary factor to promote coalescence of complex petroemulsions, the demulsifier inflow is a critical variable to incorporate into a control model for gravity separation of petroemulsions.

Keywords: Gravity separator, oil and gas, emulsion layer, modeling, control

1. Introduction

The inflows of dispersed droplets into the new emulsion layers are obtained from a simplified interfacial coalescence relationship presented by Grimes (2012). In this formulation, the product Γ of the interfacial tension σ and retarded Hamaker constant Ha are directly related to the influence of the demulsifier. Realistic combinations of the interfacial tension and Hamaker constant that provide increasing and decreasing emulsion layer thicknesses were determined and a feedback control model was formulated to adjust the demulsifier inflow based on measurement of the thickness of the emulsion layers through the parameter σ. The controllers of simple PI-type were tuned according to the SIMC rule, see e.g. Skogestad (2003), and the theoretical developments of this work are demonstrated in a simulation case study.

2. Mathematical model

The basic principles for the gravity separator model without emulsion layers have been introduced in Backi and Skogestad (2017) and Backi et al. (2018). Three dynamic state variables were defined, where the first describes the dynamics of the overall liquid level (oil plus water), h_L,

$$\frac{dh_L}{dt} = \frac{dV_L}{dt} \frac{1}{2L\sqrt{h_L(2r - h_L)}} \tag{1}$$

where r is the separator's radius and L is the length of the active separation zone. Furthermore, $\frac{dV_L}{dt} = q_{L,in} - q_{L,out}$ is the volumetric change in overall liquid, where $q_{L,out} = q_{W,out} + q_{O,out}$. The manipulated variables are the outflows of water $q_{W,out}$ and oil $q_{O,out}$, respectively.

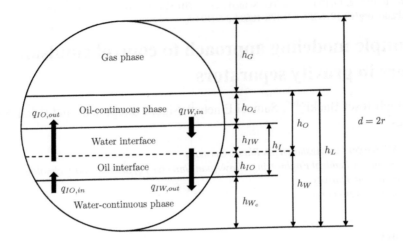

Figure 1: Schematic over the different variables in a cross-sectional view of the gravity separator

The second variable introduces the change in the water-continuous level, h_W,

$$\frac{dh_W}{dt} = \frac{dV_W}{dt} \frac{1}{2L\sqrt{h_W(2r - h_W)}} \tag{2}$$

with $\frac{dV_W}{dt} = q_{W,in} - q_{W,out}$, where $q_{W,in} = \gamma q_{L,in}$ is a fraction γ of the total liquid inflow $q_{L,in}$ entering the water-continuous phase. Thereby, $\gamma = \alpha \phi_{ww} + (1 - \alpha)(1 - \phi_{oo})$, where α is the water cut of the inflow, ϕ_{ww} describes the fraction of inflowing water entering the water-continuous phase, and ϕ_{oo} is the fraction of inflowing oil entering the oil-continuous phase.

The third variable describes the pressure dynamics derived from the ideal gas law with temperature T constant

$$\frac{dp}{dt} = \frac{RT \frac{\rho_G}{M_G}(q_{G,in} - q_{G,out}) + p(q_{L,in} - q_{L,out})}{V_{Sep} - A_L L}, \tag{3}$$

where R denotes the universal gas constant, M_G is the molar mass of the gas, ρ_G describes the gas density, and $q_{G,in}$ and $q_{G,out}$ give the in- and outflows of gas, where latter is a manipulated variable.

In addition to the dynamic part of the original model, an algebraic part is introduced. In short, the algebraic part compares horizontal to vertical residence times of different droplet size classes. Based on this calculation, a droplet size class is either separated, not separated, or partially separated. Thereby, several assumptions hold, for example that Stokes' law is the main driving force for separation and that no coalescence or breakage are considered. For further details, the authors refer to Backi and Skogestad (2017) and Backi et al. (2018).

As an extension to the model, two new state variables are introduced in order to represent the emulsion layer between the oil- and the water-continuous layers in a gravity separator. In Figure 1, a schematic of the cross-sectional areas and the variables in such a separator are given.

2.1. Oil interface layer h_{IO}

We derive a differential equation for the oil interface layer, which builds up on top of the water-continuous layer. We write the cross-sectional area of the oil interface layer (index IO) as

$$A_{IO} = A_W - A_{W-IO}, \tag{4}$$

with the two cross-sectional areas A_W (index W for water layer) and A_{W-IO} (index $W-IO$ for water-continuous layer minus oil interface layer) defined as

$$A_W = \frac{r^2}{2}\left(2\arccos\left(\frac{r-h_W}{r}\right) - \sin\left(2\arccos\left(\frac{r-h_W}{r}\right)\right)\right), \tag{5}$$

$$A_{W-IO} = \frac{r^2}{2}\left(2\arccos\left(\frac{r-(h_W-h_{IO})}{r}\right) - \sin\left(2\arccos\left(\frac{r-(h_W-h_{IO})}{r}\right)\right)\right). \tag{6}$$

After differentiating (4)–(6), we receive the expression

$$\frac{dA_{IO}}{dt} = \frac{dh_{IO}}{dt}\left[2\sqrt{(h_W-h_{IO})(2r-(h_W-h_{IO}))}\right]$$
$$+ \frac{dh_W}{dt}\left[2\left(\sqrt{h_W(2r-h_W)} - \sqrt{(h_W-h_{IO})(2r-(h_W-h_{IO}))}\right)\right], \tag{7}$$

where $\frac{dh_W}{dt}$ is introduced in (2) leading to

$$\frac{dA_{IO}}{dt} = \frac{dh_{IO}}{dt}\left[2\sqrt{(h_{IO}+h_W)(2r-(h_{IO}+h_W))}\right]$$
$$+ \frac{dV_W}{dt}\frac{1}{L}\left(1 - \sqrt{\frac{(h_{IO}+h_W)(2r-(h_{IO}+h_W))}{h_W(2r-h_W)}}\right). \tag{8}$$

Ultimately, we receive the differential equation for the oil interface layer

$$\frac{dh_{IO}}{dt} = \frac{dV_{IO}}{dt}\frac{1}{2L}\frac{1}{\sqrt{(h_W-h_{IO})(2r-(h_W-h_{IO}))}}$$
$$- \frac{dV_W}{dt}\frac{1}{2L}\left(\frac{1}{\sqrt{(h_W-h_{IO})(2r-(h_W-h_{IO}))}} - \frac{1}{\sqrt{h_W(2r-h_W)}}\right), \tag{9}$$

with $\frac{dV_W}{dt} = q_{W,in} - q_{W,out}$ and $\frac{dV_{IO}}{dt} = q_{IO,in} - q_{IO,out}$; latter is calculated using the algebraic part of the model.

2.2. Water interface layer h_{IW}

For the water interface layer (index IW) forming on the bottom of the oil-continuous layer, it holds

$$A_{IW} = A_{O+G} - A_{O+G-IW} \tag{10}$$

with the two cross-sectional areas A_{O+G} (index $O+G$ for the sum of the oil-continuous layer plus the gas phase) and A_{O+G-IW} (index $O+G-IW$ for oil-continuous layer plus gas phase minus water interface layer) defined as

$$A_{O+G} = \frac{r^2}{2}\left(2\arccos\left(\frac{r-h_{O+G}}{r}\right) - \sin\left(2\arccos\left(\frac{r-h_{O+G}}{r}\right)\right)\right), \tag{11}$$

$$A_{O+G-IW} = \frac{r^2}{2}\left(2\arccos\left(\frac{r-(h_{O+G}-h_{IW})}{r}\right) - \sin\left(2\arccos\left(\frac{r-(h_{O+G}-h_{IW})}{r}\right)\right)\right). \tag{12}$$

Differentiating (10), (11), and (12) and defining $h_{O+G} = 2r - h_W$, we obtain

$$\frac{dA_{IW}}{dt} = \frac{dh_{IW}}{dt}\left[2\sqrt{(h_{IW}+h_W)(2r-(h_{IW}+h_W))}\right]$$
$$+ \frac{dh_W}{dt}\left[2\left(\sqrt{(h_{IW}+h_W)(2r-(h_{IW}+h_W))} - \sqrt{h_W(2r-h_W)}\right)\right] \tag{13}$$

where, again, $\frac{dh_W}{dt}$ is presented in (2) and hence

$$
\frac{dA_{IW}}{dt} = \frac{dh_{IW}}{dt} \left[2\sqrt{(h_{IW} + h_W)(2r - (h_{IW} + h_W))} \right]
$$
$$
+ \frac{dV_W}{dt} \frac{1}{L} \left(\sqrt{\frac{(h_{IW} + h_W)(2r - (h_{IW} + h_W))}{h_W(2r - h_W)}} - 1 \right). \tag{14}
$$

Finally, we receive the differential equation for the water interface layer

$$
\frac{dh_{IW}}{dt} = \frac{dV_{IW}}{dt} \frac{1}{2L} \frac{1}{\sqrt{(h_{IW} + h_W)(2r - (h_{IW} + h_W))}}
$$
$$
- \frac{dV_W}{dt} \frac{1}{2L} \left(\frac{1}{\sqrt{h_W(2r - h_W)}} - \frac{1}{\sqrt{(h_{IW} + h_W)(2r - (h_{IW} + h_W))}} \right) \tag{15}
$$

with $\frac{dV_W}{dt} = q_{W,in} - q_{W,out}$ and $\frac{dV_{IW}}{dt} = q_{IW,in} - q_{IW,out}$. It must be pointed out that here $\frac{dV_W}{dt} = q_{W,in} - q_{W,out}$ excludes the oil leaving the water into the interface. Furthermore, $\frac{dV_{IW}}{dt} = q_{IW,in} - q_{IW,out}$ is calculated using the algebraic part of the model.

3. Controller Design

The controllers of PI type in the shape $K(s) = K_P + T_I \frac{1}{s}$ have the aim to regulate the gas pressure and the levels of overall liquid and water as well as the oil and the water interface layer levels. The first three controllers have already been introduced and developed in Backi and Skogestad (2017) and Backi et al. (2018), where the manipulated variables (MVs) are the outflows of water $q_{W,out}$, oil $q_{O,out}$ and gas $q_{G,out}$. Hence, in this work the focus lies on the design of the controllers for the oil and water interface layers.

3.1. Theoretical developments

In order to control the layer thickness of emulsions in gravity separators, the addition of demulsifiers to the inlet stream is necessary. Depending on the type of oil and the thickness of the emulsion, more or less demulsifier has to be added. We derive a relationship between the outflow of oil and water from the respective interface layers into their bulk phases and the coalescence time t_c for droplets with the volumetric outflow given by

$$
q_{I,out} = A_I \int_0^\infty \frac{r_d}{3t_c} f(r_d) dr_d \cong \frac{A_I r_1}{3t_c}, \tag{16}
$$

where $A_I = 2L\sqrt{h_W(r - h_W)}$ is the interfacial area, r_d are droplet radii, r_1 is a lumped, unified, theoretic droplet radius and t_c is the coalescence time

$$
t_c = 1.046 \frac{\mu_c(\rho_d - \rho_c) g_z r_1^{9/2}}{\sigma^{3/2} Ha^{1/2}}, \tag{17}
$$

where μ_c is the dynamic viscosity of the continuous phase, ρ_d and ρ_c are the densities of the dispersed and continuous phases, respectively, g_z is the gravitational acceleration, and finally Ha describes the retarded Hamaker constant and σ is the interfacial tension. Latter is the manipulated variable for the controller, hence by changing σ we can affect the volumetric outflow (16). In a first step, we aim to find a relation between the retarded Hamaker constant and the interfacial tension in order to affect the coalescence time, which affects the outflow of oil or water from the respective interfaces. Since the retarded Hamaker constant and the interfacial tension are related by the function

$$
\Gamma = \sigma^{3/2} Ha^{1/2} \tag{18}
$$

we first investigate pairings of these values that give reasonable values for Γ and then ultimately find the theoretic, lumped droplet radii r_1 for water and oil droplets in the respective interfaces. Furthermore, we can reduce the parameter space by one if we find an expression that links the retarded Hamaker constant Ha with the interfacial tension σ, namely $Ha = f(\sigma)$, and hence $\Gamma = \sigma^{3/2} [f(\sigma)]^{1/2}$. A function that satisfies the property that the retarded Hamaker constant is decreasing as the interfacial tension is rising is

$$f(\sigma) = \sigma \left[1.7564 \cdot 10^{-\left(20 + \frac{\sigma}{\sigma_{min}}\right)} \right] ln \left(1 + \frac{0.0001}{\sigma} \right), \tag{19}$$

where $\sigma_{min} = 0.005$ N m^{-1}.

Expressions for the outflows can be obtained by combining (16), (17) and (18)

$$q_{IO,out} = \frac{A_I r_{IO,1}}{3 t_{IO,c}} = \frac{2 L \sqrt{h_W (r - h_W)}}{3.138 \, \mu_{IO,c} (\rho_d - \rho_c) g_z} \frac{\Gamma_{IO}}{r_{IO,1}^{7/2}},$$

$$q_{IW,out} = \frac{A_I r_{IW,1}}{3 t_{IW,c}} = \frac{2 L \sqrt{h_W (r - h_W)}}{3.138 \, \mu_{IW,c} (\rho_d - \rho_c) g_z} \frac{\Gamma_{IW}}{r_{IW,1}^{7/2}}, \tag{20}$$

and it becomes apparent that the only tuning variables affecting the outflow are the water level h_W and furthermore the combined parameters Γ_{IO} and Γ_{IW} as well as the theoretic, lumped droplet radii $r_{IO,1}$ and $r_{IW,1}$.

As can be seen from (19) and ultimately (20), the interfacial tensions σ_{IW} and σ_{IO}, which are the MVs for the two controllers, can never be zero or negative. From these equations it can be inferred that large σ give small outflows and vice versa.

4. Results

Figure 2 shows simulation results incorporating the model and the controllers presented in sections 2 and 3, respectively. The subplots $S_{i,i}$ are thereby numerated according to entries in a matrix, hence $S_{1,1}$ is the top left plot, etc.

$S_{1,1}$ and $S_{2,1}$ show the liquid level h_L and the water level h_W in solid lines with their respective reference values in dashed lines. The water level is increased at time $t = 200$ s from initially 1.2 m to 1.4 m while the liquid level is held constant at 2.5 m. This is done to demonstrate that all controllers can handle this occurrence. $S_{1,2}$ and $S_{2,2}$ depict the outflows from the oil- and water-continuous phases, $q_{W,out}$ and $q_{O,out}$, respectively.

$S_{3,1}$ presents the oil interface level with its outflow $q_{IO,out}$ in $S_{3,2}$. One can see the that $q_{IO,out}$ is initially held constant at a very low flow rate resulting in the oil interface level to increase. The controller and hence the addition of demulsifier is activated at $t = 100$ s causing the oil interface level to be brought back to zero. Accordingly, the water interface level is depicted in $S_{4,1}$ with its outflow $q_{IW,out}$ in $S_{4,2}$. Like for the oil interface level described above, initially $q_{IW,out}$ is constant and the water interface level increases. Again, the controller is turned on at $t = 100$ s, which leads to a decrease in the water interface level.

For completeness, the interfacial tensions for the two interface layers over time are presented in $S_{3,3}$ and $S_{4,3}$, respectively. Furthermore, the oil-continuous level is shown in $S_{1,3}$ and $S_{2,3}$ depicts the water-continuous level.

Due to space limitations we refer to Backi et al. (2018), where the majority of simulation parameters are presented. The only simulation parameters specific for this work are $r_{IO,1} = 90$ µm and $r_{IW,1} = 110$ µm as well as the tuning parameters for the interface layer controllers, $K_{P,IW} = 1.84$, $T_{I,IW} = 0.092$, $K_{P,IO} = 2.19$ and $T_{I,IO} = 0.11$.

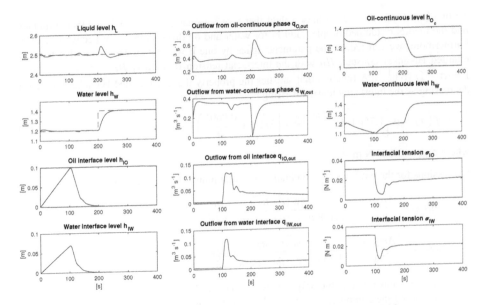

Figure 2: The different state variables with the manipulated variables

5. Conclusion

We presented a simple modeling approach in order to regulate the thickness of emulsion layers in gravity separators. The simplicity of the model makes it usable for easy controller design, as is demonstrated in the Results section. Although the approach describing two interface layers on top of each other does not strictly fulfill physical principles, the model showed itself useful. However, it must be noted that from a strict physical viewpoint, the water interface forming on top of the oil interface will not happen in reality. It rather is a way to describe the emulsion interface between the water- and oil-continuous phases in a systematic manner. We refer to Hartland and Jeelani (1987), who describe sedimentation and dense-packed zones, which inspired our approach.

6. Acknowledgement

The authors gratefully acknowledge the financial support provided by the Norwegian Research Council in the project SUBPRO (Subsea production and processing).

References

C. J. Backi, B. A. Grimes, S. Skogestad, 2018. A Control- and Estimation-Oriented Gravity Separator Model for Oil and Gas Applications Based upon First-Principles. Ind. Eng. Chem. Res. 57 (21), 7201–7217.

C. J. Backi, S. Skogestad, May 24–26 2017. A simple dynamic gravity separator model for separation efficiency evaluation incorporating level and pressure control. In: Proceedings of the American Control Conference. Seattle, USA, pp. 2823–2828.

B. A. Grimes, 2012. Population Balance Model for Batch Gravity Separation of Crude Oil and Water Emulsions. Part I: Model Formulation. J. Dispersion Sci. Technol. 33 (4), 578–590.

S. Hartland, S. A. K. Jeelani, 1987. Choice of model for predicting the dispersion height in liquid/liquid gravity settlers from batch settling data. Chem. Eng. Sci. 42 (8), 1927–1938.

S. Skogestad, 2003. Simple analytic rules for model reduction and PID controller tuning. J. Process Control 13, 291–309.

Anton A. Kiss, Edwin Zondervan, Richard Lakerveld, Leyla Özkan (Eds.)
Proceedings of the 29th European Symposium on Computer Aided Process Engineering
June 16th to 19th, 2019, Eindhoven, The Netherlands. © 2019 Elsevier B.V. All rights reserved.
http://dx.doi.org/10.1016/B978-0-128-18634-3.50195-8

Improving Waste Water Treatment Plant Operation by Ammonia Based Aeration and Return Activated Sludge Control

Melinda Várhelyi, Vasile Mircea Cristea*, Marius Brehar

Babeş-Bolyai University, Faculty of Chemistry and Chemical Engineering, 11 Arany Janos Street, Cluj-Napoca 400028, Romania

mcristea@chem.ubbcluj.ro

Abstract

Wastewater Treatment Plant (WWTP) is a very important part of every urban infrastructure, which has to achieve the desired effluent quality with cost-effective operation. Modelling and simulation help to find better operation scenarios and design alternatives, but also to investigate different control strategies. A dynamic WWTP simulator implemented in Matlab and Simulink was first calibrated with the configuration and process measured data from the local municipal WWTP. Then, the calibrated model was used for the investigation of different control strategies. This paper presents the investigation of five different ammonia based aeration control (ABAC) structures applying feedback and feedforward control. The latter is based on the ammonia influent disturbance. Five different return activated sludge (RAS) ratio control structures, coupled to the influent flow rate of wastewater, are also investigated in association to the nitrate recirculation (NR) control. According to the performed simulations, when applying ABAC with a feedforward component added to the setpoint of the dissolved oxygen (DO) controller, the air flow rate and the aeration energy can be reduced up to 45% and the effluent quality can be improved. Following the investigation of the RAS ratio-control, it was shown that flow rate of RAS and the pumping energy can be reduced up to 37% while the effluent quality remains high. This research demonstrated the benefits of applying ABAC and RAS control.

Keywords: wastewater treatment, modelling, simulation, ammonia based aeration control, return activated sludge control.

1. Introduction

Control of the Wastewater Treatment Plants (WWTPs) is a complex task due to the disturbances consisting in the wastewater influent concentrations and flow rates, the complicated behaviour of the microorganisms and the prescribed effluent quality requirements. WWTPs confront with the problem of high operational costs represented by the aeration and pumping energy, asking for control strategies to reduce them. Control strategies may be investigated by modelling and simulation, before their implementation in the WWTPs (Ostace et al., 2012).

Ammonia based aeration control (ABAC) is a cascade control in which the dissolved oxygen (DO) controller is combined with an ammonia controller (Rieger et al., 2014). The reason to apply ABAC is to reduce the aeration costs due to the implied aeration energy and to react efficiently to the ammonia disturbances appearing in the influent wastewater and at the end of the aerobic zone (Ozturk et al., 2016; Zhu et al., 2017).

Return activated sludge (RAS) flow rate is essential for the biological wastewater treatment because RAS has the role to return the activated sludge from the secondary settlers to the input of the biodegradation basins ensuring the necessary concentration of the biomass in the anaerobic and aerobic tanks (Olsson, 2012).

This paper focuses on two main control strategies: ammonia based aeration control in which the flow rate of air entering the aerated bioreactors is computed by ammonia controller in cascade with three dissolved oxygen controllers and the RAS control which is based on the flow rate of the wastewater entering the WWTP.

2. Development of the Dynamic WWTP Simulator

The WWTP model used in this study is an extension of Benchmark Simulation Model No. 1 (BSM1) (Copp, 2002) with a nonreactive primary clarifier based on Otterpohl model. The model was calibrated with the reactors and flows configuration of the investigated municipal WWTP and the measured data collected during May 2016 from this plant. The municipal WWTP under study has an anaerobic-anoxic-aerobic (A2O) configuration, according to which the nitrate (internal) recirculation from the end of the aerobic zone is entering between the anaerobic and anoxic zones of the basin. The dynamic WWTP Simulator was implemented in Matlab and Simulink. The equations of the primary clarifier, the bioreactors according to the Activated Sludge Model No. 1 (ASM1) and the nonreactive secondary settler based on Takács clarifier model were written in C programming language and introduced in the Simulink environment by S-function blocks. The model was calibrated in steady state using different optimisation methods and then tested by dynamic simulations. This calibrated model was used in the further presented studies.

3. Implemented Control Strategies

In the study of the ammonia based aeration control, the calibrated model was complemented with 3 Proportional-Integral (PI) feedback dissolved oxygen (DO) controllers in cascade configuration, as slave controllers, associated to an ammonia master controller. This control structure was augmented with the ratio based feedforward control of both the RAS and the NR, based on the WWTP leading influent flow rate. The configuration of this control strategy is presented in Figure 1. Ammonia concentration is measured at the aerated zone outlet and is used by the ammonia controller to keep the 1 g NH_3-N/m^3 setpoint value. The ammonia master controller computes the reference value for the first DO controller, which can be supplemented by an extra value based on the ammonia disturbances in the influent. This added value can be a constant, proportional or in linear relationship with the influent ammonia concentration. In this research 5 cases were tested as presented in Table 1: switching the added DO setpoint value from 0 to 0.5 g O_2/m^3 when the ammonia concentration in the influent exceeds 30 g NH_3-N/m^3 (case A), respectively 20 g NH_3-N/ m^3 (case E). Based on the 30 g NH_3-N/m^3 ammonia concentration limit, it was studied a proportional (case B) and a linear (case C) dependency of the supplement. It was also defined a linear function for the DO setpoint supplement, depending on the influent ammonia concentration, the coefficients of the linear equation being determined by optimisation. This latter approach was taking into consideration the quality of the effluent and the necessary air flow rate entering the aerobic zone (case D). The third DO controller computes the flow rate of the air entering equally into the 3 aerated reactors. The flow rate of the RAS is considered to be equal with the influent flow rate; the NR is controlled by a nitrate controller combined with a 0.7 factor of the wastewater flow rate. The simulations were performed for 66 days (three times the collected measured data of 22 days).

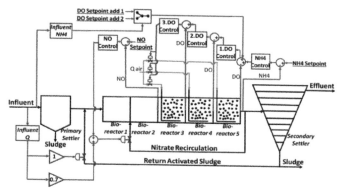

Figure 1. ABAC with supplementing the dissolved oxygen setpoint computed by the ammonium controller, control of RAS and control of NR based on the wastewater influents

Table 1. Investigated cases of ammonia based aeration control

CASE	DESCRIPTION	
ABAC Case A	If $C_{NH\_inf} \geq 30 gN/m^3$ then $y_{ref\_DO\_R5} = y_{c\_NHcontroller} + 0.5$ else $y_{ref\_DO\_R5} = y_{c\_NHcontroller}$	(1)
ABAC Case B	If $C_{NH\_inf} \geq 30 gN/m^3$ then $y_{ref\_DO\_R5} = y_{c\_NHcontroller} + 0.03 \cdot c_{NH\_inf}$ else $y_{ref\_DO\_R5} = y_{c\_NHcontroller}$	(2)
ABAC Case C	If $C_{NH\_inf} \geq 30 gN/m^3$ then $y_{ref\_DO\_R5} = y_{c\_NHcontroller} + (0.074 \cdot C_{NH\_inf} - 2.217)$ else $y_{ref\_DO\_R5} = y_{c\_NHcontroller}$	(3)
ABAC Case D	$y_{ref\_DO\_R5} = y_{c\_NHcontroller} + (0.0307 \cdot C_{NH\_inf} + 0.0101)$	(4)
ABAC Case E	If $C_{NH\_inf} \geq 20 gN/m^3$ then $y_{ref\_DO\_R5} = y_{c\_NHcontroller} + 0.5$ else $y_{ref\_DO\_R5} = y_{c\_NHcontroller}$	(5)

In the second part of this research, the DO setpoint supplement based on the ammonia influent concentration in the ABAC structure was eliminated; the configuration of the calibrated model with the investigated control structures is presented in Figure 2. Considering the RAS (Q_{RAS}) and NR (Q_{NR}) control, a study was carried out to find the optimal values of the factors k_{RAS} and k_{NR} for the associated flow rates, as described by Eq. (6) and (7). The investigated cases are presented in Table 2. The optimized values were found and simulations were performed the for 22 days period.

$$Q_{RAS} = k_{RAS} \cdot Q_{Influent} \tag{6}$$

$$Q_{NR} = k_{NR} \cdot Q_{Influent} + y_{c\_NOcontroller} \tag{7}$$

Table 2. The investigated cases with RAS control and basic ABAC structure

CASE	DESCRIPTION
RAS Case A	k_{RAS}=0.8323, k_{NR}=0.5509 optimized values for the whole period of time
RAS Case B	k_{RAS} and k_{NR} changing optimized values from day to day – I.
RAS Case C	k_{RAS} and k_{NR} changing optimized values from day to day – II.
RAS Case D	k_{RAS} changing optimized values from day to day, k_{NR}=0.7
RAS Case E	k_{RAS}=0.7761 optimized value for the whole period, k_{NR}=0.7

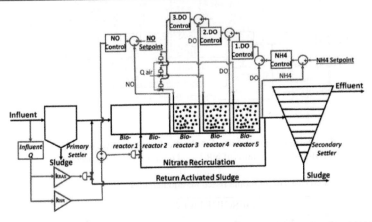

Figure 2. Return activated sludge and nitrate recirculation control associated to ABAC

4. Results and discussions

4.1. Ammonia Based Aeration Control with variable supplement of the DO setpoint

This control system design proposes the reduction of aeration energy, and consequently the operational costs at the aerated basins and the improvement of effluent quality by applying ABAC. To the ABAC structure found in literature (Rieger et al., 2014), i.e. to the control signal of the ammonia controller, it was added a supplement based on the influent ammonia disturbance. Table 3 shows the comparison between the average values obtained from simulations of the effluent concentrations and the RAS, NR and air flow rates, when the supplemented DO setpoint value was designed based on different principles. In all investigated cases it may be observed the reduction of nitrates and total nitrogen effluent concentration and the reduction of the air flow rate comparatively to the municipal WWTP.

Considering the average effluent concentrations and flow rate values, it can be stated that the most efficient case is ABAC – Case D. In Figure 3 and Figure 4 it is represented the comparison between the municipal WWTP and the simulated WWTP with ABAC – Case D.

Table 3. Comparison between the municipal WWTP measured data and the results of the simulated ABAC cases with supplemented DO setpoint

	Municipal WWTP	ABAC Case A	ABAC Case B	ABAC Case C	ABAC Case D	ABAC Case E
COD Effluent [g/m³]	19.36	22.18	22.14	22.18	22.05	22.17
NH Effluent [g N/m³]	0.17	1.00	1.00	1.00	0.99	1.00
NO Effluent [g N/m³]	3.76	0.60	0.60	0.59	0.57	0.61
N total Effluent [g N/m³]	5.70	3.60	3.61	3.60	3.57	3.62
Air flow rate [m³/day]	127,770	70,791	71,133	70,738	70,575	71,115
RAS flow rate [m³/day]	112,530	119,220	119,220	119,220	119,220	119,220
NR flow rate [m³/day]	138,350	136,390	133,260	136,380	139,340	134,180

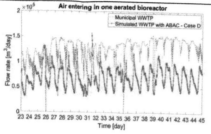

Figure 3. Comparison between the flow rate of air entering one bioreactor measured at the municipal WWTP and the simulated WWTP with ABAC – Case D

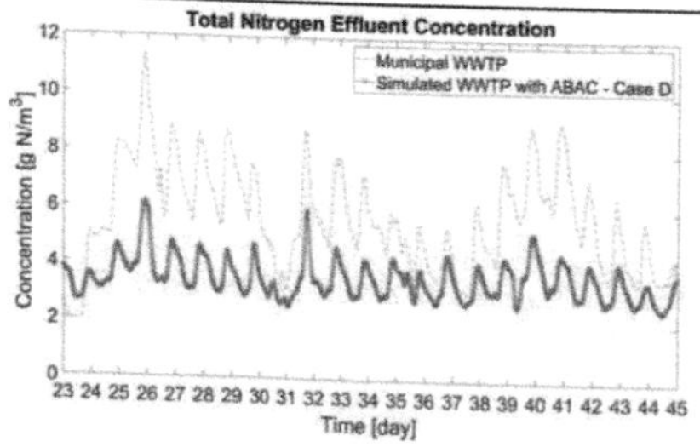

Figure 4. Comparison between the concentration of total nitrogen in effluent measured at the municipal WWTP and the simulated WWTP with ABAC – Case D

4.2. Return Activated Sludge Control

The flow rate of the RAS is dependent on the influent flow rate. An investigation, for the different RAS designs of Table 2, was performed in order to find the best factor value for RAS and, in some cases, for NR too. Table 4 presents the obtained results.

Table 4. Comparison between the municipal WWTP measured data and the results of the simulated WWTP with the different RAS cases

	Municipal WWTP	RAS Case A	RAS Case B	RAS Case C	RAS Case D	RAS Case E
COD Effluent [g/m3]	19.36	20.66	21.17	20.07	19.02	20.39
NH Effluent [g N/m3]	0.17	0.99	0.99	0.99	0.99	0.99
NO Effluent [g N/m3]	3.76	0.79	0.69	0.90	1.08	0.83
N total Effluent [g N/m3]	5.70	3.71	3.62	3.80	3.92	3.73
Air flow rate [m³/day]	127,770	71,178	71,417	70,739	70,317	71,114
RAS flow rate [m³/day]	112,530	99,231	118,860	86,601	71,336	92,534
NR flow rate [m³/day]	138,350	114,140	127,930	99,606	113,230	123,330

Considering the cases of the RAS flow rate designs, the case of changing the factor value for RAS every day and setting 0.7 for the k_{NR} (RAS – Case D) led to the flow rate of RAS reduced by 36.6%, associated to pumping energy and cost reduction. RAS – Case D also presents a reduction of the NR flow rate by 18.2%. Figure 5 and Figure 6 show the benefits of applying RAS – Case D. Figure 7 shows reduction of air flow rate entering one bioreactor and Figure 8 presents the effluent total nitrogen concentration.

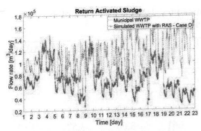

Figure 5. Comparison between RAS flow rate of measured at the municipal WWTP and simulated WWTP with RAS – Case D

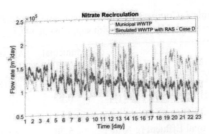

Figure 6. Comparison between the NR flow rate measured at the municipal WWTP and simulated WWTP with RAS – Case D

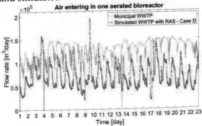

Figure 7. Comparison between the air entering one aerobic bioreactor measured at the municipal WWTP and simulated WWTP with RAS – Case D

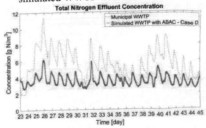

Figure 8. Comparison between total nitrogen effluent concentration measured at the municipal WWTP and simulated WWTP with RAS – Case D

5. Conclusions

New ammonia based aeration control designs associated to the return activated sludge control were investigated by simulation for WWTP control. In the ABAC structure based on the ammonia inlet concentration disturbance, the DO setpoint value handled by the ammonia controller was supplemented with a component calculated based on different principles. The RAS ratio control was coupled with the inlet wastewater flow rate and the implied factor values were computed by optimization. The investigated control strategies demonstrated benefits for WWTPs, reducing the aeration and pumping energies and improving the effluent quality.

References

J. B. Copp, 2002, The COST Simulation Benchmark: Description and Simulator Manual

G. Olsson, 2012, ICA and me – A Subjective Review, Water Research, 46, 6, 1585-1624

G. S. Ostace, V. M. Cristea, P. S. Agachi, 2012, Evaluation of different control strategies of the waste water treatment plant based on a modified activated sludge model no. 3, Environmental Engineering and Management Journal, 11, 1, 147-164

M. C. Ozturk, F. M. Serrat, F. Teymour, 2016, Optimization of aeration profiles in the activated sludge process, Chemical Engineering Science, 139, 1-14

L. Rieger, R. M. Jones, P. L. Dold, C. B. Bott, 2014, Ammonia-based feedforward and feedback aeration control in activated sludge prRiegercesses, Water Environment Research, 86, 1, 63-73

Z. Zhu, R. Wang, Y. Li, 2017, Evaluation of the control strategy for aeration energy reduction in a nutrient removing wastewater treatment plant based on the coupling of ASM1 to an aeration model, Biochemical Engineering Journal, 124, 44-53

Anton A. Kiss, Edwin Zondervan, Richard Lakerveld, Leyla Özkan (Eds.)
Proceedings of the 29th European Symposium on Computer Aided Process Engineering
June 16th to 19th, 2019, Eindhoven, The Netherlands. © 2019 Elsevier B.V. All rights reserved.
http://dx.doi.org/10.1016/B978-0-128-18634-3.50196-X

Combining the Advantages of Discrete- and Continuous-time MIP Scheduling Models

Hojae Lee[a] and Christos T. Maravelias[a]

[a]*Department of Chemical and Biological Engineering, University of Wisconsin-Madison, 1415 Engineering Dr., Madison, WI 53706, USA*
christos.maravelias@wisc.edu

Abstract

A solution algorithm that harnesses the strengths of discrete- and continuous-time scheduling models is proposed. It is a three stage algorithm consisting of (i) a discrete-time MIP scheduling model, (ii) a mapping algorithm, and (iii) a continuous-time LP model. It enables fast and accurate solution of scheduling problems, by quickly obtaining approximate solutions and improving their accuracy, while maintaining feasibility. Through a computational study, it is shown that the proposed method is capable of finding high quality solutions up to 4 orders of magnitude faster, compared to stand-alone discrete- and continuous-time models.

Keywords: chemical production scheduling, solution algorithm, continuous-time model, discrete-time model

1. Introduction

All existing grid-based mixed-integer programming (MIP) scheduling models can be classified into two categories based on the type of grids they employ: discrete- and continuous-time models. While continuous-time models provide more accurate solutions, discrete-time models are, in general, faster in large-scale instances and can accommodate various process features without additional computational cost (Floudas and Lin, 2004; Sundaramoorthy and Maravelias, 2011). In this work, we propose a solution algorithm, namely the Discrete-Continuous Algorithm, that combines the advantages of both discrete- and continuous-time scheduling models, while overcoming their limitations.

We consider a short-term chemical production scheduling problem in network environments. Given the facility data, resource availability and demand information, the objective is to find a schedule that satisfies the customer demand, while minimizing makespan. We use the state-task network (STN) representation throughout the paper (Kondili et al., 1993). We use set $\mathbf{I} \ni i$ to denote tasks, $\mathbf{J} \ni j$ to denote units, and $\mathbf{K} \ni k$ to denote materials. We also define the following subsets: $\mathbf{I}_j$ to denote tasks that can be executed in unit j, $\mathbf{I}_k^+/\mathbf{I}_k^-$ to denote tasks that produce/consume material k, and $\mathbf{J}_i$ to denote units that can process task i.

2. Discrete-continuous algorithm

The Discrete-Continuous Algorithm (DCA) consists of three components: (i) a discrete-time MIP scheduling model, (ii) a mapping algorithm, and (iii) a continuous-time LP model. In the first stage, an approximate solution is obtained with a relatively large discretization step length (δ), which gets mapped on to two new types of continuous-time grids in the second stage. Finally, in the third stage, some components of the first stage solution (e.g. timing of events, batch sizes, etc.) are recomputed to obtain a more accurate, and potentially better, solution. More details can be found in Lee and Maravelias (2018). Although not the scope of this work, we note that extending the algorithm to an iterative procedure may provide guaranteed bounds on the solution quality.

2.1. Stage 1: Discrete-time MIP scheduling model

While the algorithm can employ any discrete-time scheduling model, the STN model proposed by Shah et al. (1993) is used in this study. We divide the horizon into intervals of equal length, δ. The resulting time points are represented by $n \in \mathbf{N}$. We round-up the processing times (i.e. $\bar{\tau}_{ij} = \lceil (\tau_{ij}^F + \tau_{ij}^V \beta_j^{max})/\delta \rceil$), and round-down the horizon (i.e. $\bar{H} = \lfloor H/\delta \rfloor$) to ensure feasibility, where $\tau_{ij}^F / \tau_{ij}^V$ are the fixed/variable processing time of task i performed on unit j, and H is the horizon. The key decision variable for the first stage model is the *start of task* binary, X_{ijn}, which is equal to 1 if task i starts in unit j at time point n. The readers are encouraged to refer to Shah et al. (1993) for the full model. In the first stage, the model is solved with a chosen δ to obtain an approximate solution fast.

2.2. Stage 2: Mapping algorithm

In the second stage, we map the first stage solution onto two continuous-time grids that we introduce: (i) material-, and (ii) unit-specific grids. The batches in the first stage solution that produce/consume each material are mapped onto the corresponding material-specific grid, collecting the information regarding the sequencing of these batches. The batches that are executed in each unit are mapped onto the corresponding unit-specific grid, collecting information regarding the batch-unit assignment and the batch sequencing within each unit. By denoting the set of grid points as $\mathbf{M} \ni m$, each grid point on a given grid can be represented using an index pair (e.g. 2nd grid point on unit U1 grid is represented as (U1,$m2$)).

We introduce the following binary algorithmic parameters to store the information collected: $f_{ikm}^1 = 1$ if task i is mapped to grid point (k,m); $f_{jkm}^2 = 1$ if a task mapped to grid point (k,m) is performed in unit j; $f_{kmm'}^3 = 1$ if grid point where material k is produced (i.e. (k,m)) is before where it is consumed (i.e. (k,m')); $f_{ijkmm'}^4 = 1$ if task i is mapped to grid points (j,m) and (k,m'). The mapping algorithm is given in Table 1.

2.3. Stage 3: Continuous-time LP model

Based on the algorithmic parameters obtained in the previous stage, the timing of events, batch sizes and inventory levels are recomputed to obtain a more accurate solution. We define the following nonnegative continuous variables: batch size of the task mapped to grid point (k,m), B_{km}^C; inventory level of material k at grid point (k,m), S_{km}^C; time of grid point (j,m), T_{jm}^U; time of grid point (k,m), T_{km}^M.

Table 1: Mapping algorithm

0:	**Initialize:** $M_j^U = 0 \ \forall j,\ M_k^M = 0 \ \forall k,\ f^{1 \sim 4} = 0,\ \mathbf{M}_k^P = \mathbf{M}_k^C = \emptyset \ \forall k$
1:	**Loop** $n \in \mathbf{N}$
2:	**Loop** $k \in \mathbf{K}$
3:	**Loop** $i \in \mathbf{I}_k^+$
4:	**Loop** $j \in \mathbf{J}_i$
5:	**If** $X_{ij(n-\bar{\tau}_{ij})} = 1$
6:	$M_k^M \leftarrow M_k^M + 1;\ M_j^U \leftarrow \sum_{n' < n} \sum_{i' \in \mathbf{I}_j} X_{i' jn'}$
7:	$\mathbf{M}_k^P \leftarrow \mathbf{M}_k^P \cup \{M_k^M\}$
8:	$f^1_{ik,m(=M_k^M)} \leftarrow 1;\ f^2_{jk,m(=M_k^M)} \leftarrow 1;\ f^4_{ijk,m(=M_j^U),m'(=M_k^M)} \leftarrow 1$
9:	**EndIf**
10:	**EndLoop x2**
12:	**Loop** $i \in \mathbf{I}_k^-$
13:	**Loop** $j \in \mathbf{J}_i$
14:	**If** $X_{ijn} = 1$
15:	$M_k^M \leftarrow M_k^M + 1;\ M_j^U \leftarrow \sum_{n' \leq n} \sum_{i' \in \mathbf{I}_j} X_{i' jn'}$
16:	$\mathbf{M}_k^C \leftarrow \mathbf{M}_k^C \cup \{M_k^M\}$
17:	$f^1_{ik,m(=M_k^M)} \leftarrow 1;\ f^2_{jk,m(=M_k^M)} \leftarrow 1;\ f^4_{ijk,m(=M_j^U),m'(=M_k^M)} \leftarrow 1$
18:	$f^3_{kmm'} \leftarrow 1 \ \forall m < M_k^M,\ m \in \mathbf{M}_k^P,\ m' = M_k^M$
19:	**EndIf**
20:	**EndLoop x4**

We ensure the sequencing of batches on material- and unit-specific grids.

$$T_{km}^M \leq T_{km'}^M \quad \forall k, m, m' : f^3_{kmm'} = 1 \tag{1}$$

$$T_{jm}^U + (\tau_{ij}^F + \tau_{ij}^V B_{km'}^C) \leq T_{j(m+1)}^U \quad \forall i,j,k,m,m' : f^4_{ijkmm'} = 1 \tag{2}$$

Time matching constraints are enforced to ensure that the timing of grid points that originate from the same batch are the same.

$$T_{jm}^U = T_{km'}^M \quad \forall i,j,k,m,m' \in \mathbf{M}_k^C : f^4_{ijkmm'} = 1 \tag{3}$$

$$T_{jm}^U = T_{km'}^M - (\tau_{ij}^F + \tau_{ij}^V B_{km'}^C) \quad \forall i,j,k,m,m' \in \mathbf{M}_k^P : f^4_{ijkmm'} = 1 \tag{4}$$

We enforce unit capacity constraints, as well as batch size matching constraints to ensure that the batch sizes that correspond to the same batch are identical.

$$\beta_j^{min} \leq B_{km}^C \leq \beta_j^{max} \quad \forall j,k,m : f^2_{jkm} = 1 \tag{5}$$

$$B_{km'}^C = B_{k'm''}^C \quad \forall i,j,k,k',m,m',m'' : f^4_{ijkmm'} = f^4_{ijk'mm''} = 1,\ k < k' \tag{6}$$

The inventory level of material k at grid point m is equal to the sum of initial inventory, γ_k, and the amount produced/consumed up to that point. We make sure the demand for material k, ξ_k is satisfied.

$$S_{km}^C = \gamma_k + \sum_{m > m' \in \mathbf{M}_k^P} \sum_{i \in \mathbf{I}_k^+ : f^1_{ikm'} = 1} \rho_{ik} B_{km'}^C + \sum_{m \geq m' \in \mathbf{M}_k^C} \sum_{i \in \mathbf{I}_k^- : f^1_{ikm'} = 1} \rho_{ik} B_{km'}^C \quad \forall k, m \in \mathbf{M}_k^C \tag{7}$$

$$S_{k,M_k^M}^C \geq \xi_k \quad \forall k \tag{8}$$

We minimize makespan, MS:

$$minMS : MS \geq T_{km}^M \quad \forall k, m \in \mathbf{M}_k^P \tag{9}$$

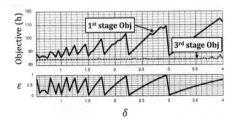 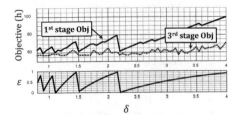

Figure 1: Using the error evaluation function, $\varepsilon(\delta)$, as a metric to determine δ's, illustrated for (a) instance 1, and (b) instance 2

3. Determining the key model parameter, δ

As mentioned earlier, DCA requires the user to select the discretization step length, δ, in the first stage model. If chosen carefully, the speed and accuracy of DCA can be improved significantly. However, selecting δ's that bring such results is non trivial.

In order to find the best performing δ, we introduce an error evaluation function denoted as $\varepsilon(\delta)$. For a given instance, $\varepsilon(\delta)$ measures the amount of discretization error introduced by the chosen δ, considering how much the shortage of each unit is constraining the objective value.

$$\varepsilon(\delta) = \frac{1}{\delta} \sum_i \sum_{j \in \mathbf{J}_i} \omega_{ij} |\tau_{ij} - \delta \bar{\tau}_{ij}| , \quad \omega_{ij} = \left(\frac{N_{ij}}{\sum_i N_{ij}}\right)\left(\frac{D_j}{\sum_j D_j}\right) \tag{10}$$

$$D_j = \sum_n \lambda_{jn}^* \ \forall j \tag{11}$$

$$N_{ij} = \sum_n X_{ijn}^* \ \forall i, j \in \mathbf{J}_i \tag{12}$$

where, X_{ijn}^* is the solution of the LP relaxed model of the first stage, and λ_{jn}^* is the dual variable associated with the allocation constraints (Eq. 9 in Shah et al. (1993)) in the LP relaxed model.

As shown in Figure 1, $\varepsilon(\delta)$ is an accurate metric to quantify the amount of discretization error introduced. Specifically, DCA obtains high quality first and third stage solutions, whenever δ's that lead to low values of $\varepsilon(\delta)$ are used. Although only shown for two instances, we observed similar results for all other instances tested.

4. Generating multiple solutions

In industrial applications, it is desired to obtain multiple solutions rather than just one. By having multiple options, the operators can choose a schedule that minimizes the potential risks that may arise from the discrepancy between the problem data and the actual plant operation conditions.

DCA offers a systematic way of obtaining multiple solutions with varying degrees of accuracy in problem data approximation. This can be achieved by running DCA with different δ's that are identified by our method (see Figure 1). More importantly, depending on the δ chosen, the users have the flexibility to decide which problem data (e.g. processing time of a specific task) to approximate more accurately.

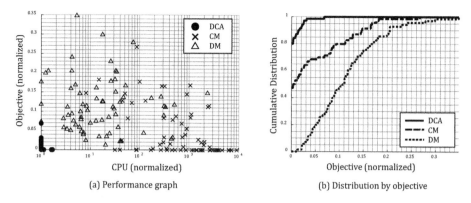

(a) Performance graph   (b) Distribution by objective

Figure 2: Performance comparison between the three approaches, DCA, DM and CM: (a) performance graph, and (b) distribution by objective value

5. Computational study

5.1. Performance test

We compare the performance of DCA with systematically determined δ against a discrete-time model (Shah et al., 1993) and one of the best performing continuous-time model (Sundaramoorthy and Karimi, 2005), denoted as DM and CM, respectively. We consider 7 networks from the literature with 10 variations each (total of 70 instances).

Figure 2a shows the performance of the three approaches. The abscissa of the graph represents the CPU time normalized with respect to the fastest approach, while the ordinate represents the objective value normalized with respect to the best. In terms of computational speed, DCA outperforms the other two approaches by up to 4 orders of magnitude. Figure 2b shows the cumulative distribution of instances as a function of normalized objective value. DCA finds the best solutions in 80% of the instances, and high quality solutions (i.e. 0~3% inferior than best found) in 98% of the instances. In general, DCA yields substantially better and often rather different schedules compared to the other approaches.

5.2. Case study

A network from (Papageorgiou and Pantelides, 1996) that consists of 19 tasks performed on 8 units producing 4 products is considered (Figure 3a). The objective is to minimize makespan while satisfying a given demand until the end of 120 h horizon. The example is solved using three different approaches, and the results are given in Table 2. DCA with δ =1.85 h finds the best solution of 56.4 h in less than 4 seconds, being the fastest approach (see Figure 3b).

Table 2: Computational results

	DM	CM	DCA
δ or #Slots	1.00	25	1.85
Constraints	8,550	5,732	4,598
Disc. Var.	2,156	742	1,166
Cont. Var.	4,782	4,829	2,577
LP Relax.	5.9	43.2	10.3
Obj (h)	62	57	59.2
(Refined)			56.4
CPU (s)	35.1	18,000.1	3.8
Nodes	1,872	>10^5	2,291
Gap (%)	-	13.5	-

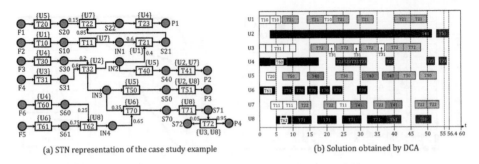

(a) STN representation of the case study example   (b) Solution obtained by DCA

Figure 3: Case study: (a) STN representation, and (b) solution obtained by DCA

6. Conclusion

The two types of grids adopted in grid-based mathematical programming models for chemical production scheduling exhibit distinctive advantages. Discrete-time models are generally faster and have higher modeling flexibility, while continuous-time models provide more accurate solutions.

In this work, we proposed a new solution algorithm, namely the Discrete-Continuous Algorithm, that enable us to exploit the advantages of both time grids. DCA is capable of obtaining accurate solutions fast, while also allowing us to model different process characteristics, such as intermediate shipments and deliveries, limited shared utilities, etc. In addition, we proposed a systematic method to choose the discretization step length, δ, used in the first stage model, which has a significant impact on the performance of DCA. Furthermore, DCA offers a systematic way to generate multiple alternative solutions with varying degrees of accuracy in problem data approximation.

Through computational study, it was shown that DCA is capable of finding high quality solutions much faster than stand-alone discrete- and continuous-time models. Specifically, we showed that it finds the best solution in 80% of test instances while being up to 4 orders of magnitude faster.

References

C. A. Floudas, X. X. Lin, 2004. Continuous-time versus discrete-time approaches for scheduling of chemical processes: a review. Computers & Chemical Engineering 28 (11), 2109–2129.

E. Kondili, C. C. Pantelides, R. W. H. Sargent, 1993. A general algorithm for short-term scheduling of batch-operations – I. MILP formulation. Computers & Chemical Engineering 17 (2), 211–227.

H. Lee, C. T. Maravelias, 2018. Combining the advantages of discrete- and continuous-time scheduling models: Part 1. framework and mathematical formulations. Computers & Chemical Engineering 116, 176 – 190.

L. G. Papageorgiou, C. C. Pantelides, 1996. Optimal campaign planning scheduling of multipurpose batch semicontinuous plants .2. A mathematical decomposition approach. Industrial & Engineering Chemistry Research 35 (2), 510–529.

N. Shah, C. C. Pantelides, R. W. H. Sargent, 1993. A general algorithm for short-term scheduling of batch-operations – II. computational issues. Computers & Chemical Engineering 17 (2), 229–244.

A. Sundaramoorthy, I. A. Karimi, 2005. A simpler better slot-based continuous-time formulation for short-term scheduling in multipurpose batch plants. Chemical Engineering Science 60 (10), 2679–2702.

A. Sundaramoorthy, C. T. Maravelias, 2011. Computational study of network-based mixed-integer programming approaches for chemical production scheduling. Industrial & Engineering Chemistry Research 50 (9), 5023–5040.

Anton A. Kiss, Edwin Zondervan, Richard Lakerveld, Leyla Özkan (Eds.)
Proceedings of the 29th European Symposium on Computer Aided Process Engineering
June 16th to 19th, 2019, Eindhoven, The Netherlands. © 2019 Elsevier B.V. All rights reserved.
http://dx.doi.org/10.1016/B978-0-128-18634-3.50197-1

Decision-making of online rescheduling procedures using neuroevolution of augmenting topologies

Teemu J. Ikonen[a] and Iiro Harjunkoski[a,b,*]

[a]*Aalto University, School of Chemical Engineering, Kemistintie 1, 02150 Espoo, Finland*
[b]*ABB Corporate Research, Wallstadter Str. 59, 68526 Ladenburg, Germany*
*iiro.harjunkoski@aalto.fi

Abstract

Online scheduling requires appropriate timing of rescheduling procedures, as well as the determination of relevant horizon length. Optimal choices of these quantities are highly dependent on the uncertainty of the scheduling environment and may vary over time. We propose an approach where a neural network is trained to make online decisions on these quantities, as well as on the choice of the rescheduling method (mathematical programming or metaheuristics). In our approach, the neural network is trained using neuroevolution of augmenting topologies (NEAT) in a simulated environment. In this paper, we also optimize the rescheduling interval and horizon length of a conventional periodically occurring rescheduling on a dynamic routing problem. The resulting approach is the baseline for the development of the proposed neural network approach.

Keywords: online scheduling, horizon length, scheduling interval, neural network, NEAT

1. Introduction

Online scheduling of industrial processes is a demanding task, mainly due to a large number of solution candidates and uncertainty in scheduling parameters. Considering batch processes, the most common optimization approach in offline scheduling is mixed-integer programming (MIP), using modeling formalisms, e.g. resource-task network (Pantelides, 1994), state-task network (Kondili et al., 1993) or generalized disjunctive programming (Raman and Grossmann, 1994). Given a reasonably large amount of computational time, these approaches can often find the optimal solution. However, in online scheduling, the duration of the optimization process is constrained due to the moving time horizon, which significantly compromises the optimality of the solution. Another aspect to consider is that the later the optimization process is initiated, the more accurate are its scheduling parameters. This raises questions, such as what is the optimal rescheduling interval and what is a relevant length for the scheduling horizon. For a recent review of online scheduling, the reader may wish to consult the paper by Gupta et al. (2016).

We propose an approach where an artificial neural network trained to decide when to initiate a rescheduling process, what length to define for the time horizon and which optimization strategy (mathematical programming or metaheuristics) to use. In many previous studies, neural networks have been used to learn dispatching rules from the historical data of scheduling decisions (Priore et al., 2014). However, this approach can typically only mimic the historical decisions, and therefore the optimality of the scheduling solutions is highly dependent on the quality of the given historical data. Our neural network approach operates at a higher level; it exploits the strength of MIP at finding the optimal, or near-optimal solution, and focuses on the allocation of the computational resources in the changing scheduling environment.

More specifically, we propose to use neuroevolution of augmenting topologies (NEAT), introduced by Stanley and Miikkulainen (2002), to train the neural network. NEAT is a genetic algorithm that simultaneously evolves the topology and weighting parameters of the neural network. Such et al. (2017) report the performance of the evolutionary neuroevolution algorithms to compare well against the gradient-based backpropagation algorithms. A key feature of NEAT is that the evolution process is initiated from very simple neural networks, the topology of which is then incrementally increased complexity during the evolution. The feature reduces unnecessary complexity of the final neural network, and is not possible using gradient-based algorithms. Recently, Hausknecht et al. (2014) applied NEAT to train a neural network to play 61 different Atari 2600 games, where the controls for the player are to move a joystick and press a button. The controls for our neural network are also of a similar low level of complexity (see the second paragraph), but the critical aspect is their timely execution.

In this paper, we first define a rescheduling problem (Section 2), suitable to be used as a test case for our neural network-based rescheduling approach. For the sake of simplicity, we define the underlying optimization problem to be a routing problem. However, our approach is also applicable to rescheduling of other industrial processes, e.g. batch processes. In Section 3, we tune the horizon length and rescheduling interval of the conventional periodically occurring rescheduling approach. This approach is a benchmark for our neural network-based approach. Finally, in Section, 4, we describe the intended input and output signals for the neural network.

2. Dynamic routing problem

Let us consider a square region, having dimensions 1000×1000 m, and a vehicle traveling in the region at a constant speed of $v = 10$ m/s. The purpose of the vehicle is to visit n sites in the region before site-specific due dates. Each site has an order date, at which the vehicle receives the location and due date of the site. The objective of the optimization problem is to minimize the delay sum of visiting the n sites. Since scheduling decision must be made with limited information, finding the optimal solution to the problem requires rescheduling.

As an optimizer, we here use ant colony optimization (ACO) (Dorigo and Gambardella, 1997), which is a probabilistic metaheuristic search method of finding good paths in a graph. In the method, ants communicate by laying pheromone on paths between nodes. The laid pheromone concentration of an ant is proportional to the objective function value of the route the ant traveled. When a new ant is at node i, it chooses the path to node j with probability

$$p_{ij} = \frac{\phi_{ij}^{\alpha} d_{ij}^{\beta}}{\sum_{i,j=1}^{n} \phi_{ij}^{\alpha} d_{ij}^{\beta}}, \tag{1}$$

where ϕ_{ij} and d_{ij} are the pheromone level and desirability[1] of the path (i, j). where α and β are influence parameters[2]. We use a population size of 200. We have implement ACO using the Python module ACOpy (Grant, 2018).

In this and the next section, we study periodic rescheduling with fixed interval and horizon length. At $t = 0$, the vehicle starts to follow an initial path, determined by a greedy algorithm (which is defined to always choose the node lying closest to the current node). The path is then updated when the first rescheduling procedure is finished. The computational budget of all the rescheduling procedures during the time span is restricted to 50 CPU seconds. The computational budget is distributed evenly between the rescheduling processes. Therefore, a frequent rescheduling will have a small computation budget per rescheduling process, and vice versa. Each rescheduling procedure is associated with times t_{info} and t_{exe}, the difference of which is the duration of the procedure. The rescheduling is conducted using the information available at t_{info}, whereas t_{exe} is the planned start time of the new schedule.

2.1. A simple numerical example

Let us next present a solution to a simple routing problem of nine sites. The sites are randomly distributed inside the region (Figure 1(a)), and their due dates are randomly selected from the time interval of [0, 500s] (Table 1). Further, the order dates of sites 1 to 7 are set to 0 s, while the order dates of sites 8 and 9 are randomly selected from the time interval of [0, t_{due}], where t_{due} is the due date of the site (Table 1).

Table 1: Order and due dates for the sites in the simple numerical example.

site [-]	order date [s]	due date [s]	site [-]	order date [s]	due date [s]
1	0	46	6	0	400
2	0	198	7	0	346
3	0	343	8	4	43
4	0	14	9	185	439
5	0	279			

In this numerical example, we choose to use a scheduling interval of 50 s and horizon length of 500 s (i.e. the entire time span of the problem). Figure 1(a) shows the initial route (determined by a greedy search) for the vehicle. Further, Figures 1(b)-1(d) visualize intermediate states of the process. At $t_{exe} = 5$ s, the optimizer schedules the visit to the most critical site 1 before visiting sites 6 and 2. At $t_{exe} = 55$ s and $t_{exe} = 205$ s, the optimizer includes visits to sites 8 and 9, respectively, in the schedule. The locations and due dates of these sites were not known at the start of the time span. Figure 1(e) shows the final (realized) route of the vehicle. The delay sum of the final route is 235.30 s, which is caused by the vehicle failing to meet the due dates at sites 1, 4 and 8 (see Figure 1(f)).

3. Tuning of the rescheduling interval and horizon length

The purpose of the numerical example in the previous section was to provide an introduction to the dynamic routing problem, which we intend to study with the neural network

---

[1] In this work, we use the distance of a path as its desirability measure.
[2] In this work, we use values $\alpha = 1$ and $\beta = 3$.

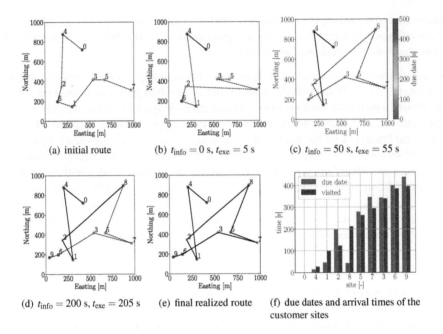

(a) initial route   (b) $t_{\text{info}} = 0$ s, $t_{\text{exe}} = 5$ s   (c) $t_{\text{info}} = 50$ s, $t_{\text{exe}} = 55$ s

(d) $t_{\text{info}} = 200$ s, $t_{\text{exe}} = 205$ s   (e) final realized route   (f) due dates and arrival times of the customer sites

Figure 1: Representative states and the final results of the rescheduling procedure with nine sites to be visited. The locations and due dates of sites 8 and 9 are received at $t = 4$ s and 185 s, respectively, whereas the other orders are already known at $t = 0$. Continuous path represents realized route of the vehicle, and dashed line the scheduled route.

based approach. In this section, we examine a similar, but larger, routing problem with a total of 50 sites, and a time span of 1000 s. The locations and due dates of 40 sites are known at $t = 0$, whereas for the remaining 10 sites the information is obtained again at a random time point in the interval of $[0, t_{\text{due}}]$, where t_{due} is the due date of the site.

When using the periodically occurring rescheduling procedures, the optimal values for the rescheduling interval and horizon length are problem-dependent. Thus, in this section, we also use a grid search to systematically seek an optimized combination of the two parameters. The results of the grid search are shown in Figure 3. Out of the tested parameter combinations, the best final objective value (i.e. the delay sum) is obtained by the scheduling interval of 120 s and the horizon length of 500 s. We omit those parameter combinations, in which the scheduling interval is larger than the scheduling horizon (see the bottom right corner of Figure 3), as the vehicle would need to wait at certain sites for the schedule to be updated. Based on the results, the performance of the rescheduling procedure is poor for combinations lying close to this constraint, but also for combinations with very frequent rescheduling and long horizon length. In the former, the algorithm does not exploit enough information it has about the future, where as in the latter the algorithm does not have enough computational time to find good solutions (to relatively large scheduling problems).

Finally, let us examine the rescheduling procedure with the optimized rescheduling interval and horizon length. Figures 2 and 4 visualize the initial route, the intermediate state at

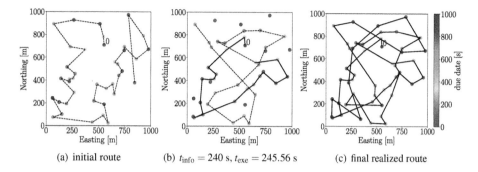

(a) initial route   (b) $t_{\text{info}} = 240$ s, $t_{\text{exe}} = 245.56$ s   (c) final realized route

Figure 2: Representative states and the final results of the rescheduling procedure with 50 sites.

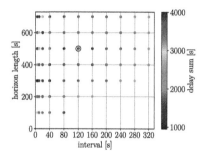

Figure 3: A grid search of best-performing rescheduling interval and horizon length. The best parameter combination is highlighted by a circle. Darkest red points exceed the scale.

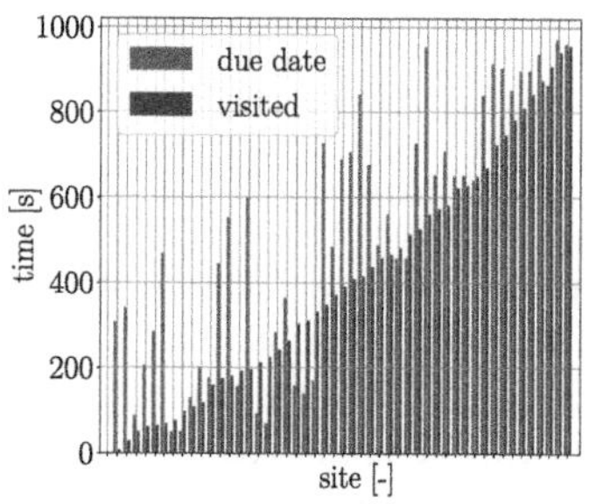

Figure 4: Due dates and arrival times of the final route (Figure 2(c)).

$t_{\text{exe}} = 245.56$ s and the final realized route of the rescheduling procedure. The delay sum of the final route is 992.89 s. Despite being a solution to a larger scale problem, the same features are also present here as in the simple example (Section 2.1): sites with urgent due dates are prioritized and, as the time proceeds, new orders are included in the schedule.

4. Proposed input and output signals

The periodically occurring rescheduling with tuned interval and horizon length seems to work reasonably well for these relatively small-scale scheduling problems. Another possible approach would be to trigger the rescheduling always when new information is obtained. However, the rescheduling decisions made by these approaches may be compromised, especially in larger problems, or if new information is obtained more frequently.

Thus, we propose the scheduling decisions to be made using a neural network that is trained by the NEAT algorithm. Figure 5 shows the input nodes we propose for the neural network. The input signals represent changes in the scheduling environment, the status of the ongoing rescheduling procedure, and the remaining computational resources.

At each evaluation, the neural network decides to either run a rescheduling procedure, using metaheuristics or mathematical programming, or to do nothing (depending which of the three top output nodes receives the highest signal). In addition, the neural network includes signals for the horizon length and allocated computational time for the possible rescheduling procedure.

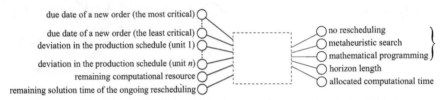

Figure 5: Proposed input and output signals for the neural network. Here, the neural network is depicted as a box as in NEAT its topology is decided during the training procedure.

5. Conclusions

We propose the rescheduling decision in online scheduling to be made by a neural network, trained using the NEAT algorithm. As a first step, this paper presents a dynamic routing problem suitable to be used as a demonstration and benchmarking problem for our approach. We optimized the rescheduling interval and horizon length of the conventional periodically occurring rescheduling procedure, in order to obtain a relevant baseline approach. The future work involves training and evaluation of the NEAT algorithm for the presented routing problem.

Acknowledgment: Financial support is gratefully acknowledged from the Academy of Finland project "SINGPRO", Decision No. 313466.

References

M. Dorigo, L. M. Gambardella, 1997. Ant colony system: a cooperative learning approach to the traveling salesman problem. IEEE Transactions on evolutionary computation 1 (1), 53–66.

R. Grant, 2018. Acopy [accessed on the 15th of october, 2018]. https://github.com/rhgrant10/acopy.

D. Gupta, C. T. Maravelias, J. M. Wassick, 2016. From rescheduling to online scheduling. Chemical Engineering Research and Design 116, 83–97.

M. Hausknecht, J. Lehman, R. Miikkulainen, P. Stone, 2014. A neuroevolution approach to general atari game playing. IEEE Transactions on Computational Intelligence and AI in Games 6 (4), 355–366.

E. Kondili, C. C. Pantelides, R. W. H. Sargent, 1993. A general algorithm for short-term scheduling of batch operations–I. MILP formulation. Computers & Chemical Engineering 17 (2), 211–227.

C. C. Pantelides, 1994. Unified frameworks for optimal process planning and scheduling. In: Proceedings on the second conference on foundations of computer aided operations. Cache Publications New York, pp. 253–274.

P. Priore, A. Gómez, R. Pino, R. Rosillo, 2014. Dynamic scheduling of manufacturing systems using machine learning: An updated review. AI EDAM 28 (1), 83–97.

R. Raman, I. E. Grossmann, 1994. Modelling and computational techniques for logic based integer programming. Computers & Chemical Engineering 18 (7), 563–578.

K. O. Stanley, R. Miikkulainen, 2002. Evolving neural networks through augmenting topologies. Evolutionary computation 10 (2), 99–127.

F. P. Such, V. Madhavan, E. Conti, J. Lehman, K. O. Stanley, J. Clune, 2017. Deep neuroevolution: genetic algorithms are a competitive alternative for training deep neural networks for reinforcement learning. arXiv preprint arXiv:1712.06567.

Anton A. Kiss, Edwin Zondervan, Richard Lakerveld, Leyla Özkan (Eds.)
Proceedings of the 29th European Symposium on Computer Aided Process Engineering
June 16th to 19th, 2019, Eindhoven, The Netherlands. © 2019 Elsevier B.V. All rights reserved.
http://dx.doi.org/10.1016/B978-0-128-18634-3.50198-3

Optimal short-term Scheduling of Industrial Packing Facilities

Apostolos P. Elekidis [a,b], Francesc Corominas [c], Michael C. Georgiadis [a,b,*]

[a] *Department of Chemical Engineering, Aristotle University of Thessaloniki, University Campus, Thessaloniki, 54124, Greece*

[b] *Chemical Process and Energy Resource Institute (CPERI), Centre for Research and Technology Hellas (CERTH), PO Box 60361, 57001, Thessaloniki, Greece*

[c] *Procter & Gamble, Temselaan 100, 1853 Strombeek-Bever, Belgium*

mgeorg@auth.gr.

Abstract

The simultaneous lot-sizing and production scheduling problem of a real-life large-scale industrial facility of packaged consumer goods is considered in this work. The problem under consideration is mainly focused on the packing stage which constitutes the major production bottleneck. Several packing lines, illustrating different design and operational characteristics, operate continuously in parallel. An efficient solution strategy is implemented to reduce the high computational cost. A Mixed-Integer Linear Programming model (MILP) is used in parallel with a decomposition-based algorithm. Appropriate constraints, referring to the production/formulation stage of the plant, are also imposed in order to ensure the generation of feasible production schedules. The model relies on tight timing and sequencing and products allocation constraints. The main objective is the minimization of the makespan. A number of different case studies have been considered and detailed optimal production schedules have been generated for over 130 products scheduled weekly. The obtained results lead to nearly-optimal scheduling solutions in reasonable computation times. The proposed model can assist decision makers towards rigorous scheduling plans in a dynamic production environment under realistic uncertainty.

Keywords: production scheduling, consumer goods industry, mixed integer linear programming, decomposition technique

1. Introduction

Modern multi-product, multipurpose plants play a key role within the overall current climate of business globalization, aiming to produce highly diversified products that can address the needs and demands of customers spread over wide geographical areas. The inherent size and diversity of these processes gives rise to the need for planning and scheduling problem, of large-scale industrial production operations. Over the past 20 years the literature illustrates a large-number of scheduling models, which have been mostly applied on generic but relatively small problem instances (Méndez et al., 2006). However, current real-world industrial applications include hundreds of different final products, produced in flexible multi-purposes facilities, under several tight design and operating constraints (Harjunkoski et al., 2014). A few approaches have been used to solve large-scale industrial scheduling problems, utilizing advances in Mixed-Integer optimisation (Kopanos, Puigjaner and Georgiadis, 2011). Furthermore, hybrid methods for large scale industrial problems have been proposed. Kopanos et al. (2010), proposed a decomposition strategy for large-scale scheduling problems in multiproduct multistage pharmaceutical batch plants. Baumann and Trautmann (2014) proposed a hybrid method for large-scale short-term scheduling of make-and-pack production processes. In this work an MILP-based decomposition algorithm is proposed for large scale scheduling problems in a multiproduct continuous plant. The model focuses on the packing stage, taking also into account constraints referring to the production formulation stage in order to ensure the generation of feasible production schedules.

2. Problem Statement

This work considers the scheduling of packaged consumer goods in a real-life industry. More than 300 products can be produced continuously in parallel packing lines. The production process consists of the formulation/production and the packing stage. In the formulation stage a number of intermediate products are produced. In most cases, more than one final product can be produced from the same intermediate product in the packing stage. Each packing line is connected to its own production/formulation unit. Sequence dependent changeovers take place in both stages. The changeover times differ among the various sequences, depending on the package size, the package color, the intermediate product etc. All changeovers, in the two stages, take place simultaneously and therefore the most time-consuming changeover determines the total changeover time for a product sequence. In addition, in the formulation stage, due to technical plant restrictions, the total number of intermediate products' changeovers should not exceed an upper limit. The short-term scheduling horizon of interest is one week, and both the packing and the formulation units are available 24 hours per day. Products' due dates are taken into account along with the necessary planned maintenance activities. The main objective is the minimization of the makespan and the minimization of products' changeover times.

3. MILP-based Decomposition Algorithm

For the aforementioned problem, an MILP-based decomposition algorithm has been developed. The original large-scale problem is decomposed into smaller scheduling sub-problems in an iterative mode. At each iteration, a subset of the involved product orders is scheduled. The main model decisions are a) the allocation of products to the packing lines, taking into account underlying equipment technical constraints, b) the relative sequence of products in the packing lines and c) the starting and completion time of product orders in the packing stage. The proposed MILP-based decomposition algorithm consists of: a) the insertion policy, b) the MILP model and c) a decomposition technique. If maintenance periods are to be taken into account, extra "maintenance-product" orders are scheduled, with processing time equal to the maintenance time. The ending times of these product orders are fixed to the ending times of the associated maintenance activities.

3.1. Insertion policy

In order to solve the original problem iteratively, the number and sequencing of the inserted products should be decided. Two insertion criteria are used in order to avoid the generation of infeasible schedules. The products with the earliest due dates are inserted first. Furthermore, according to the plant's technical restrictions, the products with the same recipe (intermediate product) are inserted first, in order to minimize the total number of intermediate products' changeovers. Several real test cases have illustrated, that a 5-by-5 product insertion policy is the optimal one, as by increasing the number of products the solution is not improved while the computational cost is increased. In case that maintenance periods have to be taken into account these "maintenance-product" orders are inserted first.

3.2. MILP model

The MILP model applied in this case study is an extension of a general-immediate precedence framework as developed by Kopanos, Mendez and Puigjaner (2010). A brief description of the model is presented below:

$$\sum_{j\in i_j} Y_{i,j} = 1 \quad \forall i \tag{1}$$

$$X_{i',i,j} + X_{i,i',j} + 1 \geq Y_{i',j} + Y_{i,j} \qquad \forall i, i' > i, j \in \left(i_j \cap i'_j\right) \tag{2}$$

$$C_{i'} \geq C_i + T_{i'} + XX_{i,i',j} changeover_{i,i'} - M\left(1 - X_{i,i',j}\right) \tag{3}$$

$$\forall i, i' \neq i, j \in \left(i_j \cap i'_j\right)$$

$$2(X_{i',i,j} + X_{i,i',j}) \leq Y_{i',j} + Y_{i,j} \qquad \forall i, i' > i, j \in \left(i_j \cap i'_j\right) \tag{4}$$

$$Z_{i,i',j} + XX_{i,i',j} \geq X_{i,i',j} \quad \forall i, i', j \in \left(i_j \cap i'_j\right) \tag{5}$$

$$Z_{i,i',j} = \sum_{i'' \neq i,i', j\in(i_j\cap i'_j)} \left(X_{i,i'',j} + X_{i',i'',j}\right) + M\left(1 - XX_{i,i',j}\right) \tag{6}$$

$$\forall i, i', j \in \left(i_j \cap i'_j\right)$$

$$\sum_{i} \sum_{\substack{i' \neq i, \\ formula_i \neq formula_{i'}}} \left(XX_{i',i,j}\right) \leq Limit_j \qquad \forall j \tag{7}$$

$$\min C_{max} \geq C_i \qquad \forall i, j \tag{8}$$

Constraint (1) forces that every product order i goes through one packing line j via the allocation binary variable $Y_{i,j}$. Constraint (2), (3) and (4) give the relative sequencing of product orders. The big-M constraint (3) determines the completion time $C_{i'}$ of a product order i' to be greater that the completion time and the processing time $T_{i'}$ of whichever product i is produced beforehand at the same unit, and greater than the changeover time, *changeover*$_{i,i'}$, only if the binary variable $X_{i,i',j}$, is active. The binary variable $X_{i,i',j}$ is active only if product i' is produced after product i. The constraints (2) and (4) state that when two products are produced at the same unit, only one global sequencing binary variable has to be active and when one of the binary variable $X_{i',i,j}$s and $X_{i,i',j}$ is active at least one of the $Y_{i,j}$ and $Y_{i',j}$ has to be active as well. The variable $Z_{i,i',j}$ signifies the position difference among two products produced in the same packing line. When $Z_{i,i',j}$ is equal to 0, the product i is produced exactly before the i'. The variable $Z_{i,i',j}$ is calculated in equation (6). As a result, from equation (5) the immediate precedence binary variable $XX_{i,i',j}$ takes the value 1 only when the variable $Z_{i,i',j}$ is equal to 0. The binary variable $XX_{i,i',j}$ takes the value 1 when the product i' is produced exactly after the product i. The constraint (7) ensures that the number of sequences between products with different recipe ($formula_i$) does not exceed an upper limit ($Limit_j$) which is determined by the plants' technical restrictions. In order this constraint to be included in the model both the immediate and the general precedence binary variables are used. Finally, the objective function of the model is expressed by constraints (8), which is the minimization of the total production makespan, C_{max}, hence it also considers the minimization of changeovers and unnecessary idle times.

3.3. Decomposition Algorithm

The large-scale industrial scheduling problem is decomposed into smaller problems in an iterative fashion. In each iteration a subset of the product orders is scheduled. In this way the MILP subproblems are solved much easier while the computational time is significantly decreased. A number of inserted products are scheduled in each iteration until all product orders are scheduled. After the resolution of the MILP model at each iteration, the global sequencing variables, as well as the allocation variables of the inserted products are fixed. However, the timing variables (ending time) and the immediate precedence binary variables remain free. When all products are inserted, the final schedule is constructed. Each iteration, aims to find a solution with a 0% integrality gap. However, industrial requirements impose an upper bound on the total computational time. Therefore, a time limit of 10 minutes, has been set for the solution of each subproblem. Figure 1 illustrates the proposed decomposition technique.

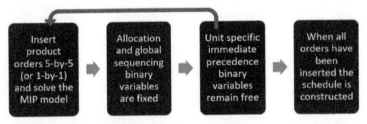

Figure 1: Decomposition-based solution algorithm

4. Results

A representative industrial case study of 178 final products and 6 packing lines is used to illustrate the applicability and efficiency of the proposed solution strategy. All data have been provided from a real-life, large-scale consumer goods industry. Technical packing lines restrictions do not allow full flexibility of the products' allocation. The first 80 products can be solely allocated to the first 3 packing lines, as the remaining 98 products can be processed only in the next 3 packing lines. Utilizing the aforementioned decomposition algorithm 5 products are scheduled at each iteration, except for the last one, where the remaining 3 products are scheduled. The decomposition technique, was implemented in GAMS and solved using the CPLEX 12.0 solver.. Figure 2. illustrates the final schedule, of the study under consideration. Maintenance activities have also been taken into account.

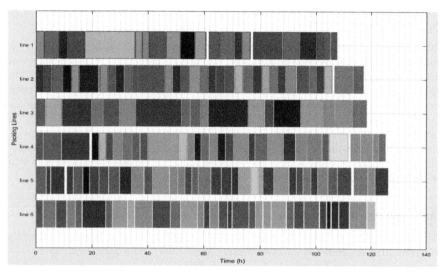

Figure 2: Gantt chart of the packing lines referring to the examined case study

Table 1: Comparison between the exact MILP model and the decomposition approach applying different insertion policies

	Exact MILP Approach	Decomposition Approach		
		Insertion policy 1-by-1	Insertion policy 5-by-5	Insertion policy 10-by-10
Makespan (hrs)	118.9	122.6	120.2	120.8
Computational CPU Time (minutes)	15	1.6	2.46	31.4

Another medium-scale problem instance including 80 products and 3 packing lines, has been also studied, in order to assess the solutions of the proposed decomposition technique. The study is solved using both the decomposition algorithm and the exact MILP model, described above. The results are illustrated in the Table 1. It is observed that the decomposition algorithm leads to nearly optimal solutions. In addition, the computation time is significantly decreased using the decomposition algorithm, when the products are inserted 1-by-1, or 5-by-5. The application of the 10-by-10 insertion policy, increases the complexity of the subproblems and the zero integrality gap cannot be achieved in several of the iterations without exceeding the underlying solution time limit.

As a result, the computational time is increased and higher makespan values are obtained. The aforementioned exact MILP model can only be applied to medium problem instances. For larger problem instances the computational cost becomes prohibitively high and even a feasible solution can cannot be guaranteed.

5. Conclusions

The main contribution of this work is the application of an MILP-based decomposition technique, in large scale, industrial packing facilities. A continuous, real-life, industrial facility is considered. More specifically, the packing stage of a consumer goods facility is scheduled, taking also into account constraints related to the formulation/production stage. The proposed decomposition-based approach leads to nearly optimal solutions for large-scale problem instances. On the other hand, the exact MILP model can only be applied to medium-sized problem instances. The solution strategy can assist decision makers to take rigorous scheduling plans in a dynamic production environment under realistic uncertainty. Current work considers an additional improvement algorithm step to further improve the efficiency of the proposed scheduling technique.

Acknowledgements

The work leading to this publication has received funding from the European Union's Horizon 2020 research and innovation programme under grant agreement No 723575 (Project CoPro) in the framework of the SPIRE PPP.

References

G.M. Kopanos, L. Puigjaner, M.C. Georgiadis, 2010, Production scheduling and lot-Sizing in dairy plants: The yoghurt production line, Industrial & Engineering Chemical Research, 49, 2, 701-718

C.A. Méndez, J. Cerdá, I.E. Grossmann, I. Harjunkoski, M. Fahl, 2006, State-of-the-art review of optimization methods for short-term scheduling of batch processes, Computer & Chemical Engineering, 30, 6-7, 913-94.

I. Harjunkoski, C.T. Maravelias, P. Bongers, P. M. Castro, S. Engell, I. E. Grossmann, J. Hooker, C.Méndez, G. Sand, J. Wassick, 2014, Scope for industrial applications of production scheduling models and solution methods, Computers & Chemical Engineering, 62, 161-193

P. Baumann, N. Trautmann, 2014, A hybrid method for large-scale short-term scheduling of make-and-pack production processes, European Journal of Operational Research, 236, 2, 718-735

G.M. Kopanos, C.A. Méndez, L. Puigjaner, 2010, MIP-based decomposition strategies for large-scale scheduling problems in multiproduct multistage batch plants: A benchmark scheduling problem of the pharmaceutical industry, European Journal of Operational Research, 207, 2, 718-735

Anton A. Kiss, Edwin Zondervan, Richard Lakerveld, Leyla Özkan (Eds.)
Proceedings of the 29th European Symposium on Computer Aided Process Engineering
June 16th to 19th, 2019, Eindhoven, The Netherlands. © 2019 Elsevier B.V. All rights reserved.
http://dx.doi.org/10.1016/B978-0-128-18634-3.50199-5

Deciphering Latent Uncertainty Sources with Principal Component Analysis for Adaptive Robust Optimization

Chao Ning, Fengqi You[*]

Cornell University, Ithaca, New York, 14853, USA

fengqi.you@cornell.edu

Abstract

This paper proposes a novel data-driven robust optimization framework that leverages the power of machine learning for decision-making under uncertainty. By performing principal component analysis on uncertainty data, the correlations among uncertain parameters are effectively captured, and latent uncertainty sources are identified. Uncertainty data are then projected onto each principal component to facilitate extracting distributional information of latent uncertainties with kernel density estimation technique. To explicitly account for asymmetric uncertainties, we introduce forward and backward deviation vectors in an uncertainty set. The resulting data-driven uncertainty set is general enough to be employed in adaptive robust optimization model. The proposed framework not only significantly ameliorates the conservatism of robust optimization but also enjoys computational efficiency and wide applicability. An application of optimization under uncertainty on batch process scheduling is presented to demonstrate the effectiveness of the proposed general framework. We also investigate a data-driven uncertainty set in a low-dimensional subspace and derive a theoretical bound on the performance gap between ARO solutions due to the dimension reduction of uncertainties.

Keywords: big data, optimization under uncertainty, principal component analysis

1. Introduction

As a promising alternative paradigm, robust optimization characterizes uncertain parameters using an uncertainty set (Mulvey et al. 1995; Ben-Tal et al., 2009;). Uncertainty set-induced robust optimization can be divided broadly into two categories, namely static robust optimization and adaptive robust optimization (ARO). In static robust optimization, all the decisions are made "here-and-now" before uncertainty realizations. By introducing "wait-and-see" decisions, ARO approach well models the sequential nature of decision-making processes, and ameliorates the conservatism issue (Ben-Tal and Nemirovski, 2002). In recent years, robust optimization methodology has found a fruitful of applications, such as biomass processing network (Gong et al., 2016), production scheduling (Shi et al., 2016), supply chain (Yue et al., 2016), energy systems (Gong et al., 2017), and model-based control (Shang et al., 2019).

Despite the burgeoning popularity of robust optimization, conventional robust optimization methods have some potential limitations (Ning and You, 2017a). First, they do not take advantage of the power of machine learning and big data analytics to leverage uncertainty data for decision-making under uncertainty (Shang and You,

2017). Moreover, their adopted uncertainty sets fail to account for the correlations or asymmetry among uncertainties. Neglecting relevant information embedded in uncertainty data, such as correlations and asymmetry, could render the induced robust optimization solution conservative. With the explosion in uncertainty data and great advances in machine learning techniques, data-driven optimization paves a new way to decision-making under uncertainty (Ning and You, 2017b). Based on all the concerns above, the research objective of our work is to develop a general data-driven robust optimization framework that exploits uncertainty data for decision-making under uncertainty. In this paper, we propose a general data-driven decision-making framework that bridges machine learning with robust optimization methodology directly. A theoretical analysis on the impact of uncertainty dimension reduction on the quality of ARO solutions is also performed within the proposed framework.

2. Data-Driven Uncertainty Set Construction

Consider an uncertainty data matrix $\mathbf{X} = \left[\mathbf{u}^{(1)}, \dots, \mathbf{u}^{(N)} \right]^{T} \in R^{N \times m}$, in which each row represents an uncertainty data point in m-dimensional space. There are a total number of N uncertainty data points. We first scale matrix X to zero-mean in (4) for principal component analysis (PCA) modelling (Murphy, 2012; Friedman et al. 2001), as shown in Eq. (1).

$$\mathbf{X}_0 = \mathbf{X} - \mathbf{e}\boldsymbol{\mu}_0^T \tag{1}$$

where $\mathbf{X}_0$ is an uncertainty data matrix after scaling, $\mathbf{e}$ denotes a column vector of all ones, and $\boldsymbol{\mu}_0$ represents the mean vector of uncertainty data.

The covariance matrix of uncertain parameters can be approximated with the sample covariance matrix $\mathbf{S}$, as shown in $\mathbf{S} = 1/(N-1)\mathbf{X}_0^T\mathbf{X}_0$. Through the eigenvalue decomposition, we can obtain $\mathbf{S} = \mathbf{P}\boldsymbol{\Lambda}\mathbf{P}^T$. The square matrix $\mathbf{P} = \left[\mathbf{p}_1, \dots, \mathbf{p}_m\right] \in R^{m \times m}$ consists of all the m eigenvectors (corresponding to the principal components), and $\boldsymbol{\Lambda} = diag\left\{\lambda_1, \mathrm{K}, \lambda_m\right\}$ is a diagonal matrix consisting of all the eigenvalues.

Consider the projection of an uncertainty data point $\mathbf{u}^{(i)}$ onto the k-th principal component, shown as follows,

$$\mathbf{t}_k^{(i)} = \mathbf{p}_k^T \left[\mathbf{u}^{(i)} - \boldsymbol{\mu}_0 \right] \tag{2}$$

Let ξ_k be a latent uncertainty along the k-th principal component, $\hat{f}_{KDE}^{(k)}\left(\xi_k\right)$ denotes the estimated probability density function of ξ_k by using the kernel density estimation approach, and $\hat{F}_{KDE}^{(k)}\left(\xi_k\right)$ be the corresponding cumulative density function. The corresponding quantile function can be expressed as follows:

$$\hat{F}_{KDE}^{(i) \, -1}\left(\alpha\right) = \min\left\{\xi_k \in R \middle| \hat{F}_{KDE}^{(k)}\left(\xi_k\right) \geq \alpha\right\} \tag{3}$$

which provides the minimum value of ξ_k based on a specific parameter α.

We propose the data-driven uncertainty set with uncertainty budget, as shown below.

$$U = \left\{ \mathbf{u} \;\middle|\; \begin{array}{l} \mathbf{u} = \boldsymbol{\mu}_0 + \mathbf{P}\boldsymbol{\xi}, \; \boldsymbol{\xi} = \underline{\boldsymbol{\xi}} \circ \mathbf{z}^- + \overline{\boldsymbol{\xi}} \circ \mathbf{z}^+, \\ \mathbf{0} \le \mathbf{z}^-, \; \mathbf{z}^+ \le \mathbf{e}, \; \mathbf{z}^- + \mathbf{z}^+ \le \mathbf{e}, \; \mathbf{e}^T \left(\mathbf{z}^- + \mathbf{z}^+ \right) \le \Phi \end{array} \right\} \tag{4}$$

where $\mathbf{z}^-$ is a backward deviation vector, $\mathbf{z}^+$ is a forward deviation vector, $\mathbf{e}$ is a column vector of all ones, and Φ is an uncertainty budget. $\boldsymbol{\xi}$ is the latent uncertainty vector, its lower bound vector $\underline{\boldsymbol{\xi}}$ is defined in Eq. (5), and the upper bound vector $\overline{\boldsymbol{\xi}}$ is defined in Eq. (6).

$$\underline{\boldsymbol{\xi}} = \left[\hat{F}_{KDE}^{(1)\,-1}(\alpha), \dots, \hat{F}_{KDE}^{(m)\,-1}(\alpha) \right]^T \tag{5}$$

$$\overline{\boldsymbol{\xi}} = \left[\hat{F}_{KDE}^{(1)\,-1}(1-\alpha), \dots, \hat{F}_{KDE}^{(m)\,-1}(1-\alpha) \right]^T \tag{6}$$

A novel feature of the data-driven uncertainty set in Eq. (4) is that PCA and kernel density estimation are utilized to accurately capture the principal components and distributions embedded within the uncertainty data. Additionally, the proposed uncertainty set U does not necessarily be symmetric by introducing the positive and negative deviations separately.

3. Data-Driven Adaptive Robust Optimization Framework

In this subsection, we propose a data-driven two-stage ARO model to organically integrate uncertainty data with adaptive optimization model, as shown in Eq. (7).

$$\min_{\mathbf{x}} \; \mathbf{c}^T \mathbf{x} + \max_{\mathbf{u} \in U} \min_{\mathbf{y} \in \Omega(\mathbf{x}, \mathbf{u})} \mathbf{b}^T \mathbf{y}$$
$$\text{s.t.} \;\; \mathbf{A}\mathbf{x} \ge \mathbf{d}, \;\; \mathbf{x} \in R_+^{n_1} \times Z^{n_2} \tag{7}$$
$$\Omega(\mathbf{x}, \mathbf{u}) = \left\{ \mathbf{y} \in R_+^{n_3} : \mathbf{W}\mathbf{y} \ge \mathbf{h} - \mathbf{T}\mathbf{x} - \mathbf{M}\mathbf{u} \right\}$$

where $\mathbf{x}$ is the first-stage decision made "here-and-now" before the uncertainty $\mathbf{u}$ is realized, while the second-stage decision or recourse decision $\mathbf{y}$ is postponed in a "wait-and-see" mode after the uncertainties reveal themselves. $\mathbf{x}$ includes both continuous and integer variables, while $\mathbf{y}$ only includes continuous variables. $\mathbf{c}$ and $\mathbf{b}$ are the vectors of the cost coefficients. U is the data-driven uncertainty set in Eq. (4).

This data-driven adaptive robust optimization model is a tri-level optimization problem, which cannot be solved directly by any off-the-shelf optimization solvers. The key idea is to decompose the multi-level MILP shown in (7) into a master problem and a sub-problem. The master problem (8) contains only a part of extreme points, so it provides a relaxation of the original problem.

$$\min_{\mathbf{x}, \eta, \mathbf{y}^k} \; \mathbf{c}^T \mathbf{x} + \eta$$
$$\text{s.t.} \;\; \mathbf{A}\mathbf{x} \ge \mathbf{d}$$
$$\eta \ge \mathbf{b}^T \mathbf{y}^k, \;\; k = 1, \dots, r$$
$$\mathbf{T}\mathbf{x} + \mathbf{W}\mathbf{y}^k \ge \mathbf{h} - \mathbf{M}\mathbf{u}^k, \;\; k = 1, \dots, r \tag{8}$$
$$\mathbf{x} \in R_+^{n_1} \times Z^{n_2}, \;\; \mathbf{y}^k \in R_+^{n_3}, \;\; k = 1, \dots, r$$

where $\mathbf{u}^k$ is the enumerated uncertainty realization at the kth iteration, $\mathbf{y}^k$ is its corresponding recourse variable, and r represents the current number of iterations.

4. Application on Batch Process Scheduling

Planning and scheduling are typical operations in process systems engineering. Batch process scheduling plays a critical role in manufacturing industries (Chu et al., 2015). In this section, an industrial multipurpose batch process is presented to demonstrate the advantages of the proposed data-driven ARO approach. Figure 1 depicts the state-task network of this batch process (Chu et al., 2013). In this application, the product demands are uncertain.

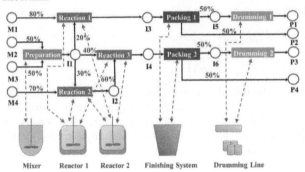

Figure 1. State task network of a multipurpose batch process.

The conventional method only utilizes the bounds of demand data, whereas the proposed approach leverages machine learning to dig out the correlation information among different product demands for scheduling decisions. As a result, the proposed approach is less conservative, and returns $299 more profits than the conventional ARO method, and the Gantt chart of the proposed approach is displayed in Figure 2.

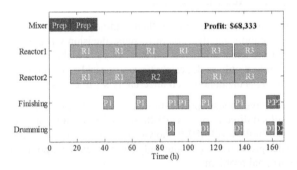

Figure 2. Optimal schedule decisions determined by the proposed data-driven approach.

5. Theoretical Analysis

High-dimensional uncertainty data often lies close to a low-dimensional subspace (Ferreira et al., 2012). Therefore, an extension of the proposed approach is to decipher the low-dimensional structure to perform "uncertainty dimension reduction" for ARO. Moreover, the discovered low-dimensional structure holds the potential to provide insights into the management of uncertainty. Only a small number of principal components can be used to retain much uncertainty information (Xu et al., 2016). The ARO problem without performing uncertainty data compression is shown as (ARO-H).

(ARO-H)
$$\min_{\mathbf{x} \in X} \left\{ \mathbf{c}^T \mathbf{x} + \max_{\mathbf{u} \in U} Q(\mathbf{x}, \mathbf{u}) \right\}$$

where $Q(\mathbf{x}, \mathbf{u})$ is the recourse function. We define problem (ARO-L) as follows:

(ARO-L)
$$\min_{\mathbf{x} \in X} \left\{ \mathbf{c}^T \mathbf{x} + \max_{\mathbf{u} \in \hat{U}} Q(\mathbf{x}, \mathbf{u}) \right\}$$

where the uncertainty set $\hat{U} = \left\{ \hat{\mathbf{u}} \mid \exists \mathbf{u} \in U \ \ \hat{\mathbf{u}} = \Pi(\mathbf{u}) \right\}$.

Note that $\Pi(g)$ is an operator of projecting uncertainty onto a low-dimensional subspace. The projected vector has the same dimension as the original vector. Suppose that $\mathbf{x}^*$ and $\hat{\mathbf{x}}$ are the optimal solutions of problems (ARO-H) and (ARO-L), respectively. We are concerned with the difference in their worst-case performance in an uncertain environment where the real uncertainty realization is still in a high-dimensional space. The following theorem provides a theoretical bound on the solution quality due to the uncertainty data compression for the ARO problem.

Theorem. Given that $\mathbf{x}^*$ and $\hat{\mathbf{x}}$ are the optimal solutions of problems (ARO-H) and (ARO-L), respectively. We have

$$0 \le \left[\mathbf{c}^T \hat{\mathbf{x}} + \max_{\mathbf{u} \in U} Q(\hat{\mathbf{x}}, \mathbf{u}) \right] - \left[\mathbf{c}^T \mathbf{x}^* + \max_{\mathbf{u} \in U} Q(\mathbf{x}^*, \mathbf{u}) \right] \le \alpha \cdot \beta \tag{9}$$

where parameter $\alpha = \max_{\mathbf{u} \in U} \left\| \mathbf{u} - \Pi(\mathbf{u}) \right\|_2$, parameter $\beta = \max_{\boldsymbol{\varphi} \in \Phi} \left\| \mathbf{M}^T \boldsymbol{\varphi} \right\|_2$ and set $\Phi = \left\{ \boldsymbol{\varphi} \mid \mathbf{W}^T \boldsymbol{\varphi} \le \mathbf{b}, \ \boldsymbol{\varphi} \ge \mathbf{0} \right\}$.

Proof of Theorem. Since $\mathbf{x}^*$ is an optimal solution of problem (ARO-H), and $\hat{\mathbf{x}}$ is a feasible solution of problem (ARO-H), we can arrive at the following inequality: $0 \le \left[\mathbf{c}^T \hat{\mathbf{x}} + \max_{\mathbf{u} \in U} Q(\hat{\mathbf{x}}, \mathbf{u}) \right] - \left[\mathbf{c}^T \mathbf{x}^* + \max_{\mathbf{u} \in U} Q(\mathbf{x}^*, \mathbf{u}) \right]$. Similarly, because $\hat{\mathbf{x}}$ is an optimal solution of problem (ARO-L), and $\mathbf{x}^*$ is a feasible solution of problem (ARO-H), we can arrive at the following inequality: $\mathbf{c}^T \hat{\mathbf{x}} + \max_{\mathbf{u} \in \hat{U}} Q(\hat{\mathbf{x}}, \mathbf{u}) \le \mathbf{c}^T \mathbf{x}^* + \max_{\mathbf{u} \in \hat{U}} Q(\mathbf{x}^*, \mathbf{u})$. Based on strong duality, we can have $Q(\mathbf{x}, \mathbf{u}) = \max_{\boldsymbol{\varphi} \in \Phi} \left[\mathbf{h} - \mathbf{Tx} - \mathbf{M}(\bar{\mathbf{u}} + \mathbf{u}) \right]^T \boldsymbol{\varphi}$, where dual feasible region is $\Phi = \left\{ \boldsymbol{\varphi} \mid \mathbf{W}^T \boldsymbol{\varphi} \le \mathbf{b}, \ \boldsymbol{\varphi} \ge \mathbf{0} \right\}$. Based on $\hat{U} \subseteq U$, we have $\mathbf{c}^T \mathbf{x}^* + \max_{\mathbf{u} \in \hat{U}} Q(\mathbf{x}^*, \mathbf{u}) \le \mathbf{c}^T \mathbf{x}^* + \max_{\mathbf{u} \in U} Q(\mathbf{x}^*, \mathbf{u})$. Next, we can have:

$$\begin{aligned} &\mathbf{c}^T \hat{\mathbf{x}} + \max_{\mathbf{u} \in \hat{U}} Q(\hat{\mathbf{x}}, \mathbf{u}) \\ &= \mathbf{c}^T \hat{\mathbf{x}} + \max_{\mathbf{u} \in U} \max_{\boldsymbol{\varphi} \in \Phi} \left[\mathbf{h} - \mathbf{Tx} - \mathbf{M}(\bar{\mathbf{u}} + \Pi(\mathbf{u})) \right]^T \boldsymbol{\varphi} \\ &\ge \mathbf{c}^T \hat{\mathbf{x}} + \max_{\substack{\mathbf{u} \in U \\ \boldsymbol{\varphi} \in \Phi}} \left[\mathbf{h} - \mathbf{Tx} - \mathbf{M}(\bar{\mathbf{u}} + \mathbf{u}) \right]^T \boldsymbol{\varphi} + \min_{\substack{\mathbf{u} \in U \\ \boldsymbol{\varphi} \in \Phi}} \left[\mathbf{M}(\mathbf{u} - \Pi(\mathbf{u})) \right]^T \boldsymbol{\varphi} \\ &= \mathbf{c}^T \hat{\mathbf{x}} + \max_{\mathbf{u} \in U} Q(\hat{\mathbf{x}}, \mathbf{u}) - \max_{\substack{\mathbf{u} \in U \\ \boldsymbol{\varphi} \in \Phi}} \left[(-1) \cdot \mathbf{M}(\mathbf{u} - \Pi(\mathbf{u})) \right]^T \boldsymbol{\varphi} \\ &\ge \mathbf{c}^T \hat{\mathbf{x}} + \max_{\mathbf{u} \in U} Q(\hat{\mathbf{x}}, \mathbf{u}) - \max_{\substack{\mathbf{u} \in U \\ \boldsymbol{\varphi} \in \Phi}} \left\| \mathbf{u} - \Pi(\mathbf{u}) \right\|_2 \cdot \left\| \mathbf{M}^T \boldsymbol{\varphi} \right\|_2 \end{aligned} \tag{10}$$

which concludes the proof. □

6. Conclusions

In this paper, we proposed a novel data-driven ARO framework based on PCA. By employing PCA on uncertainty data matrix, latent uncertainty sources were identified. The forward and backward deviation vectors were introduced into the data-driven uncertainty set. An application on batch scheduling was presented. The results showed that the proposed data-driven ARO approach enjoyed a less conservative solution compared with other methods. Additionally, a theoretical bound was also derived.

References

A. Ben-Tal, L. El Ghaoui, A. Nemirovski, 2009, Robust optimization: Princeton University Press.

A. Ben-Tal, A. Nemirovski, 2002, Robust optimization - methodology and applications. Mathematical Programming, 92, 453-480.

A. Ben-Tal, D. den Hertog, J.-P. Vial, 2015, Deriving robust counterparts of nonlinear uncertain inequalities. Mathematical Programming, 149, 265-299.

D. Bertsimas, V. Gupta, N. Kallus, 2017, Data-driven robust optimization. Mathematical Programming, 1-58.

Y. Chu, J.M. Wassick, F. You, 2013, Efficient scheduling method of complex batch processes with general network structure via agent-based modeling, AIChE Journal, 59, 2884-2906.

Y. Chu, F. You, 2015, Model-based integration of control and operations: Overview, challenges, advances, and opportunities, Computers & Chemical Engineering, 83, 2-20.

R.S. Ferreira, L.A. Barroso, M.M. Carvalho, 2012, Demand response models with correlated price data: A robust optimization approach, Applied Energy, 96, 133-149.

J. Friedman, T. Hastie, R. Tibshirani, 2001, The elements of statistical learning, Springer, Berlin.

J. Gong, D.J. Garcia, F. You, 2016, Unraveling optimal biomass processing routes from bioconversion product and process networks under uncertainty: an adaptive robust optimization approach, ACS Sustainable Chemistry & Engineering, 4, 3160-3173.

J. Gong, F. You, 2018, Resilient design and operations of process systems: Nonlinear adaptive robust optimization model and algorithm for resilience analysis and enhancement. Computers & Chemical Engineering, 116, 231-252.

J. M. Mulvey, R. J. Vanderbei, S. A. Zenios, 1995, Robust Optimization of Large-Scale Systems. Operations Research, 43, 264-281.

K. P. Murphy, *Machine learning: A probabilistic perspective*: MIT press, 2012.

C. Ning, F. You, 2017a, A data-driven multistage adaptive robust optimization framework for planning and scheduling under uncertainty. AIChE Journal, 63, 4343-4369.

C. Ning, F. You, 2017b, Data-driven adaptive nested robust optimization: General modeling framework and efficient computational algorithm for decision making under uncertainty. AIChE Journal, 63, 3790-3817.

C. Ning, F. You, 2018, Data-driven stochastic robust optimization: General computational framework and algorithm leveraging machine learning for optimization under uncertainty in the big data era. Computers & Chemical Engineering, 111, 115-133.

C. Shang, X. Huang, F. You, 2017, Data-driven robust optimization based on kernel learning. Computers & Chemical Engineering, 106, 464-479.

C. Shang, F. You, 2019, A data-driven robust optimization approach to scenario-based stochastic model predictive control. Journal of Process Control, 75, 24-39.

H. Shi, F. You, 2016, A computational framework and solution algorithms for two-stage adaptive robust scheduling of batch manufacturing processes under uncertainty, AIChE Journal, 62, 687-703.

H. Xu, C. Caramanis, and S. Mannor, 2016, Statistical optimization in high dimensions, Operations Research, 64, 958-979.

D. Yue, F. You, 2016, Optimal supply chain design and operations under multi-scale uncertainties: Nested stochastic robust optimization modeling framework and solution algorithm, AIChE Journal, 62, 3041-3055.

Anton A. Kiss, Edwin Zondervan, Richard Lakerveld, Leyla Özkan (Eds.)
Proceedings of the 29th European Symposium on Computer Aided Process Engineering
June 16th to 19th, 2019, Eindhoven, The Netherlands. © 2019 Elsevier B.V. All rights reserved.
http://dx.doi.org/10.1016/B978-0-128-18634-3.50200-9

Incipient Fault Detection, Diagnosis, and Prognosis using Canonical Variate Dissimilarity Analysis

Karl Ezra S. Pilario[a,b,*], Yi Cao[c] and Mahmood Shafiee[a]

[a]*Department of Energy and Power, Cranfield University, College Road, Bedfordshire, MK43 0AL, United Kingdom*
[b]*Department of Chemical Engineering, University of the Philippines, Diliman, Quezon City 1101, Philippines*
[c]*College of Chemical and Biological Engineering, Zhejiang University, Hangzhou, China*
k.pilario@cranfield.ac.uk

Abstract

Industrial process monitoring deals with three main activities, namely, fault detection, fault diagnosis, and fault prognosis. Respectively, these activities seek to answer three questions: 'Has a fault occurred?', 'Where did it occur and how large?', and 'How will it progress in the future?' As opposed to abrupt faults, incipient faults are those that slowly develop in time, leading ultimately to process failure or an emergency situation. A recently developed multivariate statistical tool for early detection of incipient faults under varying operating conditions is the Canonical Variate Dissimilarity Analysis (CVDA). In CVDA, a dissimilarity-based statistical index was derived to improve the detection sensitivity upon the traditional canonical variate analysis (CVA) indices. This study aims to extend the CVDA detection framework towards diagnosis and prognosis of process conditions. For diagnosis, contribution maps are used to convey the magnitude and location of the incipient fault effects, as well as their evolution in time. For prognosis, CVA state-space prediction and Kalman filtering during faulty conditions are proposed in this work. By covering the three main process monitoring activities in one framework, our work can serve as a baseline strategy for future application to large process industries.

Keywords: canonical variate analysis (CVA), incipient fault, Kalman filter (KF), dynamic process monitoring

1. Introduction

Data-driven methods for industrial process monitoring, known in general as multivariate statistical process monitoring (MSPM) techniques, have attained significant development in the last few decades (Reis and Gins, 2017). By addressing challenges such as high-dimensionality, temporal correlation, nonlinearity, non-Gaussianity, etc. in the plant data, MSPM tools have become more reliable for the automated detection of various faults in the plant (Ge et al., 2013).

However, in the larger perspective, fault detection must be followed by fault diagnosis and prognosis (Chiang et al., 2005). Fault diagnosis aims to determine the location and magnitude of a detected fault, while fault prognosis aims to predict its evolution in time. Indeed, Reis and Gins (2017) noted an increasing trend of research focus in the area of fault diagnosis and prognosis in the process industries. By developing more reliable diagnostic and prognostic tools, well-informed decisions can be made during production and maintenance operations planning, even when the fault continues to degrade the process performance. In addition, the ability to address the aforementioned challenges in handling plant data must be present in all three process monitoring activities. Hence, it is beneficial to establish them under an integrated framework.

Among the various types of faults, the incipient fault is deemed as that which requires prognosis (Li et al., 2010). As opposed to abrupt faults, incipient faults are those that start at small magnitudes, but slowly worsen in time (Isermann, 2005). If an incipient fault is left to degrade the process, it may ultimately lead to process failure or an emergency situation (Pilario and Cao, 2017). Hence, incipient fault monitoring is an important issue that needs to be addressed by more advanced diagnostic and prognostic tools.

For the sensitive detection of incipient faults under dynamically varying conditions, the Canonical Variate Dissimilarity Analysis (CVDA) method was recently developed (Pilario and Cao, 2018). This method is based on the well-known Canonical Variate Analysis (CVA) (Odiowei and Cao, 2010) but uses a dissimilarity-based statistical index for enhanced sensitivity by measuring the predictability of the hidden states of the process. In this work, the CVDA fault detection methodology is extended to fault diagnosis and prognosis. Specifically, we aim to: (i) formulate a contributions-based approach for diagnosis using the canonical variate dissimilarity index; and, (ii) propose a prognostic tool using CVA state-space modelling and Kalman filtering under faulty conditions.

This paper is structured as follows. The CVDA methodology is revisited in Section 2. The proposed diagnosis and prognosis procedures are given in Section 3. Section 4 demonstrates the monitoring performance in a simulated CSTR case study. Lastly, concluding remarks are given.

2. CVDA for Fault Detection

The CVDA detection method proceeds by performing CVA to calculate the state and residual variables from the input-output process data, and then computing the statistical indices T^2, Q, and D, which serve as health indicators of the process.

Given N samples of input and output data at normal conditions, denoted as $\mathbf{u}_k \in \mathbb{R}^{m_u}$ and $\mathbf{y}_k \in \mathbb{R}^{m_y}$ at the kth sampling instant, respectively, the past and future column vectors are formed as:

$$\mathbf{p}_k = [\mathbf{u}_{k-1}^T \quad \mathbf{u}_{k-2}^T \quad \cdots \quad \mathbf{u}_{k-p}^T \quad \mathbf{y}_{k-1}^T \quad \mathbf{y}_{k-2}^T \quad \cdots \quad \mathbf{y}_{k-p}^T]^T \in \mathbb{R}^{mp} \tag{1}$$

$$\mathbf{f}_k = [\mathbf{y}_k^T \quad \mathbf{y}_{k+1}^T \quad \mathbf{y}_{k+2}^T \quad \cdots \quad \mathbf{y}_{k+f-1}^T]^T \in \mathbb{R}^{m_y f} \tag{2}$$

where p and f are the number of lags in the past and future windows of data, respectively, and $m = m_u + m_y$. The number of lags are chosen large enough to capture autocorrelation in the output data (Odiowei and Cao, 2010).

The sample covariance matrices are then obtained using $\boldsymbol{\Sigma}_{pp} = \frac{1}{M-1} \mathbf{Y}_p \mathbf{Y}_p^T$, $\boldsymbol{\Sigma}_{ff} = \frac{1}{M-1} \mathbf{Y}_f \mathbf{Y}_f^T$, and $\boldsymbol{\Sigma}_{fp} = \frac{1}{M-1} \mathbf{Y}_f \mathbf{Y}_p^T$, where $\mathbf{Y}_p = [\mathbf{p}_{p+1} \ \mathbf{p}_{p+2} \ \cdots \ \mathbf{p}_{p+M}]$ and $\mathbf{Y}_f = [\mathbf{f}_{p+1} \ \mathbf{f}_{p+2} \ \cdots \ \mathbf{f}_{p+M}]$ are the past and future Hankel matrices, respectively, and $M = N - p - f + 1$.

CVA aims to find vectors $\mathbf{a} \in \mathbb{R}^{m_y f}$ and $\mathbf{b} \in \mathbb{R}^{mp}$ so that the correlation between the linear combinations $\mathbf{a}^T \mathbf{f}_k$ and $\mathbf{b}^T \mathbf{p}_k$ are maximized (Odiowei and Cao, 2010). The algebraic solution is given by the singular value decomposition (SVD) of the scaled Hankel matrix:

$$\mathbf{H} = \boldsymbol{\Sigma}_{ff}^{-1/2} \boldsymbol{\Sigma}_{fp} \boldsymbol{\Sigma}_{pp}^{-1/2} = \mathbf{U} \boldsymbol{\Sigma} \mathbf{V}^T \tag{3}$$

where $\mathbf{U} = [\boldsymbol{\upsilon}_1, \boldsymbol{\upsilon}_2, \ldots, \boldsymbol{\upsilon}_r]$ and $\mathbf{V} = [\mathbf{v}_1, \mathbf{v}_2, \ldots, \mathbf{v}_r]$ are the left and right singular matrices, $\boldsymbol{\Sigma} = \mathrm{diag}(\sigma_1, \sigma_2, \ldots, \sigma_r)$ is the diagonal matrix of descending non-zero singular values, and r is the rank of $\mathbf{H}$. The singular values, σ_i, are the maximum solutions of Eq. (3), which are also the canonical correlations between projected past and future data. Taking only the first n projections, corresponding to the n largest canonical correlations, the states, residuals, and dissimilarity features at the kth sampling instant are computed as:

$$\mathbf{z}_k = \mathbf{J}_n \mathbf{p}_k \in \mathbb{R}^n \tag{4}$$

$$\mathbf{e}_k = \mathbf{F}_n \mathbf{p}_k \in \mathbb{R}^{mp} \tag{5}$$

$$\mathbf{d}_k = \mathbf{L}_n \mathbf{f}_{k-f+1} - \boldsymbol{\Sigma}_n \mathbf{J}_n \mathbf{p}_{k-f+1} \in \mathbb{R}^n \tag{6}$$

where $\mathbf{J}_n = \mathbf{V}_n^T \boldsymbol{\Sigma}_{pp}^{-1/2} \in \mathbb{R}^{n \times mp}$, $\mathbf{F} = (\mathbf{I} - \mathbf{V}_n \mathbf{V}_n^T) \boldsymbol{\Sigma}_{pp}^{-1/2} \in \mathbb{R}^{mp \times mp}$, $\mathbf{L}_n = \mathbf{U}_n^T \boldsymbol{\Sigma}_{ff}^{-1/2} \in \mathbb{R}^{n \times m_y f}$, $\mathbf{V}_n$ and $\mathbf{U}_n$ are the first n columns of $\mathbf{V}$ and $\mathbf{U}$, respectively, and $\boldsymbol{\Sigma}_n$ is a diagonal matrix of the n largest singular values in Eq. (3). Finally, the statistical indices at the kth sampling instant are given by:

$$T_k^2 = \mathbf{z}_k^T \mathbf{z}_k \tag{7}$$

$$Q_k = \mathbf{e}_k^T \mathbf{e}_k \tag{8}$$

$$D_k = \mathbf{d}_k^T (\mathbf{I} - \boldsymbol{\Sigma}_n^2)^{-1} \mathbf{d}_k \tag{9}$$

which measure departures from the usual state subspace, residual subspace, and predictability of future states from the past states, respectively.

In the CVDA training phase, the distributions of T^2, Q, and D, are estimated using kernel density estimation (KDE) without any assumption of Gaussianity in the data. Upper control limits (UCLs) are computed as the values of T_{UCL}^2, Q_{UCL}, and D_{UCL} at which $P(T^2 < T_{\text{UCL}}^2) = \alpha$, $P(Q < Q_{\text{UCL}}) = \alpha$, and $P(D < D_{\text{UCL}}) = \alpha$, respectively. Throughout this paper, a significance limit of $\alpha = 99\ \%$ is adopted. During online monitoring, the process condition is deemed faulty when either of the indices exceeded their respective UCLs.

3. CVDA for Fault Diagnosis and Prognosis

3.1. Fault Diagnosis

To achieve fault diagnosis, the value of a statistical index at time point k can be decomposed into parts contributed by each measured variable. Hence, the measured variable/s with the largest contributions can be identified as most associated with the fault that occurred. Incipient fault diagnosis results must be reported using contribution maps instead of the traditional contribution bar plots, so that the fault evolution and propagation across time can be visually illustrated.

Contributions from the traditional CVA indices were already given by Jiang et al. (2015). In this paper, we derive the contributions for the dissimilarity index, D, as well. Altogether, the contributions of the ith measured variable ($i = 1, \ldots, m$) to statistical index J at time point k, denoted by $C_{i,k}^J$ with $J \in \{T^2, Q, D\}$, are given by:

$$C_{i,k}^{T^2} = \sum_{j=0}^{p-1} \left| \mathbf{z}_k^T \left(\mathbf{J}_n^{(i+mj)} \mathbf{p}_k^{(i+mj)} \right) \right| \tag{10}$$

$$C_{i,k}^{Q} = \sum_{j=0}^{p-1} \left| \mathbf{e}_k^T \left(\mathbf{F}_n^{(i+mj)} \mathbf{p}_k^{(i+mj)} \right) \right| \tag{11}$$

$$C_{i,k}^{D} = \begin{cases} \sum_{j=0}^{p-1} \left| \mathbf{d}_k^T (\mathbf{I} - \boldsymbol{\Sigma}^2)^{-1} \left(\mathbf{L}_n^{(i-m_u+m_y j)} \mathbf{f}_{k-f+1}^{(i-m_u+m_y j)} - \boldsymbol{\Sigma}_n \mathbf{J}_n^{(i+mj)} \mathbf{p}_{k-f+1}^{(i+mj)} \right) \right| & i > m_u \\ \sum_{j=0}^{p-1} \left| \mathbf{d}_k^T (\mathbf{I} - \boldsymbol{\Sigma}^2)^{-1} \left(-\boldsymbol{\Sigma}_n \mathbf{J}_n^{(i+mj)} \mathbf{p}_{k-f+1}^{(i+mj)} \right) \right| & i \leq m_u \end{cases} \tag{12}$$

where $\mathbf{J}_n^{(l)}$, $\mathbf{F}_n^{(l)}$, $\mathbf{L}_n^{(l)}$ denote the lth columns in matrices $\mathbf{J}_n$, $\mathbf{F}_n$, $\mathbf{L}_n$, respectively, and $\mathbf{p}_k^{(l)}$, $\mathbf{f}_k^{(l)}$ denote the lth element in column vectors $\mathbf{p}_k$, $\mathbf{f}_k$, respectively. For Eq. (12) to apply, we assume that $p = f$ and that measured variables are sorted as all inputs followed by all outputs.

Using Eqs. (10)-(12), contributions from a fault-free training data set are first taken. For online use, relative contributions are computed by removing the mean normal contribution from $C_{i,k}^{J}$, and scaling by the standard deviation of the normal contributions (Deng and Tian, 2011):

$$C_{i,k}^{J,\text{rel}} = \frac{C_{i,k}^{J} - \text{mean}(C_{i,k}^{J}|\text{normal})}{\text{std}(C_{i,k}^{J}|\text{normal})}. \tag{13}$$

3.2. Fault Prognosis

CVA is distinguished from other process monitoring methods in that it is also a system identification method (Larimore, 1990). This benefit is useful for prognosis under the CVDA framework. In CVA, the system is assumed to be represented by the following state-space model:

$$\mathbf{x}_{k+1} = \mathbf{A}\mathbf{x}_k + \mathbf{B}\mathbf{u}_k + \mathbf{w}_k \tag{14}$$
$$\mathbf{y}_k = \mathbf{C}\mathbf{x}_k + \mathbf{D}\mathbf{u}_k + \mathbf{v}_k \tag{15}$$

where $\mathbf{A}, \mathbf{B}, \mathbf{C}, \mathbf{D}$ are the state-space matrices, $\mathbf{w}, \mathbf{v}$ are the process and measurement noise, respectively, and $\mathbf{x}, \mathbf{y}, \mathbf{u}$ are the states, outputs, and inputs, respectively. After Eq. (4), CVA identification is achieved by letting $\mathbf{x}_k := \mathbf{z}_k$ and performing multivariate regression to estimate $\hat{\mathbf{A}}, \hat{\mathbf{B}}, \hat{\mathbf{C}}, \hat{\mathbf{D}}$ as follows (Larimore, 1990):

$$\begin{bmatrix} \hat{\mathbf{A}} & \hat{\mathbf{B}} \\ \hat{\mathbf{C}} & \hat{\mathbf{D}} \end{bmatrix} = \mathbf{\Sigma}\left[\begin{pmatrix} \mathbf{x}_{k+1} \\ \mathbf{y}_k \end{pmatrix}, \begin{pmatrix} \mathbf{x}_k \\ \mathbf{u}_k \end{pmatrix} \right] \cdot \mathbf{\Sigma}^{-1}\left[\begin{pmatrix} \mathbf{x}_k \\ \mathbf{u}_k \end{pmatrix}, \begin{pmatrix} \mathbf{x}_k \\ \mathbf{u}_k \end{pmatrix} \right] \tag{16}$$

where $\mathbf{\Sigma}[\cdot, \cdot]$ denotes the sample covariance operation. Moreover, since the noise can be recovered as $\hat{\mathbf{w}}_k = \mathbf{x}_{k+1} - \hat{\mathbf{A}}\mathbf{x}_k - \hat{\mathbf{B}}\mathbf{u}_k$ and $\hat{\mathbf{v}}_k = \mathbf{y}_k - \hat{\mathbf{C}}\mathbf{x}_k - \hat{\mathbf{D}}\mathbf{u}_k$, then the noise covariances can be computed as $\hat{\mathbf{Q}} = \mathbf{\Sigma}[\hat{\mathbf{w}}_k, \hat{\mathbf{w}}_k]$ and $\hat{\mathbf{R}} = \mathbf{\Sigma}[\hat{\mathbf{v}}_k, \hat{\mathbf{v}}_k]$. Meanwhile, $\hat{\mathbf{P}}_0 = \mathbf{\Sigma}[\hat{\mathbf{x}}_k, \hat{\mathbf{x}}_k]$ is the initial state covariance.

Given a future input sequence, the estimated state-space model can be used for predicting the output variables during normal operation. When a fault occurs, the state-space model must be re-trained by applying Eq. (16) to the samples obtained during the early stages of degradation (Ruiz-Cárcel et al., 2016). However, an incipient fault can bring changes in both the state-space model parameters and the states $\mathbf{x}_k$ themselves. Re-training only corrects the model parameters. Hence, in this paper, we propose the use of Kalman filtering (KF) to correct changes in the states in conjunction with model re-training. In KF, the pertinent equations are:

$$\mathbf{x}_{k|k-1} = \hat{\mathbf{A}}\mathbf{x}_{k-1|k-1} + \hat{\mathbf{B}}\mathbf{u}_k \tag{17}$$
$$\hat{\mathbf{P}}_{k|k-1} = \hat{\mathbf{A}}\hat{\mathbf{P}}_{k-1|k-1}\hat{\mathbf{A}}^T + \hat{\mathbf{Q}} \tag{18}$$
$$\mathbf{x}_{k|k} = \mathbf{x}_{k|k-1} + \mathbf{K}_k(\mathbf{y}_k - \hat{\mathbf{C}}\mathbf{x}_{k|k-1}) \tag{19}$$
$$\hat{\mathbf{P}}_{k|k} = (\mathbf{I} - \mathbf{K}_k\hat{\mathbf{C}})\hat{\mathbf{P}}_{k|k-1} \tag{20}$$

where $\mathbf{K}_k = \hat{\mathbf{P}}_{k|k-1}\hat{\mathbf{C}}^T(\hat{\mathbf{C}}\hat{\mathbf{P}}_{k|k-1}\hat{\mathbf{C}}^T + \hat{\mathbf{R}})^{-1}$ is the Kalman gain.

To apply Eq. (17)-(20), the initial states are first estimated using Eq. (4) on the first p samples. Assuming that $\mathbf{x}_k$, $\hat{\mathbf{w}}_k$ and $\hat{\mathbf{v}}_k$ are Gaussian-distributed, KF is used to update $\mathbf{x}_k$ at every time step. Note that changes due to other incipient faults, such as drifts in the noise statistics, may be difficult to track upon using the Kalman filter. This is a topic for future work.

4. Case Study

To evaluate the performance of CVDA for fault diagnosis and prognosis, a closed-loop CSTR case study is used, which is available in (Pilario, 2017). Data are simulated from a system of 4 state

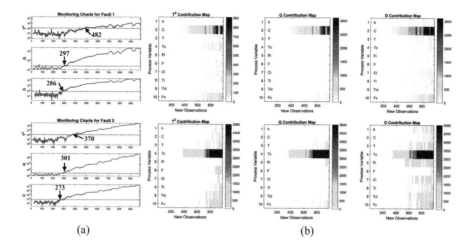

Figure 1: Monitoring results for: (a) fault detection (Arrows - detection time, mins; Red dash - UCLs; Blue solid - detection index); and (b) fault diagnosis. Top: Catalyst decay, Bottom: Fouling. For (b), the contributions maps are for T^2, Q, and D from left to right.

equations, with 2 cascade control loops to maintain liquid level, h, and reactor temperature, T. The measured variables are $\mathbf{u} = [F_i, C_i, T_i, T_{ci}]^T$ and $\mathbf{y} = [h, C, T, T_c, F, F_c]^T$. The input variables are being perturbed every 30 sampling times to simulate disturbance changes.

Using a data set of 1000 samples at normal operation, CVDA was trained so as to generate the projection matrices, the UCLs, and the state-space model. The number of lags and number of states are selected as 14 and 8, respectively. These are chosen using autocorrelation analysis and the dominant singular values method (Odiowei and Cao, 2010). Two incipient faults are studied in the CSTR, namely, catalyst decay (Fault 1) and heat transfer fouling (Fault 2). In the faulty data sets, both faults are slow drifts introduced after an initial 200 min of normal operation.

Fig. 1(a) shows detection results for the faulty data sets. As shown, the D index incurs the earliest detection time for both faults. To locate and track the magnitude of the detected fault, contribution maps are generated using Eqs. (10)-(13), and are shown in Fig. 1(b). The most associated variables to catalyst decay and fouling faults are shown to be the outlet concentration, C, and outlet temperature of the coolant, T_c. These are, indeed, the expected results for fault diagnosis. Moreover, the maps were able to illustrate how the fault effects propagate to other variables further in time. Notably, the D contribution maps revealed not only the fault-affected variables, but also the input variables that contributed to the dissimilarity between past and future states. Hence, we suggest to use all three contribution maps in conjunction to better capture incipient fault signatures.

The 1-step ahead prediction of the top 2 faulty variables in each fault case is given in Fig. 2. The estimated data are Kalman-filter predictions from a state-space model re-trained every 20 sampling times using the latest 400 samples while under faulty conditions. Table 1 gives the R^2 measure of fit of the predictions, showing that without KF, the faulty variables cannot be tracked accurately. Hence, KF improves CVDA fault prognosis in this case study.

5. Conclusion

In this paper, an incipient fault detection, diagnosis, and prognosis methodology is presented under an integrated framework, namely, Canonical Variate Dissimilarity Analysis (CVDA). Using a CSTR case study involving 2 parametric incipient fault scenarios, the method was shown to

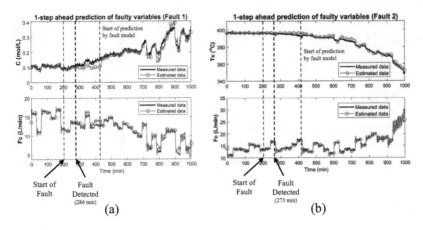

Figure 2: Fault prognosis results for the top 2 faulty variables in: (a) Catalyst decay fault; (b) Fouling fault. See Table 1 for R^2 fitness on all the charts.

Table 1: R^2 fitness values of the predictions of the faulty variables

Fault 1	with KF	without KF	Fault 2	with KF	without KF
C	90.57 %	−135.0 %	T_c	98.32 %	−35.3 %
F_c	94.41 %	27.30 %	F_c	96.52 %	12.93 %

provide early detection, reliable diagnosis, and accurate tracking of faulty variables. Hence, the framework constitutes a baseline strategy for industrial process monitoring.

References

L. H. Chiang, E. L. Russell, R. D. Braatz, 2005. Fault Detection and Diagnosis in Industrial Systems. Springer-Verlag, London.

X. Deng, X. Tian, 2011. A new fault isolation method based on unified contribution plots. Proceedings of the 30th Chinese Control Conference, CCC 2011 (10), 4280–4285.

Z. Ge, Z. Song, F. Gao, 2013. Review of recent research on data-based process monitoring. Industrial and Engineering Chemistry Research 52, 3543–3562.

R. Isermann, 2005. Model-based fault-detection and diagnosis - status and applications. Annual Reviews in Control 29 (1), 71–85.

B. Jiang, D. Huang, X. Zhu, F. Yang, R. D. Braatz, 2015. Canonical variate analysis-based contributions for fault identification. Journal of Process Control 26 (2015), 17–25.

W. E. Larimore, 1990. Canonical variate analysis in identification, filtering, and adaptive control. Proceedings of the IEEE Conference on Decision and Control 2, 596–604.

G. Li, S. J. Qin, Y. Ji, D. Zhou, 2010. Reconstruction based fault prognosis for continuous processes. Control Engineering Practice 18 (10), 1211–1219.

P.-E. Odiowei, Y. Cao, 2010. Nonlinear dynamic process monitoring using canonical variate analysis and kernel density estimations. IEEE Transactions on Industrial Informatics 6 (1), 36–45.

K. E. Pilario, 2017. Cascade-controlled CSTR for Fault Simulation.
URL https://www.mathworks.com/matlabcentral/fileexchange/65091-cascade-controlled-cstr-for-fault-simulation

K. E. Pilario, Y. Cao, sep 2017. Process incipient fault detection using canonical variate analysis. In: 2017 23rd International Conference on Automation and Computing (ICAC). IEEE, pp. 1–6.

K. E. S. Pilario, Y. Cao, 2018. Canonical variate dissimilarity analysis for process incipient fault detection. IEEE Transactions on Industrial Informatics.

M. Reis, G. Gins, 2017. Industrial Process Monitoring in the Big Data/Industry 4.0 Era: from Detection, to Diagnosis, to Prognosis. Processes 5 (3), 35.

C. Ruiz-Cárcel, L. Lao, Y. Cao, D. Mba, 2016. Canonical variate analysis for performance degradation under faulty conditions. Control Engineering Practice 54, 70–80.

Anton A. Kiss, Edwin Zondervan, Richard Lakerveld, Leyla Özkan (Eds.)
Proceedings of the 29th European Symposium on Computer Aided Process Engineering
June 16th to 19th, 2019, Eindhoven, The Netherlands. © 2019 Elsevier B.V. All rights reserved.
http://dx.doi.org/10.1016/B978-0-128-18634-3.50201-0

One-point temperature control of reactive distillation: A thermodynamics-based assessment

Mihai Daniel Moraru,[a,b,*] Costin Sorin Bildea[b]

[a]Department of Process Technology and Development, Hexion, Seattleweg 17, 3195 ND Pernis, The Netherlands

[b]Department of Chemical and Biochemical Engineering, University Politehnica of Bucharest, Str. Gh. Polizu 1-7, 011061 Bucharest, Romania

mihai.moraru@hexion.com

Abstract

A recent study introduced a new class of control structures using one-point temperature control applicable to heterogeneous reactive distillation. The feed rate of one reactant sets the plant capacity. The inventory of the other reactant is inferred from measuring the reflux rate or reflux ratio and is managed by adjusting the second feed flow. This idea can be implemented in different ways. Here, the applicability of these structures to various esterification systems of industrial importance is assessed. The analysis is based on key thermodynamic data: boiling points, azeotropy and liquid-liquid equilibria. As a rule-of-thumb, if water is the lightest and the ester the heaviest boiler in the system, and if at least the alcohol forms a minimum boiling heterogeneous azeotrope with water, then the new class of control structures may be applicable.

Keywords: acetates, acrylates, esterification, propionates, process control

1. Introduction

Reactive distillation (RD) offers significant advantages such as reduced investment and operating costs. However, the controllability of the process is as important as the economics, in order to meet important objectives such as process stability, production rate and product purity. Recently, we proposed a new class of control structures using one-point temperature control that are applicable to heterogeneous reactive distillation (Moraru et al., 2018). The *heterogeneous* attribute refers to systems that present minimum boiling heterogeneous azeotropy making possible a liquid-liquid split after vapor condensation. Hence, these control structures are appropriate for industrially important chemical systems which share this particular thermodynamic property.

Many acetates, acrylates and propionates are esters of industrial importance which have in common the minimum boiling heterogeneous azeotropy. They are produced in large quantities and used in a variety of applications. These esters are usually obtained in the reaction between an alcohol and the respective acid, which is typically performed in liquid phase and in the presence of a homogeneous catalyst. Whenever feasible, the RD technology using solid catalysts is preferred as alternative to the conventional processes (i.e., reactors followed by distillation systems).

In the perspective of showing the broad applicability of these new control structures using one-point temperature control, we study the basic thermodynamic properties of six important esterification systems: two acetates, two acrylates and two propionates;

the two acrylates were already used to test the concept of the new control structures and serve here as comparison for the other esterification systems. In the first part, the basic control idea applicable to the heterogeneous reactive distillation is presented; dynamic simulation results proving the control performance are also briefly presented. In the second part of the study, the common key properties that suggest the applicability of the new control structures are described. The azeotropy and liquid-liquid equilibria of each system are compared and discussed in detail. Aspen is used as computer-aided process engineering tool to make the analysis.

2. Basic control idea

The basic idea behind the new one-point temperature control structures addresses three basic control objectives: setting plant capacity, achieving the required product purity, and maintaining the component inventory. The latter is the key element of the concept.

To better describe the concept, we start by describing the RD process. A general esterification system which can be efficiently performed by RD comprises two reactants (A, B) and two products (C, D). Often, one product is the high-boiling species and is obtained as bottom stream, while the other is involved in a low-boiling azeotrope and is obtained as distillate. This azeotrope has the beneficial effect of displacing the chemical equilibrium, but also the disadvantage of an impure distillate. However, when the heterogeneity of the azeotrope makes possible a liquid–liquid split of the condensed distillate, the reactant-rich phase can be returned to the column as reflux, while the product-rich phase has the composition on the side of the distillation boundary which allows obtaining the top-product with a relatively high-purity. Figure 1 (left) shows a typical example. The heavy reactant B (acid or alcohol) is fed as liquid at the top of the reactive zone of the RD column. The light reactant A (acid or alcohol) is fed as vapor at the bottom of the reactive zone. The reactants flow in counter-current and reaction takes place in the liquid phase. The heavy product C (ester) is recovered in the stripping zone of the column, being obtained as high purity bottom stream. The light product D (water) forms several binary and ternary low-boiling heterogeneous azeotropes. Thus, water is removed from the liquid mixture with the result of increased reaction rate due to equilibrium displacement. The distillate is condensed, cooled, and sent to liquid–liquid separation. The organic phase is refluxed. The aqueous phase is the top product. If required, a flash can remove the small amounts of organics that are still dissolved.

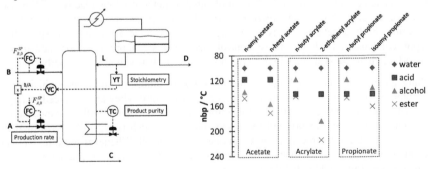

Figure 1. Basic control idea for RD process (left); different types of esterification systems (right)

During operation, the feed flow rate of one reactant sets the plant capacity. The requirement to feed the reactants in the stoichiometric ratio is addressed by feedback control. The inventory of the other reactant is inferred from the reflux rate or reflux ratio, which both can be easily measured. When this reactant is not fed in the correct stoichiometric ratio, it will either accumulate in the system or get depleted, which will be reflected by a change of the reflux rate. Thus, corrective action (change of the fresh feed rate) can be taken. This simple solution does not require concentration measurements and does not use any temperature control loop. Consequently, the reboiler duty becomes available to be used as manipulated variable in a quality-control loop. Several control strategies are possible based on this idea.

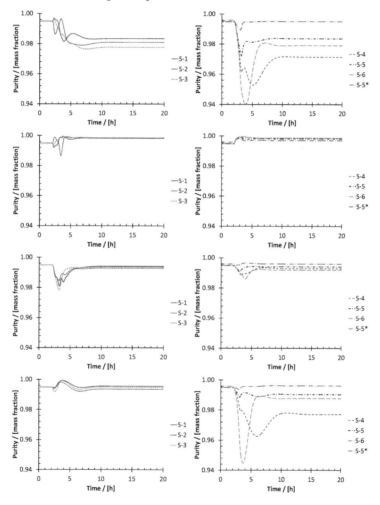

Figure 2. Process dynamics associated with each control structure (left column: literature structures S-1, S-2, S-3; right column: novel structures S-4, S-5, S-6 (S-5* uses pressure-compensated control and can be implemented to any of these six structures)) at various perturbations: top diagrams: +25% production; middle-top diagrams: -25% production; middle-bottom diagrams: 5 %mass water in fresh acid; bottom diagrams: 5 %mass water in fresh alcohol.

Figure 2 compares the performance of several control structures from the literature (S-1, S-2, (Luyben and Yu, 2008) and S-3 (Zeng et al, 2004)) with control structures based on the idea presented above (S-4, S-5, S-6). S-1 sets the production rate by the reboiler duty and uses two temperature measurements for inferential control of reactants inventory: the top feed controls a temperature in the stripping section while the bottom feed controls a temperature on the top of the reactive zone. S-2 sets the production rate directly, by changing the feed rate of the light reactant. The second reactant and the reboiler duty are used to control two temperatures in the stripping and reactive sections, respectively. S-3 uses only one temperature control loop, fixing the ratio between the reboiler duty and the flow rates of fresh reactants. In S-4 and S-5, the accumulation / depletion of one reactant is inferred from the reflux ratio and the reflux rate, respectively. S-6 takes a different approach: the reflux ratio is fixed and the alcohol is fed at two different locations: at the top of the reactive section, and directly in the decanter. This second feed provides an additional degree of freedom which is used to control the level of the organic phase, maintaining in this way the component balance. Among the new control structures, S-5 is the best, showing comparable performance with S-1 in terms of speed of disturbance rejection and offset in product purity.

3. Common key properties

The applicability of the new control concept to reactive distillation processes of industrial importance can be assessed based on key physical properties. This perspective is discussed in the remaining of this chapter.

3.1. Thermodynamic model

The model parameters for calculating the physical properties of pure components and mixtures are available in the Aspen database. The azeotropes and phase equilibria are calculated using the UNIQUAC or NRTL liquid activity coefficient models. The HOC equation of state is also considered to account for the dimerization of acids in the vapor phase. The binary interaction parameters are taken from published literature or retrieved from the Aspen databanks, including the parameters for the HOC equation.

3.2. Boiling points of pure components

The order of boiling points of the pure components in the system determines the structure of the RD column: addition of feeds and withdrawal of products. Figure 1 (right) orders by their normal boiling points the components (alcohol, acid, water and ester) of each of the esterification system considered in this work. In all systems, the ester is the highest boiler, and is obtained as bottoms; the heterogeneous azeotropes are the lightest, making possible to obtain the water as distillate (after decantation); the reactants are intermediates (acids lighter or heavier than alcohols), and are fed to the column at the bottom and top of the catalytic section. In addition, these six systems present various characteristics which shows that the proposed control structures are applicable to a wide range of esterification processes: (i) combination between 3 acids and 5 alcohols, (ii) wide boiling range for acids (118-141 °C), alcohols (118-184 °C) and esters (145-214 °C).

Note that other esterification systems (not presented in this study) such as *n*-butyl acetate, *n*-decyl acrylate and cyclohexyl acrylate, can be included in this list of industrially important esters.

3.3. Azeotropy and liquid-liquid equilibria

The formation of heterogeneous azeotropes and formation of two liquid phases after condensation are the key elements on the control concept. This allows to control the inventory of one component by monitoring the refluxed organic phase and adjusting one of the fresh feeds. Table 1 shows the azeotropes calculated at 1.013 bar for all esterification systems. Each system presents at least two binary heterogeneous azeotropes: water/alcohol and water/ester. With the exception of the 2-ethylhexyl acrylate, all the other systems also present a ternary heterogeneous azeotrope: water/alcohol/ester. All these azeotropes have a boiling point very close to that of water and will be found in the vapor distillate. Excepting the *n*-hexyl acetate system, the acid is involved in homogeneous azeotropy either with water (*n*-butyl propionate) or with the alcohol or ester. These azeotropes have a high boiling point, close to that of the ester in some cases. They are important since they may contaminate the ester product.

Figure 3 shows that all systems have large immiscibility areas, meaning that the water product is obtained with a relatively high purity; when this is not the case (n-butyl acrylate), an additional flash can recover the alcohol and obtain high-purity water. This is possible due to the minimum boiling heterogeneous azeotrope water / alcohol.

Table 1. Azeotropy[a] (mass based) at 1.013 bar of the six esterification systems

#	type[b]	T_B / [°C]	water	acid	alcohol	ester
n-amyl acetate						
1	het.	94.7	0.411		0.107	0.481
2	het.	94.9	0.404			0.596
3	het.	95.9	0.541		0.459	
4	hom.	140.0		0.154	0.572	0.284
5	hom.	140.3		0.193	0.807	
n-hexyl acetate						
1	het.	97.4	0.586		0.139	0.276
2	het.	97.6	0.579			0.421
3	het.	98.0	0.690		0.310	
n-butyl acrylate						
1	het.	92.3	0.397		0.498	0.106
2	het.	92.4	0.417		0.583	
3	het.	95.4	0.435			0.868
4	hom.	139.8		0.868	0.132	
5	hom.	148.5		0.047		0.953
2-ethylhexyl acrylate						
1	het.	99.1	0.809		0.191	
2	het.	99.6	0.855			0.145
n-butyl propionate						
1	het.	92.9	0.411		0.436	0.153
2	het.	93.0	0.422		0.558	
3	het.	94.5	0.388			0.612
4	hom.	99.9	0.834	0.166		
isoamyl propionate						
1	het.	95.3	0.472		0.286	0.242
2	het.	95.7	0.541		0.459	
3	het.	96.2	0.466			0.534
7	hom.	142.1		0.754	0.246	

[a]the calculations are performed using the *Distillation Synthesis* tool in Aspen Properties
[b]azeotrope type: heterogeneous (het.) or homogeneous (hom.)

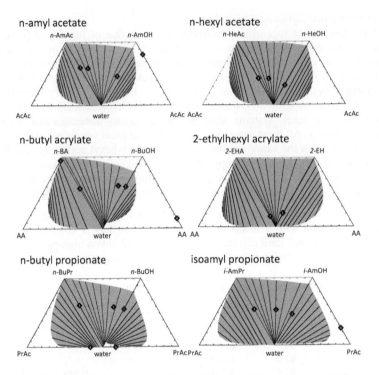

Figure 3. Liquid-liquid equilibria (mass based) at 1.013 bar and 35 C of the six esterification systems; numbering of the azeotropes as in Table 1

4. Conclusions

The applicability to a given esterification system of the basic control concept can be assessed based on key physical property data. As a rule-of-thumb, if the pure components have the right nbp order, and if at least the alcohol forms a minimum boiling heterogeneous azeotrope with water, then the new class of control structures may be applicable.

References

M.D. Moraru, E. Zaharia, C.S. Bildea, 2018, Novel Control Structures for Heterogeneous Reactive Distillation, Chem. Eng. Trans. 69, 535-540.

M.D. Moraru, C.S. Bildea, 2017, Process for n-butyl acrylate production using reactive distillation: Design, control and economic evaluation. Chem. Eng. Res. Des. 125, 130-145.

M.D. Moraru, C.S. Bildea, 2018, Process for 2-Ethylhexyl Acrylate Production Using Reactive Distillation: Design, Control and Economic Evaluation, Ind. Eng. Chem. Res. 57, 15773–15784.

W.L. Luyben, C.C. Yu, 2008, Reactive Distillation Design and Control, Wiley-AIChE.

K.L. Zeng, C.L. Kuo, I.L. Chien, 2006, Design and control of butylacrylate reactive distillation column system, Chem. Eng. Sci. 61, 4417–4431.

Anton A. Kiss, Edwin Zondervan, Richard Lakerveld, Leyla Özkan (Eds.)
Proceedings of the 29th European Symposium on Computer Aided Process Engineering
June 16th to 19th, 2019, Eindhoven, The Netherlands. © 2019 Elsevier B.V. All rights reserved.
http://dx.doi.org/10.1016/B978-0-128-18634-3.50202-2

The Effect of Indirect GHG Emissions Costs on the Optimal Water and Energy Supply Systems

Negar Vakilifard, Parisa A. Bahri[*], Martin Anda, Goen Ho

School of Engineering and Information Technology, Murdoch University, 90 South Street, Murdoch, Perth, Western Australia 6150, Australia

P.Bahri@murdoch.edu.au

Abstract

This study investigates the effect of indirect greenhouse gas (GHG) emissions on the optimal long-term planning and short-term operational scheduling of a desalination-based water supply system. The system was driven by grid-electricity and surplus output from residential rooftop photovoltaics to deliver water and energy to urban areas. The interactive two-level mixed integer linear programming model took into account demands, system configurations, resources capacities and electricity tariffs as well as GHG emission factor associated with the source of grid electricity. Both system and carbon abatement costs were considered in the formulation of the objective function. The optimal decisions for Perth (Australia) resulted in \$47,449,276 higher discounted total cost but 51,301.3 tCO$_2$eq less GHG emissions over 15 years planning horizon compared to when only system costs were minimised. Finally, the predominant effect of the indirect GHG emissions costs over system costs on the optimal solutions indicated their high sensitivity towards the source of purchased grid electricity.

Keywords: Grid electricity, Photovoltaics, GHG emissions, Desalination, Optimisation.

1. Introduction

Climate change and increasing water demand in urban areas have made it inevitable to incorporate drought-proof technologies such as desalination in water supply systems. However, meeting their intensive energy demand from fossil-fuel sources leads to higher indirect greenhouse gas (GHG) emissions, which adversely affects the existing water resources and therefore adds further complexities to sustainable supply. Considering renewable energy sources in the energy mix of this water supply option, therefore, could be a potential solution to decrease this effect.

In (Vakilifard et al., 2017), we proposed the idea of employing surplus residential grid-connected photovoltaics (PVs) output in conjunction with grid electricity to drive urban desalination-based water supply system. Using this source of energy not only assists in sustainably meeting the water-related energy demand but also mitigates the barrier of increasing the PVs installation to the existing electrical grid. In (Vakilifard et al., 2018), we developed a mixed integer linear programming (MILP) model to provide optimal strategic decisions of such water supply system incorporating short-term operational scheduling considering PV installation density as a parameter for any given year. In this paper, we extended the model to also investigate the effect of indirect GHG emissions costs associated with purchasing grid electricity on the optimal solutions and to determine to what extent they vary by the source of this energy. Additionally, the interactive effect of added water-related energy demand and installation density (as variable) was addressed through the two-run solving strategy. The results for an urban area located in the north-western corridor of Perth (Australia) were then discussed.

2. Problem statement

The problem was defined in three time frames (yearly, seasonal and hourly) for the planning horizon of 15 years (beginning from 2017). Water and energy needs were determined in 4 distinct zones in the studied area based on the demands per capita, annual population growth and service area of zone substations using ArcGIS 10 integrated with Excel analysis. It was assumed that a decentralised water supply system consisting of desalination plants, storage tanks, and a pipeline network delivers water to the zones. Plants were presumed to be operated in flexible (hourly) mode. The extra water could be desalted when renewable energy was available and could be stored for later use. Plant capacities were selected from 6 discrete values (20,000-120,000 m³/day) considering the plant factor of 0.85. For storage tanks, the capacities were chosen from 10,000 and 20,000 m³. Two pipeline capacities associated with the diameters of 30 and 54 in were also taken into account. The potential locations of water supply components in each zone, energy consumption per unit of water produced, stored and distributed as well as the capital and operational and maintenance (O&M) costs were according to (Vakilifard et al., 2018). Grid electricity and PV output supplied residential and water-related electricity demands. The maximum grid electricity that could be delivered to each zone was ascertained based on the associated substations capacities. The maximum capacity of available renewable energy was achieved based on the performance analysis of a 4 kW PV system output conducted in system advisor model (Vakilifard et al., 2017) and the installation density. The latter is the number of households equipped with PV systems in each zone divided by the total number of households and was determined by the model considering the economic-subjectivity index.

This index, in fact, is the product of economic and subjectivity indexes accounting for the economic preference of PV uptake and households' free will, respectively. To determine the economic index, Excel analysis was done based on the methodology described in (Miranda et al., 2015). This binary index is considered to be 1 if it is beneficial to install a PV system. The subjectivity index could be determined by the decision maker and could get any value between 0 and 1. It is the ratio of the households who decide to uptake PV systems (in case it is economically beneficial) to the total households. In this study, it was assumed the economic-subjectivity index is 1 meaning that in case it is economical to install a PV system, all households would decide to be equipped with one. Grid electricity price tariffs as well as the net feed-in tariff were taken from Synergy, the electricity retailer of Perth. The real discount rates for residential and business sectors were adopted from (AEC, 2017; ERA, 2017). The GHG emission factor of 0.7 for purchasing grid electricity from the south west interconnected system (SWIS), the electricity network in Perth, was adopted from (DEE, 2017). It is to be noted that this emission factor is only associated with the environmental impact of the fuels combustion in stationary sources (operational stage). The cost of carbon abatement was considered $ 40/tCO₂eq, taken from (WSAA, 2012). All cost data was converted to 2017 real Australian dollar using appropriate exchange rates from (RBA, 2017).

3. Optimisation strategy

The model was formulated as a two-level MILP model. In the first level of optimisation, the objective function maximised the economic benefits for the households equipped with PV systems (z1). This included savings from avoiding purchasing grid electricity as well as revenues from feeding surplus PV output back into the grid. The outcome of this level of optimisation, namely PV installation density, share of grid electricity in supplying residential energy demand and surplus PV output, were introduced to the level-two optimisation where the optimal decisions for the water supply system were achieved. The objective function at this stage concerned minimisation of the discounted total cost of the water supply system including indirect GHG emissions costs associated with purchasing grid electricity for water supply (z2). The model constraints of the level-one optimisation are presented in Eqs. (1)-(10). The constraints of level-two optimisation were based on our previous study (Vakilifard et al., 2017, 2018).

$$z_1 : Max \sum_t DF_t^r \cdot \sum_s nd_s \cdot \left[\sum_i \sum_b Ce_{t,s,b}^r . RE_{t,i,s,b}^r + Cr_t . Surp_{t,i,s,b} \right]$$

$$z_2 : Min \sum_t DF_t^{bi} \cdot \sum_s nd_s \cdot \left[Cfe_t^{bi} + \sum_i \sum_b Cr_t^{bi} . RE_{t,i,s,b}^w + Ce_{t,s,b}^{bi} . P_{t,i,s,b}^w + COM_t . Q_{t,i,s,b} + Cs_t . V_{t,i,s,b} \right]$$

$$+ \sum_t DF_t^{bi} \cdot \sum_i \left[\sum_c Cap_{t,c}^{DP} . XW_{t,i,c} + \sum_m Cap_{t,m}^{STT} . X_{t,i,m} \right]$$

$$+ \sum_t DF_t^{bi} \cdot \left[\sum_{(i,j) \in \{AL_{i,j} | i = j\}} Cap_t^{PI} . np_{t,i} . L_{i,j} . convf_2 + \sum_{(i,j) \in \{AL_{i,j} | i \neq j\}} Cap_t^{PI} . npIJ_{t,i,j} . L_{i,j} . convf_2 \right]$$

$$+ \sum_t DF_t^{bi} . CTax . f^{GHG} . \sum_i \sum_s nd_s . \sum_b p_{t,i,s,b}^w$$

Residential energy balance (with PVs): $\quad P_{t,i,s,b}^r + RE_{t,i,s,b}^r = k_{t,i} . D_{t,i,s,b}$ (1)

Added share of each energy source: $\quad P_{t,i,s,b}^r = P_{t-1,i,s,b}^r + addP_{t,i,s,b}^r$ (2)

$\qquad\qquad\qquad\qquad\qquad\qquad RE_{t,i,s,b}^r = RE_{t-1,i,s,b}^r + addRE_{t,i,s,b}^r$ (3)

Residential energy balance (without PVs): $\quad P_{t,i,s,b}^{rn} + RE_{t,i,s,b}^{rn} = (1 - k_{t,i}) . D_{t,i,s,b}$ (4)

Max. potential PV output: $\quad MaxR_{t,i,s,b} = k_{t,i} . dur_b . PV_{s,b} . exisR_{t,i}$ (5)

$\qquad MaxR_{t,i,s,b} = MaxR_{t-1,i,s,b} + addk_{t,i} . kp_t . dur_b . PV_{s,b} . (exisR_{t,i} - exisR_{t-1,i})$ (6)

Surplus PV output: $\quad Surp_{t,i,s,b} = MaxR_{t,i,s,b} - RE_{t,i,s,b}^r$ (7)

PV share constraints: $\quad RE_{t,i,s,b}^r \leq MaxR_{t,i,s,b}$ (8)

Grid share constraint: $\quad P_{t,i,s,b}^r + P_{t,i,s,b}^{rn} + P_{t,i,s,b}^w \leq dur_b . MaxPS_{t,i}$ (9)

Unused surplus energy: $\quad Surp_{t,i,s,b} - RE_{t,i,s,b}^{rn} - RE_{t,i,s,b}^w \cong 0$ (10)

The optimal solution was achieved in two runs using an interactive approach. In the first run, the initial estimation of the optimal water supply system was obtained regardless of remaining unused surplus PV output (relaxation of Eq. (10)). This led to the maximum PV installation density in the area and the initial estimation of the optimal decisions for water supply system while there was the highest access to the renewable energy. Considering this constraint (Eq. (10)), the initial estimation was then applied in the second run to adjust the installation density and achieve the final optimal solution. Each run included both levels of optimisation.

4. Optimal strategic and operational decisions

The model was coded into GAMS 24.3.1 software and solved by CPLEX 12.6. Two scenarios were considered. In the "minimum system and GHG costs" scenario, the objective function minimised the discounted cost of the system including capital and O&M costs as well as the carbon abatement cost over the planning horizon. This scenario was then compared with "minimum system costs" scenario, which only concerned the minimisation of discounted capital and O&M costs of the system. The optimal results for the minimum system and GHG costs scenario led to $ 2,291,309,369 discounted total cost, around $ 47,449,276 higher than minimum system costs scenario. It also resulted in 51,301.3 tCO$_2$eq less GHG emissions over 15 years of system operation. The optimal results are presented in Figures 1 and 2.

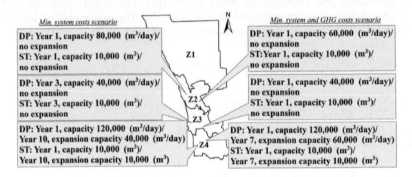

Figure 1: Optimal desalination plants (DP) and storage tanks (ST) capacities located in 4 discrete zones (z) in Min. system costs and Min. system and GHG costs scenarios

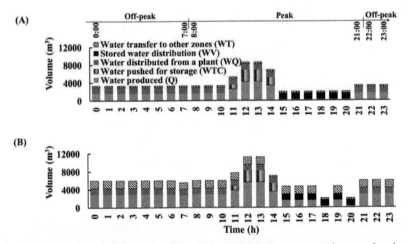

Figure 2: Operational scheduling of the water supply system in the representative zone 2 and year 2018 in: (A) Min. system costs and (B) Min. system and GHG costs scenarios

In both scenarios, three zones of 2, 3 and 4 were equipped with desalination plants (Figure 1); however, in terms of capacities and the timing of construction/expansion of the water supply components, they were different. In minimum system and GHG costs scenario, the model equipped both zones of 2 and 3 with desalination plants and associated storage tanks from the beginning of the planning horizon. Thus, despite the lack of economies of scale of the smaller desalination plants, it located a desalination plant with the capacity of 60,000 m^3/day in zone 2 as opposed to the capacity of 80,000 m^3/day in minimum system costs scenario. Instead, it placed a desalination plant with the capacity of 40,000 m^3/day in zone 3 in year 1 versus year 3 in minimum system costs scenario. The reason is to reduce the energy consumption of water transfer among allowable zones (around 3,640 MWh over 2 years) and decrease indirect GHG emissions costs. It is worth mentioning that although the expansion capacity of the desalination plant in zone 4 in the minimum system and GHG costs scenario was larger than minimum system cost scenario, the earlier time of the expansion neutralised the economic benefits of the larger scale expansion capacity.

Figures 2(A) and 2(B) show the optimal operational scheduling of the water supply system in both scenarios for a representative zone 2 and year 2018. Given flexible mode of operation for water supply system, the paradigm of the daily operational scheduling was achieved relatively the same for both scenarios. Accordingly, the highest water production and storage occurred when renewable energy was available, although it was concurrent with the peak electricity pricing hours. The stored water was then used for providing the water demand in the same zone when it was still during the peak electricity hours but no surplus PV output was available that could be assigned to the water-related electricity demand. The level of water production in zone 2 during peak electricity hours when there was no access to the renewable energy was limited to the demand of its adjacent zones (1 and/or 3). In off-peak electricity hours, the water demand was supplied directly from the desalination plant located in this zone.

5. Sensitivity analysis

Figure 3 depicts the sensitivity of the optimal solution towards purchasing grid electricity from 6 different sources in the minimum system and GHG costs scenario. The associated data was achieved from (Gifford, 2011). The results indicate a relatively high sensitivity towards emission factors higher than 0.148 (municipal waste). In fact, in higher emission factors, the effect of environmental impact was more significant and thus the optimal solution was mainly driven by indirect GHG emissions costs, which led to higher discounted total cost. By decreasing the emission factor, the effect of indirect GHG emissions costs reduced and from a certain point, it did not change the optimal decisions. Thus, the system cost turned to the predominant factor affecting the optimal results. This also explains the relatively same discounted total costs of the system with the minimum system costs scenario in lower emission factors.

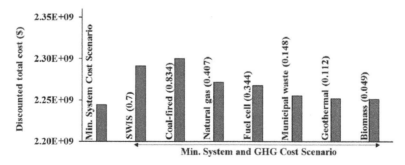

Figure 3: Sensitivity analysis towards purchasing electricity generated from different sources

6. Conclusions

In this study, we proposed an interactive optimisation model for the strategic and operational decisions of an urban water supply system driven by surplus PV output and grid electricity considering both system and carbon abatement costs in the formulation of the objective function. The optimal solutions for an urban area located in the north-western corridor of Perth (Australia) led to less GHG emissions but higher discounted total cost compared to the case where the system costs were the only components of the objective function. Finally, the results of the sensitivity analysis towards purchasing grid electricity from different sources showed the predominant effect of indirect GHG emissions costs over system costs on the optimal solutions in higher emission factors.

Nomenclatures

Sets:

AL= allowable zones for water transfer

c= discrete points of plant capacities

i, j= zone

m= discrete points of storage tank capacities

t, s, b= planning horizon, season and time block, respectively

Continuous variables:

$addP$= added share of grid electricity (kWh)

$addRE$= added share of PV output

$addk$= added PV installation density (%)

k= PV installation density (%)

$MaxR$= Max. PV output (kWh)

P= share of grid electricity (kWh)

Q= desalinated water produced (m³)

RE= share of renewable energy (kWh)

V= existing water storage (m³)

Parameters:

Cap = capital cost for plant ($) and for pipeline ($/km)

Ce= variable grid electricity cost ($/kWh)

Cfe= fixed grid electricity cost ($/day)

COM= plants O&M cost ($/m³)

$convf2$= conversion factor (km/m)

Cr= renewable electricity cost ($/kWh)

Cs= O&M cost of water storage ($/m³)

D= residential energy demand (kWh)

$CTax$= Carbon abatement cost ($/kgCO₂)

$exisR$= number of existing residential rooftops

f^{GHG}= GHG emission factor (kgCO₂/kWh)

DF= discount factor

dur= duration of the time block (h)

kp= economic-subjectivity index

L= distance (m)

$MaxPS$= Max. substation capacity (kW)

nd= number of days (day)

PV= PV system output (kW)

$Surp$= surplus PV output (kWh)

Binary variables:

$np, npIJ$= decisions for construction/ expansion capacity of the pipeline

X, XW= decisions for storage tank size and plant capacity, respectively

Superscripts associated with:

bi= business sector; w= water

DP= plant; PI= pipeline; STT= storage tank

r, rn= households with and without PVs, respectively

References

AEC, 2017, Australian Energy Council, Solar report, https://www.energycouncil.com.au, [accessed 11.01.18].

DEE, 2017, Department of the Environment and Energy, National greenhouse accounts factors, https://www.environment.gov.au, [accessed 02.02.18].

ERA, 2017, Economic Regulation Authority, The efficient costs and tariffs of the Water Corporation, Aqwest and Busselton Water, https://www.erawa.com.au, [accessed 17.02.18].

J. Gifford, 2011, Survey and sustainability of energy technologies [M.Sc. Thesis], Iowa State University.

R. F. C. Miranda, A. Szklo, R. Schaeffer, 2015, Technical-economic potential of PV systems on Brazilian rooftops, Renewable Energy, 75, 694-713.

RBA, 2017, Reserve Bank of Australia Exchange rate data, mountly data, http://www.rba.gov.au/statistics/historical-data.html#exchange-rates, [accessed 02.01.17].

N. Vakilifard, P. A. Bahri, M. Anda, G. Ho, 2017, Water security and clean energy, co-benefits of an integrated water and energy management, In: Computer aided process engineering, Elsevier, 40, 1363-1368.

N. Vakilifard, P. A. Bahri, M. Anda, G. Ho, 2018, Integrating real-time operational constraints in planning of water and energy supply, In: Computer aided process engineering, Elsevier, 43, 313-318.

WSAA, 2012, Water Services Association of Australia Ltd. , Occasional paper 28: Cost of carbon abatement in the urban water industry, https://www.wsaa.asn.au/publications?page=13, [accessed 18.01.18].

Anton A. Kiss, Edwin Zondervan, Richard Lakerveld, Leyla Özkan (Eds.)
Proceedings of the 29th European Symposium on Computer Aided Process Engineering
June 16th to 19th, 2019, Eindhoven, The Netherlands. © 2019 Elsevier B.V. All rights reserved.
http://dx.doi.org/10.1016/B978-0-128-18634-3.50203-4

Short-term Scheduling of a Multipurpose Batch Plant Considering Degradation Effects

Ouyang Wu[a,b], Giancarlo Dalle Ave[c,d], Iiro Harjunkoski[c], Lars Imsland[b], Stefan Marco Schneider[a], Ala E.F. Bouaswaig[a] and Matthias Roth[a]

[a]*Automation Technology, BASF SE, 67056 Ludwigshafen, Germany*
[b]*Department of Engineering Cybernetics, NTNU, 7491 Trondheim, Norway*
[c]*ABB Corporate Research Germany, 68526 Ladenburg, Germany*
[d]*Dept. of Biochemical & Chemical Engineering, TU Dortmund, 44221 Dortmund, Germany*
ouyang.wu@ntnu.no

Abstract

Fouling is a typical type of degradation in the process industries, which results in significant negative effects on the efficiency of plants. This paper considers a case study, where production tasks with different recipes contribute to the batch-to-batch fouling evolution depending on their sequence. The scheduling of the production tasks is improved by explicit consideration of the degradation effects, resulting in an optimized production sequence. A set of continuous-time MILP-based scheduling models integrated with degradation models are employed to formulate this batch scheduling problem. The proposed formulations are tested for different problem sizes of the case study illustrating the effectiveness of each of the proposed approaches.

Keywords: multipurpose batch scheduling, sequence-dependent degradation, continuous-time MILP formulation, disjunctive model, precedence model

1. Introduction

Multipurpose chemical batch plants are often a challenge to schedule. These processes typically produce a large number of products requiring several processing steps. When scheduling such processes several factors must be taken into account including, but not limited to, process throughput, storage constraints, equipment condition, and maintenance concerns. When creating scheduling models, one of the major decisions is the time representation chosen for the problem. The time representation can either be continuous or discrete. Continuous time problems in general result in far fewer variables than their discrete counterparts and provide an explicit representation of timing decisions. Continuous time scheduling formulations are broadly classified into precedence, single, or multiple time grid based models (Méndez et al., 2006). Continuous time models have been applied to many different scenarios. Castro and Grossmann (2012) used generalized disjunctive programming to derive generic continuous-time scheduling models.

One aspect of scheduling that has been gaining attention in literature lately is that of integrating equipment condition into production scheduling. Vieira et al. (2017) studied the optimal planning of a continuous biopharmaceutical process with decaying yield. They formulated the problem as a continuous single-time grid formulation making both production scheduling and maintenance scheduling decisions. Other applications of combined maintenance and production scheduling can be found in Biondi et al. (2017) and Dalle Ave et al. (2019) who studied the joint production and maintenance problem of steel plant scheduling under different conditions.

The goal of this work is to investigate the production scheduling of a chemical batch plant with unit degradation. A batch reactor case study considering fouling is presented in which the fouling prolongs the time needed to complete a batch and is highly influenced by the sequence of products produced in the reactor. The problem is formulated using a set of continuous-time MILP-based scheduling models which will be described and compared in the remainder of this work.

2. Problem description

This case study focuses on a batch polymerization process. Firstly, the product monomers are mixed in a single tank before being homogenized and dispatched to the reactors. Once the monomers have reacted they are dispatched via a shared piping system to storage. The major cause of degradation in this process is fouling in the reactors; polymer residues are accumulated on the inner surfaces of the reactors, heat exchangers, and pipes. This causes reduced heat transfer from the product to the coolant thereby decreasing cooling efficiency and prolonging batch duration. Moreover, the flow resistance due to fouling leads to an increased pressure drop over the heat exchanger. A pressure-based key performance indicator (KPI) was developed to indicate the degree of fouling for each batch run using a state estimation approach, and in this approach many interfering factors due to batch production are excluded from the fouling KPI (Wu et al., 2019). The batch-to-batch behaviors of the fouling evolution discussed in Wu et al. (2018, 2019) indicate that different batch recipes contribute to the fouling evolution differently. Hence, a linear model based the fouling KPI is developed to describe the effects of batch sequences on the batch-to-batch fouling growth, given as:

$$f_{k,j} = A_{R_{k-1,j}\, j} \cdot f_{k-1,j} + B_{R_{k-1,j}\, j} \tag{1}$$

where, $R_{k,j} = r$ indicates Recipe r at the kth batch of Unit j; $f_{k,j}$ is the fouling KPI for the kth batch of Unit j, which is assumed to follow recipe-based unit-specific linear dynamics $\{(A_{rj}, B_{rj}) \mid r = 1, 2, ..., |R|, \ j = 1, 2, ..., |J|\}$.

The recipe determines the total processing time at each unit, and the reaction duration, which is also affected by the coolant temperature and the fouling. The degree of fouling is approximated using eq. (1), and the coolant temperature is considered as an external disturbance due to uncertainties among the cooling capacities and the demands. Moreover, the reactors and other units are not identical, which also leads to differences in the processing time. A linear model structure is applied to fit the processing time as a function of the fouling KPI and the coolant temperature given the recipe $R_{k,j}$ in Unit j, the coolant temperature is assumed to be fixed for the scheduling scenario. Therefore, the processing time is calculated using only the fouling KPI as eq. (2) shows:

$$Du_{k,j} = AD_{R_{k,j}\, j} \cdot f_{k,j} + BD_{R_{k,j}\, j} \tag{2}$$

where, $Du_{k,j}$ is the reaction processing time at the kth batch of Unit j; $\{AD_{rj}, BD_{rj}\}$ are the model parameters for Recipe r at Unit j. The degradation impairs the batch production capacity and therefore needs to be explicitly considered in production scheduling. Provided the quantitative models in eqs. (1) and (2), a new scheduling problem considering degradation effects is formulated by integrating the degradation models into existing scheduling optimization approaches. In the next section, continuous-time MILP approaches using precedence concepts are presented for the aforementioned batch scheduling problem.

3. Methodology

In this section, four continuous-time MILP scheduling formulations are presented which integrate the sequence-dependent degradation model using precedence-based constraints. Furthermore, no changeover time and no intermediate storage are considered for the process specifications. Note that while this work focuses on a particular case study, the methodology presented here is generic and could be adapted to other scenarios.

3.1. Nomenclature

I, J, L, R	Sets of batch orders ($i = 1, 2, ..., \lvert I \rvert$), units ($j = 1, 2, ..., \lvert J \rvert$), stages ($l = 1, 2, ...,$ $\lvert L \rvert$) and batch recipes ($r = 1, 2, ..., \lvert R \rvert$),
J_l, J_{de}, L_{de}	Sets of units belonging to Stage l, units suffered from degradation and stages containing units suffered from degradation,
Ts_{il}, Te_{il}	Start time and end time of Order i in Stage l,
Tp_{ij}, Tr_l	Processing time of Order i at Unit j, transfer time in Stage l,
Ta_{ij}	Additional processing time of Order i at Unit j due to degradation,
f_{ij}, FI_j, R_{ri}	Degradation (fouling) KPI for Order i at Unit j, initial fouling KPI at Unit j, recipe binary indicator for Order i using Recipe r,
A_{rj}, B_{rj}	Degradation model parameters for Recipe r at Unit j, see eq. (1),
AD_{rj}, BD_{rj}	Duration model parameters for Recipe r at Unit j, see eq. (2),
$X_{ii'j}^{ui}, W_{ij}$	Sequencing binary variables for a unit-specific immediate precedence model,
$X_{ii'l}^{im}, Y_{ij}, W_{ij}$	Sequencing binary variables for an immediate precedence model,
$X_{ii'l}^{g}, Y_{ij}$	Sequencing binary variables for a general precedence model,
S_{ij}	Absolute position of Order i in the sequence of Unit j,

3.2. Sequence-dependent degradation

The continuous-time immediate precedence fits the form of the sequence-dependent degradation model as shown in eq. (1). Using a unit-specific immediate precedence model, the degradation initialization and propagation are presented in the form of disjunctive constraints as eqs. (3) and (4) show: if Order i is immediately processed after Order i' at Unit j ($X_{i'ij}^{ui} = 1$), then the batch degradation KPI (f_{ij}) is calculated from its previous value ($f_{i'j}$) using eq. (1); if Order i is processed first in the sequence of unit j ($W_{ij} = 1$), then f_{ij} equals an initial value of the fouling KPI. To avoid nonlinear constraints, the processing time of Order i at Unit j is further divided into a varying duration due to fouling (Ta_{ij}) and a constant duration (Tp_{ij}) according to eq. (2), where the calculations of Ta_{ij} and Tp_{ij} are presented in eq. (4).

$$\bigvee_{i, i' \in I: i \neq i', j \in J_{de}} \begin{bmatrix} X_{i'ij}^{ui} \\ f_{ij} \geqslant AS_{ij} \cdot f_{i'j} + BS_{ij} \\ f_{ij} \leqslant AS_{ij} \cdot f_{i'j} + BS_{ij} \end{bmatrix}, \quad \bigvee_{i \in I, j \in J_{de}} \begin{bmatrix} W_{ij} \\ f_{ij} \geqslant FI_j \\ f_{ij} \leqslant FI_j \end{bmatrix} \tag{3}$$

$$Ta_{ij} = \sum_{r \in R} R_{ri} \cdot AD_{rj} \cdot f_{ij}, \quad Tp_{ij} = \sum_{r \in R} R_{ri} \cdot BD_{rj}, \; \forall i \in I, j \in J_{de} \tag{4}$$

where, $AS_{ij} = \sum_{r \in R} R_{ri} A_{rj}$, and $BS_{ij} = \sum_{r \in R} R_{ri} B_{rj}$.

3.3. Unit-specific immediate precedence model [M1]

A single-stage unit-specific immediate precedence model is proposed in Cerdá et al. (1997), and an adapted version for multistage problems considering unit degradation is presented as follows: given the sequence binaries ($X_{i'ij}^{ui} = 1$, if Order i is processed immediate after Order i' at Unit j, $W_{ij} = 1$ if Order i is processed in the first place of the sequence at Unit j), the unit-specific immediate precedence constraints are presented in eqs. (5) and (6); the order timing constraints for a single order are presented in eq. (7); the order sequencing and timing constraints for two orders in a single unit are presented using disjunctive constraints as shown in eq. (8). The order sequencing and timing constraints for two orders in neighboring stages considering no immediate storage are presented in eq. (9) (the transfer procedure occupies both units in the two stages); the degradation and duration model constraints are presented in eqs. (3) and (4).

$$\sum_{i \in I} W_{ij} = 1, \; \forall j \in J \tag{5}$$

$$\sum_{j' \in J_l} W_{ij'} + \sum_{i' \in I: i' \neq i, j' \in J_l} X_{i'ij'}^{ui} = 1, \; \sum_{i' \in I: i' \neq i, j' \in J_l} X_{ii'j'}^{ui} \leqslant 1, \; \forall i \in I, l \in L \tag{6}$$

$$Te_{il} = Ts_{il} + \sum_{j \in J_l} [Tp_{ij}(W_{ij} + \sum_{i' \in I: i' \neq i} X_{i'ij}^{ui}) + Ta_{ij}], \; Ts_{i(l+1)} \geqslant Te_{il} + Tr_l, \; \forall i \in I, l \in L \tag{7}$$

$$\bigvee_{i,i'\in I: i\neq i',\, l\in L} \left[\begin{array}{c} \sum_{j\in J_l} X^{ui}_{i'ij} \\ Te_{i'l} + Tr_l \leqslant Ts_{il} - Tr_{(l-1)} \end{array} \right] \quad (8) \qquad \bigvee_{i,i'\in I: i\neq i',\, l\in L: l<|I|} \left[\begin{array}{c} \sum_{j\in J_l} X^{ui}_{i'ij} \\ Ts_{il} \geqslant Ts_{i'(l+1)} + Tr_l \end{array} \right] \quad (9)$$

3.4. Immediate precedence model [M2]

In contrast to M1, the immediate precedence model (Gupta and Karimi, 2003) uses two types of binary variables $\{X^{im}_{i'il}, Y_{ij}\}$ ($X^{im}_{i'il} = 1$ indicates Order i is immediately processed after Order i' in some units of Stage l; $Y_{ij} = 1$ indicates Order i is assigned to Unit j), which fits into the sequence-dependent degradation in eq. (3) by replacing $X^{ui}_{i'il}$ with $X^{im}_{i'il} \wedge Y_{ij} \wedge Y_{i'j}$. The order assignment and immediate precedence constraints are presented in eqs. (5), (10) and (11), and the order timing constraints for a single order are presented in eq. (12). Lastly, the order sequencing and timing constraints for two orders in a single unit and neighboring stages are presented in eq. (13).

$$\sum_{j\in J_l} Y_{ij} = 1,\; \sum_{j\in J_l} W_{ij} + \sum_{i'\in I: i'\neq i} X^{im}_{i'il} = 1,\; \sum_{i'\in I: i'\neq i, l\in L} X^{im}_{ii'l} \leqslant 1,\; \forall\, i\in I,\, l\in L \qquad (10)$$

$$Y_{ij} \leqslant Y_{i'j} + 1 - X^{im}_{ii'l} - X^{im}_{i'il},\; \forall\, i,i'\in I: i\neq i',\, j\in J_l, l\in L \qquad (11)$$

$$Te_{il} = Ts_{il} + \sum_{j\in J_l}(Tp_{ij}Y_{ij} + Ta_{ij}),\; Ts_{i(l+1)} \geqslant Te_{il} + Tr_l,\; \forall\, i\in I, l\in L \qquad (12)$$

$$\bigvee_{i,i'\in I: i\neq i',\, l\in L} \left[\begin{array}{c} X^{im}_{i'il} \\ Te_{i'l} + Tr_l \leqslant Ts_{il} - \\ Tr_{(l-1)} \end{array} \right],\quad \bigvee_{i,i'\in I: i\neq i', l\in L: l<|I|} \left[\begin{array}{c} X^{im}_{i'il} \\ Ts_{il} \geqslant Ts_{i'(l+1)} + Tr_l \end{array} \right] \qquad (13)$$

3.5. Hybrid precedence model [M3]

When comparing to M1 and M2, general precedence models proposed by Méndez and Cerdá (2003) generally prevail with less computation cost. These models use binary variables $X^g_{i'il} = 1$, $Y_{ij} = 1$, $Y_{i'j} = 1$ to represent that Order i is processed after i' at Unit j in Stage l. But unlike M1 and M2, it does not provide immediate precedence relations. Given that the sequence-dependent degradation only occurs in certain stages, a hybrid precedence model is developed by replacing M2 with general precedence constraints in the stages requiring no immediate precedence information ($l \in L_n = L \wedge \neg L_{de}$). The precedence constraints are presented in eq. (14), and the order timing constraints for a single order are the same as M2 as eq. (12) shows. The order sequencing and timing constraints for two orders in the same stage or neighboring stages are presented in eq. (15).

$$\sum_{j\in J_l} Y_{ij} = 1,\; X^g_{ii'l} + X^g_{i'il} = 1, \forall i,i'\in I: i\neq i', l\in L - L_{de} \qquad (14)$$

$$\bigvee_{\substack{i,i'\in I: i'\neq i,\\ j\in J_l, l\in L_n}} \left[\begin{array}{c} X^g_{i'il} \wedge Y_{ij} \wedge Y_{i'j} \\ Te_{i'l} + Tr_l \leqslant Ts_{il} - Tr_{(l-1)} \\ Ts_{il} \geqslant Ts_{i'(l+1)} + Tr_l \end{array} \right],\quad \bigvee_{i,i'\in I: i'\neq i, l\in L_n} \left[\begin{array}{c} X^g_{i'il} \\ Ts_{i'l} + Tr_l \leqslant Ts_{il} \end{array} \right] \qquad (15)$$

3.6. General precedence based model [M4]

A general precedence based model for sequence-dependent changeovers is proposed in Aguirre et al. (2012, 2017), which also can be adapted to the scheduling problem considering sequence-dependent degradation. The general precedence model presented in M3 is extended to all stages, and the immediate precedence binary variables $X^{im}_{i'il}$, W_{ij} are introduced for the stages and units having sequence-dependent behaviors ($j \in J_{de}, l \in L_{de}$). By introducing a continuous variable S_{ij}, defined as the absolute position of batch i in the sequence of Unit j, the immediate precedence variables are connected to the general precedence model using eq. (16).

$$\bigvee_{i,i'\in I: i\neq i', j\in J_l, l\in L_{de}} \left[\begin{array}{c} X^g_{i'il} \wedge Y_{ij} \wedge Y_{i'j} \\ S_{ij} \geqslant S_{i'j} + 1 \\ X^{im}_{i'il} + S_{ij} - 1 \geqslant S_{i'j} + 1 \end{array} \right],\; Y_{ij} \leqslant S_{ij} \leqslant \sum_{i\in I} Y_{ij},\; \forall i\in I, j\in J_{de} \qquad (16)$$

3.7. Big-M reformulation and objective function

The disjunctive programming constraints in the precedence models are reformulated into the MILP frameworks using the big-M approach. More details can be found in Castro and Grossmann (2012). This objective function considered in this work is the minimization of makespan considering fouling-influenced batch time.

4. Case study: computational results

Size	M1	M2	M3	M4
12(6,6)	3495.02 (53%)	**3495.02(74sec)**	3495.02(37%)	3495.02(37%)
15(7,8)	4315.79 (70%)	**4266.73(979sec)**	4282.61(42%)	4266.73(42%)
18(7,8,3)	5164.13 (79%)	5589.43(84%)	5116.97(46%)	5164.13(47%)
25(6,5,4,5,5)	7237.93 (91%)	13339.37(94%)	7034.91(51%)	7375.89(53%)

Table 1: Computational results of different formulations: objective function and optimality gap

Referring to the aforementioned case studies, the monomer make-up section and the reaction section are taken as the two production stages with one monomer vessel (U1) in Stage L1 and two non-identical reactors (U2 and U3) in Stage L2. The process topology can be viewed in Figure 1. Due to the lack of intermediate storage, the homogenization section (L1-Tr) is modeled as a transfer procedure between the stages. This transfer occupies both the discharging mixer as well as the reactor that is being charged, which is modeled using eq. (9) for model M1. Additionally, the only pipeline (L2-Tr) transfers product from either U2 or U3 into storage. To prevent multiple orders overlapping in L2-Tr, the ordering and timing constraints are used in a similar way as shown in eq. (8) by introducing an artificial precedence binary for this unit. In this case study, when minimizing makespan, it is not possible for jobs to overtake one another in L2 and therefore the precedence binary variable for L2-Tr same one for L1. In addition, an operational threshold is added on the degradation indicator.

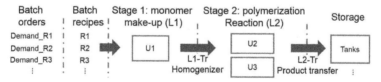

Figure 1: The topology of of the batch process

Each of the four models defined earlier were tested for varying problem sizes. Take 15(7,8) for example, 15 refers to the total amount of product demand, and the values inside the parentheses (7,8) refer to the demands for different recipes. The parameters of the scheduling problem are obtained from the case study and the historical data. The models are implemented in GAMS 25.1 and solved using CPLEX 12.8, and the time limit is set to 1200 CPU seconds. The Gantt chart and fouling curves for the 25-order case using M3 are illustrated in fig. 2; the normalized fouling KPI data are presented for two units (U2 and U3) with different symbols denoting the recipes from R1 to R5, and the threshold for the normalized fouling KPI is one. Furthermore, the computational results are presented in table 1 with objective values and relative optimality gaps. Results show that the small-size problems using M2 are solved to zero gap before 20 minutes but the computation cost increases dramatically with larger problem size; M3 has better computational performance (smaller value of objective function and smaller relative optimality gap) for the problem sizes 18(7,8,3) and 25(6,5,4,5,5).

5. Conclusions

A short-term batch scheduling problem considering sequence-dependent degradation was formulated using different continuous-time MILP methods. Taking an industrial case study as example,

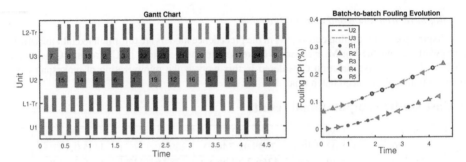

Figure 2: Result of the 25-order problem using M3

the recipe-based degradation models are integrated with the precedence models to improve the scheduling of batch production. Four integrated MILP formulations are compared by running the scheduling models with different problem sizes, and the hybrid precedence model is shown to have the best computation performance for relatively large-size problems. However, the investigated formulations do not find a provably optimal solution within 20 minutes; future work involves improving the computational performance of such models.

Acknowledgments: Financial support is gratefully acknowledged from the Marie Skłodowska Curie Horizon 2020 EID-ITN project "PRONTO", Grant agreement No 675215.

References

A. M. Aguirre, S. Liu, L. G. Papageorgiou, 2017. Mixed integer linear programming based approaches for medium-term planning and scheduling in multiproduct multistage continuous plants. Industrial & Engineering Chemistry Research 56 (19), 5636–5651.

A. M. Aguirre, C. A. Méndez, P. M. Castro, C. De Prada, 2012. Milp-based approach for the scheduling of automated manufacturing system with sequence-dependent transferring times. In: Computer Aided Chemical Engineering. Vol. 30. Elsevier, pp. 477–481.

M. Biondi, G. Sand, I. Harjunkoski, 2017. Optimization of multipurpose process plant operations: A multi-time-scale maintenance and production scheduling approach. Computers & Chemical Engineering 99, 325 – 339.

P. M. Castro, I. E. Grossmann, 2012. Generalized disjunctive programming as a systematic modeling framework to derive scheduling formulations. Industrial & Engineering Chemistry Research 51 (16), 5781–5792.

J. Cerdá, G. P. Henning, I. E. Grossmann, 1997. A mixed-integer linear programming model for short-term scheduling of single-stage multiproduct batch plants with parallel lines. Industrial & Engineering Chemistry Research 36 (5), 1695–1707.

G. Dalle Ave, J. Hernandez, I. Harjunkoski, L. Onofri, S. Engell, 2019. Demand side management scheduling formulation for a steel plant considering electrode degradation. In: IFAC International Symposium on Dynamics and Control of Process Systems (DYCOPS).

S. Gupta, I. Karimi, 2003. An improved milp formulation for scheduling multiproduct, multistage batch plants. Industrial & engineering chemistry research 42 (11), 2365–2380.

C. A. Méndez, J. Cerdá, 2003. An milp continuous-time framework for short-term scheduling of multipurpose batch processes under different operation strategies. Optimization and Engineering 4 (1-2), 7–22.

C. A. Méndez, J. Cerdá, I. E. Grossmann, I. Harjunkoski, M. Fahl, 2006. State-of-the-art review of optimization methods for short-term scheduling of batch processes. Computers & Chemical Engineering 30 (6-7), 913–946.

M. Vieira, T. Pinto-Varela, A. P. Barbosa-Pvoa, 2017. Production and maintenance planning optimisation in biopharmaceutical processes under performance decay using a continuous-time formulation: A multi-objective approach. Computers & Chemical Engineering 107, 111 – 139.

O. Wu, A. E. F. Bouaswaig, S. M. Schneider, F. Moreno Leira, L. Imsland, M. Roth, 2018. Data-driven degradation model for batch processes: a case study on heat exchanger fouling. In: Computer Aided Chemical Engineering. Vol. 43. Elsevier, pp. 139–144.

O. Wu, L. Imsland, E. Brekke, S. M. Schneider, A. E. F. Bouaswaig, M. Roth, 2019. Robust state estimation for fouling evolution in batch processes using the em algorithm. In: IFAC International Symposium on Dynamics and Control of Process Systems (DYCOPS).

Anton A. Kiss, Edwin Zondervan, Richard Lakerveld, Leyla Özkan (Eds.)
Proceedings of the 29th European Symposium on Computer Aided Process Engineering
June 16th to 19th, 2019, Eindhoven, The Netherlands. © 2019 Elsevier B.V. All rights reserved.
http://dx.doi.org/10.1016/B978-0-128-18634-3.50204-6

Real-time determination of optimal switching times for a H_2 production process with CO_2 capture using Gaussian Process Regression models

Luca Zanella[a], Marcella Porru[b,*], Giulio Bottegal[b], Fausto Gallucci[b], Martin van Sint Annaland[b] and Leyla Özkan[b]

[a]*University of Padova, Via Marzolo 9, Padova, 35131, Italy*
[b]*Eindhoven University of Technology, Eindhoven, 5600 MB, The Netherlands*
* m.porru@tue.nl

Abstract

This work presents a systematic methodology to determine in real-time the optimal durations of the three stages of a new Ca-Cu looping process for H_2 production with integrated CO_2 capture. Economic and quality criteria are proposed to determine the appropriate time to switch between the stages. These criteria rely on the time-profiles of some key variables, such as product concentrations. Given the delayed nature of hardware sensor measurements, the real-time determination of such variables is based on soft-sensors. For this purpose, Gaussian Process Regression models are employed. The predictive capabilities of these models are tested on several datasets, yielding reliable predictions in most of the cases. The values of the optimal switching times computed with the proposed method differ from the actual values by 4 %, at most.

Keywords: machine learning, gaussian process regression models, CO_2 capture, process intensification, process monitoring.

1. Introduction

The calcium-copper (Ca-Cu) looping technology is a cyclic, three-stages process for H_2 production with integrated CO_2 capture conducted in dynamically operated packed bed reactors (Abanades et al., 2010). While a satisfactory understanding of the underlying phenomena is available, investigations in the field of process automation are needed to efficiently upscale the process. Since the behaviour of each stage is inherently dynamic, one issue is the assessment of the optimal duration. The main aim of this work is to develop a methodology to identify the appropriate time instant to switch between stages. This is crucial to match satisfactory operating targets, including economic profitability and product quality specifications. Once suitable switching criteria have been defined, the optimal duration of the stages can be either scheduled or computed in real-time. The second approach is preferable, because it allows more flexibility in dealing with disturbances and process uncertainties. In practice, not all the variables needed for the evaluation of the switching criteria can be sensed in real-time. This is the case of composition measurements obtained via hardware analysers, which are often affected by substantial time-delays. Instead, soft-sensors can be employed as virtual analysers. Soft-sensors are models that allow the estimation of hard-to-measure, primary variables, making use of

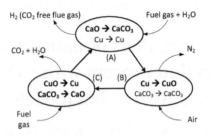

Figure 1: The three stages Ca-Cu chemical looping process for H_2 production.

secondary variables, that are easily sensed online. Several soft-sensing techniques have been reported in literature. A common classification is based on the type of models employed, either first principle or data-driven. First principle models usually allow reliable extrapolation (Porru et al., 2013), but lack of knowledge in some aspects of the underlying physics might result in poor performance (Porru and Özkan, 2017). Moreover, model complexity might make their solution too computationally demanding for real-time applications (Shokry et al., 2017). Conversely, data-driven models may suffer from limited extrapolation capabilities, since they only encode information available in the training data (Bonvin et al., 2016). Hence, the amount of data required for their training is usually large. A major advantage compared to first-principle models is that a deep knowledge of the process is not necessary. Hence, data-driven soft-sensors are particularly suitable to monitor complex and recursive processes, such as the Ca-Cu looping for H_2 production. Among a plethora of data-driven models, Gaussian Process Regression (GPR) is a method for supervised learning (Rasmussen and Williams, 2006). It allows performing regression without giving explicit parametrization to the underlying function, instead determining a distribution over the possible functions that are consistent with the observed data. The application of GPR has several advantages over popular machine learning methods used in process engineering, such as artificial neural networks. These include limited number of parameters to be estimated, the capability of automatically providing confidence intervals for the predictions and that of dealing with noisy measurements, by acting as *smoothing devices*. Although GPR has received great interest in recent years, its use for soft-sensing is still rather limited (Mei et al., 2016). Here, GPR models are proposed as soft-sensors to evaluate the conditions that trigger the switch from one stage to the next.

2. The Ca-Cu looping process

The Ca-Cu looping process is a novel concept for high purity H_2 production from CH_4 with integrated CO_2 capture. The process has been gaining increasing attention as a low-CO_2 emission solution for both power generation and H_2 production (Martini et al., 2016, 2017). The process consists of three stages, conducted cyclically in packed bed reactors. Switching between stages is performed by alternating the gaseous reactants fed to the reactor and changing the operating conditions in the fixed bed, mainly temperature and pressure. Figure 1 schematically shows the inlet and outlet flows of the stages and the main reactions involved in the solid phase. Data sets employed in the present work are collected via an experimental campaign. The solid bed contains three different solids: a Ni-based catalyst, to catalyse the reforming reactions, CaO, employed as CO_2-sorbent material and Cu which is involved in the regeneration of CaO. The available set up allows the measurement of the following variables: temperature of the solid bed along the axial profile, $T_s(z)$, inlet and outlet gas temperatures, T_G^0 and T_G^{out}, inlet and outlet gas pressure, P_G^0 and P_G^{out}, outlet (dry) volume fractions, x_i^{out} and outlet (dry) volume flow rate, $\dot{V}_{tot}^{out}$.

In Stage A, sorption enhanced reforming takes place. The reactor is fed with a mixture of CH_4 and steam with a given steam-to-carbon ratio (S/C). The three main reactions are steam-methane reforming (SMR), water-gas shift (WGS) and carbonation of CaO. The last reaction allows the removal of the CO_2 formed, increases H_2 purity, CH_4 and CO conversions by shifting the SMR and WGS equilibria toward the products. In addition, the combination of these reactions makes the overall stage exothermic, so that the only energy duty is the one needed to compensate for the heat losses. This stage is considered terminated when the CaO is fully converted to $CaCO_3$, indicating that no more CO_2 can be captured by the fixed bed. To allow continuous operation, the $CaCO_3$ formed in this stages is regenerated to CaO by the Cu/CuO chemical loop carried out in stages B and C. In stage B, Cu is oxidized to CuO by feeding pressurized diluted air to the reactor. This stage is considered concluded when most of the Cu has been oxidised. The high pressure and low O_2 content prevent the calcination of $CaCO_3$.

Switching to stage C is performed by changing the feed to H_2 or other reducing species at lower pressure, such as syngas or CH_4. The reduction of CuO to Cu provides the energy required for the calcination of $CaCO_3$ that regenerates the CO_2 sorbent. When most of the CuO has been reduced, switching to stage A is performed again.

3. Switching criteria

Economic and quality criteria are proposed to determine the appropriate time to switch (t_s) to the next stage. In the previous section the definition of end times was based on the state assumed by the solid bed. However, measurements of the solid bed composition are not available, nor easy to perform in-situ at industrial scale. Therefore, the course of each stage should be monitored employing other variables, such as temperature or gas composition. In this work, t_s is determined via an optimization-based approach. Table 1 reports the objective functions to be maximized and constraints. P (€/time) is a profit

Table 1: Switching criteria employed for stages A, B and C.

Stage	Objective Function	Constraint
A	$\int_0^t P(\tau)\,d\tau$	$x_{H_2}^{out}(t) \geq 0.85$
B	$T_s^{out}(t)$	$\dot{M}_{O_2}^{out}(t)/\dot{M}_{O_2}^0(t) \leq 0.35$
C	$T_s^{out}(t)$	$\dot{M}_{H_2}^{out}(t)/\dot{M}_{H_2}^0(t) \leq 0.35$

function, depending on several variables, including V_{tot}^{out}, x_i^{out}, prices of chemicals, carbon tax on CO_2 emissions etc. $\dot{M}_i$ is the mass flow rate of species i, computed with the ideal gas law. T_s^{out} is the outlet bed temperature. Although quite simple, the selected criteria aim to fulfil multiple requirements. For stage A, operation should last until it is economically profitable while the required H_2 purity constraint is not violated. For stages B and C, the proposed criteria aim to maximize the amount of Cu oxidation and reduction, respectively.

The constraint is motivated by the fact that a release of considerable amounts of O_2 and H_2 is an indication of the two stages coming to an end. The behaviour of the objective functions has been studied offline (i.e. after completion of the batches, and collection of the data), finding that they admit a unique maximum. These criteria are applied to all the datasets collected, considering the stages separately, resulting in a wide dispersion of the optimal durations, even at fixed operating conditions. Such differences are mainly linked to run-to-run variations of hardly controllable variables, such as the solid bed state. This makes it necessary to implement the criteria in real-time (i.e. during the batch run, as soon as a new data point is collected), to enable timely switching. The application of these criteria lies on the availability of the time-profiles of concentrations of several components with delayed measurements. Therefore, an inferential model is needed to allow their estimation in real-time. This is achieved by employing GPR.

4. Method: Gaussian Process Regression models

Let $\mathscr{D} = \{(\mathbf{x}_i, y_i) \mid i = 1, \ldots n\}$ be a training set of input-output pairs, where n is the number of measurement points, $\mathbf{x}_i$ is a vector collecting the inputs (often collected in a matrix X) and y_i is the measured output at point i (often collected in a vector $\mathbf{y}$). The goal of GPR is to perform nonlinear regression without giving explicit parametrization of the underlying function. A prior probability is assigned to any possible function, such that higher probabilities are given to those functions which are considered more likely to describe the input-output observations (Mackay, 1998). The functions providing the best representation of the data are found by performing Bayesian inference collectively at the training points. Training of GPR models consists of two main steps:

a. Model selection. It consists of the choice of a kernel, k, which is a function of two arguments that set qualitative properties of the unknown function, such as smoothness or periodicity. The kernel measures the similarity between two training points x_i and x_j.

b. Model training. This step involves the tuning of a few hyperparameters, denoted by $\boldsymbol{\theta}$, through which k is parametrized in order to give flexibility to the selected model. In this work, the estimation of $\boldsymbol{\theta}$ is achieved by maximization of marginal likelihood (we refer to Rasmussen and Williams (2006), Ch. 5, for details on marginal likelihood estimation).

Once the model is trained, the prediction of the output y_* at any point x_* that does not belong to the training set is computed assuming that its conditional probability given the training data follows a Gaussian distribution, namely $y_* | \mathbf{y} \sim \mathcal{N}(K_*(K + \gamma I)^{-1}\mathbf{y}, K_{**} - K_*(K + \gamma I)^{-1}K_*^{\top})$, where K is a covariance matrix containing kernel evaluation of each pair of points that belong to the training set, γ is a hyperparameter tuning the smoothness of the function, K_* is the vector containing the kernel evaluations between all training points and the test point and K_{**} is the variance on the test point. The predicted value y_* is computed as the mean of the previous distribution: $y_* = \bar{y}_* = K_*(K + \gamma I)^{-1}\mathbf{y}$. The method automatically provides the confidence interval of the predicted value of the function as the posterior variance of the Gaussian distribution through the simple expression: $var(y_*) = K_{**} - K_*(K + \gamma I)^{-1}K_*^{\top}$.

5. Online determination of the switching time

The use of criteria in Table 1 makes it necessary to predict in real-time the following primary variables, $\mathbf{y}$: $\dot{V}_{tot}^{out}$, $x_{H_2}^{out}$, $x_{CO_2}^{out}$ during stage A, $\dot{V}_{tot}^{out}$, $x_{O_2}^{out}$ during stage B and $\dot{V}_{tot}^{out}$, $x_{H_2}^{out}$

during C. Separate GPR models have been trained for each of them, because the method holds for scalar outputs. Both training and test sets are a subset of the data sets employed in the previous section. As the computational time needed to train a model is $\mathcal{O}(n^3)$, model training is performed offline. The MATLAB® routine fitrgp is employed for this purpose. Among all the variables sensed in the set up, the predictors are selected as the ones satisfying two requirements: (i) being influential on the outputs, since employing non-influential variables is observed to cause overfitting and large prediction variances; (ii) being available in real-time. Based on these considerations, t, P_G^0 and inlet compositions, specifically S/C, $x_{O_2}^0$ and $x_{H_2}^0$ in stages A, B and C, respectively, are employed as predictors to target all the outputs. Other variables are expected to be influential on the outputs. However, their variability in the available datasets is rather limited, hence including them would be of little use. The predictor matrices, X_A, X_B and X_C for the three stages are constructed as $X_A = [\mathbf{t} \quad \mathbf{P_G^0} \quad \mathbf{S/C}]$, $X_B = [\mathbf{t} \quad \mathbf{P_G^0} \quad \mathbf{x_{O_2}^0}]$ and $X_C = [\mathbf{t} \quad \mathbf{P_G^0} \quad \mathbf{x_{H_2}^0}]$, respectively. Bold symbols identify vectors, whose entries are the values of the variable along the time series, $\mathbf{t}$. Due to the computational demand of GPR, the number of training data is set equal to a few thousand, building matrices of dimensions $\dim X_A = 1464 \times 3$, $\dim X_B = 3927 \times 3$ and $\dim X_C = 1791 \times 3$, respectively. The observations included in X_A belong to 24 datasets, 4 for each set of nominal operating conditions (S/C = 3, 4, 5 and $P_G^0 = 2$ and 7 bar), those in X_B belong to 49 datasets, 7 for each set of operating conditions ($x_{O_2}^0 = 3$ %, 5 %, 10 % and P_G^0 variable in the interval 3-8 bar) and the ones in X_C belong to 36 datasets, 12 for each value of $x_{O_2}^0$ (equal to 20 %, 39 % and 58 %). Predictors are standardised to reduce complexity during calculation and to avoid those having greater order of magnitude dominating those of a smaller one. The squared exponential (k_{se}) and rational quadratic (k_{rq}) kernels are employed to model $\dot{V}_{tot}^{out}$ and x_i^{out}, respectively.

The predictive capabilities of the models are tested on 38, 97 and 221 datasets of the three stages. Figure 2 shows the predicted profiles of two key variables employed in stage A. These plots are representative of most of the results. GPR provides very reliable predictions, the underlying functions are quite smooth and the confidence intervals are sufficiently tight.

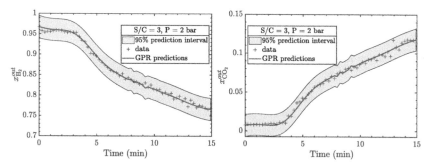

Figure 2: Time-profiles of: (a) $x_{H_2}^{out}$ and (b) $x_{CO_2}^{out}$ for stage A.

The real-time determination of t_s is simulated for several runs of each stage, by employing the GPR-based soft-sensors. At each sampling time (1 s), GPR soft-sensors use the measured values of the secondary variables to infer the values of the primary variables.

This way, the value of the objective function and constraints in Table 1 are computed. When the value of the objective function at the current time is smaller than the previously calculated one or the constraint is being hit, then switching is triggered (assuming that the constraint has a unique maximum). The use of GPR-based soft sensors, even in critical experimental runs, results in satisfactory performance. As an example, the actual value of t_s computed for one experimental run of stage A at S/C = 5, $P_G^0 = 5$ bar, results in 9.43 min. Conversely, scheduling the duration of t_s as the average of the values computed offline for all the data sets at the same operating conditions would result in $t_s = 11.59$ min, corresponding to an error of 23 %. In addition, such a time duration would not have allowed satisfying the constraint. The use of the GPR-based soft-sensors allows a real-time estimation of $t_s = 9.80$ min, which corresponds to a reduction in the error to only 3.89 %, which is a satisfactory performance for this application.

6. Conclusions

This work deals with the real-time determination of the switching instants of a cyclic process for H_2 production with integrated CO_2 capture. The proposed methodology is based on online evaluation of switching criteria, aiming to maximize economic and quality indicators. GPR-based soft sensors have been developed to make online predictions of these criteria. The use of GPR-based soft-sensors allows a real-time estimation of the optimal times to switch with a maximum error of only 4 % in the investigated runs. Future work should be devoted to identify reliable principles for a quantitative assessment of model performance and to improve model update techniques. This fact is crucial to abate process-model mismatches.

References

J. Abanades, R. Murillo, J. Fernandez, G. Grasa, I. Martínez, 2010. New CO_2 capture process for hydrogen production combining Ca and Cu chemical loops. Environ. Sci. Technol. 44, 6901–6904.

D. Bonvin, C. Georgakis, C. C. Pantelides, M. Barolo, M. A. Grover, D. Rodrigues, R. Schneider, D. Dochain, 2016. Linking models and experiments. Ind. Eng. Chem. Res. 55, 6891–6903.

D. J. C. Mackay, 1998. Introduction to gaussian processes. NATO ASI Series F Computer and Systems Sciences 168, 133–166.

M. Martini, I. Martínez, M. Romano, P. Chiesa, F. Gallucci, M. van Sint Annaland, 2017. Increasing the carbon capture efficiency of the Ca-Cu looping process for power production with advanced process schemes. Chem. Eng. J. 328, 304 – 319.

M. Martini, A. van den Berg, F. Gallucci, M. van Sint Annaland, 2016. Investigation of the process operability windows for Ca-Cu looping for hydrogen production with CO_2 capture. Chem. Eng. J. 303, 73 – 88.

C. Mei, M. Yang, D. Shu, H. Jiang, G. Liu, Z. Liao, 2016. Soft sensor based on gaussian process regression and its application in erythromycin fermentation process. Chem. Ind. Chem. Eng. Q. 22, 127–135.

M. Porru, J. Alvarez, R. Baratti, 2013. A distillate composition estimator for an industrial multicomponent ic4-nc4 splitter with experimental temperature measurements. IFAC Proceedings volumes 46 (32), 391–396.

M. Porru, L. Özkan, 2017. Monitoring of batch industrial crystallization with growth, nucleation, and agglomeration. part 1: Modeling with method of characteristics. Ind. Eng. Chem. Res. 56 (20), 5980–5992.

C. Rasmussen, C. Williams, 2006. Gaussian Processes for Machine Learning. MIT Press, Cambridge, Massachussets.

A. Shokry, M. Pérez-Moya, M. Graells, A. E. na, 2017. Data-driven dynamic modeling of batch processes having different initial conditions and missing measurements. Comput. Aided Chem. Eng. 40, 433–438.

Anton A. Kiss, Edwin Zondervan, Richard Lakerveld, Leyla Özkan (Eds.)
Proceedings of the 29th European Symposium on Computer Aided Process Engineering
June 16th to 19th, 2019, Eindhoven, The Netherlands. © 2019 Elsevier B.V. All rights reserved.
http://dx.doi.org/10.1016/B978-0-128-18634-3.50205-8

An Advanced Data-Centric Multi-Granularity Platform for Industrial Data Analysis

Marco S Reis,[a,*] Tiago J. Rato[a]

[a]*CIEPQPF, Department of Chemical Engineering, University of Coimbra, Rua Sílvio Lima, 3030-790, Coimbra, Portugal*
marco@eq.uc.pt

Abstract

Data collected in industrial processes are often high-dimensional, with dynamic and multi-granular characteristics that need to be properly accounted for in order to extract useful information for high level tasks, such as process monitoring, control and optimization. While the first two characteristics are already well addressed by state-of-the-art analytic methods, multi-granularity has received far less attention from the scientific community. Furthermore, data aggregation offers a suitable solution to handle the data deluge in Big Data scenarios, avoiding the loss of information of subsampling and multirate approaches. In this paper, we highlight the main aspects of multi-granularity data and present a novel multi-granularity framework that successfully addresses two fundamental problems: (i) handling multi-granularity data in a rigorous and consistent way; (ii) creating optimal multi-granularity structures for predictive modeling, even when raw data are available at a single granularity.

Keywords: Multi-granularity, Soft sensors, Batch processes, Continuous processes, Process Analytical Technology.

1. Introduction

Among the defining characteristics of industrial data, high-dimensionality and process dynamics are the most well-known and widely explored. They are typically handled through the use of projection-based (or latent variable) methods such as Principal Component Analysis (PCA) (Jolliffe, 2002) and Partial Least Squares (PLS) (Geladi et al., 1986) and their dynamic extensions with time-shifted variables (Ku et al., 1995, Kaspar et al., 1993, Rato et al., 2013). Alongside with these characteristics, multi-granular (or multiresolution) structures are often present in data. However, multi-granularity is often ignored and wrongly treated as a multirate problem, which is not the case. In the multirate case the recorded values regard instantaneous (pointwise) observations taken at regular sampling times. Conversely, in the multi-granularity case the recorded values contain information about the process with different levels of granularity (different resolutions) due to the aggregation of data during successive periods of time where high resolution data are merged and combined in order to produce a single low resolution observation per aggregation period. Another source of granularity can be found in situations where variables represent measurements made on composite samples collected during a certain time period (e.g., production lot, working shift, etc.), and therefore the recorded values regard specific windows of time. These windows of time are here defined as the variable's "time support". This characteristic makes multi-granular data sets fundamentally different from their multirate counterparts. Thus, they require the use of dedicated analytical methodologies that simultaneously account for high-dimensional, dynamic and multi-granular characteristics, which, until now, were not available.

To achieve the aforementioned goal, two main classes of multi-granular (or multiresolution; these terms are here used interchangeably) modelling approaches must be defined. These classes, as well as their application scope are summarized in Figure 1. The first class of methodologies addresses the case where data already presents a multi-granularity structure. Consequently, the goal of these methodologies is to properly account for the relationships caused by the multi-granularity structure and obtain models that are more parsimonious and able to accommodate this reality. On the other hand, the second class of methodologies relates to the case where raw data is available at a single-granularity (i.e., all variables have the same resolution). In this case, the goal is to change the native granularity of each variable to an optimal granularity level that maximizes the data analysis performance. By doing so, the modelling framework also becomes more flexible, as well as able to smooth noisy variables and to remove redundant information. For these reasons, multi-granularity models are also more parsimonious, accurate and interpretable than single-granularity models.

For the sake of space, only the second class of methodologies are discussed in this paper. As for the first class of methodologies (handling multi-granularity data structures), we refer to reader to (Reis et al., 2006a, Reis et al., 2006b, Rato et al., 2017), where applications to process monitoring and predictive modelling are covered.

The rest of this paper is organized as follows. In Section 2, the modelling framework used for selecting optimal multi-granular structures is described. Afterwards, in Section 3, the application of this methodology is exemplified for (i) continuous processes, (ii) batch processes and (iii) spectroscopy data. Finally, the main conclusions of this work are summarized in Section 4.

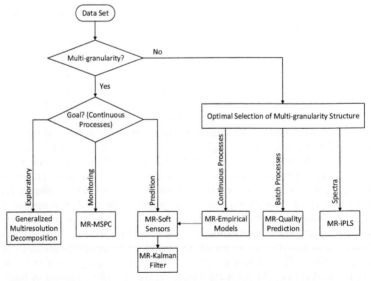

Figure 1 Organogram of the multi-granularity framework and its application scope.

2. Multi-granularity parsimonious models

To introduce the concepts behind the multi-granularity models, let us consider that a variable at a given granularity is represented by $x_{i,t}^{(q_i)}$, where i is the variable index, t is the time index and q_i is an integer defining the variable's granularity level (note that different variables may be at distinct granularities). Without loss of generality, the multi-granularity structure is assumed to follow a dyadic tree with a time support of 2^{q_i}. That is, variables at the granularity level q_i are set to be an average over 2^{q_i} time instants in the finest time domain.

For the class of modelling methodologies discussed in this paper, it is considered that all predictor and response variables are originally available at level 0 (finest granularity or highest resolution). Therefore, multi-granularity is introduced by the modelling methodology, being the optimal granularity level determined by the selection algorithm described below. Furthermore, the approach used to change the predictors' granularity can be adapted to the specificities of each type of data. For instance, in the case of continuous processes, the changes in granularity are performed over moving windows (to preserve process dynamics), while for batch processes they are applied over non-overlapping windows (to accommodate for asynchronous batches and reduce the number of observations).

The multi-granularity selection algorithm used to find the optimal granularity of the predictors is based on the standard stepwise forward algorithm for variable selection with an additional search over the granularity dimension. This search is constrained to fall between the native granularity of the variables (0) and a maximum granularity level defined by the user (q_{max}). From our experience, a maximum granularity level of $q_{max} = 5$ is usually a good starting point, when no further information about the process is available.

In general terms, the multi-granularity stepwise forward selection algorithm starts by choosing an initial model by fitting tentative models for each combination of variables and granularities (i.e., for each combination of i and q_i). In each case, a vector of Cross-Validation Root Mean Squared Error (CV-RMSE) is computed using Monte-Carlo sampling. Afterwards, the combination leading to the lowest median CV-RMSE is selected to build the initial reference model. In the following step, the algorithm proceeds to an "adding" stage, where each combination of variables and granularities (except for the variables already in the model) is tentatively added to the reference model and compared against it using the CV-RMSE values, through a pairwise Wilcoxon rank sum statistical test (Wilcoxon, 1945). Finally, the algorithm performs a "removal" stage to tentatively remove variables from the model that become redundant after the "inclusion" stage. These "inclusion" and "removal" stages are repeated until no further improvement in the CV-RMSE is achieved. Following these steps, the output of the algorithm is a multi-granularity model with the best subset of predictors at their optimal granularity.

3. Applications

3.1. Continuous processes

Based on the algorithm shortly described in Section 2, Rato et al. (2018a) proposed a modelling framework for building multiresolution empirical models for continuous processes (MR-EMC). To illustrate the advantages of MR-EMC, the results obtained for a simulated distillation column using the model described in (Skogestad et al., 2005) will be presented. This system is composed by a column with 41 stages and returns the compositions and temperatures in all stages. For this comparison, standard PLS, Dynamic PLS (DPLS) and MR-EMC models were used to estimate the distillate composition based on the temperature readings. To assess the consistency of the results, 100 replicates were made. Each replicate was composed by 1,000 observations for training the models and another 1,000 observations for testing their performance using the Mean Squared Error (MSE). In Figure 2, the pairwise differences between the MSEs obtained by PLS and DPLS are compared against those of MR-EMC. From this figure, it is clear that PLS presents the worst performance as it only uses the native granularity of the variables and has no information about process dynamics. When time-shifted variables are added, a significant improvement is observed (DPLS). Nevertheless, MR-EMC is still the best methodology since its MSEs are generally lower than those obtained with PLS and DPLS. These results demonstrate that selecting the optimal granularity for the predictors plays a major role in model accuracy. Furthermore, it was also observed that the MR-EMC models were easier to interpret since the selected combinations of variables and granularities were closely related to the true dynamic dependencies of the process.

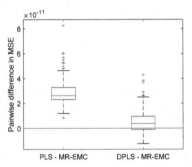

Figure 2 Pairwise differences of the MSE over 100 replicates of the distillation column case study. For reference, the variance of the response is 1.37×10^{-9}.

3.2. Batch processes

A multiresolution quality prediction (MRQP) framework for predicting the end-batch quality by exploiting the structured correlation in both the time and variables dimensions was originally proposed by Gins et al. (2018) and later on extended by Rato et al. (2018c) to account for processes with different operational phases. The selection algorithm is similar to the one presented in Section 2, but now observations are also included as predictors (batch-wise unfolding). Furthermore, the batch is divided in stages in order to allow for a finer modelling of the variables importance. Thus, this methodology can effectively select the key process variables and the prominent operational phases affecting the end-batch quality. Adjusting the granularity of the variables also reduces the number of observations, which can significantly decrease the number of predictors when batch-wise unfolding is applied. This approach was tested with several systems, including simulated and real world processes. For the real world case, regarding an industrial batch polymerization process, the improvement achieved in prediction (*PRESS*) with respect to the standard Multiway PLS approach (Nomikos et al., 1995) was of 54 % (Gins et al., 2018).

3.3. Spectroscopy

The interval PLS (iPLS) algorithm proposed by Nørgaard et al. (2000) is typically applied to compare the performance of a full-spectrum model to that of local models based on subintervals of equal length. However, as this approach does not directly select the most relevant wavebands, sequential approaches based on forward (FiPLS) or backward (BiPLS) algorithms have been suggested (Xiaobo et al., 2007). Still, these methodologies are unable to optimally aggregate the original signal and denoise/merge redundant information within the interval, which ultimately reduces their prediction capabilities. To overcome this limitation, Rato et al. (2018b) proposed a multiresolution interval PLS (MR-iPLS) methodology that also selects the optimal granularity for each interval. Again, the selection algorithm is similar to that presented in Section 2, but since there is no time dimension, the changes in granularity are applied over non-overlapping intervals of wavelengths.

To exemplify the advantages of MR-iPLS against the standard iPLS models, the data set given in (Kalivas, 1997) is here used for comparison purposes. This data set is composed by near infrared (NIR) spectra (with a total of 401 wavelengths) for 60 gasoline samples with known octane numbers. In this study, the samples were randomly divided into a training data set with 40 samples and a testing data set with the remaining 20 samples. The models were fit on the training data set using 12 intervals with 32 wavelengths and one interval with 17 wavelengths. The independent testing data set was then used to evaluate the prediction performance of each model through the MSE, leading to the results presented in Table 1. These results clearly show that the FiPLS model has the worst performance due to its inability to overcome local optima. On the other hand, BiPLS provides better results due to the exclusion of irrelevant intervals. Nevertheless, it is noticeable that using MR-iPLS to optimize the granularity of the selected intervals, a model with even lower MSE is obtained. In this case, the 9 intervals selected by MR-iPLS were all placed at a coarser granularity level ($q = 5$). Consequently, instead of using the original 288 wavelengths in these intervals, only their averages are included in the model, which reduces the number of predictors to 9.

Table 1 MSE and number of predictors included in each model for the spectroscopy case study.

	PLS	**FiPLS**	**BiPLS**	**MR-iPLS**
MSE	0.0409	0.0451	0.0410	0.0381
Number of predictors	401	160	256	9

4. Conclusions

The presence of multi-granularity structures in industrial data limits the application of current data analysis methodologies since they are neither able to accommodate for this aspect, nor to take advantage of the intrinsic relationships. Furthermore, even when data is available at a single-granularity, it is not guaranteed that the native granularity of the predictors is the most appropriate one for process modelling. To address this situation, a set of multi-granularity modelling frameworks that simultaneously select the best subset of predictors to be included in the model along with the optimal granularity for each predictor have been proposed for several types of industrial data. In all the data scenarios considered, results show that the multi-granularity methodologies perform consistently better than their single-granularity counterparts, while leading at the same time to more parsimonious and interpretable models.

Acknowledgements

The authors acknowledge financial support through project 016658 (references PTDC/QEQ-EPS/1323/2014, POCI-01-0145-FEDER-016658) co-financed by the Portuguese FCT and European Union's FEDER through the program "COMPETE 2020".

References

P. Geladi, B. R. Kowalski, 1986, Partial Least-Squares Regression: a Tutorial, Analytica Chimica Acta, 185, 1-17.

G. Gins, J. F. M. Van Impe, M. S. Reis, 2018, Finding the optimal time resolution for batch-end quality prediction: MRQP – A framework for multi-resolution quality prediction, Chemometrics and Intelligent Laboratory Systems, 172, 150-158.

I. T. Jolliffe, 2002, Principal Component Analysis, New York, Springer.

J. H. Kalivas, 1997, Two data sets of near infrared spectra, Chemometrics and Intelligent Laboratory Systems, 37, 2, 255-259.

M. H. Kaspar, W. Harmon Ray, 1993, Dynamic PLS modelling for process control, Chemical Engineering Science, 48, 20, 3447-3461.

W. Ku, R. H. Storer, C. Georgakis, 1995, Disturbance detection and isolation by dynamic principal component analysis, Chemometrics and Intelligent Laboratory Systems, 30, 1, 179-196.

P. Nomikos, J. F. MacGregor, 1995, Multivariate SPC Chart for Monitoring Batch Processes, Technometrics, 37, 1, 41-59.

L. Nørgaard, A. Saudland, J. Wagner, J. P. Nielsen, L. Munck, S. B. Engelsen, 2000, Interval Partial Least-Squares Regression (iPLS): A Comparative Chemometric Study with an Example from Near-Infrared Spectroscopy, Applied Spectroscopy, 54, 3, 413-419.

T. J. Rato, M. S. Reis, 2013, Fault detection in the Tennessee Eastman benchmark process using dynamic principal components analysis based on decorrelated residuals (DPCA-DR), Chemometrics and Intelligent Laboratory Systems, 125, 15, 101-108.

T. J. Rato, M. S. Reis, 2017, Multiresolution Soft Sensors (MR-SS): A New Class of Model Structures for Handling Multiresolution Data, Industrial & Engineering Chemistry Research, 56, 13, 3640–3654.

T. J. Rato, M. S. Reis, 2018a, Building Optimal Multiresolution Soft Sensors for Continuous Processes, Industrial & Engineering Chemistry Research, 57, 30, 9750–9765.

T. J. Rato, M. S. Reis, 2018b, Multiresolution Interval Partial Least Squares: A Framework for Waveband Selection and Resolution Optimization, Under Submission,

T. J. Rato, M. S. Reis, 2018c, Optimal selection of time resolution for batch data analysis. Part I: Predictive modeling, AIChE Journal, 64, 3923-3933.

M. S. Reis, P. M. Saraiva, 2006a, Generalized Multiresolution Decomposition Frameworks for the Analysis of Industrial Data with Uncertainty and Missing Values, Industrial & Engineering Chemistry Research, 45, 6330-6338.

M. S. Reis, P. M. Saraiva, 2006b, Multiscale Statistical Process Control with Multiresolution Data, AIChE Journal, 52, 6, 2107-2119.

S. Skogestad, I. Postlethwaite, 2005, Multivariable Feedback Control: Analysis and design, Chichester, Wiley.

F. Wilcoxon, 1945, Individual Comparisons by Ranking Methods, Biometrics Bulletin, 1, 6, 80-83.

Z. Xiaobo, Z. Jiewen, L. Yanxiao, 2007, Selection of the efficient wavelength regions in FT-NIR spectroscopy for determination of SSC of 'Fuji' apple based on BiPLS and FiPLS models, Vibrational Spectroscopy, 44, 2, 220-227.

Anton A. Kiss, Edwin Zondervan, Richard Lakerveld, Leyla Özkan (Eds.)
Proceedings of the 29th European Symposium on Computer Aided Process Engineering
June 16th to 19th, 2019, Eindhoven, The Netherlands. © 2019 Elsevier B.V. All rights reserved.
http://dx.doi.org/10.1016/B978-0-128-18634-3.50206-X

Optimal Operation and Control of Fluidized Bed Membrane Reactors for Steam Methane Reforming

Alejandro Marquez-Ruiz[a*], Jiaen Wu[b], Leyla Özkan[a], Fausto Gallucci[b], Martin Van Sint Annaland[b]

[a] Department of Electrical Engineering , Eindhoven University of Technology, 5612 AJ, Eindhoven, The Netherlands

[b] Department of Chemical Engineering and Chemistry , Eindhoven University of Technology, 5612 AZ, Eindhoven, The Netherlands

a.marquez.ruiz@tue.nl

Abstract

This work presents the optimal operation and control of a Fluidized Bed Membrane Reactor (FBMR) for Steam Methane Reforming (SMR). First, a nonlinear distributed parameter dynamic model is developed. Next, the optimal operation of the system is studied by solving a dynamic optimization problem that maximizes the conversion and separation in the reactor. Based on the optimization result, reduced order linear models are developed and used in the design of conventional and model based controllers. The performance of these controllers are tested considering the variation in the inlet concentration of the feed to the reactor and the initial conditions.

Keywords: Fluidized Bed Membrane Reactors, Steam Methane Reforming, Optimal Operation, Model Based Controllers.

1. Introduction

Hydrogen is one of the most important chemicals used in the chemical industry, and lately, it has gained particular attention due to the wide range of applications also as an energy carrier. There are several technologies available for the production of hydrogen; however, the conventional steam methane reforming (SMR) is currently still the most common and economical way to produce hydrogen. The conventional SMR process consists of two main reactions given by:

$$CH_4 + H_2O \rightleftharpoons CO + 3H_2, \Delta H_{298\,K} = +206\,\text{kJmol}^{-1}$$

$$CO + H_2O \rightleftharpoons CO_2 + H_2, \Delta H_{298\,K} = -41\,\text{kJmol}^{-1}$$

In order to intensify this process and enhance the yield of hydrogen and shift the reaction equilibrium in the direction of producing hydrogen, new reactor concepts have been developed and investigated. The Fluidized Bed Membrane Reactor (FBMR) is one of the most promising membrane reactors for integrated reforming/dehydrogenation, separation and purification (Rahimpour et al. 2017). In the FMBR, a bundle of hydrogen-selective membranes are immersed into the fluidized catalytic bed in order to carry out the separation and reaction steps simultaneously. Such reactor concepts can reduce capital costs and improve process efficiency, but at the cost of losing degrees of freedom in the operation. Despite the substantial amount of work done on process design based on steady state modelling of FBMR's the study of operational aspects such as controllability, stability, and operational feasibility have not been investigated as extensively. Only a limited number of works is available with the application of control theory for a fluidized bed reactor (FBR). However, these controllers cannot always be applied to an FBMR, because the intensified FBMR exhibits a quite different dynamic behaviour due to small equipment but same or larger process efficiency. Considering the impact of the intensification on the operation, the control design of a FBMR is very relevant.

In this work, we have studied the optimal operation and control of an FBMR for SMR. First, a dynamic model extending the steady-state models proposed by Medrano et al. (2018) is described. Next, the optimal operation of the system is studied. To this end, a dynamic optimization problem that maximizes conversion and separation in the reactor is proposed. Based on the result obtained from the optimization problem, and the dynamic model, linear models of the system have been developed. However, as the linear model is infinite dimensional (the dynamic model is given by a set of Nonlinear Partial Differential Equations), and control design based on these models is challenging, model reduction techniques have been applied. Finally, with the reduced linear model, low-level and Model Predictive controllers have been designed. The performance of these controllers has been tested in simulation in case of variations in the inlet composition of the feed to the reactor.

2. Dynamic Modelling of FBMR

Consider a FBMR where pure hydrogen is recovered via palladium-based membranes inserted into the fluidized bed (Gallucci et al. 2008). The dynamic model obtained in this work is an extension of the model proposed by Medrano et al. (2018) by including accumulation terms in the gas phase mass (bubble/wake and emulsion phases) and energy balances yielding the following set of Partial Differential Equations (PDEs),

$$\frac{\partial}{\partial t}\begin{bmatrix} \left(f_{bw}C_{i,bw}\right) \\ \left(f_e C_{i,e}\right) \\ \left(\rho C_p f_{bew}T\right) \end{bmatrix} = -\frac{\partial}{\partial z}\begin{bmatrix} \left(f_{bw}u_b C_{i,bw}\right) \\ \left(f_e u_e C_{i,e}\right) \\ \left(\rho C_p f_{bew}uT\right) \end{bmatrix} + \begin{bmatrix} \xi_i + r_{i,bw}f_{bw}(1-\varepsilon_{mf}) + S_{mb} \\ -\xi_i + r_{i,e}f_e(1-\varepsilon_{mf}) + S_{me} \\ \sum_{k=1}^{p}\sum_{i=1}^{m}(-\Delta H_{r,k})r_{i,k}f_k(1-\varepsilon_{mf}) \end{bmatrix} + \begin{bmatrix} 0 \\ 0 \\ \frac{4}{d}H_J(T_J - T) \end{bmatrix} \quad (1)$$

$$\forall i = \{CH_4, CO_2, CO, H_2O, H_2\}$$

where ξ_i, $S_{\{mb,me\}}$ are the mass transfer rates between the bubble and emulsion phases, and the membrane, and $r_{i,\{bw,e\}}$ are the reaction rates. The model given by Eq. (1) is subject to a set of algebraic equations which are not included in this paper due to space limitations. However, a summary of the hydrodynamics and mass transfer correlations used in this work is reported in Medrano et al. (2018).

2.1. Discrete Model

In order to solve the dynamic model, the spatial derivatives of Eq. (1) are replaced by a backward differences approximation such that the model can be written as a set of ODEs,

$$\frac{d}{dt}\begin{bmatrix} (f_{bw}C_{i,bw}^j) \\ (f_e C_{i,e}^j) \\ (\rho C_p f_{bew} T^j) \end{bmatrix} = -\frac{1}{\Delta z}\begin{bmatrix} \Delta(f_{bw}u_b C_{i,bw}^j) \\ \Delta(f_e u_e C_{i,e}^j) \\ \Delta(\rho C_p f_{bew} u T^j) \end{bmatrix} + M\begin{bmatrix} \xi_i^j \\ S_i^j \\ r_i^j \end{bmatrix} + \begin{bmatrix} 0 \\ 0 \\ \frac{4H_J}{d}(T_J - T) \end{bmatrix}, \forall j = 1,...,N \quad (2)$$

where Δ is defined as $\Delta C_i^j = C_i^j - C_i^{j-1}$, N *is* the number of sections in which the reactor is divided, and M is a matrix related to the mass transfer and reaction rates. After some algebraic manipulations Eq. (2) can be expressed as: $\dot{x} = f(x,u,d)$, $y = g(x,u)$, where

$x = \left[C_{i,bw}^j, C_{i,e}^j, T^j \right]^T \in \mathbb{R}^{11N}$, $y = \left[X_{CH_4,out}, C_{H_2,out} \right]^T \in \mathbb{R}^2$ and $u = [F, T_J]^T \in \mathbb{R}^2$ are the

states, manipulated, and controlled variables respectively, with F the feed flow of the reactor, T_J the jacket temperature, and $X_{CH_4,out} = C_{CH_4,in}^{-1}(C_{CH_4,in} - C_{CH_4,out})$ the methane

conversion. In addition, $d = \left[C_{i,bw}^{in}, C_{i,e}^{in}, T^{in} \right]^T \in \mathbb{R}^{11}$ are the disturbances of the system.

3. Optimal Operation of FBMR

The main objective of the FBMR is to maximize both the conversion of methane and the separation of the hydrogen throughout the membrane. By defining $X_{CH_4}^j \in [0,1]$ as the conversion of methane and $S_{H_2}^j \in [0,1]$ as the hydrogen separation factor along the reactor, we formulate the following optimization problem to achieve these objectives:

$$u_{opt} = \arg\min_u \int_{t_0}^{t_f}\left(\sum_{j=1}^N \omega \left\| 1 - X_{CH_4}^j \right\|_2^2 + (1-\omega)\left\| 1 - S_{H_2}^j \right\|_2^2 \right) dt \quad (3)$$

subject to: $\dot{x} = f(x,u,d)$, $y = g(x,u)$, $x \in \mathbf{X}$, $u \in \mathbf{U}$

Where $\omega \in [0,1]$ is a weighting parameter. In this optimization problem, the feed flow F and the jacket temperature T_j are the decision variables. From the practical point of view, the objective function of problem (3) is selected because it takes a minimum value when the raw material and the hydrogen produced are completely consumed and separated. Finally, the problem (3) is solved by parametrizing u and converting Eq. (3) into a Nonlinear Programming problem (NLP). The results of the optimization problem are shown in Figure 1.

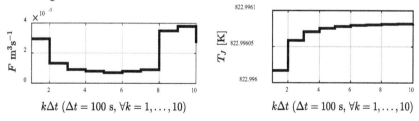

Figure 1. Optimal Trajectories for the inlet feed flow and the jacket temperature.

Notice that in Figure 1, T_J does not change significantly. That means, during the normal operation (no disturbances), this variable must remain constant around 823 K. However, it is important to clarify that the controllers can still manipulate T_J, to reject disturbances and to keep the system in the optimal trajectory.

4. Control Design for FBMR

In this paper, two control techniques are used to keep the operation of the FBMR around the optimal trajectory presented in Section 3, MPC and PID. To design these controllers, a linear model of the process must be obtained.

4.1. Linear Model and Model Reduction

The linear model of the FBMR is obtained by linearizing $\dot{x} = f(x,u,d)$, $y = g(x,u)$ around the optimal operating profiles using Taylor series expansion and considering only the first order terms as follows,

$$\dot{x} = Ax + Bu + B_d d, A = f_x(x,u)\big|_{x^*,u^*}, B = f_u(x,u)\big|_{x^*,u^*}, B_d = f_d(x,u)\big|_{x^*,u^*}$$

$$y = Cx + Du, C = g_x(x,u)\big|_{x^*,u^*}, D = g_u(x,u)\big|_{x^*,u^*} \tag{5}$$

where $x \in \mathbb{R}^{11N}$, $y \in \mathbb{R}^2$, and $u \in \mathbb{R}^2$. In this paper, the spatial domain of the reactor is divided into 100 sections, resulting in 1100 number of states. The controllability analysis shows the system has uncontrollable states. These two facts, a large number of states and uncontrollable states, make the design and implementation of model-based controllers for the FBMR very difficult. A good solution to this issue is to eliminate the uncontrollable states. This also reduces the number of states . To this end, a model reduction technique called balance truncation (Gugercin and Antoulas, 2004) is implemented in the FBMR model, and briefly described in Table 1.

Table1: Mathematical formulation of the reduced order model

Full order Model	Reduced Order Model
$G(s): \begin{cases} \dot{x} = Ax + Bu + B_d d \\ y = Cx + Du \end{cases}, x \in \mathbb{R}^{11N}$	$\tilde{G}(s): \begin{cases} \dot{\tilde{x}} = \tilde{A}\tilde{x} + \tilde{B}u + \tilde{B}_d d \\ y = \tilde{C}x + \tilde{D}u \end{cases}, \tilde{x} \in \mathbb{R}^n, n \le 11N$
$\tilde{x} = Tx, \tilde{A} = TAT^{-1}, \tilde{B} = TB, \tilde{B}_d = TB, \tilde{C} = T^{-1}C, \tilde{D} = D$, where T is computed using reachability and observability Gramians of $G(s)$. It is important to highlight that $\tilde{x}$ does not preserve the physical meaning of x, because $G(s)$ is not unique.	

Figure 2 shows the steady state concentration profile of the nonlinear, linear and the reduced order models for CH_4, H_2O, CO, CO_2, and H_2 in the emulsion phase. In Figure 2, it can be observed that the reduced order model with only 11 states provides a good approximation of the linear and the nonlinear model.

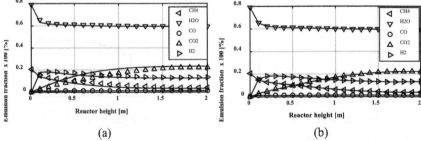

(a)   (b)

Figure 2: Models comparison: concentration of CH_4, H_2O, CO, CO_2, and H_2 in the emulsion phase in steady state for: (a) the nonlinear (–) and linear ($\triangle,\triangledown,\triangleright,\triangleleft,\circ$) models. (b) Linear (–) and reduced order ($\triangle,\triangledown,\triangleright,\triangleleft,\circ$) models.

4.2. PID Control

We present the PID control design in Table 2. As the PID controller is a SISO control technique, the Relative Gain Array ($\Lambda(j\omega)$) method is used in this work to calculate the best input-output pairing for the process.

Table 2: Mathematical formulation of the PID including RGA

Model	PID
$y(s) = \tilde{G}(s)u(s)$ $\Lambda(j\omega) = \tilde{G}(j\omega) * \tilde{G}(j\omega)^{-T} = \begin{bmatrix} -0.0028 & 1.0028 \\ 1.0028 & -0.0028 \end{bmatrix}$	$u_i(s) = \left(K_p + \dfrac{K_i}{s} + K_d s \right) e_i(s)$ $e_i(s) = y_{ref} - y_i(s)$

Based on the RGA the input-output pairing must be $T_J - X_{CH_4,out}$ and $F - C_{H_2,out}$. Additionally, the transfer function $\tilde{G}(s)$ is identified based on the step response of the reduced order model. The results of this controller are presented in Section 4.4.

4.3. Model Predictive Control (MPC)

MPC has been widely adopted by the industrial process control community and has been implemented successfully in many applications. MPC can handle constraints, which often have a significant impact on the quality and safety in process operations. To design predictive controllers, a discrete time model of the process must be obtained. To this end, a Zero-Order Hold discretization method is used. In Table 3, the discrete time version of the reduced order model is described.

Table 3: Mathematical formulation of the discrete time model

Reduced Order Model	Discrete Time Model
$\tilde{G}(s): \begin{cases} \dot{\tilde{x}} = \tilde{A}\tilde{x} + \tilde{B}u + \tilde{B}_d d \\ y = \tilde{C}x + \tilde{D}u \end{cases}$ $\xrightarrow{\ ZOH\ }$	$\tilde{G}(z): \begin{cases} \tilde{x}_{x+j+1} = \tilde{A}_k \tilde{x}_{k+j} + \tilde{B}_k u_{k+j} + \tilde{B}_{k,d} d_{k+j} \\ y_{k+j} = \tilde{C}_k \tilde{x}_{k+j} + \tilde{D}_k u_{k+j} \end{cases}$
The notation $k + j \triangleq (k + j)\Delta t$ where Δt is the sampling time	

The MPC control problem can be written as,

$$u_{MPC} = \arg \min_{\{u_{k+j}\}_{j=0}^{j=H_p-1}} \left\| x_{k+H_p} \right\|_P^2 + \sum_{j=0}^{H_p-1} \left\| x_{k+j} \right\|_Q^2 + \left\| u_{k+j} - u_{opt} \right\|_R^2$$

Subject to: (6)

$$\tilde{x}_{x+j+1} = \tilde{A}_k \tilde{x}_{k+j} + \tilde{B}_k u_{k+j} + \tilde{B}_{k,d} d_{k+j}, \ y_{k+j} = \tilde{C}_k \tilde{x}_{k+j} + \tilde{D}_k u_{k+j}, \ x_{k+j} \in \mathbf{X}_k, u_{k+j} \in \mathbf{U}_k$$

where H_p is the prediction horizon, and $\mathbf{X}_k$ and $\mathbf{U}_k$ are convex polyhedrons. The optimization problem (6) can be solved efficiently converting Eq. (6) into a Quadratic Programming (QP) problem.

4.4. MPC and PID comparison

In order to test the performance of the controllers for the FBMR, two experiments have been carried out: step changes in the disturbances of the process at $t = 220s$, and different initial conditions. Figure 3 shows the comparison between the MPC and PID controllers.

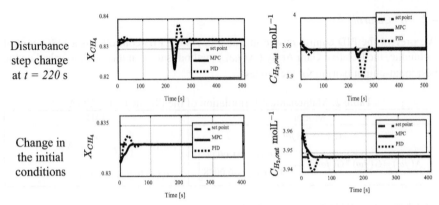

Figure 3: Controllers Comparison: Closed-loop trajectories of the outputs, MPC (—)
and PID (··) controllers.

Clearly, from Figure 3, the performance of the MPC is better than the one of PID. Of course, solving an MPC problem in real time is more complex; however, the constraint handling is an advantage. In this case, the inlet flow rate should be kept above minimum fluidization velocity.

5. Conclusions

In this paper, the optimal operation and control of an FBMR for SMR are studied. A dynamic model of the process is obtained, and an offline dynamic optimization problem is proposed to maximize the conversion of methane and separation factor in the FBMR. Based on the optimal operation, and the dynamic model, a reduced order linear model is developed using balance truncation methods. Finally, PID and MPC controllers are designed, and their performance tested in simulations considering variation in the inlet concentration of the feed to the reactor and initial conditions.

References

S. Gugercin, A. Antoulas, 2004, A survey of model reduction by balanced truncation and some new results, International Journal of Control 77.8, 748-766.

F. Gallucci, M.V.S. Annaland, J.A.M. Kuipers, 2008. Autothermal reforming of methane with integrated CO2 capture in a novel fluidized bed membrane reactor. Part 1: experimental demonstration. Topics in catalysis, 51(1-4), p.133.

M. R. Rahimpour, F. Samimi, A. Babapoor, T. Tohidian, S. Mohebi, 2017, Palladium membranes applications in reaction systems for hydrogen separation and purification: A review. Chemical Engineering and Processing: Process Intensification, 121, 24-49.

J. A. Medrano, I. Potdar, J. Melendez, V. Spallina, D.A. Pacheco-Tanaka, M. van Sint Annaland, F. Gallucci, 2018, The membrane-assisted chemical looping reforming concept for efficient H2 production with inherent CO2 capture: Experimental demonstration and model validation. Applied Energy, 215, 75-86.

Anton A. Kiss, Edwin Zondervan, Richard Lakerveld, Leyla Özkan (Eds.)
Proceedings of the 29th European Symposium on Computer Aided Process Engineering
June 16th to 19th, 2019, Eindhoven, The Netherlands. © 2019 Elsevier B.V. All rights reserved.
http://dx.doi.org/10.1016/B978-0-128-18634-3.50207-1

Stochastic nonlinear model predictive control of a batch fermentation process

Eric Bradford[a*] and Lars Imsland[a]

[a]*Engineering Cybernetics; NTNU; O. S. Bragstads plass 2D, Trondheim 7034, Norway*
eric.bradford@ntnu.no

Abstract

Nonlinear model predictive control (NMPC) is an attractive control approach to regulate batch processes reliant on an accurate dynamic model. Most dynamic models however are affected by significant uncertainties, which may lead to worse control performance and infeasibilities, considering the tendency of NMPC to drive the system to its constraints. This paper proposes a novel NMPC framework to mitigate this issue by explicitly taking into account time-invariant stochastic uncertainties. Parametric uncertainties are assumed to be given by so-called polynomial chaos expansions (PCE), which constitutes a flexible approach to depict arbitrary probability distributions. It is assumed that at each sampling time only noisy output measurements are available. The proposed procedure uses a sparse Gauss-Hermite sampling rule to formulate an efficient scenario-based NMPC algorithm based on the PCE, while a stochastic nonlinear filter is employed to update the PCE given the available measurements. The framework is shown to be effective on a challenging semi-batch fermentation process simulation case study.

Keywords: Chemical process control, Polynomial chaos, Nonlinear filters, Model-based control

1. Introduction

Batch processes are commonly used in many chemical sectors, including pharmaceuticals, bulk chemicals and biotechnology. Batch processes are operated at unsteady state and are highly non-linear, which motivates the use of nonlinear model predictive control (NMPC). The performance of the NMPC algorithm depends strongly on the accuracy of the dynamic model used and inherent uncertainties may lead to constraint violations and worse control actions. If we assume these uncertainties to be given by known probability distributions, stochastic NMPC (SNMPC) methods can be used (Mesbah, 2016). The main difficulty in SNMPC lies in propagating stochastic uncertainties through nonlinear system models. Several SNMPC algorithms have been proposed using different methods to propagate stochastic uncertainties:

- Unscented transformation sampling (Bradford and Imsland, 2018a)
- Polynomial chaos expansions (Fagiano and Khammash, 2012)
- Markov Chain Monte Carlo (Maciejowski et al., 2007)
- Gaussian processes (Bradford and Imsland, 2018b)
- Quasi Monte Carlo methods (Bradford and Imsland, 2017)
- Particle filters (Sehr and Bitmead, 2017)

Most work in SNMPC assumes full state feedback, which is uncommon for real processes. Instead, the measurements made at each sampling time are both noisy and incomplete. In this paper we therefore propose to use a nonlinear filter to update the stochastic uncertainties at each sampling time. The uncertainties are represented by so-called polynomial chaos expansions (PCE), which allow for complex probability distributions to be given by polynomials of simpler stochastic variables. In addition, we suggest to efficiently formulate the SNMPC problem using a sparse Gauss-Hermite quadrature rule. The framework is verified on a fermentation case study.

2. Problem formulation

We aim to control a discrete-time nonlinear equation system with stochastic uncertainties:

$$\mathbf{x}(t+1) = \mathbf{f}(\mathbf{x}(t), \mathbf{u}(t), \boldsymbol{\theta}(\boldsymbol{\xi})), \qquad \mathbf{x}(0) = \mathbf{x}_0(\boldsymbol{\theta}(\boldsymbol{\xi})) \tag{1}$$

$$\mathbf{y}(t) = \mathbf{h}(\mathbf{x}(t), \boldsymbol{\theta}(\boldsymbol{\xi})) + \mathbf{v} \tag{2}$$

where k is the discrete time, $\mathbf{x} \in \mathbb{R}^{n_x}$ are the system states, $\mathbf{u} \in \mathbb{R}^{n_u}$ denote the control inputs, $\boldsymbol{\theta}(\boldsymbol{\xi}) \in \mathbb{R}^{n_\theta}$ are time-invariant uncertainties, $\boldsymbol{\xi} \in \mathbb{R}^{n_\theta} \sim \mathcal{N}(\mathbf{0}, \mathbf{I})$ describe standard normally distributed random variables parametrizing the PCE of $\boldsymbol{\theta}$, $\mathbf{f} : \mathbb{R}^{n_x} \times \mathbb{R}^{n_u} \times \mathbb{R}^{n_\theta} \to \mathbb{R}^{n_x}$ represents the nonlinear dynamic system, $\mathbf{y} \in \mathbb{R}^{n_y}$ denote the measurements, $\mathbf{h} : \mathbb{R}^{n_x} \times \mathbb{R}^{n_\theta}$ are the output equations and $\mathbf{v} \in \mathbb{R}^{n_y} \sim \mathcal{N}(\mathbf{0}, \boldsymbol{\Sigma}_\mathbf{v})$ is the measurement noise assumed to be zero mean multivariate normally distributed with known covariance matrix $\boldsymbol{\Sigma}_\mathbf{v}$. The initial condition $\mathbf{x}(0)$ may also be uncertain and hence is a function of $\boldsymbol{\theta}(\boldsymbol{\xi})$.

The time-invariant uncertainty described by $\boldsymbol{\theta}$ are assumed to be given by a known truncated PCE:

$$\theta_i(\boldsymbol{\xi}) = \sum_{0 \le |\alpha| \le m} a_j^{(i)} \phi_{\alpha_j}(\boldsymbol{\xi}) = \mathbf{a}_i^T \boldsymbol{\phi}(\boldsymbol{\xi}) \tag{3}$$

where each component $\theta_i(\boldsymbol{\xi})$ is given by an individual polynomial series with multivariate polynomials $\phi_{\alpha_j}(\boldsymbol{\xi})$ with coefficients $a_j^{(i)}$. We summarise the coefficients as $\mathbf{A} = [\mathbf{a}_1, \ldots, \mathbf{a}_{n_\theta}]$. The multivariate polynomials are given by products of univariate polynomials $\phi_{\alpha_j} = \prod_{i=1}^{n_\xi} \phi_{\alpha_{j_i}}(\xi_i)$ with $\phi_{\alpha_{j_i}}(\xi_i)$ being univariate polynomials of ξ_i of degree α_{j_i}. The vector $\boldsymbol{\phi}(\cdot) = [\phi_1(\cdot), \ldots, \phi_L(\cdot)]^T$ contains the multivariate polynomials of the expansions, m denotes the order of truncation and $|\alpha| = \sum_{i=1}^{n_\xi} \alpha_i$. Each polynomial series consists of $L = \frac{(n_\xi + m)!}{n_\xi! m!}$ terms and $\mathbf{a}_i \in \mathbb{R}^L$ represents a vector of coefficients of these terms. The univariate polynomials are given by Hermite polynomials:

$$\phi_j(\xi_i) = (-1)^j \exp\left(\frac{1}{2}\xi_i^2\right) \frac{d^j}{d\xi_i^j} \exp\left(-\frac{1}{2}\xi_i^2\right) \tag{4}$$

3. Gauss-Hermite nonlinear model predictive control

Once the uncertainties are defined as PCE as shown in section 2, we aim to exploit this information together with the dynamic equation system in Eq.(1) to formulate an optimal control problem (OCP) to be solved iteratively. To achieve this we use Gauss-Hermite quadrature rules to create several realizations of $\boldsymbol{\theta}$ from its PCE representation to formulate a scenario-based MPC problem. The Gauss-Hermite quadrature rules can be seen to give an approximation to the integral:

$$\mathbb{E}[f(\boldsymbol{\xi})] = \int_{-\infty}^{\infty} f(\boldsymbol{\xi}) p(\boldsymbol{\xi}) d\boldsymbol{\xi}, \qquad p(\boldsymbol{\xi}) = \prod_i \exp(-\xi_i^2)/\sqrt{\pi} \tag{5}$$

where $\boldsymbol{\xi}$ is a standard normally distributed random variable and $\mathbb{E}[\cdot]$ is the expectation operator.

The Gauss-Hermite quadrature rules accomplish this by creating deterministic samples of $\boldsymbol{\xi}$ with corresponding weights. The approximations of expectation and variances of a function $f(\boldsymbol{\xi})$ are:

$$\mu_f = \mathbb{E}[f(\boldsymbol{\xi})] \approx \sum_{q=1}^{N_q} w_q f(\boldsymbol{\xi}_q), \quad \sigma_f^2 = \mathbb{E}\left[(f(\boldsymbol{\xi}) - \mu_f)^2\right] \approx \sum_{q=1}^{N_q} w_q \left(f(\boldsymbol{\xi}_q) - \mu_f\right)^2 \tag{6}$$

where $\boldsymbol{\xi}_q$ and w_q are given by the quadrature rule with overall N_q points. The Gauss-Hermite rule used in this work was taken from Jia et al. (2012), which is a sparse Gauss-Hermite quadrature rule. These require less samples than the full Gauss-Hermite rules for the same order of accuracy.

Using Chebyshev's inequality probability constraints can be robustly reformulated involving only the mean and variance of the random variable. Let γ be a generic random variable, then

$$\mathbb{P}(\gamma \leq 0) \geq 1 - \varepsilon \implies \kappa_\varepsilon \sigma_\gamma + \hat{\gamma} \leq 0, \quad \kappa_\varepsilon = \sqrt{(1-\varepsilon)/\varepsilon} \tag{7}$$

where $\varepsilon \in (0,1) \subset \mathbb{R}$ is the probability that γ exceeds 0, $\hat{\gamma}$ and σ_γ^2 are the mean and variance of γ respectively.

The optimal control problem to be solved can subsequently be stated as follows using the Gauss-Hermite quadrature rule with N_q points, the dynamic system in Eq.(1), the PCE representation of $\theta(\cdot)$ and the initial condition at time t given as a function of $\theta(\cdot)$ as $\mathbf{x}_t(\theta(\cdot))$:

$$\underset{\mathbf{U}}{\text{minimize}} \quad \sum_{q=1}^{N_q} w_q J(\mathbf{x}^{(q)}(0), \mathbf{U}, \xi_q)$$

subject to

$$\mathbf{x}^{(q)}(k+1) = \mathbf{f}(\mathbf{x}^{(q)}(k), \mathbf{u}(k), \theta(\xi_q)) \qquad \forall (k,q) \in \mathbb{N}_k \times \mathbb{N}_q$$

$$\mu_{g_{jk}} + \kappa_\varepsilon \sigma_{g_{jk}} \leq 0, \quad \kappa_\varepsilon = \sqrt{(1-\varepsilon)/\varepsilon} \qquad \forall (j,k) \in \mathbb{N}_g^{(k)} \times \mathbb{N}_{k+1}$$

$$\mu_{g_{jk}} = \sum_{q=1}^{N_q} w_q g_j^{(k)}(\mathbf{x}^{(q)}(k), \theta(\xi_q)) \qquad \forall (j,k) \in \mathbb{N}_g^{(k)} \times \mathbb{N}_{k+1} \tag{8}$$

$$\sigma_{g_{jk}} = \sum_{q=1}^{N_q} w_q \left(g_j^{(k)}(\mathbf{x}^{(q)}(k), \theta(\xi_q)) - \mu_{g_{jk}} \right)^2 \qquad \forall (j,k) \in \mathbb{N}_g^{(k)} \times \mathbb{N}_{k+1}$$

$$\mathbf{u}(k) \in \mathbb{U} \qquad \forall k \in \mathbb{N}_k$$

$$\mathbf{x}^{(q)}(0) = \mathbf{x}_t(\theta(\xi_q)) \qquad \forall q \in \mathbb{N}_q$$

where $\mathbb{N}_k = \{0, \ldots, N-1\}$, $\mathbb{N}_{k+1} = \{1, \ldots, N\}$, $\mathbb{N}_g = \{1, \ldots, n_g^{(k)}\}$, $\mathbb{N}_q = \{1, \ldots, N_q\}$, N is the time horizon, $\mathbf{U} = \{\mathbf{u}(0), \ldots, \mathbf{u}(N-1)\}$, the objective is given by the expectation of a nonlinear function $J(\mathbf{x}_q(0), \mathbf{U}, \xi_q)$ approximated by the Gauss-Hermite rule, $\mathbb{U}$ represents the constraints on $\mathbf{u}(k)$ and $\mathbf{x}^{(q)}$ represents the state vector for each sampling point q. The chance constraints are approximated by Chebyshev's inequality given in Eq.(7) as nonlinear functions $g_j^{(k)}(\mathbf{x}^{(q)}(k), \theta(\xi_q))$ constrained robustly to be less than 0 with a probability of ε.

4. Polynomial chaos expansion filter

The PCE filter updates θ given the noisy measurements available from Eq.(2) and was first proposed in Madankan et al. (2013). It has further been applied for linear stochastic MPC in Mühlpfordt et al. (2016). Let $D_t = \{\mathbf{y}(1), \ldots, \mathbf{y}(t)\}$ be the measurements collected up to time t. Bayes's rule can be employed to update θ recursively using the previous PCE of θ:

$$p(\theta|D_t) = \frac{p(\mathbf{y}(t)|\theta) p(\theta|D_{t-1})}{p(\mathbf{y}(t)|D_{t-1})} \tag{9}$$

where $p(\theta|D_{t-1})$ is the prior distribution of θ given observations up to time $t-1$, $p(\mathbf{y}(t)|\theta)$ is the likelihood $\mathbf{y}(t)$ is observed given θ at time t. We defined $p(\mathbf{y}(t)|\theta) = \mathcal{N}(\mathbf{h}(\mathbf{x}(t), \theta), \Sigma_\mathbf{v})$ as standard normal distribution in Eq.(2) with mean $\mathbf{h}(\mathbf{x}(t), \theta)$ and covariance $\Sigma_\mathbf{v}$.

If we take both sides of Eq.(9) times $\prod_{j=1}^{n_\theta} \theta_j^{r_j}$ and integrate over both sides we obtain:

$$M_\mathbf{r}^+ = \frac{\int \prod_{j=1}^{n_\theta} \theta_j^{r_j} p(\mathbf{y}(t)|\theta) p(\theta|D_{t-1}) d\theta}{p(\mathbf{y}(t)|D_{t-1})} \tag{10}$$

where $M_\mathbf{r}^+ = \int \prod_{j=1}^{n_\theta} \theta_j^{r_j} p(\theta|D_t) d\theta$ and let $k = \sum_{j=1}^{n_\theta} r_j$. Now $M_\mathbf{r}^+$ refers to the various k-th order moments with respect to the updated distribution of θ, $p(\theta|D_t)$.

In our case the distribution of $\theta(\xi)$ depends on its PCE coefficients $\mathbf{A}$ and hence these need to be updated, see Eq.(3). We can approximate the distribution of $p(\theta(\xi)|D_{t-1})$ using sampling, since it is assumed that ξ follows a standard normal distribution. Let $\theta_t(\xi)$ denote the PCE of θ at time t with coefficients $\mathbf{A}_t$. Applying a sample estimate to Eq.(10) with a sample design of size N_s given by $\{\xi_1, \ldots, \xi_{N_s}\}$, where $\xi_i \sim \mathcal{N}(\mathbf{0}, \mathbf{I})$ we obtain the following:

$$M_{\mathbf{r}}^{(s)+} = \frac{1}{\alpha N_s} \sum_{i=1}^{N_s} \prod_{j=1}^{n_\theta} (\theta_{t-1}(\xi_i))_j^{r_j} p(\mathbf{y}(t)|\theta_{t-1}(\xi_i)) \tag{11}$$

where $M_{\mathbf{r}}^{(s)+}$ is the sample approximation of the RHS of Eq.(10) of the posterior moments and $\alpha = \frac{1}{N_s} \sum_{i=1}^{N_s} p(\mathbf{y}(t)|\theta(\xi_i))$. The sample design was created using Latin hypercube sampling and the inverse transform of the standard normal distribution.

To update the coefficients $\mathbf{A}_{t-1}$ of θ_{t-1} to $\mathbf{A}_t$ to represent θ_t we use moment matching. It is possible to determine closed-form expressions of moments of PCE expansions as given in Eq.(3) in terms of their coefficient, see Dutta and Bhattacharya (2010). The difference between these and the sample estimate of the posterior moments is then minimized to update the coefficients:

$$\hat{\mathbf{A}}_t = \arg\min_{\mathbf{A}_t} \sum_{k \leq m} ||M_{\mathbf{r}}^+(\mathbf{A}_t) - M_{\mathbf{r}}^{(s)+}||_2^2 \tag{12}$$

where $k = \sum_{j=1}^{n_\theta} r_j$ was defined as the order of the moments and hence m defines the total order of moments we want to match. $M_{\mathbf{r}}^+(\mathbf{A}_t)$ denotes the moments of the PCE expansion as a function of the coefficients $\mathbf{A}_t$ and $\hat{\mathbf{A}}_t$ are the updated coefficients to match the posterior moments.

5. Case study

The overall framework is summarised in Algorithm 1 for receding horizon SNMPC. First it is initialized by specifying the problem, including initial coefficients of the PCE expansion of θ.

Algorithm 1: Output feedback SNMPC

Input : $\hat{\mathbf{A}}_0, \Sigma_\mathbf{v}, \mathbf{f}(\cdot), \mathbf{h}(\cdot), \mathbf{x}_0(\theta)$

for *each sampling time* $t = 0, 1, 2, \ldots$ **do**

1. Determine $\mathbf{x}_t(\theta_t(\xi))$ using $\mathbf{f}(\cdot, \cdot, \theta_t(\xi))$ recursively from an updated initial condition $\mathbf{x}(0) = \mathbf{x}_0(\theta_t(\xi))$.

2. Solve SNMPC problem with $\theta_t(\xi)$ and $\mathbf{x}_t(\theta_t(\xi))$ and obtain optimal control actions.

3. Apply first part of the control actions to the plant.

4. Measure $\mathbf{y}(t+1)$.

5. Apply the PCE filter to update $\theta_t(\xi)$ to $\theta_{t+1}(\xi)$ by determining the coefficients $\hat{\mathbf{A}}_{t+1}$.

end

The case study aims to control a fermentation bioreactor using Algorithm 1. Fermentation is an important process in the biochemical and pharmaceutical industries. Uncertainties are often considerable and disregarding these may lead to inadequate performance. In this paper we consider a semi-batch bioreactor with the inlet substrate flowrate as the control variable. The two variables that are considered uncertain are the inlet concentration of the substrate $C_{S,in}$ and the kinetic parameter μ_{max}. The dynamic model was taken from Petersen and Jørgensen (2014) and describes the fermentation of a single cell protein using Methylococcus Capsulatus:

$$\dot{V} = F \tag{13a}$$

$$\dot{C}_X = -FC_X/V + \exp[\mu_{max}(\xi)] \frac{C_S}{K_S + C_S + C_S^2/K_I} \tag{13b}$$

$$\dot{C}_S = -F(C_S - \exp[C_{S,in}(\xi)])/V - \gamma_s \exp[\mu_{max}(\xi)] \frac{C_S}{K_S + C_S + C_S^2/K_I} \tag{13c}$$

where V is the volume of the reactor in m^3, F is the feed rate of substrate in m^3/h, C_X and C_S are the concentrations of biomass and substrate respectively in kg/m^3. The uncertain parameters are assumed to be given as a PCE in terms of standard normally distributed variables ξ, which were log-transformed to ensure positiveness. This defines $\mathbf{f}(\cdot)$ in Eq.(1) with $\mathbf{x} = [V, C_X, C_S]^T$, $u = F$ and $\theta(\xi) = [\mu_{max}(\xi), C_{S,in}(\xi)]^T$. The parameter values are $\gamma_S = 1.777$, $K_S = 0.021 kg/m^3$ and $K_I = 0.38 kg/m^3$. The corresponding output equation is given as follows:

$$\mathbf{y} = \mathbf{Kx} + \mathbf{v}, \quad \mathbf{K} = \text{diag}([1,0,1]), \quad \mathbf{v} \sim \mathcal{N}\left(\mathbf{0}, \text{diag}\left([3 \times 10^{-4}, 3 \times 10^{-4}]\right)\right) \tag{14}$$

The respective initial PCE expansions are given as, which defines $\hat{\mathbf{A}}_0$ and $\theta_0(\xi)$:

$$\mu_{max}(\xi) = \log(0.37) + 0.04\xi_1 + 0.02(\xi_1^2 - 1) + 0.0067(\xi_1^3 - 3\xi_1) \tag{15}$$

$$C_{S,in}(\xi) = \log(1.00) + 0.04\xi_2 + 0.02(\xi_2^2 - 1) + 0.0067(\xi_2^3 - 3\xi_2) \tag{16}$$

The objective was set to minimize the batch time with a chance constraint to produce a minimum concentration of biomass of $10 kg/m^3$ with a probability of 0.05 with 10 control intervals in a shrinking horizon implementation. The input F was constrained between $0 kg/h$ and $10 kg/h$.

6. Results, discussion and conclusions

Firstly the simulation was run by setting the parameters of the plant model to $[\mu_{max}, C_{S,in}] = [0.41, 1.05]$. The results of this are shown in Fig. 1. Next the initial parameter PCE of μ_{max} and $C_{S,in}$ given in Eq.(15) and Eq.(16) were sampled 100 times randomly and used to simulate the "true" system according to Eq.(13), for which the results are given in Fig. 2. In each case Algorithm 1 was then used to control these systems in closed-loop to verify the performance given the objective and constraints outlined in the previous section. $N_s = 200$ samples were used for the PCE filter, see section 4.

In Fig. 1 it can be seen that given the measurement available at $t = 1$ the initial distribution at $t = 0$ moves towards the correct value. With several more updates it then converges to a sharp distribution at $t = 10$ due to the relatively low measurement noise. The last row highlights the working of the algorithm: The lower left plot shows that less and less time is necessary to reach the required biomass concentration due to the better estimates available of the uncertain parameters, i.e. it becomes possible to reduce the sampling times as it becomes less and less conservative. In the second graph we see that the biomass reaches nearly exactly $10 kg/m^3$ at the final time with the control inputs shown in the last graph.

In Fig. 2 we see that the batch times of the Monte Carlo simulations vary significantly with batch times ranging from 70h to 140h, with most batch times around 110h. In the last two graphs it can be seen that the required biomass of $10 kg/m^3$ was reached in most simulations despite the uncertainties present, however in about 7% of the scenarios this was not achieved. This is due to the parameter update being overconfident from the limited number of samples used. In particular, it can be seen that two scenarios do not reach even $4 kg/m^3$, which happens if the parameters converge to the wrong value due to the limited number of samples used.

In conclusion a novel framework has been proposed by employing PCE to describe parametric uncertainties and exploiting this uncertainty description in a sparse Gauss-Hermite MPC formulation taking into account these uncertainties efficiently. Noisy measurements were used at each sampling time to update the PCE of the uncertain parameters and reduce the inherent uncertainties present significantly. It could be shown that the algorithm is able to achieve the required biomass in 93% of the scenarios, however the parameter update may be overconfident using 200 samples.

Acknowledgements

This project has received funding from the European Union's Horizon 2020 research and innovation programme under the Marie Sklodowska-Curie grant agreement No 675215.

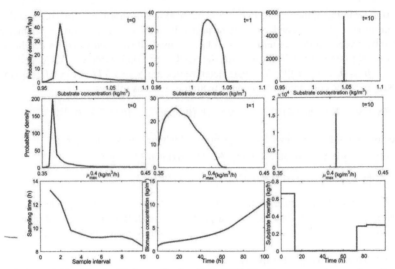

Figure 1: Run of Algorithm 1 with $[\mu_{max}, C_{S,in}] = [0.41, 1.05]$. The first two graph rows show the evolution of the marginal distribution of the inlet substrate concentration and μ_{max} at $t = 0$, $t = 1$ and $t = 10$. The last graph row shows the trajectories of the sampling time, biomass concentration and substrate flowrate.

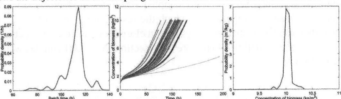

Figure 2: From left to right: Probability density function of batch times, biomass concentration trajectories and probability density function of final biomass concentrations

References

E. Bradford, L. Imsland, 2017. Expectation constrained stochastic nonlinear model predictive control of a batch bioreactor. Computer Aided Chemical Engineering 40, 1621–1626.

E. Bradford, L. Imsland, 2018a. Economic Stochastic Model Predictive Control Using the Unscented Kalman Filter. IFAC-PapersOnLine 51 (18), 417–422.

E. Bradford, L. Imsland, 2018b. Stochastic Nonlinear Model Predictive Control Using Gaussian Processes. In: 2018 European Control Conference (ECC). IEEE, pp. 1027–1034.

P. Dutta, R. Bhattacharya, 2010. Nonlinear estimation with polynomial chaos and higher order moment updates. In: American Control Conference (ACC), 2010. IEEE, pp. 3142–3147.

L. Fagiano, M. Khammash, 2012. Nonlinear stochastic model predictive control via regularized polynomial chaos expansions. In: 51st IEEE Conference on Decision and Control (CDC). IEEE, pp. 142–147.

B. Jia, M. Xin, Y. Cheng, 2012. Sparse-grid quadrature nonlinear filtering. Automatica 48 (2), 327–341.

J. M. Maciejowski, A. L. Visintini, J. Lygeros, 2007. NMPC for complex stochastic systems using a Markov chain Monte Carlo approach. In: Assessment and Future Directions of Nonlinear Model Predictive Control. Springer, pp. 269–281.

R. Madankan, P. Singla, T. Singh, P. D. Scott, 2013. Polynomial-chaos-based Bayesian approach for state and parameter estimations. Journal of Guidance, Control, and Dynamics 36 (4), 1058–1074.

A. Mesbah, 2016. Stochastic model predictive control: An overview and perspectives for future research.

T. Mühlpfordt, J. A. Paulson, R. D. Braatz, R. Findeisen, 2016. Output feedback model predictive control with probabilistic uncertainties for linear systems. In: American Control Conference (ACC), 2016. IEEE, pp. 2035–2040.

L. N. Petersen, J. B. Jørgensen, 2014. Real-time economic optimization for a fermentation process using Model Predictive Control. In: 13th European Control Conference (ECC). IEEE, pp. 1831–1836.

M. A. Sehr, R. R. Bitmead, 2017. Particle model predictive control: Tractable stochastic nonlinear output-feedback MPC. IFAC-PapersOnLine 50 (1), 15361–15366.

Anton A. Kiss, Edwin Zondervan, Richard Lakerveld, Leyla Özkan (Eds.)
Proceedings of the 29th European Symposium on Computer Aided Process Engineering
June 16th to 19th, 2019, Eindhoven, The Netherlands. © 2019 Elsevier B.V. All rights reserved.
http://dx.doi.org/10.1016/B978-0-128-18634-3.50208-3

Control structure design for a CO_2- refrigeration system with heat recovery

Adriana Reyes-Lúa[a], Glenn Andreasen[b], Lars F. S. Larsen[c], Jakob Stoustrup[b] and Sigurd Skogestad[a*]

[a]*Norwegian University of Science and Technology, Department of Chemical Engineering (NTNU), 7491 Trondheim, Norway*
[b]*Aalborg University, Department of Electronic Systems, Fr. Bajers Vej 7C, 9220 Aalborg, Denmark*
[c]*Danfoss A/S, Refrigeration and Air conditioning, 6430 Nordborg, Denmark*
*sigurd.skogestad@ntnu.no

Abstract

In this work, we analyze a generic supercritical CO_2-refrigeration system with parallel compression, based on systems used for supermarket use. In order to maximize energy efficiency, this system has a "heat-recovery" function, in which part of the heat rejected at high pressure and temperature can be recovered to provide heating. Operating conditions and active constraints are strongly affected by seasonal requirements and ambient temperature. Thus, it is necessary to find a control structure that satisfies operational constraints and maintains (near-)optimal operation with different sets of active constraints. In this paper, we use a systematic procedure to define such control structure.

Keywords: constrained operation, self-optimizing control, PID control, control structure

1. Introduction

An appropriately designed control structure should maintain (near-) optimal operation, also when there are disturbances which cause the system to operate under conditions different than the design point. Optimal operation of a process in the presence of disturbances could be maintained using optimization-based control. However, in some cases, it is possible to design and implement advanced PI(D)-based control structures that also maintain optimal operation when constraints are reached (Skogestad, 2000; Reyes-Lúa et al., 2018). The advantage of such a PI(D)-based control structure compared to optimization-based control is simpler tuning and independence of an explicit model for every system (Forbes et al., 2015).

CO_2-refrigeration systems with parallel compression is environmentally attractive. Finding optimal design and operating conditions is an ongoing area of research (Gullo et al., 2018). In order to maximize energy efficiency, some systems have a "heat recovery" function, in which part of the heat rejected at high pressure and temperature can be recovered to provide heating (e.g. district heating or tap water) (Sawalha, 2013). The available energy can be increased by operating the cooler at a higher pressure, at the expense of a higher compression work. In this work, we design a PI(D)-based control structure for the studied CO_2-refrigeration system.

2. Description of the CO_2-refrigeration system with heat recovery

A flow diagram of the analyzed CO_2-refrigeration cycle with parallel compression and heat recovery is shown in Fig. 1, and the pressure-enthalpy diagram is shown in Fig. 2. The main function of this system is to provide cooling ($\dot{Q}_{ev}$) and maintain the desired cabinet temperature (T_{cab}) via heat exchange in the evaporator, which operates at low pressure (P_l). Low-pressure CO_2 in vapor phase is compressed to high-pressure (P_h) and temperature supercritical CO_2, which may be used to heat tap water in the heat recovery section. Excess heat ($\dot{Q}_{gc}$) is rejected to the ambient air in the gas cooler.

High-pressure CO_2 is expanded to an intermediate (sub-critical) pressure (P_{IP}) in the high-pressure valve (V_{hp}). Vapor and liquid CO_2 are separated in the liquid receiver. The evaporator valve (V_{ev}) regulates the flow of liquid CO_2 from the receiver to the evaporator. By opening and closing V_{hp} and V_{ev}, we regulate the refrigerant charge (mass) at the high and low pressures. Vapor CO_2 from the liquid receiver is recycled to the high-pressure side either via parallel compression ($K2$) or the intermediate pressure valve (V_{IP}) and the main compressor ($K1$). The total compression work can be reduced by utilizing the parallel compressor instead of the intermediate pressure valve and the main compressor.

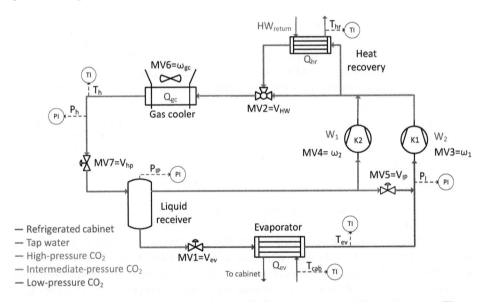

Figure 1: CO_2-refrigeration system with parallel compression and heat recovery. There are seven available manipulated variables (MV).

2.1. High-side pressure

In the supercritical region, there is no saturation condition and the pressure is independent of the temperature. From the control point of view, this means that it is necessary to control the high pressure (P_h), since it influences the gas cooler exit enthalpy (and evaporator inlet enthalpy). In other words, P_h will determine specific refrigeration capacity. As P_h is determined by the relationship between refrigerant charge, inside volume and temperature in the high-pressure side, we can actively control it using V_{hp} (Kim et al., 2004).

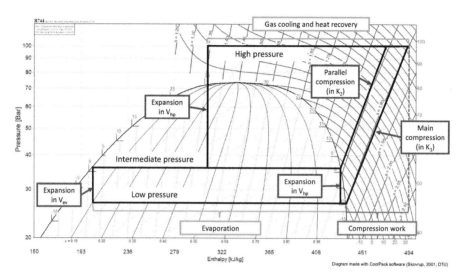

Figure 2: Pressure-enthalpy diagram of CO_2-refrigeration system with parallel compression.

It is relevant to analyze the effect of high pressure on the coefficient of performance (COP). In the case of a refrigeration system, it is defined as the ratio between cooling and compression work ($COP = \dot{Q}_{ev}/W_s$). As the isentropic compression line in the pressure-enthalpy diagram (blue lines in Fig. 2) is linear, compression work will linearly increase as P_h increases. On the other hand, in the supercritical region the isotherm (red lines in Fig. 2) becomes steeper with pressure, reducing the capacity enhancement from a given increase in pressure. For this reason, the COP reaches a maximum above which the added capacity no longer fully compensates for the additional work of compression. Thus, there is an optimal high pressure that maximizes COP (Nekså, 2002).

We should also note that in the supercritical region, at a fixed pressure, a small change in refrigerant exit temperature can produce a large change in gas cooler exit enthalpy (and evaporator inlet enthalpy), making COP very sensitive to the gas cooler refrigerant exit temperature. Previous studies (Liao et al., 2000; Jensen, 2008; Sawalha, 2013) are in line with this and have shown that the optimal set-point for the high pressure (P_h) should be corrected by the outlet temperature of the gas cooler (T_h).

2.2. Heat-recovery section

Part of the heat rejected at high pressure can be recovered to provide hot water in the heat recovery section. Heat is rejected at gliding temperature, as supercritical CO_2 is cooled. This way, the temperature profile of the CO_2 matches the heating-up curve of water, giving reduced thermodynamic losses and high efficiency (Kim et al., 2004). As it can be deduced from Fig. 2, increasing the high pressure increases the available heat for recovery, at the expense of a higher compression work. Additionally, the available heat for recovery in the supercritical region is much higher than with sub-critical CO_2.

3. Design of the PI(D)-based control structure

In this section, we apply part of the systematic plantwide control procedure proposed by Skogestad (2000) to design a self-optimizing control structure which maintains (near-) optimal operation, also in the presence of disturbances. The first step of the procedure is to define the operational objective. Here, we want to maximize the coefficient of performance (COP), subject to the system itself and operational constraints:

$$\min_{u} \ -COP(u,x,d) = -(\dot{Q}_{ev} + \dot{Q}_{hr})/(W_1 + W_2)$$

$$\text{s.t.} \quad f(u,x,d) = 0 \qquad \text{system equations (model)} \qquad (1a)$$

$$g(u,x,d) \leq 0 \qquad \text{operational and physical constraints} \qquad (1b)$$

$$e(u,x,d) = 0 \qquad \text{set-points} \qquad (1c)$$

where x are the internal states, u are the degrees of freedom, and d are the disturbances.

Physical constraints in Eq. (1b) are related to pressure (P_i), motor velocities (ω_i) and valve openings (z_i), specifically: $P_i \leq P_i^{max} \ \forall i$, $(P_{IP} - P_l)^{min} \leq (P_{IP} - P_l)$, $\omega_j^{min} \leq \omega_j \leq \omega_j^{max} \ \forall j$, and $z_k^{min} \leq z_k \leq z_k^{max} \ \forall k$. The most important set-point in Eq. (1c) is to supply enough cooling ($\dot{Q}_{ev}$) to maintain $T_{cab} = T_{cab}^{sp}$. Additionally, we would like to supply enough heating ($\dot{Q}_{hr}$) to maintain $T_{hr} = T_{hr}^{sp}$.

The next step is to determine the steady-state optimal operation. In order to do this we:

- *Identify steady-state degrees of freedom:* The analyzed system has seven available manipulated variables, MVs in Fig. 1: $u = [\omega_1, \omega_2, \omega_{gc}, z_{Vev}, z_{Vhp}, z_{VHW}, z_{VIP}]^T$. These degrees of freedom can be used to achieve optimal operation. Note that ω_2 and z_{VIP} are not independent, as either would have a similar effect in P_{IP}.

- *Identify important disturbances and their expected range:* In this case study, important disturbances (d) are cooling demand ($\dot{Q}_{ev}$, corresponding to T_{cab}^{sp}), and heating demand ($\dot{Q}_{hr}$, corresponding to T_{hr}^{sp}). The range for both is $\dot{Q}_i^{min} \leq \dot{Q}_i \leq \dot{Q}_i^{max}$.

- *Identify active constraints regions:* Once that the disturbances and their range are specified, the active constraints regions are found. This can be done by optimization or using engineering insight (Jacobsen and Skogestad, 2011). There are three relevant operating regions:

  1. *"Unconstrained" case:* corresponding to spring/fall operation.
  2. *Maximum heating:* corresponding to winter, when $\dot{Q}_{hr} = \dot{Q}_{hr}^{max}$.
  3. *Maximum cooling:* corresponding to summer, when $\dot{Q}_{ev} = \dot{Q}_{ev}^{max}$.

In every case, cooling requirements ($T_{cab} = T_{cab}^{sp}$) must be met. If possible, heating requirements ($T_{hr} = T_{hr}^{sp}$) should also be met. We do not consider $\dot{Q}_{ev} = \dot{Q}_{ev}^{min}$, as it corresponds to shut-down. $\dot{Q}_{hr} = \dot{Q}_{hr}^{min}$ is included in the "unconstrained" and maximum cooling cases. Then, we will design a control structure that works for the three relevant cases mentioned above. Fig. 3 shows the proposed control structure. The procedure to design this control structure is explained below.

For each region, each steady-state degree of freedom (MV) needs to be paired with a controlled variable. First, we pair active constraints. Then, for the remaining degrees of freedom, we identify self-optimizing controlled variables. These are usually a combination of measurements found by optimization. When designing control structures for systems with changing active constraint regions, it is useful to organize constraints in a priority list (Reyes-Lúa et al., 2018). Physical constraints (Eq. (1b)) have the highest priority. Regarding set-points (Eq. (1c)), $T_{cab} = T_{cab}^{sp}$ has a higher priority than $T_{hr} = T_{hr}^{sp}$.

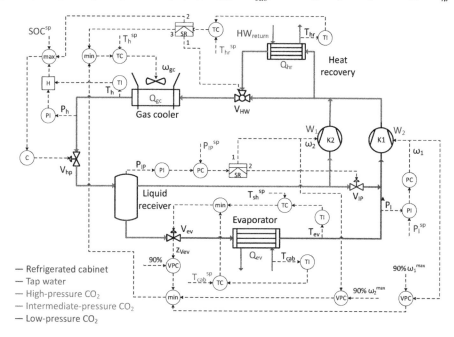

Figure 3: Control structure for the CO_2-refrigeration system with heat recovery.

3.1. "Unconstrained" case

This is the base case and we can satisfy every constraint. We use MV1=V_{ev} to control T_{cab}, and MV2=V_{HW} to control T_{hr}. The set-points are given by the operator. In order to assure that the evaporator is not over-flooded, we include a controller for the evaporator outlet temperature (T_{sh}). We have five remaining unconstrained degrees of freedom, two of which are not independent (ω_2 and V_{IP}). We pair these degrees of freedom as follows:

1. MV3=ω_1 controls P_l. P_l^{sp} is found by optimization (self-optimizing variable).

2. The parallel compressor (MV4=ω_2) and MV5=V_{IP} are used to control the pressure in the liquid receiver (P_{IP}). The set-point defined by optimization and may be a self-optimizing variable. Normal operation is using ω_2, but when the flow is too low, we use V_{IP}. We can implement this with a split-range controller.

3. MV6=ω_{gc} controls T_h (outlet of the gas cooler). T_h^{sp} is defined by optimization.

4. MV7=V_{hp} controls P_h. As explained in Section 2.1, the set-point is a linear combination (H) of P_h and T_h, which is a self-optimizing variable (Jensen, 2008).

3.2. Maximum heating

When V_{HW} becomes fully open, we must switch the manipulated variable to continue controlling T_{hr}. This is handled using split-range control with selectors. First, we switch to V_{hp} as manipulated variable and increment the available heat for recovery by increasing P_h. To implement this, we include a selector for the set-point of the high-pressure controller. Once we reach P_h^{max}, we get additional capacity for the heat-recovery section by increasing T_h, using ω_{gc} as manipulated variable. This will increase mass flow through the compressors and, as consequence, the discharge temperature.

If we continue to increase T_h, at some point liquid in the low-pressure section may be insufficient and V_{ev} will reach its maximum opening. Alternatively, the compressors could reach maximum capacity due to the increased mass flow. To prevent this, we implement valve-positioning controllers (VPC) with a *min* selector, which will prevent T_h from increasing in such a way that either the valve or the compressors ($z_{V_{ev}}$, ω_1 or ω_2) saturate.

3.3. Maximum cooling

As cooling requirements increase, z_{Vev} will open and reach z_{Vev}^{max}. The valve positioning controller for V_{ev} will adjust T_h (and indirectly P_h) such that the system reaches Q_{ev}^{max}.

4. Final remarks

Using a systematic procedure, we designed a PI(D)-based control structure for a CO_2-refrigeration system, that maintains (near-)optimal steady-state operation, also with changes in the set of active constraints. We should point out that pairing on the low-pressure side could be different (e.g. controlling T_{cab} with the main compressor, and P_l with V_{ev}). The final decision would consider system dynamics. It is important to mention that we can usually reach the same steady-state control objectives we reach with split-range controllers by using valve-positioning control or different controllers with different set-points.

References

M. G. Forbes, R. S. Patwardhan, H. Hamadah, R. B. Gopaluni, Jan 2015. Model Predictive Control in Industry: Challenges and Opportunities. IFAC-PapersOnLine 48 (8), 531–538.

P. Gullo, K. M. Tsamos, A. Hafner, K. Banasiak, Y. T. Ge, S. A. Tassou, Dec 2018. Crossing CO_2 equator with the aid of multi-ejector concept: A comprehensive energy and environmental comparative study. Energy 164, 236–263.

M. G. Jacobsen, S. Skogestad, Oct 2011. Active Constraint Regions for Optimal Operation of Chemical Processes. Industrial & Engineering Chemistry Research 50 (19), 11226–11236.

J. B. Jensen, 2008. Optimal operation of refrigeration cycles. Phd thesis, NTNU, Norway.

M.-H. Kim, J. Pettersen, C. W. Bullard, 2004. Fundamental process and system design issues in CO_2 vapor compression systems. Progress in Energy and Combustion Science 30 (2), 119–174.

S. Liao, T. Zhao, A. Jakobsen, Jun 2000. A correlation of optimal heat rejection pressures in transcritical carbon dioxide cycles. Applied Thermal Engineering 20 (9), 831–841.

P. Nekså, Jun 2002. CO_2 heat pump systems. International Journal of Refrigeration 25 (4), 421–427.

A. Reyes-Lúa, C. Zotică, S. Skogestad, Jul 2018. Optimal Operation with Changing Active Constraint Regions using Classical Advanced Control. In: 10th ADCHEM. IFAC, Shenyang, China.

S. Sawalha, Jan 2013. Investigation of heat recovery in CO_2 trans-critical solution for supermarket refrigeration. International Journal of Refrigeration 36 (1), 145–156.

S. Skogestad, Oct 2000. Plantwide control: the search for the self-optimizing control structure. Journal of Process Control 10 (5), 487–507.

Anton A. Kiss, Edwin Zondervan, Richard Lakerveld, Leyla Özkan (Eds.)
Proceedings of the 29th European Symposium on Computer Aided Process Engineering
June 16th to 19th, 2019, Eindhoven, The Netherlands. © 2019 Elsevier B.V. All rights reserved.
http://dx.doi.org/10.1016/B978-0-128-18634-3.50209-5

Real-Time Optimisation of Closed-Loop Processes Using Transient Measurements

Jack Speakman* and Grégory François

School of Engineering, The University of Edinburgh, Edinburgh EH9 3FB, UK
jack.speakman@ed.ac.uk

Abstract

Real-time optimisation (RTO) has the ability to boost the performance of a process whilst satisfying a set of constraints by using process measurements to refine the model-based optimal operating conditions towards plant optimality. Modifier adaptation (MA) is a methodology of RTO which can find the optimal operating point of a process even in the presence of plant-model mismatch. This work presents an extension to MA through the combination of two established frameworks. This new combined framework allows for the optimisation of a controlled process using transient measurements whilst using a steady-state open-loop model. In conjunction, an investigation into model-based gradient estimation methods of the controlled process using the open-loop model was undertaken and a correction has been proposed, which allows for the use of limited controller information in gradient estimation using the neighbouring extremals method.

Keywords: Real-time optimisation, Modifier adaptation, Transient measurements, Plant-model mismatch, Model-based gradient estimation

1. Introduction

The optimisation of chemical processes is vital for improving the economic profile, whilst ensuring the meeting of safety and environmental objectives, in the face of increasing global competition and tightening regulations. Process optimisation is typically carried out using a model to calculate the operating conditions which maximise the performance of the process. Fully accurate models are not possible to obtain due to uncertainties, noise, and simplifications, resulting in sub-optimal operation. Real-time optimisation (RTO) solves this through the use of measurements from the process to adjust the operating point towards the true optimal solution.

Modifier adaptation (MA) is an approach to RTO which adds input-affine modifier terms to the cost and constraints (Marchetti et al., 2009), and measurements of the process are used to iteratively update these modifiers. The main advantage of MA lies in the mathematically proven capacity to converge to a KKT point of the process, even in the presence of plant-model mismatch, a case for which the standard two-step approach (Jang et al., 1987) typically fails (Forbes et al., 1994). Recently, an extension to MA has been proposed allowing for the optimisation of a controlled plant using an open-loop model (François et al., 2016), which is mathematically proven to reach a plant KKT point upon convergence. Such cases occur for complex plants using built-in control systems while the model has been developed in the lab. This article aims to extend this framework to allow for the use of transient measurements, and for the use of model-based gradient estimation methods when there is a difference in degrees of freedom between the model and the plant. This paper will state the mathematical optimisation problem, with a brief overview of MA; both extensions will be described, followed by a proposition of the KKT converging nature of the combined framework and associated proof. Finally this new framework will be demonstrated in a case study of a CSTR.

2. Problem formulation

2.1. Steady-State Optimisation Problem

The optimisation of a continuous process consists of finding the inputs (u) which minimise the cost (ϕ) whilst satisfying a set of constraints (g_j). These functions are assumed to be known for a measured output (y_p), in practice this is modelled by $\bar{y} = H(u, \theta)$, resulting in two problems:

Plant Optimisation Problem	Model Optimisation Problem
$u_p^* = \arg\min_u \Phi_p(u) := \phi(u, y_p)$	$u^* = \arg\min_u \Phi(u, \theta) := \phi(u, \bar{y}(u, \theta))$
$s.t. \quad G_{p,j}(u) := g_j(u, y_p) \leq 0$	$s.t. \quad G_j(u, \theta) := g_j(u, \bar{y}(u, \theta)) \leq 0$

$$(1) \qquad (2)$$

Where θ are the uncertain model parameters, $(.)_p$ refers to the plant and $j = 1, \ldots, n_g$. If there is plant-model mismatch (i.e. $\bar{y} \neq y_p$), the solution to Problem 2 will not match that of Problem 1.

2.2. Standard Modifier Adaptation

Standard MA adds correction terms to the model cost and constraints to reconcile the gradient of the plant and model problems with respect to the inputs at the current operating point, which requires the plant cost and constraints to be measured or estimated from the plant input and outputs.

$$u_{k+1}^* = \arg\min_u \Phi_m(u, \theta) := \Phi(u, \theta) + \lambda_k^{\Phi, u}(u - u_k^*)$$

$$s.t. \quad G_{m,j}(u, \theta) := G_j(u, \theta) + \varepsilon_k^{G_j} + \lambda_k^{G_j, u}(u - u_k^*) \leq 0, \quad j = 1, \ldots, n_g \tag{3}$$

Where ε and λ are the zeroth and first order correction terms, defined as follows:

$$\varepsilon_k^{G_j} = G_{p,j}(u_k^*) - G_j(u_k^*, \theta) \tag{4}$$

$$\lambda_k^{\Phi, u} = \nabla_u \Phi_p(u_k^*) - \nabla_u \Phi(u_k^*, \theta) \tag{5}$$

$$\lambda_k^{G_j, u} = \nabla_u G_{p,j}(u_k^*) - \nabla_u G_j(u_k^*, \theta), \quad j = 1, \ldots, n_g \tag{6}$$

This problem is shown to only converge at a KKT point of the plant (Marchetti et al., 2009). This is a key result from MA and is highly sought after from the extensions to MA.

3. Recent Advances

3.1. Controlled Plant, Open-loop model

Standard MA requires the same degrees of freedom between the model and the process. An extension to MA was proposed which allows for the use of an open-loop model in the optimisation of a process which operates in closed-loop with different degrees of freedom (r) than the model (François et al., 2016), as illustrated in Figure 1a. This framework has three possible methods of implementation depending on whether the optimisation and modifiers are in terms of u or r:

UR:
$$u_{k+1}^* = \arg\min_u \Phi_{m,k}(u) := \Phi(u) + \lambda_k^{\Phi, r}(y(u) - r_k)$$

$$s.t. \quad G_{m,j,k}(u) := G_j(u) + \varepsilon_k^{G_j} + \lambda_k^{G_j, r}(y(u) - r_k) \leq 0, \quad j = 1, \ldots, n_g \tag{7}$$

UU:
$$u_{k+1}^* = \arg\min_u \Phi_{m,k}(u) := \Phi(u) + \lambda_k^{\Phi, u}(u - u_k)$$

$$s.t. \quad G_{m,j,k}(u) := G_j(u) + \varepsilon_k^{G_j} + \lambda_k^{G_j, u}(u - u_k) \leq 0, \quad j = 1, \ldots, n_g \tag{8}$$

RR:
$$r_{k+1}^* = \arg\min_r \Phi_{m,k}(r) := \Phi(r) + \lambda_k^{\Phi, r}(r - r_k)$$

$$s.t. \quad G_{m,j,k}(r) := G_j(r) + \varepsilon_k^{G_j} + \lambda_k^{G_j, r}(r - r_k) \leq 0, \quad j = 1, \ldots, n_g \tag{9}$$

Where $\lambda_k^{\Phi, r}$ and $\lambda_k^{G_j, r}$ are the difference between the plant and model gradients with respect to r

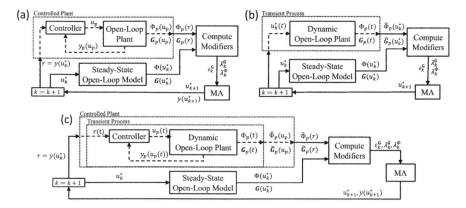

Figure 1: Framework schematics. (a) Closed-loop. (b) Transient. (c) Combined.

3.2. Transient Measurements

Standard MA, and the framework described above in Section 3.1, wait until steady-state before calculating the next operating point, ignoring all the transient information between operating points. Therefore, an extension to standard MA was developed that uses these transient measurements to update the operating points before steady-state is achieved (François and Bonvin, 2014), as illustrated in Figure 1b, resulting in a faster convergence to the process optimum. This method estimates the steady-state value as the current measured value ($\hat{y}_p = y_p(t)$) and performs MA with these values, which was proven by François and Bonvin to reach a KKT point of the process upon convergence, *but only when the process is operated in open-loop.*

3.3. Gradient Estimation Methods

The gradient of the process cost and constraints with respect to the degrees of freedom at u_k^* is required to calculate $\boldsymbol{\lambda}$. This cannot be directly measured, but must be inferred from process measurements. This paper will implement two different gradient estimation techniques and extend these to work in the case of a difference in degrees of freedom between the model and the process. The first technique, multiple units (MU) (Srinivasan, 2007), uses the finite differences method by running $n_u + 1$ identical units in parallel, offsetting each unit from the current operating point.

The second method, neighbouring extremals (NE) (Gros et al., 2009), is a model-driven approach which takes the difference between the model and process outputs, and relates it to the parametric uncertainty of the model. This can be used when structural plant-model mismatch is present with imperfect accuracy. This can be written mathematically as follows, where $f = \Phi, G_1, \ldots, G_j$:

$$\nabla_{u_p} f_p \approx \nabla_{u\theta}^2 f (\nabla_\theta H)^+ (y_p - y_0) + (\nabla_{uu}^2 f - \nabla_{u\theta}^2 f (\nabla_\theta H)^+ \nabla_u H)(u_p - u_0) \tag{10}$$

4. Transient MA of Controlled Processes using Open-Loop Models

The frameworks outlined in the previous section are combined hereafter into a single concise framework, allowing for the use of transient measurements in the optimisation of a controlled process using a steady-state open-loop model, as shown in Figure 1c. A proposition and proof using method UR is given, but similar arguments can be made using method UU and method RR.

Proposition 1. *If upon convergence, the estimates of the process* $(\widehat{\phi}_p, \widehat{g}_p, \widehat{\nabla_r \Phi_p}, \widehat{\nabla_r G_p})$, *estimated via a function of the transient measurements* $(u_p(t), y_p(t))$, *approach their true values, then the operating point* $r_\infty = y(u_\infty)$ *is a KKT point of the process.*

Proof. If the algorithm converges, i.e. $\boldsymbol{u}_\infty = \lim_{k\to\infty} \boldsymbol{u}_k$, then $\boldsymbol{u}_\infty$ must be steady. Hence, $\boldsymbol{r}_\infty = \boldsymbol{y}(\boldsymbol{u}_\infty)$, and the plant must also be steady. Therefore, the steady-state estimates approach their true values, as stated in the proposition. The gradient of the process, along with steady state estimation of the modifier $\boldsymbol{\lambda}_\infty^{\phi,r}$, can be combined and rearranged into the following:

$$\frac{\partial \phi_{m,\infty}^r}{\partial \boldsymbol{u}}\left(\left(\frac{\partial \boldsymbol{y}}{\partial \boldsymbol{u}}\right)^+ \frac{\partial \boldsymbol{y}}{\partial \boldsymbol{u}}\right) = \frac{\partial \phi_p}{\partial \boldsymbol{r}}\left(\frac{\partial \boldsymbol{y}}{\partial \boldsymbol{u}}\right) \tag{11}$$

The same can be applied to $\frac{\partial g_{m,j,\infty}^r}{\partial \boldsymbol{u}}$. From the definition of the zeroth order term $\varepsilon_\infty^{g_j,r}$, the following can be stated: $g_{m,j,\infty}^r(\boldsymbol{u}_\infty) = G_{p,j}(\boldsymbol{r}_\infty)$. Hence, as the modified problem is optimised such that $\boldsymbol{u}_\infty$ is a KKT point, satisfying the following:

$$\frac{\partial \phi_{m,\infty}^r}{\partial \boldsymbol{u}} + \boldsymbol{v}\frac{\partial \boldsymbol{g}_{m,\infty}^r}{\partial \boldsymbol{u}} = 0 \tag{12}$$

It follows that the operating point of the plant must be a KKT point since the following must be valid:

$$\frac{\partial \Phi_p}{\partial \boldsymbol{r}} + \boldsymbol{v}\frac{\partial \boldsymbol{G}_p}{\partial \boldsymbol{r}} = 0 \tag{13}$$

$$\boldsymbol{G}_p(\boldsymbol{r}_\infty) \leq 0 \tag{14}$$

$$\boldsymbol{v}\boldsymbol{G}_p(\boldsymbol{r}_\infty) = 0 \tag{15}$$

Hence, $\boldsymbol{r}_\infty$ is a KKT point of the closed-loop plant. $\qquad\square$

The above proposition and proof are similar to those from the original frameworks, an in-depth proof of each can be found in the respective papers (François and Bonvin, 2014; François et al., 2016). The estimates must be correct at steady state, this is achieved in the same manner as in (François and Bonvin, 2014), by setting them equal to the current transient value, i.e. $\hat{\phi}_p = \phi_p(t)$.

Model-driven gradient estimation methods, such as NE, applied to an open-loop model will obtain an estimate for the open-loop process ($\nabla_{\boldsymbol{u}_p}\phi_p$). The method described above requires an estimate of the gradient of the process cost and constraints with respect to the controller setpoints ($\nabla_{\boldsymbol{r}}\phi_p$). Therefore, the gradients must be remapped by multiplying by $\partial \boldsymbol{u}_p/\partial \boldsymbol{r}$. However, this relationship is often unknown for complex plants using built-in control systems (whereby the supplier does not provide full information) or when advanced controllers are implemented (e.g. MPC), hence an estimation must be made using some limited information about the controller.

5. Case Study

The methods discussed above have been simulated on the Williams-Otto reactor (Williams and Otto, 1960), which is a benchmark case for MA problems. The open-loop plant follows a three reaction system ($A+B \to C, B+C \to P+E, C+P \to G$). Whilst the open-loop model follows a 2 reaction system ($A+2B \to P+E, A+B+P \to G$). These are illustrated in Figure 2 and the systems of dynamic equations are described in Table 1.

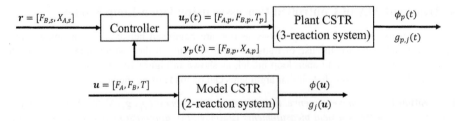

Figure 2: Block flow diagram of controlled plant and open-loop model

Table 1: Dynamic systems for the plant and the model

Plant	Model (steady)
$\frac{dX_A}{dt} = \frac{F_{A,in}}{\rho V} - k_1 X_A X_B - \frac{F}{\rho V} X_A$	$\frac{F_{A,in}}{\rho V} - k_1^* X_A X_B^2 - k_2^* X_A X_B X_P - \frac{F}{\rho V} X_A = 0$
$\frac{dX_B}{dt} = \frac{F_{B,in}}{\rho V} - k_1 X_A X_B - \frac{F}{\rho V} X_B$	$\frac{F_{B,in}}{\rho V} - 2k_1^* X_A X_B^2 - k_2^* X_A X_B X_P - \frac{F}{\rho V} X_B = 0$
$\frac{dX_C}{dt} = 2k_1 X_A X_B - 2k_2 X_B X_C - k_3 X_C X_P - \frac{F}{\rho V} X_B$	–
$\frac{dX_P}{dt} = k_2 X_B X_C - \frac{1}{2} k_3 X_C X_P - \frac{F}{\rho V} X_P$	$k_1^* X_A X_B^2 - k_2^* X_A X_B X_P - \frac{F}{\rho V} X_P = 0$
$\frac{dX_E}{dt} = 2k_2 X_B X_C - \frac{F}{\rho V} X_E$	$2k_1^* X_A X_B^2 - \frac{F}{\rho V} X_E = 0$
$\frac{dX_G}{dt} = \frac{3}{2} k_3 X_C X_P - \frac{F}{\rho V} X_G$	$3k_2^* X_A X_B X_P - \frac{F}{\rho V} X_G = 0$

Where the X_i is the mass fraction of component i, F_i is the mass flowrate of component i, k are the reaction rates, and $\rho V = 2105$ kg is the mass holdup. The values for the parameters of the model are the same as in François et al. (2016). There are 2 cases for the model, with differing parameters, case I and case II, both described in the above paper.

The inputs to the open-loop plant and model are the flowrates of both A and B, and the temperature ($\boldsymbol{u} = [F_A, F_B, T]$). The controller on the open-loop plant has been designed to reduce and change the nature of the degrees of freedom of the system, with the setpoints being $\boldsymbol{r} = [F_{B,s}, X_{A,s}]$. This is done by introducing an unmodelled ratio controller on the two input flowrates such that $F_{B,p}/F_{A,p} = 2.4$. Further, the temperature is controlled via a proportional controller of the concentration in component A, $T = T_0 + K_P(X_{A,s} - X_{A,p})$, with $T_0 = 120\,°C$ and $K_P = -1000$. Also, there is an offset in the flowrate of B, $F_{B,p} = F_{B,s} + 2$, introducing additional uncertainty. In practice better controllers would certainly be designed and implemented but the main point of this article is about when the actual controller is unknown and not about its performances. The objective and constraint functions can be written as follows:

$$\phi(\boldsymbol{u}, \boldsymbol{y}(\boldsymbol{u})) = -(1143 X_P F + 25.92 X_E F - 76.23 F_{A,in} - 114.34 F_{B,in}) \tag{16}$$

$$g_1(\boldsymbol{u}, \boldsymbol{y}(\boldsymbol{u})) = X_A - 0.09 \tag{17}$$

$$g_2(\boldsymbol{u}, \boldsymbol{y}(\boldsymbol{u})) = X_G - 0.6 \tag{18}$$

The model optimum, ϕ, for the two cases are -140 and -161 respectively, whilst the process optimum, ϕ_p, is -212. Initially the algorithms were tested using the MU gradient estimation method, which provides perfect estimations with identical units. Figures 3 and 4 show that the new framework approaches the plant optimum more rapidly than for steady-state to steady-state operation.

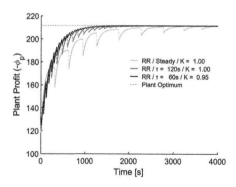

Figure 3: Plant profit for *RR* Algorithm with *MU* gradient estimation

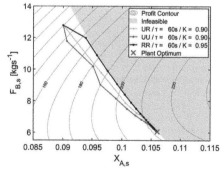

Figure 4: Controller setpoints for *MU* gradient estimation

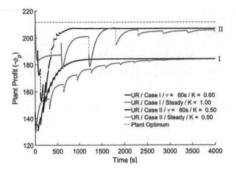
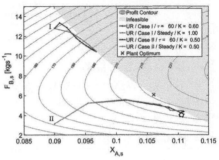

Figure 5: Plant profit for *UR* Algorithm with *NE* gradient estimation

Figure 6: Controller setpoints *UR* Algorithm with *NE* gradient estimation

The algorithms were then tested using NE gradient estimation with some controller knowledge. The assumed knowledge is that the flowrates have a fixed, but unknown, ratio control, which is approximated as the flow ratio of the model's previous operating point $\partial F_{A,p}/\partial F_{B,s} = F^*_{A,k}/F^*_{B,k}$. Also, the mass fraction of A is purely controlled by T, however the control law is unknown. This was approximated as $\partial T/\partial x_{A,s} = \left(\partial x_A/\partial T\right)^+$, at the previous operating point. As can be seen in Figures 5 and 6, both cases converge to sub-optimal points on the edge of the feasible region, at a ϕ value of -184 and -207 respectively. This is expected for a model-based gradient estimation technique, which is based on a linearisation of the steady-state nominal model.

6. Conclusion

This paper has proposed that the combination of two established frameworks within modifier adaptation, namely the use of transient measurements and the use of a open-loop model in optimising a controlled process, will reach a KKT point of the process if the algorithm converges. This is backed up with a mathematical proof and case study, resulting in a method which can use transient measurements to optimise a controlled plant with only a steady-state open-loop model. An extension to a model-based gradient estimation technique was proposed to allow for use of the open-loop model to estimate the gradient of the controlled process by transforming the estimate to be with respect to the setpoints using limited controller knowledge. This did not converge to a KKT point of the plant, since NE is a model- linearisation-based gradient estimation technique. However, this did approach a point which is on the edge of the feasible region, while significantly reducing the loss in plant optimality with minimal information about the structure of the controller.

References

J. Forbes, T. Marlin, J. F. MacGregor, 1994. Model adequacy requirements for optimizing plant operations. Comp. Chem. Eng. 18, 497–510.

G. François, D. Bonvin, 2014. Use of transient measurements for the optimization of steady-state performance via modifier adaptation. Ind. Eng. Chem. Res. 53 (13), 5148–5159.

G. François, S. Costello, A. Marchetti, D. Bonvin, 2016. Extension of modifier adaptation for controlled plants using static open-loop models. Comp. Chem. Eng. 93, 361 – 371.

S. Gros, B. Srinivasan, D. Bonvin, 2009. Optimizing control based on output feedback. Comp. Chem. Eng. 33 (1), 191 – 198.

S.-S. Jang, B. Joseph, H. Mukai, 1987. On-line optimization of constrained multivariable chemical processes. AIChE J. 33 (1), 26–35.

A. Marchetti, B. Chachuat, D. Bonvin, 2009. Modifier-adaptation methodology for real-time optimization. Ind. Eng. Chem. Res. 48 (13), 6022–6033.

B. Srinivasan, 2007. Real-time optimization of dynamic systems using multiple units. Int. J. of Rob. and Nonlin. Cont. 17 (13), 1183–1193.

T. Williams, R. Otto, 1960. A generalized chemical processing model for the investigation of computer control. AIEE Trans. 79, 483–473.

Anton A. Kiss, Edwin Zondervan, Richard Lakerveld, Leyla Özkan (Eds.)
Proceedings of the 29th European Symposium on Computer Aided Process Engineering
June 16th to 19th, 2019, Eindhoven, The Netherlands.
http://dx.doi.org/10.1016/B978-0-128-18634-3.50210-1

Process Dynamic Analysis and Control Strategy for COGEN option Used for Flare Utilization

Monzure-Khoda Kazi,[a] Fadwa Eljack,[a,*] Saad Ali Al-Sobhi,[a] Nikolaos Kazantzis,[b] Vasiliki Kazantzi[c]

[a]Qatar University, Department of Chemical Engineering, College of Engineering, Doha P.O. Box-2713, Qatar

Fadwa.Eljack@qu.edu.qa

[b]Worcester Polytechnic Institute, Department of Chemical Engineering and Center for Resource Recovery and Recycling, 100 Institute Road, Worcester, MA 01609, USA

[c]University of Applied Sciences / TEI of Thessaly, Department of Business Administration, Business School, Larissa 41110, Greece

Abstract

The aim of this work is to develop suitable process dynamic analysis and control strategies for implementing cogeneration unit (COGEN) as a flare utilization alternative to handle unexpected disturbances from unknown and uncertain flare sources. The developed steady state and dynamic simulations using Aspen Plus and Aspen Plus Dynamics allow the design of pressure and temperature controllers that demonstrate good performance. The simulation software and its built-in controller tuning option are used to create the dynamic modelling framework, and to perform controller installation and design. Finally, the proposed process dynamic analysis and control strategies are evaluated using an illustrative regulatory case study. It included disturbance rejection capability assessment in the presence of unexpected pressure load disturbances as well as temperature excursions from desirable operating conditions. The simple designed controller and its automated monitoring systems offered means of evaluating COGEN as a flare utilization alternative, while providing insights in to the techno-economic performance of the system.

Keywords: COGEN systems, Flare utilization, Dynamic process analysis, Simulation and control.

1. Introduction

Emissions from industrial flaring contribute to the accumulation of GHG in the environment and consequently to global warming. Thus, there is a need to develop an effective flare mitigation strategy and operational implementation plan that could reliably contribute to a systematic reduction of flare emissions while at the same time allow a further usage of otherwise unutilized flared streams as potentially valuable energy sources. Previous experience and research findings have demonstrated the utility of systematic process systems engineering (PSE) approaches for recovering and utilizing hydrocarbon streams typically flared during abnormal process operations (Kamrava et al., 2015, Kazi et al., 2018). These approaches have shown promising results and opportunities for substantial emissions reduction, enhanced energy recovery and cost-effective performance through the use of modular systems, such as cogeneration (COGEN) unit and thermal membrane distillation (TMD) systems (Elsayed et al., 2014, Kazi et al., 2016).

However, the prevailing design philosophy and most of the associated mathematical models for flare recovery/utilization in the pertinent literature rely on a set of design steady state conditions. Therefore, PSE research studies for flare recovery system analysis and control purposes conducted with the aid of a comprehensive, insightful and computationally tractable dynamic process modelling framework remain relatively limited and deserve further attention (Dinh et al., 2016, Eljack et al., 2014, Wang et al., 2016).

It should be pointed out, that abnormal situations encountered in process industries represent major sources of flaring when the process of interest deviates significantly from what is deemed the normal and acceptable operating conditions range. Consequently, these process deviations from the desirable performance profile are quite concerning since they often lead to financial losses as well as poor environmental records. The key challenge in abnormal situation management is to provide the operating team with tools that will enable them to avoid or minimize the impact of abnormal conditions. At the same time, if some portion of the unburned hydrocarbon streams is recovered and utilized under stable and robust conditions, significant techno-economic-environmental benefits could emerge by appropriately minimizing the associated waste streams. In this regard, the employment of classical yet comprehensive process control strategies could offer a potentially interesting solution option since satisfactory performance characteristics can be complemented by insightful cost-effective practical implementation strategies. Indeed, such strategies tend to be well received by plant personnel and engineering management teams resulting in greater plant operational efficiency through robust abnormal situation management approaches.

Based on preliminary results reported in Kazi et al. (2016), the present research work considers a COGEN unit as a flare mitigation alternative technology where by the dual process requirement of power and heating can be satisfied using a mixture of fresh fuel (natural gas) and unburned hydrocarbon streams. In particular, a systematic dynamic modelling approach is followed that leads to the development of a process analysis framework and control strategies focusing on the flare utilization section/component of the COGEN unit and the partial replacement of fresh fuel feed in order to enhance process system overall techno-economic-environmental performance. The primary objective of the proposed control strategies is the design, tuning and use of PI and PID controllers for a standard regulation problem, namely for smoothly driving key pressure and temperature variables in the combustion unit of the flare utilization section to desirable set-point values in the presence of various unexpectedly occurring disturbances that cause excursion patterns from operationally desirable steady states in flare systems. Therefore, within such a context, the COGEN unit can be used to recover some of the unburned hydrocarbon streams and utilize them to produce some value added products thus enhancing overall environmental and economic performance.

2. Methodology

The Aspen Plus V9 and Aspen Plus Dynamics software packages have been used for the development of a dynamic modelling framework focusing on the behaviour of the whole COGEN unit that encompasses the flare utilization section/component. First, open-loop (manual mode) tune-up is introduced based on an appropriately validated First Order Plus Dead Time (FOPDT) dynamic process modelling framework. Then, comprehensive closed-loop dynamic response tests are carried out for controller tuning and stability characterization purposes. The required steps for controller installation and controller design have been performed using the same software and its built-in controller tuning option. A succinct step-by-step graphic depiction of the methodology followed in this manuscript is presented in Fig.1.

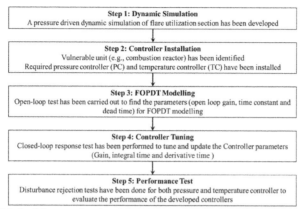

Figure 1. Methodology to develop PC and TC strategies using Aspen Plus Dynamics

3. Development of steady state and dynamic models

In this work, functional baseline steady state and dynamic models have been developed for the entire COGEN unit (see Fig. 2). Special focus was placed on the development of a pressure driven dynamic model for the flare utilization section in order to identify the vulnerable unit and the requisite action plan for reliable pressure or temperature monitoring (see dashed box in Fig. 2). Here, in the pressure controller case the pressure of the combustion reactor (CR) represents the process variable (PV), the set point value is set at 50 bar while the operating variable (OP) is the pressure maintained by the combustion stream valve (CSV). In the temperature controller case, PV is the temperature of CR, the set point value is the steady state temperature 2104 °C and the OP variable is the flow of the cooling medium.

4. Controller design and performance

4.1. Open-loop (manual mode) tune up for FOPDT dynamic modelling

A rather standard open-loop test has been performed to identify and quantitatively capture key unknown complex process dynamics through a first-order plus dead time model-postulate. The response profile of the main process variables is used to estimate meaningful open-loop gain (K_P), time constant (τ_P), and dead time (θ_P) values for the given process. The estimated model parameter values of the installed pressure and temperature controllers are shown in Fig. 3. Specifically, a step up by 5% of the output range method has been used which is a built-in method/feature in Aspen Plus Dynamics software.

4.2. Closed-loop dynamic response tests and controller tuning

The closed-loop Auto Tuning Variation (ATV) test better describes the controlled process and offers more reliable estimates of the tuned-up controller parameters (Kamal et al., 2016). In the present work, using the "Tyreus–Luyben" tuning rule (Luyben et al., 1997), a set of controller parameter values: K_C =6.71, τ_I =1.98 min, and τ_D =0.0 min were calculated for a "PI"-type pressure controller. Furthermore, using the "Ziegler–Nichols" tuning rule, the following set of controller parameter values: K_C =1586.209, τ_I =3.75 min, and τ_D =0.6 min were calculated for a "PID"-type temperature controller. It should be noted that temperature control loops within the specific controlled process system's structure typically generate oscillatory behaviour. Therefore, different controller tuning methods were tested to induce stable responses and appealing controlled process economic and environmental performance profiles. Moreover, temperature measurements can be noisy and the derivative action can lead to a deterioration of the noise-to-signal ratio and process control quality. To remedy this situation, a filtered PID controller or cascade control loop option is advisable.

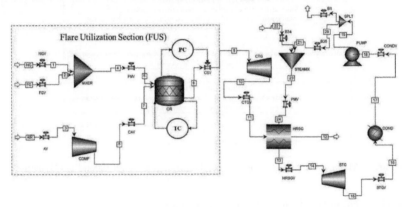

Figure 2. Developed process flow diagram of COGEN unit and its flare utilization section

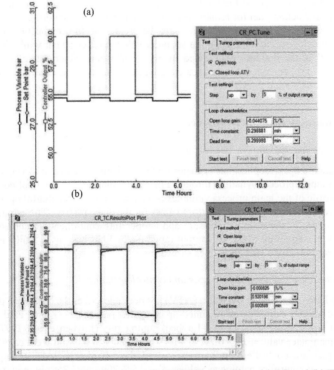

Figure 3. Open-loop response test for FOPDT modelling: (a) PC and (b) TC

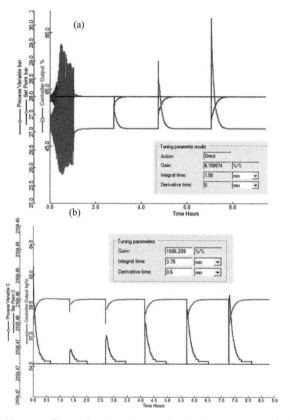

Figure 4. Closed-loop profiles under disturbance rejection (process regulation) test: (a) PC response and (b) TC response

4.3. Regulation Problem: Disturbance rejection test and controller performance evaluation

The regulatory capabilities of the aforementioned pressure and temperature controllers were tested under various disturbances in an effort to evaluate their respective performance during abnormal process operations and upsets. For testing the PI pressure controller, pressure has been changed deliberately from 50 bar to 60 bar, 75 bar and 90 bar at certain time instants to generate a pattern of unexpected occurring output disturbances. From Fig. 4a, it can be inferred that the PI controller successfully brought the pressure variable to its set point value in an offset-less, smooth and stable manner thus rejecting the disturbance effect while inducing desirable dynamic characteristics in the closed-loop response profile. Please note that unfiltered strong fluctuation behavior was observed in the beginning as the controller was not activated during that period. These fluctuations may have occurred due to coupling effects of the installed controllers, an issue which was later adequately addressed by the specific controller's action. Similar satisfactory behavior was observed in the PID temperature controller case where the cooling fluid flow rates were varied to create a pattern of occurrence of sudden disturbances, suggesting that each time the proposed PID controller was able to mitigate the disturbance effect and accomplish the primary regulatory objectives as delineated in the previous case (see Fig. 4b).

5. Techno-economic-environmental performance analysis

Within the context of the present study and the steady state process simulation framework, standard heat/energy integration allows the calculation of the heating utility of the COGEN system which was found to be 52.18 MW as well as power utility that was found to be 105.50 MW. Furthermore techno-economic-environmental performance assessment for the designed COGEN unit with the proposed controllers installed was conducted within the framework presented in (Kazi et al., 2016) where the optimal sizing of the COGEN as a flare mitigation alternative was also pursued. It was found that the proposed configuration of the COGEN unit can lead to savings of 0.18 million USD annually realized by the associated carbon tax savings and can also generate 0.25 million USD annual income from power/heat generation by utilizing the flare during process upsets.

6. Conclusion

A dynamic modelling framework for the flare utilization component of a COGEN system was first developed on the basis of which process dynamic analysis, techno-economic-environmental performance assessment as well as controller design and tuning were also pursued. Simple and practical control strategies were proposed capable of successfully regulating pressure and temperature in the presence of various disturbances as well as inducing appealing performance characteristics for the controlled process. Finally, within the above context, it was demonstrated that there is indeed ample scope to use COGEN as a potentially viable flare mitigation technique by utilizing unburned hydrocarbon streams while generating heat (52.18 MW), power (105.50 MW) as well as annual tax savings (0.18 million USD).

Acknowledgment

This paper was made possible by NPRP grant No 10-0205-170347 from the Qatar National Research Fund (a member of Qatar Foundation). The statements made herein are solely the responsibility of the author[s].

References

H. Dinh, F. Eljack, S. Wang, Q. Xu, 2016, Dynamic simulation and optimization targeting emission source reduction during an ethylene plant start-up operations, *Journal of Cleaner Production*, 135, 771-783.

F. Eljack, M. M. El-Halwagi, Q. Xu, 2014, An Integrated approach to the simultaneous design and operation of industrial facilities for abnormal situation management, In *Computer Aided Chemical Engineering*, Mario R. Eden, J. D. S.; Gavin, P. T., Eds. Elsevier, 34, 771-776.

N. A. Elsayed, M. A. Barrufet, M. M. El-Halwagi, 2014, Integration of thermal membrane distillation networks with processing facilities, *Industrial & Engineering Chemistry Research*, 53, (13), 5284-5298.

K. Al-Malah, 2016, Apen Plus®: Chemical Engineering Applicaitons, Wiley & Sons, Inc., DOI: 10.1002/9781119293644.

S. Kamrava, K. J. Gabriel, M. M. El-Halwagi, F. Eljack, , 2015, Managing abnormal operation through process integration and cogeneration systems, *Clean Technologies and Environmental Policy*, 17, (1), 119-128.

M.-K. Kazi, F. Eljack, N. A. Elsayed, M. M. El-Halwagi, 2016, Integration of energy and wastewater treatment alternatives with process facilities to manage industrial flares during normal and abnormal operations: Multiobjective extendible optimization framework, *Industrial & Engineering Chemistry Research*, 55, (7), 2020-2034.

M.-K. Kazi, F. Eljack, M. Amanullah, A. M.N. AlNouss, V. Kazantzi, 2018, A process design approach to manage the uncertainty of industrial flaring during abnormal operations. *Computers & Chemical Engineering*, 117, 191-208.

S. Wang, J. Zhang, S. Wang, Q. Xu, , 2016, Dynamic simulation for flare minimization in chemical process industry under abnormal operations. *Current Opinion in Chemical Engineering*, 14, 26-34.

M. Luyben, W. Luyben, 1997, Essentials of process control, McGraw-Hill.

Anton A. Kiss, Edwin Zondervan, Richard Lakerveld, Leyla Özkan (Eds.)
Proceedings of the 29th European Symposium on Computer Aided Process Engineering
June 16th to 19th, 2019, Eindhoven, The Netherlands. © 2019 Elsevier B.V. All rights reserved.
http://dx.doi.org/10.1016/B978-0-128-18634-3.50211-3

Efficient robust nonlinear model predictive control via approximate multi-stage programming: A neural networks based approach

Wachira Daosud[a], Paisan Kittisupakorn[b], Miroslav Fikar[c], Sergio Lucia[d] and Radoslav Paulen[c,*]

[a]*Department of Chemical Engineering, Faculty of Engineering, Burapha University, Chonburi 20131, Thailand*
[b]*Department of Chemical Engineering, Faculty of Engineering, Chulalongkorn University, Bangkok 10330, Thailand*
[c]*Faculty of Chemical and Food Technology, Slovak University of Technology in Bratislava, Bratislava 812 37, Slovakia*
[d]*Laboratory of Internet of Things for Smart Buildings, Technische Universität Berlin, Einstein Center Digital Future, Berlin 10587, Germany*
radoslav.paulen@stuba.sk

Abstract

Multi-stage nonlinear model predictive control (msNMPC) is a robust control strategy based on the description of the uncertainty propagation through a dynamic system via a scenario tree and is one of the least conservative approaches to robust control. The computational complexity of the msNMPC, however, grows with respect to the number of uncertainties and with respect to the length of the prediction horizon. This paper presents a new approach, where the optimal cost-to-go function is approximated after a specific point in time, here in particular by neural networks, so the independent branches do not have to be optimized but are approximated. The optimization might be casted over the robust horizon only, which reduces the computational burden, but still guarantees robust satisfaction of the constraints. Moreover, this approach allows to consider any length of the prediction horizon for the same computational cost. The neural network models are trained offline using the optimal profiles in all branches of the scenario tree. The potential of the proposed approach is demonstrated by simulation studies on a semi-batch reactor.

Keywords: Nonlinear model predictive control, robust control, neural networks.

1. Introduction

Robust nonlinear model predictive control (NMPC) methods aim at overcoming the limitations of conventional NMPC w.r.t. the influence of model errors and presence of uncertainty. Very popular are min-max approaches (Witsenhausen, 1968) despite that these approaches are quite conservative because they ignore the fact that new measurements will be available in the future and that the future control actions can be adapted accordingly. Another possibility to formulate the NMPC controller within the framework of stochastic optimization. Such a formulation leads to a multi-stage NMPC (msNMPC) (Lucia et al., 2013), which has shown very promising results

Acknowledgements: Support from the SAIA, n.o., the national scholarship programme of the Slovak Republic and the Slovak Research and Development Agency under the project APVV 15-0007 are gratefully acknowledged.

for challenging nonlinear examples. Nevertheless, the main disadvantage of the msNMPC lies in solving a large-scale optimization problem. The size of the optimization problem grows exponentially with the length of the prediction horizon and with the number of uncertainties. To overcome this problem, deep neural networks (NNs) (LeCun et al., 2015) were proposed to approximate the NMPC policy (Lee and Lee, 2005) or the msNMPC policy (Karg and Lucia, 2018).

A neural networks-based (NN-based) msNMPC is investigated in this paper. The NNs are used to approximate the optimal cost-to-go functions of different scenarios. Then, after a specific point in time, the independent branches do not have to be optimized but are approximated using the NNs, which results in the reduction of the computational burden, that the approach retains the same robustness guarantees as the standard msNMPC. To demonstrate the potential of the proposed strategy, simulation tests are performed and the control performance of the approach is compared to standard NMPC and msNMPC approaches.

2. Multi-stage NMPC

Multi-stage NMPC is a robust NMPC strategy that is based on the description of the uncertainty propagation through a dynamic system by a tree of discrete scenarios (Lucia et al., 2013). Each branch of the tree represents an evolution of the system states under a certain realization of the uncertainty. The main advantage of this formulation is that the availability of information provided by the future measurements is taken into account, so that the future control inputs depend on the future knowledge and can be adapted w.r.t. the expected realizations of the uncertainty.

The optimization problem to be solved at each sampling instant using the msNMPC reads as:

$$
\min_{x_{k+1}^j, u_k^j, \forall (j,k) \in I_{N_p}} \sum_{i=1}^{N} \omega_i \overbrace{\left(\sum_{k=0}^{N_p-1} L(x_{k+1}^j, u_k^j) + \phi(x_{N_p}^j) \right)}^{J_i(x_0)}, \qquad \forall x_{k+1}^j, u_k^j \in S_i, \quad (1a)
$$

$$
\text{s.t. } x_{k+1}^j = f(x_k^{p(j)}, u_k^j, d_k^{r(j)}), \qquad \forall (j, k+1) \in I_{N_p}, \quad (1b)
$$

$$
g(x_{k+1}^j, u_k^j) \leq 0, \qquad \forall (j, k+1) \in I_{N_p}, \quad (1c)
$$

$$
u_k^j = u_k^l \text{ if } x_k^{p(j)} = x_k^{p(l)}, \qquad \forall (j,k), (l,k) \in I_{N_p}, \quad (1d)
$$

where, at stage $k+1$ and position in the tree j, the state vector x_{k+1}^j depends on $x_k^{p(j)}$, u_k^j and $d_k^{r(j)}$, i.e., the parent state vector, the vector of control inputs, and the realization of the uncertainty, respectively, via the system dynamics equations (1b). The scenario tree originates at the root node x_0 and is defined with the same number (s) of branches at each node, given by $d_k^{r(j)} \in \{d_k^1, d_k^2, \ldots, d_k^s\}$, and with the index set I_{N_p}. Constants N, N_p are the number of scenarios and the length of the prediction horizon, respectively. The functions $L(\cdot)$ and $\phi(\cdot)$ denote stage and terminal costs, respectively. The constraints on inputs and states are denoted by $g(\cdot)$. Parameter $\omega_i \geq 0$ is a weighting coefficient of the ith scenario (S_i) and is chosen based on the relative importance of the scenario while $\sum_{i=1}^N \omega_i = 1$. The constraints (1d) are non-anticipativity constraints, which denote that all the control inputs that branch at the same parent node must be equal. The branching of the tree stops at a certain point, which is denoted as the robust horizon of the length N_r, which can be set to 1 or 2 in practice (Lucia et al., 2013). Note that after N_r, all the scenarios become independent.

3. Neural networks-based multi-stage NMPC

The size of the optimization problem grows exponentially with the length of the prediction horizon and with the number of uncertainties. The main idea of this work is to use NNs (LeCun et al., 2015; Daosud et al., 2017; Kittisupakorn et al., 2017) to approximate the optimal cost-to-go function of the individual scenarios such that the contribution of the independent scenarios can be approximated after the robust horizon when solving (1). This allows simultaneously for efficient real-time execution, for robust constraint satisfaction, and for an efficient training of NNs, since only an optimization over a single branch is needed therein. The NN is trained to give an approximate cost $\tilde{J}_i^*(x_{k,p}, W, b)$ of the scenario S_i, where $x_{k,p}$ is a training point for which (1) is solved. The optimal weights (W) and biases (b) for each neuron are found via

$$\min_{W,b} \sum_{p=1}^{N_t} (J_i^*(x_{k,p}) - \tilde{J}_i^*(x_{k,p}, W, b))^2, \tag{2}$$

where N_t is the number of data pairs $(J_i^*(x_{k,p}), x_{k,p})$.

The NN-based msNMPC then approximates the problem (1) as:

$$\min_{x_{k+1}^j, u_k^j, \forall (j,k) \in I_{N_c}} \sum_{i=1}^{N} \omega_i \left(\sum_{k=0}^{N_c-1} L(x_{k+1}^j, u_k^j) + \tilde{J}(x_{N_c}^j) \right), \qquad x_{k+1}^j, u_k^j \in S_i, \tag{3a}$$

$$\text{s.t. Eqs. (1b)–(1d),} \qquad \forall (j, k+1) \in I_{N_c}, \tag{3b}$$

$$g_{SS}(x_{k+1}^j) \leq 0, \qquad \forall (j, k+1) \in I_{N_c}, \tag{3c}$$

where $N_c \in \{N_r, \ldots, N_p - 1\}$ is the length of the reduced prediction horizon and functions $g_{SS}(\cdot)$ represent the so-called safe sets (Rosolia and Borrelli, 2017), which constrain the states such that these are in the range of the data in the training set of the NNs. In practice, construction of suitable safe sets can be cumbersome, so one can then decide to drop the safe sets entirely from the formulation (3) or replace them with tightened $g(\cdot)$. This however requires tuning of N_c by increasing it to guarantee recursive feasibility of the problem. This path is examined in this study.

4. Case study

We use a challenging case study from the chemical engineering domain, i.e., a semi-batch reactor that is equipped with a cooling jacket, where the cooling jacket inlet temperature is controlled by a thermostat (Thangavel et al., 2015). An exothermic reaction of A and B produces C. The model includes mass balances for the reactor content (expressed through reactor volume V and species concentrations c_A, c_B, and c_C) and energy balances of the reaction mixture, the reactor jacket and the thermostat (expressed using the temperature of the reaction mixture T, cooling jacket T_J, and jacket inlet temperature $T_{J,in}$. The control inputs are the feed rate of the component B and the set point of jacket inlet temperature, summarized in $u = (\dot{V}_{in}, T_{J,in,set})^T$. The detailed mathematical model with all parameters and considered constraints can be found in Thangavel et al. (2015).

The control goal is to maximize the concentration of C in the fixed batch time $t_f = 1{,}800\,\text{s}$. Thus, a shrinking-horizon strategy is used and the sampling time is taken as $120\,\text{s}$. The constraints on T and V must be respected despite a $\pm 30\%$ uncertainty in the reaction enthalpy H and in the reaction rate k, with nominal values $H_{nom} = -355\,\text{kJ}\,\text{mol}^{-1}$ and $k_{nom} = 3.35 \times 10^{-7}\,\text{m}^3\,\text{mol}^{-1}\,\text{s}^{-1}$.

5. Results

The dynamic optimization problems are solved using orthogonal collocation on finite elements using CasADi (Andersson et al., 2012). The nonlinear programs are solved via Ipopt (Wächter and Biegler, 2006). The plant response is simulated using a 4th-order Runge-Kutta scheme. All the results are obtained on a workstation with Intel i7 running at 2.4 GHz with 8 GB RAM.

Neural networks with several hidden layers (deep NNs) are employed due to the theoretical evidence of outperforming the shallow NNs (Safran and Shamir, 2016). For the approximation of

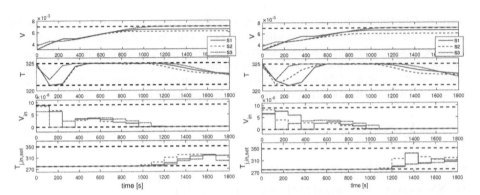

Figure 1: Input/state trajectories for $[S_1, S_2, S_3]$; Left: standard NMPC, right: NNs-based NMPC.

Table 1: Performance comparison standard NMPC vs. NNs-based NMPC for $[S_1, S_2, S_3]$.

Algorithm	$c_C(t_f) \, [\mathrm{mol\,m^{-3}}]$	Cons. viol. $[\mathrm{m^3\,s^{-1}}]$	Cons. viol. $[\mathrm{K\,s^{-1}}]$
standard NMPC	$[379.1, 470.8, 525.9]$	$[0, 0, 1.03] \times 10^{-6}$	$[0, 6.48, 0] \times 10^{-8}$
NNs-based NMPC	$[378.3, 470.0, 526.5]$	$[0, 0, 1.64] \times 10^{-5}$	$[0, 0, 9.39] \times 10^{-4}$

cost-to-go functions, a state vector $x = (V, c_A, T, T_j, T_{j,in}, t)^T$ is used as an input to the NN, which represents the minimal number of states (all other states can be obtained as combinations of these states) and the current time t is considered due to the problem nature (batch, shrinking horizon).

The NNs are trained with the data generated for 360 batches, where the initial conditions are randomly varied, w.r.t. the ones in Thangavel et al. (2015), between $\pm 5\,\mathrm{K}$, $\pm 5 \times 10^{-4}\,\mathrm{m^3}$, and $\pm 50°C\,\mathrm{mol\,m^{-3}}$ for the temperatures, volume, and concentration of A, respectively. 75 % of the 34,560 data points is used for training and the remaining 25 % for testing. The structure of NN is selected based on the MSE criterion (Kittisupakorn et al., 2009). The activation function is the tanh function. The number of hidden nodes is varied between 2–34. The resulting structure of the NNs consists of 2 hidden layers with different number of nodes for different scenario (S_i).

We compare the performance of the standard NMPC and NNs-based NMPC for some selected scenarios. Figure 1 shows trajectories of constrained states and control inputs in $S_1 = (H_{min}, k_{min})$, $S_2 = (H_{nom}, k_{nom})$, and $S_3 = (H_{max}, k_{max})$. The optimal solution tries to feed B as much as possible at the beginning of the batch. Then the feed rate is adjusted to respect the temperature constraints. At these phases cooling is maximal. Towards the end of the batch, feeding is stopped since either maximal volume is reached or any further feeding only dilutes the reaction mixture and decreases the final concentration of C. At this phase, cooling is adjusted to respect the constraints on T. It can be seen that the use of the NNs-based NMPC results in similar state and input trajectories as the standard NMPC. Both control methods result in similar final concentrations of C (see Tab 1), while the constraint violations occur only marginally, due to numerical reasons. The average CPU times per sampling time are 0.35 s and 0.14 s for standard and NN-based NMPC, respectively.

For the msNMPC, we use $N_r = 1$. Nine scenarios are considered with all the combinations of extremal and nominal values of the uncertain parameters. As we do not use safe sets in this work, a tuning has to be done for the value of N_c. The tuning procedure is illustrated in Fig. 2, where state and input trajectories resulting from the use of NN-based msNMPC are shown for different values of N_c under the most extreme scenario (H_{min}, k_{max}), i.e., the most exothermic and the fastest reaction. As the NN-based msNMPC is myopic from its nature, large constraint violations can occur if N_c is too small. While the msNMPC would be initialized with $N_p = 15$, the NN-based

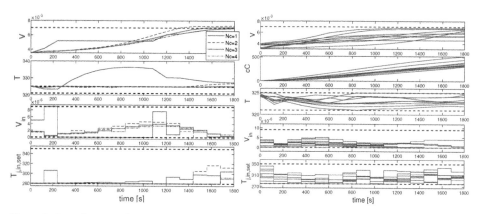

Figure 2: Input/state profiles using the NN-based msNMPC for (H_{min}, k_{max}) and different N_c.

Figure 3: Input/state profiles using the NN-based msNMPC for the 9 scenarios.

Table 2: Performance comparison msNMPC vs. NN-based msNMPC averaged over 9 scenarios.

Algorithm	$c_C(t_f)$ [mol m^{-3}]	Cons. viol. [m^3 s^{-1}]	Cons. viol. [K s^{-1}]
msNMPC	354.4	3.72×10^{-10}	1.36×10^{-2}
NN-based msNMPC	371.5	3.50×10^{-4}	0

msNMPC is able to robustly satisfy the constraints with $N_c = 4$.

Figure 3 shows the state and input trajectories resulting from the use of the NN-based msNMPC in all 9 scenarios. The trajectories show great similarity with the trajectories obtained by msNMPC.

The performance comparison averaged over the 9 scenarios considered is shown in Tab. 2 for msNMPC and NN-based msNMPC. Use of both algorithms results in certain minor constraint violations while the control performance is better when using NN-based msNMPC, which is attributed to slightly greater constraints violations. The optimization problems solved by the msNMPC have at most 5,686 optimization variables and are solved in average in 2.19 s, while NN-based msNMPC requires 0.95 s of CPU time in average for solving optimization problems with at most 1,528 variables.

Figure 4 shows input and state trajectories obtained using NN-based msNMPC for 40 batches with different values of the uncertain parameters being randomly generated with uniform distribution ($\pm 30\%$) around the nominal values. It can be seen that the performance is satisfactory for all batches and that the constraints are consistently respected, though with minor violations. The msNMPC and NN-based msNMPC achieve the maximum concentration of product C as 313.6 mol m^{-3} and 297.6 mol m^{-3}, respectively. The 5 % performance loss of NN-based NMPC is attributed to the interpolation inaccuracy of the neural networks and represents the price to pay for the computational efficiency.

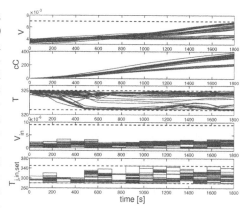

Figure 4: Input/state trajectories for 40 batches with random uncertainties (NN-based msNMPC).

The developed NN-based msNMPC scheme shows a great promise regarding an efficient implementation of the robust NMPC schemes. The results obtained on the challenging case study are encouraging, despite the fact that certain portion of performance must be given up for robust constraint satisfaction. This aspect can be further improved when safe sets are used. Another way of performance improvement, which we will concentrate on in our further studies, is the reduction of the sampling time. Our preliminary results show that using NN-based msNMPC and safe sets, one can reduce the sampling time up to 12 s and N_c to 1, while the CPU time reduces to 0.8 s in the worst case. Note that the use of msNMPC here requires 15 s for execution in the worst case.

6. Conclusions

Non-conservative robust NMPC control methods, such as msNMPC, are challenged by a large computation burden. To deal with this problem, an NN-based msNMPC is proposed. The idea pursued is to simplify the optimization problems of the msNMPC by approximating the cost-to-go functions of independent scenarios (after the robust horizon of the msNMPC). In this work, neural networks are used for this purpose. The results show that the proposed approach obtains similar results compared to the msNMPC, which is demonstrated on a challenging case study from chemical engineering domain. Most importantly, the NN-based msNMPC shows superior performance in terms of computational time. Thus the presented approach is a promising method of real-time implementation of msNMPC in case of complex control problems with long prediction horizons, large number of parametric uncertainties and short sampling times. Our future work will consist in improving the closed-loop performance of the proposed approach by the use of safe sets.

References

J. Andersson, J. Åkesson, M. Diehl, 2012. Casadi: A symbolic package for automatic differentiation and optimal control. In: Recent advances in algorithmic differentiation. Springer, pp. 297–307.

W. Daosud, J. Thampasato, P. Kittisupakorn, 2017. Neural network based modeling and control for a batch heating/cooling evaporative crystallization process. Eng J 21 (1), 127–144.

B. Karg, S. Lucia, 2018. Efficient representation and approximation of model predictive control laws via deep learning. arXiv preprint arXiv:1806.10644.

P. Kittisupakorn, P. Somsong, M. A. Hussain, W. Daosud, 2017. Improving of crystal size distribution control based on neural network-based hybrid model for purified terephthalic acid batch crystallizer. Eng J 21 (7), 319–331.

P. Kittisupakorn, P. Thitiyasook, M. Hussain, W. Daosud, 2009. Neural network based model predictive control for a steel pickling process. Journal of Process Control 19 (4), 579–590.

Y. LeCun, Y. Bengio, G. Hinton, 2015. Deep learning. Nature 521 (7553), 436.

J. M. Lee, J. H. Lee, 2005. Approximate dynamic programming-based approaches for inputoutput data-driven control of nonlinear processes. Automatica 41 (7), 1281–1288.

S. Lucia, T. Finkler, S. Engell, 2013. Multi-stage nonlinear model predictive control applied to a semi-batch polymerization reactor under uncertainty. Journal of Process Control 23 (9), 1306–1319.

U. Rosolia, F. Borrelli, 2017. Learning model predictive control for iterative tasks: a computationally efficient approach for linear system. IFAC-PapersOnLine 50 (1), 3142–3147.

I. Safran, O. Shamir, 2016. Depth-width tradeoffs in approximating natural functions with neural networks. arXiv preprint arXiv:1610.09887.

S. Thangavel, S. Lucia, R. Paulen, S. Engell, 2015. Towards dual robust nonlinear model predictive control: A multi-stage approach. In: American Control Conference 2015. pp. 428–433.

A. Wächter, L. T. Biegler, 2006. On the implementation of an interior-point filter line-search algorithm for large-scale nonlinear programming. Mathematical programming 106 (1), 25–57.

H. Witsenhausen, 1968. A minimax control problem for sampled linear systems. IEEE Transactions on Automatic Control 13 (1), 5–21.

Anton A. Kiss, Edwin Zondervan, Richard Lakerveld, Leyla Özkan (Eds.)
Proceedings of the 29th European Symposium on Computer Aided Process Engineering
June 16th to 19th, 2019, Eindhoven, The Netherlands. © 2019 Elsevier B.V. All rights reserved.
http://dx.doi.org/10.1016/B978-0-128-18634-3.50212-5

Filter-based Constraints to Easier Plant Feasibility in Modifier Adaptation Schemes

A. Papasavvas[a] and G. Francois[a]

[a]*School of Engineering, Institute for Material and Processes, The University of Edinburgh, Edinburgh EH93FB*
A.Papasavvas@sms.ed.ac.uk

Abstract

Output modifier adaptation (MAy) is an iterative model-based real-time optimization (RTO) method with the proven ability to reach, upon converge, the unknown plant optimal steady-state operating conditions despite structural and parametric plant-model mismatch and disturbances. However, as such, feasibility of the iterates cannot be guaranteed before convergence is achieved. This issue can be mitigated with additional modelling and/or experimental efforts, which is costly and sometimes cannot be envisaged, especially for large-scale systems. In this article, MAy is compared to one of its most recent extensions, namely "KMAy", which is shown to have more favorable sufficient conditions for feasibility of the iterates. A simulation study, performed on the Williams-Otto reactor – a standard benchmark case study for RTO algorithms – illustrates that KMAy is safer for practical applications.

Keywords: Real-time optimization, modifier adaptation, feasibility.

1. Introduction

Plant operators often manipulate the degrees of freedom of industrial processes to maximize performances, while trying to enforce the satisfaction of safety and quality constraints. However, such experience-based decision methods often lead to sub-optimal operations. Real-time optimization (RTO) methods use both the available model and measurements in the decision-making process. Output modifier-adaptation (MAy-Marchetti et al. (2016), Papasavvas et al. (2019)) has been recently proposed as an extension of Modifier Adaptation (MA) schemes (Marchetti et al., 2016). Basically, MAy uses all real-time plant measurements to implement input-affine corrections on the modelled input-output mapping to improve the prediction of the cost and constraints, while preserving the proven ability of MA schemes to reach upon convergence the true plant optimal inputs.

One challenge to push MAy further is to provide, at least, the same level of feasibility guarantees as experience-based decision methods. Indeed, forcing the satisfaction of the plant constraints typically requires global knowledge of the plant behaviour (Bunin et al., 2013; Marchetti et al., 2017), which is unfortunately rarely the case. In this paper, the focus is on a recent extension of MAy, i.e. "KMAy' '(Papasavvas and François, 2019). This method is shown to offer better feasibility guarantees than other MA algorithms without requiring more data or knowledge. On the basis of this study, a simple method for improving most MA algorithms is proposed.

The paper is organized as follows. After a brief review of MAy and KMAy in Section 2, these schemes are compared in Section 3 w.r.t. their ability to enforce plant feasibility. These results are illustrated by means of simulated chemical reactor in Section 4, and Section 5 concludes the paper.

2. Real-Time Optimization via Modifier Adaptation

Plant optimal operating conditions $u_p^\star \in \mathbb{R}^{n_u}$ are such that they minimize the operating cost $\phi(u, y_p(u)) \in \mathbb{R}$ while satisfying the constraints $G(u, y_p(u)) \leq 0 \in \mathbb{R}^{n_g}$, where $y_p(u) \in \mathbb{R}^{n_y}$ is the input-output mapping of the plant, i.e. the relationship between the operating conditions and the steady-state measured outputs of the plant. Finding $u_p^\star$ when only an approximate model of the plant $y(u)$ is available, is generally not possible with standard model-based optimization methods but becomes handy with RTO algorithms like MAy. The main idea behind MAy is to use plant measurements to iteratively modify the model-based optimization problem so that its first order conditions of optimality match those of the plant optimization problem at each operating point u_k. It is proven that the true plant optimum is the only possible fixed point of MAy despite structural plant-model mismatch (Papasavvas et al., 2019). MAy can be summarized as follows:

Output Modifier Adaptation (MAy): At the kth iteration, u_k is applied to the plant until steady state is reached, and the modified input-output mapping is modified: $y_{i,m,k}(u) := y_i(u) + \varepsilon_k^{y_i} + (\lambda_k^{y_i})^\mathsf{T}(u - u_k)$, where $\varepsilon_k^{y_i} := y_{i,p}(u_k) - y_i(u_k)$ and $\lambda_k^{y_i} := \nabla_u y_{i,p}\big|_{u_k} - \nabla_u y_i\big|_{u_k}$ are the 0th- and 1st-order modifiers, respectively, and $\nabla_u(\cdot)$ is the gradient operator w.r.t. u. Then, the following optimization problem is solved to derive the model-based optimal inputs $u_{k+1}^\star$:

$$u_{k+1}^\star := \quad \arg\min_{u} \quad \Phi_{\text{MAy},k}(u) := \phi(u, y_{m,k}(u)) \tag{2.1}$$

$$\text{s.t.} \quad G_{\text{MAy},k}(u) := g(u, y_{m,k}(u)) \leq 0, \tag{2.2}$$

The next iterate u_{k+1} is obtained by applying the following filter:

$$u_{k+1} := \quad u_k + K(u_{k+1}^\star - u_k), \tag{2.3}$$

with $K \in \mathbb{R}^{n_u \times n_u}$ a filter gain matrix, typically diagonal with $K_{i,i} \in (0,1], \forall i \in [1, n_g]$.

The filter of (2.3) is used here to enable asymptotic stability and to avoid excessive corrections (Marchetti et al. (2016)) but its implementation is indeed risky: *Figure 1 illustrates that this filter can lead MAy to generate the next inputs where the model predicts infeasibility! This is due to the fact that the filter of (2.3) picks u_{k+1} in the segment $[u_k, u_{k+1}^\star]$, while the model predicts that the constraint is not satisfied in the red-shaded area. Of note is that this is a major common weakness to standard MA algorithms using a similar filter.* To avoid this, the following extension of MAy can be used:

Output Modifier Adaptation Extension (KMAy): Apply MAy until equation (2.1), and solve the following problem instead of (2.1)-(2.3):

$$u_{k+1} := \quad \arg\min_{u} \quad \Phi_{\text{KMAy},k}(u) := \Phi_{\text{MAy},k}(v_k(u)) \tag{2.4}$$

$$\text{s.t.} \quad G_{\text{KMAy},k}(u) := \left[G_{\text{MAy},k}(v_k(u))^\mathsf{T}, G_{\text{MAy},k}(u)^\mathsf{T}\right]^\mathsf{T} \leq 0, \tag{2.5}$$

$$v_k(u) := K^{-1}(u - u_k) + u_k. \tag{2.6}$$

As seen in (2.5), KMAy duplicates the model-based constraints, which are in turn also checked at what would be the next iterate as the filter (2.3) is here implemented by means of equation (2.6). Note that loss of constraint qualification upon convergence could be expected with the duplication of the constraints, but it has been recently shown that if linear constraint qualification holds for MAy, then Mangasarian Fromowitz qualification holds for KMAy, while KMAy also leads to model adequacy conditions that are less difficult to satisfy compared to MAy (Papasavvas and François, 2019). Since model-based feasibility is checked at both at $v_k(u_{k+1})$ and u_{k+1}, KMAy will generate a u_{k+1} where modelled constraints are satisfied, while MAy will not, as illustrated in Figure 1. Selecting u_{k+1} such that the modified model predicts feasibility also at the next iterate does not guarantee, per se, plant feasibility. However, KMAy should outperform MAy in most cases, especially when the model provides a reasonable prediction of the plant contraints.

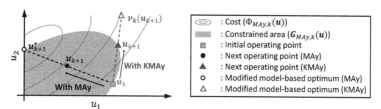

Figure 1: A 2D optimization problem solved with MAy and KMAy.

3. Comparison between MAy and KMAy from a Feasibility Perspective

In this section, we compare the ability of MAy and KMAy to generate feasible iterates or sequences of feasible iterates. To do so, the following definitions are required:

Definitions:
- $\mathcal{U} \subseteq \mathbb{R}^{n_u}$ is the set of feasible inputs for the subset of known constraints (i.e., not subject to modelling errors). The *known* constraints functions only depend of u, e.g., input bounds: $u^L \leq u \leq u^U$. $\mathcal{F}_p \subseteq \mathcal{U}$ denotes the set of feasible inputs for the plant.
- A *model* is an object m in the universe of models $\mathcal{M}$: $m \in \mathcal{M}$.
- $\forall k$, $\mathcal{C}_{\text{KMAy}}^k$ denotes the set of inputs that satisfy the constraints (2.5) and $\mathcal{C}_{\text{MAy}}^k$ denotes the set of inputs generated by the application of the filter of (2.3) to any input satisfying the constraint (2.2). This can be rewritten as: $\mathcal{C}_{\text{KMAy}}^k := \{u \in \mathbb{R}^{n_u} | G_{\text{KMAy},k}(u) \leq 0\}$ and $\mathcal{C}_{\text{MAy}}^k := \{u \in \mathbb{R}^{n_u} | G_{\text{MAy},k}(K^{-1}(u - u_k) + u_k) \leq 0\}$, respectively. In other words, $\mathcal{C}_X^k$ are the sets of candidate inputs u_{k+1} that can be generated by MAy or KMAy for $X \in \{\text{MAy}, \text{KMAy}\}$.
- $\forall k$ and for each $X \in \{\text{MAy}, \text{KMAy}\}$, $\mathcal{S}_X^k \subseteq \mathcal{M}$ are the sets of models m such that the candidate inputs u_{k+1} are feasible for the plant, i.e., $\mathcal{S}_X^k := \{m \in \mathcal{M} | \mathcal{C}_X^k \subseteq \mathcal{F}_p\}$. From the definition of $\mathcal{S}_X^k$, it is clear that $m \in \mathcal{S}_X^k$ is a *sufficient condition for the plant feasibility at the next iterate.*

For any given $u_k \in \mathcal{F}_p$ and given a model $m \in \mathcal{M}$, the following proposition states that KMAy is more likely to lead to $u_{k+1} \in \mathcal{F}_p$ than MAy:

Proposition 1 $\forall k$, *the set of models satisfying the sufficient condition for plant feasibility at $k+1$ with MAy is a subset of the set of models satisfying it with KMAy, i.e.,* $\mathcal{S}_{\text{MAy}}^k \subseteq \mathcal{S}_{\text{KMAy}}^k$.

Proof. u_{k+1} (generated by $X \in \{\text{MAy}, \text{KMAy}\}$) is in $\mathcal{C}_X^k$ and is feasible for the plant when $m \in \mathcal{S}_X^k$, since, as said before, $m \in \mathcal{S}_X^k$ is a sufficient condition for plant feasibility at the next iterate. Since KMAy includes the same constraints as MAy *together with additional constraints*, $\mathcal{C}_{\text{KMAy}}^k \subseteq \mathcal{C}_{\text{MAy}}^k$. In other words, the duplication of the constraints with KMAy shrinks the set of candidate inputs $\boldsymbol{u}_{k+1}$ compared to MAy. Finally, $\mathcal{C}_{\text{MAy}}^k \subseteq \mathcal{F}_p \Rightarrow \mathcal{C}_{\text{KMAy}}^k \subseteq \mathcal{F}_p$ while $\mathcal{C}_{\text{KMAy}}^k \subseteq \mathcal{F}_p \not\Rightarrow \mathcal{C}_{\text{MAy}}^k \subseteq \mathcal{F}_p$. Thus, $\mathcal{S}_{\text{MAy}}^k \subseteq \mathcal{S}_{\text{KMAy}}^k$. $\square$

Indeed, Proposition 1 implies that any model satisfying the aforementioned sufficient condition for feasibility with MAy also satisfies it for KMAy, while the opposite does not hold. Thus, from the point of view of the feasibility of the iterates, KMAy should be preferred to MAy, since the chances that $\boldsymbol{u}_{k+1}$ is in $\mathcal{F}_p$ are higher.

Next, we go one step further and show that $\mathcal{C}_{\text{KMAy}}^k$ can be manipulated through the filter gain K in a way that plant feasibility of $\boldsymbol{u}_{k+1}$ is guaranteed if a *local assumptions* about the model m holds. The observation that a reduction of the value of K leads to the aforementioned shrinkage of $\mathcal{C}_{\text{KMAy}}^k$ on $\boldsymbol{u}_k$ is key to this result.

Assumption 1: Assume that at iteration k, there exists a subset $\mathcal{D}_k$ of $\mathcal{U}$, i.e., $\mathcal{D}_k \subseteq \mathcal{U}$, in which the modified model provides pessimistic predictions of the plant feasibility, i.e., the constraints of the modified model upper-bound the constraints of the plant: $\mathcal{D}_k := \left\{ \boldsymbol{u} \in \mathcal{U} | G_{i,\text{MAy},k}(\boldsymbol{u}) \geq G_{i,p}(\boldsymbol{u}), \forall i \in [1, n_g] \right\}$. Assume that $\boldsymbol{u}_k \in \mathcal{D}_k^o$ the interior of $\mathcal{D}_k$.

These assumptions are likely to hold since at $\boldsymbol{u}_k$ the modified-model and the plant have identical local first-order properties thanks to the affine corrections of the model. Note that one way to enforce the satisfaction of this assumption over a larger area around $\boldsymbol{u}_k$ would be to amend the model so that the modelled constraints are local upper bounds of the plant constraints, similarly to what has been proposed in Marchetti et al. (2017). Contrary to what is suggested in (Marchetti et al., 2017), such modifications would only be performed here at a *local* level and will not necessitate more than the available measurements. Pessimistic predictions of the plant constraints should always be a favorable case, since pessimistic predictions are such that $G_{i,\text{MAy},k}(\boldsymbol{u}) \geq G_{i,p}(\boldsymbol{u})$, while the optimization problems aims at enforcing $G_{i,\text{MAy},k}(\boldsymbol{u}) \leq \boldsymbol{0}$, but it is shown hereafter that only KMAy can take advantage of this situation.

Theorem 1 *If (i) Assumption 1 holds, and (ii) $\mathcal{U}$ is bounded, then there exist diagonal filters $\boldsymbol{K} = K\boldsymbol{I}_{n_u}$ such that $\mathcal{C}_{\text{KMAy}}^k \subseteq \mathcal{F}_p$ and, in turn, such that $\boldsymbol{u}_{k+1} \in \mathcal{F}_p$.*

Proof. Assumption 1 allows to define the biggest ball $\mathcal{B}(\boldsymbol{u}_k, r) \subseteq \mathcal{D}_k$, centered at $\boldsymbol{u}_k$ of radius r. Notice that $r > 0$ because $\boldsymbol{u}_k \in \mathcal{D}_k^o$. Now we show that the distance $||\boldsymbol{u}_{k+1} - \boldsymbol{u}_k||_2$ can be manipulated through K such that $\boldsymbol{u}_{k+1} \in \mathcal{B}(\boldsymbol{u}_k, r)$. By definition, $\boldsymbol{u}_{k+1} \in \mathcal{U}$, $\boldsymbol{v}_k(\boldsymbol{u}_{k+1}) \in \mathcal{U}$, and $||\boldsymbol{u}_{k+1} - \boldsymbol{u}_k||_2 := ||\boldsymbol{u}_k + K(\boldsymbol{v}_k(\boldsymbol{u}_{k+1}) - \boldsymbol{u}_k) - \boldsymbol{u}_k||_2 = K||\boldsymbol{v}_k(\boldsymbol{u}_{k+1}) - \boldsymbol{u}_k||_2 \leq K \max_{\boldsymbol{u} \in \mathcal{U}}\{||\boldsymbol{u} - \boldsymbol{u}_k||_2\}$. According to (ii) $\mathcal{U}$ is bounded, so $\max_{\boldsymbol{u} \in \mathcal{U}}\{||\boldsymbol{u} - \boldsymbol{u}_k||_2\} \in \mathbb{R}$. Therefore, $\forall K$ such that $K \leq r/(\max_{\boldsymbol{u} \in \mathcal{U}}\{||\boldsymbol{u} - \boldsymbol{u}_k||_2\})^*$, the inequality $||\boldsymbol{u}_{k+1} - \boldsymbol{u}_k||_2 \leq r$ holds, and $\boldsymbol{u}_{k+1} \in \mathcal{B}(\boldsymbol{u}_k, r) \subseteq \mathcal{D}_k$. Given that the feasibility is checked** at $\boldsymbol{u}_{k+1} \in \mathcal{D}_k$ and that the model provides pessimistic predictions of the plant constraints in $\mathcal{D}_k$, it follows that $\boldsymbol{u}_{k+1} \in \mathcal{F}_p$. $\square$

Despite the fact that the set $\mathcal{D}_k$ is unknown in practice, Theorem 1 shows that reducing K in KMAy is a very simple way to increase the chances to have $\boldsymbol{u}_{k+1} \in \mathcal{D}_k$ and, thus,

---

*Since $r > 0$, satisfying this inequality does not requires to have $K \to 0$.

**This difference between MAy and KMAy is the reason why a similar theorem can not be stated for MAy.

to make RTO iterations safer. Interesting enough is that in (Papasavvas and François, 2019), it is shown that *reducing K* is also the way to relax the model adequacy condition for KMAy. Also, it has to be noticed that a similar result cannot be obtained from MAy, mainly because model-based feasibility is not checked at the filtered inputs. In fact, when the conditions (i) and (ii) hold, the iterate $\boldsymbol{u}_{k+1}$ can be brought arbitrarily close to $\boldsymbol{u}_k$ if K is taken small enough. However, with MAy, the feasibility being not checked at $\boldsymbol{u}_{k+1}$, the fact that the model is pessimistic in $\mathcal{D}_k$ is ignored and changing K so that $\boldsymbol{u}_{k+1} \in \mathcal{D}_k$ is not helpful for feasibility with MAy. As illustrated on Figure 1, reducing K with MAy would also bring $\boldsymbol{u}_{k+1}$ closer to $\boldsymbol{u}_k$, but on the segment $[\boldsymbol{u}_{k+1}^{\star}, \boldsymbol{u}_k]$, while it is ignored that this segment lies in set the of infeasible inputs (red-shaded area) for the modified-model.

4. Case study

Williams-Otto Reactor: The continuous stirred-tank reactor described in (Williams and Otto, 1960) is considered. 3 reactions take place in reality: (i) $A + B \rightarrow C$, (ii) $C + B \rightarrow P + E$, and (iii) $P + C \rightarrow G$; but the models only considers two (different) reactions (Marchetti et al., 2017; Forbes et al., 1994). So there is structural plant-model mismatch.

A and B are fed separately, with mass flowrates of F_A and F_B ([kg/s]), respectively. P and E are the desired products, C is an intermediate product and G is an undesired by-product. The reactor is operated isothermally at T_R ([$^\circ C$]). Steady-state mass balances can be found in Zhang and Forbes (2000). The model-based optimization problem of (Marchetti et al., 2017) is considered, where $\boldsymbol{u} = [F_A, F_B, T_R]^\mathsf{T}$ and $\boldsymbol{y} = [X_E, X_P, X_G]^\mathsf{T}$ are the inputs and outputs, respectively, and X_i denotes the concentration of species i. The goal is to maximize profit at steady state, and both inputs and X_G are constrained:

$$\max_{\boldsymbol{u}} \quad \phi(\boldsymbol{u}, \boldsymbol{y}) = (1143.38 X_P + 25.92 X_E)(F_A + F_B) - 76.23 \alpha F_A - 114.34 \beta F_B \quad (4.1)$$

$$\text{s.t.} \quad g(\boldsymbol{y}) = X_G - 0.08 \le 0, \quad F_A \in [3, 4.5], \quad F_B \in [6, 11], \quad T_R \in [80, 105]. \quad (4.2)$$

The two parameters $\alpha \in \mathbb{R}$ and $\beta \in \mathbb{R}$ are used to simulate market fluctuations. Also, note that $g(\boldsymbol{y})$ is a concave constraint (Marchetti et al., 2017).

Three different market regimes are considered: mode 1: (1.1, 1.1), mode 2: (0.85, 1), and mode 3: (1, 0.9), with different pairs of α and β values (in brackets). For the two mode transitions mode 1 $\rightarrow$ mode 2 and mode 2 $\rightarrow$ mode 3, MAy and KMAy have been applied and constraints violations have been analyzed for different filter values of the diagonal filter $K = K I_{n_u}$. Other transitions have been simulated with similar conclusions.

Simulation Results: Note that the optimal operating conditions activating the concave constraint (4.2) for each mode, the segment that joins the optimal operating points of two different modes always lies in the infeasible part with MAy. Therefore, constraint violations will clearly be likely to occur with MAy similarly to what is illustrated in Figure 1. Both problems are indeed comparable, i.e., "sliding around a concave constraint" with KMAy or going potentially through it with MAy. Figures 2 and 3 show a large difference between the maximal constraint violations observed with MAy and KMAy for different K values. As expected, using KMAy always leads to less violations.

5. Conclusion

In this article, a weakness that is common to all MA-inspired RTO algorithms using a filter has been identified. Two recent MA extensions are compared and it is argued that KMAy

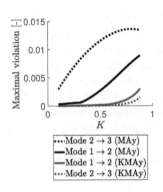

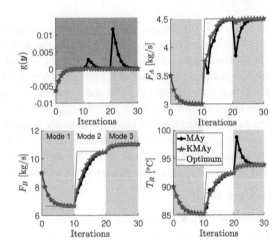

Figure 2: Maximal constraint violations with MAy and KMAy, for two different market variations.

Figure 3: Simulation results for $K = 0.5$.

is safer than MAy, in the sense that less, or smaller, constraints violations can be expected. Sufficient conditions for plant feasibility have been compared for the two algorithms and it has been shown that they are in favor of KMAy. It has been proven that, under certain conditions, a filter value can always be found such that plant feasibility can be enforced at the next iteration with KMAy, while this is not the case with MAy and it indeed appears that most MA algorithms would benefit from a formal implementation of the filter at the level of the optimization problem. Future research could include an analysis of the potential effects of such an integration to most MA-inspired RTO methods. Also, the conditions for the existence of $\mathcal{D}_k$ in Assumption 1, though beyond the scope of this article, should be investigated. Finally, the main benefit of the integration of the filter into the KMAy formulation (with the relaxation of model adequacy conditions, (Papasavvas and François, 2019)) being to prevent the generation of operating conditions where the modified model predicts infeasibility, future research should investigate how this double model-based feasibility check could, in practice, enforce plant feasibility under moderate assumptions about the nature of plant-model mismatch.

References

G. A. Bunin, G. François, D. Bonvin, 2013. Sufficient conditions for feasibility and optimality of real-time optimization schemes - I. Theoretical foundations. ArXiv:1308.2620.

J. F. Forbes, T. E. Marlin, J. F. MacGregor, 1994. Model adequacy requirements for optimizing plant operations. Comp. Chem. Eng. 18 (6), 497–510.

A. G. Marchetti, T. Faulwasser, D. Bonvin, 2017. A feasible-side globally convergent modifier-adaptation scheme. J. Process Contr. 54, 38–46.

A. G. Marchetti, G. François, T. Faulwasser, D. Bonvin, 2016. Modifier adaptation for real-time optimization – Methods and applications. Processes 4 (55), 1–35.

A. Papasavvas, G. François, 2019. Filter-based additional constraints to relax the model adequacy conditions in modifier adaptation. (Accepted) In Proc. IFAC International Symp. DYCOPS. Florianopolis.

A. Papasavvas, A. Marchetti, T. de Avila Ferreira, D. Bonvin, 2019. Analysis of output modifier adaptation for real-time optimization. Comp. Chem. Eng. 121, 285–293.

T. J. Williams, R. E. Otto, 1960. A generalized chemical processing model for the investigation of computer control. AIEE Trans. 79, 458.

Y. Zhang, J. F. Forbes, 2000. Extended design cost: A performance criterion for real-time optimization systems. Comp. Chem. Eng. 24, 1829–1841.

Anton A. Kiss, Edwin Zondervan, Richard Lakerveld, Leyla Özkan (Eds.)
Proceedings of the 29th European Symposium on Computer Aided Process Engineering
June 16th to 19th, 2019, Eindhoven, The Netherlands. © 2019 Elsevier B.V. All rights reserved.
http://dx.doi.org/10.1016/B978-0-128-18634-3.50213-7

Real-time feasible model-based crystal size and shape control of crystallization processes

Botond Szilagyi [a], Zoltan K. Nagy [a,b*]

[a] *Davidson School of Chemical Engineering, Purdue University, West Lafayette, IN 47907-2100, USA*

[b] *Department of Chemical Engineering, Loughborough University, Loughborough, LE11 3TU, UK;*

zknagy@purdue.edu

Abstract

The simultaneous control of crystal size and shape is particularly important in fine chemical and pharmaceutical crystallization. These two quantities influence the dissolution rate and bioavailability of final drug products, and also contribute to the manufacturability and efficiency of downstream operations. Numerous practical issues are associated with the implementation of an industrially relevant crystal size and shape control algorithm, such as the limited number of commercially available measurement devices and tight productivity constraints. Model-based control algorithm is required to address these problems, which, however, brings an additional challenge: the high computational demand of the applicable process simulation.

In this work we show that the model predictive control, relying on measurements coming from routinely applied, commercially available process analytical technology (PAT) tools may be feasible. The algorithm is based on GPU accelerated full 2D population balance model (PBM) solution, which does not require external computation units (i.e. cloud computing). The state estimation, which is a crucial part of any robust model-based control system, is carried out by fitting the model, through the re-adjustment of some model parameters, on the measured FBRM count, PVM based mean aspect ratio (AR) and solute concentration, in real time. Artificial neural network (ANN) based soft-sensor is employed to simulate the most likely mean AR of the bivariate crystal size distribution (2D CSD) calculated by the PBM, which is compared then to the mean AR measured by the *in-situ* imaging tool, as a part of real-time parameter re-adjustment.

Keywords: crystallization, model predictive control, real time control, crystal shape

1. Introduction

The size and shape of crystalline active pharmaceutical ingredients (APIs), in addition to the purity and polymorphic form, impact both the downstream operations (filtration, drying *etc.*) and some of physical properties (dissolution rate, flowability *etc.*). Consequently, simultaneous size and shape control of particles during solution crystallization has gained more attention from both academic and industrial perspectives (Nagy and Braatz, 2012). The size control in crystallization processes is a well-studied field, but simultaneous size and shape control has gained more attention only in the recent years, due to two major factors: (1) the lack of real-time crystal shape measurement tools and (2) limitations by computational power for the (often required) model-based control.

Numerous variables can be manipulated to control crystal shape, including the supersaturation (through temperature or antisolvent addition), facet-specific growth rate inhibitor concentration, stirring power or internal/external milling (breakage). The most convenient is to manipulate supersaturation, which influences the crystal shape through the different supersaturation dependencies of facet-specific growth and dissolution rates. Although model free shape control was already successfully achieved by these simple principles, the productivity and yield is far under the industrially desired levels. Model-based shape control approaches, which can directly involve productivity and yield constraints, were not implemented so far, mainly due to the aforementioned limitations.

In this work we propose a real-time feasible nonlinear model predictive control (NMPC) algorithm, with growing horizon state estimator (GHE). The system is based on full 2D PBM (with nucleation, growth and dissolution) to describe the entire 2D CSD, not only its moments. ANN based soft-sensor model is applied to quickly approximate the most likely measurable AR distribution (ARD) in conjunction with a GPU accelerated bi-dimensional FVM solver, already presented in the literature (Szilágyi and Nagy, 2018). In this paper we show that the NMPC is real-time feasible and it has no practical limitations from the standpoint of measurement system, since it relies on routinely applied concentration, chord-length distribution (CLD) and ARD measurements, which is a significant novelty of this work. For brevity, we will analyse the performance and calculation time requirement of GHE and NMPC individually. In the feasibility analysis we assume 4 hours batch crystallization of rod-like crystals and 5 minutes sampling time. The timings were obtained on a Dell Alienware machine with Intel i7-7700 CPU, 2400 MHz main memory clock speed, and nVidia GeForce GTX 1060m GPU.

2. The process model and the soft-sensor

The process model and simulation procedure was described in details in the literature (Szilágyi and Nagy, 2018), thus here we give only a brief overview. Assuming that the crystals are characterized by the length (L_1) and width (L_2) size dimensions, denoted by the $(L_1, L_2) \rightarrow \mathbf{L}$ size vector, and the crystal population is described by the $n(\mathbf{L}, t)$ bivariate size density function, the 2D PBE governing the variation of 2D CSD is:

$$\frac{\partial n(\mathbf{L}, t)}{\partial t} + \sum_{i=1}^{2} G_i \frac{\partial n(\mathbf{L}, t)}{\partial L_i} = B\delta(\mathbf{L} - \mathbf{L}_n) \qquad (1)$$

B is the nucleation rate, and G_i denotes the size-independent growth rates along the length and width axes. Eq.(1) has $n(\mathbf{L}, t=0) = n_0(\mathbf{L})$ initial and $n(\mathbf{L} = \infty, t) = 0$ boundary conditions. The initial condition is given by the seed distribution, whereas the boundary condition expresses that the crystals must have finite size. The first term in Eq. (1) describes the temporal evolution of the 2D CSD, the second term is for the growth. The right hand side is the nucleation term. The solute mass balance is written as:

$$\frac{dc}{dt} = -k_v \rho_c \left(\iint G_1 L_2^2 n(\mathbf{L}, t) d\mathbf{L} + 2 \iint G_2 L_1 L_2 n(\mathbf{L}, t) d\mathbf{L} \right) \qquad (2)$$

with the $c(t=0) = c_0$ initial condition. Here, k_v is the volume shape factor, whereas ρ_c is the crystal density. Since the temperature is manipulated variable, energy balance is not required for model closure.

The applied kinetic rate equations are listed in Table 1. The equations were compiled based on the literature data for potassium dihydrogen phosphate (KDP) (Eisenschmidt et al., 2015; Gunawan et al., 2002). Since the activation energies and supersaturation exponents are from separate works, the rate constants were modified so to bring the growth and dissolution rates in reasonable domain. The nucleation parameters were modified to give reasonable nucleation rate with the applied growth rate expressions.

Table 1. Kinetic rate equations applied in this work. σ is the absolute supersaturation.

Mechanisms	Expression (for length)	Expression (for width)
Nucleation [#/m^{3}s]	$B = 3 \cdot 10^{12} \sigma^{5.24}$	
Growth [µm/s]	$G_1 = 6.04 \cdot 10^6 \sigma^{1.74} \exp\left(-\dfrac{39100}{RT}\right)$	$G_2 = 2.42 \cdot 10^5 \sigma^{1.48} exp\left(-\dfrac{37100}{RT}\right)$
Dissolution [µm/s]	$D_1 = 2.72 \cdot 10^5 \sigma^1 \exp\left(-\dfrac{26800}{RT}\right)$	$D_2 = 8.18 \cdot 10^5 \sigma^1 exp\left(-\dfrac{29700}{RT}\right)$

The model equations were solved with a high-resolution finite volume method (HR-FVM) on non-uniform grid involving GPU acceleration. This allows up to two orders of magnitude speedup compared to a standard serial C implementation, which makes the simulation capable of real-time optimization (for timing details see the literature Szilágyi & Nagy, 2018). The advantage of the GPU acceleration, beyond enabling NMPC, is that it provides the computational power on site for low investment cost, eliminating the need for supercomputers or cloud-services, which are both undesired in industrial environment.

It is known that the *in-situ* imaging based shape measurement underestimates the real AR due to random spatial orientation of the crystals. To calculate the most likely measurable AR of a crystal, a geometrical model-based technique was proposed, that constructs the ARD of the single 2D crystal by mapping all possible two dimensional projections of the crystal. Then, using large number of $ARD = f(L)$ pairs, an ANN is trained to give the most likely ARD of an arbitrary sized (L) crystal. The most likely ARD of the 2D CSD is then approximated in real time as a weighted sum of individual ARDs corresponding to the HR-FVM grid sizes, returned by the ANN, where the weighting factor is the $n(\mathbf{L}, t)$.

3. Growing horizon state estimator (GHE)

There are two major reasons that require the utilization of already available measurement data. Firstly, the system states (2D CSD and concentration) are required in NMPC optimization as initial condition of the simulation. While the concentration can be measured, the 2D CSD generally not. Therefore, the 2D CSD must be estimated from the available measurements, but for robustness (e.g. with respect to sensor communication errors), the concentration is often estimated, too. Secondly, there are often uncertainties related to seed loading, and the crystallization kinetics is known to be sensitive to process conditions (through disturbances from impurities, mixing energy *etc.*), which is a typical plant-model mismatches (PMMs) that leads to sub-optimal control.

The basic idea of the NMPC is that if the system states are accurate and the model describes well the process, the process output can be predicted, hence, optimized. Our GHE is based on the same principle: the parameters that capture the main sources of PMM are identified, then used as decision variables for fitting the model to the measured data in the estimation horizon (from zero to the actual process time) in real-time. Once the updated model parameters become available, process simulation is carried out from the initial time to the actual process time, which gives the system states.

In a batch crystallization process, the parameters related to nucleation and growth kinetics as well as the seed loading generally exhibit the most batch-to-batch variability, thus are ideal candidates for decision variables in the GHE optimization. Denoting the vector of model parameters by ψ, listed in Table 2, the following GHE objective is proposed:

$$O_{GHE}(\psi) = \sum_i \left(c_{m,i} - c_{s,i}\right)^2 + w \sum_i \left(AR_{m,i} - AR_{s,i}\right)^2 + u \sum_i \frac{\left(N_{m,i} - N_{s,i}\right)^2 \left(N_{m,i}^2 + N_{s,i}^2\right)}{N_{m,i}^2 N_{s,i}^2} \quad (3)$$

Subscript "m" and "s" are for measured and simulated quantities, N_m is the FBRM count and N_s is the simulated number density normalized to $N_m(t = 0)$. The values w and u are weighting factors. The first part of the objective is for the fitting of concentrations, the second is for the ARs. It can be shown that the count term is considerably more sensitive to the correct capturing of the nucleation point than to the actual value of final count. This enables to reliably use the FBRM for nucleation detection.

Table 2. Initial values and bounds of decision variables of GHE optimization as well as the performance of state estimator in parameter re-adjustment

ψ_i	Seed v.f.	$k_b{}^*$	b	$k_{g1}{}^*$	g_1	$k_{g2}{}^*$	g_2
Starting point	$1.21 \cdot 10^{-3}$	$3.0 \cdot 10^{12}$	5.24	$6.04 \cdot 10^6$	1.74	$2.42 \cdot 10^5$	1.48
Actual parameters	$1.43 \cdot 10^{-3}$	$1.1 \cdot 10^{12}$	5.30	$2.91 \cdot 10^6$	1.72	$1.38 \cdot 10^5$	1.44
Lower bound	$8.5 \cdot 10^{-4}$	$7.1 \cdot 10^{11}$	5.08	$2.72 \cdot 10^6$	1.68	$1.32 \cdot 10^5$	1.43
Upper bound	$1.6 \cdot 10^{-3}$	$1.2 \cdot 10^{13}$	5.40	$1.31 \cdot 10^7$	1.80	$4.49 \cdot 10^5$	1.52
Readjusted parameters	$1.38 \cdot 10^{-3}$	$7.8 \cdot 10^{11}$	5.15	$2.94 \cdot 10^6$	1.71	$1.38 \cdot 10^5$	1.44

*In the GHE the log(k) was readjusted rather than the nominal value, to improve the optimization performance

A recent study suggested that in such a GHE, the kinetic parameters move within the confidence intervals of the off-line parameter estimation (Szilágyi et al., 2018). Hence, the bounds for kinetic parameters are set to the 95 % confidence interval limits available from the off-line parameter estimation. The bounds of the seed loading may be chosen arbitrarily, based on the accuracy and precision of the balance, for instance. The nominal values of ψ, as well as the (assumed) bounds are listed in Table 2.

To improve the convergence, the decision variables were scaled ($\psi \rightarrow \psi_S$), using scale parameters (χ) that normalizes the initial values to 1. The ith element of vector χ is:

$$\chi_i = \frac{1}{\psi_i} \rightarrow \psi_{S,i} = \psi_i \times \chi_i \tag{4}$$

The optimization was solved in Matlab using the interior-point algorithm (implemented in the *fmincon* inner optimization routine). The iteration limit was set to 13. It worth

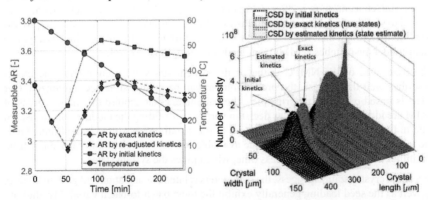

Figure 1. State estimator performance: fitting the model on measured ARs (the concentrations and counts are not plotted for brevity) as well as the estimated and exact 2D CSD. The calculation time was 115 s.

noticing that in the demonstration (Figure 1) we assumed that the estimation horizon is the same as the batch time. Therefore, this GHE operates on the longest possible time frame, so the 115 s timing gives the maximum calculation time requirement. According to Table 1 and Figure 1, the estimator readjusted successfully the model parameters, and was able to estimate the actual system states with acceptable accuracy.

4. The real time optimization problem and provisionally results

The NMPC is a repeated real-time optimization of the control signal in every sampling instance, using the estimated system states and the updated model parameters. Hence, the calculation time, i.e. the NMPC optimization and the state estimation time, must fit into the sampling time. The 5 minutes sampling time is a reasonable choice for the dynamics of a usual crystallization processes.

For the sake of simplicity, the batch NMPC problem is formulated as end-point (product property) optimization. The decision variables are the vector of temperatures in the control horizon, i.e. the current time to the final time, which time frame is shrinking as the batch evolves. We defined the temperature profiles by 12 discrete temperature points, which translates to 10 decision variables (the initial and final temperature are fixed).

$$O_{NMPC}(T) = w_1 AR_{prod.}^2 + w_2 F_{idx}^2 + w_3 \left(\frac{n_{t_{final}}}{n_{t_{initial}}}\right)^2 + w_4 \sum_i (t_{i+1} - t_i)\frac{(T_{i+1} - T_i)^2}{(t_{i+1} - t_i)^2} \qquad (5)$$

The first term of objective function is to minimize the product AR, the second term minimizes the fine index (i.e. volume fraction of crystals under a threshold sphere-equivalent diameters, here set to $L_f = 100$ μm), the third term is for the ratio of final to initial number density (zero moment), whereas the last term is to ensure smooth temperature profile. The values of weight factors (w_i, i=1,...,4) define the relative importance of sub-objectives. Setting all values to 0, except w_1, leads to pure AR minimization NMPC, w_2 controls minimization of fines in the product, whereas w_3 is for the nucleation minimization, or, indirectly crystal size maximization.

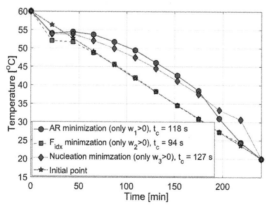

Figure 2. Optimal temperature profiles calculated for different objectives and the corresponding. All timing is an average of three run.

Although there is an AR term in the Eq.(5), the AR can be manipulated only within the attainable AR domain. The attainable size and shape domains are delimited by the crystallization kinetics and process parameters (temperature profile, seed properties *etc.*) (Vetter et al., 2014; Acevedo and Nagy).

In the optimization constraints are imposed on the heating and cooling rates (0.5 °C/min both) and for minimal and maximal temperatures (60 and 15 °C). The initial and final temperatures are fixed for yield to 60 and 20 °C. The optimization was solved using a sequential quadratic programming approach in the *fmincon* routine of Matlab. Provisional optimization results and timings are presented in Figure 2. The development of the NMPC control algorithm from its main components is associated with numerous challenges to ensure robustness and stability (Mayne et al., 2000).

5. Conclusions

In this paper a full 2D PBM based NMPC was presented for batch crystallization processes. In the study it was shown that the system is real-time feasible and implementable using commercially available monitoring systems and locally available, low cost GPUs as efficient accelerators for improved optimization times. Another enabling invention used in this work was the fast, approximate, ANN based soft-sensor, which takes the 2D CSD calculated by the PBM and transforms it to the most likely measurable ARD (mean AR), which is measured by the most common, commercially available imaging based PAT tools. Due to brevity, we analyzed separately the feasibility and performance of the NMPC and GHE optimizations only; both simulations revealed that, using the accelerated computation, fast soft-sensor and the search-space delimited by the constraints, all the calculations can be done within the 5 minutes sampling time on a standard computer, which is reasonable for many industrial crystallization processes.

6. Acknowledgements

The financial support of the International Fine Particle Research Institution is acknowledged gratefully.

7. References

Acevedo, D., Nagy, Z.K., 2014. Systematic classification of unseeded batch crystallization systems for achievable shape and size analysis, J. Cry. Gro., 394, 97-105.

Eisenschmidt, H., Voigt, A., Sundmacher, K., 2015. Face-specific growth and dissolution kinetics of potassium dihydrogen phosphate crystals from batch crystallization experiments. Cryst. Growth Des. 15, 219–227.

Gunawan, R., Ma, D.L., Fujiwara, M., Braatz, R.D., 2002. Identification of Kinetic Parameters in Multidimensional Crystallization Processes. Int. J. Mod. Phys. B 16, 367–374.

Mayne, D.Q., Rawlings, J.B., Rao, C. V., Scokaert, P.O.M., 2000. Constrained model predictive control: Stability and optimality. Automatica 36, 789–814.

Nagy, Z.K., Braatz, R.D., 2012. Advances and New Directions in Crystallization Control. Annu. Rev. Chem. Biomol. Eng. 3, 55–75.

Szilágyi, B., Borsos, Á., Pal, K., Nagy, Z.K., 2018. Experimental implementation of a Quality-by-Control (QbC) framework using a mechanistic PBM-based nonlinear model predictive control involving chord length distribution measurement for the batch cooling crystallization of l-ascorbic acid. Chem. Eng. Sci. doi: 10.1016/J.CES.2018.09.032.

Szilágyi, B., Nagy, Z.K., 2018. Aspect Ratio Distribution and Chord Length Distribution Driven Modeling of Crystallization of Two-Dimensional Crystals for Real-Time Model-Based Applications. Cryst. Growth Des. doi:10.1021/acs.cgd.8b00758.

Vetter, T., Burcham, C.L., Doherty, M.F., 2014. Regions of attainable particle sizes in continuous and batch crystallization processes. Chem. Eng. Sci. 106, 167–180.

Anton A. Kiss, Edwin Zondervan, Richard Lakerveld, Leyla Özkan (Eds.)
Proceedings of the 29th European Symposium on Computer Aided Process Engineering
June 16th to 19th, 2019, Eindhoven, The Netherlands. © 2019 Elsevier B.V. All rights reserved.
http://dx.doi.org/10.1016/B978-0-128-18634-3.50214-9

Integration of Max-Plus-Linear Scheduling and Control

Risvan Dirza[a], Alejandro Marquez-Ruiz[a], Leyla Özkan[a] and Carlos S. Mendez-Blanco[a]

[a]*Department of Electrical Engineering, Eindhoven University of Technology, 5612 AJ, Eindhoven, The Netherlands*
A.Marquez.Ruiz@tue.nl

Abstract

In this paper, we investigate the Max-Plus-Linear (MPL) representation for the integration of scheduling and control problem. This is an attractive approach because the MPL representation can result in a convex scheduling problem. Using this representation, we have formulated the integration of scheduling and control problem considering the corresponding transitions time for each processing time. Furthermore, the MPL representation is combined with the sequential decomposition method (SDM) (Chu and You (2012)) in order to solve the integrated problem. The final result is a formulation that preserves the convexity of the original scheduling problem. Finally, the performance of the new formulation is tested on a simple case study and compared with the traditional method.

Keywords: Integration, Scheduling, Control, Max-Plus-Linear Systems

1. Introduction

Scheduling and control are complementary tasks for optimizing chemical process operations. Traditionally, these tasks have been implemented separately in a sequential way. Scheduling problems, whose solutions provide the optimal production sequence and production times, do not in general consider the dynamic behavior of the processes. The inclusion of such dynamics in the scheduling problem is called the integration of scheduling and control, and it has shown important economic benefits for the overall performance of the process. Despite the advantages of the integration, several challenges have been also reported. Nevertheless, the most critical issues are the modeling approach used in the scheduling problem and the computational aspects. Both topics are strongly related.

There are several approaches to model the scheduling problem i.e. the state-task network (STN) or resource-task network (RTN). These modeling options lead to transforming the integration of scheduling and control into an MINLP problem which is relatively computationally expensive (Chu and You (2012), Subramanian et al. (2012)). An attractive approach to avoid the MINLP formulation is modeling the scheduling problem using the MPL representation (Schutter and van den Boom (2001), van den Boom et al. (2018)). There are a lot of advantages of the MPL representation. The most remarkable one is that the scheduling problem can be transformed into a convex optimization problem. In addition, the MPL representation has not been investigated for the integration of scheduling and control problem.

In this paper, we assume a perishable goods scheduling problem and no disturbances affecting the completions of the tasks. Based on these assumptions, we propose an approach to formulate an integrated problem of MPL scheduling and control, and methods to solve the problem.

This paper is organized as follows. In Section 2, preliminaries related to the control problem and the MPL system are presented. Section 3 presents the formulation of the integration of scheduling and control using the MPL representation. The details on the new proposed methods to solve the integrated problem are presented in Section 4. The proposed formulation and solutions are demonstrated on a simple case study and compared in Section 5. Finally, the conclusion is given in Section 6.

2. Preliminaries

2.1. Max-Plus Linear (MPL) system

The MPL utilizes timed-discrete event system (timed-DES) in the modeling of time (Schutter and van den Boom (2001)). A max-plus-algebraic (MPA) model, called MPL system, can represent timed-DES with no concurrency, where we assume no capacity constraint and each unit starts working as soon as all parts or conditions are available. The general form of MPL scheduling model is as follows:

$$\check{x}_i(k) = \check{A}_i \otimes \check{x}_i(k-1) \oplus \check{B}_i \otimes \check{u}_i(k); \quad \check{y}_i(k) = \check{C}_i \otimes \check{x}_i(k) \tag{1}$$

where $\check{A}_i \in \mathbb{R}_{\varepsilon}^{n_{x_i} \times n_{x_i}}, \check{B}_i \in \mathbb{R}_{\varepsilon}^{n_{x_i} \times n_{u_i}}$, and $\check{C}_i \in \mathbb{R}_{\varepsilon}^{n_{y_i} \times n_{x_i}}$ are system matrices configured by all units contributing in production line i. $\otimes$ is equal to operator "+" and $\oplus$ is equal to operator "max". In Eq.(1), the components of the input, the output, and the state are event times, and the counter k is an event cycle counter. For a process system, $\check{u}_i(k)$ would typically represent the time instants at which raw material is fed to the system for the $(k+1)^{th}$ time, $\check{x}_i(k)$ represent the time instants at which the unit start processing the kth batch of intermediate products, and $\check{y}_i(k)$ represent the time instants at which the kth batch of finished products leaves the system (completion time).

2.2. Control Problem

We consider a batch reactor as a plant model equipped with controllers to reach an optimum set point. The controller(s) of each unit, i.e. MPC or PID can be expressed in general as: $u = K(x, t_s)$, where u, x and t_s are the input, states and the transition time of a processing unit, respectively. For example, a PI controller can be expressed as: $u = K_P(Cx - y^{sp}) + K_I \int (Cx - y^{sp})$, where y^{sp} is the optimal set point of a processing unit given by the Real-Time Optimization (RTO). The parametrization of the controller(s) in terms of the transition time can be done because the tuning parameters are a function of t_s. In the planning layer, where the planning decisions are taken, a set of transitions times has been determined based on the historical data. This set is expressed as: $\widetilde{t_s}^{(1)} = \{t_s^{(1)}, \dots, t_s^{(n_l)}\}$, where l is the index of the reachable transition time.

The presence of transition time influences the accuracy of the scheduling policy. Thus, the controller is also responsible for reaching and tracking the given transition time that is regulated by controller parameters $\Omega_{i,ii}^{(l)}$ (e.g. $\Omega_{i,ii}^{(l)} = \{K_P, K_{I_{i,ii}}^{(l)}\}$ for PI controller), where

i is the production line and ii is the index of the unit contributing in line i. The controller parameters can minimize the difference between given transition time and the transition time obtained by an off-line simulation $\tilde{t}^l_{s_{i,ii}}$. This simulation is only executed once at the beginning and it requires the dynamic behavior of the closed-loop control system that depends on three parameters: the initial condition value $y_{0i,ii}$, the set point $y^{sp}_{i,ii}$, and the controller parameters $\Omega^{(l)}_{i,ii}$ (Chu and You (2012)). Moreover, it also depends on three numerical parameters: the tolerable bound width of steady output condition $b_{i,ii}$ [in %], sampling time T_s, and the prediction horizon $N_{MPC_{i,ii}}$ for MPC controller. Thus, $\tilde{t}^{(l)}_{s_{i,ii}} = f(\Omega^{(l)}_{i,ii}, y^{sp}_{i,ii}, y_{0i,ii}, b_{i,ii}, T_s, N_{MPC_{i,ii}})$. After minimizing this difference, we assign the optimal controller parameters $\Omega^{(l)\star}_{i,ii} \rightarrow \Omega^{(l)}_{i,ii}$. Thus, we have the following set of controller parameters $\widetilde{\Omega}^{(l)}_{i,ii} = \{\Omega^{(1)}_{i,ii}, \ldots, \Omega^{(n_l)}_{i,ii}\}$. From the same calculation, we can also estimate the set of transition cost $\widetilde{\delta}^{(l)}_{i,ii} = \{\delta^{(1)}_{i,ii}, \ldots, \delta^{(n_l)}_{i,ii}\}$. This computation is part of Algorithm (1).

Data: Historical data, Closed-loop system dynamic model
Result: $u^\star_{i,\mathbf{ii}}(t)$
Get transition set: $\{\widetilde{t}^{(\mathbf{l})}_{s_{i,\mathbf{ii}}}, \widetilde{\Omega}^{(\mathbf{l})}_{i,\mathbf{ii}}, \widetilde{\delta}^{(\mathbf{l})}_{i,\mathbf{ii}}\} \leftarrow$ Execute simulation (Off-line computation) ;
$k = 1$;
while $k \leq$ *the total amount of production cycles* **do**

> **procedure**: Execute selection algorithm for all units ii (see Tab. (1));
>
>> Identify the relation of $\{\widetilde{t}^{(\mathbf{l})}_{s_{i,ii}}, \widetilde{\delta}^{(\mathbf{l})}_{i,ii}\}$;
>>
>> Get $\{t^{(l)\star}_{s_{i,ii}}(k), \Omega^{(l)\star}_{i,ii}(k)\}$;
>
> Update (Integrated) MPL scheduling model (Eq. (3)): $t^{(l)}_{s_{i,\mathbf{ii}}}(k) \leftarrow t^{(l)\star}_{s_{i,\mathbf{ii}}}(k)$;
> Get $\breve{u}^\star_i(k) \leftarrow$ Solve Eq. (5);
> Controller $\leftarrow \{\breve{u}^\star_i(k), \Omega^{(l)\star}_{i,\mathbf{ii}}(k)\}$;
> Get $u^\star_{i,\mathbf{ii}}(t) \leftarrow$ Solve control problem from $t = \breve{u}^\star_i(k)$ to $t = \breve{y}^\star_i(k) - t_{ei,n_s}$;
> (i, n_s *indicates the last unit in the production line* i);
> $k = k + 1$;

end

Algorithm 1: Sequential Decomposition Method

3. Integration of Scheduling and Control using MPL Systems

3.1. Batch scheduling problem

Batch scheduling problems typically have decision variables i.e. starting time. Currently, many scheduling problems are represented in the conventional algebra. This study focuses on the MPL scheduling problem called due date perishable goods which is a convex optimization problem (Schutter and van den Boom (2001)). If the due dates for the finished products are known, every difference between the due dates and the actual output time instants has to be penalized. Hence, the suitable cost criterion takes the following form: $J(k) = \bigotimes_{j=1}^{N_p} \bigotimes_{i=1}^{n_p} |\breve{y}_i(k+j|k) - \breve{r}_i(k+j|k)|$; where $\bigotimes_{q=1}^{Q}$ is equal to $\sum_{q=1}^{Q}$, N_p is the length of prediction horizon and $\breve{r}$ is the expected completion time. The standard

MPC for the MPL systems scheduling problem is introduced as follows:

$$\min_{\mathring{\mathbf{u}}_i(k)} J(k) \tag{2}$$

where $\mathring{\mathbf{u}}_i(k) = \begin{bmatrix} \breve{u}_i(k|k) & \cdots & \breve{u}_i(k+N_p|k) \end{bmatrix}^\top$ and Eq. (2) is subject to: MPL scheduling model shown in Eq. (1) and scheduling linear constraint for N_p horizon, time as a non-decreasing input, and starting time constraint regulation. A well-developed MPC theory for MPL system has also been introduced (Schutter and van den Boom (2001)). Moreover, the study of MPL convexity has been conducted.

3.2. Integrated scheduling model

In a stand-alone scheduling problem, the transition time is assumed to be '0'. The presence of transition time cannot be ignored; therefore, the decision variables of scheduling problem should cope with the existence of the transition time. The scheduling problem requires information of the reachable transition time from all control problems, while a control problem in each cycle needs the optimal transition time and the set-points. As an additional information, the control layer interacting with scheduling layer can be a local, a supervisory control or the combination of both.

For simplicity reasons, this study considers a single product that is manufactured cyclically in a processing unit for a certain processing time $t_{p_{i,ii}}$ so that the completion time of product satisfies the due date at every cycle given by the planning layer.

In MPL model, we can modify Eq.(1) to facilitate the existence of transition, filling time $t_{f_{i,ii}}$ and emptying time $t_{e_{i,ii}}$ and represent the integrated MPL scheduling model as follows, where the filling and emptying times are usually constant:

$$\breve{x}_i(k) = \breve{A}_i(t_{s_{i,ii}}^{(l)}) \otimes \breve{x}_i(k-1) \oplus \breve{B}_i(t_{s_{i,ii}}^{(l)}) \otimes \breve{u}_i(k); \qquad \breve{y}_i(k) = \breve{C}_i(t_{s_{i,ii}}^{(l)}) \otimes \breve{x}_i(k) \tag{3}$$

To accommodate an economic objective, we introduce a total cost ω consisting of due date deviation penalty cost ω_1, total production costs ω_2, and total transition costs ω_3 related to $\delta_{i,ii}^l$. The relationship between $\widetilde{t}_{s_{i,ii}}^{(l)}$ and $\widetilde{\delta}_{i,ii}^{(l)}$ will define the approach to minimize ω.

$$\omega = \omega_1 \otimes \omega_2 \otimes \omega_3 \tag{4}$$

4. Proposed algorithms

By means of off-line computation explained in Section 2.2, we obtain a set containing achievable transition times, their associated controller parameters, and transition costs: $\{\widetilde{t}_{s_{i,ii}}^{(l)}, \widetilde{\Omega}_{i,ii}^{(l)}, \widetilde{\delta}_{i,ii}^{(l)}\}$. In general, the integrated problem can be solved either sequentially or simultaneously. In this study, we utilize SDM, illustrated in Fig. (1) and described in Algorithm (1).

SDM consists of two steps in order to solve two main problems. The first problem is determining the element inside $\widetilde{t}_{s_{i,ii}}^{(l)}$ leading to the most (economically) optimal result (see Eq. (4)) by selecting the associated controller parameters. The idea of selection algorithm which is described in Tab. (1) consist of three cases. The solution for case (#1) is Conventional Sequential Method (CSM) and (#2) is Integration Method (IM) proposed

$\widetilde{\delta}_{i,ii}^{(l)}$ vs $\widetilde{t}_{s_{i,ii}}^{(l)}$	(#1) **Monotonically non-decreasing**	(#2) **Random**	(#3) **Monotonically non-increasing**
Key appr.	$t_{s_{i,ii}}^{(l)\star} \leftarrow t_{s_{i,ii}}^{(1)}$	$t_{s_{i,ii}}^{(l)\star} \leftarrow t_{s_{i,ii}}^{(l^\star)}$	$t_{s_{i,ii}}^{(l)\star} (t_{s_{i,ii}}^{(n_l)}, \breve{r}_i, y_i, t_{f_{i,ii}}, t_{p_{i,ii}}, t_{ei,ii})$

Table 1: The key approach of selection algorithms: to determine the most economically optimal transition set based on the relation of $\widetilde{\delta}_{i,ii}^{(l)}$ vs $\widetilde{t}_{s_{i,ii}}^{(l)}$.

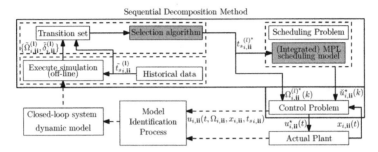

Figure 1: Illustration of SDM

in Chu and You (2012). Meanwhile, the solution for case (#3) is Constrained Deterministic Min-Max Method (CDM) inspired by Necoara et al. (2009) with some modifications incorporating threshold time to avoid penalty cost. Once we obtain the optimal transition time $t_{s_{i,ii}}^{(l)\star}$ and its index $(l)^\star$, we can determine the associated control parameters $\Omega_{i,ii}^{(l)\star}$ required for the controllers.

The second problem is determining the scheduling policy that considers the information provided by the solutions of the first step. By utilizing the SDM, where we incorporate $t_{s_{i,ii}}^{(l)\star}$ in the integrated MPL scheduling model (3), we preserve the convexity of the problem (see problem (2) as the integrated MPL scheduling problem (5)). Once we solve the integrated MPL scheduling problem, the (integrated) scheduling policy is determined.

$$\min_{\hat{\mathbf{u}}_i(k)} J(k) = \min_{\hat{\mathbf{u}}_i(k)} \bigotimes_{j=1}^{N_p} \bigotimes_{i=1}^{n_p} |\breve{y}_i(k+j|k) - \breve{r}_i(k+j|k)| \tag{5}$$

subject to (for details of notations, please see Schutter and van den Boom (2001)):
- Scheduling dynamic for N_p horizon: $\hat{\mathbf{\breve{y}}}_i(k) = \mathscr{H}(\hat{t}_{s_{i,ii}}^{(l)\star}) \otimes \hat{\mathbf{u}}_i(k) \oplus \mathscr{G}(\hat{t}_{s_{i,ii}}^{(l)\star}) \otimes \breve{x}_i(k)$
- Scheduling linear constraint for N_p horizon: $\mathscr{D}_i\hat{\mathbf{u}}_i(k) + \mathscr{F}_i\hat{\mathbf{\breve{y}}}_i(k) \le \hat{\mathbf{h}}_i(k)$
- Time as non-decreasing input: $\Delta\breve{u}_i(k+j) \ge 0$ for $j = 0, ..., N_p - 1$
- Starting time constraint regulation: $\Delta\breve{u}_{i_-} \le \Delta\breve{u}_i(k+j) \le \Delta\breve{u}_{i_+}$

5. Case Study

We consider a simple case study, comprising a single processing unit of an input refinement problem for temperature control of a nonlinear chemical batch reactor. It is assumed that the reactor has a cooling jacket whose temperature is directly manipulated (Lee et al. (2000)). The control objective of this case study is to reach and control the reactor temperature at certain fixed set-point ($y^{sp} = 30^oC$) for a certain period ($t_p = 40$ sec) by means of inlet coolant temperature. Moreover, the scheduling objective is that the heating process should be completed cyclically.

Fig. (2) shows the monotonically non-increasing relation between $\widetilde{\delta}_{i,ii}^{(\mathrm{I})}$ and $\widetilde{t}_{si,ii}^{(\mathrm{I})}$. In Fig. (3), the simulation result shows that the (integrated) MPL model requires less computation time than the traditional one in more accurate (frequent) sampling time, indicating similar result for longer processing time, $t_p \gg 40$ sec (even for less frequent sampling). Moreover, the MPL can be computationally efficient for a large-scale interconnected system. In terms of the performance, both MPL model and the traditional one yield the same trajectory. Unlike stand-alone model, Fig. (4) shows that the integrated MPL scheduling model is able to eliminate delay because it considers the transition time in the model.

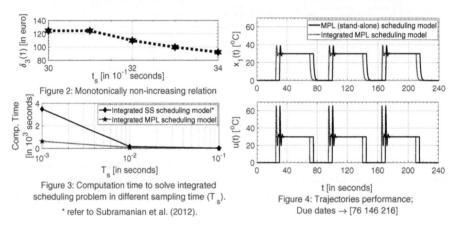

Figure 2: Monotonically non-increasing relation

Figure 3: Computation time to solve integrated scheduling problem in different sampling time (T_s).
* refer to Subramanian et al. (2012).

Figure 4: Trajectories performance;
Due dates → [76 146 216]

6. Conclusion

In this paper, a new formulation for an integrated scehduling and control problem using MPL representation is proposed. Based on this formulation, the SDM sets the optimal (integrated) scheduling policy. Finally, the proposed formulation and solution are tested in simulation on a temperature control and scheduling of a nonlinear chemical batch reactor. Future work could include incorporating uncertainties and capacity constraint, investigating simultaneous solution method and the extension to other type of scheduling problems.

References

Y. Chu, F. You, 2012. Integration of scheduling and control with online closed-loop implementation: Fast computational strategy and large-scale global optimization algorithm. Computers and Chemical Engineering 47, 248 – 268.

J. Lee, K. Lee, W. Kim, 2000. Model-based iterative learning control with a quadratic criterion for time-varying linear systems. Automatica 36, 641 – 657.

I. Necoara, B. D. Schutter, T. van den Boom, H. Hellendoorn, 2009. Robust control of constrained max-plus-linear systems. Int. J. robust and nonlinear control 19, 218 – 242.

B. D. Schutter, T. van den Boom, 2001. Model predictive control for max-plus-linear discrete event systems. Automatica 37, 1049 – 1056.

K. Subramanian, C. Maravelias, J. Rawlings, 2012. A state-space model for chemical production scheduling. Computers and Chemical Engineering 47, 97 – 110.

T. van den Boom, H. de Bruijn, B. D. Schutter, L. Özkan, 2018. The interaction between scheduling and control of semi-cyclic hybrid systems. IFAC PapersOnLine 51-7, 212 – 217.

Anton A. Kiss, Edwin Zondervan, Richard Lakerveld, Leyla Özkan (Eds.)
Proceedings of the 29th European Symposium on Computer Aided Process Engineering
June 16th to 19th, 2019, Eindhoven, The Netherlands. © 2019 Elsevier B.V. All rights reserved.
http://dx.doi.org/10.1016/B978-0-128-18634-3.50215-0

Nonlinear Model Predictive Control of Haemodialysis

Tianhao Yu[a,b], Vivek Dua[a,*]

[a] Department of Chemical Engineering, Centre for Process Systems Engineering, University College London, London WC1E 7JE, United Kingdom

[b] Current address: University of Surrey, Guildford GU2 7XH, United Kingdom

*Corresponding author: v.dua@ucl.ac.uk

Abstract

Haemodialysis is a blood treatment technique used for patients whose kidneys are not working normally. A number of Linear Model Predictive Control (LMPC) formulations for controlling haemodialysis have been proposed in the literature. In this paper, a Nonlinear Model Predictive Control (NMPC) formulation is proposed that instead of using an approximated linear model takes into account the nonlinear model for haemodialysis. One of the key advantages of the NMPC formulation is that it gives more comprehensive control of the haemodialysis process. Four manipulated variables are considered: the blood flow out of the body, the blood flow returned from the dialysis machine to the body, the dialysate rate and the sodium concentration of the inlet blood flow. By manipulating these variables, the fluid volumes and the toxin clearance are monitored and controlled. The results show that a better control performance for the blood volumes and toxin clearance is obtained while the blood and dialysate flow rates are held at a relatively low level, which is expected to reduce the risk of adverse reactions for the patients.

Keywords: Haemodialysis, Model Predictive Control, Optimisation

1. Introduction

Human kidney functions as a crucial organ which balances electrolytes and volumes of body fluid, and filters toxin content out of the blood before the blood returns to heart. Approximately 20% of the blood is processed by kidneys which amounts to 1800 litres per day (Lote 2013). Haemodialysis refers to a blood treatment process that removes toxins and waste for patients with severe renal failure. In order to achieve normal functions of the kidney, the haemodialysis treatment typically takes 3 to 5 hours per week for one patient (Pierce, Tubing, and Cassettes 2008).

The blood and dialysate flowrates are normally considered as two key variables during the treatment process. A trade-off arises when determining the two variables: increasing the two flowrates will generally shorten the required treatment time period, but will result in higher risk of adverse reactions for the patients including headache and muscle cramps. Therefore, finding the optimum operating conditions for haemodialysis becomes a complex optimisation problem. Currently the haemodialysis treatment is usually performed by nurses under empirical operating conditions.

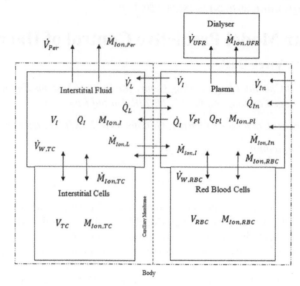

Figure 1.1: Mathematical model describing the haemodialysis process where four compartments are separated.

Eck and Dua (Eck and Dua 2016) performed simulation of the haemodialysis process based on the relevant mathematical model describing the dynamic behaviours of haemodialysis proposed by Gyenge et al.(1999) and Canete and Huang (2010). In the LMPC formulations proposed in the literature, the controlled variables include the relative blood volume (RBV), heart rate (HR) and systolic blood pressure (SBP). These variables are controlled by manipulating the ultrafiltration rate (UFR), the rate at which the fluid is being removed from the patient's body but the objectives regarding toxin clearance and excess water were not considered. In this work an NMPC approach taking into account the nonlinear model of the haemodialysis process, incorporating toxin clearance and excess water, is proposed.

2. Methodology

2.1. Mathematical model
A mathematical model describing the dynamic behaviour of the haemodialysis has been developed in order to achieve the desired optimisation. The model used in this paper is derived based on studies by Gyenge et al. (1999) and Canete and Huang (2010), as well as the modifications by Eck and Dua (2016). The proposed model divides the human body into four individual compartments and one dialyser compartment, as presented in Figure 1.1. The four compartments consist of two extracellular compartments: the red blood cells and the interstitial cells, as well as two intracellular compartments including the blood plasma and interstitial fluid, which are separated by the capillary walls. From Figure 1.1 it can be observed that the interstitial cells are connected to the interstitial fluid while the red blood cells are connected to the plasma compartment. The dialyser only connects to the plasma compartment, which indicates that the plasma compartment is the only compartment that is directly affected by external elements. V refers to the

volumes of the compartments (L), M represents the ion contents (kg), $\dot{M}$ is the ion flux (kg/h) and Q denotes the protein content (kg). The subscripts denote the different compartments and components including ions and toxins.

The development of the mathematical model takes the assumptions into consideration that the four compartments are perfectly mixed and homogeneous, and all proteins have the similar physical and chemical properties as albumin. The model contains theoretical and empirical correlations from literatures such as mass balances, ion distribution and concentration gradients. The state variables of the haemodialysis treatment process are described by simultaneous ordinary differential equations (ODEs) representing the interconnected compartments as shown in Figure 1.1. The values of the parameters are taken from literature (Chapple et al. 1993).

2.2. Orthogonal collocation on finite elements (OCFE)
When solving the mathematical model symbolic and numerical methods are considered. The symbolic method is not suitable due to the complexity of the model. Orthogonal collocation on finite elements (OCFE) is a numerical method for solving ODEs and is widely applied to chemical and biomedical engineering systems. The method minimises the difference between the Lagrange interpolation and the actual solution on predetermined collocation points. The OCFE method is used to both simulate the haemodialysis process and solve the control problem based on the proposed mathematical model.

2.3. Nonlinear model predictive controller setup
The objective function of the NMPC formulation is defined as follows:

$$\min_{u,z} E_{MPC} = \sum_{t=0}^{H_p} Q_z (z_c^{pred}(t_i) - z_c^{sp}(t_i))^2 + \sum_{t=0}^{H_c} Q_u (u(t_i) - u^{ref}(t_i))^2 \tag{1}$$

The objective function minimises the differences between the controlled predictive state variables $z_c^{pred}(t)$ and the set point $z_c^{sp}(t)$ over the prediction horizon H_p. In the objective function the volumes of plasma and interstitial cells compartments, the total body fluid volume and the urea contents are selected as controlled predictive state variables. The objective function also minimises the deviation of the input variables $u(t)$ from a reference value $u^{ref}(t)$ over the control horizon H_c. As mentioned above, four input (manipulated) variables are proposed: the blood flow out of the body, the blood flow returned from dialysis machine to the body, the dialysate rate and the sodium concentration of the inlet blood flow. Q denotes the penalty ratios which can be modified to adjust the controller performance. The total simulation time is set to 210 minutes since normal haemodialysis treatment time is 3 to 5 hours. In order to apply the OCFE method, the simulation time period is divided into 210 elements with 2 internal collocation points in each element. The prediction horizon is set to 5 timesteps while each timestep represents 1 minute. The values of the parameters and the initial conditions of the variables are taken from the literature assuming a 70-kg patient (Xie et al. 1995). The bounds and set points for the variables are given in Table 2.1.

Table 2.1: Values of variables in the NMPC set up

Variable		Units	Lower Bound	Upper Bound	Maximum Change	Set Point
Manipulated Variables	$J_{B,Out}$	ml/min	0.0	800.0	25.0	0.0
	$J_{B,In}$	ml/min	0.0	800.0	25.0	0.0
	J_D	ml/min	0.0	800.0	25.0	0.0
Controlled Variables	$[Na]_{J_{B,In}}$	mmol/l	96.6	179.4	-	139.7
	V_{Total}	L	-	-	-10; +5	40.74
	V_{Pl}	ml	320	32000	-	3200
	V_I	ml	840	84000	-	8400
	$Urea_{Ex}$	mmol	-	-	-	0.0

3. Results and discussion

3.1. Model validation

The mathematical model has to be firstly validated before implementation of the NMPC. The results of infusion simulation are compared with the experimental results obtained by Manning and Guyton (1980) who performed experiments in which dogs were infused with RS in an amount equivalent to 5%, 10, and 20% of the actual body weight. The results of dialysis simulation regarding the toxin concentrations were in good comparison with the experimental data obtained by Ziofcko (2000). The experimental data consists of the urea, uric and creatinine blood concentrations, which were obtained by measuring the dialysate and blood toxin concentrations 11 times in a 210-minute interval. The data was taken from a 58-kg bodyweight patient who experienced a dialyzing treatment lasting 3 hours and 30 minutes, 3 times a week. Two examples of comparison between the simulation results and experimental data, the blood volume change for infusion and the urea concentration change for dialysis simulation, are given in the top two plots in Figure 3.1. The comparisons show a good agreement which indicates that the model can be considered as a valid representation of the haemodialysis process.

3.2. NMPC

Figure 3.1 presents the change of the values of the objective function E_{MPC}. It can be observed that the initial value of 2.82 rapidly declines towards 0 in 18 minutes and is kept around the desired value of zero. Concerning the fluid volume change for the total body, it falls from 41.72 L to the set point of 40.73 L in 25 minutes. Afterwards, the total fluid volume is controlled around the desired level until the completion of simulation.

The graph shown in Figure 3.1 shows the change of total urea content in both the extracellular and intracellular compartments. It can be observed that the NMPC controller is able to achieve approximately 60% urea clearance, which satisfies one of the purposes of the haemodialysis process.

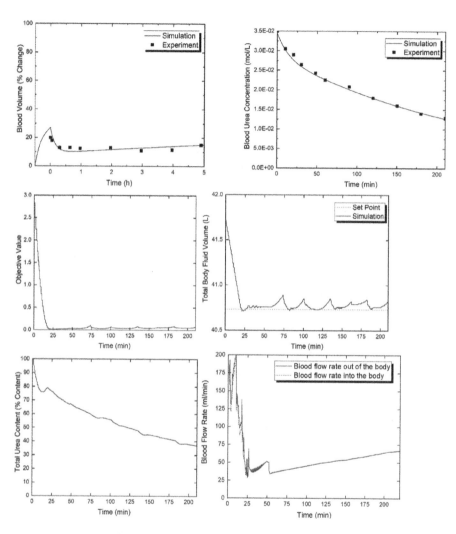

Figure 3.1: Results for NMPC of haemodialysis.

The dynamic behaviours of the blood flow rate out of the body and returned to the body are also plotted in Figure 3.1. From the graph the values of the two manipulated variables follows each other very closely during the entire horizon. This is due to the inequality constraint embedded in the NMPC: $0 \leq J_{B,Out} - J_{B,In} \leq 10$. The difference between the two blood flow rates also represent the ultrafiltration rate. The blood flow rates rapidly fall to their set points of zero in 25 minutes and from minute 50 on, the blood flowrates gradually increases and become stable eventually. In the previously developed Linear Model Predictive Controller (LMPC) by Javed et al. (2009, 2010), the systolic blood pressure (SBP) was controlled by manipulating the relative blood volume (RBV) and heart rate (HR) and the haemodialysis objectives regarding the toxin clearance and excess water content were not considered. In this paper, the NMPC controller successfully demonstrated that the desired toxin clearance can be achieved while the flowrates are maintained at a relative low level to ensure the side effects will not occur. The NMPC controller also controls the sodium concentration to avoid hypotension.

4. Conclusions

A non-linear model predictive controller (NMPC) for haemodialysis treatment is proposed in this work. The developed mathematical model which describes the dynamic behaviours of haemodialysis is validated by comparing the simulation results with previously reported experimental data. The implementation of NMPC on haemodialysis gives promising results. The total body fluid volume is maintained around its set point and the blood flowrates are kept at a low level. As mentioned above, increasing the blood and dialysate flowrates will generally enhance the efficiency of haemodialysis and shorten the treatment time period, but may result in adverse reactions. The proposed NMPC controller is able to reduce the risk of side effects by controlling the flowrates. At the same time, it achieves desired goals of haemodialysis: clearance of toxins and filtration of excess water. Comparing to the Linear Model Predictive Control (LMPC) proposed in literatures (Javed et al. 2009, 2010), the NMPC gives more comprehensive control and could be considered as a more reliable controller due to the good control performance. Future work could include personalised therapy and clinical validation of the controller.

5. References

Chapple, C., B. D. Bowen, R. K. Reed, S. L. Xie, and J. L. Bert. 1993. "A Model of Human Microvascular Exchange: Parameter Estimation Based on Normals and Nephrotics." *Computer Methods and Programs in Biomedicine* 41(1):33–54.

Eck, Thomas and Vivek Dua. 2016. "Control Relevant Modelling for Haemodialysis." Pp. 949–954 in *Computer Aided Chemical Engineering*. Vol. 38.

Fernandez de Canete, J. and P. Del Saz Huang. 2010. "First-Principles Modeling of Fluid and Solute Exchange in the Human during Normal and Hemodialysis Conditions." *Computers in Biology and Medicine* 40(9):740–750.

Gyenge, CC; Bowen, BD; Reed, RK; Bert, JL. 1999. "Transport of Fluid and Solutes in the Body II. Model Validation and Implications." *Am J Physiol* 277(3 Pt 2):H1228–H1240.

Javed, F., A. V Savkin, G. S. H. Chan, P. M. Middleton, P. Malouf, E. Steel, J. Mackie, and T. M. Cheng. 2009. "Modeling and Control of the Heart Rate and Blood Volume Responses to Hemodialysis." *2009 IEEE International Conference on Control and Automation* 625–630.

Javed, Faizan, Andrey V. Savkin, Gregory S. H. Chan, and James D. Mackie. 2010. "Modeling and Model Predictive Control of Hemodynamic Variables during Hemodialysis." Pp. 4673–4678 in *Proc. of the IEEE Conf. on Decision &Control*.

Lote, Christopher J. 2013. *Principles of Renal Physiology*.

Pierce, Thermo Scientific, Dialysis Tubing, and Dialysis Cassettes. 2008. "Dialysis : An Overview." *Thermo Scientific*.

Xie, S. L., R. K. Reed, B. D. Bowen, and J. L. Bert. 1995. "A Model of Human Microvascular Exchange." *Microvascular Research* 49(2):141–162.

Anton A. Kiss, Edwin Zondervan, Richard Lakerveld, Leyla Özkan (Eds.)
Proceedings of the 29th European Symposium on Computer Aided Process Engineering
June 16th to 19th, 2019, Eindhoven, The Netherlands. © 2019 Elsevier B.V. All rights reserved.
http://dx.doi.org/10.1016/B978-0-128-18634-3.50216-2

Development of Guidelines for Optimal Operation of a Cogeneration System

Jia-Lin Kang[a], Hsu-Hung Chang[b], Shyan-Shu Shieh[c], Shi-Shang Jang[b*],

Jing-wei Ko[d], Hsiang-Yao Sun[d]

[a] Department of Chemical and Material Engineering, Tamkang University,
New Taipei City, 25137, Taiwan, ROC
[b] Department of Chemical Engineering, Tsinghua University, Hsinchu,
30031, Taiwan, ROC
[c] Department of Occupational Safety and Health, Chang Jung Christian
University, 71101, Taiwan 30031, Taiwan, ROC
[d] Refining and Manufacturing Research Institute, CPC Corporation,
Taiwan, ROC
ssjang@mx.nthu.edu.tw

Abstract

Guidelines for optimal operation strategy of cooling water temperature are presented to find the maximum net electrical energy of a cogeneration system for turbine generators (TGs) and cooling towers (CTs). In this study, a real cogeneration plant in Taiwan, consisting of variable-frequency drive fan-based CTs and a generator, was chosen as a case study. Statistical linear models were used to build TGs and CTs. A multi-linear model was established to approach a nonlinear system in the cooling tower using a physically meaningful index, the cooling capability index, as an indicator of data clustering. The results showed that the cooling capability index was a good clustering index for the CTs. The cooling tower multi-linear model regressed by the clustering data performed well in predicting the cooling water outlet temperature. The simulated case verified that the online operating guidelines were consistent with real plant data and that it has the ability to find the optimal cooling water temperature. Two cases of plant test prove that the online operating guidelines can assist operators in operating cooling water temperature to the maximum net electrical energy.

Keywords: cooling capability index, multi-linear model, online operating guide, net electrical energy

1. Introduction

Steam turbine generators (TGs) are widely used in the chemical and energy industries, and optimization of the energy efficiency of a generator is an important issue in industrial energy conservation. In general, low-pressure vapor at the outlet of the generator is introduced into the condenser in order to increase the vacuum, thus increasing power generation. On the other hand, the supplied cooling water requires additional fan power in the cooling towers (CTs) to maintain a low water temperature.(Pan et al., 2011) Due to the fact of the complexities in TG-CT systems, the optimization of these power generation processes is still an active area among the researchers (Hajabdollahi et al., 2012; Bornman et al., 2016; Li et al., 2018).

Cooling tower and generator modeling are essential to finding the optimal operating conditions. Models of CTs can be divided into theoretical models and data-driven models. The modeling of CTs and TGs using theoretical models to investigate the potential optimization of operating conditions have been reported in many papers(Zhao et al., 2008; Zhao and Cao, 2009; Ganjehkaviri et al., 2015; Hafdhi et al., 2015). However, the construction of theoretical models requires sufficient knowledge and is time consuming. Thus, constructions of such models cannot provide optimal conditions in a short period. Data-driven models were adopted in a few studies to shorten the modeling time and to investigate the interaction between steam TGs and the CW temperature of CTs to find the optimal operating conditions. Furthermore, the optimal operating condition between steam TGs and the CW temperature of CTs was rarely discussed and verified. Pan et al.(Pan et al., 2011; Pan et al., 2013) used the LMN method to establish a multi-linear model for cogeneration systems with two-speed fans. The model results suggested adjusting the CW temperature to achieve the optimization. The multi-linear model obtained by LMN via clustering had high prediction performance for the cooling water temperature, but the physical meaning of each cluster is unknown, such that operators cannot trust the model to implement on plants. Therefore, it is desirable to use a physically meaningful method to cluster the data to guide the operator in adjusting the cooling water temperature. On the other hand, the optimum operation strategy of fans was determined by the arrangement of fan speed modes for two-speed fans but it is not the ideal optimization method for variable-frequency fans. Furthermore, with an automatic clustering approach in LMN, the model may become rather unstable and need retune frequently. According to the above previous works, we found the following problems remained unsolved. (1) there exist many disturbances in the TG-CT systems, such as steam unstable steam flow, changing humidity as well as environmental temperature. Empirical models are basically not making sense to these changes. (2) First principle models are difficult to obtain due to the lack of trustworthy physical parameters. (3) The optimal conditions are changing from time to time due to the change of the system parameters mentioned above.

The purpose of this study was to tackle the above difficulties by constructing models of TGs with variable-frequency fans using a physical clustering index. Multi-linear model for variable-frequency fans was built. Online operating guidelines based on the models of TGs and CTs were presented to find the optimal cooling water temperature and maximum net electrical energy of the cogeneration system with some physically meaningful plant tests. The rest paper is organized as follows. Section 2 presents details of the modeling and optimal approach, including the TG model, the multi-linear model of CTs, and an optimal strategy named online operating guidelines. Results of a simulated case and plant tests are discussed in Section 3 to demonstrate the feasibility of the online operating guidelines, and a summary of findings is given in the concluding section.

2. Methodology

2.1 Cooling Tower Modeling with Multi-linear Model

Wang et al. (2013) presented a cooling capability index, γ, which can describe the cooling capability of a CT in terms of effective power utilization. The definition of γ is

$$\gamma = \frac{F_{CW}C_p\Delta T_{CW}}{W_{fan}} \tag{1}$$

In this study, the cooling capability index was adopted as a clustering index for regressing the CT multi-linear model. The format of the multi-linear model is a set of the following equations:

$$T_{CW,out} = a_0 + a_1 T_{CW,in} + a_2 H_{R,in} + a_3 T_{air,in} - a_4 W_{fan} \tag{2}$$

In the multi-linear model, the reciprocal of a_4 ($1/a_4$, kW/°C) is physically representative of the required fan power consumption per unit CW outlet temperature change, $\Delta \widehat{W}_{fan}$.

2.2 Last Stage Turbine Generator Modeling

The enthalpy of the steam at the last stage of the TG can be simplified and regressed as follows:

$$H_{steam@the\ last\ stage\ of\ TG} = b_0 + b_1 T_{CW,out} + b_2 F_{steam} \tag{3}$$

2.3 Online Operating Guidelines

The procedure is shown in Figure 1. The optimal CW outlet temperature cannot be over maximum temperature, T_{max} because that will cause the vacuum in the condenser is not enough, leading the TG tripped. The procedure will give two online operating guidelines of the ΔE_{net} and W_{fan}. According to these two online operating guidelines, operators can find the optimal CW outlet temperature and corresponding fan power consumption.

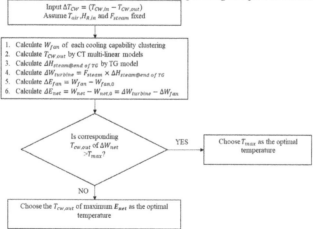

Figure 1 Procedure of operating guidelines for CW temperature

3. Results and Discussion

3.1 Plant Description

In this study, real cogeneration plant in Taiwan, consisting of a variable-frequency drive fan-based cooling tower and a generator, was chosen as a case study to demonstrate model establishment and the online operating guidelines according to the last stage TG and CT models. As Figure 2 shows, the cogeneration plant consists of a TG, a three-cell CT in which every cell has a fan, and a condenser. The range of the fan power is approximately 0–200 kW. The total supplied cooling water of the cooling towers is 5400 T/H. High-pressure steam is introduced into a TG, which in turn drives a power generator to produce electric power.

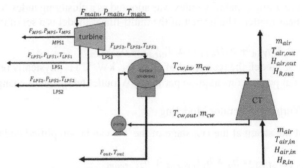

Figure 2 Schematic of a typical cogeneration plant.

3.2 Cooling Tower Model
The CT data presented in this study were divided into nine clusters ($\alpha = 1$ to 9) according to the range of the γ. Every eight intervals of γ starting from the smallest value of γ served as a cluster. The range of the minimum γ_α was from 67 to 75; the range of the maximum γ_α was from 133 to 141. These nine clusters were used to regress the multi-linear model. The R^2 of each linear model is above 0.97.

3.3 Last Stage Turbine Generator Model
The enthalpy of the steam at the last stage of the TG correlation is regressed as the following:

$$H_{\text{steam@the last stage of TG}} = 2506 + 1.78 T_{CW,out} + 0.2 F_{steam} \tag{4}$$

The R^2 of the correlation is 0.99.

3.4 Demonstration with a Simulated Case
Figure 3 shows the online operating guidelines of the relative net electrical energy, ΔE_{net}. The values are 29.70 °C and 30.25 °C for the two cases of ΔT_{CW}, respectively. The trends of the ΔE_{net} calculated by historical data were consistent with the trend of the online operating guideline of the ΔE_{net}. As Figure 3(a) shows, the optimal CW outlet temperature is approximately 32.7 °C for the CT. However, the CW outlet temperature can only be increased to 34 °C if the optimal temperature cannot be found during the increase in the cooling water outlet temperature, as Figure 3(b) shows, because the T_{max} of the CW outlet temperature is 34 °C in this case study.

3.5 Plant Test
Figure 5 shows the plant test results and online operating guidelines of the case 2. The online operating guidelines were given based on 24.9 °C of T_{air} and 59.9 % of H_{air}. The initial CW outlet temperature was 29.1 °C. However, because the cooling capability index was 155 at this condition, which beyond the range of the CT multi-linear model, the online operating guidelines at this condition were obtained by extrapolation. According to the online operating guidelines, the CW outlet temperature must be raised to increase the net electrical energy. Thus, the CW outlet temperature was raised to 31.4 °C. The finial ΔE_{net} increased by 21.75 kW. Compared with Case 1, Case 2 had greater ΔE_{net} because the

CW outlet temperature was operated at the lower temperature, which required the larger fan power consumption. As the CW outlet temperature increased, the fan power consumption dropped dramatically, leading to the energy saving benefit appear. The results of cases 1 and 2 showed that the net electrical energy can be increased by the online operating guidelines and the finial ΔE_{net} of the plant tests are close to the finial ΔE_{net} of the online operating guidelines. These results prove that the online operating guidelines using the TG model and the CT multi-linear model are feasible.

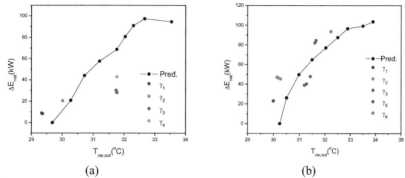

(a)   (b)

Figure 3 The agreement of the relative net electrical energy between multi-linear model and measurement using the conditions of (a) $T_{air,in}$: 28.5~29 °C, $H_{R,in}$: 78~79.5%, and F_{steam}: 30~32 (T/H), and (b) $T_{air,in}$: 30~32 °C, $H_{R,in}$: 70~73%, and F_{steam}: 35~36 (T/H)

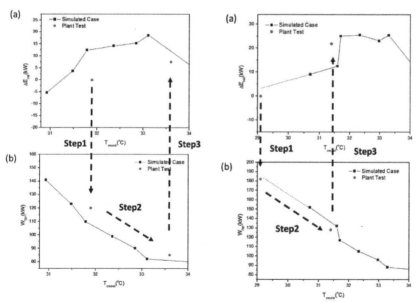

Figure 4 Measured data of (a) cooling water outlet temperature, and (b) fan power consumption for Case 1.

Figure 5 Measured data of (a) cooling water outlet temperature, and (b) fan power consumption for Case 2

4. Conclusions

In this study, we presented an optimal operation method for finding the maximum power generation for cogeneration systems by using a CT multi-linear model and a TG model. An online operating guidelines were used to determine the optimal CW outlet temperature to achieve maximum net electrical energy. In the demonstration, the online operating guidelines were consistent with historical data and that it had the ability to find the optimal CW temperature. Two cases of the plant test based on the online operating guidelines showed that the final ΔE_{net} increased by 7.5 and 21.75 kW, respectively. The difference of ΔE_{net} between Case 1 and Case 2 is mainly caused by the different initial CW outlet temperatures. The ΔE_{net} of plant test and ΔE_{net} of the online operating guidelines were close to prove that the online operating guidelines can be practically applied.

Acknowledgment

The authors acknowledged the financial support provided by Refining and Manufacturing Research Institute, CPC Corporation through the grant EEA-0615001.

References

Bornman, W., Dirker, J., Arndt, D.C., Meyer, J.P., 2016. Operational energy minimisation for forced draft, direct-contact bulk air cooling tower through a combination of forward and first-principle modelling, coupled with an optimisation platform. Energy 114, 995-1006.

Ganjehkaviri, A., Jaafar, M.M., Hosseini, S., 2015. Optimization and the effect of steam turbine outlet quality on the output power of a combined cycle power plant. Energy Conversion and Management 89, 231-243.

Hafdhi, F., Khir, T., Yahyia, A.B., Brahim, A.B., 2015. Energetic and exergetic analysis of a steam turbine power plant in an existing phosphoric acid factory. Energy conversion and management 106, 1230-1241.

Hajabdollahi, F., Hajabdollahi, Z., Hajabdollahi, H., 2012. Soft computing based multi-objective optimization of steam cycle power plant using NSGA-II and ANN. Applied Soft Computing 12, 3648-3655.

Li, X., Wang, N., Wang, L., Kantor, I., Robineau, J.-L., Yang, Y., Maréchal, F., 2018. A data-driven model for the air-cooling condenser of thermal power plants based on data reconciliation and support vector regression. Applied Thermal Engineering 129, 1496-1507.

Pan, T.-H., Shieh, S.-S., Jang, S.-S., Wu, C.-W., Ou, J.-J., 2011. Electricity gain via integrated operation of turbine generator and cooling tower using local model network. IEEE Transactions on Energy Conversion 26, 245-255.

Pan, T.-H., Xu, D., Li, Z., Shieh, S.-S., Jang, S.-S., 2013. Efficiency improvement of cogeneration system using statistical model. Energy conversion and management 68, 169-176.

Wang, J.-G., Shieh, S.-S., Jang, S.-S., Wu, C.-W., 2013. Discrete model-based operation of cooling tower based on statistical analysis. Energy conversion and management 73, 226-233.

Zhao, B., Liu, L., Zhang, W., 2008. Optimization of cold end system of steam turbine. Frontiers of Energy and Power Engineering in China 2, 348-353.

Zhao, H., Cao, L., 2009. Study on the optimal back-pressure of direct air-cooled condenser in theory, Power and Energy Engineering Conference, 2009. APPEEC 2009. Asia-Pacific. IEEE, pp. 1-4.

Anton A. Kiss, Edwin Zondervan, Richard Lakerveld, Leyla Özkan (Eds.)
Proceedings of the 29th European Symposium on Computer Aided Process Engineering
June 16th to 19th, 2019, Eindhoven, The Netherlands. © 2019 Elsevier B.V. All rights reserved.
http://dx.doi.org/10.1016/B978-0-128-18634-3.50217-4

On the Optimization of Production Scheduling in Industrial Food Processing Facilities

Georgios P. Georgiadis[a,b], Chrysovalantou Ziogou[b], Georgios Kopanos[a], Borja Mariño Pampín[c], Daniel Cabo[d], Miguel Lopez[d], Michael C. Georgiadis[a,b,*]

[a]Department of Chemical Engineering, Aristotle University of Thessaloniki, Thessaloniki 54124, Greece

[b]Chemical Process and Energy Resources Institute (CPERI), Centre for Research and Technology Hellas (CERTH), PO Box 60361, 57001, Thessaloniki, Greece

[c]Frinsa del Noroeste S.A., Avenida Ramiro Carregal Rey – Parcela 29, Ribeira, La Coruña, Spain

[d]ASM Soft S.L., Crta de Bembrive 109, Vigo, Spain

mgeorg@auth.gr

Abstract

This work presents the development and application of an efficient solution strategy for the optimal production scheduling of a real-life food industry. In particular, the case of a canned fish production facility for a large-scale Spanish industry is considered. Main goal is to develop an optimized weekly schedule, in order to minimize the total production makespan. The proposed solution strategy constitutes the basis to develop an efficient and robust approach for this complex scheduling problem. A general precedence Mixed-Integer Linear Programming (MILP) model is utilized for all scheduling-related decisions (unit allocation, timing and sequencing). To solve the scheduling problem in a computational time accepted by the industry, a two-step decomposition algorithm is employed. Salient characteristics of the canned fish industry are aptly modelled, while valid industry-specific heuristics are incorporated. The suggested solution strategy is successfully applied to a real study case, corresponding to the most demanding week of the plant under study.

Keywords: production scheduling, food industry, MILP, decomposition

1. Introduction

Market trends and competitiveness has steered food industry towards large production volumes, complex alternative recipes and an increasing product portfolio, making production scheduling a challenge. The current industrial practice imposes scheduling-related decisions to be mainly derived by managers and operators, hence the overall plant performance is subject to their experience. Computer-aided scheduling tools can significantly improve these decisions by proper consideration of all involved parameters and therefore significantly enhance production scheduling (Harjunkoski, 2016). As a result, productivity is improved, while customers remain satisfied and profits increase. Acknowledging the importance of optimized production scheduling, the scientific community has widely studied the topic over the last 30 years, introducing numerous scheduling models (Méndez et al., 2006). However, most of these works consider small-scale study cases. This is mainly attributed to the fact that production scheduling is an NP-hard problem, therefore large complex instances can become intractable. The scientific community has widely recognised the lack of applications to real industrial cases (Harjunkoski et al., 2014). Recently some attempts have been made to close the existing gap between theory and industrial practice. In (Kopanos et al., 2010), the authors studied a real-life yoghurt production facility using a novel mixed discrete and continuous MILP model. Furthermore, Baumann and Trautmann, (2014) proposed a hybrid MILP method for make-and-pack processes using a decomposition strategy and a critical-path improvement algorithm. Moreover, Aguirre et al., (2017) combined a novel MILP model that incorporates TSP (Travelling Salesman Problem) constraints, with a rolling horizon algorithm. In this work a solution strategy is proposed that deals with scheduling problems of large-scale industrial food production facilities. In particular, an MILP model is proposed to optimize the production schedule, while a two-step decomposition algorithm is utilized to solve the problem in an acceptable computational time.

2. Problem statement

In this work, the canned fish production in a real-life industrial facility is examined. Specifically, the production process of Frinsa del Noroeste S.A., located in Ribeira, Spain, is investigated using real process data. The plant is capable of producing more than 400 codes, and it is one of the largest canned fish industries in Europe. The facility comprises of multiple stages, including both batch and continuous processes (Fig. 1). The raw materials arrive in the facility in the form of frozen fish blocks, and as such they need to be unfrozen in the thawing chambers. Then, the blocks are chopped in the appropriate size and filled in the cans alongside with all other ingredients (brine, olive oil etc.) required by the recipe. In the next stage, the sealed cans are sterilized in order to ensure the microbiological quality of the final products. Finally, the cans are packaged in their final form (6-pack, 12-pack, boxes etc.) and are stored in the warehouse, to be later distributed in the market.

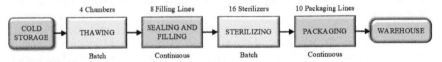

Figure 1: Facility layout

The plant under consideration can be identified as a multiproduct, multistage facility with both batch (thawing, sterilizing) and continuous (sealing and filling, packaging) processes each utilizing multiple parallel units. Additionally, the large production demand and high production flexibility increases significantly the plant's complexity. The thawing stage is overdesigned compared to the processing capacity of all other stages, therefore it is a valid assumption to omit it from this study. Unfortunately, no clear bottlenecks exist, and as such all other processing stages need to be modelled. The short-term scheduling horizon of interest is 5 days, whereas all units are available 24 hours per day. Sequence-dependent changeovers are considered. All design and operating constraints of the facility, such as a limited waiting time between stages to ensure microbiological integrity, are taken into account. The objective is to minimize the total production makespan, while ensuring demand satisfaction.

3. Mathematical framework

The key scheduling decisions to be made are related to: a) the number of product batches required to satisfy the incoming orders, b) the allocation of product batches to units in every processing stage, c) when will the process of each batch in every stage start and finish and d) in what relative sequence. A typical industrial practice in most food industries, imposes the operation of the intermediate batch processes in their maximum capacity. Utilizing the batch stage to its fullest, leads to reduction of changeovers between products and a general increase in the plant's productivity. Thus, the number of batches of each product required to satisfy the demand is calculated a priori, based on the given demand, the inventory levels and the capacity of the sterilization chambers.

3.1. MILP model

The suggested MILP model is based on the general precedence framework. Due to lack of space, only a brief description of the model is presented:

$$\sum_{j \in (SJ_{s,j} \cap PJ_{p,j})} \overline{Y}_{p,b,s,j,n} = 1 \qquad \forall p \in I_p^{in}, b \in PB_{p,b,n}, n \in I_n^{in} \tag{1}$$

$$L_{p,b,s,n} + \sum_{j \in (SJ_{s,j} \cap PJ_{p,j})} (fs_{p,b,j,n}^{time} \cdot \overline{Y}_{p,b,s,j,n}) = C_{p,b,s,n} \qquad \forall p \in I_p^{in}, b \in PB_{p,b,n}, n \in I_n^{in} : s = 1 \tag{2}$$

$$C_{p,b,s,n} + W_{p,b,s,n} = L_{p,b,s+1,n} \qquad \forall p \in I_p^{in}, b \in PB_{p,b,n}, n \in I_n^{in} : s < 3 \tag{3}$$

$$L_{p',b',s,n} \geq C_{p,b,s,n} + ch_{p,p',j}^{time} - M \cdot (1 - X_{p,p',n}) - M \cdot (2 - \overline{Y}_{p,b,s,j,n} - \overline{Y}_{p',b',s,j,n})$$
$$\forall p \in I_p^{in}, p' \in I_{p'}^{in}, b \in PB_{p,b,n}, b' \in PB_{p',b',n}, j \in (PPJ_{p,p',j} \cap SJ_{s,j}), n \in I_n^{in} : p < p', s \neq 2 \tag{4}$$

$$L_{p',b',s,n} > C_{p,b,s,n} - M * (1 - \overline{X}_{p,b,p',b',n}) - M * (2 - \overline{Y}_{p,b,s,j,n} - \overline{Y}_{p',b',s,j,n}),$$
$$\forall p \in I_p^{in}, b \in PB_{p,b,n}, p' \in I_{p'}^{in}, b' \in PB_{p',b',n}, j \in (PPJ_{p,p',j} \cap S_{s,j}), n \in I_n^{in} : p < p', s = 2 \tag{5}$$

$$C_{p,b,s+1,n} - L_{p,b,s,n} \leq Q_p \qquad \forall p \in I_p^{in}, b \in PB_{p,b,n}, n \in I_n^{in} : s = 1 \tag{6}$$

$$C_{p,b,s,n} \leq 24 \qquad \forall p \in I_p^{in}, b \in PB_{p,b,n}, n \in I_n^{in} : s = 3 \tag{7}$$

$$C^{max} \geq C_{p,b,s,n} \qquad \forall p \in I_p^{in}, b \in PB_{p,b,n}, n \in I_n^{in} : s = 3 \tag{8}$$

Constraints (1) guarantee that all product batches p,b to be scheduled on day n will be processed by exactly one unit j in every stage s, using the binary allocation variable $\overline{Y}_{p,b,j,n}$. Constraints (2) impose the timing constraints in the sealing and filling stage. More specifically, they state that the completion of the sealing and filling task for every product batch to be scheduled in every day $C_{p,b,s,n}$ is equal to the starting time of the task $L_{p,b,s,n}$ plus the required processing time $fs_{p,b,j,n}^{time}$. Similar constraints are used for the sterilization and packing stages. To synchronize the stages, constraints (3) are employed. The continuous variable $W_{p,b,s,n}$ defines the waiting time between each stage. The sequencing constraints between product batches in every stage are portrayed in constraints (4) and (5). Two general precedence variables are introduced, $X_{p,p',n}$ and $\overline{X}_{p,b,p',b',n}$, alongside a big-M parameter. The first precedence variable defines the sequencing of product batches in the continuous stages (sealing and filling and packing), while the latter in the batch stage (sterilization). Notice, that the batch sets b,b' are not used in the first precedence variable, since a single campaign policy is followed in the continuous stages. This way the binary variables are significantly decreased, thus the computational complexity of the problem is reduced. In particular, constraints (4) state that if a product p is processed prior to p' on day n ($X_{p,p',n}=1$) and both product batches are processed in the same unit j ($\overline{Y}_{p,b,j,n} = \overline{Y}_{p',b',j,n} = 1$), then the starting time of p',b' must be greater than the completion time of p,b plus any required changeover $ch_{p,p',j}^{time}$. Similarly, constraints (5) impose the sequencing constraints in the sterilization stage. Constraints (6) enforce the waiting time between the sealing and filling stage and the sterilization stage to be less than a specific limit Q_p. This limit ensures the microbiological integrity of the final product. To ensure that the daily scheduling horizon is not violated, constraints (7) are used. The objective of the model is the minimization of the total production makespan C_{max} and is expressed by constraints (8).

3.2. Decomposition algorithm

The complexity of the examined plant is such that an exact method cannot solve the scheduling problem in reasonable time. Therefore, a two-step decomposition algorithm is employed to split the initial problem into several tractable subproblems. First, the weekly scheduling problem is decomposed in a temporal manner into 5 daily scheduling subproblems. Then, an order-based decomposition is utilized to solve the daily scheduling problem for a specific number of products in each iteration. Fig. 2. illustrates the flowchart of the proposed solution strategy. At first the batching subproblem is solved to translate the product orders into batches. Afterwards, the number of orders to be scheduled in each iteration are set. Then, the MILP is solved for the specified subproblem area (day and number of products) and only the binary variables (unit allocation, sequencing) are fixed. When all orders are scheduled for a given day, all variables are fixed, and the algorithm continues to the next day. Finally, when all days are considered, the complete schedule is generated.

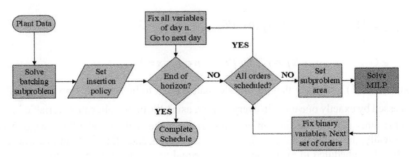

Figure 2: Solution strategy

4. Results

An industrial study case using real data from the Frinsa production plant is presented. In total 136 final products are to be scheduled, corresponding to a real weekly demand from a period with the most intensive production. To solve this complex case, the proposed solution strategy is utilized. In each iteration the daily schedule for half of the product-orders was chosen to be optimized. The MILP model was implemented in GAMS 25.1 and solved in an Intel Core i7 @3.4Gz with 16GB RAM, using CPLEX 12.0. Optimality is reached for all iterations of the suggested solution strategy. Figure 3 illustrates the complete schedule generated for all units of every processing stage. Each color corresponds to a batch or lot of a product-code.

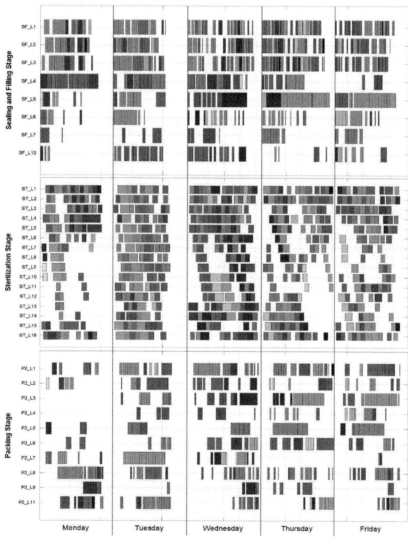

Figure 3: Gantt chart of units of all processing stages

Compared to the real weekly schedule proposed by Frinsa, the optimized schedule of the proposed strategy illustrates interesting results. To satisfy the given demand, the manually derived schedule by Frinsa, requires the addition of a shift on Sunday evening, while the optimized schedule satisfy all orders within 5 days. Moreover, the total CPU time for the solution of the problem is approximately 1 hour and it is acceptable by the company.

5. Conclusions

This work presents the optimization-based production scheduling of a large-scale real-life food industry. More specifically, all major processing stages of a canned fish production facility have been optimally scheduled. The industrial problem under consideration illustrates significant complexity, due to the mixed batch and continuous stages, each having numerous shared resources, the large number of final products and the various operational, design and quality constraints. This make-and-pack structure (one or multiple batch or continuous processes followed by a packing stage) is typically met in most food and consumer packaged goods industries, hence, the presented solution strategy can be easily implemented in other industrial problems. It has been shown that the suggested solution strategy can optimally schedule even the most demanding weeks of the examined industry in acceptable time, leading to reduction of overtime production. The proposed strategy can be the core for a computer-aided scheduling tool that can facilitate the decision-making process for the production scheduling of food industries. Current work focuses on the introduction of cost related objectives, as well as, the incorporation of uncertainty in product demands.

Acknowledgements

The work leading to this publication has received funding from the European Union's Horizon 2020 research and innovation programme under grant agreement No 723575 (Project CoPro) in the framework of the SPIRE PPP.

References

Aguirre, A.M., Liu, S., Papageorgiou, L.G., 2017. Mixed Integer Linear Programming Based Approaches for Medium-Term Planning and Scheduling in Multiproduct Multistage Continuous Plants. Ind. Eng. Chem. Res. 56, 5636–5651.

Baumann, P., Trautmann, N., 2014. A hybrid method for large-scale short-term scheduling of make-and-pack production processes. Eur. J. Oper. Res. 236, 718–735.

Harjunkoski, I., 2016. Deploying scheduling solutions in an industrial environment. Comput. Chem. Eng. 91, 127–135.

Harjunkoski, I., Maravelias, C.T., Bongers, P., Castro, P.M., Engell, S., Grossmann, I.E., Hooker, J., Méndez, C., Sand, G., Wassick, J., 2014. Scope for industrial applications of production scheduling models and solution methods. Comput. Chem. Eng. 62, 161–193.

Kopanos, G.M., Puigjaner, L., Georgiadis, M.C., 2010. Optimal Production Scheduling and Lot-Sizing in Dairy Plants : The Yogurt Production Line. Ind. Eng. Chem. Res. 49, 701–718.

Méndez, C.A., Cerdá, J., Grossmann, I.E., Harjunkoski, I., Fahl, M., 2006. State-of-the-art review of optimization methods for short-term scheduling of batch processes. Comput. Chem. Eng. 30, 913–946.

Anton A. Kiss, Edwin Zondervan, Richard Lakerveld, Leyla Özkan (Eds.)
Proceedings of the 29th European Symposium on Computer Aided Process Engineering
June 16th to 19th, 2019, Eindhoven, The Netherlands. © 2019 Elsevier B.V. All rights reserved.
http://dx.doi.org/10.1016/B978-0-128-18634-3.50218-6

A Stacked Auto-Encoder Based Fault Diagnosis Model for Chemical Process

Yi Qiu, Yiyang Dai*

College of Chemistry and Chemical Engineering, Southwest Petroleum University, Chengdu 610500, China

daiyiyang1984@163.com

Abstract

Fault detection and diagnosis (FDD) is one of the key technologies to ensure the safe operation of chemical processes. With the widespread application of automation technology in chemical plants and the era of big data, data-based methods have become a hot research topic in the field of fault diagnosis. How to effectively extract the fault characteristics from the data and determine the cause of the fault is the key to help the operator deal with the abnormal conditions. Stack auto-encoder is a deep learning model with strong feature extraction and generalization capabilities. This paper proposes a SAE-based chemical process fault diagnosis model and applies it to Tennessee Eastman process. The performance of the SAE-based model is illustrated by comparison with the results of other methods.

Keywords: fault diagnosis, chemical engineering, stacked auto-encoder, neural network.

1. Introduction

With the development of science and technology, the chemical production process has become increasingly large-scale and complicated. When an abnormal situation occurs in the system, if it cannot be discovered and dealt with in time, it will cause huge economic losses and even casualties. Therefore, fault detection and diagnosis (FDD) have important application significance for chemical process.

Since Beard (1971) proposed the concept of fault diagnosis, many scholars have conducted related research in the field of fault diagnosis. Venkatasubramanian et al. (2003) classify fault diagnosis methods into three general categories according to models: quantitative model-based methods, qualitative model-based methods, and process history data based (data-driven) methods. With the widespread use of automation technology in the process industry, researchers can easily obtain a large amount of process history data. Therefore, the data-based FDD method has become a hot research topic in the field of fault diagnosis.

There are two types of FDD methods based on historical data. One type is statistical methods, such as principal component analysis (PCA), independent component analysis (ICA), kernel principal component analysis (KPCA), partial least squares (PLS) and their derivatives. The other type is pattern recognition methods, such as artificial neural networks (ANN), support vector machines (SVM), and so on. However, with the large-scale and complicated development of chemical units, the application of statistical methods and traditional pattern recognition methods will be limited by high-dimensional data and high-correlation data. Since Hinton & Salakhutdinov (2006) proposed deep learning, the development and application of artificial intelligence technology has achieved great success in the past decade. Today, artificial intelligence is widely used in computer science, finance, and natural language processing. Zhang & Zhao (2017) proposed a chemical process fault diagnosis model based on deep belief network (DBN). Wu & Zhao (2018) proposed a chemical process fault diagnosis model based on convolutional neural network (CNN). The application of artificial intelligence technology in the field of chemical process fault diagnosis has also shown excellent performance.

In this paper, we propose a chemical process fault diagnosis model based on stack auto-encoder (SAE) and apply it to the fault diagnosis of Tennessee Eastman Process (TEP). The remainder of this paper is organized as follows: A brief introduction to SAE is given in Section 2. Section 3 shows a SAE-based fault diagnosis model. In section 4, The SAE-based model is applied in the TE process. Finally, we conclude the paper and describe some work in the future.

2. Stacked Auto-Encoder

The auto-encoder (AE) is proposed by Rumelhart et al. (1986), which is a typical single hidden layer neural network. The stack auto-encoder (SAE) is a modified model proposed by Hinton & Salakhutdinov (2006) based on the auto-encoder. SAE is an important deep learning model and it is a deep neural network essentially.

2.1. Structure of SAE

A SAE with 5 hidden layers is shown in Fig. 1. Structurally, the SAE consists of a coder layer (Layer-1 to Layer-4) and a decoder layer (Layer-4 to Layer-7), which assume a symmetrical structure. Each two adjacent layers from left to right is an AE, the weighted connections of neurons exist between each two adjacent layers.

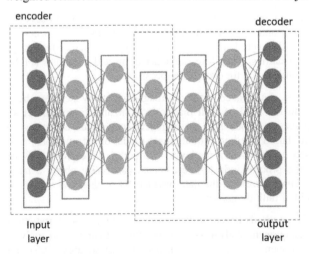

Fig. 1 Structure of a SAE with 5 hidden layers

Because the network structure of SAE has multiple hidden layers, it can well represent complex high-dimensional functions and has powerful feature extraction capabilities.

2.2. Training of SAE

The SAE training process consists of the following two stages: Pre-training stage and fine-tuning stage. The first stage is to initialize the weight of the deep neural network by means of layer-by-layer unsupervised learning (unsupervised greedy algorithm). The layer-by-layer greedy algorithm can effectively avoid the network falling into local optimum. The second stage is to fine-tune the model produced in the previous step by supervised learning (such as backpropagation algorithm).

For a training set x containing m samples, each of the data passes through the encoder to obtain a characteristic expression of the hidden layer, as shown in Eq. (1).

$$y^{(i)} = f_\theta(x^{(i)}) = s(Wx^{(i)} + b) \tag{1}$$

Where $\theta = (Wx,b)$ is the network parameter, W is the weight, b is the bias term, and s is the activation function.

The reconstructed vector is obtained by the $y^{(i)}$ input decoder, as shown in Eq. (2).

$$z^{(i)} = g_{\theta'}(y^{(i)}) = s(W'y^{(i)} + b') \tag{2}$$

Where $\theta' = (W',b')$, $W' = W^T$.

In order to prevent the model from over-fitting, a penalty term (regularization) needs to be added. The cost function is shown in Eq. (3).

$$J = \frac{1}{m}\sum_{i=1}^{m}(\frac{1}{2}\|z^{(i)} - x^{(i)}\|) + \frac{\lambda}{2}\|W\|^2 \tag{3}$$

Where λ is the regularization parameter.

Iterative calculation is performed by the gradient descent method, and the optimal weight is obtained when the cost function is the minimum. In each iteration, the update process of the weight W and the bias term b as shown in Eq. (4) and Eq. (5).

$$W \leftarrow W - \alpha\frac{\partial J}{\partial W} \tag{4}$$

$$b \leftarrow b - \eta\frac{\partial J}{\partial b} \tag{5}$$

Where η is the learning rate.

After the pre-training stage, the weight of the pre-training model is iterated and updated again using the backpropagation algorithm until the error between the predicted label and the input label is lower than the expected value. Finally, the best weights and bias terms for each layer of AE are obtained.

3. Fault diagnosis model

According to the training process of SAE, we constructed a SAE-based chemical process fault diagnosis model. The fault diagnosis model is shown in Fig. 2, which includes the offline stage and the online stage. 80% of the historical data is used as training data, and its corresponding label is generated. The model is trained using the training set and its corresponding labels, and the remaining data is entered into the model for testing. Compare the test results to the label, and apply the model to online-stage if it performs well. If the model has a low correct rate on the test set, the parameters should be changed for retraining.

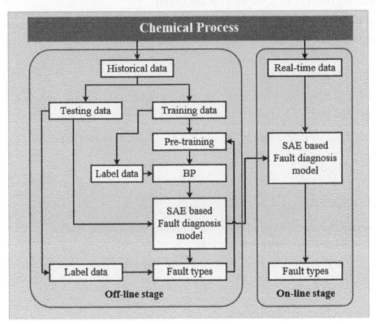

Fig 2. The framework of the SAE based fault diagnosis model

4. Application to Tennessee Eastman process

In order to verify the performance of the model which is proposed in this paper. The dataset of Tennessee Eastman (TE) process is applied to the SAE-based fault diagnosis model. The results are compared to some other FDD methods.

4.1. TE process

TE process is a simulation based on the actual chemical industry process, which was proposed by Downs & Vogel (1993). TE process includes 12 manipulated variables and 41 measured variables. Since the 12th manipulated variable is a constant, the diagnosis model involves a total of 52 variables. There are 20 faults in TE process.

All simulation data come from http://web.mit.edu/braatzgroup/links.html. There are 2 sets in this data set. Each set contains 21 sets of data, which are normal state data and 21 types of fault state data. The first dataset was first simulated in the normal state for 1 hour, and simulation was continued for 24 hours after adding the disturbance. The second dataset was first simulated in the normal state for 8 hours, and simulation was continued for 40 hours after the disturbance was added. The sampling period for both sets of data is 3 minutes, so the dataset includes 30,260 samples.

4.2. Diagnosis results and comparison

We conducted comparative experiments by setting different hyperparameters. A multivariate hyperparameter set was obtained through experiments. The SAE model includes 6 layers of neurons, and the number of neurons in each layer is 52, 100, 50, 50, 100, 21. The number of pre-trainings is 200, the size of each batch in pre-training stage and fine-tuning stage is 89, the learning rate is 1, and the regularization parameter is 0.0000005.

The performance of the model is usually evaluated by the fault diagnosis rate (FDR). p_i is the count of type i samples that are classified to type i.

$$FDR_i = \frac{p_i}{total\ count\ of\ type\ i\ samples} \tag{6}$$

The comparison of diagnosis performance with several statistical methods and traditional pattern recognition methods is shown in Table 1.

Table 1 Diagnosis performance comparison of different methods. (a) based partitioning PCA (Wang et al., 2016); (b) based dynamic ICA (Hsu et al., 2010); (c) based designing a hierarchical neural network (Eslamloueyan, 2011); (d) based support vector machines(Yélamos et al., 2009); (e) based SAE (proposed in this paper).

FDR (%)	(a)	(b)	(c)	(d)	(e)
Fault01	99.8	100.0	97.0	95.0	98.2
Fault02	98.8	99.0	98.0	100.0	99.6
Fault03	13.6	2.0	53.0	0.0	34.6
Fault04	86.5	97.0	95.0	57.0	96.3
Fault05	100.0	100.0	96.0	64.0	99.2
Fault06	99.5	100.0	100.0	93.0	100.0
Fault07	100.0	100.0	100.0	100.0	100.0
Fault08	98.3	98.0	60.0	100.0	95.3
Fault09	13.4	1.0	29.0	0.0	37.8
Fault10	64.4	82.0	47.0	53.0	89.2
Fault11	77.1	54.0	48.0	21.0	74.3
Fault12	99.3	100.0	46.0	0.0	94.7
Fault13	94.6	95.0	32.0	91.0	94.0
Fault14	100.0	100.0	67.0	0.0	95.2
Fault15	16.9	2.0	66.0	0.0	84.9
Fault16	49.0	82.0	37.0	88.0	82.6
Fault17	96.3	90.0	72.0	68.0	92.4
Fault18	91.3	90.0	94.0	82.0	90.0
Fault19	39.4	81.0	52.0	16.0	86.3
Fault20	54.0	88.0	67.0	100.0	86.4
Average	74.6	78.1	67.8	56.4	86.5

5. Conclusions and prospects

With the widespread use of automation technology in chemical processes, it has become increasingly convenient for researchers to access historical data. Many new data-based methods have been applied to the study of chemical process fault diagnosis. In the increasingly complex chemical process, SAE has strong feature extraction and generalization capabilities. This paper proposes a SAE-based chemical process fault diagnosis model, and uses the data generated by the TE process for model construction and training, the average fault diagnosis rate reaches 86.5%. Compared to traditional shallow neural networks, deep learning requires a larger training sample to present its advantages. Therefore, our next work is to simulate the TE process to get more data. Then, the key variables are divided according to the relevance of the variables.

Acknowledgement

The authors gratefully acknowledge financial support from the National Natural Science Foundation of China (NSFC) (No. 21706220).

References

Beard, Vernon, R., 1971. Failure accommodation in linear systems through self-reorganization.

Eslamloueyan, R., 2011. Designing a hierarchical neural network based on fuzzy clustering for fault diagnosis of the Tennessee – Eastman process. APPL SOFT COMPUT, 1407-1415.

GuozhuWang, JianchangLiu, YuanLi, ChengZhang, 2016. Fault diagnosis of chemical processes based on partitioning PCA and variable reasoning strategy☆. CHINESE J CHEM ENG, 869-880.

Hinton, G.E.H.C., Salakhutdinov, R.R., 2006. Reducing the Dimensionality of Data with Neural Networks. SCIENCE, 504-507.

Hsu, C.C.C.E., Chen, M., Chen, L., 2010. A novel process monitoring approach with dynamic independent component analysis. CONTROL ENG PRACT, 242-253.

Rumelhart, D.E., Hinton, G.E., Williams, R.J., 1986. Learning representations by back-propagating errors. NATURE, 533-536.

Venkatasubramanian, V.V.C.P., Rengaswamy, R.R.C.E., Yin, K., Kavuri, S.N., 2003. A review of process fault detection and diagnosis Part I: Quantitative model-based methods. Computers and Chemical Engineering, 293-311.

Vogel, J.J.D.A., 1993. A plant-wide industrial process control problem. Computers and Chemical Engineering, 245-255.

Wu, H., Zhao, J., 2018. Deep convolutional neural network model based chemical process fault diagnosis. COMPUT CHEM ENG, 185-197.

Yélamos, I., Escudero, G., Graells, M., Puigjaner, L., 2009. Performance assessment of a novel fault diagnosis system based on support vector machines. Computers and Chemical Engineering, 244-255.

Zhang, Z.Z.Z., Zhao, J.Z.J., 2017. A deep belief network based fault diagnosis model for complex chemical processes. COMPUT CHEM ENG, 395-407.

Anton A. Kiss, Edwin Zondervan, Richard Lakerveld, Leyla Özkan (Eds.)
Proceedings of the 29th European Symposium on Computer Aided Process Engineering
June 16th to 19th, 2019, Eindhoven, The Netherlands. © 2019 Elsevier B.V. All rights reserved.
http://dx.doi.org/10.1016/B978-0-128-18634-3.50219-8

Multi-objective optimization approach to design and planning hydrogen supply chain under uncertainty: A Portugal study case

Diego Câmara,[a,*] Tânia Pinto-Varela,[a] Ana Paula Barbósa-Povoa[a]

[a]*Centro de Estudos de Gestão-IST, Instituto Superior Técnico, Universidade de Lisboa, Av. Rovisco Pais, Lisboa 1049-001, Portugal*

diegofelipe@tecnico.ulisboa.pt

Abstract

The progress of the current energy infrastructure to a new pattern, where hydrogen plays an important role has been increasing the interest of industrial communities, as well as, academics. This is mainly associated with sustainability goals where reducing environmental impacts is a target while guaranteeing economic benefits. In this context, the development of hydrogen optimized global supply chains (SC) that can support industrial processes through renewable energy sources is to be pursued. In this paper, we explore this challenge and a multi-objective, multi-period, stochastic Mixed Integer Linear Programming (MILP) problem for design and planning a hydrogen SC is developed. Uncertainty over the availability of primary energy sources (PES) is considered through the definition of set scenarios. A Portuguese case is explored, from where it as concluded that the reduction in global warming potential is very small when compared to the increase in cost.
Keywords: Hydrogen supply chain, Renewable energy source, Multi-objective optimization, Uncertainty in primary energy sources.

1. Introduction

The use of fuels with low emissions has been recently highlighted as one of the trends to mitigate environmental impacts. In this set, the use of hydrogen as a fuel has been considered an alternative with significant potential to integrate a more sustainable energy matrix, especially in the transportation sector (Almansoori et al., 2011). Hydrogen has become a key element for governments under pressure of emission constraints and environmental laws. Despite this, there is still no appropriate infrastructure available and uncertainties are one of the main drawbacks to the development of the hydrogen economy as the case of primary energy sources (PES) availability.

Different approaches to formulating a hydrogen SC have been appearing in the literature, as an attempt to determine if hydrogen is a competitive option, as an energetic vector, while mitigating the high capital investment, this is the case of the study of Ogumerem et al. (2017), in which the oxygen co-produced form the electrolysis is further processed and sold to generate revenue. In the process of designing a hydrogen SC, forecasts associated with demand, price, and type of energy sources are the main uncertainty sources. But the existent literature studies have mainly focused on demand uncertainty and a few addressed cost uncertainty. To the author's knowledge, none explored uncertainty associated with the availability of PES, which are required for the hydrogen production technology.

Considering the aforementioned, a model to determine strategic and planning decisions taking into account uncertainty on the PES availability is here explored. A MILP is developed, considering a multi-period model and two objective functions: minimization of cost and minimization of global warming potential (GWP) in the SC network. A scenario-based tree approach is defined where scenarios describe the possible PES availability. The ε-constraint method with lexicography optimization is implemented and the results of several non-dominated solutions from the Pareto curve are characterized.

2. Problem definition

The hydrogen SC can include diverse echelons, depending on the specifications and needs of each application. The problem in this paper considers three mains echelons: primary energy sources, production plants, and storage facilities. The links between the different echelons are guaranteed through a multimodal transportation network to satisfy the hydrogen demand. Hydrogen production can use a wide diversity of energy sources, from this, were considered only renewables energies, and based on the hydrogen production technologies selection. The hydrogen produced is stored in storage facilities according to its physical form (gaseous or liquid). In order to satisfy the demand, importation is also allowed. In sum the problem in the study can be described as follows:
Given:

- A set of potential locations to install hydrogen production units as well as storage facilities;
- type of production and storage facilities;
- the maximum and minimum capacities for flows, production, and storage;
- primary energy source availability;
- global warming potential for each transportation mode;
- regional delivery distances;
- operating and capital costs;
- the initial existent number of production plants and storage facilities;
- customer's demand.

Determine:

- the number, location, size, capacity, and technologies of hydrogen production plants and storage facilities;
- the network planning with all flows, rates of hydrogen and primary energy sources consumption, production rates and average inventory of materials.

So as to:

- minimize the cost of the supply chain and the GWP.

GWP, stands for the global warming potential and is taken as an environmental indicator of the overall effect related to emissions of greenhouse gases (CO_2-equiv).

3. Model main characteristics

As mentioned above the model developed is a multi-objective, multi-period, stochastic Mixed Integer Linear Programming (MILP) where two main objectives are considered: minimization of cost and minimization of global warming potential (GWP). A set of constraints are defined considering: network expansion; PES availability, demand requirements, production, storage capacities, and transportation capacities; mass balances and importation needs. The demand is assumed to have a profile over the long-term planning horizon, and it follows an s-shape pattern, according to a hydrogen market penetration factor (Almansoori et al., 2009). Uncertainty in the PES availability is characterized using a dynamic scenario tree approach as detailed in the following section. A ε-constraint method with lexicography optimization is implemented to avoid weak (dominated) solutions.

3.1 The scenario tree approach

To model PES uncertainty a dynamic scenario tree is considered formed by nodes and arcs. Every node represents a state of the random parameter, in our case the PES availability, at a given time point, which evolves over the planning horizon. A scenario comprises a sequence of nodes, and arcs, from the first time point to the last one. The probability of a scenario is given by the product of the probability of each arc between the root and leaf nodes (Lima et al, 2018). Fig. 1 depicts a scenario tree example with 8 scenarios, in which two branches leave, from the root node until the last node.

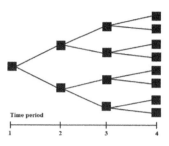

Figure 1 - Scenario tree example.

4. Case study

The design and planning of a hydrogen supply chain in Portugal is taken as case-study. An administrative segmentation of Portugal has been used to obtain a realistic path between districts with the existing main roads and truck lines and to estimate the potential demand from local population statistics (De-León Almaraz e al., 2014).

Four renewable energy sources are considered: wind power, solar power, hydroelectric and biomass. In production terms, two types of production technologies are assumed: electrolyzes and biomass gasification. Hydrogen is produced in liquefied form. Two transportation modes are used, a tanker truck and a railway tanker. Cryogenic storage is assumed.

To estimate the hydrogen demand was assumed the current number of private and light vehicles in Portugal, the average distance travelled, and fuel economy hydrogen (kg H_2/km) to calculate the total equivalent hydrogen demand. Furthermore, the demand equivalent was treated based on the penetration factor in each time period, in this case, 5 years per stage, in a horizon of 15 years of network planning. The hydrogen penetration factor starts at 5% in the first period, growing to 25% in the second period, and it reaches 50% in the third period.

To generate the scenarios considering the PES uncertainty availability an S-shape PES trajectory is considered, then a scenario tree of the form shown in Fig. 2 was obtained. Three time periods and nine distinctive scenarios are considered. The scenarios were structured from current energy data and considering variability over time, reaching a maximum growth of 11% in a scenario in the last time period compared to the initial information. The probabilities of occurrence of each scenario are given in Table 1, assuming a pessimist, a neutral and an optimistic scenario in each time period after the first one. The first time period has a deterministic PES profile. The model was solved through GAMS 25.1.1, using CPLEX 12.0, in a two Intel Xeon X5680, 3.33 GHz computer with 24 GB RAM. The time presented in table 3 is the sum of the execution times of the 10 points of the Pareto border.

Table 1 - Probabilities of occurrence.

Scenarios"w"	Probabilities
w1	6,25 %
w2	12,5 %
w3	6,25 %
w4	12,5 %
w5	25,0 %
w6	12,5 %
w7	6,25 %
w8	12,5 %
w9	6,25 %

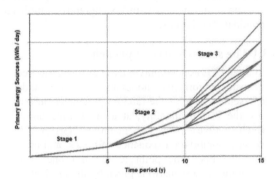

Figure 2 - S-shape scenario tree of PES.

5. Results and Discussion

The results are summarised in the Pareto frontier represented in Figure 3. In order to detail its analysis, three points of the curve are selected as identified in Figure 3. The extreme points 1 and 10 corresponds respectively to the maximum values of GWP and total daily cost. The solution from the middle, point 7, was selected to evaluate the differences with the other selected solutions.

Table 2 depicts the topology, total costs and GWP of the three solutions. Comparing the point 1 and the point 10 solutions it can be seen an increase of 17 times in cost, and a reduction of 0.94, about 6%, in the GWP impact. Moreover, by comparing the lowest cost solution, point 1, and the middle solution, point 7, there is an 8-fold increase in the total daily cost of the network, while the reduction of the GWP is about a ratio 0,96, which indicates a reduction approximately 4%. This shows that the reduction in total GWP is very small when compared to the necessary increment in costs.

Thus, despite the high variability in the supply chain cost, a small impact in the GWP value was obtained, which is justified by the type of PES selected and its respective production technology. In point 10 biomass was selected as the primary energy source, while point 1 presents the selection of two types of production technologies, electrolyzes and biomass gasification, therefore having as energy sources not only biomass but integrating the other ones, such as solar power, wind power and hydroelectric.

These choices are justified by the fact that biomass has the lowest price compared to other energy sources, but biomass gasification is a less environmental friendly production technology when compared to electrolysis.

Related to the network decisions it is possible to observe that the number of production plants increases from 8 to 17 from point 1 to 10. These results are justified by the incremental use of biomass as the energy source that has associated lower technologies capacities. Since the demand variability is unchanged, the number of storage facilities remains the same for all solutions.

In sum the obtained results suggest decentralized SC topologies for lower values of GWP (point 10), with smaller capacities, and where the demand for hydrogen is supplied locally by production plants. On the contrary, when the objective is the minimization of costs a centralized topology is obtained (point 1), where higher capacities are installed.

Multi-objective optimization approach to design and planning hydrogen
supply chain under uncertainty: A Portugal study case

1313

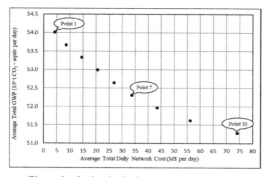

Figure 3 - Optimal solutions in the Pareto curve.

Table 2 – Comparison of multiperiod solutions topologies

Point in Pareto Frontier	1	7	10
Total number of production plants	8	13	17
Total number of storage facilities	54	54	54
Total number of transportation units	1283	200	405
Average Total Daily Network Cost (M$ per day)	4.19	33.67	74.11
Average GWP Production plants (10^3 t CO2 - equiv per day)	18.09	16.44	15.40
Average GWP Storage facilities (10^3 t CO2 - equiv per day)	35.85	35.85	35.85
Average GWP Transportation modes (10^3 t CO2 - equiv per day)	0.075	0.014	0.034
Average Total GWP (10^3 t CO2 - equiv per day)	54.02	52.31	51.28

Considering point 1 it can be analyzed through Figure 4 the topology evolution over time. As demand increases, the number of production plants and storages facilities increases accordingly, in the same regions, to satisfy the increment of local demand. Production plants are strategically centralized disseminated along Portugal territory.

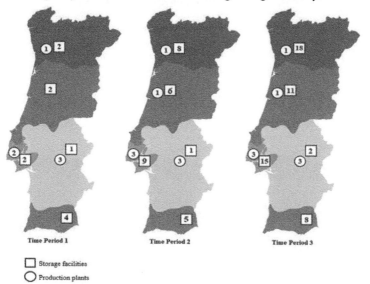

Figure 4 – SC topology evolution over the multi-period solution (point 1)

Table 3 – Computational results for the multi-objective optimization

Continuous variables	Binary variables	Constraints	CPU (s)	GAP (%)
8252	1621	10457	122	0

6. Conclusion

In this work, a MILP model to support the decision maker in the design and planning of hydrogen supply chains is developed. The model considers variability in renewable energy sources and integrates a set of constraints describing the supply chain characteristics. Economic and environmental objectives are considered.

The model is applied to the design and plan of a hydrogen supply chain in Portugal. When a cost objective is at stake a centralized network is obtained when compared to a decentralized supply chain when an environmental objective is considered.

It was also concluded that the reduction in total GWP is very small compared to the necessary increment in costs.

As future work is intended to explore other sources of uncertainty in the model and thus analyze their impact in the SC network. It is also expected to integrate the risk component as an objective function into the model.

References

Almansoori, A., & Shah, N. (2009). Design and operation of a future hydrogen supply chain : Multi-period model. *International Journal of Hydrogen Energy*, *34*(19), 7883–7897. https://doi.org/10.1016/j.ijhydene.2009.07.109

Almansoori, A., & Shah, N. (2011). Design and operation of a stochastic hydrogen supply chain network under demand uncertainty. *International Journal of Hydrogen Energy*, *37*(5), 3965–3977. https://doi.org/10.1016/j.ijhydene.2011.11.091

De-León Almaraz, S., Azzaro-Pantel, C., Montastruc, L., & Domenech, S. (2014). Hydrogen supply chain optimization for deployment scenarios in the Midi-Pyrénées region, France. *International Journal of Hydrogen Energy*, *39*(23), 11831–11845. https://doi.org/10.1016/j.ijhydene.2014.05.165

De-León Almaraz, S., Azzaro-Pantel, C., Montastruc, L., Pibouleau, L., & Senties, O. B. (2013). Assessment of mono and multi-objective optimization to design a hydrogen supply chain. *International Journal of Hydrogen Energy*, *38*(33), 14121–14145. https://doi.org/10.1016/j.ijhydene.2013.07.059

Lima, C., Relvas, S., & Barbosa-póvoa, A. (2018). Stochastic programming approach for the optimal tactical planning of the downstream oil supply chain. *Computers and Chemical Engineering*, *108*, 314–336. https://doi.org/10.1016/j.compchemeng.2017.09.012

Ogumerem, G. S., Kim, C., Kesisoglou, I., Diangelakis, N. A., & Pistikopoulos, E. N. (2017). A Multi-objective Optimization for the Design and Operation of a Hydrogen Network for Transportation Fuel. *Chemical Engineering Research and Design*, 1–14. https://doi.org/10.1016/j.cherd.2017.12.032

Anton A. Kiss, Edwin Zondervan, Richard Lakerveld, Leyla Özkan (Eds.)
Proceedings of the 29th European Symposium on Computer Aided Process Engineering
June 16th to 19th, 2019, Eindhoven, The Netherlands. © 2019 Elsevier B.V. All rights reserved.
http://dx.doi.org/10.1016/B978-0-128-18634-3.50220-4

Morris screening for FMECA of valve failure modes on offshore gas reinjection

Emil Krabbe Nielsen[a,b], Jérôme Frutiger[b] and Gürkan Sin[b,*]

[a]*Department of Electrical Engineering, Technical University of Denmark, Elektrovej 326, Kgs. Lyngby 2800, Denmark*
[b]*Department of Chemical Engineering, Technical University of Denmark, Søltofts Plads 227, Kgs. Lyngby 2800, Denmark*
ekrani@elektro.dtu.dk

Abstract

FMECA is commonly used as a tool for assessing the consequences of different failure modes of a component and the criticality of the consequences. Traditionally failure modes are generated using a one-factor-at-a-time method. In this study, we propose extensions to the FMECA procedure in which failure modes are sampled using statistical sampling techniques, and their effects are evaluated under a wide range of operating conditions. Morris's efficient sampling technique is used for generating failure mode effect analysis scenarios. The scenarios are evaluated in a K-Spice simulator with a model of an offshore gas reinjection system in the Danish North Sea. The impact of the failure modes on the process performance and process safety is evaluated under varying process conditions. The extended methodology enables fast screening of the effects of the failure mode under different realistic process conditions. This provides a more comprehensive and global assessment of the consequences for process safety and reliability in the chemical industry.

Keywords: Failure Mode Effect and Criticality Analysis, Morris screening, Gas reinjection

1. Introduction

For safety-critical systems, tools like Failure Mode, Effects and Criticality Analysis (FMECA), and Hazard and Operability Study (HAZOP) are important in reliability analysis as well as Quantitative Risk Assessment (QRA) studies to ensure an inherently safe design and safe operation of the system. For modern chemical process plants, a model is a requirement in the design phase. This model should be exploited, not just to ensure mass, energy and momentum balances of the system, and to optimize productivity, but also to ensure the safety of the system. In a previous study Enemark-Rasmussen et al. (2012), simulations are performed according to a single defined operational setpoint to evaluate failure modes . However as the economical experimental designs proposed by Morris (1991) are available and computational power increases, such evaluations should be carried out by rigorously investigating the influence of failure modes under a range of potential variation of the process conditions.

In this work we propose a quantitative approach to analyse the effects of failure modes of a control valve under a range of process conditions. Four failure modes of control valves on offshore oil and gas platforms have been identified, which will be simulated under varying process conditions. The method is more time consuming and computationally expensive than conventional approaches, especially as the number of process conditions, components and failure modes increases with system complexity. To avoid an excessively time consuming method, the process conditions, and the failure modes are sampled by using a random and economically efficient sampling method: Morris screening. Morris screening is a discrete sampling method, and allows for evaluation of different and completely unrelated failure modes in one study.

2. Process system

The approach is applied to an offshore gas reinjection system of Mærsk, that has been modelled in K-Spice based on design and operation data in Enemark-Rasmussen et al. (2012). The purpose of the gas reinjection system is to maintain the well pressure for enhanced oil recovery. The feed of gas from upstream is controlled by a control valve. In the reinjection system the gas is first cooled by a heat exchanger, supplied with sea water. The cooling rate is controlled by a control valve, at the outlet of the heat exchanger. Next, the gas enters the scrubber to avoid vapour in the gas being reinjected to the well. Production of liquids from the scrubber is undesired, however a control valve releases the liquid to a flare, if the level becomes too high. A compressor sucks the gas from the scrubber, after which the gas is either reinjected to the wells or recycled. The gas is recycled to avoid the compressor from surging. Surging is undesirable, however so is recycling the gas as it limits the productivity. An anti-surge controller controls the setpoint of four control valves, one for the gas feed, two for the gas recycling and one for the gas reinjection, to ensure a steady flow through the compressor. The system is modelled in K-Spice as shown in Figure 1. K-Spice, a dynamic process simulator for the oil and gas industry, is used for process design and engineering verification. It features parameterised process models for oil and gas processing.

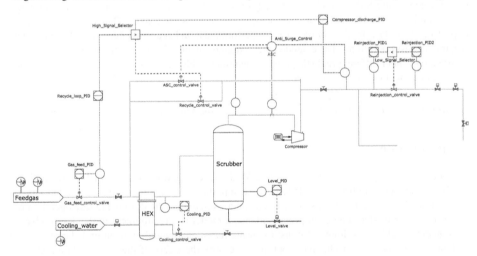

Figure 1: Offshore gas reinjection system.

Various different failure modes have previously been simulated in K-spice for this system by Enemark-Rasmussen et al. (2012). However common for all simulations was the use of only a single set of process conditions according to the design specifications. The process conditions before and after introducing the failure mode was used to calculate the sensitivity of the process to assess the criticality of each failure mode using a one-factor-at-a-time approach. In this study, we use a global sampling based approach to assess the criticality of the consequences, and focus the study on four failure modes of one control valve evaluated under a range of process conditions.

The effort to document the occurrence of failure modes for valves, pumps etc. is extensive in the oil and gas industry, however the level of detail on reported failure modes is very sparse (Management, 2002; Peters and Sharma, 2003). In this study, the valve opening P, and the valve opening and closing time t_{open} and t_{close} are used for implementing four different failure modes for the gas feed control valve. When implementing the failure mode, the manipulated parameter is fixed for the remaining simulation from t_{fm} to t_{end}. Apart from the failure modes, the sampled initial process conditions are the temperature T_{gas} and the pressure P_{gas} of the feed gas, and additionally the feed temperature T_{cool} of the sea water used for cooling.

The values of the sampled initial conditions is determined by a discrete cumulative distribution function (CDF). The variation in percentage for the conditions are as follows: $T_{gas} = 5\%$, $P_{gas} = 4\%$ and $T_{cool} = 4\%$. As described by Sin et al. (2009), it can be used to determine the minimum and maximum values of a uniform distribution discretized into $p = 4$ levels, around a mean or nominal value of operation. The nominal operation is defined here as: $T_{gas} = 44.5\,°C$, $P_{gas} = 121\,bar$ and $T_{cool} = 30\,°C$. Gaussian noise is added to the process conditions as measurement noise and varies during the simulation. The standard deviation σ_{noise} in Table 3 defines the normal distribution of the noise, around the values sampled by Morris screening for θ_{1i}, θ_{2i} and θ_{3i} as mean. The noise is randomly sampled and added by K-Spice as part of the simulation.

3. Procedure

First a set of k process conditions including the failure mode is sampled as the input $\theta_i = [\theta_{1i}, \theta_{2i}...\theta_{ki}]$ for the simulation for $i = 1, 2 ... n$ simulations. The n number of sampled sets are then simulated in K-Spice. All simulations start with a sampled set of initial process conditions θ_{1i}, θ_{2i} and θ_{3i} at time $t_{start} = 0\,s$. These conditions are used throughout the entire simulation. To allow the system to stabilize under the set of process conditions specified by the sampling, the simulation is run for 1200 seconds. At time $t_{fm} = 1200\,s$ the failure mode (FM) θ_4 is introduced. Next, the simulation is run for 7200 seconds until $t_{end} = 8400\,s$ after which the simulation is stopped, and the simulation output for every simulation $Y_{ij} = [y_{1ij}, y_{2ij} ... y_{mij}]$ for $j = [10, 20 ... t_{end}]$ is recorded by sampling every ten seconds. Here, j denotes a point in the time series of the mth output signal from the ith simulation. The steps, inputs and outputs of the procedure is shown in Table 3.

#	Step	Input	Output
1	Morris screening	r=40, k=4, p=4, $\Delta = 2/3$, CDF	θ_i
2	Load initial conditions in simulation i	$\theta_{1i}, \theta_{2i}, \theta_{3i}$	
3	Run simulation i untill $t_{fm} = 1200\,s$	$\sigma_{Noise} = [5, 2.5, 2.5]$	
4	Introduce failure mode	θ_4	
5	Run simulation i untill $t_{end} = 8400\,s$		Y_{ij}
6	Sensitivity analysis	θ_i, Y_{ij}	EE_i

Table 1: Procedure for sampling, simulating and analysing.

4. Experiment and sample design

Morris screening is a global sensitivity analysis method that employs a one-factor-at-a-time (OAT) approach for producing experimental designs. OAT analysis is performed for a number of randomly sampled nominal values in the design space. The analysis is based on the mean and standard deviation of the elementary effects, EE_i for the input i to determine how sensitive the outputs are to changes in the inputs $i = 1, 2, ...$ For these experiments $k = 4$ input parameters are sampled at $p = 4$ levels. Typically, the number of repetitions needed for Morris screening is in the range of $r = [10, 50]$ (Sin et al., 2009), and based on the required number of elementary effects F_i to be calculated. For this study $r = 40$ is used, however only five elementary effects are shown. Based on this, the number of samples are $n = r(k+1) = 40(4+1) = 200$ Sin et al. (2009). The elementary effects are calculated based on changes to the inputs, defined by the step size $\Delta = p/[2(p-1)] = 4/[2(4-1)] = 2/3$ for a uniform distribution of the sampling space. These steps are used for performing local sensitivity analysis, however when averaging the mean of the elementary effects, the method can be used in a global context. The sampled probabilities in unit hyperspace [0 1] have been converted to real values by using their inverse discrete and cumulative probability distribution function. The values of the resulting four levels are shown in table 4.

Level	T_{gas} (θ_1)	P_{gas} (θ_2)	T_{cool} (θ_3)	Failure mode (FM) (θ_4)
1	28.6°C	116.2 bar	17.9°C	Valve seized at current position P
2	39.2°C	119.4 bar	26.0°C	Valve fail open at $P = 1\%$
3	49.8°C	122.6 bar	34.0°C	Valve fail close at $P = 99\%$
4	60.4°C	125.8 bar	42.1°C	Stiction $3t_{close}$ & $3t_{open}$

Table 2: Design space.

5. Results

For each simulation i, the output Y_{ij} is a time series with j samples in time, corresponding to the sampled input θ_i. A selection of the process signals from the simulations are shown in Figure 2. The average of all 200 samples for each time step j is plotted for each output, along with a lower and upper safety limit. The limits are an average of the 200 process signals for each output from time t_{start} to t_{fm}, multiplied by $\pm 25\%$. In general the control system is capable of maintaining the temperature close to nominal operating conditions as can be seen on plot 1-5 in Figure 2. However, under some conditions, the temperature exceeds the upper limit at the gas outlet, and in the recycle loop. The plots 6-9 shows a stable pressure under the majority of process conditions, although it drops below the lower limit to nearly 0 bar under some conditions. This does however not pose a safety risk. Plot 10 shows that the amount of recycled gas varies a lot between 0 and 200 kg/h, and under some specific conditions the recyled amount approaches 300 kg/h. In general the productivity is stable, but it drops close to 0 kg/h under some conditions. The surge rate, defined as $\frac{\int \text{Time surging } dt \ [s]}{\text{Time } [s]}$, varies a lot initially, as expected, and in time splits up into four dinstinct bands, for which two of them increases over time, one remains constant and one decreases over time.

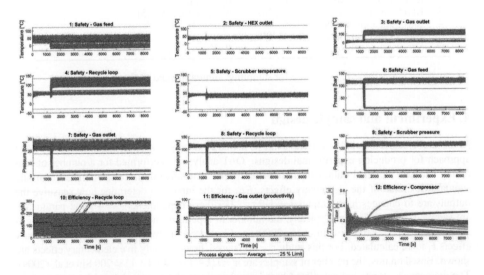

Figure 2: A selection of process signals for output Y_{ij} from the $n = 200$ samples.

The plots in Figure 3 of each output as a function of each input shows that failure mode two is the sole cause of the high temperatures for the gas outlet and recycle loop, independent of the process conditions. The plot also shows, that failure mode two is the sole reason for the pressure drop in the gas reinjection system, as only very little new gas is supplied. On the contrary, failure mode four consistently produces high pressure with almost no variation. The amount of recycled gas drops close to 0 kg/h for failure mode two as only a small amount of gas is supplied to the system. Failure mode three causes the high amount of recycled gas for a gas feed pressure of 125.8 bar. As previously mentioned, the productivity is stable, which is true for all process conditions, except those including failure mode two. Failure mode two results in three bands of surging, resulting in both the lowest and highest amount of time surged in total. The plots do however not show any unique set of conditions producing these bands. The variation in the amount of time surged is very low for both failure mode one and four as opposed to two and three.

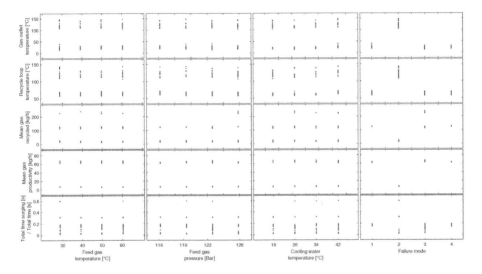

Figure 3: Matrix plot of model output Y_i as a function of model input θ_i.

The most significant elementary effect has the highest average, and a high standard deviation is an indication of interactions or non-linearity (Morris, 1991). The elementary effects for some of the critical outputs are shown in Figure 4. The wedge in the figure can be used as an indication of significance, and is calculated as the mean of samples of the elementary effects: $\hat{d}_i = \pm 2SEM_i$. The standard error of mean, SEM_i is calculated as the variance S_i of the elementary effects of input sample i divided by the square root of repetitions r: $SEM_i = \frac{S_i}{\sqrt{r}}$. If an elementary effect is outside of the wedge it is considered significant, and insignificant if inside (Morris, 1991).

A change in failure mode can be considered as the most influential input parameter on the gas outlet temperature, recycle loop temperature, amount of gas recycled and the gas productivity, but shows no significance on the amount of time surged. This result for the total surge rate contradicts the observations from Figure 3. The values in Figure 4 are averaged elementary effects based on all changes to the respective input parameter. This can lead to ambiguous results and for this reason these results cannot be interpreted alone. The gas feed pressure change should also be considered significant for the increase in gas outlet temperature, and a change in temperature of the cooling feed is significant for an increase in temperature in the recycle loop. The mean of the total surge rate shown in Fig. 4 indicates no significant inputs, as the standard error of the mean SEM_i is quite large due to the high variance S_i of elementary effects. The estimated mean μ^* of the distribution of absolute values of the elementary effects is however useful for Type II errors when the variance

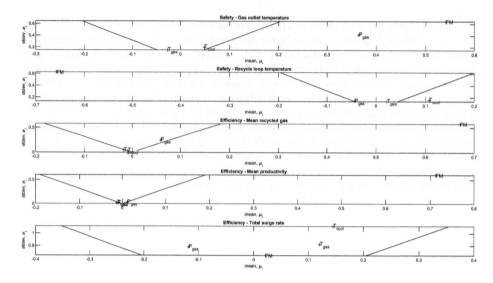

Figure 4: Elementary effects of outputs.

is both negative and positive (Campolongo et al., 2007). It reveals that the failure mode is the most significant parameter for the variation of the surge rate as $\mu^* = 5.70$ for FM, $\mu^* = 5.51$ for T_{cool}, $\mu^* = 4.02$ for T_{gas} and $\mu^* = 3.98$ for P_{gas}.

In general, the standard deviation of the elementary effects of the failure mode is quite large for all outputs except for the surge rate. However as this input parameter is not continuous, but merely four completely different parameters in itself, a high standard deviation is expected and should thus be either interpreted with caution or disregarded.

6. Conclusion

Morris screening has been used for analysing failure modes of a valve under varying process conditions in simulations of an offshore gas reinjection system. The analysis showed that only the temperature at the gas outlet and in the recycle loop could compromise the safety, caused by the gas feed valve closing. In general the results showed that within the studied range of variation the process conditions are insignificant, and that the effect of the failure modes is independent of the process conditions. The efficiency of the plant was also examined, and in general the productivity was stable, except for when the valve closes, and the gas feed is shut off. Additionally, the closed valve impacts the amount of time which the compressor surged both negatively and positively. In case of an open valve, the recycling rates increased, but only under high pressure.

References

F. Campolongo, J. Cariboni, A. Saltelli, 2007. An effective screening design for sensitivity analysis of large models. Environmental modelling & software 22 (10), 1509–1518.

R. Enemark-Rasmussen, D. Cameron, P. B. Angelo, G. Sin, 2012. A simulation based engineering method to support hazop studies. In: Computer aided chemical engineering. Vol. 31. Elsevier, pp. 1271–1275.

S. I. Management, 2002. OREDA: Offshore Reliability Data Handbook, 4th Edition. OREDA Participants.

M. D. Morris, 1991. Factorial sampling plans for preliminary computational experiments. Technometrics 33 (2), 161–174.

J. Peters, R. Sharma, 2003. Assessment of valve failures in the offshore oil & gas sector. Health and Safety Executive.

G. Sin, K. V. Gernaey, A. E. Lantz, 2009. Good modeling practice for pat applications: Propagation of input uncertainty and sensitivity analysis. Biotechnology progress 25 (4), 1043–1053.

Anton A. Kiss, Edwin Zondervan, Richard Lakerveld, Leyla Özkan (Eds.)
Proceedings of the 29th European Symposium on Computer Aided Process Engineering
June 16th to 19th, 2019, Eindhoven, The Netherlands. © 2019 Elsevier B.V. All rights reserved.
http://dx.doi.org/10.1016/B978-0-128-18634-3.50221-6

Optimal Maintenance Scheduling for Washing of Compressors to Increase Operational Efficiency

Frederik Schulze Spüntrup[a], Giancarlo Dalle Ave[b,c], Lars Imsland[a,*] and Iiro Harjunkoski[b]

[a]*Department of Engineering Cybernetics, Norwegian University of Science and Technology, 7491 Trondheim, Norway*
[b]*ABB Corporate Research Germany, 68526 Ladenburg, Germany*
[c]*Technical University Dortmund, 7491 Dortmund, Germany*
lars.imsland@ntnu.no

Abstract

The need for simultaneously decreasing carbon emissions and increasing operational profit motivates gas turbine and gas compressor operators to understand, minimize and control performance deterioration. Fouling is one of the most prevalent deterioration problems. This work addresses the causes and effects of fouling and investigates an approach to give decision-support on the questions if, how often and when compressor washing should be conducted. Integration with other maintenance actions is also considered. In this work, a discrete time-scheduling approach that follows the Resource Task Network framework is developed and formulated as a Mixed Integer Linear Program. A novel enumerator formulation makes this method simpler and easier to extend for different maintenance types than existing methods. Degradation is included in a linearized way for a case study from the Oil and Gas industry. Results indicate that washing scheduling is beneficial for the profit. This study gives the foundation for decision-support regarding additional investments in existing production systems, amortization, and supplies optimal maintenance schedules for various applied maintenance types.

Keywords: Maintenance scheduling, asset degradation, energy efficiency

1. Introduction and Background

Degradation of assets is an omnipresent phenomenon in the process industries where the lifetime of a plant easily reaches 30-50 years. Deteriorioation of assets is inevitable. Over the lifecycle of an equipment its efficiency will decrease and it will degrade up to a level where it cannot be operated any longer or the operating company decides to take countermeasures. According to Diakunchak (1992) compressor fouling constitutes 70-85% of the performance loss caused by deterioration in gas turbines. With the latest IPCC report on achieving the maximum 1.5°C average temperature increase goal it is important to avoid any unnecessary waste of energy.

While the literature identifies performance degradation and declares washing as an appropriate countermeasure (Stalder, 2001), there is not much work done on the decision when to perform the washing procedure. The contribution of this work is to include the efficiency degradation of assets into the decision on when to perform maintenance tasks.

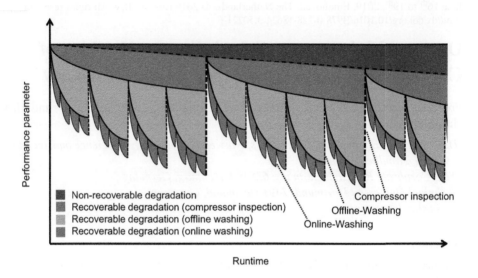

Figure 1: Example of the different types of degradation in a compressor and different types of maintenance to restore efficiency.

Kurz and Brun (2001) describe the degradation of gas turbine systems. Different mechanisms affect the degradation process: Increased tip clearance (Khalid et al., 1999) and changes in the airfoil geometry (Singh et al., 1996) are interlinked with the effect of non-recoverable degradation. Recoverable degradation, which can be (partially) reversed by specific methods (e.g. washing) is described as changes in the airfoil surface quality (Elrod et al., 1990). Kurz and Brun (2012) reported significant deterioration of different gas turbines, ranging between 2 and 12 % after one year of operation.

Hovland and Antoine (2006) optimise the power generation by applying parameter estimation via Kalman filtering combined with a hybrid dynamic model. Model predictive control is used with this hybrid dynamic model and compared to fixed washing schedules.

Xenos et al. (2016) investigate an air separation plant. Optimal operation and the maintenance of compressor networks are part of the developed MILP model. Online and offline washing are considered as part of the condition-based maintenance approach. The degradation is considered in three different cases: low, medium and high degradation. They are based on a study of industrial data. Offline-washing is not considered to be done during non-performance-related maintenance actions or regular shutdowns of the equipment. The degradation in the case study causes an increase in electrical power consumption of 2 % per month, which leads to high saving potential.

This paper addresses optimal scheduling of maintenance actions to increase operational efficiency. Section 2 presents the problem statement of this work. Even though this problem is common in many kinds of process industries, we will focus on a case study of an offshore compressor fleet of an oil and gas company, which is introduced in Section 3. A short overview of the mathematical formulation of the scheduling approach is given in Section 4. The results of the maintenance scheduling are illustrated and discussed within the different subcases in Section 5. The conclusion gives an overview of the findings of this work and shows possible directions for future work.

2. Problem Statement and Case Study

The problem considered in this paper investigates how often online and offline washing in the context of other maintenance types should be conducted to optimally countermeasure deterioration effects and efficiency loss in compressors due to fouling. Furthermore, the actual benefit of washing can not always be determined beforehand. This information is important for the decision-making process; if a washing system should be installed on existing assets or not.

The case study that is used is inspired by a real fleet of compressors. In order to assess the performance of the formulation, the case study is analyzed with varying grid coarseness and time horizon length. For the length of the scheduling horizon a duration of 1-3 months was chosen, as this resembles the need for ahead-scheduling of maintenance workers and the ordering of specific resources such as spare parts or other supplies in most companies.

Table 1: Notation of the scheduling formulation

Index/Set	Description	Continuous Variable	Description
i	Task	$C_{r_c,t}$	Enumerator
j	Unit	$G_{r_c,t}$	Goodness of unit
t	Time Interval	K	Profit/Cost term
R	Resource	$\mu_{i,t}$	Discrete interaction
R_c	Resource (asset)	**Binary Variable**	**Description**
T_S	Time (Scheduling)	$N_{i,t}$	Maintenance task i starts at t
I	Maintenance tasks	**Parameters**	**Description**
		D_{r_c}	Degradation factor
Superscript	**Description**	**Superscript**	**Description**
NonRecov	Non-recoverable degradation	*ROnW*	Recoverable degradation (Online Washing)
RInsp	Recoverable degradation (Inspection)	*ROffW*	Recoverable degradation (Offline Washing)
Min	Minimum	*Max*	Maximum

3. Mathematical Modeling

The scheduling model is designed as a discrete-time model. While the exact starting times of maintenance actions depend on other external factors such as weather conditions or the changeover between two shifts, it is not important to work with a continuous-time model. A generic approach based on the Resource-Task Network (RTN) framework (Pantelides, 1994) is applied. The notation of the model is shown in Table 1.

The central element of the scheduling approach is the inherent goodness of every asset. This goodness is affected by degradation and therefore the efficiency decreases over time.

As depicted in Figure 1 the efficiency of the compressor is decreasing over time. There are different terms for the various types of degradation and there are 5 distinctive maintenance modes that have particular effects. Different counters for each type of degradation are used to calculate the time passed since the last maintenance has taken place. Every conducted maintenance resets one or multiple counters, which are used to model the degradation in a linear fashion. The goodness degradation is defined as follows:

$$
\begin{aligned}
G_{r_c,t} = G_{r_c}^{initial} &- D_{r_c}^{NonRecov}(C_{r_c,t}^{GN} - 1) - D_{r_c}^{RInsp}(C_{r_c,t}^{RInsp} - 1) \\
&- D_{r_c}^{ROffW}(C_{r_c,t}^{ROfW} - 1) - D_{r_c}^{ROnW}(C_{r_c,t}^{ROnW} - 1), \forall r_c \in R_c, t \in T_S
\end{aligned}
\tag{1}
$$

The enumerators are each formulated in a comparable way. As an example, the enumerator $C_{r_c,t}^{ROnW}$ for counting the time intervals since the last online washing is introduced:

$$
C_{r_c,t}^{ROnW} \geq C_{r_c,t-1}^{ROnW} + 1 - M \sum_{i_{r_c}}^{I_{OnW}} N_{i,t}, \forall r_c \in R_c, t \in T_S
\tag{2}
$$

$$
C_{r_c,t}^{ROnW} \leq C_{r_c,t-1}^{ROnW} + 1, \forall r_c \in R_c, t \in T_S
\tag{3}
$$

$$
C_{r_c,t}^{ROnW} \leq 1 + M(1 - \sum_{i_{r_c}}^{I_{OnW}} N_{i,t}), \forall r_c \in R_c, t \in T_S
\tag{4}
$$

$$
C_{r_c,t}^{ROnW} \geq 1, \forall r_c \in R_c, t \in T_S
\tag{5}
$$

The enumerator for the Remaining Useful Life of the assets works in a similar fashion. However, it is reset by a short-term maintenance I_{SM} that does not restore the goodness but fixes operational problems with the asset. The formulation is not shown in this paper. In a real-world application a continuous input from a Condition Monitoring System would be required.

The resource balance describes that every active maintenance task $N_{i,t}$ is consuming a specific amount of resources μ_i, t from the initial amount of resources R_0, including the compressors and the maintenance personnel. When a task is concluded, these resources are released again for the following time interval.

$$
R_{r,t} = R_{initial} - \sum_i^I N_{i,t} \mu_{i,t}, \forall r \in R, t \in T_P
\tag{6}
$$

The target of the optimization is to maximize the profit, calculated by the production profit from each compressor, multiplied by the goodness of the compressor in each time interval minus the cost for maintenance.

$$
\max z = K_{profit} \sum_{plat}^{r_c} G_{r_c,t} - K_i \sum_i \sum_t^{T_S} N_{i,t}
\tag{7}
$$

Table 2: Results of the washing scheduling for a fleet size of 10 compressors. (Legend: H - Time Horizon, G - Grid Size, Off - Offline Washing, On - Online Washing)

Case	H: 1 month, G: 4 hours			H: 2 month, G: 4 hours		
	On	Off	On+Off	On	Off	On+Off
Improvement[1]	0.97%	0.13%	1.07%	2.06%	1.50%	3.30%
MIP-Gap	0.72%	0.46%	0.61%	7.03%	4.45%	2.24%
CPU-s[2]	1800	1800	1800	1800	1800	1800

Case	H: 1 month, G: 8 hours			H: 2 months, G: 8 hours		
	On	Off	On+Off	On	Off	On+Off
Improvement[1]	1.29%	0.38%	1.28%	1.88%	0.84%	2.71%
MIP-Gap	0.26%	0.01%	0.28%	3.42%	2.91%	1.19%
CPU-s[2]	1800	1411	1800	1800	1800	1800

[1] Improvement is calculated as the ratio between the case with solely regular maintenance types, but without any washing.

[2] The computation was terminated after a maximum of 1800 s or when the MIP-Gap was below a threshold of 0.01%.

4. Results and Discussion

Four different cases with fixed fleet size, but varying time horizon and grid size were solved using GAMS 24.8.4/CPLEX 12.7.1.0. The calculations are carried out as follows: Each case has the information that two different maintenance actions (long overhaul and short maintenance, e.g. change of a broken seal) must be conducted within the time horizon. This information could come in a real-world implementation e.g. from a higher level planning system. Furthermore, a specific combination of washing actions (online, offline or combined) can be conducted. The optimization problem tries to optimize production profit by recovering degradation of the compressors at the right time.

The improvement is calculated by comparing the three different scenarios for each case against the base-scenario in which no washing is performed, just the basic maintenance such as long overhauls and short-term maintenance. The quotient of the objective function for a specific scenario and the base scenario indicate the improvement.

The results of four different cases with three scenarios each are shown in Table 2. While the specific values of the objective function represent the entire production profit within the time horizon (see Eq. 7), the actual values of the production profit are not insightful. As the size of the optimization problem is larger the longer the time horizon is and the finer the grid size is, the scheduling algorithm is terminated prematurely in all cases at 1800 seconds. Therefore, the result shows the minimum improvement, as further calculation might improve the objective function. More exhaustive calculations may be conducted in a follow-up study.

The results show that the algorithm is able to improve the operating profit. The improvement ranges between 0.4% and 3.3%. Several trends can be observed: The longer the time horizon is, the more improvement can be observed. As the overall degradation (non-recoverable plus all recoverable degradation) is set to 5% per year, this makes sense, as smaller decreases in the efficiency do not necessarily justify conduction of washing, while on a longer time horizon washing is more economically viable. With fixed computational time, the gap between the best bound of the problem and the integer solution is dependent on the time horizon and the grid size and is thus scaling with the problem size.

5. Conclusion

A novel maintenance scheduling approach was presented in this work. The model differs from other work, as it combines the information about the asset degradation via an enumerator-approach into the decision-making for several types of maintenance actions, especially washing actions.

Degradation is considered to be a combination of recoverable degradation, non-recoverable degradation and the decreasing remaining useful lifetime of an asset. Results show that the scheduling of the proposed compressor washing in addition to other maintenance types is able to improve the production profit by up to 3.3% compared to operations without the compressor washing. Next to the fact that the scheduling model can give help to optimize the operation of process networks, it can be used as a decision-support tool. By comparing the possible saving introduced through new maintenance types, e.g. washing, a maximum capital investment for the amortization within a specific timeframe can be calculated.

Future work will investigate a multi-time-scale model that integrates both long-term planning and short-term scheduling of maintenance.

6. Acknowledgment

Financial support is gratefully acknowledged from the Marie Skłodowska Curie Horizon 2020 EID-ITN project "PROcess NeTwork Optimization (PRONTO)", Grant agreement No 675215.

References

Diakunchak, I. S., 1992. Performance Deterioration in Industrial Gas Turbines. Journal of Engineering for Gas Turbines and Power 114 (2), 161.

Elrod, W. C., King, P. I., Poniatowski, E. M., 1990. Effects of Surface Roughness, Freestream Turbulence, and Incidence Angle on the Performance of a 2-D Compressor Cascade. Volume 1: Turbomachinery, V001T01A061.

Hovland, G., Antoine, M., 2006. Scheduling of gas turbine compressor washing. Intelligent Automation and Soft Computing 12 (1), 63–73.

Khalid, S. A., Khalsa, A. S., Waitz, I. A., Tan, C. S., Greitzer, E. M., Cumpsty, N. A., Adamczyk, J. J., Marble, F. E., 1999. Endwall Blockage in Axial Compressors. Journal of Turbomachinery 121 (3), 499–509.

Kurz, R., Brun, K., 2001. Degradation in Gas Turbine Systems. Journal of Engineering for Gas Turbines and Power 123 (1), 70.

Kurz, R., Brun, K., 2012. Fouling Mechanisms in Axial Compressors. Journal of Engineering for Gas Turbines and Power 134 (3), 032401–032401.

Pantelides, C. C., 1994. Unified framework for optimal process planning and scheduling. In: Proc. Second Conf. on Foundations of Computer Aided Operations. pp. 253–274.

Singh, D., Tabakoff, W., Mechanics, E., 1996. Simulation of Performance Deterioration in Eroded Compressors. ASME. Turbo Expo: Power for Land, Sea, and Air Volume 1:, 1–8.

Stalder, J.-P., 2001. Gas Turbine Compressor Washing State of the Art: Field Experiences. Journal of Engineering for Gas Turbines and Power 123 (2), 363.

Xenos, D. P., Kopanos, G. M., Cicciotti, M., Thornhill, N. F., 2016. Operational optimization of networks of compressors considering condition-based maintenance. Computers and Chemical Engineering 84, 117–131.

Anton A. Kiss, Edwin Zondervan, Richard Lakerveld, Leyla Özkan (Eds.)
Proceedings of the 29th European Symposium on Computer Aided Process Engineering
June 16th to 19th, 2019, Eindhoven, The Netherlands. © 2019 Elsevier B.V. All rights reserved.
http://dx.doi.org/10.1016/B978-0-128-18634-3.50222-8

A Quality-by-Control Approach in Pharmaceutical Continuous Manufacturing of Oral Solid Dosage via Direct Compaction

Qinglin Su,[a] Sudarshan Ganesh,[a] Dan Bao Le Vo,[a] Anushaa Nukala,[a] Yasasvi Bommireddy,[b] Marcial Gonzalez,[b] Gintaras V. Reklaitis,[a] Zoltan K. Nagy[a],*

[a]*Davidson School of Chemical Engineering, Purdue University, West Lafayette, IN 47907, United States*

[b]*School of Mechanical Engineering, Purdue University, West Lafayette, IN 47907, United States*

znagy@purdue.edu

Abstract

The pharmaceutical industry has been undergoing a paradigm shift towards continuous manufacturing, under which novel approaches to real-time product quality assurance have been investigated. A new perspective, entitled Quality-by-Control (QbC), has recently been proposed as an important extension and complementary approach to enable comprehensive Quality-by-Design (QbD) implementation. In this study, a QbC approach was demonstrated for a commercial scale tablet press in a continuous direct compaction process. First, the necessary understanding of the compressibility of a model formulation was obtained under QbD guidance using a pilot scale tablet press, Natoli BLP-16. Second, a data reconciliation strategy was used to reconcile the tablet weight measurement based on this understanding on a commercial scale tablet press, Natoli NP-400. Parameter estimation to monitor and update the material property variance was also considered. Third, a hierarchical three-level control strategy, which addressed the fast process dynamics of the commercial scale tablet press was designed. The strategy consisted of the Level 0 built-in machine control, Level 1 decoupled Proportional Integral Derivative (PID) control loops for tablet weight, pre-compression force, main compression force, and production rate control, and Level 2 data reconciliation of sensor measurements. The effective and reliable performance, which could be demonstrated on the rotary tablet press, confirmed that a QbC approach, based on product and process knowledge and advanced model-based techniques, can ensure robustness and efficiency in pharmaceutical continuous manufacturing.

Keywords: Quality-by-Design, Quality-by-Control, Continuous manufacturing, Process control, pharmaceutical

1. Introduction

The recent approval of four drug products using continuous manufacturing technologies, e.g., Orkambi (lumacaftor/ivacaftor) in 2015, Prezista (darunavir) in 2016, Verzenio (abemaciclib) and Symdeko (tezacaftor/ivacaftor) in 2018, by United States Food and Drug Administration (US FDA) is a strong evidence of the on-going paradigm shift from batch to continuous manufacturing in the pharmaceutical industry. The US FDA Quality-by-Design guidance has also been widely acknowledged in providing directions with respect to product and process knowledge development, such as identifying critical process parameters (CPPs) in process design and linking critical material attributes (CMAs) to critical quality attributes (CQAs) (Yu et al., 2014). Control strategies that include specification of the drug substance, excipients, and drug products as well as controls for each step of the manufacturing process are also considered as important elements of the QbD concept. In addition to designing quality into the product during the early stages of drug development, the quality attributes must also be automatically and consistently controlled, in the presence of process uncertainties and disturbances, during drug manufacturing (Lee et al., 2015). This recognition has led to a call for Quality-by-Control, particularly in continuous manufacturing, consistent with the current Industry 4.0 or smart manufacturing approaches. The proposed QbC idea consists of the design and operation of a robust manufacturing system that is achieved through an active process control system designed in accordance with hierarchical process automation principles, based on a high-degree of quantitative and predictive understanding of product and process.

Our previous work has investigated the characterization of the compressibility of a model formulation, consisting of Acetaminophen (API, 10.0%), Avicel Microcrystalline Cellulose PH-200 (excipient, 89.8%), and SiO_2 (lubricant, 0.2%), using a Natoli BLP-16 tablet press (Su et al., 2018). System dynamics and hierarchical process control development for a direct compression line were also undertaken for this pilot scale tablet press. In this study, we demonstrate the use of the proposed QbC approach in transferring the product and process understanding generated with the Natoli BLP-16 to the commercial scale, Natoli NP-400, tablet press. The rest of the manuscript will first discuss the QbC concept, followed by a brief introduction of the features of the tablet press in a continuous direct compression process. The advantages of a QbC approach in continuous tablet manufacturing will be presented in the result and discussion section. Concluding remarks and considerations for future work are given at the end of the manuscript.

2. Quality-by-Control

In traditional batch manufacturing quality attributes of products are tested at the end of each batch manufacturing step, following the so-called Quality-by-Testing (QbT) approach, as shown in Figure 1. Under the QbD guidance, systematic understanding of drug quality, including identification of CMAs and CPPs, is developed and monitoring is employed to assure that the quality attributes are met. The QbD approach is also very important to advancing the adoption of continuous manufacturing. However, assurance of quality in the continuous manufacturing mode in addition requires the use of QbC concepts to actively drive reduction in the variance of quality attributes, in the presence of process disturbances, raw material property variations, or uncertainties introduced as a result of scale up or technology transfer. The QbC approach builds on QbD by employing quantitative and predictive product and process knowledge in the form of models of appropriate fidelity, together with process analytical technology (PAT), to actively and robustly control the CQAs at the specified levels by adjusting CPPs, thus, achieving real-time quality assurance. The on-going trend towards Industry 4.0 or smart manufacturing also demands a high-level automation in process operation and quality control, which is fully consistent with the QbC approach.

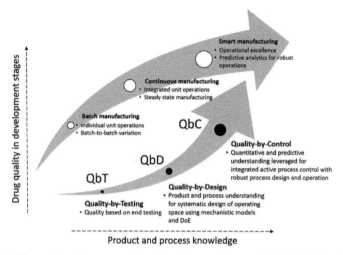

Figure 1. The systematic progression in quality assurance via QbT, QbD, and QbC.

3. Continuous tablet manufacturing

3.1. Continuous direct compaction

The continuous direct compression process under study consists of two Schenck AccuRate PureFeed® AP-300 loss-in-weight feeders which continuously feed the API and excipient ingredients into a Gericke GCM-250 continuous blender. A Schenck AccuRate DP4 micro feeder adds the lubricant into the powder blend exiting the continuous blender, which is then conveyed directly to a rotary tablet press (Su et al., 2017). The tablet press is a multi-stage process, in which each station undergoes the recurring major steps of die filling, metering, pre-compression, main-compression, tablet ejection and take-off, as shown in Figure 2. The tablet weight can be controlled by changing the dosing position subject to the variation of powder bulk density or filling time due to change in turret speed and/or feeder rotation speed. A Natoli BLP-16 tablet press (16 stations with flat-head punches) was employed in our previous work to characterize the formulation compressbility while a commerical scale Natoli NP-400 (22 stations with concave-head punches) was used in the present work for continuous tablet manufacturing. The two tablet presses are also different in size and design of hopper and feed frame, which may result in differences in powder bulk density at the die table.

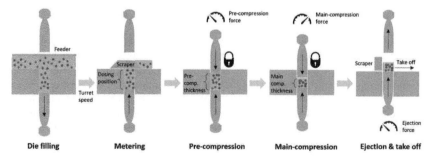

Figure 2. Major steps in a Natoli rotary tablet press.

3.2. Formulation compressibility

The classical Kawakita model was employed to characterize the formulation compressibility in the Natoli BLP-16 tablet press (Su et al., 2018), as shown in Figure 3,

$$\frac{CF}{1 - \rho_c/\rho_r} = \frac{CF}{a} + \frac{\pi D^2/4}{ab} \tag{1}$$

$$\rho_r = \frac{4W_t}{\pi D^2 \rho_t T} \tag{2}$$

where CF is the main compression force, kN; ρ_c is the critical relative density, -, ρ_r the calculated in-die relative density from tablet weight W_t, and ρ_t the known a prior true density of the powder, g/cm³; parameters a and b (MPa) are interpreted as the maximum degree of compression and the reciprocal of the pressure applied to attain the maximum degree of compression, respectively; D is the diameter of the die (mm) and T (mm) is the minimum in-die tablet thickness pre-set by the main compression thickness for B tooling punches with flat cylindrical punch surfaces.

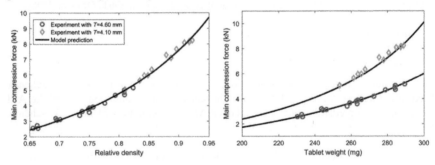

Figure 3. Compressibility characterization by a classical Kawakita model.

3.3. Hierarchical control system

A three-level hierarchical control system, shown in Figure 4, was developed based on the product and process understanding developed using the Natoli BLP-16 tablet press (Su et al., 2018) and was transferred to the NP-400 tablet press. The CQA and CPP measurement data were collected using an Emerson DeltaV DCS system to support the process control system design following the ISA 95 standard (Su et al., 2017). Specifically, the vendor built-in machine control at Level 0 manipulates the dosing position, pre-compression thickness, main compression thickness, and turret speed. At Level 1, the DCS system employs four PID controllers, controlling the tablet weight, pre-compression force, main compression force, and production rate by manipulating the above four Level 0 variables, respectively. A Level 2 data reconciliation module was implemented which serves to reconcile the tablet weight measured by an in-house adapted load cell with the main compression force measurement using the constraints imposed by the Kawakita model, Eqs. (1) and (2) (see Su et al., 2018). Specifically, the model parameter ρ_c was continuously re-estimated and updated during data reconciliation to account for variation in the powder bulk density due to material property changes (particle size, water content) or differences in equipment scale (hopper and feed frame, etc.) that also results in changes in powder bulk density at die table.

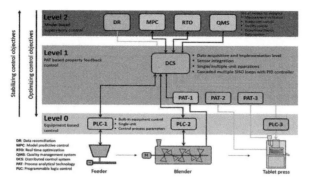

Figure 4. A hierarchical three-level process control for direct compaction.

4. Results and discussion

With a QbC approach based on a quantitative model of compressibility, the Level 2 data reconciliation approach was able to reduce the uncertainty in real-time tablet weight measurement from the load cell for both tablet presses (see the reconciled measurement matched the at-line measurement of tablet weight), as shown in Figure 5. More importantly, the model parameter of critical density (which is related to powder bulk density) in the Kawakita equation (1), which was first estimated from BLP-16 runs, was readily updated from NP-400 tablet press real-time data. The shift in the critical density which was observed in the transfer from BLP-16 to NP-400 may be a result of changes in the bulk density at the die table due to different scale of hopper and different feed frame design. Assurance of model validity and parameter updating are the common concerns in most technology transfers and these results demonstrate that these concerns can be managed within a QbC approach. The high-level understanding of the process dynamics of the tablet press and resulting control structure design, such as the pairing of the control input and output variables, were readily transferable from the pilot to a larger scale press. Specifically, the control structure in which tablet weight was controlled by manipulating the dosing position and the production rate was controlled by adjusting turret speed (the feed frame stirrer speed also changes accordingly), resulted in effective control performance of both tablet presses. Both tablet presses could reach their tablet weight set points steadily. For instance, as shown in Figure 5, when the production rate was increased at 600 s from 3 kg/hr to 5 kg/hr after start-up of the NP-400 tablet press or reduced at 1000 s from 5 kg/hr to 4 kg/hr, the tablet weight was maintained at nearly constant levels.

5. Conclusions and future work

A Quality-by-Control approach was implemented in a continuous tablet manufacturing process based on the product and process understanding gained through the previous Quality-by-Design studies on a pilot unit. Compared to rigid process operation within a predefined design space, active process control response to common process variations, disturbances, or uncertainties can be automatically achieved under the QbC paradigm in a quantitative and predictive way to maintain consistent product quality. The systematic implementation of a hierarchical process control system in continuous direct compression process was highlighted, leveraging QbD understanding of the product and process to achieve robust and efficient process operations and real-time quality control of oral solid dosages. Future efforts in systematic sensor network maintenance, control performance monitoring and continuous improvement should be pursued to further advance QbC implementation.

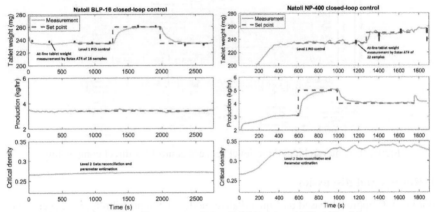

Figure 5. Control performance of Natoli BLP-16 and NP-400 tablet press.

Acknowledgement

Funding for this project was made possible, in part, by the Food and Drug Administration through grant U01FD005535. Views expressed by authors do not necessarily reflect the official policies of the Department of Health and Human Services; nor does any mention of trade names, commercial practices, or organization imply endorsement by the United States Government. This work was also supported in part by the National Science Foundation under grant EEC-0540855 through the Engineering Research Center for Structure Organic Particulate Systems. Purdue Process Safety and Assurance Centre (P2SAC) and technical support from Douglas Voss of Natoli are also appreciated.

References

L. X. Yu, G. Amidon, M. A. Khan, S. W. Hoag, J. Polli, G. K. Raju, J. Woodcock, 2014, Understanding pharmaceutical quality by design. The AAPS Journal, 16(4), 771-783.

S. L. Lee, T. F. O'Connor, X. Yang, C. N. Cruz, S. Chatterjee, R. D. Madurawe, J. Woodcock, 2015, Modernizing pharmaceutical manufacturing: from batch to continuous production. Journal of Pharmaceutical Innovation, 10(3), 191-199.

Q. Su, S. Ganesh, M. Moreno, Y. Bommireddy, M. Gonzalez, G. V. Reklaitis, Z. K. Nagy, 2018, A perspective on Quality-by-Control (QbC) in pharmaceutical manufacturing. Computer and Chemical Engineering. Under review.

Q. Su, Y. Bommireddy, M. Gonzalez, G. V. Reklaitis, Z. K. Nagy, 2018, Variation and risk analysis in tablet press control for continuous manufacturing of solid dosage via direct compaction, Computer Aided Chemical Engineering, 44, 679-684.

Q. Su, M. Moreno, A. Giridhar, G. V. Reklaitis, Z. K. Nagy, 2017, A systematic framework for process control design and risk analysis in continuous pharmaceutical solid-dosage manufacturing, Journal of Pharmaceutical Innovation, 12, 327-346.

Q. Su, Y., Bommireddy, Y. Shah, S. Ganesh, M. Moreno, J. Liu, M. Gonzalez, N. Yazdanpanah, T. O'Connor, G. V. Reklaitis, Z. K. Nagy, 2018, Data reconciliation in the Quality-by-Design (QbD) implemention of pharmaceutical continuous tablet manufacturing. Submitting.

Anton A. Kiss, Edwin Zondervan, Richard Lakerveld, Leyla Özkan (Eds.)
Proceedings of the 29th European Symposium on Computer Aided Process Engineering
June 16th to 19th, 2019, Eindhoven, The Netherlands. © 2019 Elsevier B.V. All rights reserved.
http://dx.doi.org/10.1016/B978-0-128-18634-3.50223-X

Dynamic transitions in a reactive distillation column for the production of silicon precursors

Salvador Tututi-Avila[a*], Nancy Medina-Herrera[b], Luis Ricardez-Sandoval[c] and Arturo Jiménez-Gutiérrez[d]

[a] *Facultad de Ciencias Químicas, Universidad Autónoma de Nuevo León, Av. Universidad S/N Ciudad Universitaria, 66455, San Nicolás de los Garza, N.L., México.*

[b] *Facultad de Agronomía, Universidad Autónoma de Nuevo León, Francisco Villa S/N, ExHacienda el Canadá, General Escobedo, N.L., México.*

[c] *Department of Chemical Engineering, University of Waterloo, 2NL 3G1, Waterloo, Ontario, Canada.*

[d] *Departamento de Ingeniería Química, Instituto Tecnológico de Celaya, Av. Tecnológico y García Cubas S/N, 38010, Celaya, Gto., México.*

salvador.tututivl@uanl.edu.mx

Abstract

The production of energy from solar cells is being considered as one key technology of the future. The remarkable growth in this market is primarily based upon solar cells made from polycrystalline silicon, which is typically produced via chemical vapor deposition of dichlorosilane or silane. Different technologies including reactive distillation have been considered for the production of polycrystalline silicon precursors. In the reactive distillation process, the trichlorosilane disproportionation reaction takes place producing monochlorosilane, dichlorosilane and silane. Steady state studies have reported that different silane products can be obtained by adjusting the operating condition of the reactive column. However, dynamic transitions from one product to another that can meet product quality specifications during operation still remains as a key challenge. This work investigates a feasible strategy for the dynamic transition and production of three silanes in a reactive distillation column. Results from this study show that there is a feasible dynamic transition from silane to monochlorosilane, then to dichlorosilane, and then back to monochlorosilane and silane. Temperature control and equipment sizing were identified as key design and operating variables to accomplish those transitions.

Keywords: reactive distillation; silane; solar cell; optimal transitions; dynamic transition; process control

1. Introduction

The need for additional renewable energy resources has led to the development of new green technologies that can aid to meet global energy demands in the coming decades. Solar energy technology has received special attention from industry and the scientific community mainly because of its positive impact on the environment. In this context, solar cells have emerged as one of the preferred technologies to produce clean energy; thus, the global market for this technology has been growing every year. Semiconductor industries use ultra-high purity polysilicon as the main raw material to produce solar cells. Polysilicon is synthetized using volatile silicon hydride. The most common silicon hydride is silane; however, dichlorosilane or monochlorosilane can also be employed (Ceccaroli and Lohne, 2010). The first process used to produce polysilicon was the Siemens process, which is based on thermal decomposition of trichlorosilane; however, this process has been recently replaced by others because of the higher energy requirements of the Siemens process. Two industrial-scale alternative technologies used nowadays are the Union Carbide process and the Ethyl corporation process. In the latter, silane is used as the main precursor for polysilicon production (Ceccaroli and Lohne, 2010).

The trichlorosilane disproportionation process conveys three reversible reactions, which involve five components including dichlorosilane and monochlorosilane. The latter compounds are of particular interest because of their wide application in the semiconductor industry as well as the thin-film solar photovoltaic industry (Mochalov, 2016). The intensified reactive distillation process has been considered as an attractive process to produce silane, dichlorosilane and monochlorosilane with low energy consumption (Medina-Herrera et al., 2017). The pressure, distillate to feed ratio and reflux ratio in the reactive column have been reported as key variables that directly determine the desired silane product specifications. Optimal steady-state operation for the production of the three different silane products has also been reported (Medina-Herrera et al., 2017). Similarly, controllability studies showing that temperature control is a key feature for this process have also been reported (Medina-Herrera et al., 2017). On the other hand, studies on the dynamic transitions in the reactive column that can maintain dynamic feasibility and that comply with the product specifications are very limited (Ramírez-Márquez et al., 2016). Previous studies have shown that dynamic transitions impact process performance and the logistics of multi-product systems (Koller and Ricardez-Sandoval, 2017; Patil et al. 2015). The aim of this work is to present a systematic method that specifies the way in which the reactive distillation column needs to be operated in closed-loop to achieve dynamically feasible transitions for three trichlorosilane products, i.e. silane, monochlorosilane and dichlorosilane.

2. Steady-State: Reactive Distillation Column

The trichlorosilane decomposition process consists of three reversible reactions, i.e.

$$2SiHCl_3 \xleftrightarrow{cat} SiCl_4 + SiH_2Cl_2$$
$$2SiH_2Cl_2 \xleftrightarrow{cat} SiHCl_3 + SiH_3Cl \tag{1}$$
$$2SiH_3Cl \xleftrightarrow{cat} SiH_2Cl_2 + SiH_4$$

This reaction mechanism, along with the kinetic parameters reported by Huang et al. (Huang et al., 2013), were used in this work to carry out both steady state and dynamic the simulations. The Radfrac rigorous distillation model of AspenPlus® process simulator was used for modelling the reactive distillation column.

In a previous study, a sensitivity analysis was performed to identify suitable conditions for the production of the silane products considered in this work (Medina-Herrera et al., 2017). That study suggested that the production of the different silane products could be achieved with appropriate pressure conditions and distillate to feed (D/F) ratio. Table 1 shows the parameters specified for the optimal steady-state operation of the reactive distillation unit reported in Medina-Herrera et al. (2017). The specifications obtained from that analysis aimed to minimize the total annual cost by considering equal production of the three different products during the year.

Table 1. Optimal steady-state parameters, reactive distillation column (Medina-Herrera et al., 2017).

	Silane Production	Monochlorosilane Production	Dichlorosilane Production
Number of stages	89	89	89
Feed stage	42	42	42
Feed flowrate of SiHCl₃ (kmol/h)	10	10	10
Reactive Zone (Stage to Stage)	29-72	29-72	29-72
Purity of product in D1 (mol%)	99.5	99.5	99.5
Distillate to feed ratio	0.25	0.333	0.5
Reflux ratio	34.9	38.9	11.3
Operating pressure (atm)	4.9	2.3	1
Column diameter (m)	0.73	0.73	0.73
Holdup of Reactive Section (m³)	0.068	0.068	0.068
Qc (kW)	292	712	375
QR (kW)	333	739	387

3. Transient Analysis

In order to carry out the dynamic analysis, the reflux drum tank was sized taking into account the inventories for the different silane products. As shown in Table 1, the production of the monochlorosilane requires a higher vapor flow rate; therefore, the column's specifications for monochlorosilane were used to determine the size of the reflux drum. Once the steady state simulation converged, the simulation was exported to Aspen Dynamics. Rigorous hydraulics were considered for all simulations. Hold-up was evaluated at steady-state assuming a residence time of 2.5 second per stage (Huang et al., 2013). When dynamic transitions are considered, internal mass flows and therefore the residence time, will change accordingly, but under the assumption that hold-up on the reactive trays remains the same. The overall hold-up for the reactive zone is reported in Table 1. Ideal trays were assumed for the column.

Identification of the critical process variables in the transient domain was carried out by a degrees of freedom analysis. Due to their relevance for the dynamic operation, the pressure and D/F ratio were considered as part of the degrees of freedom in the analysis. Also, the liquid level in the reflux drum was controlled with the reflux flowrate. For this analysis, the heuristic rule of using reflux flowrate to control the level of the drum for reflux ratios higher that three was considered. Column hydraulics plays an important role on the dynamic simulations, since temperature and composition profiles inside the column are likely to change when disturbances enter into the process. Therefore, a key variable that should be taken into account for dynamic simulations is temperature. This variable needs to be controlled to promote production of the desired product. Furthermore, temperature control is required to avoid thermal degradation of the catalyst.

A sensitivity analysis was carried out to determine the most sensitive column trays for each silane product under changes in the reboiler duty (Luyben, 2006). For that purpose, changes of +/- 0.1% in the reboiler duty and in the reflux ratio were performed. The results from the temperature tray profiles allowed the identification of the trays with largest deviation in temperature. The trays that resulted with the largest temperature change were selected as controlled variables (Luyben, 2006). The results of the sensitivity analysis that show the tray selection for each individual silane product are presented in Table 2.

Table 2. Identification of the most sensitive stage for the production of silanes and PI tuning parameters for their controllers.

Production	Stage	PI tuning parameters	
Silane	79	P=4.7	I=9.2 min
Dichlorosilane	69	P=1.1	I=14.5 min
Monochlorosilane	63	P=0.9	I=11.8 min

Temperature, pressure and D/F ratio control loops were taken into account to develop the following dynamic analysis procedure for the production of the silane products. In order to meet the desired operating conditions of each silane production, pressure, D/F and controlled tray temperature should be changed progressively. It was decided to implement setpoint ramps, as opposed to step ramps, to prevent abrupt changes that could lead to process instabilities. Therefore, setpoint ramps were imposed for distillate to feed ratio, pressure and the temperature controllers simultaneously while monitoring the reflux drum level. Figure 1 shows the control configuration of the reactive column as implemented in Aspen Dynamics to carry out the dynamic transitions, which was based upon an implementation for disturbance rejection strategies, with the tuning parameters for temperature controllers reported in Table 2. It is important to note that pressure changers where configured in order to develop pressure driven simulations.

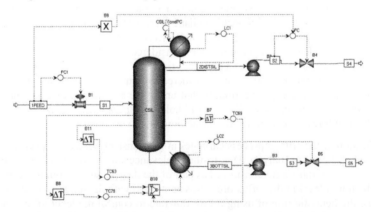

Figure 1. Control configuration of the reactive column for the dynamic transitions.

As three different products can be produced independently for each design in the reactive distillation column, six possible routes can be followed to change from the production of a silane product to another. Those possible transitions are: silane to monochlorosilane, monochlorosilane to dichlorosilane, dichlorosilane to monochlorosilane, monochlorosilane to silane, silane to dichlorosilane and dichlorosilane to silane. In order to implement the procedure described above for all the different transitions, a selector structure was implemented in the simulation. The selector model was configured so that the appropriate temperature stage in the loop was set as active for the desired transition.

4. Results

Dynamic transitions using the methodology described above were tested for all the different combinations. Figure 2 shows the liquid composition profiles for the production of monochlorosilane from dichlorosilane. Ramp set points of 20 h were configured in the column's pressure, D/F ratio and temperature controllers. As shown in Figure 2, it takes around 60 hours to obtain a high purity stream of monochlorosilane. This figure also shows that the dynamic setpoint changes imposed in pressure, and D/F ratio and temperature are tracked properly.

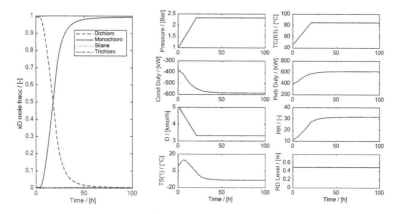

Figure 2. Column profiles for the transition of Dichlorosilane to Monochlorosilane in the distillate.

When the setpoint ramps in the controlled variables were applied for the other five transitions, it was observed that all of them were feasible in terms of reaching the desired new steady state for the corresponding silane products. Nevertheless, when the transitions silane to dichlorosilane and dichlorosilane to silane were tested, monochlorosilane was initially produced thus affecting the time to reach the final silane product (dichlorosilane). This behaviour is due to the nature of the reaction mechanism. Because of its impact on the process operating costs, another important result is that related to the column heat duty. Figure 2 also shows that higher energy requirements are needed for the dichlorosilane to monochlorosilane transition. This result agrees with the results reported in Table 1, which showed that higher energy requirements were needed for the production of monochlorosilane. Table 3 reports the overall transition times for the different silane products. Even when all transitions were successful to reach the desired set points for the products, an economic analysis should be used to assess the impact of the transition times observed for each case to quantify the impact of production losses during such time periods.

Table 3. Transition time for the different silane productions.

Transition	Time (h)	Transition	Time (h)
Dichlorosilane to monochlorosilane	62.6	Dichlorosilane to silane	94.5
Monochlorosilane to Silane	67.8	Silane to monochlorosilane	100
Silane to dichlorosilane	100	Monochlorosilane to dichlorosilane	61.2

5. Conclusions

This work explored the feasibility of implementing dynamic transitions for the production of three high-value silane products. Operating points obtained from steady-state calculations for the production of three silane products were used as target values for each transition. A control policy based on ramps for the controlled variables was implemented for the six possible transitions considered in this study. The results showed that such a control strategy was effective to yield feasible transitions from a silane product to another. The most critical transition was that from silane to dichlorosilane and viceversa. This was mostly due to the production of monochlorosilane as an intermediate product, which affected the length of the dynamic transition between these two products. An effective transition policy should avoid this condition; that is, given that the three products are desired to be produced, the production of either silane or dichlorosilane should be preceded by the production of monochlorosilane.

References

B. Ceccaroli, L. Otto, 2010, "Solar Grade Silicon Feedstock." In Handbook of Photovoltaic Science and Engineering, John Wiley & Sons, Ltd, 169–217.

X. Huang, D. Wei-Jie, Y. Jian-Min, X. Wen-De, 2013, "Reactive Distillation Column for Disproportionation of Trichlorosilane to Silane: Reducing Refrigeration Load with Intermediate Condensers." Industrial & Engineering Chemistry Research 52(18): 6211–20.

W. Luyben, 2006, "Evaluation of Criteria for Selecting Temperature Control Trays in Distillation Columns." Journal of Process Control 16(2): 115–34.

N. Medina-Herrera, S. Tututi-Avila, A. Jiménez-Gutiérrez, J. Gabriel Segovia-Hernández, 2017, "Optimal Design of a Multi-Product Reactive Distillation System for Silanes Production." Computers & Chemical Engineering: 1–10.

L. A. Mochalov, 2016, "Preparation of Silicon Thin Films of Different Phase Composition from Monochlorosilane as a Precursor by RF Capacitive Plasma Discharge." Plasma Chemistry and Plasma Processing 36(3): 849–56.

C. Ramírez-Márquez, 2016, "Dynamic Behavior of a Multi-Tasking Reactive Distillation Column for Production of Silane, Dichlorosilane and Monochlorosilane." Chemical Engineering and Processing: Process Intensification 108: 125–38.

R. W. Koller, L. A. Ricardez-Sandoval, 2017, "A dynamic optimization framework for integration of design, control and scheduling of multi-product chemical processes under disturbance and uncertainty." Computers & Chemical Engineering, 106: 147-159.

B. P. Patil, E. Maia, L. A. Ricardez-Sandoval, 2015, "Integration of scheduling, design, and control of multiproduct chemical processes under uncertainty." AIChE Journal, 61(8): 2456-2470.

Anton A. Kiss, Edwin Zondervan, Richard Lakerveld, Leyla Özkan (Eds.)
Proceedings of the 29th European Symposium on Computer Aided Process Engineering
June 16th to 19th, 2019, Eindhoven, The Netherlands. © 2019 Elsevier B.V. All rights reserved.
http://dx.doi.org/10.1016/B978-0-128-18634-3.50224-1

Dynamics and control of a heat pump assisted azeotropic dividing-wall column (HP-A-DWC) for biobutanol purification

Iulian Patraşcu,[a] Costin Sorin Bîldea,[a] Anton A. Kiss,[b,c,*]

[a]University "Politehninca" of Bucharest, Polizu 1-7, Romania

[b]School of Chemical Engineering and Analytical Science, The University of Manchester, Sackville Street, Manchester, M13 9PL, United Kingdom

[c]Sustainable Process Technology Group, Faculty of Schience and Technology, University of Twente, PO Box 217, 7500 AE Enschede, The Nethelends

tony.kiss@manchester.ac.uk

Abstract

Recently, the butanol purification from acetone-butanol-ethanol mixture was achieved in one azeotropic dividing-wall column requiring 2.7 MJ/kg butanol, which represents just 7.5% of the energy content of butanol. Compared to a conventional separation sequence, this design allows 60% energy savings due to heat integration and use of a heat pump. This work considers the dynamics and control of the process. The basic regulatory control can provide stability for small and short-time disturbances, but the process shuts down if the disturbances persist. By adding a reboiler and a condenser, better control becomes possible. As a result, the process can handle large and persistent disturbances in feed rate and composition, achieving good control of product purity.

Keywords: Dividing-wall column, optimal design, heat integration, process control.

1. Introduction

The biobutanol is considered a competitive fuel over ethanol, being characterized by low water miscibility, flammability and corrosiveness. In the upstream fermentation process, biobutanol is obtained with a concentration lower than 3%wt. Using gas stripping technology, the acetone-butanol-ethanol (ABE) mixture reaches 4.5 %wt. acetone, 18.6 %wt. butanol and 0.9 %wt. ethanol (Xue et al., 2013). In downstream processing, these components can be separated by distillation, reverse osmosis, adsorption, liquid-liquid extraction, pervaporation and others (Abdehagh et al., 2014).

For large scale plants, distillation remains the most promising technique. Significant energy savings can be achieved by employing novel configurations (such as dividing-wall column) and heat integration facilitated by heat pumps (Kiss, 2013; Luo et al., 2015; Kiss and Infante Ferreira, 2016). A novel heat pump assisted azeotropic dividing-wall column (A-DWC) was suggested for the ABE separation, allowing up to 60% energy savings compared to a conventional separation sequence (Patrascu et al., 2017, 2018). The present work considers the dynamics and controllability of this highly-integrated process. For this purpose, a rigorous pressure-driven simulation is developed in Aspen Plus Dynamics. After few design modifications, the process proves to be controllable by using only conventional control loops.

2. Problem statement

Fewer equipment units can drive the ABE separation by combining three distillation columns in a single A-DWC, while using heat integration and vapor recompression (VRC) technologies can further reduce the energy requirement. This highly integrated design delivers large energy savings of up to 60% but must be also operable. However, combining all these technologies in a single process can reduce the flexibility of the process due to fewer degrees of freedom, raising difficulties in the controllability of the process. In this paper the dynamics of the process is investigated by testing the stability and introducing minor design changes which offer two new variables to be manipulated.

3. Process design

Patrascu et al. (2018) proposed a heat-integrated VRC-assisted A-DWC for ABE separation (Figure 1). The A-DWC contains two stripping and one fractionation sections. The feed-side stripping section (DWC-L) separates the most plentiful component (water) with high purity in bottom. In the fractionation section, the light components (acetone, ethanol and some water) are obtained as distillate, while the butanol-water azeotrope is routed to a side stream, and cooled for decantation. Here, the distillation boundary is crossed and breaking the butanol-water azeotrope is achieved. The aqueous phase is sent to DWC-L for water purification and the organic phase sent to the second stripping section (DWC-R) for butanol purification. The top vapor stream is compressed from 1 bar to 5.8 bar (60 °C to 150 °C) and used to drive the reboiler from DWC-L and to preheat the fresh feed, and the organic and aqueous phases, before being returned to the column. Performing this heat integration, the energy requirement is reduced by 69% compared to a simple A-DWC column. The purity of water product is 99.8 %wt. and the purity of butanol product is 99.4 %wt. (Figure 2). A small amount of water is present in distillate due to the ethanol – water azeotrope.

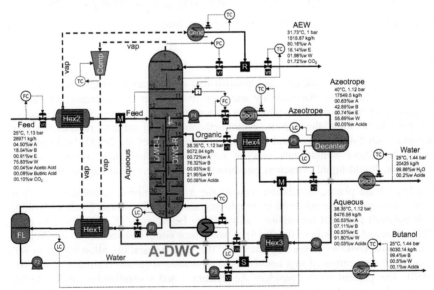

Figure 1. Heat-integrated VRC-assisted A-DWC for ABE separation

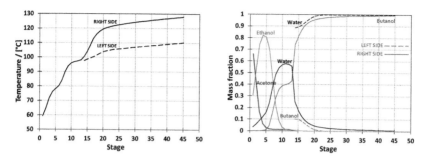

Figure 2. Temperature (left) and mass composition (right) profiles

4. Process dynamics

A dynamic simulation is built in Aspen Plus Dynamics. Two RADFRAC units are used to model the water removal section (DWC-L) and the combined fractionation (AEW) and butanol purification (DWC-R) sections. The reboiler on the water-removal side is represented as a HEATX (counter-current heat exchanger) followed by a FLASH. Several pumps and valves are provided, as required by the pressure-driven simulation. The basic control loops of flow rate, liquid level, pressure and temperature are shown in Figure 1. The column pressure is controlled with the reflux flow rate, which is subcooled at 31 °C. Besides the inventory control loops, the temperature on stage 5 of the fractionation section is controlled by manipulating the valve position of the distillate stream; the temperature on stage 40 of the butanol-purification section (DWC-R), is controlled by manipulating the reboiler duty; the temperature of the compressed vapor (to not exceed 150 °C) is controlled by the compressor brake-power.

The dynamic of the process is tested by changing the feed flow rate (by ±5% for 10 minutes and by ±5% for one hour), as shown in Figure 3. The control structure can successfully reject small and non-persistent disturbances (top diagrams). However, when the disturbance persists (bottom diagrams), the stream entering the column cools down, and the pressure and the temperature in column drop. As a result, less vapours are sent to the fractionation section and to the compressor. In the meantime, the stage 40 temperature controller located on the butanol-purification section tries to compensate the disturbance by increasing the reboiler duty. This affects the temperature of stage 5, so the top temperature controller closes the distillate stream valve. Further, the pressure controller closes the valve from the reflux stream, trying to increase the column pressure. However, the vapours flow is too small. Although the compressor brake-power is available as a variable to be manipulated, this action would increase the temperature of the compressed vapours above 150 °C, which is not allowed for safety reasons. Therefore, another manipulated variable is needed.

When the feed flow rate is decreased by 5% and the disturbance persists for a long time (Figure 3, bottom diagrams) the feed enters the column at higher temperature. In this case, more vapours are sent to fractionation section and more vapours are compressed, leading to a continually temperature increase in the water-removal section. As a result, more water is vaporized and eventually reaches the butanol product stream. Such disturbance could be rejected by controlling the temperature of the stream entering the column, but no manipulated variable is available.

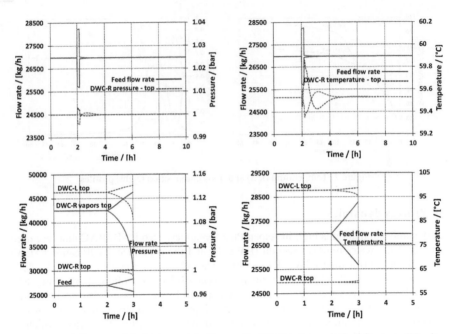

Figure 3. Stability test of the basic control structure for 5% disturbance in feed flow rate (10 min. top, 1 hour bottom)

5. Process control

To overcome these control difficulties, two new variables to be manipulated were added (Figure 4). The first one is a small additional duty on the reboiler of the water-removal section, which represents 10% of the energy required. The second one is an additional cooler placed on the compressed vapor stream, between the reboiler and the feed preheating. These variables are used in new control loops, namely: a) temperature control of the preheated feed, by manipulating the duty of the additional cooler; and b) temperature on stage 25 in water-removal section (DWC-L), by manipulating the additional reboiler duty.

In addition, a valve-position controller has the task to keep the additional duty at its steady state value by changing the set point of the temperature controller of compressed vapor stream (a low-select block ensures that the setpoint does not exceed 150 °C). This controller is essential when the feed flow rate is decreased because if less energy is required, the additional duty will remain at steady state value by reducing the temperature of the compressed vapours. The additional duty will increase only when more energy is required. By introducing these controllers, the process can handle large feed flow rate disturbances for long time.

Other dynamic simulation results (not shown here) indicate that the transient regime takes about 5 hours. As the feed rate changes, the water purity and the mass fraction of water in the distillate are practically unaffected, while the butanol purity deviates by only 0.1 %wt. from the design value. When the fraction of butanol in the feed stream changes (from 18.6 %wt. to 17 %wt. or 20 %wt.) the behaviour is similar: water and distillate composition is practically unchanged, while the concentration of the butanol product deviates by about 0.1 %wt. from the design value. The performance of the control system is excellent, considering that only temperature measurements were used.

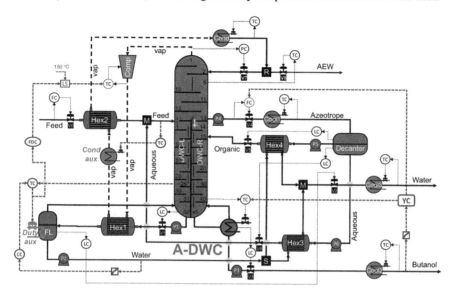

Figure 4. Control structure of the heat-pump assisted A-DWC

Finally, the performance of the new control structure can be improved by introducing two concentration controllers for maintaining the products (water and butanol) quality (Figure 4). The slow dynamics of these measurement was taken into account by using the Aspen Plus Dynamics sensor model with 5 minutes sampling period and 5 minutes dead time. Figure 6 and Figure 6 show dynamic simulation results for the control structure employing the concentration controllers, for feed flow rate and concentration disturbances, respectively.

The deviation of butanol purity from the steady state value is quite small. Eventually, 4 hours after the disturbance, the butanol purity comes back to the design value (99.4 %wt.). Water purity is practically unchanged, while the mass fraction of water in the distillate is acceptable.

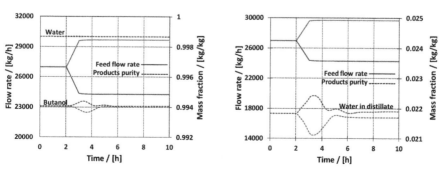

Figure 5. Performance of the control structure for 10% disturbance in feed flow rate

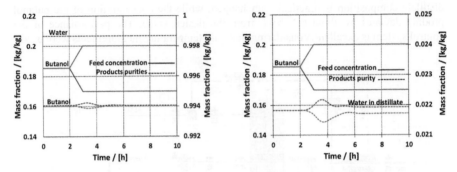

Figure 6. Performance of the control structure for feed concentration disturbances

6. Conclusions

The basic control structure of the A-DWC can reject small and short-time disturbance, but the process shuts down when the disturbance persists. Adding an additional reboiler and a condenser on the feed side stripping section provides two new variables to be manipulated. This change in the main design reduces the energy savings of 60% by only 4%. This is small penalty to pay, as the control structure performs successfully when disturbances of 10% in the flow rate and butanol concentration occur in the fresh feed.

Acknowledgements

The financial support of the European Commission through the European Regional Development Fund and of the Romanian state budget, under the grant agreement 155/25.11.2016 Project POC P-37-449 (ASPiRE), is kindly acknowledged. AAK is thankful for the Royal Society Wolfson Research Merit Award.

References

N. Abdehagh, F. H. Tezel, J. Thibault, 2014, Separation techniques in butanol production: Challenges and developments, Biomass and Bioenergy, 60, 222-246.

A. A. Kiss, Novel applications of dividing-wall column technology to biofuel production processes, 2013, Journal of Chemical Technology and Biotechnology, 88, 1387-1404.

A. A. Kiss, C. A. Infante Ferreira, 2016, Heat pumps in chemical process industry, CRC-Press, Taylor & Francis Group.

H. Luo, C. S. Bildea, A. A. Kiss, 2015, Novel heat-pump-assisted extractive distillation for bioethanol purification, Industrial & Engineering Chemistry Research, 54, 2208-2213.

I. Patrascu, C. S. Bildea, A. A. Kiss, Eco-efficient butanol separation in the ABE fermentation process, 2017, Separation and Purification Technology, 177, 49-61.

I. Patrascu, C. S. Bildea, A. A. Kiss, Eco-efficient downstream processing of biobutanol by enhanced process intensification and integration, 2018, ACS Sustainable Chemistry & Engineering, 6, 5452-5461.

C. Xue, J-B. Zhao, F-F. Liu, C-G. Lu, S-T. Yang, F-W. Bai, 2013, Two-stage in situ gas stripping for enhanced butanol fermentation and energy-saving product recovery, Bioresource Technology, 135, 396-402.

Anton A. Kiss, Edwin Zondervan, Richard Lakerveld, Leyla Özkan (Eds.)
Proceedings of the 29th European Symposium on Computer Aided Process Engineering
June 16th to 19th, 2019, Eindhoven, The Netherlands. © 2019 Elsevier B.V. All rights reserved.
http://dx.doi.org/10.1016/B978-0-128-18634-3.50225-3

Closed-loop dynamic real-time optimization of a batch graft polymerization reactor

Ryad Bousbia-Salah[a], François Lesage[a], Miroslav Fikar[b], Abderrazak Latifi [a,*]

[a]Laboratoire Réactions et Génie des Procédés, CNRS – ENSIC, Université de Lorraine, 1 rue Grandville, 54001, Nancy Cedex, France

[b]Faculty of Chemical and Food Technology, Slovak University of Technlogy in Bratislava, Radlinskeho 9, 81237 Bratislava, Slovakia

*Abderrazak.Latifi@univ-lorraine.fr

Abstract

In a recent work (Bousbia-Salah et al, 2018), dynamic real-time optimization (D-RTO) of a batch reactor where polymer grafting reactions take place is experimentally implemented within open-loop control. This paper constitutes the continuation of the work and presents the experimental implementation of closed-loop D-RTO in the same reactor. The objective is to determine and to apply the online reactor temperature profile that minimizes the batch period while meeting terminal constraints on the overall conversion rate and grafting efficiency. The computed and measured optimal profiles of temperature, overall conversion rate and grafting efficiency exhibit a good agreement and show also that the terminal constraints are satisfied.

Keywords: Dynamic real-time optimization, Closed-loop control, Polymer grafting batch reactor, Experimental implementation

1. Introduction

Dynamic real-time optimization (D-RTO) is the most suitable technology to successfully handle uncertainty and changing conditions in the operation of plants particularly those that are never in steady-state (Rawlings et al, 2011; Ellis et al, 2013 and 2014). It makes use of the online available measurements to maximize a process performance index while meeting environmental and operating constraints (e.g. terminal specifications, temperature limits…). The results from the optimization problem can be directly applied to the process or sent to a lower level where a (e.g. MPC) control law is designed to apply to the process the optimal profiles of decision variables.

In a recent study (Bousbia-Salah et al, 2018), we considered a batch reactor where polymer grafting reactions take place. The objective was to determine the reactor temperature profile that minimizes the batch period while meeting terminal constraints on the overall conversion rate and grafting efficiency. The resulting optimal temperature profile was then experimentally implemented within the reactor in open-loop control. Although the results showed a good agreement between the computed and the measured profiles, the implementation was not able to handle the possibly disturbances or model mismatch. Closed-loop control is therefore needed in order to overcome this problem.

In the present paper, the D-RTO approach developed in (Bousbia-Salah et al, 2018) is experimentally implemented within the same batch polymerization reactor, but in closed-loop control with the help of an efficient moving horizon observer designed to estimate unmeasured states of the system.

2. Grafting polymerization

When styrene is added to ground tire rubber (GTR) particles, it can be located inside and/or outside them, depending on the styrene/GTR ratio. The latter is kept below 2 so that styrene is completely located inside GTR particles where it polymerizes in two types of polymerization. The first type is the polymerization of styrene itself leading to polystyrene (PS) chains which are not linked to GTR particles. These PS chains are called free PS. The second type is the polymerization of styrene from the rubber chains and the resulting PS chains are attached (grafted) to the rubber ones. These PS chains are designated as GTR-g-PS. The resulting material is GTR particles inside which there are free PS and GTR-g-PS. The main objective is to minimize the amount of free PS in the PS/GTR-g-PS while maximizing the amount of GTR-g-PS in order to allow recycling a maximum of GTR with the best properties, especially the impact strength (Yu, 2015).

This objective is taken into account here through constraints on the grafting efficiency (GE) and the styrene conversion rate (X). GE is defined as the ratio between the amount of grafted PS and that of the total PS (grafted PS + free PS).

3. Formulation of the D-RTO problem

The dynamic optimization problem that is solved within the D-RTO approach may be formulated as

$$\min_{T(t), t_f} J = t_f \tag{1}$$

Subject to
$$\dot{x} = f(x, T) \text{ with } x(t_k) = \hat{x}_k \text{ and } t \in [t_k, t_f] \tag{2}$$
$$GE(t_f) \geq GE_f \tag{3}$$
$$X(t_f) \geq X_f \tag{4}$$
$$T \leq T_{max} \tag{5}$$

where the reactor temperature $T(t)$ and the batch period t_f are the decision variables and x is the vector of state variables. GE and X are the measured process outputs and GE_f and X_f are their desired final values. T_{max} is the reactor temperature upper bound.

The model equations are derived after obtaining the moment rates of the polymer chains. Moreover, the model consists of a system of 20 non-linear ODEs involving 24 unknown kinetic parameters.

The process model equations (2) are derived after obtaining the moment rates of the polymer chains. Moreover, the model consists of a system of 20 non-linear ODEs involving 24 unknown kinetic parameters. (Yu, 2015).

The optimization problem is solved over a shrinking horizon using the control vector parametrization (CVP) method (Goh and Teo, 1988). The latter is based on the approximation of decision variables by means of piece-wise constant functions over the optimization horizon taken equal to 3. The vector of time-independent parameters is then given by $p^T = (T_1, T_2, T_3, t_f)$.

The resulting non-linear programming problem is solved by means of a gradient-based method where the gradients are computed through the integration of sensitivities at each iteration of the optimizer. The equations of sensitivities are defined as

$$\dot{s} = \frac{\partial f}{\partial x} s + \frac{\partial f}{\partial p} \text{ with } s(t_k) = 0 \tag{6}$$

where $s(t) = \frac{\partial x(t)}{\partial p}$ are the sensitivities of the vector of state variables with respect to the vector of parameters p. Note that Eqs.(6) are integrated from t_k to t_f at each iteration of the optimization solver.

The D-RTO approach proceeds then as follows. Starting from the real process to which the optimal decision variables are applied at a sampling time t_k, the outputs as well as inputs are used to estimate the state vector $\hat{x}_k$ that will be used as the initial condition for the next optimization at the sampling time t_{k+1}. The corresponding initial condition for the sensitivities is always taken equal to zero. The estimation $\hat{x}$ of x is carried out by means of a moving horizon estimator (MHE) in a closed-loop control (Michalska and Mayne, 1995; Robertson and Lee, 1995) using input and output variables.

4. Experimental rig and measurements

4.1. Experimental rig

The polymerization reactions take place in a stirred batch reactor equipped with a condenser, a temperature control device and a sample collection system. The substances involved in the reactions are GTR particles of 800 μm in diameter, styrene (monomer) and two initiators, i.e. benzoyl peroxide (BPO) and dicumyl peroxide (DCP). Hydroquinone (HQ) and chloroform are used for the measurements of the conversion rate and the grafting efficiency.

2.2. Conversion measurement

The conversion of monomer is measured by a gravimetric method. Samples are taken from the reactor and weighed. A known amount of hydroquinone is added in to stop the polymerization. Then they are put into a Halogen Moisture Analyzer and the temperature is raised to 170°C and kept at this temperature for 10 min. The residual monomer could completely evaporate and the remaining material is composed of free PS, GTR-g-PS and HQ.

4.2. Grafting efficiency measurement

Samples taken from the reactor are put in chloroform to measure the grafting efficiency of the polymerization of styrene inside GTR particles. The free PS and HQ are soluble in chloroform, while GTR and/or GTR-g-PS are not. After at least 6 hours during which the samples are shaken many times, the solvent is removed using filter paper. The free PS and HQ are in the solvent, and the non-soluble portion in the filer paper is GTR or GTR-g-PS.

5. D-RTO implementation

The algorithm used to implement the closed-loop D-RTO is detailed below (Bousbia-Salah, 2018.

1. Initialize the iteration number and time: $k = 0$ and $t_k = 0$.
2. Solve the optimization problem (1-5) for T_1, T_2, T_3 and t_f.
3. Apply the first temperature T_1 during 60 minutes and set $t_k := t_k + 60$min and k :=k+1.
4. Take a reaction mixture sample from the reactor for conversion rate measurement.
5. Apply the second temperature T_2 while the measurement is being performed, i.e. 30 minutes (time required for measurement). Set $t_k := t_k + 30$min and k := k+1.

6. Use the measured conversion rate to estimate the state $\hat{x}_k$.
7. Repeat the steps (2) to (6) until t_f is reached and stop the reaction.

It is noteworthy that since the dynamics of the polymerization reaction is relatively slow, manual temperature control is found to be sufficient to achieve reasonable accuracy. On the other hand, as mentioned above, the conversion rate is measured in 30 minutes, whereas the measurement of *GE* needs much more time and takes about 6 hours.
The implementation is carried out for $GE_f = 60\%$ and $X_f = 85\%$.

Figures 1, 2 and 3 show the temperature, monomer conversion and grafting efficiency profiles, respectively. They compare the computed optimal profiles and the measured ones in the reactor. The temperature profile exhibits a regular increase in order to fulfill the required monomer conversion rate and grafting efficiency. This regular increase is meaningful since at constant temperature, the conversion rate increases with time whereas the grafting efficiency decreases. Therefore the temperature should increase with time in order to achieve the desired conversion rate but not too much in order to guarantee the specified terminal value of grafting efficiency.

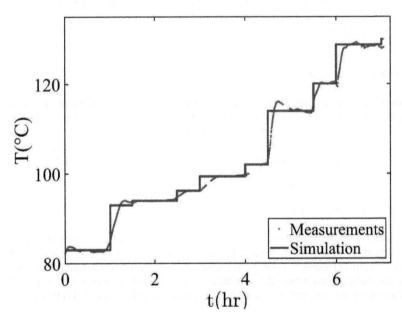

Figure 1: Optimal profiles of reactor temperature for $GE_f = 60\%$ and $X_f = 85\%$.

Figures 2 and 3 show the time-varying process outputs, i.e. conversion rate and grafting efficiency, respectively, with two different profiles. The estimated profile using the moving horizon state observer, and the measured one. It can be seen that the computed terminal inequality constraints on both process outputs are satisfied and are almost the same as the measured ones. This very good agreement between the computed and measured profiles can be explained by the absence of disturbances and a very good quality of the process model.

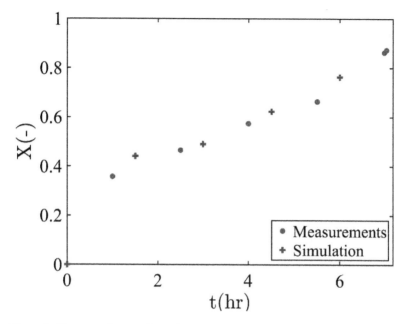

Figure 2: Time-varying profiles of the conversion rate for $GE_f = 60\%$ and $X_f = 85\%$.

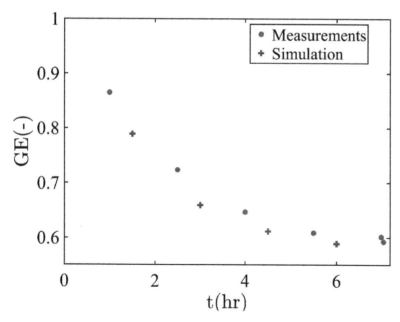

Figure 3: Time-varying profile of the grafting efficiency for $GE_f = 60\%$ and $X_f = 85\%$.

R. Bousbia-Salah et al.

6. Conclusions

The closed-loop control results presented in this paper showed that dynamic real-time optimization is a suitable technology for online dynamic optimization of the grafting polymerization reactor considered here. It allowed to determine and implement online the temperature profile that minimizes the batch period under terminal constraints on monomer conversion and grafting efficiency using an optimization horizon of 3. The results show very good agreement between the computed and measured profiles of temperature, monomer conversion and grafting efficiency. Moreover, the computed and measured terminal constraints are fully satisfied. On the other hand, the stability of the computed optimal temperature is guaranteed since the process considered is of the batch type and the shrinking optimization horizon covers the whole operational time. The current works deal with the development of Raman spectroscopy to measure online the monomer conversion rate in order to improve the experimental implementation of the closed-loop control at the laboratory scale prior to its implementation within a reactor at industrial scale.

References

R. Bousbia-Salah, F.Lesage, G.H. Hu, M.A.Latifi, 2018, Experimental implementation of dynamic real-time optimization in a graft polymerization reactor, Computer - Aided Chemical Engineering, 43, 829 – 834, 2018.

R. Bousbia-Salah, 2018, Optimisation dynamique en temps-réel d'un réacteur de polymérisation par greffage, PhD Thesis, Université de Lorraine, Nancy, France.

M.Ellis, H.Durand, P.D. Christofides, 2014, A tutorial review of economic model predictive control methods, Journal of Process Control, 24, 1156 – 1178.

M.Ellis, M.Heidarinejad, P.D.Christofides, 2013, Economic model predictive of nonlinear singularly perturbedsystems, Journal of Process Control, 23, 743 – 754.

C.J. Goh, K.L. Teo, 1988, Control parametrization: A unified approach to optimal control problems with general constraints, Automatica, 24, 3-18.

H. Michalska, D. Q. Mayne, 1995, Moving horizon observers and observer-based control, IEEE Transactions on Automatic Control},40,995-1006.

J.B.Rawlings, D.Angeli, C.N. Bates, 2012, Fundamentals of economic model predictive control, In Proceedings of the 51st IEEE Conference on Decision and Control, 3851 – 3861, Manui, Hawaii, USA.

D.G. Robertson, J.H. Lee, 1995, A least squares formulation for state estimation, Journal of Process Control, 5, 291-299.

N.Yu, 2015, Etude de la cinétique de polymérisation radicalaire du styrène dans un réseau tridimensionnel et application à la valorisation de pneus usagés, PhD Thesis, Université de Lorraine, Nancy, France.

Anton A. Kiss, Edwin Zondervan, Richard Lakerveld, Leyla Özkan (Eds.)
Proceedings of the 29th European Symposium on Computer Aided Process Engineering
June 16th to 19th, 2019, Eindhoven, The Netherlands. © 2019 Elsevier B.V. All rights reserved.
http://dx.doi.org/10.1016/B978-0-128-18634-3.50226-5

Application of cyclic operation to acetic / water separation

Catalin Patrut[a], Elena Catalina Udrea[a], Costin Sorin Bildea[a*]

aUniversity "Politehnica" of Bucharest, Polizu 1-7, 011061 Bucharest, Romania

s_bildea@upb.ro

Abstract

Acetic acid is an important bulk chemical, used as raw material in the production of vinyl acetate, terephtalic acid, acetate esters and acetic anhydride. All these processes include a step for separating the acetic acid from water, both high-purity and high-recovery of acetic acid being required. Although acetic acid and water do not form an azeotrope, the separation is difficult due to the tangent pinch present on the pure water end. A common industrial practice to overcome the tangent pinch is to add an entrainer and, therefore, to use heterogeneous azeotropic distillation. This work investigates the applicability of cyclic distillation to this difficult separation. A mathematical model is presented and used for sizing and optimization considering the total annual cost as objective function and the number of trays, feed tray location, and the duration of vapor-flow period as decision variables.

Keywords: Water, Acetic Acid, Cyclic distillation.

1. Introduction

Acetic acid is an important chemical, produced by methanol carbonylation, acetaldehyde oxidation and the oxidation of butane and/or naphtha fractions. All these processes contain a step for separating the acetic acid from water. Although acetic acid and water do not form an azeotrope, the separation is difficult due to the tangent pinch present on the pure water end. In this contribution, the applicability of cyclic distillation for separation of 10780 kg/h mixture of water (71%wt) and acetic acid (29%wt) is investigated. The required separation performance is given as purities of the acetic acid (99.3 %wt) and water (99.4 %wt) products.

Cyclic operation of distillation columns is attractive for difficult separations, being already used for kerosene fractionation (petrochemical industry) and ethanol concentration (food industry). Cyclic operation of distillation columns (Patrut et al., 2014) consists of a vapour flow period and a liquid flow period, as shown in Figure 1. During the vapor flow period, the liquid remains stationary on the tray while the vapour is flowing upwards through the column. During the liquid flow period, the vapor flow is stopped, reflux and feed are supplied to the column, and the liquid holdup is dropped from each tray to the tray below. This mode of operation can be achieved by using perforated trays, without downcomers, combined with sluice chambers located under each tray.

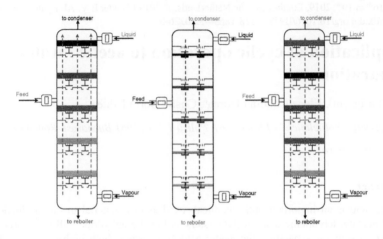

Figure 1. Schematics illustrating the working principle of cyclic distillation: (a) vapor-flow period; (b) liquid flow-period; (c) beginning of a new vapor-flow period.

2. Mathematical model

The model of the cyclic distillation column (Patrut et al., 2014) is derived under the following assumptions: ideal stages (vapor-liquid equilibrium is reached); equal heat of vaporization; perfect mixing on each stage; negligible vapor holdup; saturated liquid feed. No assumptions are made here regarding the linearity of vapor-liquid equilibrium, negligible or constant pressure drop, infinite reboiler holdup or zero condenser holdup.

The vapor-flow period: For each component, j = 1, NC, the following equations describe the evolution of species holdup on each tray:

Condenser:
$$\frac{dM_{1,j}}{dt} = V \cdot y_{2,j} \tag{1}$$

Trays:
$$\frac{dM_{k,j}}{dt} = V\left(y_{k+1,j} - y_{k,j}\right); \qquad k = 2, NT-1 \tag{2}$$

Reboiler:
$$\frac{dM_{NT,j}}{dt} = -V \cdot y_{NT,j} \tag{3}$$

For each tray, k = 1,*NT*, the following relationships describe the liquid-vapor equilibrium:

$$P_k \cdot y_{k,j} - x_{k,j} \cdot \gamma_{k,j}\left(x_1,\dots x_{NC}, T_k\right) \cdot P_j^{vap}\left(T_k\right) = 0 \tag{4}$$

$$x_{k,j} = \frac{M_{k,j}}{\sum_j M_{k,j}} \; ; \qquad \sum_j y_{k,j} - 1 = 0 \tag{5}$$

For each tray, k = 3, *NT*, the pressure is calculated based on the vapor flow rate and the amount of liquid on the tray above:

$$P_2 = P_{cond} + \Delta P_{cond} \; ; \qquad P_k = P_{k-1} + \Delta P\left(M_{k-1}, V\right) \tag{6}$$

The state of the system at the beginning of the vapor-flow period is the same as the state at the end of the liquid-flow period:

$$\mathbf{M}\left(t = 0\right) = \mathbf{M}^{(L)} \tag{7}$$

The state of the system at the end of the vapor flow period is found by integrating the equations (1)-(7)

$$\mathbf{M}^{(V)} = \mathbf{M}\left(t = t_{vap}\right) \tag{8}$$

The liquid-flow period: for each component, $j = 1$, NC, the following equations give the stage holdup at the end of the liquid-phase period:

Condenser: $$M_{1,j}^{(L)} = M_{1,j}^{(V)} - \left(D + L\right) \cdot x_{1,j}^{(V)} \tag{9}$$

Trays, rectifying section: $$M_{2,j}^{(L)} = L \cdot x_{1,j}^{(V)} \; ; \; M_{k,j}^{(L)} = M_{k-1,j}^{(V)} \; , k = 2, NF \tag{10}$$

Feed tray: $$M_{NF+1,j}^{(L)} = M_{NF,j}^{(V)} + F \cdot x_{F,j} \tag{11}$$

Trays, stripping section: $$M_{k,j}^{(L)} = M_{k-1,j}^{(V)} \; , \quad k = NF + 2, NT - 1 \tag{12}$$

Reboiler: $$M_{NT,j}^{(L)} = M_{NT,j}^{(V)} - B \cdot x_{NT,j}^{(V)} + M_{NT-1,j}^{(V)} \tag{13}$$

Equations (1) - (6) and (9) - (13) can be written in the following condensed form, where $\Phi^{(V)}$ and $\Phi^{(L)}$ are mappings relating the state at the start and the end of the vapor- and liquid-flow periods, respectively.

$$\left(M^{(V)}, x^{(V)}\right) = \Phi^{(V)}\left(M, x\right) \tag{14}$$

$$\left(M^{(L)}, x^{(L)}\right) = \Phi^{(L)}\left(M, x\right) \tag{15}$$

Periodicity condition requires:

$$\left(M^{(L)}, x^{(L)}\right) = \Phi^{(L)} \circ \Phi^{(V)}\left(M^{(L)}, x^{(L)}\right) \tag{16}$$

A straightforward solution of equation (16) can be obtained by considering an initial state and applying relationships (14) and (15) until the difference between two iterations becomes small. However, the convergence can be accelerated by applying numerical methods suitable for algebraic equations (for example Newton).

The fugacity and activity coefficients have been calculated using the NRTL activity model for the vapor-liquid phase and Hayden O'Connell (HOC) fugacity coefficient model for the vapor phase. The parameters for the NRTL and HOC models have been regressed from the NIST TDE database using Aspen Plus.

3. Results and discussion

In order to size and to determine the most efficient column from an economical point of view, the operating characteristics of the cyclic distillation systems have been studied. It is observed that, for a fixed number of trays, the required vapor flow-rate (which is directly related to the energy requirements) depends on the chosen feed tray. The variation of the required vapor flow-rate with the chosen feed tray for a column with 20 trays is presented in Figure 2.

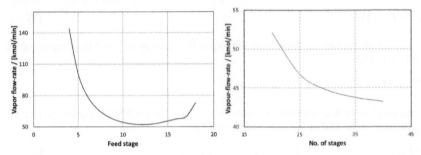

Figure 2. Dependence of the vapor flow-rate with the feed tray for a 20 trays column (left);
Variation of the vapor flow-rate with the number of trays (right)

The influence of the number of trays on the vapor flow-rate has been investigated by calculating the vapor flow-rate for columns with 20, 25, 30, 35 and 40 trays, fed at the optimal feed tray. The variation of the vapor flow-rate with the number of trays is represented in Figure 2 (right). The vapor flow-rate decreases with the increase of the number of trays, which indicates the existence of an optimum number of trays. The flow-rate decreases sharply between 20 and 25 trays, followed by a slower decrease between 25 and 40 trays. The variation of the total annual cost (calculated with well-known correlations) with the number of trays is shown in Figure 3.

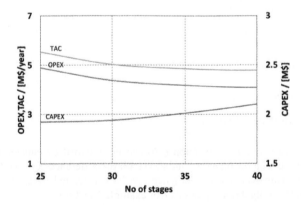

Figure 3. Total annual cost for the water acetic acid through cyclic distillation column

The total annual cost decreases with the increase of the number of trays. However, the decrease between 35 and 40 trays is very small. For practical reasons, a column with 35 trays is further studied. The variation of the required vapor flow-rate with the bottoms product purity is shown in Figure 4.

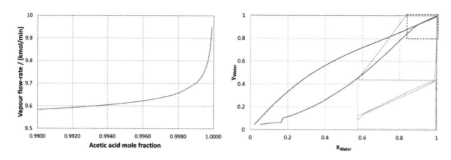

Figure 4. Variation of the vapour-flow-rate with the bottoms stream purity (left); Equilibrium and operating curves for a 35 tray cyclic distillation column (right)

The equilibrium and operating curves for the 35 trays column are shown in Figure 4 (right). In the tangent pinch region, the equilibrium and operation curves are very close to each other, as in conventional distillation. The mass balance of the cyclic distillation column is shown in Table 1. The characteristics of the column are shown in Table 2.

Table 1. Mass balance of the water - acetic acid cyclic distillation column

Stream	Feed	Distillate	Bottoms
Flow-rate / [kmole/h]	477.31	424.76	52.55
Water mole fraction	0.8919	0.9982	0.0230
Acetic acid mole fraction	0.1081	0.0018	0.9770
Water mass fraction	0.7100	0.9940	0.0070
Acetic acid mass fraction	0.2900	0.0060	0.9930

Table 2. Characteristics of the water – acetic acid cyclic distillation column

Number of trays	35
Feed tray	25
Diameter / [m]	3.7
Height / [m]	22.2
Reboiler duty / [kW]	17037
Condenser duty / [kW]	12957

It should be noted that the water / acetic acid separation can be also performed by azeotropic distillation, using i-butyl acetate as entrainer. The acetic acid / water mixture and i-butyl acetate are fed to the first column. High-purity acetic acid is obtained as bottoms product, while the water – i-butyl acetate azeotrope is obtained in the distillate. The 2-phase azeotrope mixture is separated in a decanter. The organic phase is returned into the first column as reflux. The aqueous phase is fed into the second column, where high-purity pure water is produced as bottoms product and water – i-butyl acetate azeotrope is produced as distillate (Li et al., 2014).

Table 3 compares the economics of cyclic distillation, conventional distillation and azeotropic distillation alternatives, for an operating time of 8000 h/year and a payback period of 3 years.

Table 3. Cost summary for the water / acetic acid separation by different methods

	Cyclic distillation	Conventional distillation	Azeotropic distillation	
			Acetic acid column	Water column
Column / [k$]	806	1318	305	19.6
Trays / [k$]	142	280	45	1.2
Reboiler / [k$]	758	1129	515	64.7
Condenser / [k$]	397	521	605	10.1
CAPEX / [k$]	2102	3248	1470	95.6
Cooling water / [k$/y]	269	412	178	0.3
Steam / [k$/y]	3817	4527	1914	125.6
OPEX / [k$/y]	4087	4938	2092	125.9
Total annual cost / [k$/y]	4787	6021	2582	158

4. Conclusion

Separation of the acetic acid / water mixture in a distillation column operated in a cyclic manner appears to be economically competitive for the cases when impurification with a solvent is not acceptable. For comparison, when the separation is performed by conventional distillation, the optimal column has 70 trays, requires 6.03 kWh/kg acetic acid, for a total annual cost of 6200 k$/y. Azeotropic distillation (for example, using isobutyl acetate as entrainer) is cheaper (2.9 kWh/kg acetic acid, 2800 k$/y). Even though possible from a theoretical point of view, separation of water and acetic acid through binary distillation is unpractical. A very large number of trays is required in the rectifying section, to overcome the tangent pinch formed by the two components. Through cyclic operation, the number of trays is reduced to half and slightly lower energy consumption (therefore a lower total annual cost) is required for the same distillate purity and a higher purity of the bottoms product.

References

J. Hayden, J. P. O'Connell, 1975, A generalized method for predicting second virial coefficients, Ind. Eng. Chem. Process Des. Dev., vol. 14, no. 3, pp. 209-216.

K.L. Li, I.L. Chien, C.L. Chen, Design and optimization of acetic acid dehydration process, The 5th international symposium of advanced control industrial processes, Hiroshima (2014).

C. Patrut, C. S. Bildea, I. Lita, A. A. Kiss, 2014, Cyclic distillation - Design, control and applications, Separation and purification technology, vol. 125, pp. 326-336.

H. Renon, J. Prausnitz, 1968, Local compositions in thermodynamic excess functions for liquid mixtures, AIChE Journal, vol. 14, no. 1, pp. 135-144.

Anton A. Kiss, Edwin Zondervan, Richard Lakerveld, Leyla Özkan (Eds.)
Proceedings of the 29th European Symposium on Computer Aided Process Engineering
June 16th to 19th, 2019, Eindhoven, The Netherlands. © 2019 Elsevier B.V. All rights reserved.
http://dx.doi.org/10.1016/B978-0-128-18634-3.50227-7

Control analysis of batch reactive distillation column with intermittent fed

C F. Rodriguez-Robles,[a,*] S. Hernandez-Castro,[a] H. Hernandez-Escoto,[a] J. Cabrera-Ruiz,[a] F. O. Barroso-Muñoz,[a] J. E. Terrazas-Rodriguez[b]

[a]Universidad de Guanajuato, Noria Alta s/n Col. Noria Alta, Guanajuato, Guanajuato.

[b]Universidad Veracruzana, Av. Universidad Veracruzana, Paraiso, Coatzacoalcos, Veracruz, 96538, Mexico.

cf.rodriguezrobles@ugto.mx

Abstract

This work addresses control issues of a batch reactive distillation for the production of methyl lactate via esterification reaction. The particular challenge of this classic process intensification case lies on that reactant is lighter than product and is easier to be taken away from the reaction, in such a way the conversion of loaded reactant is shrunk; to overcome this drawback, excess of reactant is typically added in the reactive zone. In order to guarantee a high conversion, a practical control approach is proposed: maintaining the maximum allowed temperature of the reaction through reboiler duty, and intermittently feeding the reactant excess, which is driven by maintaining a fixed amount of reactant in reboiler. The controllers for this two-point control configuration are conventional PI linear. The results in a simulation framework show that a reactant conversion greater than 90 % is reached with ease.

When the product is the lightest component in the mixture, the equilibrium reactions are favoured because the products are removed as the reaction proceeds; as a result, conversion is increased; however, when a reactant is the lightest component there are problems causing low conversion, and to overcome the problem, excess of such a reactant is fed in the reactive zone. Control strategy (one and two-point control) is proposed such a practical solution and novel conversion control is explored to ensure a 90% conversion. Also, the maximum temperature of the process is 393,15 K. The temperature in reboiler is satisfactorily controlled using the reboiler duty as the input control. The conversion set point (90 %) is reached with an intermittent fed of excess reactant, in this loop the control input is the feed.

Keywords: batch reactive distillation, batch distillation control, lactic acid.

1. Introduction

Reactive distillation is one of the most representative cases of process intensification (reaction and separation in one equipment), where equilibrium reactions are favoured because the products are removed as the reaction proceeds; as a result, conversions are increased and energy is saved. Nevertheless, in some reactions, certain reactants can be the lighter components, leading to low conversion because they are taken out from the reaction zone. In order to overcome this problem, it is necessary to use an excess of reactant to support the advance of the reaction; one example of this case come up in the

the esterification reaction between alcohols and carboxylic acids, such as methanol and lactic acid yielding methyl acetate and water.

A typical control structure of a continuous reactive distillation column uses the reactants feed to control the temperature in a selected stage. However, in a batch reactive distillation, the control strategy is different, one-point control can regulate temperature, but is not sufficient to guarantee conversion; then, using a two-points control configuration is resorted to (Sørensen and Skogestad, 1994). Moreover, the reactive system is another important factor due the physical properties.

Methyl acetate can be produced by reactive distillation. Figure 1 shows the reactive batch distillation with an inlet to allow intermittent reactant feed. The reaction is carried out in the reboiler, during which the methanol (MeOH) and water (W) are evaporated but condensed in the top accumulator, while the methyl acetate (ML) remain in the reboiler with the lactic acid (LA) and the methanol flow is fed into the reboiler, and the initial load of reactants as well. Aqar et al. (2016) studied the production of ML in a batch reactive distillation to optimize the LA conversion following an optimal control approach: When the column operates in total reflux, the maximum LA conversion achieved 62 %, and with the control system implemented, using the reflux ratio (RR) and the methanol flow as control inputs, the achieved conversion was greater than 90%. The result showed that the RR must be low to extract the water from the reboiler; however, the temperature profile was not shown in such study.

In the open literature there are a few works about control of batch reactive distillation: Sørensen and Skogestad (1994) demonstrated that one-point control strategy is enough to have a good control in the column for a typical case of reactive batch distillation where the products are the lighter components and reaction zone is in the reboiler. On this framework of designing a control system to guarantee a high conversion of reactant, controllability analysis in batch distillation is possible by linearized model, but if the model has high nonlinear behaviour large deviations could be presented and multiple points are necessary to know how controllability changes during the process (Biller and Sørensen, 2002). In this work, the nonlinear model of a batch reactive distillation column is rather used and the controllability is set by evaluating the performance of the proposed control system using one and two-points control strategy.

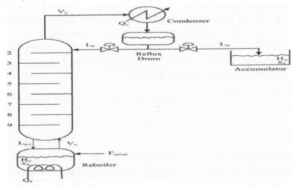

Figure 1. Batch reactive distillation with intermittent fed.

2. Method

Two control configuration are considered: (i) one-point control, and (ii) two-points control. In the one-point one, methanol inflow is manipulated to control the reboiler level; in this control scheme, the reboiler duty is kept constant at its maximum value. In the case of two-points control, two pairs were analysed. In the first structure, the reboiler duty was manipulated to control the reboiler temperature and the methanol feed was selected as input to control the bottom level. In the second structure, the heat duty supplied to the reboiler was used to control the reboiler temperature and the methanol feed flow was the input to control the LA conversion.

2.1. Reactive case study

In this work, the esterification of methanol with LA in a batch reactive distillation column was considered (Saenz et al. 2004). The initial load of LA and methanol was 100 mole (H_{N0}) in the reboiler, an equimolar mixture. Saenz et al. (2004) studied the kinetic of reaction LA and methanol to produce ML and water: they proposed different kinetics using Amberlyst 15 (Supelco) resin. The maximum temperature was 393,15 K to avoid desulfonization.

2.2. Model

An equilibrium model (total mass balances, component mass balances, equilibrium, summation, energy balances) is used for the distillation column and the condenser holdup is constant assuming a perfect control through flow distillate. The number of stages (NT) in the column was 10 including the condenser and reboiler and accumulator (Figure 1). The maximum reboiler duty was fixed at 7200 kJ h^{-1}, the maximum reboiler level was 157 moles (around 0,009 m^3). The process configuration and process features were recalled from Aqar et al. (2016), and the dynamic model followed Sørensen and Skogestad (1994) with a lineal tray hydraulic. The thermodynamically model UNIQUAC was used to calculate the Vapour–Liquid Equilibrium. A summary of column process conditions is shown in Table 1. The amount of catalyst used was 2% of the total grams of the mixture.

Table 1. Operation parameters in the column.

Maximum reboiler duty	7200 kJ h^{-1}
Maximum temperature in reboiler	373.15 K
Column pressure	100,325 N m^2
Hydraulic time constant	100(RR) s
Initial holdup on the trays	[0.25(H_{N0})]/(NT-2)
Initial liquid flow	10 % of initial charge (H_{N0})
Condenser holdup	0.25(H_{N0})

Hydraulic time constant was a function of RR and the condenser holdup was 2.5 % of the initial charge of the mixture. The vapour flow V2 is equal to the vapour flow VN.

2.3. Control strategies

To ensure reactant (methanol) in excess during the whole process, the initial reboiler level was kept constant and this will be the one-point control strategy. Control strategies summary are given in Table 2. Sørensen and Skogestad (1994) used the reboiler heat duty to control the reboiler temperature. So, it is possible to use the methanol reactant and the reboiler temperature in the next two-point control strategy.

Table 2. Control Strategies Summary.

Strategy	Control Input	Control Output
One-point Control	Methanol Flow	Reboiler Level
Two-point Control (a)	Reboiler Duty	Reboiler Temperature
	Methanol Flow	Reboiler Level
Two-point Control (b)	Reboiler Duty	Reboiler Temperature
	Methanol Flow	LA Conversion

In the first two-point control configuration (a), the level both and temperature in reboiler will be controlled. The temperature is the control output to avoid desulfonization of the resin and oligomerization of LA (Thotla and Mahajani, 2009). Another way to ensure the LA conversion is to control it directly with a SP equal or greater of 90 %. The two-point control strategy (b) used the methanol flow to control the LA conversion. In the two-point control strategies (a) and (b), the reboiler temperature is controlled with the reboiler duty. Proportional Integral (PI) controllers were used. A parametric tuning was employed through the equation (1), selecting the parameters that requires less control effort for the system. Where u_0 is the initial value of the control input at time zero.

$$effort = \int_{t_0}^{t_f} (u-u_0)^2 dt \tag{1}$$

3. Results

In this section, the simulation results will be presented. The control parameters for the PI-controllers used in the simulations are shown in the Table 3. The reboiler level is RL. For the one-point control, the reboiler duty was kept to its maximum value.

Table 3. Control parameters used in the PI controllers.

One-point Control	$K_p = 1\ kmol^{-1}$	$\tau_I = 10\ h$	$RL \rightarrow F_{MeOH}$
Two-point Control (a)	$K_p = 0.1\ kmol^{-1}$	$\tau_I = 10\ h$	$RL \rightarrow F_{MeOH}$
	$K_p = 100\ kmol^{-1}$	$\tau_I = 10\ h$	$T_{NT} \rightarrow Q_R$

Table 2. Control parameters used in the PI controllers.

Two-point Control (b)	$K_p = 100 \ kmol^{-1}$ $\tau_I = 10 \ h$	$X_{LA} \rightarrow F_{MeOH}$
	$K_p = 200 \ kmol^{-1}$ $\tau_I = 10 \ h$	$T_{NT} \rightarrow Q_R$

The reboiler level control presented a poor performance caused by the use of maximum value of reboiler duty in the whole operation; this is observed in Figure 2. Reboiler level SP was the initial charge of the mixture (100 mol).

In the one-point control, the final LA conversion is 63 % because the system reached the equilibrium as in the reflux total operation; the RR was 0.3 for this case. In the two-point control strategy (a), the level and temperature in reboiler were controlled. The RR value was 0.3. The system has good control performance still the LA conversion is 65.2%. The control output response and reboiler composition profile are given in Figure 3.

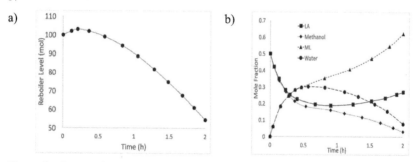

Figure 2. a) control output response SP: 100 mol. b) Reboiler composition profiles for the one-point control strategy.

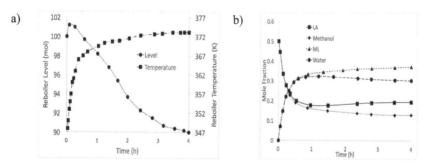

Figure 3. a) control output response; SP1= 100 mol, SP2= 373.15 K. b) Reboiler composition profiles for the two-point control (a) strategy.

In the two-point control strategy (b), the reboiler temperature was controlled satisfactorily while the LA conversion reaches the 96 % at the 6 h of operation with a RR = 0.1 and SP of 99 %. The ML mole composition at the final time (15 h) was 0.7, the methanol composition remains constant throughout the operation and the total amount of methanol flow was 397. 75 mol.

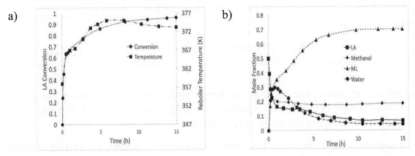

Figure 4. a) control output response b) Reboiler composition profiles for the two-point control (a) strategy.

4. Conclusions

The reboiler temperature in the reactive batch distillation was effectively controlled with two-point control (a), but it did not improve the LA conversion compared with the total reflux operation condition. Although considering temperature and reboiler level as control outputs did not raise LA conversion, reboiler temperature and LA conversion as control output did. The temperature specification was achieved using the two-point control strategy (b). Conversion control is a good option to improve the conversion in a batch reactive distillation. Even though the PI controller has shown good performance, in order to reduce the time operating a nonlinear controller is needed.

Acknowledgements

C. F. Rodriguez-Robles acknowledges the scholarship granted by CONACYT for the accomplishment of their graduate studies.

References

D. Y. Aqar, N. Rahmanian, and I. M. Mujtaba, 2016, Methyl lactate synthesis using batch reactive distillation: Operational challenges and strategy for enhanced performance, Separation and Purification Technology 158, 193-203.

N. C. T. Biller, E. Sørensen, 2002, Controllability of reactive batch distillation columns, Computer Aided Chemical Engineering 10, 445-450.

M. T. Sanz, , R. Murga, S. Beltrán, J. L. Cabezas, J. Coca, 2004, Kinetic study for the reactive system of lactic acid esterification with methanol: methyl lactate hydrolysis reaction. Industrial & engineering chemistry research 43 (9), 2049-2053.

E. Sørensen, S. Skogestad, 1994,Control strategies for reactive batch distillation, Journal of Process Control 4 (4), 205-217.

S. Thotla, S. Mahajani, 2009, Reactive distillation with side draw, Chemical Engineering and Processing: Process Intensification 48 (4), 927-937.

Anton A. Kiss, Edwin Zondervan, Richard Lakerveld, Leyla Özkan (Eds.)
Proceedings of the 29th European Symposium on Computer Aided Process Engineering
June 16th to 19th, 2019, Eindhoven, The Netherlands. © 2019 Elsevier B.V. All rights reserved.
http://dx.doi.org/10.1016/B978-0-128-18634-3.50228-9

New Methodology for Bias Identification and Estimation – Application to Nuclear Fuel Recycling Process

Amandine Duterme[a*], Marc Montuir[a], Binh Dinh[a], Julia Bisson[b], Nicolas Vigier[c], Pascal Floquet[d], Xavier Joulia[d]

[a]CEA Marcoule, Nuclear Energy Division, Research Department on Processes for Mining and Fuel Recycling, 30207 Bagnols-sur-Cèze, France

[b]ORANO, La Hague Plant, Technical Department, 50444 Beaumont-Hague, France

[c]ORANO, La Défense, Technical Department – R&D, 92400 Courbevoie, France

[d]Laboratoire de Génie Chimique, Université de Toulouse, CNRS, INP, UPS, Toulouse, France

amandine.duterme@cea.fr

Abstract

This paper focuses on the data reconciliation technique (DR) in case of numerous biases. DR improves the degree of confidence in available information and generates consistent data. The inventory and analysis of the plant data (position and type of sensors …) enable an evaluation of the process redundancy. Classical Gross Error Detection and Identification (GEDI) techniques delete the biased variables, decreasing the redundancy. This leads to information loss and possibly an inability to apply DR. The methodology proposed here combines DR, based on a reduced model, and rigorous simulations to locate and estimate multiple biases and to make data consistent in case of inter-connected flows. This methodology is applied to the nuclear fuel recycling process within the scope of a state estimation tool built on a process simulation code.

Keywords: simulation, nuclear fuel treatment, data reconciliation, bias estimation.

1. Introduction

A measurement intrinsically possess uncertainty that prevents straightforward closure of mass and energy balances. In the data reconciliation (DR) methodology, accuracy is given to the measurements by exploiting redundancies in process data and physical constraints, from steady-state mass balances (Simpson et al., 1991) to nonlinear dynamic constraints (Liebman et al., 1992).

There are two main approaches to dealing with gross errors that impact DR. The first uses Gross Error Detection and Identification (GEDI) methods (Narasimhan and Jordache, 2000) and sequentially deletes the biased variables from the DR. The redundancy, which implies the ability of DR to correct the measurements in order to satisfy the process constraints, is reduced. However, performances of DR and GEDI are still limited in disrupted cases, such as multiple flows between two units, numerous gross errors, and the position and magnitude of gross errors (Corderio do Valle et al. 2018). The second approach, not discussed in this study, modifies the objective function of DR to mitigate the effect of gross errors (Fuente, M.J. et al. 2015).

A new methodology for a nonlinear system is proposed here, combining the DR approach, based on a simplified model, and simulations, based on a first-principle model. It prevents the removal of the biased variables from the measurement set. The bias estimation is performed by the rigorous model, which enables the maximum redundancy to be kept. With a set of consistent input data generated by DR, the simulation can precisely estimate key indicators.

2. Methodology

Graph theory can be used to classify data in order to distinguish observable (measured or calculable) data from non-observable data. Among observable data, three categories can be defined: redundant data (deleting this measurement does not change the system observability), non-redundant and measured data, non-measured data. The redundant data are reconciled.

The n measurement vector X^B is linked to the true value of the measured variables X^T, the random error ε^B (assumed to be independent, with a zero mean and normally distributed), and the gross error B, here, the bias, by the following equation:

$$X^B = X^T + \varepsilon^B + B \qquad (1)$$

Data reconciliation consists of minimizing an objective function constrained by a set of constraints f:

$$\underset{X^R,B}{Min}((X^B - B) - X^R)^T V^{-1}((X^B - B) - X^R) \qquad (2)$$

$$s.t. f(X^R, \theta) = 0$$

where X^R is the n reconciled values vector, V the (n,n) covariance matrix of the measured data, and θ the parameters of the system. X^R are the best estimates of process variables, in the sense of the maximum likelihood. A study of the redundant variables, depending on the process topology and the number of independent equations, enables the determination of the ability of the DR to calculate a consistent set of reconciled data.

A global approach is to solve the DR with the rigorous model as constraints and simultaneously estimates the reconciled values X^R and the biases B. In most industrial applications, the entire first-principle model of the process, named rigorous model thereafter, cannot be directly used as the constraints f for the DR. This is generally due to implementation difficulties, such as code interfaces, and complex numerical estimation of the gradient of the constraints. The methodology proposed performs bias estimation outside the DR by an iterative strategy. It divides the problem into two sub problems: the rigorous model estimates the biases while the DR solves Eq. (2), in regards of X^R only, with a reduced model as constraints. This simplified model is made up of a selection of linear and nonlinear equations specially chosen in order to exploit all information available from the measurements. In particular, it contains the total mass balance and, depending on the case study, some partial mass balances and equations for the calculation of physical properties and fluid phase equilibria.

Figure 1 displays the new methodology. First, a map of the process (list of fluxes, units, sensors, uncertainties etc.) is built offline to generate the redundancy graph. The second step makes use of process expertise to identify bias locations, concerning measurements on internal or output fluxes. An initial simulation, based on the rigorous model, with input fluxes raw measurements $X_{in}{}^B$ as input data, gives a first estimation for the biases $B^{C(0)}$. As regards the vector of the calculated bias B^C, each element is null except for the identified biased output variables. These elements are equal to the difference between the biased measurements $X_{out}{}^B$ and the rigorous simulation calculated outputs $X_{out}{}^C$.

The DR is then solved iteratively with respect to X^R only, the values of biases $B^{C(i)}$ being considered fixed. Therefore, the process redundancy is unreduced by biases. To solve the nonlinear steady-state DR problem, this study uses the Fmincon function of Scilab software. The uncertainty for reconciled values is estimated at each DR solution by uncertainty propagation (Narasimhan and Jordache, 2000).

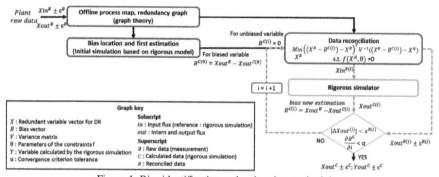

Figure 1: Bias identification and estimation methodology

At iteration i, the reconciled values of the input fluxes $X_{in}{}^{R(i)}$ are transferred to the rigorous simulator. The reconciled and calculated output flow information, $X_{out}{}^{R(i)}$ and $X_{out}{}^{C(i)}$ respectively, are compared. If the difference $|\Delta X_{out}{}^{C-R(i)}|$ between them is smaller than the uncertainties of reconciled data, the DR gives consistent values for rigorous model equations in the case of Lipschitz continuity around the solution. New bias values $B^{C(i)}$ are estimated with the last rigorous calculation. The best estimation of the bias values is reached when the biases between two iterations are constant. If these two criteria are not respected, bias information given to the DR is not satisfactory. The new bias values $B^{C(i)}$ are given to the DR for the next step. The iterations continue until the bias value estimation enables consistent data to be reached. The final DR is performed with fully known bias information, and has a minimal objective function value.

3. Case study

The methodology proposed in this paper is applied to the PUREX process and uses the PAREX simulation code developed and validated by the CEA (Dinh et al., 2008). This process carries out the treatment of spent nuclear fuel. Spent fuel contains the elements of interest, uranium and plutonium, and the waste, i.e. fission products. TBP (tributyl-phosphate) is the extractive molecule used to recover and purify uranium and plutonium through interconnected liquid-liquid extraction steps. For the final products, very specific features in terms of purity as well as extraction efficiency are required. In order to reach the necessary high performances, the metal loading of the solvent (metal mass flowrate in the solvent for a specific TBP mass flowrate) must be precisely controlled. This ratio is a sensitive parameter which deeply impacts the process state (Bisson et al., 2016). Therefore, DR aims to reduce uncertainty on this key process indicator by giving reliable input data to the rigorous simulator PAREX. It is based on first-principle models notably taking into account the partitioning of the species, the transfer, and chemical kinetics.

This study deals with an extraction-stripping step of the PUREX process where many sensors are implemented, and can be separated into two categories. Major consideration is given to a specific set of sensors essential for operation, control, and to respect the safety regulations (multiple sensors, regular checking, preventive maintenance, etc.). They are listed as reference information for the industrial plant. The secondary sensors are not used for process control or for industrial safety. They give additional information, increasing redundancy, which can help process state estimation. Some of this additional data can have biases non-detectable with previously-acquired data. A scenario is defined in order to encounter identified causes of GEDI performance loss (Corderio do Valle et al., 2018): the biases concern flows connecting the same two units, and their suppression leads to the system being non-redundant.

The initial graph of the PUREX process (Figure 2a) contains information about flows (directed arcs) and units (nodes). The redundancy graph (Figure 2b) is free of internal non-measured physical quantities: arcs depict constraints linking measurements of interest from one unit to the other. The measured variables used in the DR problem are volumetric flowrates, densities for each arc, temperature, and uranium, plutonium, nitric acid and TBP composition for specific arcs. The six identified biases are all located on internal and output flowrates measured by secondary sensors.

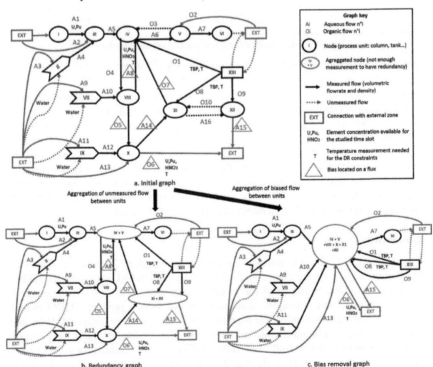

Figure 2: Graphs of an extraction-stripping step of the PUREX process.

For the classical GEDI methodology, each time a bias is detected, the redundancy decreases (Narasimhan and Jordache, 2000). The bias removal graph (Figure 2c) shows this loss of redundancy: DR cannot be applied on the aggregated node (IV+V+VIII+X+XI+XII), as only the calculation of the biases on output flowrates is possible in this scenario. In addition, classical GEDI techniques cannot locate a bias within two-way arcs between two units, such as between IV and VIII.

4. Results

The first bias estimation $B^{C(0)}$ was obtained by the comparison between measured internal and output flow-rates and the initial PAREX calculation. Four iterations were needed to obtain consistent bias estimations B^C and the minimum of the objective function. As soon as all biases can be considered constant, the iterations stop ($|B^{C(4)} - B^{C(3)}| < \alpha$, with $\alpha = 10^{-4}$ the tolerance of the convergence criterion). Note that the bias values are considerably higher than the measurement uncertainty in this scenario; therefore their contribution must be isolated.

All redundant data (80 variables), biased and unbiased, are reconciled. Figure 3 displays the differences ΔX_{out}^{B-R} between some of the measured X_{out}^B and reconciled X_{out}^R values and their corresponding uncertainties. The differences $\Delta X_{out}^{B-C(0)}$ and ΔX_{out}^{B-C} between the measured X_{out}^B and, respectively, initial $X_{out}^{C(0)}$ and final X_{out}^C PAREX values, are also laid out.

The differences $\Delta X_{out}^{B-C(0)}$ result from input measurement uncertainty. For flowrate A7, flowrate O4, and density A8, the reconciled values are closer to the PAREX calculations than the measured values, highlighting the consistency of the final data set. Moreover, the uncertainty of the reconciled values is smaller than the measurement uncertainty.

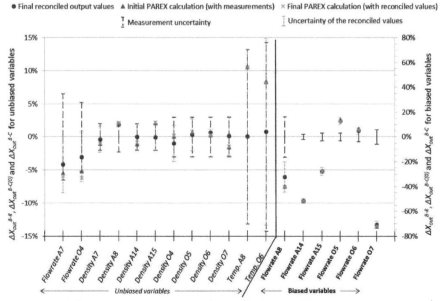

Figure 3: Comparison between measured, reconciled, and PAREX calculated outputs.

Concerning density A8 and density O6, the uncertainties of the reconciled values are very small and surround the reconciled and calculated values. These densities are linked to uranium and plutonium concentrations through a density equation. The analytical concentration measurement methods are more precise than the density sensors. Thus, the uncertainty propagation through the constraints enables the X_{out}^R to be more precise.

Concerning the other unbiased variables, mostly linked by mass balances, the reconciled X_{out}^R values are only slightly different from the measured X_{out}^B values. The two PAREX simulations give very similar results. This reflects the low sensitivity of these physical quantities to the change in the inputs from measured to reconciled values.

As PAREX input data are reconciled, accuracy is given to process indicators estimation. For instance, the uncertainty of the TBP mass flowrate is reduced by half (measurement uncertainty: 5.26 %, uncertainty of the reconciled value: 2.81 %). The uncertainties of uranium and plutonium mass flowrates are also reduced (from 2.8 % to 2.0 %), which leads to a better estimation of the metal loading in the solvent.

5. Conclusions

In classical GEDI methods, each bias decreases the redundancy of the system. The new methodology is based on nonlinear DR in which the biases are fixed and estimated by a rigorous model, with the reconciled values as input data. Therefore, the redundancy is not modified. For the bias estimation to be precise, the rigorous code and the DR iterate until the bias values offer a consistent set of reconciled data. This methodology enables explicit and inexplicit constraints for a DR problem to be addressed.

The proposed methodology was applied with success to a spent nuclear fuel treatment process. As a tool to reduce uncertainty in nuclear matter management within the plant, combining data reconciliation and the PAREX code could help in process monitoring and control.

References

J. Bisson, B. Dinh, P. Huron, C. Huel, 2016, PAREX, a numerical code in the service of La Hague plant operations, Procedia Chemistry, 32, 117-124.

E. Corderio do Valle, R. De Araùjo Kalid, A. Resende Secchi, A. Kiperstok, 2018, Collection of benchmark test problems for data reconciliation and gross error detection and idetnficaition, Computers and Chemical Engineering, 111, 134-148.

B. Dinh, P. Baron, J. Duhamet, 2008, Treatment and recycling of spent nuclear fuel, monograph of CEA DEN, the PUREX process separation and purification operations, 55-70.

M. Fuente, G. Gutierrez, E. Gomez, D. Sarabia, C. de Prada, 2015, Gross error management in data reconciliation, IFAC-PapersOnLine, 48, Issue 8, 623-628.

M. Liebman, T. Edgar, L. Lasdon, 1992, Efficient data reconciliation and estimation for dynamic processes using nonlinear programming techniques, Computers & chemical engineering, 16, 963-986.

S. Narasimhan, C. Jordache, 2000, Data reconciliation and gross error detection: an intelligent use of process data, Gulf Publishing Co.

C. Pantelides, J. Renfro, 2013, The online use of first-principles models in process operations: review, current status and future needs, Computers & chemical engineering, 51, 136-148.

D. Simpson, V. Voller, M. Everett, 1991, An efficient algorithm for mineral processing data adjustment, International Journal of Mineral Processing, 31, 73-96.

Anton A. Kiss, Edwin Zondervan, Richard Lakerveld, Leyla Özkan (Eds.)
Proceedings of the 29th European Symposium on Computer Aided Process Engineering
June 16th to 19th, 2019, Eindhoven, The Netherlands. © 2019 Elsevier B.V. All rights reserved.
http://dx.doi.org/10.1016/B978-0-128-18634-3.50229-0

A Blockchain Framework for Containerized Food Supply Chains

Dimitrios Bechtsis, [a,d,e] * Naoum Tsolakis, [a,b] Apostolos Bizakis, [a,c] Dimitrios Vlachos [a,e]

[a]*Laboratory of Statistics and Quantitative Analysis Methods, Industrial Management Division, Department of Mechanical Engineering, AUTh, Greece*

[b]*Centre for International Manufacturing, Institute for Manufacturing, Department of Engineering, School of Technology, University of Cambridge, United Kingdom*

[c]*TREDIT SA Engineering/Consulting, Thessaloniki, Greece*

[d]*Department of Automation Engineering, Alexander Technological Educational Institute of Thessaloniki, Greece*

[e]*Center for Interdisciplinary Research and Innovation (CIRI-AUTH), Balkan Center, Buildings A & B, Thessaloniki,Greece.*

dimbec@autom.teithe.gr

Abstract

Global agricultural trade flows demonstrated a three-fold growth during the past decade particularly in emerging economies. At the same time, global food scandals in conjunction with the spillover effects in economy and society highlight the unreliability of existing food tracking systems and the inefficiency in monitoring food quality and fraud incidents across global food supply chains (SCs). Blockchain technology (BCT), which has already been successfully applied on the financial industry to validate critical transactions, seems to be a promising option. This research investigates a two-stage containerized food SC by implementing a demonstrator application at the Hyperledger Fabric framework. The study findings indicates that on one hand BCT has entered its maturity phase while on the other hand its adoption in food SC operations could add significant value by authenticating critical parameters and providing enhanced traceability . At the same time, BCT enabled by other digital technologies could allow for the optimization of global food SCs. Thus, BCT constitutes a promising digital technology that provides the capability to food SC stakeholders to securely share information, enhance process control and traceability and prevent potential risks.

Keywords: container handling, food supply chain, blockchain, traceability.

1. Introduction

Global agricultural trade flows are valued at about USD 1.7 trillion annually (FAO, 2017), demonstrating a three-fold growth during the past decade, particularly in emerging economies. At the same time, wastage/losses of perishable products along with global food scandals and in conjunction with associated spillover effects in economy and society, highlight the unreliability of existing food tracking systems and the inefficiency in monitoring food quality and fraud incidents across farm-to-fork food supply chains (SCs). To that end, blockchain constitutes a promising digital technology that provides food SC stakeholders the capability to securely share information and prevent potential risks. To this effect, the financial technology (FinTech) industry provides solutions, such as the blockchain technology (BCT), that could (Christidis and Devetsikiotis, 2016): (i) provide a sustainable, robust and secure option for promoting synergies and reliable information sharing across SCs; (ii) minimize time consuming actions that comply with continuously changing legislation; (iii) reduce the need for trusted third parties; and (iv) introduce the use of smart contracts. Despite the prospects of BCT in end-to-end SC operations, a framework describing data requirements for containerized food SCs, particularly in case of maritime logistics, is still lacking.

In this regard, this research first provides a systematic literature review for mapping the blockchain landscape and proposes a blockchain framework for containerized food SCs. The framework contributes to the SC management research by establishing a blockchain infrastructure that employs smart interconnected devices across network echelons and further provides the capability to food SC stakeholders to securely share information, enhance process control and traceability, and prevent potential risks to containerized food SCs. A pilot case study in Greece is presented for the shipping container industry transporting perishable food products in refrigerated containers. The smart containers enable the uninterrupted communication with the blockchain backbone and allow the continuous and reliable storage of critical information. The case particularly focuses on critical maritime logistics processes involved in, namely: (i) preannouncement phase; (ii) port of origin storage yard; (iii) origin vessel; (iv) sea transport; (v) destination vessel; (vi) port of destination storage yard; and (vii) last mile logistics. To that end, our analysis captures key data requirements from key involved stakeholders, namely line shipping companies, terminal operators and port authorities (Lee and Song, 2017).

2. Blockchain Technology in Food Supply Networks

Global food trade has increased the total transportation time and distance from the farmer to the end consumer. In addition, fresh and processed food is vulnerable to hygiene and safety risks at all food SC echelons, thus rendering the farm-to-fork assessment of the products' quality very challenging. Aung and Chang (2013) stated that the transition of the food industry to a global value network necessitates traceability systems and faster response times to prevent food scandals. Wahyuni et al. (2018) identified risks across end-to-end food SCs by analyzing a case study for the fish industry. Across the entire spectrum of food SC operations, all stakeholders are responsible for preventing any biological and chemical contamination. Stakeholders are encouraged to use Information and Communication Technologies (ICTs) for advanced control, and monitor all 'source-make-deliver' processes to safeguard perishable products and eliminate quality variations from the stringent food safety requirements.

BCT could signal a digital transformation to the food SC landscape, affect key SC objectives (e.g. cost, quality, speed, risk reduction, flexibility) and increase transparency and accountability (Kshetri, 2018). Caro et al. (2018) presented AgriBlockIoT, a practical implementation of BCT using the Ethereum and the Hyperledger. Internet of Things (IoTs) devices capture and directly transmit data to the BCT ecosystem, while sensors ensure the automated execution of smart contracts. Tse et al. (2018) introduced BCT to the food industry in China to tackle food safety issues, as it provides a permanent record for each transaction and a replacement for traditional paper tracking systems. The authors also included an Administration and Regulator level at their implementation in order to help government track, monitor and audit the food SC. Finally, Tian et al. (2017) identified the emerging role of ICTs to the SC ecosystem and focused on an innovative decentralized BCT for implementing a real-time food tracing system based on Hazard Analysis and Critical Control Points and the IoTs.

3. Business Processes in Containerized Food Supply Chains

In order to analyze the critical business processes in containerized food SCs, this research explores the illustrative real-world case study of an end-to-end food SC. In particular, we study a Greek Third-Party-Logistics provider, located in the city of Thessaloniki, offering services for containerized food. The investigated case study takes into consideration the following business processes – farming-processing-distribution – where all retrieved critical information is stored in a BCT structure to ensure the product quality. The main focus of the Third-Party-Logistics provider is on logistics processes involved in maritime containerized global transportation through using reefers.

The end-to-end perspective of the investigated case study enables integrated data gathering and ensures visibility and traceability across all food SC stages. We recognize that the gathered data across the food SC under investigation can be segmented in two categories: (i) critical data that are considered as input to a BCT framework; and (ii) supporting data that are considered as informative metadata. We claim that the identification of these data categories supports the viable development of BCT software as the required resources (e.g. storage capacity, energy consumption) are minimized.

3.1. Farming and Processing

In a typical food SC ecosystem, the agricultural commodities are being collected by the farmer and are then forwarded downstream the SC to the retailer and the final consumers. Every farmer is characterized by primary production focus (Farmer ID and metadata) at a specific geolocation of a field (Field ID and metadata) in the farm (Farm ID and metadata). Furthermore, the farming processes involved include sowing, cultivation and harvesting of agricultural commodities. At each process it is necessary to retain critical information and transactions, including: quality of the seeds (Seed ID and metadata); cultivation practices (Cultivation Type IDs and metadata); used fertilizers (Fertilizer ID and metadata); environmental conditions during the growth of the product (environmental metadata). IoTs devices are vital for the uninterrupted communication of the SC software and BCT with the farming infrastructure. At a second stage, food and beverages processing industry utilizes the inbound agricultural materials as input to manufacture processed foods and distribute the final products to wholesalers. Critical data include: industry's ID; food processing methods; ingredients and preservatives; processed food ID and all relevant metadata. Industry 4.0 technologies enable the direct communication of the manufacturing environment (equipment and machinery) with the infrastructure and BCT.

3.2. Distribution

Wholesalers are responsible for the distribution of the final products to retailers. At the distribution process transportation conditions are critical as they could tamper the quality of food products, particularly in the case of perishable goods. Refrigerated containers store agricultural products and with the use of smart sensors a 24/7 link for informing the ICTs and BCT of the SC is ensured. The distribution of agri-commodities from the wholesaler to retailers involves three major entities: (i) port of origin; (ii) shipping line for maritime freight; and (iii) port of destination.

3.3. Port of Origin

At the port of origin, the process begins with the preannouncement phase. The list of export containers to enter the Container Terminal is required for the communication between the shipping agent who handles the container and the port authority, and it must be submitted before the arrival of each container (48h in advance). Port authorities use this list to plan the transport of the container by examining the origin, content, port of destination, Verified Gross Mass as well as space availability at the port, and the required temperature limits for preserving the agricultural products. The port's handling equipment (e.g. forklift, and straddle carrier) is used for transferring the container boxes close to the dock in order to be positioned in the vessels by Gantry Cranes based on specific EDI messages (container loading list). Containers are categorized according to: (i) port of destination; (ii) agent responsible for transportation (i.e. primary or secondary); (iii) container's properties (i.e. Verified Gross Mass, container size – 20, 40, 45, high-cube); (iv) load (e.g. barrels, parcels, boxes); (v) container type (e.g. reefer); (vi) hazard type (e.g. non-flammable gas, organic peroxides, oxidizing substances, substances demonstrating fire hazard, mass explosion hazard); and (vii) payment responsible. Once the refrigerated container is at the storage yard it is plugged at the power supply in order to maintain appropriate temperature while at the same time it communicates its status to the information communication infrastructure. The final responsibility of the port of origin is the ship loading from the origin vessel with the use of straddle carriers. Port employees are equipped with personal digital assistant devices to confirm the execution of every individual task. It is critical to continuously monitor containers, in an automated manner, to track container flows.

3.3.1. Shipping Line for Maritime Freight

Once a container is correctly loaded at the ship, the port has no further obligations and the ship has to fulfil the consignment. Provided that a refrigerated container is safely loaded, it must be plugged in to the ship's power supply. The ship's operator manually establishes the correct temperature at the reefer and ensures that it properly operates during shipment. The Electronic Data Interchange message that presents the exact position of every container in the ship is generated in order to inform all stakeholders. The reverse process then takes place as the container arrives at the port of destination. The CORPAR order for unloading containers, along with the determination of the containers' exact position, is executed to complete the unloading process.

3.3.2. Port of Destination

At the destination port, and prior to the arrival of the ship, port authorities use the bill of lading in order to plan the handling of the inbound containers by examining their origin, content and destination as well as space availability at the port and temperature requirements for preserving the transported food products. Bills of lading are submitted in either Electronic Data Interchange or Extensible Markup Language formats and contain attributes such as: quantity; weight; type of hazards; type of container (e.g. size, reefer); state information (e.g. temperature tolerance limits). Port authorities review the bill of lading and proceed to the appropriate preparatory activities. In case a specific product is on the bill of lading, the port receives the container and temporarily stores it at the destination warehouse. Containers are stored according to their load, destination or transportation agent. In the case of reefer containers, a power supply is necessary for ensuring appropriate storage conditions. The port operator is responsible for the reefer and the temperature settings which should be clearly stated at the bill of lading. Selected containers could be moved for customs control where all the obligatory custom declaration forms are checked and then the container is moved for a short-term storage to the destination storage yard. All transport information during last mile logistics must be monitored to safeguard the quality of the delivered products.

4. Digital Transformation: Blockchain Framework

Modeling the containerized food SC processes enables the digital transformation of the shipping industry and provides added value to the SC (Scholliers, 2016). Food value networks are sensitive to time critical parameters and to the transportation and storage conditions of the agricultural commodities. Perishable products should be delivered to the consumers without losing their organoleptic and nutritional properties and this could be safeguarded by the use of innovative information technology-based solutions. The proposed framework enables the integration of BCT from a farm-to-fork perspective, with special focus on the business model of containerized food SCs. Every step of the process is designed in order to provide all the obligatory and critical data to the sequential blocks of the BCT framework. The blockchain architecture could follow a public (Ethereum) or a private (Hyperledger) scenario and blockchain nodes are responsible for safeguarding the information flow of the SC and ensure data integrity. In tandem with IoTs devices (e.g. radio-frequency identification) and autonomous vehicles in industrial environments that can operate on a 24/7 basis (Bechtsis et al., 2018), communication with the blockchain allows real-time information and data sharing among all stakeholders, hence increasing traceability and trust. Finally, a demonstrator working model is implemented at the Hyperledger Fabric framework. The use case scenario involves a two-stage Distributor and Retailer SC business model that includes Assets (Product, Driver, Container, Truck), Transactions (Product Transportation Steps) and Events (Distributor and Retailer Updates). The Business model for all entities was implemented at the Hyperledger Composer and was tested using the Hyperledger's Playground interface. Every participant could receive credentials for joining the network either by simply communicating with other peers or by creating a distinct node. The proof of concept scenario used the levelDB for the immutable transaction log and CouchDB for storing metadata. Further research is needed for developing a full scale SC tool that will implement the proposed containerized food SC framework (Figure 1).

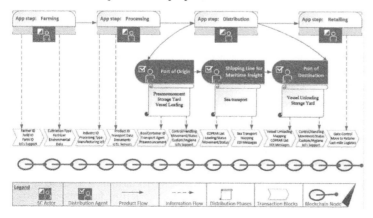

Figure 1. Blockchain Framework for Containerized Food SCs

5. Conclusions

The study findings indicate that BCT is in its maturity phase as it has been successfully applied on the FinTech processes. Digital technologies and the blockchain framework focus on critical transactions provide added value to SC operations and promote optimization techniques to global organizational challenges. The proposed framework integrates operations of the various partners across a global containerized food supply chain on a single secure information sharing process implemented on a farm-to-fork blockchain platform. A working model of a two stage SC was implemented on Hyperledger Fabric as a proof of concept demonstrator. The cost of developing/operating or leasing this platform, which is closely related to the volume of exchanged data, as well as the additional costs (sensors, IoTs devices) for collecting the necessary information will be the key parameter for documenting the feasibility (under business terms) of this system for a supply network.

References

M. M., Aung, and Y. S., Chang, 2014, Traceability in a food supply chain: Safety and quality perspectives, Food Control. Elsevier Ltd, 39(1), pp. 172–184

D., Bechtsis, N., Tsolakis, D., Vlachos, and J.S., Srai, 2018, Intelligent Autonomous Vehicles in digital supply chains: A framework for integrating innovations towards sustainable value networks, Journal of Cleaner Production, 181, pp. 60-71

M. P., Caro, M.S., Ali, M., Vecchio, R., Giaffreda, 2018, Blockchain-based traceability in Agri-Food supply chain management: A practical implementation', in 2018 IoT Vertical and Topical Summit on Agriculture - Tuscany, IOT Tuscany 2018, pp. 1–4

K. ,Christidis, and M., Devetsikiotis, 2016, Blockchains and Smart Contracts for the Internet of Things, IEEE Access, 4, pp. 2292–2303

FAO, 2017, Trade and Food Standards. Rome: Food and Agriculture Organization of the United Nations and the World Trade Organization

J. F., Galvez, J. C., Mejuto, and J., Simal-Gandara, 2018, Future challenges on the use of blockchain for food traceability analysis, TrAC - Trends in Analytical Chemistry. Elsevier Ltd, 107, pp. 222–232

N., Kshetri, 2018, 1 Blockchain's roles in meeting key supply chain management objectives, International Journal of Information Management. Elsevier, 39(December 2017), pp. 80–89

C.Y., Lee, and D.P, Song, 2017, Ocean container transport in global supply chains: Overview and research opportunities, Transportation Research Part B: Methodological, 95, pp. 442-474

J., Scholliers, A., Permala, S., Toivonen, H., Salmela, 2016, Improving the Security of Containers in Port Related Supply Chains, Transportation Research Procedia. Elsevier B.V., 14, pp. 1374–1383

F., Tian, 2017, A supply chain traceability system for food safety based on HACCP, blockchain and Internet of Things, in Proc. of the ICSSSM, pp. 1–6

D., Tse, B. , Zhang, Y., Yang, C., Cheng, H., Mu, 2017, Blockchain application in food supply information security, in 2017 IEEE International Conference on Industrial Engineering and Engineering Management (IEEM), pp. 1357–1361

H. C., Wahyuni, W. Sumarmi, and I. A., Saidi, 2018, Food safety risk analysis of food supply chain in small and medium enterprises (case study: Supply chain of fish), International Journal of Engineering and Technology(UAE), 7(2.14 Speci), pp. 229–233

Anton A. Kiss, Edwin Zondervan, Richard Lakerveld, Leyla Özkan (Eds.)
Proceedings of the 29th European Symposium on Computer Aided Process Engineering
June 16th to 19th, 2019, Eindhoven, The Netherlands. © 2019 Elsevier B.V. All rights reserved.
http://dx.doi.org/10.1016/B978-0-128-18634-3.50230-7

Data-Based Robust Model Predictive Control Under Conditional Uncertainty

Chao Shang, Wei-Han Chen, Fengqi You

Cornell University, Ithaca, New York, 14853, USA

Abstract

In this work, a novel data-driven robust model predictive control (RMPC) framework is outlined for optimal operations and control of energy systems, where uncertainty in predictions of energy intensities enters into the process in an additive manner. However, the distribution of prediction errors may be time-varying and depend on other external variables. To appropriately describe the distribution of uncertainty, a novel concept of conditional uncertainty as well as the conditional uncertainty set is proposed, which disentangles the dependence of distribution on external variables and hence reduces the conservatism. In general, the conditional uncertainty set can be modelled by integrating domain-specific knowledge and data collected from previous experience. An example arising from agricultural irrigation control is presented to illustrate of the effectiveness of the proposed methodology.

Keywords: Model predictive control, robust optimization, data-based decision-making

1. Introduction

In the past two decades, model predictive control (MPC) has established itself as an enabling technology for optimal control and operation of dynamic complex systems, with uncountable applications in process industries (Morari and Lee, 1999). In the MPC framework, a dynamic process model is adopted to describe future system behaviour, based on which the input sequence is optimized by various MPC algorithms. Despite of its significant success, MPC technology commonly requires heavy manual workload for maintenance, mainly because of model mismatch and random disturbances that drive the system to deviate from its trajectory and affect the control performance (Chu et al. 2015).

To address this issue, robust MPC (RMPC) has been extensively investigated in both academia and industries (Saltık et al., 2018), with representative applications in building control (Zhang et al., 2017), operations of smart grid (Zong et al., 2012), and irrigation systems (Delgoda et al., 2016), where process systems are influenced by various weather and climatic factors in an additive manner, such as solar radiation, wind energy, rainfall, etc. Fortunately, thanks to the development of information technology, forecasts of many climatic factors have become more and more easy to attain nowadays. However, these forecasts themselves are not sufficiently accurate, thereby bringing uncertainties in forecast errors. Consequently, in the associated RMPC scheme, the external disturbance can be decomposed into a known deterministic input arising from forecasts, and the uncertainty representing forecast errors. In existing studies, norm-based uncertainty sets, as a basic ingredient in robust optimization and RMPC, have been adopted to characterize the distribution of uncertain forecast errors.

In this work, we point out the limitations of using traditional norm-based uncertainty sets to deal with forecast errors of energy intensities. In fact, their distributions can be heavily dependent on some other external variables, e.g. forecast values, thereby showing significant time-varying characteristics. Motivated by this, we make clear such dependence with some concrete examples, and define a new conditional uncertainty set to disentangle the dependence of forecast error on external variables. In this way, primitive uncertainties independent from external variables can be effectively extracted, based on which the time-varying uncertainty distribution can be expressed in a unified manner, thereby significantly reducing the conservatism of RMPC. To determine the distribution of primitive uncertainties in a data-driven way, the use of support vector clustering (SVC)-based uncertainty set (Shang et al., 2017) is recommended. We adopt an example from irrigation control and carry out closed-loop simulations based on real weather prediction and measurement data. Case studies show that the proposed approach can help reducing conservatism of RMPC significantly and safely maintain soil moisture above a certain level with reduced water consumptions.

2. Robust Model Predictive Control under Conditional Uncertainty

2.1 Robust Model Predictive Control

We consider optimal control problems of the following linear system:

$$x_{t+1} = Ax_t + Bu_t + Ev_t + Hw_t, \tag{1}$$

here x_t and u_t are the system state and control input, v_t and w_t are deterministic disturbance and random disturbance, and $\{A, B, E, H\}$ are system matrices of appropriate dimensions (Tatjewski, 2007; Seborg et al. 2010). Given the control horizon N, Compact expression of system dynamics given the control horizon N:

$$x = Ax_0 + Bu + Ev + Hw, \tag{2}$$

where $x = [x_1^T, L, x_N^T]^T$ is the stacked sequence of future states, and $\{u, v, w\}$ can be defined in a similar way. System constraints can be generally expressed as $Cx + Du \leq f$. In the presence of uncertainty, future control inputs u are essentially functions of past disturbance w, which can be approximately parameterized by affine disturbance feedback (ADF) policy (Ben-Tal et al., 2004; Goulart et al., 2006; Shang et al., 2018):

$$\pi(w) = Pw + q. \tag{3}$$

Here P has a lower-triangular structure to respect the causality of ADF policy. By optimizing over coefficient matrix P and vector q, the formulation of traditional RMPC reads as follows (Saltık et al., 2018):

$$\min_{\pi} J_H(x_0, \pi)$$
$$\text{s.t. } x = Ax_0 + B\pi(w) + Ev + Hw \tag{4}$$
$$Cx + D\pi(w) \leq f, \ \forall w \in \Omega$$

where $J_H(x_0, \pi)$ denotes the loss function to be minimized, and Ω is the uncertainty set to describe the region of uncertainty. The most-used formulations of Ω are typically based on l_1, l_2 and l_∞ norms, e.g., $\Omega = \{w \mid \|w\|_1 \leq \Gamma\}$ (polyhedral), $\{w \mid \|w\|_2 \leq \Gamma\}$

(ellipsoidal), and $\{w \mid \| w \|_\infty \leq \Gamma\}$ (box), where Γ stands for the size parameter (Bertsimas and Sim, 2004). By adopting techniques from robust optimization, one can translate (4) into a tractable optimization problem by introducing auxiliary variables (Ben-Tal et al., 2004).

2.2 *Description of Conditional Uncertainty*

A potential limitation of (4) lies in that, unnecessarily conservative solutions may be attained because the structure of Ω cannot well describe the distribution of w. Specifically, in the optimal operations and control of energy systems, it is common that w and v represents future predictions and prediction errors of energy intensities. For example, in operations of micro-grid, predictions of wind and solar energies are typically available based on weather forecasts (Zong et al., 2012). In this case, the distribution of prediction errors may exhibit time-varying characteristics. Consider two different cases where predictions of solar energies are made in the daytime and at night. In the daytime, due to varying cloudiness, predictions tend to be uncertain, while at night predictions are much more accurate in the absence of sunlight. Hence, prediction errors in the day time will be more pronounced than those at night. It implies that one cannot use generic uncertainty sets to appropriately tackle such uncertainty, and hence research attentions are required in this vein.

We state that, such uncertainty can be understood as being dependent on some external variables, which can be prediction values in the case of solar energy predictions. This is because prediction values of solar energy will be high in the daytime and be low at night, which embody time-varying information themselves. Our idea is to disentangle the dependence the time-varying uncertainty on the external variables, thereby obtaining the underlying primitive uncertainty whose distribution tends to be time-invariant. Formally, such dependence can be described in the following generative manner:

$$w = g(\boldsymbol{\delta}, z), \tag{5}$$

where $\boldsymbol{\delta}$ and z denote external variables and the primitive uncertainty, respectively. $g(\cdot, \cdot)$ is a specific function to be determined. We further assume that with $\boldsymbol{\delta}$ given, there is a one-to-one mapping from w to z, which is denoted as:

$$z = g^{-1}(\boldsymbol{\delta}, w). \tag{6}$$

Because of the effect of decoupling $\boldsymbol{\delta}$ from z, which has a homogeneous distribution, a constant uncertainty set D can be adopted. Recently, a popular and efficient way to parameterize D is to utilize machine learning techniques (Ning and You, 2017, 2018), where SVC induced by the weighted generalized intersection kernel (WGIK) is an appealing one due to its compact representation capability, ease of implementation and robustness to outliers. The typical expression of the SVC-based uncertainty set is given by (Shang et al., 2017):

$$D = \left\{ z \mid \sum_i \alpha_i \| \mathbf{Q}(z - z^{(i)}) \|_1 \leq \theta \right\}, \tag{7}$$

where $\mathbf{Q}$ is a weighting matrix, and $\{z^{(i)}\}$ are historical samples of z. Parameters $\{\alpha_i\}$ and θ can be obtained by solving a dual quadratic program (Ben-Hur et al., 2001). With the distribution of z described by D, we can further arrive at the expression of Ω:

$$\Omega = \{w \mid w = g(\boldsymbol{\delta}, z), \ z \in D\}. \tag{8}$$

We refer to (8) as *conditional uncertainty set*, which is established based on the data-driven uncertainty set D of primitive uncertainty z.

2.3 *Robust Optimal Control with Conditional Uncertainty Set*

Because w can be expressed in terms of z, the ADF policy can be defined directly on z, which is $\pi(z) = \mathbf{P}z + q$. Up to now, we can derive the following optimal control problem:

$$\min_{\pi} J_{H}(x_0, \pi)$$

$$\text{s.t. } x = \mathbf{A}x_0 + \mathbf{B}\pi(z) + \mathbf{E}v + \mathbf{H}g(\delta, z) \tag{9}$$

$$\mathbf{C}x + \mathbf{D}\pi(z) \le f, \ \forall z \in D$$

With $g(\cdot, \cdot)$ appropriately defined, (9) can be regarded as a traditional robust optimization problem and is amenable to a dual reformulation (Shang and You, 2019). Next, an irrigation control example is presented to demonstrate its generality.

3. An Irrigation Control Example

In the context of irrigation control, the irrigation amount is manipulated to keep soil moisture level within a proper range. The water balance in the root zone soil can be mathematically expressed as $x_{t+1} = (1-c) \cdot x_t + u_t + v_t + w_t$, where x_t is the soil moisture level, u_t is the irrigation amount, and $c \cdot x_t$ stands for the amount of runoff and water percolation ($0 < c < 1$) (Delgoda et al., 2016). We assume that $v_t = -e_t + y_t$, where e_t denotes the amount of crop evapotranspiration that is precisely known, and y_t is the predicted rainfall amount. w_t is the rainfall prediction error at stage t. The control goal is to minimize water usage subject to limited water supply $u_t \le 10$ mm while maintaining soil moisture above a safety level $x_t \ge 30$ mm, because the deficiency in soil moisture can probably lead to crop devastations. Such goal can be easily interpreted by choosing appropriate loss functions and constraints.

To devise a conditional uncertainty set for w, we first plot in Figure 1 the joint empirical distribution of w_t and y_t ($t = 1$) based on real weather data collected at Des Moines, Iowa. It can be observed that the distribution of w_t is apparently asymmetric and varies with the value of y_t. In particular, when the value of w_t is negative, the ground truth rainfall amount p_t is smaller than predicted, and hence we have $w_t = p_t - y_t \ge -y_t$. By the same token, when the value of w_t is non-negative, the ground truth rainfall amount p_t is larger than predicted but smaller than a maximum $p^{\max}$, and hence we have $w_t = p_t - y_t \le p^{\max} - y_t$. Therefore, the ranges of positive and negative parts of w_t rely on the predicted value y_t, which highlight the requirement of employing the conditional uncertainty set. Based on such an observation, we propose to parameterize Ω as follows:

$$\Omega = \left\{ w \,\middle|\, w = \mathbf{G}(y)^+ w^+ - \mathbf{G}(y)^- w^-, \ w^+ - w^- \in D, \ 0 \le w^+, w^- \le 1 \right\}. \tag{10}$$

Therefore, $z = \{w^+, w^-\}$ stands for the primitive uncertainty that explains variations within w. Diagonal coefficient matrices are given by:

$$\mathbf{G}(\mathbf{y})^{+} = \mathrm{diag}\left\{ p^{\max} - y_1, \mathrm{L}, p^{\max} - y_N \right\}, \; \mathbf{G}(\mathbf{y})^{-} = \mathrm{diag}\left\{ y_1, \mathrm{L}, y_N \right\}, \tag{11}$$

which help normalizing the primitive uncertainty. To construct D, historical prediction and measurement data have been collected at Des Moines, Iowa from Jun 2016 to Oct. 2016. Since D is essentially a polytope, the above problem is a typical robust optimization problem that allows for a tractable dual reformulation. Here the external variables $\boldsymbol{\delta}$ become future rainfall predictions $\mathbf{y}$. Finally, the optimal control problem (9) becomes:

$$\min_{\boldsymbol{\pi}} \; J_H(x_0, \boldsymbol{\pi})$$

$$\text{s.t. } \mathbf{x} = \mathbf{A}x_0 + \mathbf{B}u + \mathbf{E}v + \mathbf{H}\mathbf{G}(\mathbf{y})^{+}\mathbf{w}^{+} - \mathbf{H}\mathbf{G}(\mathbf{y})^{-}\mathbf{w}^{-} \tag{12}$$

$$\mathbf{C}\mathbf{x} + \mathbf{D}\boldsymbol{\pi}(\mathbf{w}^{+}, \mathbf{w}^{-}) \le \mathbf{f}, \; \forall \mathbf{w}^{+} - \mathbf{w}^{-} \in D, \; \mathbf{0} \le \mathbf{w}^{+}, \mathbf{w}^{-} \le 1$$

where $\boldsymbol{\pi}(\mathbf{w}^{+}, \mathbf{w}^{-}) = \mathbf{P}^{+}\mathbf{w}^{+} + \mathbf{P}^{-}\mathbf{w}^{-} + \mathbf{q}$ stands for the ADF policy. Closed-loop simulations have been executed from May 2017 to Oct. 2017 under open-loop control, certainty equivalent MPC (CEMPC), and the proposed method. The results with respect to monthly irrigation amounts and constraint violation probabilities have been summarized in Tables 1 and 2.

Table 1. Monthly Irrigation Amounts in 2017

	Jun.	Jul.	Aug.	Sep.	Oct.
Open-Loop Ctrl. (mm)	300.00	283.00	226.00	268.00	92.00
CEMPC (mm)	196.64	209.03	141.11	131.12	47.32
RMPC (mm)	198.87	209.55	142.39	132.87	50.11

Table 2. Monthly Violations Probabilities of Soil Moisture Levels in 2017

	Jun.	Jul.	Aug.	Sep.	Oct.
Open-Loop Ctrl. (%)	0.00	0.00	0.00	0.00	0.00
CEMPC (%)	21.67	23.38	15.32	17.50	5.98
RMPC (%)	0.00	0.00	0.00	0.00	0.00

It can be observed that, the open-loop control strategy that determines weekly irrigation schedule in advance, leads to zero probability of soil moisture deficiency along with the greatest water consumptions. In contrast, CEMPC yields the most aggressive control actions, since the least amount of water is used. However, the price is that state constraints are frequently violated, which may heavily compromise crop productivity. The proposed data-based RMPC achieves a satisfactory performance. On one hand, it reduces the water consumption by about 40% compared to open-loop control, which is the state-of-the-art technology. On the other hand, it successfully safeguards soil moisture level constraints. This owes to an appropriate characterization of the uncertainty distribution based on the proposed conditional uncertainty set constructed with historical data. The proposed method can achieve a desirable balance between saving water, and hedging against uncertainty in forecast errors with the aim to reliably avoid potential harm to crop.

4. Conclusions

This paper features a new framework for RMPC under additive conditional uncertainty, where the dependence of original uncertainty on primitive uncertainty is clearly described, based on which data-driven uncertainty sets are constructed to indirectly characterize the distribution of original uncertainty. The proposed method applies to optimal operations and control of energy systems where predictions of energy intensities are available but inevitably induce uncertain prediction errors. By compactly describing the distribution of uncertainty with the proposed conditional uncertainty set, the conservatism of control actions can be efficiently reduced, which uncovers the value of leveraging big data. A real-world case study based on agricultural irrigation systems is performed to show the effectiveness and merits of the proposed method.

References

A. Ben-Hur, D. Horn, H. T. Siegelmann, V. Vapnik, 2001, Support vector clustering. Journal of Machine Learning Research, 2(Dec), 125–137.

A. Ben-Tal, A. Goryashko, E. Guslitzer, A. Nemirovski, 2004, Adjustable robust solutions of uncertain linear programs, Mathematical Programming, 99, 351–376.

D. Bertsimas, M. Sim, 2004, The price of robustness. Operations Research, 52(1), 35–53.

Y. Chu, F. You, 2015, Model-based integration of control and operations: Overview, challenges, advances, and opportunities. Computers & Chemical Engineering, 83, 2-20.

D. Delgoda, H. Malano, S. K. Saleem, M. N. Halgamuge, 2016, Irrigation control based on model predictive control. Environmental Modelling & Software, 78, 40–53.

P. J. Goulart, E. C. Kerrigan, J. M. Maciejowski, 2006, Optimization over state feedback policies for robust control with constraints. Automatica, 42(4), 523–533.

M. Morari, J. H. Lee, 1999, Model predictive control: Past, present and future. Computers & Chemical Engineering, 23(4-5), 667–682.

C. Ning, F. You, 2017, Data-Driven Adaptive Nested Robust Optimization: General Modeling Framework and Efficient Computational Algorithm for Decision Making Under Uncertainty. AIChE Journal, 63, 3790-3817.

C. Ning, F. You, 2017, A data-driven multistage adaptive robust optimization framework for planning and scheduling under uncertainty. AIChE Journal, 63, 4343-4369.

C. Ning, F. You, 2018, Data-driven stochastic robust optimization: General computational framework and algorithm leveraging machine learning for optimization under uncertainty in the big data era. Computers & Chemical Engineering, 111, 115-133.

C. Ning, F. You, 2018, Data-driven decision making under uncertainty integrating robust optimization with principal component analysis and kernel smoothing methods. Computers & Chemical Engineering, 112, 190-210.

M. B. Saltık, L. Özkan, J. H. Ludlage, S. Weiland, P. M. Van den Hof, 2018, An outlook on robust model predictive control algorithms: Reflections on performance and computational aspects. Journal of Process Control, 61, 77–102.

D.E. Seborg, D.A. Mellichamp, T.F. Edgar, et al. (2010). Process Dynamics and Control: John Wiley & Sons.

C. Shang, X. Huang, F. You, 2017, Data-driven robust optimization based on kernel learning. Computers & Chemical Engineering, 106, 464-479.

C. Shang, F. You, 2018, Distributionally robust optimization for planning and scheduling under uncertainty. Computers & Chemical Engineering, 110, 53-68.

C. Shang, F. You, 2019, A data-driven robust optimization approach to scenario-based stochastic model predictive control. Journal of Process Control, 75, 24-39.

P. Tatjewski. (2007). Advanced control of industrial processes: structures and algorithms: Springer.

X. Zhang, M. Kamgarpour, A. Georghiou, P. Goulart, J. Lygeros, 2017, Robust optimal control with adjustable uncertainty sets. Automatica, 75, 249–259.

S. Zhao, F. You, 2019, Resilient supply chain design and operations with decision-dependent uncertainty using a data-driven robust optimization approach. AIChE Journal, 65, 1006-1021.

Y. Zong, D. Kullmann, A. Thavlov, O. Gehrke, H. W. Bindner, 2012, Application of model predictive control for active load management in a distributed power system with high wind penetration. IEEE Transactions on Smart Grid, 3(2), 1055–1062.

Anton A. Kiss, Edwin Zondervan, Richard Lakerveld, Leyla Özkan (Eds.)
Proceedings of the 29th European Symposium on Computer Aided Process Engineering
June 16th to 19th, 2019, Eindhoven, The Netherlands. © 2019 Elsevier B.V. All rights reserved.
http://dx.doi.org/10.1016/B978-0-128-18634-3.50231-9

Centralised versus localised supply chain management using a flow configuration model

Bogdan Dorneanu[a], Elliot Masham[a], Evgenia Mechleri[a], Harvey Arellano-Garcia[*a,b]

[a]*Department of Chemical & Process Engineering, University of Surrey, Guildford, United Kingdom*

[b]*LS Prozess- und Anlagentechnik, Brandenburgische Technische Universität Cottbus-Senftenberg, Cottbus, Germany*

h.arellano-garcia@surrey.ac.uk

Abstract

Traditional food supply chains are often centralised and global in nature. Moreover they require a large amount of resource which is an issue in a time with increasing need for more sustainable food supply chains. A solution is to use localised food supply chains, an option theorised to be more sustainable, yet not proven. Therefore, this paper compares the two systems to investigate which one is more environmentally friendly, cost efficient and resilient to disruption risks. This comparison between the two types of supply chains, is performed using MILP models for an ice cream supply chain for the whole of England over the period of a year. The results obtained from the models show that the localised model performs best environmentally and economically, whilst the traditional, centralised supply chain performs best for resilience.

Keywords: supply chain management, flow configuration model, ice cream, centralised versus localised, MILP.

1. Introduction

Food supply chains (FSCs) are complex due to the large amount of actors, but are also unique amongst the supply chains (SCs) due to seasonal production and demand, as well as requiring to operate at refrigerated temperatures to ensure food safety and quality. The traditional FSC require a large amount of resources, which is an issue in a time with increasing need for more sustainable food supply [1, 2]. The majority of the FSC are very global and centralised, a solution designed to minimise the cost of the whole chain and therefore to increase the profits of a company. However, these are not considered sustainable due to the large amount of resource they consume [3]. A theoretical approach has been proposed to make the FSC more sustainable by having it localised [4, 5], a solution where all actors function within the same area. These FSC are seen to be more sustainable since the local customers are using local ingredients and a decrease in CO_2 emissions and resource demand can be achieved [6]. Thus there are significant business trade-offs to consider in deciding on an operational strategy – and some of these trade-offs may be entirely valid between food chains and food transport. Although the localised FSC has been theorised to be more sustainable in environmental terms, this has not been proven through a model, as most existing studies optimise without considering all the chain stages [7, 8]. For comparison of centralised and localised SCs, MILP models have been developed. To perform this comparison, an ice cream SC model is developed for England, incorporating the use of the 8N configuration neighborhood flow approach [9].

2. The neighbourhood flow approach and the FSC

The model is built for the ice cream SC over a period of one year and consists of 4 different actors. The first set of actors are represented by a total of 16 different suppliers, all with different supply capacities, providing two ingredients. Half of the suppliers will simulate dairy farms, supplying only milk, while the other half will simulate sugar factories, providing the sugar. The second set of actors are the factories where both milk and sugar from the suppliers will use. The milk is transported from suppliers to each factory, where ingredient storage is available, before manufacturing. In the centralised model there is 1 large factory while the localised model considers 8 small factories. All factories make 4 different types of ice cream (SKUs). The third set of actors is represented by the SKU warehouses, which is the location where products are stored during the time the factories are producing more than the demand. In the localised model there will be 8 small warehouses, while the centralised model assumes 3 large ones. Only the product that leaves the factory and needs to be stored will be transported to the warehouse, while the rest is sent directly to the demand centre. The demand centre is the final actor and the customer of the FSC. There is a total of 8 different demand centres where the demand and locations are the same for both the localised and the centralised model. The demand at the centres varies for each type of SKU and across the individual 52 weeks of a year. The movement between the actors of the FSC is summarised in Fig.1.

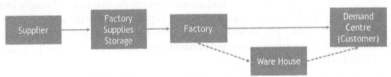

Figure 1: The flow between the actors of the ice cream supply chain

To simulate the locations, the transport of ingredients and ice cream between the actors all the models will use a neighbourhood flow approach with the use of the 8N configuration as set out by [9]. The neighbourhood flow approach requires the use of a grid system which consists of multiple different regions. A region is specified to contain an actor and the ice cream or the ingredients can be transported from one actor to another through flow from one region to another neighbouring region through 4N or 8N flows. If the corresponding actors that send and receive the transported goods are not in neighbouring regions the simulation model will move the goods from one neighbouring region to another and repeat this process until the goods are delivered to the required actors. The size of each region is taken as being 50x50 km, the distance travelled by the goods through 4N neighbourhood flow is 50 km, while for the 8N flow is 70.7 km. the grid system that the neighbourhood flow approach uses has been set up to simulate the whole England in both models, which resulted in a total of 69 different regions of 50x50 km size (Fig.2). In the centralised model all actors required for the SC can be in any region, and the ingredients and products can be transported through any region shown in Fig.2. For the localised models, different local area sizes and location are selected where the different regions of England are designated by double lines in Fig.2. Moreover, in the localised models London and the SE of England are considered to be the same region due to Greater London's small size, and the practical assumption that some actors may not be present here, such as the farm, the warehouse or the factory.

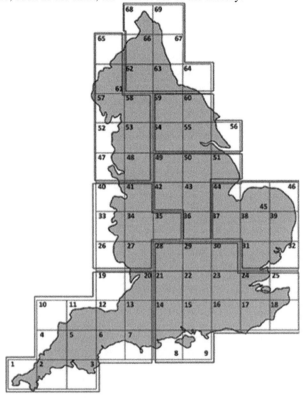

Figure2: Grid system for England, used to simulate the FSC models

3. Mathematical formulation

In all the MILP models developed, the constraints remain the same and only the objective equations differ.

3.1. Model constraints

The amount of ingredients h, being taken from the supplier in region g in week t, cannot exceed the amount of ingredients that are available weekly from the supplier:

$$SupplyUse_{h,g,t} \leq MaxSupply_{h,g} \qquad \forall h, g, t \qquad (1)$$

The total amount of ingredients h, stored in the factory in region g in week t cannot exceed the factory maximum ingredient storage.

$$INVIng_{h,g,t} \leq INVIngCap_{h,g} \qquad \forall h, g, t \qquad (2)$$

The amount of ingredients h in the factory storage in region g in week t is equal to the amount of ingredients h that is in the factory storage in the previous week, plus the ingredient delivered from the suppliers minus the amount of ingredients used to produce the SKU in the factory in week t. The set up time and the production time of every SKU in a factory in region g in week t cannot exceed the maximum weekly production time allowed for each SKU. The ingredients and the SKU should satisfy mass balances for each region g and week t.

3.2. Economic objective equation

The objective of the centralised and localised economic models is to minimise the total economic cost of the whole SC. The costs consists of the factory production costs, *ProdC*, the ingredients purchase, transport, *IngTransC* and storage, *ISC*, *WHC* costs, the SKU storage and transport, *SKUTransC* costs and a penalty if demand is not met at the demand centres, *DNMPen*:

$$TEC = ISC + ProdC + WHC + IngTransC + SKUTransC + DNMPen \qquad (3)$$

3.3. Resilience objective equation

In the resilience model, localised and the centralised SC are tested under disruptive scenario of the supplier not being able to provide ingredients over an 8 week period of high demand. The objective is to minimise the total demand that is not met at the demand centres.

$$DNM = \sum_i \sum_g \sum_t (Demand_{i,g,t} - Retailer_{i,g,t}) \qquad (4)$$

To this end, an extra equation is added to the model to calculate the rate of fulfilment for every SKU in every week t for all the demand centres.

3.4. Model parameters

The value of the parameters is chosen to closely match realistic values of an ice cream SC. The total yearly demand is split into weekly demands that match the typical seasonal demand (Fig.3). The weekly demand split into the 4 SKUs, and split between the 8 different demand centres across England. The selection of the regions is based on the population for the localised area. The demand centres and their proportion are localised in regions 6 (10%), 23 (32%), 34 (10%), 37 (11%), 43 (8%), 53 (13%), 55 (10%), 67 (5%). The location and maximum supply available from each supplier is the same for both the localised and centralised models. The availability of supply is assumed constant over the year. Each localised region must have a milk and a sugar supplier. The region with milk suppliers are assumed in locations where the density of dairy farms is high. The milk suppliers are in regions 5 (600 t/wk), 16 (1,550 t/wk), 39 (600 t/wk), 41 (600 t/wk), 49 (500 t/wk), 60 (600 t/wk), 61 (700 t/wk), 63 (300 t/wk), while the sugar supplier are in regions 3 (450 t/wk), 9 (800 t/wk), 31 (350 t/wk), 40 (350 t/wk), 43 (300 t/wk), 56 (300 t/wk), 57 (350 t/wk), 67 (150 t/wk). For the centralised model there will be only one large factory located in region 42, the most central region.

For the localised model there are 8 small factories in regions 25, 52, 44, 35, 13, 54, 51, 62. It is assumed that ice cream consists of 75% of milk and 25% of sugar. The factory maximum ingredients capacity differs for all the 8 different factories in the localised model, and the factory maximum ingredients capacity for the single centralised factory in the centralised model is the sum of all the 8 different localised factories maximum ingredients capacities. The warehouses are located in regions 21 (4,000 t), 34 (2,000 t), and 54 (1,170 t) for the centralised model, and in regions 17 (3,000 t), 57 (1,100 t), 45 (620 t), 26 (590 t), 7 (560 t), 56 (560 t), 50 (475 t), 50 (475 t) and 64 (260 t).

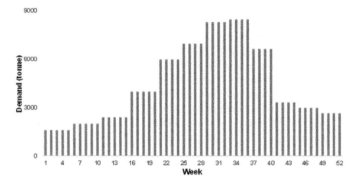

Figure3: The weekly demand profile

Moreover, all the cost parameters (ingredients, transport, manufacturing, storage) are expressed according to the values for the UK. If the SC does not meet the demand there is a penalty defined as the price the ice cream would sell at the demand centre.

4. Results

4.1. Economic results

When the centralised economic model is optimised to minimise the total SC cost, the results was that it would cost a total of £214.3 million over all of the 52 weeks. Since over the 52 weeks the SC handles 231,600 tonnes of ice cream, the cost in terms of every unit of ice cream is £925.25 per tonne of ice cream. When the localised economic model is optimised to minimise the total SC costs, it would cost £213 million over the 52 week period. Since the localised SC handles overall the same amount of ice cream as the centralised one, the cost in terms of unit of ice cream is £919.50 per tonne of ice cream. Fig.4 shows the layout of the economic model for the two types of SC, for week 33.

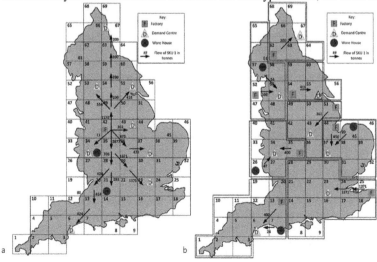

Figure4: Flow of SKU1 in week 33 for the a) centralised SC and b) localised SC

4.2. Resilience results

The disruption scenario assumes that the milk supplier in region 16 is not able to provide any milk over an 8 week period in the high season. In the centralised SC the minimum of the demand not being met achieved is 8,582 tonne of ice cream, with the model specifying the overall weekly fulfilment rate. The resilience index for the centralised SC is 0.932. For the localised SC the minimum demand not being met achieved from the optimised model is 16,485 tonnes of ice cream, with the optimised model specifying the overall weekly fulfilment rates. The resilience index in this case is 0.869. Fig.5 shows the demand profile fulfilment during the SC disruption scenario.

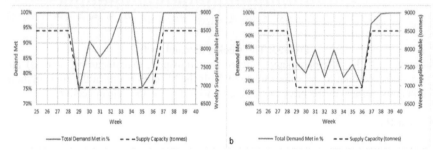

Figure5: Demand profile fulfilment during the SC disruption scenario for the a) centralised and b) localised model;

5. Conclusions

This work describes the development of two MILP models to simulate a centralised and a localised ice cream SC in England over 52 weeks, using the 8N neighbourhood flow configuration. The two SC are compared at economical and resilience level. The localised SC is £1.377 million cheaper than the centralised one due to the shorter transportation distances between the various actors. At resilience level, the centralised SC has a higher resilience index, due to the possibility to being supplied from other regions when one of the suppliers fails to fulfil the demand. Although centralised distribution shows efficiency benefits, the distributed system may have the ability to reduce temporal, geographical and administrative gap between decision makers and the impact of their decisions, reduce the possibility of widespread impacts if production systems fail and avoid the need for long-distance transport infrastructure. Rather than considering the two configurations as alternatives to each other, a sustainability assessment considering economic, environmental and societal situations where complementarities and synergies between centralised and localised FSC occur should be investigated.

References

[1]. Akkerman, et al., 2010, *OR Spectrum 32*, 863; [2]. Kirwan, et al., 2017, *Journal of Rural Studies 52*, 21; [3]. Mundler & Rumpus, 2012, *Food Policy 37*, 609; [4]. Rothwell, et al., 2016, *Journal of Cleaner Production 114*, 420; [5]. Rahimifad, et al., 2017, *Sustainable Design and Manufacturing*, Springer, 13; [6]. Rauch, et al., 2016, *Journal of Cleaner Production 135*, 127; [7]. Esteso, et al., 2018, *International Journal of Production Research 56*, 4418; [8]. Zhu, et al., 2018, *International Journal of Production Research 56*, 5700; [9]. Akgul, et al., 2012, *Biomass and Bioenergy 41*, 55

Anton A. Kiss, Edwin Zondervan, Richard Lakerveld, Leyla Özkan (Eds.)
Proceedings of the 29th European Symposium on Computer Aided Process Engineering
June 16th to 19th, 2019, Eindhoven, The Netherlands. © 2019 Elsevier B.V. All rights reserved.
http://dx.doi.org/10.1016/B978-0-128-18634-3.50232-0

An improved approach to scheduling multipurpose batch processes with conditional sequencing

Nikolaos Rakovitis, Jie Li,* Nan Zhang

School of Chemical Enginneering and Analytical Science, University of Manchester, Manchester M13 9PL, UK

Jie.li-2@manchester.ac.uk

Abstract

Scheduling of multipurpose batch processes has gained much attention in the past decades. Numerous models using different mathematical modelling approaches have been proposed. The model size and computational performance largely depend on the number of time points, slots or event points required. Most existing models still require a great number of time points, slots or event points mainly because a consumption task always takes place after its related production tasks regardless of whether it consumes materials from the related production tasks. Although two existing models have been developed to overcome this limitation, they can either lead to real time violation or generate suboptimal solutions in some cases. In this work, we develop an improved unit-specific event-based model in which a consumption task takes place after its related production task only if it consumes materials from the production task. A consumption task starts immediately after its related production task completes only if there is no enough storage for materials produced from the producing task. We also allow production and consumption tasks related to the same state to take place at the same event points. The results show that the proposed model generates same or better solutions than existing models with less number of event points and less computational time.

Keywords: Scheduling, Multipurpose batch process, Mixed-integer linear programming.

1. Introduction

Scheduling of multipurpose batch processes has received much attention during the past three decades (Harjunkoski et. al. 2014). A number of mathematical models have been developed using discrete-time and continuous-time modelling approaches. The continuous-time modelling approaches include slot-based, global event-based, sequence based, and unit-specific event-based approaches. The model size and the computational performance largely depend on the number of time points, slots or event points required. Most existing models still require a great number of time points, slots, or event points to generate the optimal solution since a consuming task must start after the related producing tasks even if it does not consume materials from the producing tasks. In addition, production and consumption tasks related to the same state are not allowed to take place at the same time points, slots, or event points. To tackle these issues, Seid and Majozi (2012) imposed consuming tasks to take place after related production tasks if materials produced from the production units are used by any related consuming task. However, their model could lead to storage violations in real time. Vooradi and Shaik

(2013) enforced a consuming task to take place after a related production task only if this consuming task uses materials from the specific production task with introduction of a high number of additional binary variables, leading to an intractable model size. Furthermore, materials are not allowed to be stored in a processing unit if it is idle, leading to suboptimality in some cases. These two efforts do not allow production and consumption tasks related to the same state to take place at the same time points, slots, or event points yet. Recently, Shaik and Vooradi (2017) and Rakovitis et al. (2018) did allow related production and consumption tasks to take place at the same event points. They still impose a consuming task must always start after the related producing tasks.

In this work, we develop an improved unit-specific event-based approach for this scheduling problem, which requires less number of event points than existing formulations. We use the definition of recycling tasks from Rakovitis et al. (2018) and allow related non-recycling producing and consumption tasks to take place at the same event points. Furthermore, the proposed model sequence a processing unit which process a consumption task with its related production task, only if it consumes materials from the unit that process that task, and it force the finish time of a consuming task to be equal to the start time of the production task, if materials produced by the producing task are not able to be stored. Finally, we allow processing units to store the materials that were produced during the events that the unit is inactive. The computational results demonstrate that the proposed model generates same or better solutions using less number of event points and less computational time.

2. Problem description

Figure 1 illustrates a multipurpose batch process facility involving S ($s = 1, 2, ..., S$) states, J ($j = 1, 2, ..., J$) processing units and I ($i = 1, 2, ..., I$) tasks. The states include feed materials S^{feed}, intermediates S^{IN} and final products S^{FP}. Each processing unit can process I_j tasks, but process at most one task at a time. Two storage policies including unlimited or finite intermediate storage (UIS or FIS) for intermediate states are considered, while unlimited storage is available for feed materials and products. Unlimited amount of feed materials is assumed during the scheduling horizon. Finally, unlimited wait policy (UW) is considered for all batches. Given production recipes, minimum/maximum unit capacities, processing times, storage policy for each intermediate and the scheduling horizon, the scheduling problem is to determine the optimal allocation of tasks to units, the start and end times of each task, task sequences in a processing unit as well as the inventory profiles in order to maximize productivity.

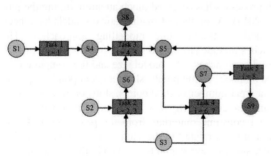

Figure 1 STN representation of a multipurpose batch process facility

3. Mathematical formulation

The mathematical formulation is developed using unit-specific event-based modelling approach because the advantages of this modelling approach have been well established in the literature. In the model, we have similar allocation constraints, capacity constraints, material balance constraints, and duration constraints to those of Rakovitis et al. (2018), which are not presented here. Next, we introduce new features of the proposed model.

3.1. Different tasks in different units

We define continuous variables $bc_{i,i',s,n}$ as material s produced from task i consumed by task i' at event n. Thus, the material consumed should not exceed total available materials.

$$-\sum_{j'}\sum_{i'\in I_s^C, i'\in I_{f}}\left(\rho_{s,i'}\cdot\sum_{n\leq n'\leq n+\Delta n}b_{i',j',n,n'}\right)\leq ST_{s,n-1}+\sum_{i\in I_s^P}\sum_{i'\in I_s^C}bc_{i,i',s,n}\qquad \forall s\in \mathbf{S}^{IN},n \qquad (1)$$

The amount of materials from i to i' (i.e., $bc_{i,i',s,n}$) should not exceed the production amount of task i or consumption amount of task i'.

$$\sum_{i'\in I_s^C}bc_{i,i',s,n}\leq \rho_{s,i}\cdot\sum_{n-\Delta n\leq n'\leq n}b_{i,j,n',n}\qquad \forall s\in \mathbf{S}^{IN}, i\in I_s^P, i\notin \mathbf{I}^{Re}, j, n \qquad (2)$$

$$\sum_{i\in I_s^P}bc_{i,i',s,n}\leq -\rho_{s,i'}\cdot\sum_{n\leq n'\leq n+\Delta n}b_{i',j',n,n'}\qquad \forall s\in \mathbf{S}^{IN}, i'\in I_s^C, j, n \qquad (3)$$

$$bc_{i,i',s,n}\leq z_{j,j',s,n}\cdot \rho_{s,i}\cdot B_{i,j}^{\max}\qquad \forall s\in \mathbf{S}^{IN}, i\in I_s^P, i\in I_j, i'\in I_s^C, i'\in I_{j'}, j, j', j\neq j', n \qquad (4)$$

$$T_{s,j,n}\geq Tf_{j,n}-M\cdot\left(1-\sum_{i\in I_s^P, i\in I_j}\sum_{n-\Delta n\leq n'\leq n}w_{i,j,n',n}\right)\qquad \forall s\in \mathbf{S}^{IN}, j, n \qquad (5)$$

$$T_{s,j,n}\leq Ts_{j',n}+M\cdot\left(2-\sum_{i'\in I_s^C, i\in I_j}\sum_{n\leq n'\leq n+\Delta n}w_{i',j',n,n'}-z_{j',j,s,n}\right)$$
$$\forall s\in \mathbf{S}^{IN}, j, j', i\in I_j, i\in I_s^P, i\notin \mathbf{I}^{Re}, n \qquad (6)$$

An active consumption task i' should be processed in a unit after the time that state s is available at the previous event n.

$$T_{s,j,n}\leq Tf_{j',n+1}+M\cdot\left(1-\sum_{i'\in I_s^C, i'\in I_{f}}\sum_{n+1\leq n'\leq n+1+\Delta n}w_{i',j',n+1,n'}\right)$$
$$\forall s\in \mathbf{S}^{IN}, j, j'\in I_s^P, i\in I_j, i\notin \mathbf{I}^{Re}, n<N \qquad (7)$$

3.2. Sequencing constraints for limited intermediate storage policies

For FIS states, it should be examined whether the materials produced can be stored.

$$\sum_{i\in I_s^P, i\in I_j, i\notin \mathbf{I}^{Re}}\sum_{j}\rho_{i,s}\sum_{n-\Delta n\leq n'\leq n}b_{i,j,n',n}+ST_{s,n}\leq ST_s^{\max}+\sum_{i\in I_s^P}\sum_{i'\in I_s^C}bs_{i,i',s,n}\qquad \forall s\in \mathbf{S}^{FIS},n \qquad (8)$$

If they cannot be stored then they either have to be consumed immediately or stored in the unit, processing consuming task i' at event n, must take a non-zero value.

$$bs_{i,i',s,n} \leq B_{i,j}^{\max} \left(x_{i,i',s,n} + u_{i,i',s,n} \right) \qquad \forall s \in \mathbf{S}^{FIS}, j, j', i \in \mathbf{I}_s^P, i' \in \mathbf{I}_s^C, n \tag{9}$$

$$x_{i,i',s,n} + u_{i,i',s,n} \leq 1 \qquad \forall s \in \mathbf{S}^{FIS}, j, j', i \in \mathbf{I}_s^P, i' \in \mathbf{I}_s^C, n \tag{10}$$

$$\rho_{s,i} \cdot \sum_{n-\Delta n \leq n' \leq n} b_{i,j,n',n} \geq \sum_{i' \in \mathbf{I}_s^C} bs_{i,i',s,n} \qquad \forall s \in \mathbf{S}^{FIS}, j, i \in \mathbf{I}_s^P, i \in \mathbf{I}_j, i \notin \mathbf{I}^{Re}, n \tag{11}$$

$$-\rho_{s,i'} \cdot \sum_{n \leq n' \leq n+\Delta n} b_{i',j',n,n'} \geq \sum_{i \in \mathbf{I}_s^P} bs_{i,i',s,n} \qquad \forall s \in \mathbf{S}^{FIS}, i' \in \mathbf{I}_s^C, j', n \tag{12}$$

If $bs_{i,i',s,n}$ takes a non-zero value, then the start time of consuming task as well as the finish time of the producing task are enforced to be equal.

$$T_{s,j,n} \leq Tf_{j,n} + M \cdot \left(1 - \sum_{i \in \mathbf{I}_s^P, i \in \mathbf{I}_j, i \notin \mathbf{I}^{Re}} x_{i,i',s,n} \right) \qquad \forall s \in \mathbf{S}^{FIS}, j, n \tag{13}$$

$$T_{s,j,n} \geq Ts_{j',n} - M \cdot \left(1 - \sum_{i' \in \mathbf{I}_s^C} x_{i,i',s,n} \right) \qquad \forall s \in \mathbf{S}^{FIS}, j, j', i \in \mathbf{I}_s^P, i \in \mathbf{I}_j, i \notin \mathbf{I}^{Re}, n \tag{14}$$

In order to avoid real time storage violations, between producing and consumption tasks occurring at the previous event the following constraint is introduced.

$$Tf_{j',n+1} \geq Ts_{j,n} - M \cdot \left(1 - \sum_{n+1-\Delta n \leq n' \leq n+1} w_{i',j',n',n+1} \right) \forall s \in \mathbf{S}^{FIS}, j, j', i' \in \mathbf{I}_s^C, i' \in \mathbf{I}_j, n < N \tag{15}$$

3.3. Allow production units to hold materials

We define $us_{j,n}$ as the extra amount of materials stored in a processing unit j at n.

$$ST_{s,n} \leq ST_s^{\max} + \sum_{j} \sum_{i \in \mathbf{I}_s^P, i \in \mathbf{I}_j} us_{j,n} \qquad \forall s \in \mathbf{S}^{FIS}, n \tag{16}$$

The amount of materials stored in a unit at n cannot exceed the amount produced at $n-1$.

$$us_{j,n} \leq \sum_{s \in \mathbf{S}^{FIS}} \sum_{i \in \mathbf{I}_j, i \in \mathbf{I}_s^P} \sum_{n-1-\Delta n \leq n' \leq n} \rho_{i,s} b_{i,j,n',n-1} + us_{j,n-1} \qquad \forall j, n > 1 \tag{17}$$

If a unit j holds some material at n, then it cannot process any task at this event.

$$us_{j,n} \leq \left[\max_{i \in \mathbf{I}_j} \left(B_{i,j}^{\max} \right) \right] \cdot \left(1 - \sum_{i \in \mathbf{I}_j} \sum_{n-\Delta n \leq n' \leq n} \sum_{n \leq n' \leq n'+\Delta n} w_{i,j,n',n''} \right) \qquad \forall j, n \tag{18}$$

We define binary variables $xs_{j,n}$ to denote if a unit j holds materials.

$$us_{j,n} \leq \left[\max_{i \in \mathbf{I}_j} \left(B_{i,j}^{\max} \right) \right] \cdot xs_{j,n} \qquad \forall j, n \tag{19}$$

Finally, if a unit holds some material at n, the finish time of all consuming tasks must be equal to the start time of that unit.

$$T_{s,j,n} \leq Tf_{j,n} + M \cdot \left(1 - xs_{j,n} \right) \qquad \forall s \in \mathbf{S}^{FIS}, j, i \in \mathbf{I}_s^P, i \in \mathbf{I}_j, n \tag{20}$$

$$T_{s,j,n} \geq Ts_{j',n} - M \cdot \left(2 - xs_{j,n} - \sum_{n \leq n' \leq n + \Delta n} w_{i',j',n,n'} \right)$$

$$\forall s \in \mathbf{S}^{FIS}, j, j', i \in \mathbf{I}_s^P, i \in I_j, i \notin \mathbf{I}^{Re}, i' \in \mathbf{I}_s^C, i' \in I_{j'}, n \quad (21)$$

4. Computational results

We solve five examples including four well-established examples in the literature (Shaik and Floudas, 2009; Li and Floudas, 2010) to illustrate the capability of the proposed model. Both UIS and FIS policies are considered. We also compare the performance of the proposed model with the models of Li and Floudas (2010) (denoted as LF) and Vooradi and Shaik (2013) (denoted as VS). The computational results are given in Tables 1-2. From Tables 1-2, it can be observed that the proposed model reduces the total number of event points required for all examples compared to the LF model. This is because as it is depicted in figure 2 unit J1 is able to process 100 units of intermediate state using the proposed formulation since 50 of them can be temporary stored in the processing unit before being processed by J2, while by using existing formulations only 60 units can be processed since they cannot be stored in the processing unit and the storage capacity is 10. Compared to the VS model, the proposed model reduces the number of event points required for most examples except Examples 2 and 3 because all tasks in these two examples have to be treated as recycling task. The proposed model required less computational time for all examples compared to the LF and VS models especially for Example 3 which is reduced by 55% with UIS policy and 15% with FIS policy. More significantly, the proposed model generates a better solution than LF and VS models for Example 5. This is because a production unit is allowed to store materials produced. Therefore, the amount of materials produced can exceed the storage capacity. The optimal schedule for the proposed model is illustrated in Figure 2.

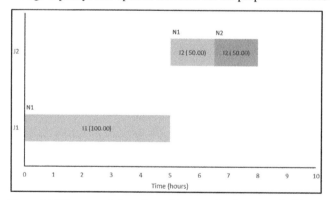

Figure 2 STN representation of a multipurpose batch process facility

5. Conclusions

In this work, an improved unit-specific event-based formulation for multipurpose batch processes has been presented. The proposed model allows related non-recycling production and consumption units to take place at the same event. Furthermore, a unit processing a consuming task is sequenced with a related production task, only if the unit processing the consuming task consumes materials from the unit processing the producing task. Additionally, a consumption task is allowed not to start immediately after the finish of the related production task if storage is available. Finally, a processing unit is allowed to store materials if it is idle. The results demonstrate that the proposed model is able to generate same or better solutions, using less events, which can significantly reduce the computational time, especially in computationally expensive problems.

Table 1 Comparative results for examples 1-5 with UIS and FIS policy

Ex.	Model	UIS				FIS			
		Event points	CPU time (s)	RMILP	MILP	Event points	CPU time (s)	RMILP	MILP
1	LF[a]	5	0.1	3000.00	2628.19	6	0.2	3973.92	2628.19
	VS[b]	5	0.3	3000.00	2628.19	5	0.2	3000.00	2628.19
	U-S[c]	3	0.1	3000.00	2628.19	3	0.2	3000.00	2628.19
2	LF	6[d]	3.7	2730.66	1962.69	6[d]	4.4	2730.66	1962.69
	VS	5	0.2	2436.69	1962.69	5	0.3	2436.69	1962.69
	U-S	5	0.2	2436.69	1962.69	5	0.4	2436.69	1962.69
3	LF	8[d]	1244	3618.64	2358.20	9[d]	3600[e]	3618.64	2345.31[f]
	VS	7	1179	3369.69	2358.20	7	950	3369.69	2358.20
	U-S	7	546	3369.69	2358.20	7	821	3369.69	2358.20
4	LF	5	0.1	80.0000	58.9870	5	0.1	80.0000	58.9870
	VS	5	0.2	80.0000	58.9870	5	0.2	80.0000	58.9870
	U-S	2	0.2	80.0000	58.9870	2	0.1	80.0000	58.9870
5	LF	3	0.1	500.000	500.000	3	0.1	500.000	300.000
	VS	3	0.1	500.000	500.000	3	0.1	500.000	300.000
	U-S	2	0.1	500.000	500.000	2	0.1	500.000	500.000

[a] Li et. al. 2010. [b] Vooradi and Shaik 2013. [c] Proposed unit-specific model [d] $\Delta n = 1$ [e] Relative gap 7.11%. [f] Suboptimum solution.

References

I. Harjunkoski, C. Maravelias, P. Bongers, P. Castro, S. Engell, I. Grossmann, J. Hooker, C. Méndez, G. Sand, J. Wassick, 2014, Scope for industrial application of production scheduling models and solution methods, Computers and Chemical Engineering, 62(5), 161-193

J. Li, C. Floudas, 2010, Optimal Event Point Determination for Short-Term Scheduling of Multipurpose Batch Plants via Unit-Specific Event-Based Continuous-Time Approaches, Industrial & Engineering Chemistry Reshearch, 49(16), 7446-7469

N. Rakovitis, J. Li, N. Zhang, 2018, A novel modelling approach to scheduling of multipurpose batch processes, Computer Aided Chemical Engineering, 44, 1333-1338

R. Seid , T. Majozi, 2012, A robust mathematical formulation for multipurpose batch plants, Chemical Engineering Science, 68(1), 36-53

M. Shaik, C. Floudas, 2009, Novel Unified Modeling Approach for Short-Term Scheduling, Industrial & Engineering Chemistry Research, 48(6), 2947-2964

M. Shaik M., R. Vooradi, 2017, Short-term scheduling of batch plants: Reformulation for handling material transfer at the same event, Industrial & Engineering Chemistry 56(39), 11175-11185

R. Vooradi, M. Shaik, 2013, Rigorous unit-specific event based model for short term scheduling of batch plants using conditional sequencing and unit-wait times, Industrial and Engineering research, 52(36), 12950-12792

Anton A. Kiss, Edwin Zondervan, Richard Lakerveld, Leyla Özkan (Eds.)
Proceedings of the 29th European Symposium on Computer Aided Process Engineering
June 16th to 19th, 2019, Eindhoven, The Netherlands. © 2019 Elsevier B.V. All rights reserved.
http://dx.doi.org/10.1016/B978-0-128-18634-3.50233-2

Electroencephalogram based Biomarkers for Tracking the Cognitive Workload of Operators in Process Industries

Mohd Umair Iqbal,[a] Babji Srinivasan,[a] Rajagopalan Srinivasan[,b,*]

[a]*Indian Institute of Technology Gandhinagar, Gujarat, 382355, India*

[b]*Indian Institute of Technology Madras, Tamil Nadu, 600036, India*

babji.srinivasan@iitgn.ac.in, raj@iitm.ac.in

Abstract

Human errors are a root cause of majority of accidents occurring in the process industry. These errors are often a result of excessive workload on operators, especially during abnormal situations. Understanding and measurement of cognitive workload (overload), experienced by human operators while performing key safety critical tasks, is thus important to the understanding of human errors. Subjective measurements of workload are often not reliable and there is a need for physiological based parameters of workload. In this work, we propose a methodology to measure cognitive load of a control room operator in terms of a biomarker, specifically theta/alpha ratio, obtained from a single electrode EEG signal. Real-time detection of the biomarker can enable minimize errors and improve safety.

Keywords: EEG, human errors, cognitive workload, safety

1. Introduction

With the advancement of technology and enhanced automation, the role of human operators has been shifting more towards monitoring the process operations. However, the disturbances in the process often call for human intervention. It often happens that humans commit errors in either planning or execution of actions; human errors have been a root cause of over 70% of accidents in process industries (Mannan, 2005). This situation is similar in other safety critical domains like aviation, health and nuclear industries. The reason for such errors is often attributed to the high cognitive workload experienced by the operator(s) during abnormal process events (Bhavsar et al., 2015). During these events, the operator uses their mental model of the process (developed during training, prior knowledge about the process, heuristic rules learnt from experience with the process, etc.,) to initiate and execute a sequence of steps that would bring the process back to normal operating conditions. However, any mismatch between the plant behaviour and the mental model would result in higher cognitive workload, a common cause for human errors. Therefore, it is imperative to understand the nature and evolution of the mental model of the operators.

Mental models of an operator can be broadly divided into three categories: skill based, rule based and knowledge-based (Reason, 1990). Operators use skill-based models to perform repetitive tasks which do not require any mental effort. For instance, an experienced driver on a traffic free road subconsciously performs driving related actions. An if-then rule-based mode would be used to tackle some abnormal situations in the process, specifically those that they are well-versed with a priori. Operators resort to a knowledge-based mode when they face a new situation where skill-based and rule-based modes are not available; they therefore need to apply their knowledge to identify suitable actions. However, operators traverse between the three mental modes (Reason, 1990) in the course of handling an abnormal situation. The resulting cognitive workload is different in each of the mental states. Therefore, it is essential to understand, explain and predict the transition among the various mental modes based on cognitive load so that the nature of the human errors can be uncovered and interventions taken to minimize the same. This is the ultimate objective of the research described here.

2. Literature Survey

Cognitive workload is experienced because of the limited working memory capacity of humans (Parasuraman, 2008). A mismatch between this capacity and demands of the process leads to errors which may transform into catastrophic accidents. To understand the underlying cognitive constructs during such mismatch, the nature and extent of cognitive workload needs to be quantified. Physiological parameters offer a potent approach to understand the underpinnings of human behaviour and provide information about the internal state of an individual. One such is Electroencephalogram (EEG) which has been used in diverse fields to understand the mental states of humans and predict stress levels and cognitive load.

EEG involves measurement of electrical activity of the brain using electrodes which are placed on skin. The electrical activity is the result of neural-oscillations occurring in brain due to fluctuations in excitability of neurons in response to stimuli. A raw EEG signal is comprised of various frequency components which are often clubbed into different frequency bands for the purpose of drawing meaningful observations. The commonly used frequency bands are delta (1-4 Hz), theta (4-7 Hz), alpha (8-12 Hz), beta (12-30 Hz), and gamma (30-80 Hz). Alpha and theta band activity have been extensively used for cognitive load measurement. Specifically, alpha activity is associated with idling, arousal and workload. A decrease in alpha activity is associated with increase in arousal, mental load, stress and anxiety (Neurosky, 2015). Conversely, an increase in alpha activity signifies idling or calmness (Pfurtscheller et al., 1996). Lim et. al, (2014) showed a significant increase in alpha activity when the eyes were closed, reflecting relative calm. Theta activity is also associated with attention and internal mental processing (Harmony et al., 1996); hence, an increase in theta activity means more mental effort. The power spectral density of delta waves has been reported to show a positive correlation with mental stress (Awang et al., 2011). Other frequency bands also have some significance e.g. beta waves in optimum condition can help with memory, conscious focus and problem solving (Abhang, 2016); delta waves are associated with adult sleep; and gamma waves with hypertension and high mental activity.

The potential of EEG to predict cognitive workload for dynamic processes has, however, been limited. Nevertheless, these frequency bands in the EEG signal have been used as biomarkers, for instance in the aviation sector involving dynamic tasks. A study by Gentili et. al, (2014) highlight the potential of theta by alpha ratio to measure cognitive workload and degradation of pilot performance when carrying out complex tasks involving manoeuvring. Theta by alpha ratio is reported to increase with task complexity. A study by Wilson (2002) revealed that the alpha band relative power decreased while delta band relative power increased for pilots, when compared at the same electrode locations in comparison to their pre-flight baseline values, for the majority of cases. These studies highlight the potential of EEG to offer insight into task difficulty and workload. There is, hence, a need to carry out cognitive workload studies in chemical process industries which like aviation offer highly dynamic situations involving high-risk and potential for disasters. The cognitive workload of the plant operators, in the process industry domain, depends on their ability to develop a mental model in a given situation, which is distinctly different from the time space representation of airplane pilots and drivers. Previously, the authors have explored eye-tracking based approaches to understand the cognitive behaviour of control room operators in process plants (Bhavsar et al., 2015). However, to the best of authors' knowledge, there is hardly any work, in this domain, to understand the cognitive workload of operators using EEG. In this paper, we seek to evaluate the potential of EEG biomarkers to discriminate between levels of cognitive load (mental states) and task difficulty.

3. Experimental Methodology

The experiment consists of monitoring a simulated ethanol production plant via a Human Machine Interface (HMI). The process involves an exothermic reaction between ethene and water taking place in a jacketed Continuous Stirred Tank Reactor (CSTR) whose output goes to a distillation column for separation of ethanol product. In case of any disturbance in the plant, the operator needs to take action(s) to bring the process variables to within acceptable limits, using various flow control sliders operable from HMI, as deemed appropriate by the participant. The reader is referred to Iqbal and Srinivasan (2017) for a detailed description of the process.

The study involved 6 post graduate chemical engineering students who performed 8 disturbance rejection tasks after receiving proper training (reading handouts and video training). During the course of the experiment the EEG data was recorded using Neurosky Mindwave+ single electrode headset. Different tasks involved dealing with different abnormalities. Tasks 1 and 2 involved a disturbance in coolant flow rate which resulted in the coolant flow rate crossing the upper alarm limit. Tasks 3 and 4 involved tackling the disturbance in coolant flow rate going outside the lower shutdown limits. Tasks 5 and 6 demanded bringing back the process to within alarm/acceptable limits in response to feed flow rate to CSTR breaching the upper alarm limit. The tasks differed by the magnitude of disturbance which in turn accordingly affected other process variables. Task 7 and 8 involved a disturbance in the feed flowrate going outside the lower alarm limits.

4. Results

An overview of the methodology employed to study the cognitive workload and related constructs is shown in Figure 1. We recorded the operators' EEG during the course of different tasks using an EEG headset and then carried out task-wise Short Time Fourier Transform (STFT) of the recorded raw EEG signal to obtain information in frequency domain. Among various possible biomarkers that have been reported in literature, theta (4-7 Hz) and alpha activity (8-12 Hz) showed promising results, which

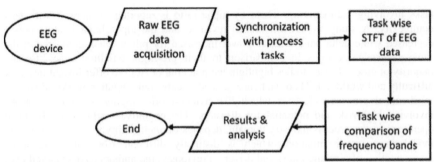

Figure 1: Overview of the methodology

are reported after the detailed description of a typical task in the following paragraph.

A typical task involves a disturbance resulting in a process variable(s) going outside its acceptable limits designated by upper and lower alarm limits. The operator is notified about the disturbance by an alarm entry in the alarm log and by change in colour of the tag(s), representing that variable(s) in the process schematic to red. The operator needs to assess the situation, decide on the appropriate recovery action and then take the action by moving appropriate slider(s) to bring the process variable(s) back to within acceptable limits. If the operator is unsuccessful to bring the process within normal bounds within a short time, the disturbance will result in the process variables crossing the shutdown limits (which is considered as a failure). The operator can see the trend of any process variable on HMI by clicking on the tag corresponding to that variable. The tasks have been designed in such a way that different aspects of cognitive behaviour and workload of operators could be studied; the tasks involve abnormal events (depending on the nature and magnitude of initial disturbance) with single alarm triggered, multiple alarms triggered in slow/quick succession and similar tasks i.e. tasks of varied difficulty as well as tasks of similar difficulty are included in the study.

Consider tasks 1 and 8, the simplest and the most difficult tasks, respectively; Task 1 involves an abnormality in coolant flow rate (through jacket) which does not affect other process variables immediately whereas during Task 8 an initial disturbance in the feed flowrate results in triggering of other alarms (temperature and concentration of CSTR) in quick succession. The comparison of theta/alpha activity of participants for these tasks reveals that the ratio is lower for a majority of simple tasks compared to Task 8, as shown in Figure 2. Out of 6 participants, 4 participants showed higher theta/alpha ratio for Task 8. For those 4 participants the average value is 1.05 for Task 8 compared to 1.03 for Task 1. Only one showed a lower score for Task 8 and for 1 participant it showed a negligible change. We therefore evaluated if this ratio has the potential to differentiate task difficulty and the consequent mental workload.

A comparison of two tasks - 7 and 8 – was therefore performed. Task 8 is a more difficult task compared to Task 7 as it involves a larger disturbance in the feed flowrate. This larger disturbance in feed flowrate in turn results in triggering of multiple alarms (temperature and concentration of CSTR) in quick succession compared to Task 7 (where the deviations are relatively slow), and also involves rapid evolution of process variables to outside shutdown limits. A relative comparison of these two tasks, for all

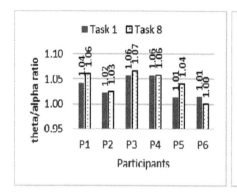
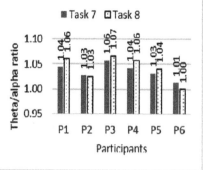

Figure 2 Theta by alpha ratio comparison of Task 1 vs Task 8

Figure 3 Theta by alpha ratio comparison of Task 7 vs Task 8

the 6 participants revealed that out of the 6 participants, 4 participants showed a higher theta by alpha ratio for task 8. For those 4 participants the average value was 1.05 for Task 8 compared to 1.04 for Task 7. One participant showing negligible change. Thus, theta by alpha ratio can act as a marker to predict the cognitive load as imposed by the task difficulty and as perceived by the participant.

The trend shown by the theta by alpha ratio in our study is in line with the results reported by Gentili et. al, (2014) during their pilot cognitive workload assessment studies. The alpha by theta ratio changes between easy and medium difficulty tasks and between medium & highest difficulty tasks in their study was around 0.05 and 0.08, respectively. The reason for the smaller difference in the values observed in our study can be attributed to the different nature and level of the complexity of the tasks in our study compared to theirs; the nature of the EEG device especially the electrode employed is also known to lead to differences. Our tasks involved control room tasks which would last 2-3 minutes while in their case the operator would carry out each task for 10 minutes. The persistent workload for longer duration experienced by pilots in their study might have invoked different levels of theta and alpha activity compared to our study. Our study is also consistent with the pilot-workload study by Wilson (2002) in terms of alpha activity; the study revealed that alpha activity corresponding to the different electrodes decreases for majority of the electrode positions in comparison to pre-flight base line values, though the differences were not always significant for all the electrodes; alpha activity alters the theta by alpha ratio. Nevertheless, in real control room settings where the disturbances and the remedial procedures can persist for longer time, our biomarker can show more significant differences. Further, more studies need to be carried out to identify the regions of brain which are more important for the cognitive workload determination involving different nature of dynamic tasks involved in process industries. This will help in a holistic understanding of cognitive workload determined using EEG based bio-markers. The tracing of the workload of the operators using biomarkers can help us minimize human errors due to high cognitive workload. If the workload during an unknown disturbance crosses a particular threshold/limit, automatic cues (may be in the form of additional information about the disturbed segment of the process) can be provide to the operator accordingly to minimize the likelihood of error.

5. Conclusions

In this work, we have found that EEG based bio-markers have the potential to help quantify the task difficulty and the consequent cognitive workload on the participants. Our results show that theta by alpha ratio is higher for tasks which are more difficult and impose higher cognitive workload on the participants. This marker can be used to assess training level of the operators before they are asked to monitor a plant. These will ensure their preparedness and feasibility for tasks involving monitoring and control of the processes in highly hazardous chemical process industries. In the future, we plan to measure the cognitive workload using EEG in real-time and also correlate them with markers obtained from our eye-tracking studies. Real time gauging of cognitive workload can help create systems which provide relevant cues to check the workload of operators by providing appropriate support in the form of certain information or otherwise. Multivariate analysis, involving various physiological parameters obtained from EEG and eye-tracking will help arrive at robust indices of cognitive load which will go a long way in understanding and obviating human errors.

References

Abhang, P. A., Gawali, B. W., & Mehrotra, S. C. (2016). Chapter 3—technical aspects of brain rhythms and speech parameters. Introduction to EEG-and Speech-Based Emotion Recognition, 51-79.

AlZu'bi, H. S., Al-Nuaimy, W., & Al-Zubi, N. S. (2013, December). EEG-based driver fatigue detection. In Developments in eSystems Engineering (DeSE), 2013 Sixth International Conference on (pp. 111-114). IEEE.

Awang, S. A., Pandiyan, P. M., Yaacob, S., Ali, Y. M., Ramidi, F., & Mat, F. (2011). Spectral density analysis: theta wave as mental stress indicator. In Signal Processing, Image Processing and Pattern Recognition (pp. 103-112). Springer, Berlin, Heidelberg.

Bhavsar, P., Srinivasan, B., & Srinivasan, R. (2015). Pupillometry based real-time monitoring of operator's cognitive workload to prevent human error during abnormal situations. Industrial & Engineering Chemistry Research, 55(12), 3372-3382.

Harmony, T., Fernández, T., Silva, J., Bernal, J., Díaz-Comas, L., Reyes, A., ... & Rodríguez, M. (1996). EEG delta activity: an indicator of attention to internal processing during performance of mental tasks. International journal of psychophysiology, 24(1-2), 161-171.

Iqbal, M. U., & Srinivasan, R. (2017). Simulator based performance metrics to estimate reliability of control room operators. Journal of Loss Prevention in the Process Industries.

Lim, C. K. A., Chia, W. C., & Chin, S. W. (2014, August). A mobile driver safety system: Analysis of single-channel EEG on drowsiness detection. In Computational Science and Technology (ICCST), 2014 International Conference on (pp. 1-5). IEEE.

Mannan, S. (Ed.). (2005). Lees' Loss Prevention in the Process Industries: Hazard Identification, Assessment and Control (Vol. 1). Butterworth-Heinemann.

Neurosky. (2015). Greek Alphabet Soup – Making Sense of EEG Bands (Last assessed: Nov 10, 2018) http://neurosky.com/2015/05/greek-alphabet-soup-making-sense-of-eeg-bands/

Parasuraman, R., Sheridan, T. B., & Wickens, C. D. (2008). Situation awareness, mental workload, and trust in automation: Viable, empirically supported cognitive engineering constructs. Journal of Cognitive Engineering and Decision Making, 2(2), 140-160.

Pfurtscheller, G., Stancak Jr, A., & Neuper, C. (1996). Event-related synchronization (ERS) in the alpha band—an electrophysiological correlate of cortical idling: a review. International journal of psychophysiology, 24(1-2), 39-46.

Reason, J. (1990). Human error. Cambridge university press.

Wilson, G. F. (2002). An analysis of mental workload in pilots during flight using multiple psychophysiological measures. The International Journal of Aviation Psychology, 12(1), 3-18.

Anton A. Kiss, Edwin Zondervan, Richard Lakerveld, Leyla Özkan (Eds.)
Proceedings of the 29th European Symposium on Computer Aided Process Engineering
June 16th to 19th, 2019, Eindhoven, The Netherlands. © 2019 Elsevier B.V. All rights reserved.
http://dx.doi.org/10.1016/B978-0-128-18634-3.50234-4

A Data-Driven Robust Optimization Approach to Operational Optimization of Industrial Steam Systems under Uncertainty

Liang Zhao, Chao Ning, Fengqi You

Cornell University, Ithaca, New York 14853, USA

Abstract

This paper proposed a data-driven adaptive robust optimization approach to deal with operational optimization problem of industrial steam systems under uncertainty. Uncertain parameters of the proposed steam turbine model are derived from the semi-empirical model and historical process data. A robust kernel density estimation method is employed to construct the uncertainty sets for modeling these uncertain parameters. The data-driven uncertainty sets are incorporated into a two-stage adaptive robust mixed-integer linear programming (MILP) framework for operational optimization of steam systems. By applying the affine decision rule, the proposed multi-level optimization model is transformed into its robust counterpart, which is a single-level MILP problem. To demonstrate the applicability of the proposed method, the case study of an industrial steam system from a real-world ethylene plant is presented.

Keywords: data-driven, adaptive robust optimization, industrial steam system, operational optimization, uncertainty

1. Background

In recent years, system modelling and optimization methods are employed to improve the efficiency and economic benefits of industrial steam systems. In steam systems, it is an important issue to accurately model the steam turbines to improve the steam system efficiency (Petroulas and Reklaitis, 1984; Li et al. 2014). Turbine models based on thermodynamics were proposed to optimize the performance of steam system (Aguilar et al. 2007). Hybrid models were developed to improve the performances of steam turbine models (Luo et al. 2011). Building upon the developed steam turbine models, several optimization models for synthesis and design of steam systems were proposed (Tveit et al. 2006, Li et al. 2013, Beangstrom et al. 2016). A stochastic programming approach was employed to deal with uncertainties in the site utility system (Sun et al. 2017). However, the optimization results of stochastic programming are highly sensitive to the pre-defined probability distributions, which could deviate significantly from the "true" uncertainty information. The static robust optimization (SRO) approach might make the resulting solutions very conservative (Soyster, 1973; Bertsimas et al. 2011; Ben-Tal and Nemirovski, 1998). More recently, data-driven robust optimization has emerged as an effective tool for optimization under uncertainty (Ning et al. 2017, 2018, Shang et al. 2017). In industrial steam systems, large amounts of historical data are collected and stored in a database that records all operational information of the devices and their connected networks. A semi-empirical model of steam turbine is first developed based on process mechanism and operational data. Then, the uncertain parameters are collected from process historical data. A data-driven two-stage ARO method for operational optimization of steam systems under uncertainty is presented. The robust counterpart of the proposed data-driven ARO model is derived as a single-level MILP model to facilitate the solution process. A case study of the steam system in a real-world ethylene plant is presented to demonstrate the effectiveness of the proposed method.

2. Deterministic MILP model of industrial steam system

Mixed-integer optimization models were developed for operational optimization of the steam systems, which rely on the linear or nonlinear model of the basic components (Varbanov et al. 2004; Chu and You, 2015). A decomposition method is employed to develop extraction steam turbine model (Aguilar et al. 2007). The semi-empirical model of steam turbine is presented as follows,

$$m_t^{in} = a_t \cdot m_t^{ext} + b_t, \ \forall t \tag{1}$$

where a_t and b_t, are parameters of the model and can be calculated by process data.

The deterministic MILP model optimization model of steam system (DOSS) is given by,

$$\min \quad c = c^{SS} \sum_{t \in TSS} m_t^{in} + c^e \sum_j y_j^e + c^{water} \sum_{lv} (l_{lv} - 1) m_{lv}^{in} \tag{2}$$

$$\text{s.t.} \quad \sum_t (a_{sh,t}^{in} m_t^{in} + a_{sh,t}^{ext} m_t^{ext}) + \sum_j a_{sh,j}^M (1 - y_j) F_j^S + \sum_{lv} (a_{sh,lv}^{in} m_{lv}^{in} + a_{sh,lv}^{out} m_{lv}^{out}) \geq md^{sh} \tag{3}$$

$$e_t + W_t \geq WD_t \tag{4}$$

$$m^{min} \leq m \leq m^{max} \tag{5}$$

Equations (2)-(4) are steam demand constraints, mechanical power demand constraints and variables range constraints, respectively.

3. Data-Driven ARO Model for Steam System

3.1 Deriving uncertain parameters of semi-empirical model for steam turbines

The parameters of the steam turbine models are derived from process operation parameters according to equation (1). The schematic diagram of uncertain parameter derivation process is given in Figure 1.

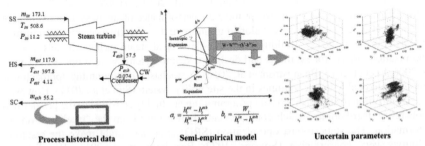

The procedure of deriving uncertainty parameters from historical operation data.

The process historical data of steam system can typically be collected conveniently. A couple of parameters, a_t and b_t, in the proposed semi-empirical model can be calculated by process data, including temperatures, pressures, and flow rates of inlet, extraction and exhausting steam of turbines. Then, the uncertain parameters are derived.

3.2 Data-driven uncertainty sets for ARO

RKDE is employed to construct data-driven uncertainty sets for the ARO framework (Kim et al. 2012). The correlation information is also considered in the construction of uncertainty sets (Ning et al. 2017). The data-driven uncertainty set is reformulated as follows,

$$U^{cor} = \left\{ \zeta \left| \begin{array}{l} \hat{F}_{RKDE}^{i-1}(\alpha) \leq \zeta_i \leq \hat{F}_{RKDE}^{i-1}(1-\alpha), \ \forall i \\ \sum_i \zeta_i^0 (1-r_i h) \leq \sum_i \zeta_i \leq \sum_i \zeta_i^0 (1+r_i h) \\ \left\| M(\zeta - \zeta^0) \right\|_1 \leq \beta \end{array} \right. \right\} \tag{6}$$

where α is a predefined parameter to denote the confidence level (1-2α), h is a predefined uncertainty budget to control the level of conservatism as the parameter α. re $M=\Sigma^{-1/2}$, Σ is the covariance matrix of the uncertain parameters, and β is a parameter used to adjust the size of l_1 norm ball in $\left\| M(\zeta - \zeta^0) \right\|_1 \leq \beta$.

3.3 Data-driven robust MILP model of steam system

A novel data-driven adaptive robust MILP model for operational optimization of steam systems is then developed based on the uncertainty sets (6). The multi-level data-driven adaptive robust steam system model based on RKDE incorporating correlation information (DDARSS-RKDECI) is developed as follows. Noted that the model is also subjected to constraint (4) in model DOSS.

$$\min_{m_t^{in}(\cdot), m_t^{ext}, m_{lv}^{in}, y_j} \max_{\zeta \in U^{cor}} c_e \sum_j y_j^e + c_{water} \sum_{lv} (l_{lv} - 1)m_{lv}^{in} + c_{ss} \sum_{t \in TSS} m_t^{in}(\zeta) \tag{7}$$

$$\text{s.t.} \quad \begin{array}{l} \sum_t (a_{sh,t}^{in} m_t^{in}(\zeta) + a_{sh,t}^{ext} m_t^{ext}) + \sum_j a_{sh,j}^M (1 - y_j) F_j^S \\ + \sum_{lv} (a_{sh,lv}^{in} m_{lv}^{in} + a_{sh,lv}^{out} m_{lv}^{out}) \geq md^{sh}, \ \forall sh \in SH, \ \zeta \in U^{cor} \end{array} \tag{8}$$

$$m_t^{in,min} \leq m_t^{in}(\zeta) \leq m_t^{in,max}, \ \forall t \in T, \ \zeta \in U^{cor} \tag{9}$$

$$m_t^{exh,min} \leq m_t^{in}(\zeta) - m_t^{ext} \leq m_t^{exh,max}, \ \forall t \in T, \ \zeta \in U^{cor} \tag{10}$$

3.4 Robust counterpart reformulation of data-driven ARO model for steam system

Because there are uncertain parameters in the objective function, the epigraph reformulation of objective function (7) is presented by,

$$\min_{m_t^{in}(\cdot), m_t^{ext}, m_{lv}^{in}, y_j, z} z \tag{11}$$

$$\text{s.t.} \quad c_e \sum_j y_j^e + c_{water} \sum_{lv} (l_{lv} - 1)m_{lv}^{in} + c_{ss} \sum_{t \in SS} m_t^{in}(\zeta) \leq z, \ \forall \zeta \in U^{cor} \tag{12}$$

By applying the affine decision rule $m_t^{in}(\zeta) = P_t \zeta + q_t$, constraint (12) is reformulated as follows,

$$\max_{\zeta \in U^{cor}} c_{ss} \sum_t P_t \zeta \leq z - c_e \sum_j y_j^e - c_{water} \sum_{lv} (l_{lv} - 1)m_{lv}^{in} - c_{ss} \sum_t q_t \tag{13}$$

Because the left-hand side of constraint (13) is a linear program, dual variables λ_i, γ_i, π, Ω, $C1$, $C2_k$ and $C3_k$ are introduced. Constraint (13) is then reformulated as follows,

$$\sum_i \zeta_i^{\max} \lambda_i - \sum_i \zeta_i^{\min} \gamma_i + \zeta^{sum,\max} \pi - \zeta^{sum,\min} \Omega + \beta C1 + \zeta^{sum,cor} C2_k - \zeta^{sum,cor} C3_k$$
$$\leq z - c_e \sum_j y_j^e + c_w \sum_{lv} (l_{lv} - 1) m_{lv}^{in} - c_{ss} \sum_t q_t \tag{14}$$

$$\lambda_i - \gamma_i + \pi - \Omega + \sum_k M_{k,j} \cdot C2_k - \sum_k M_{k,j} \cdot C3_k = c_{ss} \sum_t P_t, \ \forall i \tag{15}$$

$$C1 + C2_k - C3_k = 0, \ \forall k \tag{16}$$

$$\lambda_i, \ \gamma_i, \ \pi, \ \Omega, \ C1, \ C2_k, \ C3_k \geq 0, \ \forall i, k \tag{17}$$

Similar to constraint (13), constraints (8)-(10) can be transformed to new constraints by dual variables. Then, the data-driven adaptive robust counterpart for steam system operational optimization (DDARCSS-RKDECI) is formulated as a single-level optimization problem, which can be solved by optimization solvers like CPLEX directly.

4. Applications

To demonstrate the applicability of the proposed approach, a case study on the steam system of an ethylene plant is presented. There are four steam grades (SS, HS, MS, and LS), four extractions turbines (T1, T2, T3, and T4), 29 backpressure turbines (T5 to T33), three let-down valves in the given steam system. The prices of SS, water and electricity are set to be 180 m.u./t, 15 m.u./h and 1.5 m.u./kwh, respectively (where m.u. stands for "monetary unit"). In RKDE, the parameter α is set as 0.02, and the parameter h is set as 0.8. The parameter β in uncertainty set U^{cor} is set as 3.2.

The proposed method is compared with the deterministic and RKDE-based method (DDARCSS-RKDE). All optimization problems are coded in GAMS 24.8.3. The CPLEX 12.7.0.0 with an optimality tolerance of 0% is used to solve the single-level MILP problems. The computational results of different methods are given in Table 1.

Table 1 Problem sizes and solution comparisons for different methods.

	DOSS	DDARCSS-RKDE	DDARCSS-RKDECI
Binary variables	29	29	29
Continuous variables	11	397	732
Constraints	13	196	369
Min. operating cost(m.u./h)	88,720	96,750	96,366

The optimal operating cost obtained by the deterministic method is 88,720 m.u./h, and it is obviously less than the values determined by the data-driven ARO methods, because the deterministic model uses fixed parameters during the optimization process. Due to the consideration of correlation information among uncertain parameters, the solution of DDARCSS-RKDECI method is less conservative than that of the DDARCSS-RKDE approach.

The effects of parameters α and h on the solution of the proposed method are investigated by changing their values. α is set as 0.02, 0.05, 0.1, 0.2 and 0.4, and h is set as 0.05, 0.1, 0.2, 0.4 and 0.8. The heat map of operational cost is shown in Figure 2. When the value of α decreases from 0.4 to 0.02 for any fixed h, the operational cost increases by 6.3% on average; when h increases from 0.05 to 0.8 for any fixed α, the operational cost increases by 0.9% on average. For a fixed value of parameter h, the operational cost decreases obviously with the increase of α, because the confidence interval becomes smaller with a large value of α. Since h is used to control the effect of uncertainty deviations, the objective value increases slowly as the value of parameter h increases.

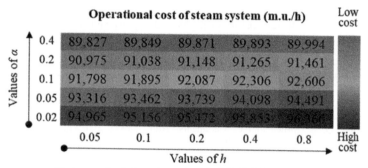

Heat map of solutions determined by DDARCSS-RKDECI.

The breakdown of the optimal operational costs determined by different methods in Table 1 are shown in Figure 3. The percentage of electricity obtained by the deterministic method is the largest among the three methods. The parameters of the semi-empirical model are fixed as nominal values, meaning the steam demands from processes can be satisfied easily. As a result, more electrical motors will be turned on, replacing the steam turbines as the method of power generation for the pumps. The data-driven two-stage ARO methods have similar breakdowns of operational costs, because the consumptions of electricity and water are determined at the first stage. The minor differences are caused by the difference of their uncertainty sets.

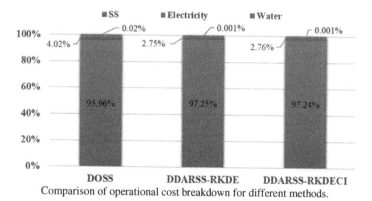

Comparison of operational cost breakdown for different methods.

5. Conclusion

A novel operational optimization model of steam system under uncertainty using data-driven two-stage ARO was proposed. The uncertain parameters of semi-empirical models for steam turbines were collected from historical process data of an industrial plant. The RKDE method was then employed to construct the uncertainty sets from the derived uncertain parameters. The correlation information was considered in the construction of uncertainty sets through the covariance matrix of uncertain parameters. The data-driven two-stage ARO model of steam system was formulated based on the constructed uncertainty sets, and the corresponding robust counterpart was developed using affine decision rule. A real-world case study on the steam system of an ethylene plant was presented to demonstrate the effectiveness of the proposed methods.

References

O. Aguilar, S. Perry, J.-K. Kim and R. Smith, 2007. Design and optimization of flexible utility systems subject to variable conditions: Part 1: Modelling framework, Chemical Engineering Research and Design 85(8): 1136-1148.

S. G. Beangstrom, and T. Majozi, 2016, Steam system network synthesis with hot liquid reuse: II. Incorporating shaft work and optimum steam levels, Computers & Chemical Engineering 85: 202-209.

A. Ben-Tal, A. Nemirovski, 1998, Robust convex optimization. Mathematics of Operations Research, 23, 769-805.

D. Bertsimas, D. B. Brown and C. Caramanis, 2011, Theory and Applications of Robust Optimization, SIAM Review 53(3): 464-501.

Y. Chu, F. You, 2015, Model-based integration of control and operations: Overview, challenges, advances, and opportunities. Computers & Chemical Engineering, 83, 2-20.

J. Kim, and C. D. Scott, 2012, Robust kernel density estimation, Journal of Machine Learning Research 13(Sep): 2529-2565.

Z. Li, W. Du, L. Zhao and F. Qian, 2014, Modeling and Optimization of a Steam System in a Chemical Plant Containing Multiple Direct Drive Steam Turbines, Industrial & Engineering Chemistry Research 53(27): 11021-11032.

Z. Li, L. Zhao, W. Du and F. Qian, 2013, Modeling and Optimization of the Steam Turbine Network of an Ethylene Plant, Chinese Journal of Chemical Engineering 21(5): 520-528.

X. Luo, B. Zhang, Y. Chen and S. Mo, 2011, Modeling and optimization of a utility system containing multiple extractions steam turbines, Energy 36(5): 3501-3512.

C. Ning, F. You, 2017, Data-Driven Adaptive Nested Robust Optimization: General Modeling Framework and Efficient Computational Algorithm for Decision Making Under Uncertainty. AIChE Journal, 63, 3790-3817.

C. Ning, F. You, 2017, A data-driven multistage adaptive robust optimization framework for planning and scheduling under uncertainty. AIChE Journal, 63, 4343-4369.

C. Ning, F. You, 2018, Data-driven stochastic robust optimization: General computational framework and algorithm leveraging machine learning for optimization under uncertainty in the big data era. Computers & Chemical Engineering, 111, 115-133.

C. Ning, F. You, 2018, Data-driven decision making under uncertainty integrating robust optimization with principal component analysis and kernel smoothing methods. Computers & Chemical Engineering, 112, 190-210.

T. Petroulas, G.J.A.j. Reklaitis, 1984, Computer-aided synthesis and design of plant utility systems. 30, 69-78.

C. Shang, X. Huang, F. You, 2017, Data-driven robust optimization based on kernel learning. Computers & Chemical Engineering, 106, 464-479.

C. Shang, F. You, 2019, A data-driven robust optimization approach to scenario-based stochastic model predictive control. Journal of Process Control, 75, 24-39.

A.L. Soyster, 1973, Convex programming with set-inclusive constraints and applications to inexact linear programming. Operations Research, 21, 1154-1157.

L. Sun, L. Gai and R. Smith, 2017, Site utility system optimization with operation adjustment under uncertainty, Applied Energy 186: 450-456.

T.-M. Tveit, and C.-J. Fogelholm, 2006, Multi-period steam turbine network optimisation. Part II: Development of a multi-period MINLP model of a utility system, Applied Thermal Engineering 26(14-15): 1730-1736.

P. S. P. S.,arbanov, S. Doyle and R. Smith, 2004, Modelling and Optimization of Utility Systems, Chemical Engineering Research and Design 82(5): 561-578.

Anton A. Kiss, Edwin Zondervan, Richard Lakerveld, Leyla Özkan (Eds.)
Proceedings of the 29th European Symposium on Computer Aided Process Engineering
June 16th to 19th, 2019, Eindhoven, The Netherlands. © 2019 Elsevier B.V. All rights reserved.
http://dx.doi.org/10.1016/B978-0-128-18634-3.50235-6

Synthesis technology for failure analysis and corrective actions in process systems engineering

Ákos Orosz[a], Ferenc Friedler[b,*]

[a]*Department of Computer Science and Systems Technology, University of Pannonia, Veszprém H-8200, Hungary*
[b]*Institute for Process Systems Engineering & Sustainability, Pázmány Péter Catholic University, Budapest H-1088, Hungary*
friedler.ferenc@ppke.hu

Abstract

The ever increasing complexity of processing systems and the requirements to be satisfied by the processes make process design and process operation more and more complex. Because of that, the reliability/availability is to be taken into account together with other requirements in process design/synthesis. Similarly, failure analysis is also an important part of process operation. In the current work, it has been shown that in addition to the widely used unit level redundancy, process level redundancy must also be considered in synthesizing a process for satisfying high level availability. Therefore, in case of failure, the determination of the best corrective action must be based on more general process networks than considered before. The proposed procedure determines the best or n-best corrective actions for the most general types of highly complex process networks, e.g., including process level redundancy. The P-graph based procedure is effective in generating corrective action suitable for industrial processes.

Keywords: process synthesis, process availability, failure analysis, corrective action.

1. Introduction

Availability, the rate of operability, is an important indicator of a process since it directly influences the profitability. A high level availability can usually be achieved if redundancy is built in the process. If this is the case, the failure of an equipment unit of the process does not necessarily mean that the process becomes non-operational. To keep the process operational in an event of failure of equipment units, a corrective action is to be done. Since a process may have a large number of alternative feasible operating modes with different operating costs, the determination of the corrective action should be systematic and based on optimization if possible.

Corrective actions are, especially, widely used in power distribution networks. Because of the need of an almost immediate action, decisions are usually based on heuristics and genetic algorithms with no guarantee of the optimality. Ela and Spea (2009) applied genetic algorithms to minimize the losses in transmission lines. Korad and Hedman, (2013) developed a hybrid method for taking into account uncertainties in determining the corrective switching action.

Distefano and Puliafito (2007) solved computer system failure problem by two different approaches, one is based on reliability block diagram and the other on fault tree analysis. Though the results were similar for them, the typical issue is that the mathematical model is not generated algorithmically for either method. Goel et al. (2002) simultaneously considered availability and profitability in synthesizing a chemical process in a mixed integer programming model. The method considers only specific flowsheets, e.g., no redundancy is allowed in the flowsheet of the synthesized process.

Stated formally, if an equipment unit is capable of performing its designated job on a sufficient level, it is called as functional, otherwise, non-functional. Similarly, an operating unit is functional if there are functional equipment units assigned to this operating unit that can perform the transformation identified by the operating unit.

A process is operational if it is capable of generating all products in the required amounts and qualities. An operational process may have several different modes of operations, i.e., the operation can be performed under different process networks. The best of the operating modes is called as on-design operation, all the others are the off-design operations. The mode of operations that is currently working is called as active operating mode. Not necessarily all functional equipment units assigned to the operating units of an active operating mode are in service, consequently, the equipment units of this operating mode can be either active or non-active.

For practical reason, the procedure for determining the corrective action must systematically satisfy the following four criteria:

(i)   Keep the system operational as far as it is possible;

(ii) Keep the operating cost of the selected operating mode as low as possible, i.e., it is based on optimization;

(iii) Robust, i.e., it determines the corrective action for any possible flowsheet;

(iv) Generates the corrective action promptly.

To fulfil criterion (iii), the types of flowsheets (networks) that can appear in a real process must be determined first. For example, Goel et al (2002) considered redundancy free flowsheets. However, if redundancy is allowed, the most common redundancy is realized on operating unit level, i.e., more than one equipment unit is established for an operating unit. The basic question now is if this operating unit level redundancy covers all possible options or more general types of redundancies are to be considered.

The flowsheet or network of a process is determined during its synthesis. Thus, the types of networks to be considered in making a corrective action depend on the possible flowsheets (networks) that can be the result in synthesizing a process. An example, Example 1, will illustrate that unit level redundancy does not necessarily cover all important cases for optimal decision.

In practice, the network of the most preferred processes are usually complex, highly interconnected involving several loops. There are two types of optimization tools for process design in terms of structure representation, implicit, e.g., mathematical programming approaches and explicit, e.g., the P-graph framework. Methods with explicit structure-representation provide additional opportunities on structural properties of process networks including an appropriate interface to failure analysis and to develop systematic method for determining the optimal corrective action. The P-graph framework (see, e.g., Peters et al, 2003) is formulated by combinatorial mathematics, based on the five axioms of feasible process networks and includes combinatorial algorithms, e.g., for synthesizing processes. There is a close relation between the process feasibility in process synthesis and the process operability in failure analysis, since the axioms are valid for both. Therefore, these two types of requirements can be considered simultaneously without compromising any of them. A P-graph is a directed graph and it represents the process network by two types of vertices, i.e., dots for materials and horizontal bars for operating units. Software implementations of the main algorithms are available at www.p-graph.org.

2. Structural redundancy in a process

A processing system is termed as redundancy-free, if the failure of any of its equipment units makes the system non-operational. Because of the high level of requirements on the availability, in most cases, redundancy free systems are not acceptable, redundant systems are to be designed. The redundancies in the process are determined during its synthesis. The availability of a redundancy-free system is simply the product of the availability indicators of the equipment units. The determination of the availability for systems with redundancies is complex both in theory and applications. Since the availability of a processing system can be determined similarly to its reliability, an available method can be applied for that (Kovacs et al., 2018).

Frequently, designing a process with unit-level redundancy satisfies the system's availability requirement, and has been widely used in practice. However, this doesn't necessarily give rise to the optimal process for simple reasons. The off-design operations are performed only a short period of time compared to the on-design operations. It is quite possible that the flowsheet (network) of the off-design operation is different from that of the on-design operation. For example, in total site system, there can be other resources of the intermediate products with possibly higher cost against reduced investment requirements. Another type of process level redundancy is shown in the following example where the synthesis procedure is also given in determining the optimal process.

Example 1

Figure 1 shows the maximal structure (super-structure) of a synthesis problem for producing material A, it is also supposed that the availability of the process must be at least 0.999. Figure 2 shows the optimal process with 9,312 USD/y cost, it is a redundancy free solution. The availability of this process, 0.989, does not satisfy the requirement, and there is no other redundancy free solution with the required availability. Thus, redundancy must be considered during synthesis. On Figure 3, the maximal structure is shown with assigning double equipment units to each operating unit. Figure 4 shows the cost optimal process synthesized by simultaneously taking into account the availability constraint (Kovacs et al, 2018). Bold lines give the on-design operating mode and the thin lines represent the spare part of the system. Note that the optimal solution is based on process level redundancy. Consequently, optimality can only be guaranteed if process level redundancy is taken into account in process synthesis under availability constraint. Furthermore, methods for generating optimal corrective actions must consider flowsheets (networks) with no structural limitations.

3. Systematic procedure for corrective actions in case of failure

The procedure for determining the corrective action in an event of failure is based on P-graph synthesis tools, it is called as procedure CAF. It is supposed that the flowsheet of the process is given in the form of a P-graph. Let the mathematical model of each operating unit with its cost function, the unit cost of each raw material, and the demand of each product be given. If for a running process one or more equipment units become non-functional, their identifiers are assumed to be available immediately for the CAF procedure.

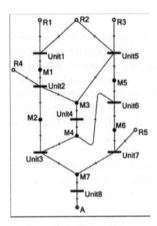

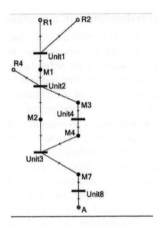

Figure1. Maximal structure of a process synthesis problem

Figure 2. Network of the optimal solution of process synthesis problem given on Figure 1, availability requirement is violated

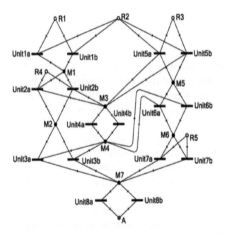

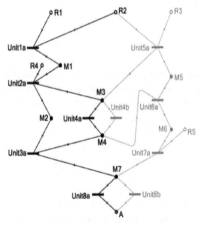

Figure 3. Maximal structure of process synthesis problem given in Figure 1 with redundant equipment units

Figure 4. The network of optimal solution of process synthesis problem given on Figure 3 satisfying the availability requirement (bold lines indicate the on-design operating mode)

First, algorithm MSG (Friedler et al., 1993) determines if the remaining functional equipment units may compose a structurally feasible process. In case of negative answer, the procedure halts, there is no feasible operating mode on the basis of the functional equipment units. Otherwise, algorithm ABB (Friedler et al., 1996) is executed to determine the optimal operating mode, if there is one. Note that a structurally feasible process is usually feasible, but not necessarily. Algorithm ABB is a branch&bound type algorithm, the bounding part can be linear or nonlinear depending on the mathematical models of the operating units. The algorithm is also able to give the n-best solutions in addition to the optimal one. Note that algorithms MSG and ABB are available at www.p-graph.org.

Example 2

One product, PMM (Perchloromethyl mercaptan), is produced by a processing system of 16 available operating units from 9 available raw materials. Each operating unit is realized by one equipment unit. The network of a process that includes process level redundancy is shown on Figure 5. The mathematical models of the operating units in this example are linear, they are available at www.p-graph.org/case-study-pmm. There are 24 feasible operating modes producing the product. The best of them in terms of cost is shown on Figure 6, it is considered as the on-design operating mode with the cost of 3,144,010 USD/y.

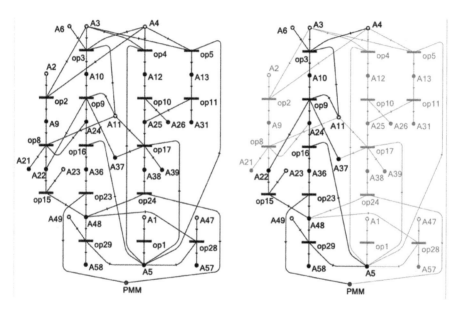

Figure 5. A process with process level redundancy for producing material PMM

Figure 6. Bold lines indicate the on-design operating mode of production PMM (cost: 3,144,010 USD/y)

Suppose that equipment unit that realizes operating unit op29 fails, then, the number of feasible operating modes is reduced to 12. Figure 7 shows the optimal operating mode under this condition, the cost of operation is increased to 3,228,140 USD/y.

If equipment units op9 and op16 fail simultaneously as a possible consequence of cascading effect, there are 8 remaining feasible operating mode, the optimal of them is shown on Figure 8, its cost is 3,819,490 USD/y. If both equipment units op16 and op17 fail, the process becomes non-operational, as no feasible operating mode exists.

4. Conclusions

It has been shown that process level redundancy must also be considered in process synthesis for optimally satisfying high level availability. Because of that, the corrective action in an event of failure must be prepared for a more general network redundancy. The procedure developed here keeps the process operational as far as it is possible, determines the most cost effective corrective operating mode practically promptly for any possible process network.

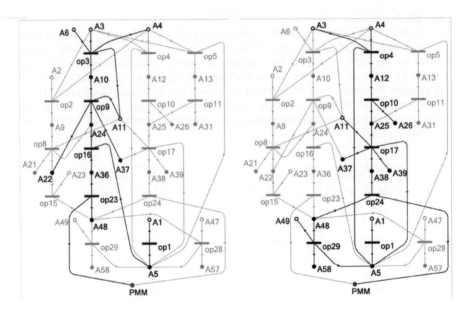

Figure 7. The optimal operating mode in case of failure of equipment unit op29 (cost: 3,228,140 USD/y)

Figure 8. The optimal operating mode in case of failure of equipment units op9 and op16 (cost: 3,819,490 USD/y)

References

S. Distefano, A. Puliafito, 2007, January. Dynamic reliability block diagrams vs dynamic fault trees, In Reliability and Maintainability Symposium, RAMS'07. Annual, pp. 71-76, IEEE

A.A.A.E. Ela, S.R. Spea, 2009, Optimal corrective actions for power systems using multi-objective genetic algorithms, Electric Power Systems Research, 79(5), pp.722-733

F. Friedler, K. Tarjan, Y.W. Huang, L.T. Fan, 1993, Graph-theoretic approach to process synthesis: polynomial algorithm for maximal structure generation, Computers & Chemical Engineering, 17(9), pp.929-942

F. Friedler, J.B. Varga, E. Feher, L.T. Fan, 1996, Combinatorially accelerated branch-and-bound method for solving the MIP model of process network synthesis, In State of the art in global optimization, pp. 609-626, Springer, Boston, MA.

H.D. Goel, J. Grievink, P.M. Herder, M.P. Weijnen, 2002, Integrating reliability optimization into chemical process synthesis, Reliability Engineering & System Safety, 78(3), pp.247-258

A.S. Korad, K.W. Hedman, 2013, Robust corrective topology control for system reliability, IEEE Transactions on Power Systems, 28(4), pp.4042-4051

Z. Kovacs, A. Orosz, F. Friedler, 2018, Synthesis algorithms for the reliability analysis of processing systems, Central European Journal of Operations Research, https://doi.org/10.1007/s10100-018-0577-0

M.S. Peters, K.D. Timmerhaus, R.E. West, 2003, Plant design and economics for chemical engineers, McGraw-Hill, New York

http://www.p-graph.org

Acknowledgment

Supports of the Pázmány Péter Catholic University under KAP Project and University of Pannonia under EFOP-3.6.1-16-2016-00015 project are acknowledged.

Anton A. Kiss, Edwin Zondervan, Richard Lakerveld, Leyla Özkan (Eds.)
Proceedings of the 29th European Symposium on Computer Aided Process Engineering
June 16th to 19th, 2019, Eindhoven, The Netherlands. © 2019 Elsevier B.V. All rights reserved.
http://dx.doi.org/10.1016/B978-0-128-18634-3.50236-8

Challenges in Decision-Making Modelling for New Product Development in the Pharmaceutical Industry

Catarina M. Marques[a,b,*], Samuel Moniz[c], Jorge Pinho de Sousa[a,b],

[a]*Faculdade de Engenharia da Universidade do Porto, Portugal*

[b]*INESC TEC, Porto, Portugal*

[c]*Departamento de Engenharia Mecânica da Universidade de Coimbra, Portugal*

eq98022@fe.up.pt

Abstract

This study presents an assessment of the main research problems addressed in the literature on New Product Development (NPD) and its methodologies, for the pharmaceutical industry. The work is particularly focused on the establishment of an evolutionary perspective of the relevant modelling approaches, and on identifying the main current research challenges, considering the fast-changing business context of the industry. Main findings suggest a generalized misalignment of recent studies with today's technological and market trends, highlighting the need for new modelling strategies.

Keywords: New product development, capacity planning, portfolio management, modelling, decision-making.

1. Introduction

The world is currently experiencing unprecedented fast changes, with major technological breakthroughs that are transforming the way companies operate and manage their systems. Pharmaceutical companies are constantly being challenged to become ever more cost efficient and responsive in the delivery of drugs. Consistently launching new drugs into the market is, therefore, critical to support the industry's economic sustainability. However, the traditional NPD process still imposes considerable challenges for practitioners and researchers. In this sector, the Operations Research (OR) and the Process Systems Engineering (PSE) communities have been developing reliable and effective model-based decision-making tools. Understanding how these developments have been impacting the industry is crucial for establishing new management directions that take into account the paradigm changes currently faced by the industry. The main goal of this work is, therefore, to present a research review with two perspectives: first, identifying the main problems on the pharmaceutical NPD process and the corresponding modelling approaches; and second, analyzing how research has evolved along the years, assessing the methodological main progresses and envision future directions.

2. Literature Review

In order to clearly define the scope of the present study the following research question was formulated: "*What have been the main research concerns and modelling approaches in the pharmaceutical NPD process?*" Based on this question, the following criteria were used to perform the review: (i) studies specifically addressing decision-making problems in the pharmaceutical sector; (ii) works within OR, PSE, and related communities; and (iii) full research papers, written in English. The search was performed considering the boolean combination of keywords ((*"planning" OR "scheduling" OR "decision-making" OR "supply chain") AND pharma**), complemented by the analysis of several review papers. After excluding duplicates, review papers, conference papers (with some limited exceptions), and works at a hospital/pharmacy level, a total of 113 research works was obtained covering the complete pharma industry management problems. A second round of analysis was then performed to keep only works related to the development of new products. Finally, a set of 53 papers was considered for a complete analysis.

3. Analysis

The 53 papers selected for full analysis cover the years from 1996 to early 2018 and are distributed in 16 scientific journals and 2 conference proceedings. "Computers and Chemical Engineering" appears as the most represented journal, with 18 published papers. Two broad research areas emerge from the selected papers, namely: (i) the portfolio management area; and (ii) the clinical trial supply chain management area. Portfolio management is the most developed area, being addressed in 31 papers and representing about 58% of the total selected works, while the clinical trial supply chain management area is tackled in the remaining 22 papers. Each one of these areas is described in the following sections.

3.1. Portfolio Management (PM)

The main goal of PM is to determine the portfolio of new products to be developed and the schedule of the associated testing tasks (clinical trials) under uncertainty in the outcomes of the tests. Therefore, two decision problems are usually found in the literature, namely: i) the portfolio selection (PS), and ii) the scheduling of the testing tasks (ST). In the first case, the main decisions include the determination of which targets should be selected for further clinical development. These decisions rely on two issues: i) the assessment of the value of the portfolio, based on its probability of success and on its potential returns; ii) and on the company market strategy.

The ST problem is closely related to the typical Resource-Constrained Project Scheduling (RCPS) problem, but with the possibility of task failing during the development process. The main decisions involved here are related to the determination of the tasks starting and completion times. Uncertainty in trials' outcomes is addressed in 28 of the 31 selected papers, thus highlighting its importance in these problems. In both cases (PS and ST) there are also resource-related decisions, in order to ensure the necessary resources availability. Moreover, most of the studies deal with the two problems simultaneously (17 papers). The modelling details and progresses on portfolio management in the pharmaceutical industry are presented in the next section.

3.1.1. Modelling approaches
Analysing the complete set of the selected papers, the prevailing modelling approaches can be broadly divided in optimization (18 papers), simulation-optimization (11 papers), and simulation (2 papers) approaches. In the optimization case, problems are formulated as deterministic MILP models or as stochastic versions of RCPS problems (two-/multi-stage stochastic programming MILP models). In the simulation-optimization case, there is a combination of combinatorial optimization models with a discrete-event simulation.

Simulation is used to capture uncertainty more realistically and to evaluate the optimization results, thus giving insights to the determination of the optimized solution, in an iterative way. In the simulation case, models are used to perform sensitivity analysis and capture the risk/reward of the portfolio value. Figure 1 depicts the evolutionary path of the most relevant contributions to the PM problem. The first work addressing this problem (Schmidt and Grossmann (1996)) was based on a MILP model and a two-stage stochastic programming (TSSP) approach, to account for uncertainties in task costs, durations, and trials outcomes. The goal was to determine the schedule of testing tasks that maximizes the expected net present value (eNPV), considering task precedence and timing constraints. This model, however, does not account for important features such as product interdependencies or resources limitations.

Later, an evolution has been observed (Figure 1), with works incorporating ever more sophisticated modelling features and broadening the problem scope. It is, therefore possible to outline the basic structure of the MILP formulation that is common to most of the works addressing ST/PS decisions. This formulation can be generalized as follows. We are given: i) a set of time periods $t \in \mathbf{T}$; ii) a set of potential drugs $i \in \mathbf{I}$ to develop; iii) a set of clinical tests $j \in \mathbf{J}$; iv) a set of resource types $r \in \mathbf{R}$; and v) a set of scenarios $s \in \mathbf{S}$. The goal is to determine which tests should be selected and when (X_{ijts} binary variables) to maximize the eNPV, subject to the following general constraints: i) assignment constraints of clinical tests to starting times; ii) test precedence constraints; and iii) resource constraints. The main evolution here was on the resource constraints representation, by considering more industry-specific decisions, such as investment in new resources, outsourcing resources or out-licensing tests.

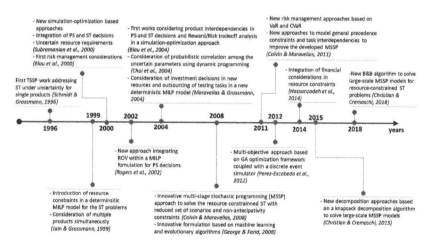

Figure 1. Timeline of the most relevant contributions in Portfolio Management.

3.2. Clinical Trials Supply Chain (CTSC) management

At CTSC management, two broader areas were identified. The first area *(CTSC operations)* is related to management of all the activities required to undertake the clinical trials, including: product manufacturing, packaging, storage, and distribution to clinical sites. In this type of supply chains (SC), special attention needs to be given to aspects such as: the limited time-horizon by trials duration; leftovers at the end of each trial that cannot be reused; delivery failures (seriously compromising time-to-market); and the significant uncertainty affecting the process. The main challenge is, therefore, to efficiently manage this unique SC, guaranteeing the timely fulfilment of the clinical trials demand. Despite their relevance for the industry, only 8 papers explicitly addressed the CTSC operational problems, showing a very limited attention paid by researchers to this issue. These works span from SC/plant design to production/ inventory planning, with most of the papers integrating these two decision-levels at some extent.

In a second area, the Product Launch planning problem is related to the important link between product development and the capacity planning needed to accommodate both existing and new products. 14 works addressing this link were considered with decisions comprising portfolio selection, capacity and production planning.

3.2.1. Modelling approaches

CTSC operations

CTSC operations are clearly a new area in the OR and the PSE communities, being still at its infancy of development. Selected works can be divided into simulation (2 papers) or simulation-optimization (6 papers) approaches. The early works addressing this problem, particularly regarding inventory management, were based on simulation approaches for assessing different scenarios, and decisions were taken mainly based on sensitivity analysis. The introduction of optimization approaches starts to emerge only recently, and works are still very scarce. The most relevant contributions were by Chen et al. (2012) with an integrated approach to manage the multi-echelon CTSC under uncertainty on patient demand, and the work by Marques et al. (2017), integrating capacity planning and process design decisions under uncertainty with an on time supply of the clinical trials. Both approaches integrate simulation methods to tackle uncertainty with MILP optimization models.

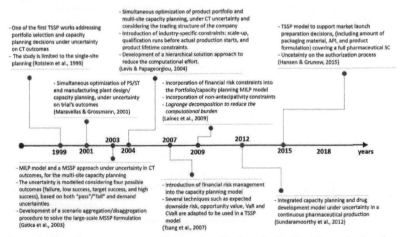

Figure 2. Timeline of the most relevant contributions in Product Launch planning.

Product Launch planning (PL)

Regarding the product launch planning problem, all the selected papers developed optimization approaches based on MILP formulations, and only 3 presented deterministic models. The remaining 11 papers addressed the uncertainty on trials' outcomes through multi-stage stochastic programming (MSSP) approaches. Most of the works deal with the decisions such as portfolio selection and capacity/production planning, and constraints on material balance, production/manufacturing, resource allocation/investments, or sales/demand.

Figure 2 depicts the evolutionary path of the most relevant contributions to the PL problem. One of the first works addressing this problem was by Rotstein et al. (1999), proposing a MILP model under uncertainty in trial's outcomes for the selection of the most profitable products, capacity planning, and investment strategy to accommodate the new products. Subsequent works tend to deal with ever more complex features by adding problem-specific constraints to the base model, such as: scale-ups; qualification runs; product lifetime, financial, and risk constraints. It is interesting to note that the increase on the models' size/complexity has also motivated a further development of new decomposition techniques, contrasting with the lack of exploitation of alternative modelling approaches.

4. Research directions and final remarks

This work shows that PM problems have been object of significant research activities, contrasting with the CTSC operations with still very modest developments. However, current technological trends (continuous manufacturing and digital capabilities) and market (personalization) trends are still absent or only discreetly considered in research works.

Future modelling approaches should take the following new aspects into account: i) assessment of the portfolio "value", including patient-related considerations (e.g. specific unmet needs); ii) patient-centric delivery models, linking "quantity" variables to new "outcome" variables reflecting benefits for patients; iii) market segmentation constraints; iv) integrated approaches and coordination between actors; v) exploitation of the benefits of continuous manufacturing settings and digital capabilities to leverage efficiency; vi) functional integration between marketing (patient perspective), product development, and manufacturing; vii) exploitation of new performance metrics and multi-objective approaches to capture the multi-dimensionality of the problems; and finally viii) exploitation of innovative modelling approaches to tackle large-scale problems.

In short, future work should extend the analysis throughout the entire drug life-cycle, and define future modelling components (constraints, variables, ...) that are more aligned with the current context of the pharmaceutical industry.

Acknowledgement

The authors gratefully acknowledge the financial support of Fundação para a Ciência e Tecnologia (FCT), under the grant PD/BD/105987/2014.

References

Blau, G., Mehta, B., Bose, S., Pekny, J., Sinclair, G., Keunker, K., & Bunch, P. (2000). Risk management in the development of new products in highly regulated industries. Computers & Chemical Engineering, 24, 659-664.

Blau, G., Pekny, J. F., Varma, V. A., & Bunch, P. R. (2004). Managing a portfolio of interdependent new product candidates in the pharmaceutical industry. Journal of Product Innovation Management, 21, 227-245.

Choi, J., Realff, M. J., & Lee, J. H. (2004). Dynamic programming in a heuristically confined state space: a stochastic resource-constrained project scheduling application. Computers & Chemical Engineering, 28, 1039-1058.

Christian, B., & Cremaschi, S. (2015). Heuristic solution approaches to the pharmaceutical R&D pipeline management problem. Computers & Chemical Engineering, 74, 34-47.

Christian, B., & Cremaschi, S. (2018). A branch and bound algorithm to solve large-scale multistage stochastic programs with endogenous uncertainty. AIChE Journal, 64, 1262-1271.

Colvin, M., & Maravelias, C. T. (2008). A stochastic programming approach for clinical trial planning in new drug development. Computers & Chemical Engineering, 32, 2626-2642.

Colvin, M., & Maravelias, C. T. (2011). R&D pipeline management: Task interdependencies and risk management. European Journal of Operational Research, 215, 616-628.

Gatica, G., Papageorgiou, L., & Shah, N. (2003). An aggregation approach for capacity planning under uncertainty for the pharmaceutical industry. Found Comp-Aided Proc Oper, 4, 245-248.

George, E. D., & Farid, S. S. (2008). Stochastic combinatorial optimization approach to biopharmaceutical portfolio management. Industrial & Engineering Chemistry Research, 47, 8762-8774.

Hansen, K. R. N., & Grunow, M. (2015). Planning operations before market launch for balancing time-to-market and risks in pharmaceutical supply chains. International Journal of Production Economics, 161, 129-139.

Hassanzadeh, F., Modarres, M., Nemati, H. R., & Amoako-Gyampah, K. (2014). A robust R&D project portfolio optimization model for pharmaceutical contract research organizations. International Journal of Production Economics, 158, 18-27.

Jain, V., & Grossmann, I. E. (1999). Resource-constrained scheduling of tests in new product development. Industrial & Engineering Chemistry Research, 38, 3013-3026.

Laínez, J. M., Reklaitis, G. V., & Puigjaner, L. (2009). Managing financial risk in the coordination of supply chain and product development decisions. Computer Aided Chemical Engineering, 26, 1027-1032.

Levis, A. A., & Papageorgiou, L. G. (2004). A hierarchical solution approach for multi-site capacity planning under uncertainty in the pharmaceutical industry. Computers & Chemical Engineering, 28, 707-725.

Maravelias, C. T., & Grossmann, I. E. (2004). Optimal resource investment and scheduling of tests for new product development. Computers & Chemical Engineering, 28, 1021-1038.

Perez-Escobedo, J. L., Azzaro-Pantel, C., & Pibouleau, L. (2012). Multiobjective strategies for New Product Development in the pharmaceutical industry. Computers & Chemical Engineering, 37, 278-296.

Rogers, M. J., Gupta, A., & Maranas, C. D. (2002). Real options based analysis of optimal pharmaceutical research and development portfolios. Industrial & Engineering Chemistry Research, 41, 6607-6620.

Rotstein, G., Papageorgiou, L., Shah, N., Murphy, D., & Mustafa, R. (1999). A product portfolio approach in the pharmaceutical industry. Computers & Chemical Engineering, 23, S883-S886.

Schmidt, C. W., & Grossmann, I. E. (1996). A mixed integer programming model for stochastic scheduling in new product development. Computers & Chemical Engineering, 20, S1239-S1244.

Subramanian, D., Pekny, J. F., & Reklaitis, G. V. (2000). A simulation—optimization framework for addressing combinatorial and stochastic aspects of an R&D pipeline management problem. Computers & Chemical Engineering, 24, 1005-1011.

Sundaramoorthy, A., Evans, J. M., & Barton, P. I. (2012). Capacity planning under clinical trials uncertainty in continuous pharmaceutical manufacturing, 1: mathematical framework. Industrial & Engineering Chemistry Research, 51, 13692-13702.

Tsang, K., Samsatli, N., & Shah, N. (2007). Capacity investment planning for multiple vaccines under uncertainty: 2: Financial risk analysis. Food and Bioproducts Processing, 85, 129-140.

Anton A. Kiss, Edwin Zondervan, Richard Lakerveld, Leyla Özkan (Eds.)
Proceedings of the 29th European Symposium on Computer Aided Process Engineering
June 16th to 19th, 2019, Eindhoven, The Netherlands. © 2019 Elsevier B.V. All rights reserved.
http://dx.doi.org/10.1016/B978-0-128-18634-3.50237-X

Extension of a Particle Filter for Bioprocess State Estimation using Invasive and Non-Invasive IR Measurements

Julian Kager[a,b,*], Vladimir Berezhinskiy[a], Robert Zimmerleiter[c], Markus Brandstetter[c] and Christoph Herwig[a,b]

[a]*ICEBE, TU Wien, Gumpendorfer Straße 1a 166/4, 1060 Wien, Austria*
[b]*CD Laboratory on Mechanistic and Physiological Methods for Improved Bioprocesses, TU Wien, Gumpendorfer Straße 1a 166/4, 1060 Wien, Austria*
[c]*Research Center for Non Destructive Testing (RECENDT) GmbH, 4040 Linz, Austria*
julian.kager@tuwien.ac.at

Abstract

Producers of pharmaceuticals have to guarantee the quality of their products. Therefore, substantial efforts are invested in process monitoring and control. However, crucial parameters, such as the nutrient and product precursor concentrations often require time-consuming analysis of a sample. Infrared (IR) spectroscopy is a promising measurement technique for the online quantification of multianalyte solutions, such as fermentation broths. Besides the high investment cost of devices, chemometric models often are not fully transferable and provide noisy estimates which need to be treated before further usage. To increase robustness and accuracy of these measurements they can be combined with kinetics models under the usage of a state observation algorithms, such as Kalman- or Particle filters.

In this work, we present a unique combination of a transferable mechanistic process description with the real-time information derived from near and mid IR spectroscopy leading to stable, smooth and accurate state estimates. IR spectra were collected in *Penicillium chrysogenum* fed-batch processes. PLS models were trained for the prediction of nitrogen, product and product precursor concentrations, which were used as inputs for the state observer. The resulting probabilistic estimates provide a good basis for automatic control.

Keywords: bioprocess monitoring, spectroscopic measurements, state observeration, non-invasive NIR spectroscopy, MIR spectroscopy, PLS-regression

1. Introduction

To control fermentation processes, real-time measurements of critical process parameters are required. Since fermentation broths are a complex mixture of different analytes, their single quantification is difficult. Over the last decades promising technologies emerged to provide real-time information on the reaction kinetics during such processes (Vojinović et al. (2006)).

Near and mid infrared (NIR; MIR) spectroscopy combined with multivariate analysis can be used to quantify multi-analyte solutions. For the penicillin producing organism *P.*

chrysogenum the product, the product precursor and the nitrogen concentration could be successfully determined via NIR and MIR spectroscopy (Luoma et al. (2017); Koch et al. (2014)). Besides advantages like the non-destructive and possible non-invasive measurement, the multivariate models are not fully transferable and provide noisy estimates which need to be treated before further usage (Krämer and King (2017)).

In chemical engineering state observers, which are computational algorithms to estimate unmeasured state variables, are widely applied (Ali et al. (2015)). Hereby, secondary measurements provide real-time information of the ongoing process and Bayesian filters estimate the most probable states. Besides the often used extended Kalman filter, particle filters, which approximate the distribution of the estimated states, gained importance with increasing computational capacities (Goffaux and Wouwer (2005)). In recent works a particle filter configuration was compared to an extended Kalman filter for the examined penicillin production process (Stelzer et al. (2017)), and the better suited particle filter was extended to use delayed offline measurements (Kager et al. (2018)).

Although estimations already show good results the combination of spectroscopy and Bayesian state estimation offers a possibility to further improve bioprocess monitoring. This potential was shown by Golabgir and Herwig (2016) under usage of Raman spectroscopy and by Krämer and King (2017) with NIR spectroscopy for a yeast process. In this work two different IR measurements were carried out. One using an immersion probe in the MIR spectral region and the other through a glass window of the reactor in the NIR spectral region. The acquired IR data was included in a particle filter in order to estimate important process parameters such as product, product precursor, nitrogen and biomass concentration. The possibility to include redundant measurements and to combine them, to get robust and complete state estimates is of high relevance to biochemical production industry.

2. Materials and Methods

2.1. Cultivation & Reference Measurements

Fed-batch experiments were performed in a 2.7 l parallel bioreactor system (Eppendorf AG, Germany) using a spore suspension of an industrial *P. chrysogenum* strain for penicillin production. Dissolved oxygen was controlled above 40% by stirrer speed (350 - 850 rpm) and subsequent oxygen addition to pressurized air, while the reactor was aerated with 1 vvm of gas mixture. The temperature was maintained at 25 °C and the pH value was maintained at 6.5 by addition of KOH and H_2SO_4. Glucose (500 g/l), the penicillin V precursor (POX) (80 g/l phenoxyacetate) and the nitrogen source (100 g/l $(NH_4)_2SO_4$) were supplied as feeds. As glucose should be the only limiting component, nitrogen and precursor were kept at non-limiting concentrations (1 g/l and 2 g/l), by manipulating the $(NH_4)_2SO_4$ and POX feeds.

For online analytics, CO_2 and O_2 content in the off-gas were measured by a gas analyzer (DASGIP GA4, Eppendorf AG, Germany). The conversion of O_2 to CO_2 was calculated according to Aehle et al. (2011), giving the carbon evolution rate (CER) and the oxygen uptake rate (OUR) in mol/h. Reference samples were taken every 8 hours. Analysis of penicillin V (PEN) and POX concentration in the culture media was performed by HPLC using a ZORBAX C-18 Agilent column and 28% acetonitrile, 6 mM H_3PO_4 and 5 mM

KH$_2$PO$_4$ as elution buffer. Glucose and ammonia concentration were quantified by an enzymatic analyzer (Cedex Bio HT, Roche GmbH, Switzerland). Biomass concentration was determined gravimetrically by separating the cells from 5 ml culture broth via centrifugation at 4800 rpm for 10 min at 4 °C. The cell pellet was dried at 105 °C after a washing step with 5 ml of deionized water in weighted glass tubes.

2.2. Infrared Spectroscopy and Partial Least Square (PLS) Regression

Mid IR absorption (650-3000 cm^{-1}) was captured by an optical fiber immersion probe (ReactIR 45m, Mettler Toledo, USA) and near IR (5000-7500 cm^{-1}) was captured, through the glass wall of the reactor (NIRONE, Spectral Engines, Finland).

After data preprocessing (standard normal variate (SNV) transformation and Savitzky-Golay smoothing (SG) and differentiation (1st; 2nd), partial least square regression models (SIMPLS) were build in MATLAB (R2018a, MathWorks, USA). Spectra preprocessing was done according to Koch et al. (2014) for mid and Luoma et al. (2017) for near IR. Suitable wavenumber ranges were selected individually by selecting ranges with high PLS weights in the whole band including 1st and 2nd order derivatives. By an randomly sampling cross-validation procedure the number of latent variables were determined. The selected procedures for different analytes are summarized in Table 1.

Table 1: Near and mid IR PLS regression procedures for penicillin V (PEN), product precursor (POX) and nitrogen (NH3).

Model	Preprocessing	Wavelength ranges [cm^{-1}]	Latent variables	Datasets
MIR-PEN	1st	1307-1352 ; 1747-1817	4	3
MIR-POX	none	1000-1150 ; 1400-1750 ; 1900-2600	3	3
MIR-NH3	1st	850-1150 ; 1400-1520 ; 1950-2300	4	3
NIR-PEN	SNV, 2nd	6050-7500 ; 5200 -5700	4	4
NIR-POX	SNV, 2nd	5100-5500 ; 6100-6400 ; 6500-7500	4	4
NIR-NH3	1st	6200-7500	4	4

2.3. Non-linear Kinetic Model and Bayesian State Estimation

A kinetic process model taken from Paul et al. (1998) describing the states x (hyphal tips, hyphal bodies, glucose, gluconate, nitrogen, product precursor and product concentration) by ordinary differential equations was adapted to the current process (Stelzer et al. (2017)). In order to use the model during the process a state estimation algorithm, a SIR particle filter derived from Simon (2006) was used. The model states are hereby described by x and linked to the measurements y. As initialization step, a sample of N particles is drawn by the known probability density function $p(x_0)$ of the initial states x_0. Iteratively, for time steps $k \geq 1$, N particles $\tilde{x}_k^i$ are propagated based on the process model. Importance weights fro every particle, $\tilde{w}_k^i = p(y_k \mid \tilde{x}_k^i)$, $i = 1, \ldots, N$, are calculated at measurement arrival. All measurements are hereby weighted by their error. Applying multinominal resampling, N posteriori particles x_k^i, $i = 1, \ldots, N$ are resampled from $\tilde{x}_k^i$ according to their importance weights. The distribution of these particles is an approximation of the probability density function, i.e. $x_k^i \sim p(x_k \mid y_k)$, from which the actual state estimates can be derived.

3. Results and Discussion

3.1. Model Building and Calibration

In Figure 1 observed vs. predicted plots show the calibrated models, with their confidence intervals and their Root Mean Square Errors (RMSE). PLS models were built according to Table 1 and parameters for the kinetic model were estimated according to the procedures described by Ulonska et al. (2018). For the product, the kinetic model is unable to predict the product degradation properly, which can be seen from the overestimation at high concentrations. Overall the submersed MIR shows the best calibration results, whereas the kinetic model predictions and the non-invasive NIR measurements are within the same error range.

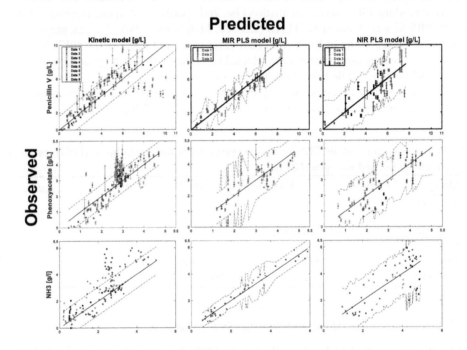

Figure 1: Observed vs. predicted of kinetic model, MIR and NIR PLS model for Penicillin V (RMSE = 1.62; 0.69; 1.33 R^2 = 0.61; 0.93; 0.71), Phenoxyacetate (RMSE = 0.74; 0.76; 0.75 R^2 = 0.59; 0.62; 0.68) and NH_3 (RMSE = 1.05; 0.41; 1.01 R^2 = 0.55; 0.93; 0.58) under usage of historical experiments. Not in all 8 fermentations, IR spectra could be recorded

3.2. Monitoring Results

In Figure 2 prediction results for product and precursor concentrations of an selected process are displayed, wereas the errors are displayed in Table 2. The state observer is able to estimates of the biomass concentration, however it fails to estimate the product and precursor concentration correctly as no information of these states are included. By applying the PLS models on the incoming IR spectra good real-time estimates of product, precursor and nitrogen can be obtained. However, results derived from spectral data are very noisy, outlier need to be removed and the signal needs to be smoothed to obtain meaningful results.

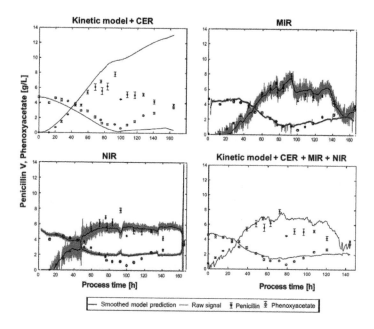

Figure 2: State estimation of validation dataset with kinetic model & CER and MIR & NIR PLS prediciton and their combination. Savitizky Golay with a window of 100 datapoints was used to smooth the raw signal.

In order to add needed real-time information to the state observer it can be extended by the IR predictions. This has the advantages that untreated results can be used and by including the measurement errors, redundant measurements derived by the submersed MIR probe and the external NIR sensor can be combined. This combination gives accurate and stable estimates for the entire process.

Table 2: Overall Root Mean Square Errors (RMSE) of different state estimation strategies

	Penicillin V [g/l]	Phenoxyacetate [g/l]	NH_3 [g/l]	Biomass [g/l]
state observer	3.97	1.14	0.90	4.01
MIR (immersion)	0.60	0.50	0.32	-
NIR (non- invasive)	1.11	0.65	0.88	-
state observer + IR	1.15	0.51	0.62	4.26

4. Conclusions

On the example of a *P. chrysogenum* fed batch process penicillin V, product precursor (Phenoxyacetate) and nitrogen (NH_3) concentration could be estimated by invasive MIR and non-invasive NIR measurements. Through a kinetic model, describing growth, nutrient consumption, product formation and the measurement of the CO_2 production rate, biomass formation during the process can be estimated. The used state estimator, which in our case was a particle filter, is a useful tool for the combination of different and redundant measurements. By including the PLS prediction of the IR spectra, the state observer is able to predict all states accurately. The particle filter hereby filters the measured and partially redundant information and estimates the most probable system states through the kinetic model, giving a solid basis for feedback control.

5. Acknowledgements

This work has been supported by the project "multimodal and in-situ characterization of inhomogeneous materials" (MiCi) by the federal government of Upper Austria and the European Regional Development Fund (EFRE) in the framework of the EU-program IWB2020. Further support was provided by the Christian Doppler Forschungsgesellschaft (Grant No. 171) and the strategic economic- research program "Innovative Upper Austria 2020" of the province of Upper Austria.

References

M. Aehle, A. Kuprijanov, S. Schaepe, R. Simutis, A. Lübbert, 2011. Simplified off-gas analyses in animal cell cultures for process monitoring and control purposes. Biotechnology letters 33 (11), 2103.

J. M. Ali, N. H. Hoang, M. A. Hussain, D. Dochain, 2015. Review and classification of recent observers applied in chemical process systems. Computers & Chemical Engineering 76, 27–41.

G. Goffaux, A. V. Wouwer, 2005. Bioprocess state estimation: some classical and less classical approaches. In: Control and Observer Design for Nonlinear Finite and Infinite Dimensional Systems. Springer, pp. 111–128.

A. Golabgir, C. Herwig, 2016. Combining mechanistic modeling and raman spectroscopy for real-time monitoring of fed-batch penicillin production. Chemie Ingenieur Technik 88 (6), 764–776.

J. Kager, C. Herwig, I. V. Stelzer, 2018. State estimation for a penicillin fed-batch process combining particle filtering methods with online and time delayed offline measurements. Chemical Engineering Science 177, 234–244.

C. Koch, A. E. Posch, H. C. Goicoechea, C. Herwig, B. Lendl, 2014. Multi-analyte quantification in bioprocesses by fourier-transform-infrared spectroscopy by partial least squares regression and multivariate curve resolution. Analytica chimica acta 807, 103–110.

D. Krämer, R. King, 2017. A hybrid approach for bioprocess state estimation using nir spectroscopy and a sigma-point kalman filter. Journal of Process Control.

P. Luoma, A. Golabgir, M. Brandstetter, J. Kasberger, C. Herwig, 2017. Workflow for multi-analyte bioprocess monitoring demonstrated on inline nir spectroscopy of p. chrysogenum fermentation. Analytical and bioanalytical chemistry 409 (3), 797–805.

G. Paul, M. Syddall, C. Kent, C. Thomas, 1998. A structured model for penicillin production on mixed substrates. Biochemical engineering journal 2 (1), 11–21.

D. Simon, 2006. Optimal state estimation: Kalman, H infinity, and nonlinear approaches. John Wiley & Sons.

I. V. Stelzer, J. Kager, C. Herwig, 2017. Comparison of particle filter and extended kalman filter algorithms for monitoring of bioprocesses. In: Computer Aided Chemical Engineering. Vol. 40. Elsevier, pp. 1483–1488.

S. Ulonska, P. Kroll, J. Fricke, C. Clemens, R. Voges, M. M. Müller, C. Herwig, 2018. Workflow for target-oriented parametrization of an enhanced mechanistic cell culture model. Biotechnology journal 13 (4), 1700395.

V. Vojinović, J. Cabral, L. Fonseca, 2006. Real-time bioprocess monitoring: Part i: In situ sensors. Sensors and Actuators B: Chemical 114 (2), 1083–1091.

Anton A. Kiss, Edwin Zondervan, Richard Lakerveld, Leyla Özkan (Eds.)
Proceedings of the 29th European Symposium on Computer Aided Process Engineering
June 16th to 19th, 2019, Eindhoven, The Netherlands. © 2019 Elsevier B.V. All rights reserved.
http://dx.doi.org/10.1016/B978-0-128-18634-3.50238-1

Design of Multi Model Fractional Controllers for Nonlinear Systems: An Experimental Investigation

G Maruthi Prasad[a], A Adithya [a] and A Seshagiri Rao[a,*]

[a]*Department of Chemical Engineering, National Institute of Technology, Waranagal - 506 004, Telangana, India*
Email: seshagiri@nitw.ac.in

Abstract

Control of nonlinear processes is challenging when compared to those of linear processes. Conventional PID controllers can be used to control nonlinear processes based on a condition that these controllers must be tuned in such a way that they should provide a very stable behaviour over the entire range of operating conditions. Inspite of tuning the PID controllers, it still results in a degradation of the control system performance. In order to stabilise nonlinear systems, multi model control approaches are found suitable. Multi-Model Approaches (MMA) rely on the problem decomposition strategy where a nonlinear system is segregated into set of many linear models based on their operating points. These reduced models are used to tune model based PI/PID controllers. In the literature, different methods are proposed to design these controllers in MMA framework and they are all integer order controllers. In this work, fractional controllers based MMA framework is developed for enhanced control of nonlinear systems. Gap metric based weighting methods are used with proper weighting functions to obtain the global controller. The weighting functions are formed by using gap metric to combine local controllers. Three different nonlinear processes (CSTR, conical tank and spherical tank processes) are considered and implemented the proposed methodology. For the purpose of comparison, MMA framework with integer order controllers is considered and it is observed that MMA framework with fractional controllers provide improved closed loop performances. Experimental investigation is also carried out to verify the applicability of the proposed method for level control on conical tank process and it is observed that the proposed method provide enhanced closed loop responses.

Keywords: Multi-model approach, integer order controller, fractional order controller, gap metric, nonlinear system

1. Introduction

In process industries, behavior of most of the systems will be nonlinear and this type of systems mostly have performance degradation by using conventional PID controller. In order to overcome this degradation there are different types of techniques available and one of the most popular techniques which has been consider by several researches is the multi model approach (MMA) Adeniran and El Ferik (2017).This approach divides nonlinear systems into multi linear models sequentially based on the operating points. Controllers are designed by using this linear model and over decades most of the researchers considered integer order controllers only. When integer order controllers are implemented practically overshoot and resonance are observed and in order to deal such effects fractional order controllers can be used which shows more promising results Podlubny (1999). In this paper an attempt has been made to apply fractional order controllers in order to design a new system called Multi model fractional order controller. This paper presents effectiveness of Multi model fractional order controller and Extensive numerical studies on nonlinear system demonstrate its performance.Section II of the paper describes multi-model approach, gap metric and fractional order controller design. In Section III nonlinear system are studied to demonstrate effectiveness of Multi model fractional order controller. Section IV numerical results and conclusions are presented.

2. Multi Model Fractional controller

2.1. Multi-model approach

The MMA deals with decomposition of nonlinear system into multi linear models based on sequentially steady states and mathematical model is developed for each of the operating range. These models now as a set, can be used as a valid representation of the nonlinear process. These models are further reduced into minimal set using gap metric.

2.2. Gap metric

Recent studies have showed that the most appropriate tool to quantify the difference between different linear systems using metrics based on norms is the gap metric Du et al. (2012); Du and Johansen (2014). The gap metric is calculated based on following formula

$$\delta_g(P_1, P_2) = max(\vec{\delta}_g(P_1, P_2), \vec{\delta}_g(P_1, P_2)) \tag{1}$$

Where as P_1 and P_2 are linear models and

$$\delta_g(P_1, P_2) = \inf_{Q \epsilon H_\infty} \left\| \begin{bmatrix} N_1 \\ M_1 \end{bmatrix} - \begin{bmatrix} N_2 \\ M_2 \end{bmatrix} Q \right\|_\infty \tag{2}$$

From eq. (1), the gap metric range is considered as $0 \le \delta_g(P_1, P_2) \le 1$. If the value of gap metric is near to 0 a single controller is enough to control two operating regions effectively else it is difficult to use single controller to control the operating regions.

2.3. Fractional Order Controller

Implementation of Fractional calculus which is generalization of Integer order calculus is making a noteworthy advancement. Its significance lies in the fact that practical systems can be better identified as fractional order differential equations instead of integer order differential equations David et al. (2011). Fractional Order PID Controller which is usually described as $PI^\lambda D^\mu$ Controller was introduced by Podlubny (1999).

$$C(s) = K_p + \frac{K_i}{s^\lambda} + k_d s^\mu \tag{3}$$

where K_p is Proportional Gain, K_i is Integral Gain, K_d is Derivative Gain, λ and μ are integral and derivative orders and can be varied between 0 to 2. Non Integer Order Controllers offer more degrees of freedom and by using these controllers for Integer Order plants, there is more flexibility in adjusting the gain and phase characteristics than using Integer Order controllers. Methods for design of fractional order controllers are discussed in Podlubny (1999); Monje et al. (2008). One such tuning method for Fractional Order PI controllers was proposed by Gude and Kahoraho (2009) in which a performance criteria (J_v) is minimized which is a measure of systems ability to handle low frequency load disturbances. Finally the normalized controller parameters are designed based on normalized dead time τ.

2.4. Multi model integer and fractional order controller

After multi model approach, minimized models are obtained using gap metric. Based on these models, PI values of integer and fractional order controllers are tuned, with the help of IMC technique and Gude method respectively. The global controller is formed by combination of local controllers. The combination is carried out using gap metric weighting function proposed by Du and Johansen (2014).

3. Nonlinear system

Spherical tank, conical tank and isothermal continuous stirred tank reactor(CSTR) are considered to study the behavior of nonlinear process.

3.1. Case study: Spherical Tank System

In a spherical tank process, level (h) is the controlled variable and input flow rate (q_i) is the manipulated variable. Applying mass balance equation, a mathematical model is obtained as

$$q_i - q_o = \pi[2rh - h^2]\frac{dh}{dt} \qquad (4)$$

where r is the radius of the spherical tank, q_o is outlet flow rate and α is valve coefficient. Here, R is the resistance of outlet flow and is found experimentally. After linearizing the above Equation by using Taylor series, the transfer function model is obtained as

$$\frac{H(s)}{Q(s)} = \frac{(-K_1/K_2)}{(-K_2)s + 1} \qquad (5)$$

where as $K_1 = (\frac{1}{\pi(2rh_s - h_s^2)})$ and $K_2 = (\frac{1}{\pi(2rh_s - h_s^2)^2})[-2q_{is}(r - h_s) - \frac{\alpha\sqrt{h_s}}{R}(1.5h_s - r)]$
Various steady state values for (h_s, q_{is}), in sequence are considered and the corresponding multi-model transfer functions are derived. Nine different multi linear transfer function models are found. Multi model control schemes states that, the number of linear models for design of controllers should be reduced. This is done by considering the gap metric value ≈ 0.1 and only 3 models are taken as shown in Table 1. Based on these three models, Integer and Fractional PI controllers are designed using the respective methods.

3.2. Case Study: Conical Tank Process

In a conical tank process, the level (h) is controlled by the input flow rate (q_i). It's mathematical model is obtained as

$$q_i - q_o = \frac{\pi D}{12H}\frac{d(h^3)}{dt} \qquad (6)$$

Table 1: Reduced models by using gap metric for the nonlinear processes

Spherical Tank		Conical Tank		Isothermal CSTR	
Operating range	Min-imised Model	Operating range	Min-imised Model	Operating range	Min-imised Model
0-15 cm	$M_2 = \frac{0.26261}{330.132s+1}$	0-15 cm	$M_2 = \frac{0.6667}{12.253s+1}$	0-0.5 mole\ L	$M_2 = \frac{0.62834}{21.84s+1}$
15-40 cm	$M_6 = \frac{0.63323}{1194.09s+1}$	15-40 cm	$M_6 = \frac{1.466}{225.68s+1}$	0.5-0.77 mole\ L	$M_6 = \frac{0.32}{11.46s+1}$
40-50 cm	$M_9 = \frac{0.7972}{563.73s+1}$	40-62 cm	$M_9 = \frac{0.9579}{1160.9s+1}$	0.77-1 mole\ L	$M_9 = \frac{0.10417}{3.662s+1}$

Linearizing eq. (6) using Taylors expansion, the transfer function model is obtained as

$$\frac{H(s)}{Q_1(s)} = \frac{\frac{4H}{\pi Dh_s^2}}{s + \frac{8q_{is}}{\pi Dh_s^2} - \frac{C_D}{2h_s^{5/2}}} \tag{7}$$

where H is height of the tank, D is diameter and C_D outlet valve coefficient. Sequentially different steady state values for (h_s, q_{is}) are considered and derived the corresponding multi-model transfer functions and 12 different multi linear transfer function models as obtained. According to multi-model control schemes, multi models are reduce to the minimum number of local models using gap metric value of 0.1 is considered and only 3 models are retained as shown in Table 1. Based on these three models, the corresponding controller parameter values are tuned.

3.3. Case Study: Isothermal CSTR

An isothermal CSTR, consider a first-order irreversible reaction takes place. The mass balance is

$$\frac{dC_A}{dt} = -KC_A + (C_{Ai} - C_A)u \tag{8}$$

Where reactant concentration$(C_A(mol/L))$ is controlled variable, input$(u = q/V(min^{-1}))$ is the manipulated variable, q $(lmin^{-1})$ is the inlet flow rate and C_{Ai} is the inlet feed concentration$(1.0$ mol/L$)$ and constant rate k is 0.028 (min^{-1}). Linearizing the above equation using Taylors expansion, the transfer function model is

$$\frac{C_A(s)}{u(s)} = \frac{1 - C_{AS}}{s + (0.028 + u_s)} \tag{9}$$

The CSTR model is divided into multi linear model based sequential steady states (u_s, C_{As}) of different operating ranges. Eighteen linear models are found and these are minimized into three model by using gap metric value of ≈ 0.1 are shown in table 1, based on these controller parameters are tuned.

4. Simulation and Experimental Results

Simulation results for spherical and conical tank processes and CSTR are carried out and are given below. Experimental implementation is also carried out for conical tank, the corresponding results are presented here.

4.1. Spherical Tank Process

Comparative results of Multi model integer and fractional order control are implemented in simulation for tracking of different set points of level and the corresponding responses are shown in Figure 1. Quantitative analysis of the spherical tank process has been done and integral absolute error (IAE) values for different set points 10, 30, 45, 35, 15 of integer order control are 32695, 14143, 8940.5, 5475.3, 16853 and fractional order control are 123.55, 2569.9, 2127.4, 579.44, 5512.7. It is observed that multimodel fractional order control approach is efficiently reducing the overshoot and the response is enhanced.

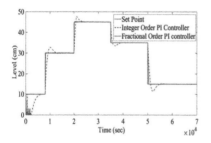 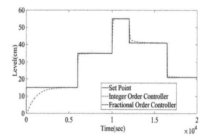

Figure 1: Compared closed loop response of spherical tank process

Figure 2: Compared closed loop response of conical tank process

4.2. Conical Tank Process

In the same way, for conical tank also the comparative results of Multi model integer and fractional order control is implemented in the simulation for tracking of different set points of level and the corresponding responses are shown in Figure 2. Quantitative analysis of the conical tank process has been done and IAE values for different set points 15, 35, 55, 41, 21 of integer order control are 1590.4, 91.70, 109.9, 231.64, 123.42 and fractional order control are 4.31, 13.29, 50.14, 37.61, 68.46. Experimental implementation is carried out for this process and response curves are plotted. The Figure 3 shows closed loop response of multi-model integer order controller and Figure 4 shows closed loop response of multi-model fractional order controller. The oscillations observed in Figure 3 (Integer order response) are very low in amplitude and in the case of fractional order controller, these oscillations become miniscule. Quantitative analysis of the conical tank process has been done and IAE values for different set points 15, 35, 55, 41, 21 of integer order control are 70995, 53654, 178050, 60935, 91032 and fractional order control are 862.05, 911.57, 4428.5, 680.81, 897.64.

4.3. Isothermal CSTR

Similarly, for CSTR also the comparative results of Multimodel integer and fractional order control is implemented in the simulation for tracking of different set points of concentration and the corresponding responses are shown in Figure 5. Quantitative analysis

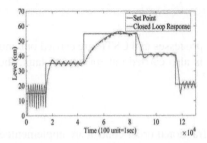

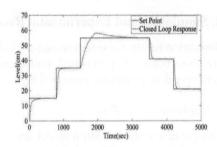

Figure 3: Experimental closed loop response of conical tank process using multi-model integer order controller

Figure 4: Experimental closed loop response of conical tank process using multi-model fractional order controller

of the CSTR process has been done and IAE values for different set points 0.4, 0.55, 0.85, 0.6, 0.35 of integer order control are 2.71, 0.0872, 0.408, 2.865, 4.3766 and fractional order control are 0.096, 0.0464, 0.2337, 2.865, 4.3766.

5. Conclusion

Multi-model fractional order controller is evaluated for control of nonlinear processes and is compared with multi-model integer order controller. Both the methods are evaluated first by the simulation and then by performing experiments on conical tank process. It is observed that multi model fractional order controller provides better performance when compared to Multi model integer order controller.

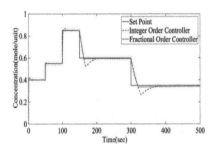

Figure 5: Closed loop response of CSTR

References

A. A. Adeniran, S. El Ferik, 2017. Modeling and identification of nonlinear systems: A review of the multi-model approach part 1. IEEE Transactions on Systems, Man, and Cybernetics: Systems 47 (7), 1149–1159.

S. David, J. Linares, E. Pallone, 2011. Fractional order calculus: historical apologia, basic concepts and some applications. Revista Brasileira de Ensino de Física 33 (4), 4302–4302.

J. Du, T. A. Johansen, 2014. Integrated multimodel control of nonlinear systems based on gap metric and stability margin. Industrial & Engineering Chemistry Research 53 (24), 10206–10215.

J. Du, C. Song, P. Li, 2012. Multimodel control of nonlinear systems: an integrated design procedure based on gap metric and hâĿd loop shaping. Industrial & Engineering Chemistry Research 51 (9), 3722–3731.

J. J. Gude, E. Kahoraho, 2009. New tuning rules for pi and fractional pi controllers. IFAC Proceedings Volumes 42 (11), 768–773.

C. A. Monje, B. M. Vinagre, V. Feliu, Y. Chen, 2008. Tuning and auto-tuning of fractional order controllers for industry applications. Control engineering practice 16 (7), 798–812.

I. Podlubny, 1999. Fractional-order systems and pi/sup/spl lambda//d/sup/spl mu//-controllers. IEEE Transactions on automatic control 44 (1), 208–214.

Anton A. Kiss, Edwin Zondervan, Richard Lakerveld, Leyla Özkan (Eds.)
Proceedings of the 29th European Symposium on Computer Aided Process Engineering
June 16th to 19th, 2019, Eindhoven, The Netherlands. © 2019 Elsevier B.V. All rights reserved.
http://dx.doi.org/10.1016/B978-0-128-18634-3.50239-3

Optimal Operation and Control of Heat-to-Power Cycles: a New Perspective using a Systematic Plantwide Control Approach

Cristina Zotică[a], Lars O. Nord[b], Jenö Kovács[c] and Sigurd Skogestad[a*]

[a]*Norwegian University of Science and Technology (NTNU), Department of Chemical Engineering, 7491, Trondheim, Norway*
[b]*Norwegian University of Science and Technology (NTNU), Department of Energy and Process Engineering, 7491, Trondheim, Norway*
[c]*University of Oulu, Intelligent Machines and Systems Research Unit, POB4300, Oulu, Finland*
sigurd.skogestad@ntnu.no

Abstract

The present control solutions for power plants[1] have been developed to a level where it is difficult to make significant performance improvements using more advanced and systematic control policies, such as model predictive control, unless one has a clear definition of the overall control problem. Hence, the objective of this work is to make a first step in the direction of systematically defining the optimal operation and control problem for power plants. We use a systematic plantwide control framework to analyze a simple heat-to-power steam cycle with a single pressure level and a drum boiler. We evaluate two typical economic modes: I) given plant load, and II) given heat input. We determine that after fulfilling controlling the active constraints there are two unconstrained degrees of freedom left that can be used to achieve optimal operation.

Keywords: Plantwide control, power plant operation, power plant control, steam cycle

1. Introduction

The objective of this work is to provide a new perspective on optimal operation and control of heat-to-power cycles by viewing it from the more general setting of plantwide control. To the best of the authors knowledge, this procedure has so far mainly been applied to chemical plants, and it is yet to be applied to the steam side of heat-to-power cycles. Power plant control systems have been developed by industrial practices over many years, and it is not obvious what are the actual specifications and degrees of freedom. Moreover, power plants contribute to the electric grid stability, which may have priority over optimal operation of the plant itself.

In this paper we analyze this systematically, in the framework of plantwide control, and the goal is to find a control policy, preferably a simple one, which stabilizes the plant on a short time scale (regulatory control), and achieves near-optimal operation on a longer time scale (supervisory control).

The rest of the paper is organized as follows: in Section 2 we present the plantwide procedure; followed by applying it to the water side of a power plant in Section 3; presenting the resulted control structures and operation modes in Section 4 and making the final remarks in Section 5.

---

[1]In this work, the terms heat-to-power cycles and power plants are used interchangeably.

2. Plantwide control procedure

The typical control hierarchy in a process plant is illustrated in Figure 1. It is decomposed based on time scale separation into several simpler layers: scheduling (weeks), site-wide optimisation (days), local optimisation (hours), supervisory control (minutes) and regulatory control (seconds). The layers communicate by sending the process measurements upwards (i.e from the lower to the upper layers), while the setpoints for each layer are given by solving an optimisation problem in the upper layer (Skogestad, 2004). The procedure consists of a top-down analysis concerning optimal steady-state operation, and a bottom-up analysis targeting the lower regulatory control layer structure. In this work, we focus on the steady-state top-down analysis which involves the following steps (Skogestad, 2004):

Step 1 Define the optimal economic operation (cost function J and constraints).

Step 2 Identify the steady-state control degrees of freedom (DOF) (i.e. manipulated variables that are the setpoints for the lower layers). Use a steady-state model to find the optimal operation for the expected disturbances.

Step 3 Based on the previous step, select the control variables (CVs). Always control the active constraints, and decide to which process variables to allocate the remaining DOF.

Step 4 Select the location of the throughput manipulator (TPM), i.e. choose where to set the production rate. This is a dynamic issue, but with economic implications.

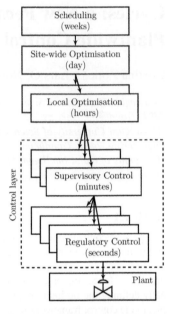

Figure 1: Typical control hierarchy in a process plant.

3. Plantwide control applied to a simple heat to power cycle

3.1. Process Description

We consider the steam side of a heat-to-power cycle with a drum boiler and a single pressure level, as shown in the process flowsheet in Figure 2. In the boiler, there are three physically independent heat exchangers (e.g. economizer (ECO), evaporator (EVAP) and superheater (SH)) dedicated to well defined regimes such as heating of liquid water to or close to boiling point, evaporation and superheating. The superheated steam is expanded in a condensing type turbine, which drives a generator connected to the electric grid. We choose the drum configuration over a once-through boiler (with a single heat exchanger) because it is most common in operating power plants. Further, the drum has the advantage of energy storage which can provide additional operational flexibility.

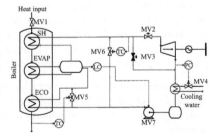

Figure 2: Flowsheet of a power plant with a drum boiler and one pressure level. MV1 and MV2 are the two remaining degrees of freedom considered in this paper. Liquid water is in blue and vapor in red.

3.2. Top-down analysis

We proceed by applying the top-down analysis to the described process.

Step 1. The plant has two operational objectives. On a slow time scale (steady-state) it should achieve the economic optimum, while on a fast time scale it contributes to the grid stability. Due to the time scale separation, these objectives are decoupled. However, the grid stability requirement may impose a back-off from the maximum power production. We define the objective cost function to be minimize the negative profit, given by Eq.1.

$$J = -(p_W W + p_Q Q - p_F F - p_U U) \quad [\$/s] \tag{1}$$

Here, W [J/s] is the produced power, Q [kg/s] is the produced steam (if any), F [J/s] is the heat input source, U [kg/s] are the utilities, and p [\$/kg] or [\$/J] is the price of each. As we consider an operating plant, capital costs, personal, and maintenance costs are not included. The cost function should be minimized subject to satisfying a set of constraints, related to products specifications, safe operation and environment. Typical constraints for a heat-to-power cycle include:

C1 Superheated steam pressure and temperature (e.g. $T \leq 550$ °C, $P \leq 200$ bar to minimize thermal and mechanical stresses and extend the operating life).

C2 Condenser pressure should be low to maximize the pressure ratio in the turbine, but also above a minimum threshold to minimize condensation on the last turbine blade and reduce erosion (e.g. $P \geq 0.01$ bar).

C3 Exhaust gas temperature should be low to minimize heat losses, but above the dew point to prevent corrosion (e.g. $T \leq 100$ °C).

C4 Drum level (stabilizaion and safety).

C5 Requirement to participate in grid frequency regulation (some plants).

C6 Fixed turbine speed (equal to the grid frequency, e.g. $n = 50$ Hz in Europe). For this reason, the turbine speed is not a degree of freedom for operation.

In addition, there are operation constraints on the combustion side (e.g. O_2, CO_2 or NO_x percentage in the flue gas or furnace pressure for combustion power-plants), but these are not relevant for the purpose of this study of the water side.

Step 2. Table 1 shows the degrees of freedom, main disturbances and control variables including the active constraints (a subset of the operational constraints from *Step1*). Here, the active constraints are determined based on engineering insight. The MVs are also shown in Figure 2.

Table 1: Manipulated, disturbances and controlled variables

Manipulated variable	Disturbance	Controlled variable
MV1: Heat (fuel) input	DV1: Quality of heat input	CV1: Power
MV2: Steam turbine valve	DV2: Grid frequency (Load)	CV2: (Live) Steam pressure
MV3: Turbine bypass[*]	DV3: Cooling water temperature	CV3: Drum level
MV4: Cooling water		CV4: Steam temperature[*]
MV5: Economizer bypass		CV5: Condenser pressure[*]
MV6: Feedwater bypass		CV6: Exhaust gas temperature[*]
MV7: Feedwater pump		

[*] Active constraint (at steady-state) controlled in Figure 2.

Step 3. We want to select control variables such that desired optimal operation is maintained even when disturbances occur. Firstly, all the active constrains need to be controlled, and one degree of freedom is used to control each of them. One possible pairing for the identified constraints is shown in Figure 2, and is: control of steam temperature with MV6, gas temperature with MV5, condenser pressure with MV4, and the bypass MV3 should be closed. Further, the drum level can be controlled with MV7. The turbine speed is usually controlled by changing the steam mass flow using MV2. In this case, the setpoint given to the speed controller becomes a new degree of freedom instead of the valve position. In practice, this is often called *droop control*. In practice all turbines have droop control (Kurth and Welfonder, 2006). Note that the pairing of MVs and CVs (here, the active constraints) can be different, and that only the number of the remaining degrees of freedom is important for this step. We now have two degrees of freedom to use for optimal operation: the heat input MV1 and the and the speed controller setpoint MV2.

3.3. Droop Control

Droop control is commonly used to proportionally allocate the load between electrical generators running in parallel, to avoid a scenario where the load is taken by only one of them (Lipták and Bálint, 2003). This is a proportional controller, with a gain of typical 3 %/% to 10 %/%. In droop control, there is a fast MV (i.e. MV2, turbine valve), and a MV with a slow effect (i.e. MV1, heat). For exam-

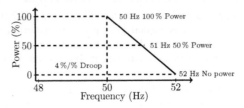

Figure 3: Droop characteristics of 4 %/%.

ple, when the turbine speed decreases, the droop controller acts by opening MV2, which increasing the steam mass flow. This can be sustained on a long term only if the heat increases as well. Figure 3 shows an example of 4 %/% droop, i.e. a change of power of 100 %/% gives $50 \cdot 0.04 = 2\,\text{Hz}$ change in turbine frequency setpoint (Anderson and Fouad, 2003).

Step 4. The location of the throughput manipulator depends on the type of power plant, and we can identify two cases (i.e. economic modes) (Skogestad, 2004):

Case I The load is the throughput manipulator and optimal operation means maximizing the fuel usage (efficiency) (i.e. the plant is required to participate in the grid frequency regulation, or to supply utilities to a downstream process).

Case II The heat input is the throughput manipulator and optimal operation implies maximizing production (load). This case applies when the heat input is cheap, or "free" (e.g. solar or heat recovery from a gas turbine).

4. Control structures and operation modes

The standard industrial control structures are boiler driven, turbine driven, floating pressure and its variation, sliding pressure (Welfonder, 1999). However, if one disregards the common industrial practices for power plant control, alternative control structures can be proposed. For example, using valve position control (VPC), or both a P and a PI controller with a MID-selector. In the following, we only show the pairing options for the two remaining degrees of freedom (MV1 and MV2), and we assume that the inner turbine speed controller is active. We start with Case I.

4.1. Floating pressure operation

One degree of freedom is used to keep the throughput manipulator to its specification. The simplest options is to use MV1 to control the power output, and have MV2 fully open to minimize throttling losses. This structure is called floating pressure because the live steam pressure is left uncontrolled, as shown in Figure 4a. However, when a constraint on the maximum allowed pressure is reached, and the set of active constraints has changed, we have to give-up another constraint that is less important (Reyes-Lúa et al., 2018). In this case, we have to use the steam turbine bypass (MV3) as a degree of freedom. The disadvantage is a slow time response for load changes.

4.2. Boiler driven operation

To improve the response time of the system, one should control another intermediate variable. The only option is to keep the live steam pressure at a given set-point. The result is the boiler driven control structure, shown in Figure 4b. The output power is controlled with MV1, while the live steam pressure is controlled with MV2. This structure has the advantage that it dynamically responds faster to a change in load by utilizing the energy stored in the system, which comes at an expense of not utilizing the heat input at the maximum. Though it has a faster response compared to the floating pressure, the system still has a large time constant from MV1 to the power output, and the overall response is slow.

4.3. Turbine driven operation

The reverse pairing is turbine driven, as illustrated in Figure 4c. The live steam pressure is controlled with MV1, while the power output is controlled with MV2. It also has the advantage of utilizing the drum and the superheater energy storage, and the response time to load changes is faster. However, it also does not efficiently utilizes the heat input due to valve throttling losses.

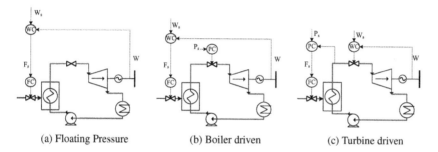

| (a) Floating Pressure | (b) Boiler driven | (c) Turbine driven |

Figure 4: Control structures for three operation modes for a simple heat-to-power cycle.

4.4. Sliding pressure operation

Sliding pressure is when the pressure setpoint is adjusted online, usually for the turbine driven operation mode. In this case, the power plant can participate in the primary frequency control, at the expense of having to back-off from optimal operation and have MV2 partially open (e.g. 90 %) (Weissbach et al., 2006). The cycle efficiency decreases due to throttling losses in the steam valve, but it allows the power plant to have a energy backup that can be dynamically used by fully opening the valve, and participate in the grid frequency control. However, assuming an isenthalpic expansion through the valve, the steady-state effect will be small. In practice, it is common to have a pressure controller with a varying setpoint given by a master controller in a cascade structure that receives the load as input, as shown in Figure 5a (Klefenz, 1986). Here, the pressure setpoint is changed based on a model that has the steam mass flowrate as input and the pressure setpoint as output.

Valve position controller. The conventional sliding pressure control structure can be improved by adding a valve position controller to bring back MV2 to its nominal opening (at steady-state) after it has dynamically contributed to grid frequency regulation, as shown in Figure 5b.

MID selector. We use both MV1 and MV2 to control the power, and the pressure is floating, as illustrated in Figure 5c. Here we use one P-controller, and one PI-controller because we can have only one integral action in a two-inputs one-output system. When pressure becomes an active constraint (i.e. at the minimum or maximum), we use a *MID* selector to control it using MV1.

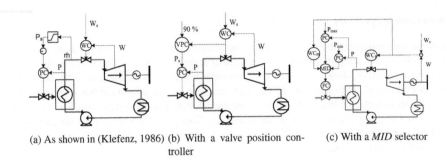

(a) As shown in (Klefenz, 1986)   (b) With a valve position con-   (c) With a *MID* selector
troller

Figure 5: Control structures for sliding pressure operation mode

4.5. Case II - the heat input is the throughput manipulator

From a steady-state point of view, there are no degrees of freedom left to control the power. However, the turbine bypass can be used to reduce the power produced by decreasing the steam mass flow. However, the efficiency of the cycles will also decrease. Ideally, one should make use of energy storage solutions of superheated steam for the situations when more steam can be produced than is needed. This is an attractive idea for power plants using solar energy, especially since the maximum solar power is at midday, when the electricity market demand is lower.

5. Conclusions and final remarks

We have systematically identified the operational objectives, operational and environmental constraints for a heat to power cycle. The degrees of freedom left are MV1, the heat input, and MV2, the steam turbine valve. Now that we have systematically defined the optimal operation and control problem for power plants, it becomes easier to make improvements using more advanced and systematic control policies, such as model predictive control.

6. Acknowledgment

This work is partly funded by HighEFF Centre for an Energy Effcient and Competitive Industry for the Future. The authors gratefully acknowledge the financial support from the Research Council of Norway and user partners of HighEFF, an 8 year Research Centre under the FME-scheme (Centre for Environment-friendly Energy Research, 257632/E20).

References

Anderson, P., Fouad, A., 2003. Power System Control and Stability, 2nd Edition. Wiley.

Klefenz, G., 1986. Automatic Control of Steam Power Plants. Bibliographisches Institut.

Kurth, M., Welfonder, E., 2006. IMPORTANCE OF THE SELFREGULATING EFFECT WITHIN POWER SYSTEMS. In: IFAC Proceedings Volumes. Vol. 39. IFAC, pp. 345–352.

Lipták, B. G., Bálint, A., 2003. Application and Selection of Control Valves. In: Lipták, B. G. (Ed.), Instrument Engineers' Handbook, 4th Edition. ISA - The Instrumentation, Systems and Automation Society, Ch. 8.38, p. 2142.

Reyes-Lúa, A., Zotică, C., Skogestad, S., 2018. Optimal Operation with Changing Active Constraint Regions using Classical Advanced Control. IFAC-PapersOnLine 51 (18), 440–445.

Skogestad, S., 2004. Control structure design for complete chemical plants. Computers and Chemical Engineering 28 (1-2), 219–234.

Weissbach, T., Kurth, M., Welfonder, E., Haake, D., Gudat, R., 2006. Control Performance of Large Scale Steam Power Plants and Improvments. In: Westwick, D. (Ed.), IFAC Symposium on Power Plants and Power Systems Control. Vol. 39. IFAC, Kananaskis, Canada, pp. 183–188.

Welfonder, E., 1999. Dynamic interactions between power plants and power systems. Control Engineering Practice 7 (1), 27–40.

Anton A. Kiss, Edwin Zondervan, Richard Lakerveld, Leyla Özkan (Eds.)
Proceedings of the 29th European Symposium on Computer Aided Process Engineering
June 16th to 19th, 2019, Eindhoven, The Netherlands. © 2019 Elsevier B.V. All rights reserved.
http://dx.doi.org/10.1016/B978-0-128-18634-3.50240-X

Novel strategies for predictive particle monitoring and control using advanced image analysis

Rasmus Fjordbak Nielsen[a], Nasrin Arjomand Kermani[b], Louise la Cour Freiesleben[c], Krist V. Gernaey[a] and Seyed Soheil Mansouri[a,*]

[a]*Process and Systems Engineering Centre (PROSYS), Department of Chemical and Biochemical Engineering,Technical University of Denmark, Søltofts Plads, Building 229, DK-2800 Kgs. Lyngby, Denmark*
[b]*Section of Thermal Energy, Department of Mechanical Engineering, Technical University of Denmark, Nils Koppels Allé, Building 403, DK-2800 Kgs. Lyngby, Denmark*
[c]*ParticleTech ApS, Hirsemarken 1, DK-3520 Farum, Denmark*
seso@kt.dtu.dk

Abstract

Processes including particles, like fermentation, flocculation, precipitation, crystallization etc. are some of the most frequently used operations in the bio-based industries. These processes are today typically monitored using sensors that measure on liquid and gas phase properties. The lack of knowledge of the particles itself has made it difficult to monitor and control these processes. Recent advances in continuous in-situ sensors, that can measure a range of particle properties using advanced image analysis, have now however opened up for implementing novel monitoring and modeling strategies, providing more process insights at a relatively low cost. In this work, an automated platform for particle microscopy imaging is proposed. Furthermore, a model based deep learning framework for predictive monitoring of particles in various bioprocesses using images is suggested, and demonstrated on a case study for crystallization of lactose.

Keywords: Bioprocess Monitoring, Advanced Image Analysis, Modeling Framework

1. Introduction

During the past two decades, bioprocesses have become an increasingly important part of many food, chemical, agrochemical and pharmaceutical industries. This has had an impact on both upstream and downstream processing. Here, fermentation, flocculation, precipitation and crystallization are frequently used processes for formation, separation and purification of products, respectively (Santos et al. (2017)). A common denominator for these processes is that they all contain particles in some form, whether it is cells, aggregates or crystals.

Despite the broad applications of these processes in industry, there is typically a problem of monitoring and controlling them to a satisfactory degree. This is evident in the common use of heuristics based control and the difficulties in obtaining consistent product qualities. Today, most of these bioprocesses are monitored and controlled using traditional sensors that measure gas and liquid phase properties. Only recently, there has been a focus on measuring particle properties continuously (Biechele et al. (2015)).

Historically, particle characterization has been been either very time-consuming or associated with relatively high capital costs of the sensors needed. To accommodate this, in the past decade, there has been a move towards developing soft-sensors, that utilize the measured gas and liquid phase properties, to estimate different process variables that relates to the particles. These have shown great potential (Damour et al. (2009); Mears et al. (2017)). However, these soft-sensors typically require a large amount of infrastructure, in terms of on-line sensors, and significant modeling efforts, as they are only indirect measures of the real process variables.

Meanwhile, there have been significant developments within the application of advanced image analysis in recent years, which has enabled reliable particle analysis in real time, providing an easy way of measuring a range of particle properties. Furthermore, with the current trends towards lab-on-a-chip solutions for industrial sensors, the use of advanced image analysis in bioprocesses can now be implemented at a relatively low cost, and add valuable insights to various bioprocesses.

In this work, a novel platform for automated particle microscopy is presented, where on-line microscopy images are taken from a flow cell. A model-based, predictive monitoring framework is proposed for a generic particle process, utilizing both quantifiable particle properties obtained from advanced image analysis and the captured raw-images themselves. Here, a deep neural network is used for estimating rates of phenomena, that can be used for prediction of process variable trajectories like particle count and size distribution. This model-based framework is subsequently demonstrated on a case study of lactose crystallization, showing the potential of this method, and finally compared to results of conventional white-box modeling.

2. Flow cell microscopy

To measure a whole range of particle properties in a continuous fashion, a robust platform is introduced to automatically transfer samples from the production tank to the flow cell under the microscope for image acquisition and on-line monitoring of the process. The automated platform,

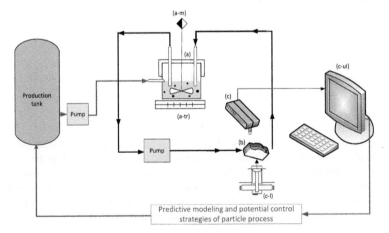

Figure 1: Illustration of proposed platform: (a) Sample unit with automated flowpath (blue and black lines) consists of mixer (a-m), temperature regulator (a-tr) and two pumps; (b) a modular flow cell; (c) Automated microscope with light adjustment (c-l) from bottom side; and a user interface (c-ui)

shown in Figure 1, consists of four core elements:

- Sample unit with automated flowpath consists of a mixer, temperature regulator and two pumps to transfer the sample from the production tank to the flow cell under controlled conditions. The sample can be diluted inside the sample unit to a convenient volume for on-line microscopy.

- A modular flow cell consist of a monolayer channel, adjustable for a wide range of particle sizes and carrier fluid properties to hold the samples under the microscope for image acquisition

- An automated microscope to capture high quality images

- A user interface to extract meaningful information from an image through several steps such as image processing, segmentation, and characterization of the identified object. The segmented 2D- image will provide valuable information such as particle size, particle shape, and particle size distribution.

The proposed configuration allows rapid and nondestructive visual monitoring of samples with various compositions and enables the extraction of valuable chemical and physical information.

3. Model-Based Framework

To benefit the most from these new real-time measurements, a predictive monitoring framework is proposed, utilizing the large amount of information that is generated from the captured images. The phenomena observed in a generic particle process include particle birth, particle growth (both positive and negative), particle agglomeration and particle breakage. To describe the overall par-

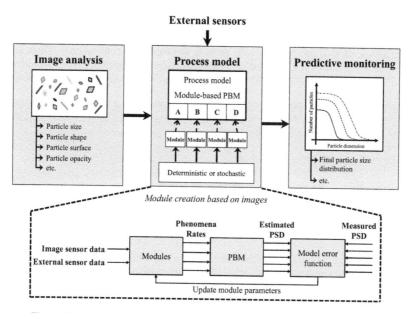

Figure 2: Proposed framework for monitoring of particles in bioprocesses

ticle processes, population balance modeling (PBM) has become a very popular and generic tool (Ramkrishna and Singh (2014)). For each type of process, a range of kinetic expressions are typically derived from experiments, describing the rate of a given phenomenon, as a function of a range of process variables like temperature, pH and composition. To determine these expressions, one has historically tried to study one phenomenon at a time, where a large fraction of these studies have only been looking at the development of a single particle as a function of time.

In this framework, we instead aim towards describing these phenomena, by analyzing sample populations during the process operation, containing several hundreds of particles. For each phenomenon that needs to be explained, a corresponding module will be included in the model. These modules can either be traditional white-box models like power-law expressions, utilizing a limited range of very specific measurements, or black box models like neural networks, random forests etc. that can take into account a much larger range of measured process variables. An overview of the framework can be seen in Figure 2.

By using this framework, with the latter black-box approach, which is combined with the white-box generic PBM, one can potentially save much time in process modeling. One will however still get an accurate description of the system dynamics, and at the same time take into account a lot more possible process disturbances and variables than in the conventional approach. Finally, this framework allows for implementing a robust model predictive control (MPC) in various particle processes.

4. Case study: Lactose crystallization

The presented framework is now demonstrated for a case study of lactose. For this study, a single crystallization experiment was carried out for the model generation and one for process model validation. A scanned 2D plane of 2288.5 μm x 1408 μm was used in both experiments.

A lactose solution was prepared by dissolving 35.0 g lactose per 100.0 g water, by heating the solution to 70 °C. Afterwards, the solution was cooled down by natural heat loss to the surroundings, which results in crystallization of lactose. Samples were taken regularly (every 3-4 minutes) using the presented flow cell microscopy, obtaining a range of particle properties, including the Feret diameters, projected area and sphericity of each particle detected on each image. Furthermore, the temperature of the crystallization process was measured continuously. Three examples of segmented images, showing the crystallization progress, can be seen in Figure 3. Note how the number of crystals is increasing and the crystals are increasing in size as a function of time.

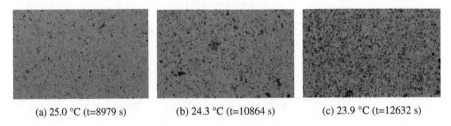

(a) 25.0 °C (t=8979 s)   (b) 24.3 °C (t=10864 s)   (c) 23.9 °C (t=12632 s)

Figure 3: Images after segmentation of crystals.

In this case study, the crystals were binned according to their mean Feret diameter, starting from 20 μm moving up to 80 μm, with a total of 40 bins. Using the presented predictive monitoring modeling framework, a rolling prediction model was generated, using a deep neural network for estimating the nucleation and growth rates, and then compared to a conventional full simulation modeling approach, using white-box power-law expressions. Both approaches utilizes a generic 1D PBM, relying on the method of classes, where the effects coming from agglomeration, breakage and dissolution have been neglected in both.

The deep neural network used here consists of 4 dense layers in total, with a total of 296,000 weights. The inputs provided to the network are: a grey-scale and down-sampled version of the raw image (64 x 64 pixels) taken at the estimation time, the current temperature, the current temperature gradient and the time-horizon Δt to predict. This results in one growth rate per bin and a single nucleation rate. By transferring these rates to the generic white-box PBM, one can obtain a grey-box running prediction of the size distribution at time $t + \Delta t$, where the initial size-distribution of the PBM is set to the measured distribution at the current time t.

In the conventional full simulation modeling approach, the solubility of lactose in water as a function of temperature was approximated by a second order polynomial fit of the datapoints measured by Hudson (1908). Furthermore, the density of lactose was set to 1.52 g/cm^3 as reported by Lewis (2007). The following three white-box rate expressions, similar to the ones presented by Szilagyi and Nagy (2016), where used in this work:

$$P = k_p \cdot \sigma^p \quad , \quad S = k_s \cdot \sigma^s \cdot V_c \quad , \quad G = k_g \cdot \sigma^g \cdot (1 + \gamma \cdot L) \tag{1}$$

where V_c is the volume of crystals per volume solute and L is the characteristic particle dimension. Thus, in the white-box model, there are seven kinetic parameters that have to be determined by experiments and a single parameter relating to the particle shape, the volumetric shape factor k_v.

The two models were fitted to the experimentally obtained data from a single batch crystallization, by minimizing the absolute error of the relative size distribution. The two fitted models can be seen in Figure 4a + 4b, showing the time-evolution of the relative size distribution. Note that only a selection of size-bins have been included in the figures for improved readability.

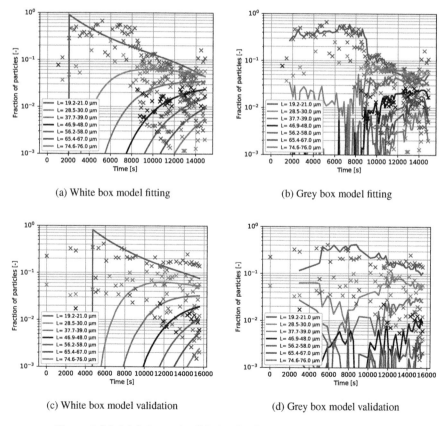

(a) White box model fitting

(b) Grey box model fitting

(c) White box model validation

(d) Grey box model validation

Figure 4: Model fitting and validation for the two modeling approaches.

The two modeling approaches have subsequently been used to predict the evolution of the size distribution as function of time in the validation experiment. These results can be seen in Figure

4c + 4d, where the predicted time-horizon for the rolling prediction is set to 10 minutes. I.e. the deep neural network estimates the appropriate nucleation and growth rates for the next 10 minutes of crystallization operation.

It can be seen that both modeling approaches predicts the trends in the size distribution evolution well. However, the rolling prediction can be seen to produce more accurate predictions. In this, the adaptive nature of the rolling prediction does play a big role, as it is utilizing size distribution data as they become available. However, even for larger predicted time-horizons Δt, the grey-box model approach performs very well, indicating that the deep neural network does estimate the actual phenomena rates to a satisfactory degree.

With the presented grey-box model approach, it is shown possible to capture some effects in the crystallization, that would not be taken into account in a conventional modeling approach. However, the grey-box modeling is also relatively prone to measurement noise in he size-distribution sampling, which can be seen in as fluctuations in Figure 4b + 4d. These sources are especially critical, as they are both used as input and output variables while training the neural network, and would therefore be beneficial to minimize these uncertainties, possibly by using MCMC methods or input noise during training, as previously reported for neural networks (Wright et al. (2000)).

5. Conclusions

In this work, a novel continuous microscopy platform has been presented, showing a simple and cheap way of taking images of particles in suspension. A model-based predictive monitoring framework has been presented, utilizing particle parameters obtained from the images and the raw images themselves. This is used for modeling a generic particle process, and predicting the future particle population, utilizing deep learning neural networks for estimating birth and growth rates. The framework has been tested on a case study of lactose crystallization, where it has been shown that the framework can provide accurate predictions, even for small amounts of training data.

6. Acknowledgments

This work partly received financial support from the Greater Copenhagen Food Innovation project (CPH-Food), Novozymes, from EU's regional fund (BIOPRO-SMV project) and from Innovation Fund Denmark through the BIOPRO2 strategic research center (Grant number 4105-00020B).

References

P. Biechele, C. Busse, D. Solle, T. Scheper, K. Reardon, 2015. Sensor systems for bioprocess monitoring. Engineering in Life Sciences 15, 469–488.

C. Damour, M. Benne, B. Grondin-perez, J. Chabriat, 2009. Model based soft-sensor for industrial crystallization: on-line mass of crystals and solubility measurement. World Academy of Science, Engineering and Technology 54, 196–200.

C. S. Hudson, 1908. Further studies on the forms of milk-sugar. Journal of the American Chemical Society 30, 1767–1783.

R. J. Lewis, 2007. Hawley's Condensed Chemical Dictionary. John Wiley & Sons, Inc.

L. Mears, S. M. Stocks, M. O. Albaek, G. Sin, K. V. Gernaey, 2017. Application of a mechanistic model as a tool for on-line monitoring of pilot scale filamentous fungal fermentation. Biotechnology and Bioengineering 114, 589–599.

D. Ramkrishna, M. R. Singh, 2014. Population Balance Modeling: Current Status and Future Prospects. Annual Review of Chemical and Biomolecular Engineering 5, 123–146.

R. Santos, A. Carvalho, A. Roque, 2017. Renaissance of protein crystallization and precipitation in biopharmaceuticals purification. Biotechnology Advances 35, 41 – 50.

B. Szilagyi, Z. Nagy, 2016. Graphical processing unit (GPU) acceleration for numerical solution of population balance models using high resolution finite volume algorithm. Computers and Chemical Engineering 91, 167–181.

W. Wright, G. Ramage, D. Cornford, I. Nabney, 2000. Neural Network Modelling with Input Uncertainty: Theory and Application. Journal of VLSI signal processing systems for signal, image and video technology 26, 169–188.

Anton A. Kiss, Edwin Zondervan, Richard Lakerveld, Leyla Özkan (Eds.)
Proceedings of the 29th European Symposium on Computer Aided Process Engineering
June 16th to 19th, 2019, Eindhoven, The Netherlands. © 2019 Elsevier B.V. All rights reserved.
http://dx.doi.org/10.1016/B978-0-128-18634-3.50241-1

Integrating Simulation and Optimization for Process Planning and Scheduling Problems

Miguel Vieira[a], Samuel Moniz[b], Bruno Gonçalves[c], Tânia Pinto-Varela[a*], Ana Paula Barbosa-Póvoa[a]

[a]CEG-IST, Universidade de Lisboa, Av. Rovisco Pais, 1049-001 Lisboa, Portugal

[b]Departamento de Engenharia Mecânica, Universidade de Coimbra, Coimbra, Portugal

[c]Departamento de Produção e Sistemas, Universidade do Minho, Guimarães, Portugal

tania.pinto.varela@tecnico.ulisboa.pt

Abstract

The development of hybrid simulation-optimization methods has allowed to explore the complexities of planning and scheduling problems. In this work, a simulation-optimization approach is developed to provide decision-support to industrial processes, combining a mathematical formulation with a detailed discrete-event simulation model. To solve industrial planning and scheduling problems, the methodology iteratively estimates the modelling parameters to maximize the output in a multistage production facility, while guaranteeing allocation constraints and main process uncertainties to satisfy the demand. The production plan is evaluated at the MILP model and the capacity-feasible schedule is generated by the dispatching rules in the simulation model. Results highlight the advantages of the hybrid approach to guarantee optimized process performance metrics in a simulated operational planning.

Keywords: simulation-optimization; production planning and scheduling; hybrid methodology

1. Introduction

Current industries are facing competitive environments that demand enhanced performance indicators, such as customer satisfaction, manufacturing/inventory holding costs, and delivery times. These challenging production environments have enabled the development of advanced planning and scheduling support tools, following the increasing digital transformation with process automation. Despite the significant developments in exact/non-exact modelling techniques in recent years, it is acknowledged that generalized mathematical formulations are limited by the inherent computational hindrance of system modelling with combinatorial complexity. As well, the development of novel simulation platforms has empowered the evaluation of scenarios that mimics real operations and explore the complexity of nonlinear and stochastic systems. Therefore, the potential hybridization of simulation with optimization approaches, as noted by Figueira and Almada-Lobo (2014), combines the advantages of the process detail representation with the ability to optimize solutions. With different abstraction levels in each model, the simulation includes features not addressed in the optimization to decrease complexity, while considering iterative parameters adjustment to allow solution convergence. This methodology is categorized as a Recursive Optimization-Simulation Approach (ROSA), which recursively evaluates performance metrics until a stopping criterion is met.

In this work, we explore this ROSA methodology to address the planning and scheduling optimization problem in multistage production process. Based in Kim and Kim (2001) approach, the proposed hybrid approach combines a mathematical model with a detailed discrete-event simulation model to provide decision-support on the production planning and the required number of allocated shared resources. As follows, we present an overview of methodology of the proposed approach and discuss the results in a case example in order to formulate a planning/scheduling solution for the industrial operational management.

2. Background

The general problem of planning/scheduling operations has become one of the most challenging problems in process industry. Due to arbitrary network configurations of resources that accounts multiple materials recipes, the inherent complexity endures researchers on finding the key modelling solution on how to optimize the sizing, sequencing and allocation while matching the expected performance of an industrial process. Since operational decisions are strongly interconnected, it is difficult to develop competent decomposition approaches to solve real-size applications with good schedules and modest computational hindrance (Castro et al., 2011). And, as note by Harjunkoski et al. (2014), novel optimization methods are required to address the pressure toward further cost-savings, efficient use of resources, regulatory aspects and new technologies.

Among the diverse solution approaches, from mathematical programming to heuristic algorithms, the combination of the best assets of different methods has been gathering significant relevance. Hybrid formulations have been consistently addressing the reduction of the solution search-space to limit the number of operational and strategic decisions to a manageable level (Vieira et al. 2018). Noteworthy, simulation and optimization approaches are widely applied as sole methodologies to address operational problems. But with modern discrete-event simulation platforms, several authors have addressed the drawbacks by exploring the iterative combination of both in the evaluation of parameters, optimization of solution variables, and sampling of scenarios (Martins et al. 2017). For example, Kim and Kim (2001) developed an extended hybrid model for the multiperiod multiproduct production planning problem, based on an iterative process where the release plan determined by a linear programming model is evaluated by the simulation model. One of planning challenges relies in the real estimate of the lead time of production, since the release of the production order to its output dispatch to satisfy the demand, accounting to the uncertainty in resources' workload capacities. When the results are not feasible (e.g. the number of orders released by the optimization exceeds the workload capacity under simulated conditions), the simulation model adjusts a specific set of parameters of the optimization model and repeats the process under more restrict conditions. When the results from the analytical method are feasible and met the stopping criteria (e.g. number of iterations, production demand met or solution convergence), a near-optimal production schedule can be generated. The iteration considers factors that correlate the capacity and workload of the resources, namely loading ratios and unit utilization. The authors verified that the method consistently converges in few iterations, even in real problems of industrial complexity (Kim and Lee, 2016). However, it is acknowledged that the right level of abstraction in each model is critical to avoid runtime becoming impracticable, since the simulation model is left to include the hard-to-model parameters, with particular significance on solution quality and convergence parameters of the method.

3. Methodology

The hybrid approach proposed in this work consists in the iterative use of optimization and simulation models to address the optimal planning and scheduling of an industrial production process. The problem defines the optimal allocation of shared resources to perform a production mix per demand period in a multistage facility considering the cost-efficient use of resources. This decomposition approach enables to generate a solution at the planning capacity level with a MILP model that is, iteratively, evaluated at the scheduling level with the visualization of the dynamic behavior of the production process. While the optimization model is formulated using GAMS/CPLEX, the detailed simulation model is implemented in SIMIO which systematically reproduces the real production facility. Besides the production mix, the capacities and workload of workstations in the manufacturing system are affected by various factors such as buffer capacity, task sequencing, operational uncertainty, input policy or dispatching rules. The main components and tools available at the SIMIO platform are used to achieve the most accurate representation of the manufacturing system, which requires advanced programming for modelling specific process features.

Based on the approach by Kim and Kim (2001), the proposed algorithm structure considers an iterative procedure to reach solution convergence at stopping criteria, i.e. if the solution objective is identical to the previous iteration. The original mathematical formulation considers four types of continuous variables expressed by product i at time period t: X_{ip} as the orders amount released at time period p, Y_{it} auxiliary variable that defines the effective proportion quantity following the release period ($t \geq p$), I_{it} as the inventory amount, and B_{it} as the backorder amount. That formulation is now extended to consider a binary variable O_{ijr} which defines the set of resources r allocation to perform the sequential operation stage j of a product i, while determining the number of shared resources R_t required to fulfil a demand plan μ_{it}.

$$Min \sum_i \sum_t (Y_{it}\alpha_{it} + I_{it}\sigma_{it} + B_{it}\pi_{it}) + \sum_t R_t\theta_t \qquad (1)$$

$$\sum_i \sum_j \sum_r \sum_{p \leq t} \varepsilon_{ipt} X_{ip} O_{ijr} \tau_{ijkr} \leq \lambda_{kt}\delta_{kt} \qquad \forall k,t \qquad (2)$$

$$I_{it} - B_{it} = Y_{it} - \mu_{it} + I_{it-1} - B_{it-1} \qquad \forall i,t \qquad (3)$$

$$Y_{it} = \sum_{p \leq t} \varepsilon_{ipt} X_{ip} \qquad \forall i,t \qquad (4)$$

$$\sum_i \sum_j \sum_k \sum_{r \in SR_A} \sum_{p \leq t} \varepsilon_{ipt} X_{ip} O_{ijr} \tau_{ijkr(A)} \leq \phi_t D_t R_t - S_t.TimeS \quad \forall t \qquad (5)$$

$$\sum_r O_{ijr} = 1 \qquad \forall i,j \qquad (6)$$

$$X_{ip}, Y_{it}, I_{it}, B_{it}, R_t \geq 0 \qquad X_{ip}, R_t \in \mathbb{Z} \qquad (7)$$

The objective function Eq. (1) considers the minimization of total costs related to production α_{it}, inventory σ_{it}, backlog π_{it} and number of shared resources θ_t. The capacity constraint of each unit stage k in Eq. (2), with the average operation time given by τ_{ijkr}, is adjusted by the recursive simulated parameters, the average utilization of stage k, λ_{kt}, of the available time per period t, δ_{kt}, and by the effective loading ratio ε_{ipt} per periods t of the amount released in p. The linearization of $X_{ip}O_{ijr} \equiv \Omega_{ijpr}$ as the released quantity to be performed by resources r was performed, considering the standard auxiliary restrictions to linearize the product of a binary and continuous variables. Eq. (4) considers the inventory balance of material flow and demand per period, and Eq. (3) enables the calculation of auxiliary variable Y_{it}. Eq. (5) considers the capacity constraint related to the operations that make use of shared resources SR_A per time period D_t, accounting the time spent ($TimeS$) with the number of changeovers S_t and the traveling between stages (ϕ_t as average operating rate). Finally, Eq. (6) enforces that all operations of product i are performed by one set of resources and Eq. (7) the non-negativity of variables (details in Kim and Kim (2001)). The first step of the algorithm consists in running the model in GAMS to determine the amount of production released X_{ip}, allocation of resources O_{ijr}, and number of shared resources required per period R_t. In the second step, the feasibility of the sequential release plan (one by one order) is evaluated in the SIMIO model and the statistics on effective loading ratios, unit stages utilization and number of changeovers/utilization of shared resources are collected. In the third step, these iteration statistics in the model updates the capacities constraints by using the recursive parameters from the simulation. The iterations procedure stops when the plan met the stopping criteria and the simulation generates the schedule based on a given dispatching rule.

4. Case study

4.1. Problem description

The example case is based on a multistage industrial process in a production plant composed by seven single-unit stages k (k1-k7) to produce five products i (i1-i5) in a U layout. The different processing stages for each 5 products follow a predefined sequence operation set j given by Table 1, where at each corresponding stage a series of tasks are executed, as displayed in Figure 1. Currently, all tasks are performed by one dedicated operator per stage (Op). Since some of the tasks at each stage can be performed in parallel, the decision-support at the operational level relies on the allocation of a mobile shared resource (Aut) able to perform one of these tasks, therefore existing the option for two sets of resources r per operation i,j. This can enable the decrease of the total processing time per stage, as shown in Table 1 (for this example, only one shared resource is allowed per stage to perform one of the simultaneous tasks.).

Table 1 – Sequence of process stages per product and corresponding processing time

Product i	Unit stage k							Resources r	Average operation time stage j (min/ton)				
	1	2	3	4	5	6	7		1	2	3	4	5
P1	●	●			●	●	●	1 Op	61,2	60,0	116,4	61,2	78,0
								1 Op+1 Aut	61,2	60,0	104,4	60,0	48,0
P2			●	●	●	●		1 Op	93,6	61,2	116,4	61,2	-
								1 Op+1 Aut	74,4	49,2	104,4	60,0	-
P3			●	●	●		●	1 Op	60,0	93,6	61,2	78,0	-
								1 Op+1 Aut	60,0	74,4	49,2	48,0	-
P4	●		●		●		●	1 Op	61,2	93,6	116,4	78,0	-
								1 Op+1 Aut	61,2	74,4	104,4	48,0	-
P5	●	●		●		●		1 Op	61,2	60,0	61,2	61,2	-
								1 Op+1 Aut	61,2	60,0	49,2	60,0	-

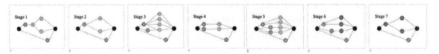

Figure 1 – Sequence of tasks per unit stage

The production follows a pull strategy, which means that the products are manufactured to assess a given demand for a time horizon of seven continuous 24 h production days. While the optimization model considers the Eqs. (1-6), the simulation model is able to accommodate the manufacturing system characteristics as close at the real facility, accounting for the routing distances, queuing, system layout and the detailed tasks sequence at each stage. The simulation inputs the determined release plan, allocation sequence and number of resources, which replicates the process operation during the planning horizon. The uncertainty in the process data is also considered, namely at the processing times of tasks performed by the operator and the changeover time of the shared resource, by assessing the corresponding variability distribution. A FIFO dispatching rule in the simulation is assumed throughout. For the minimization of operational costs, the request cost of one shared resource, θ_t, is assumed of 100 monetary units (mu), while the production α_{it}, inventory σ_{it}, backlog π_{it} costs are, respectively, 100, 25 and 400 mu.

4.2. Computational results

The results of the given example show that the methods quickly converge to a solution where the total output of the production plan is similar between models. The final results were obtained at the third iteration and the results are generated in a few seconds (optimality gap <1%), running on an Intel Core i7-7700HQ with 16GB of RAM. Whenever the optimization results were too optimistic (non-feasible), resources capacity constraints parameters were tightened to generate a more realistic production plan, given the simulation statistics (20 replications for a confidence level of 95%). For example, in the first run of the optimization model, since it considers nonrestrictive capacities of resources, the results match the total demand with only one shared resource allocation (Table 2). But when evaluated in the simulation, the output is a far different solution, due to lower effective loading ratio, percentage utilization of resources and number of setups. Iteratively the solution space is reduced, with a production release on the initial periods to reach a scheduling solution given by the simulation software. In the following iterations, the number of available shared resources per time period is increased to be able to accommodate the production, as well as the increase in inventory backorder and costs since the total demand is not satisfied on this planning horizon. The simulation model also enables to analyze the scenario results on the bottlenecks of the process, where it can be verified that stages 3 and 5 have utilization ratios between 98 to 99%, therefore restraining the fulfillment of demand (assumed to be satisfied in the following plan horizon). The reduction in the number of setups of the shared resources is verified since, with the increase of their availability with almost a full utilization, they were modelled in the simulation to preferably choose same product operations. This number is also influenced by the different allocation of resources to shared tasks performed in each iteration, which increased with the availability of the resources.

Table 2 – Iterations results of order release amount X_{ip}

| Time period | Iteration 1 | | | | | | >>> | Iteration 3 | | | | |
	P1	P2	P3	P4	P5	Shared resources		P1	P2	P3	P4	P5	Shared resources
t1	2	6	4	5		1		13	19	6	1	20	1
t2	4	4	2	7	8	1		5	8	8	9		1
t3	5	3		6	3	1		1	6		18	7	2
t4	7	2	8		2	1				10		6	2
t5	6	4	4	5	4	1							1
t6	2	6	4	5		1							2
t7	4	4	2	7	8	1							1

Opt. model	Total Output ΣYit:	140	Total Output ΣYit:	131
	Total Costs (mu):	14525	Total Costs (mu):	45117
Sim. model	Total Output:	89	Total Output:	122
	Number Setups:	132	Number Setups:	129

5. Conclusions

This work presented a hybrid methodology to provide decision-support for the planning and scheduling of multistage industrial systems, combining simulation and optimization approaches. The methodology solution potentiates the advantages of an iterative procedure, combining an upper level optimization plan, which feasibility is tested in a detailed simulation of the real production process, while considering the production mix and allocation of a number of shared resources in order to generate a scheduling solution. The iterative procedure considers recursive parameters obtained from the simulations so that the solution between the two model converges, such as the effective loading ratio, utilization capacities and number of changeovers of the shared resources.

Simulation-optimization methods have been gathering significant relevance to overcome the challenges of dealing with the computational complexity and cost of implementation in industrial planning and scheduling problems. The ability to integrate the stochastic and non-linear constraints, such as allocation of resources and variable processing times or setups, in a solution search with mathematical foundation becomes of key importance. Although further experiments are required to evaluate the quality of convergence solution and the impact of simulation dispatching rules, along with additional decisional variables as the ones proposed in this formulation. The advantages of the simulator should also be explored to provide enhanced information to the decision-maker on the efficiency of resources utilization, while being reactive to changes in the production capacity, operational ergonomic or layout. Future work will aim to explore these issues by evaluating different optimization objectives to assess operational sustainability.

Acknowledgment

The authors would like to acknowledge the financial support from Portugal 2020 project POCI-01-0145-FEDER-016418 by UE/FEDER through the program COMPETE2020.

References

P. M. Castro, A. M. Aguirre, L. J. Zeballos and C. A. Méndez, 2011, Hybrid mathematical programming discrete-event simulation approach for large-scale scheduling problems. Industrial & Engineering Chemistry Research, 50(18), 10665-10680.

G. Figueira. and B. Almada-Lobo, 2014, Hybrid simulation–optimization methods: A taxonomy and discussion, Simulation Modelling Practice and Theory, 46:118-134.

I. Harjunkoski, C. T. Maravelias, P. Bongers, P.M. Castro, S. Engell, I. E. Grossmann, J. and Wassick, J., 2014, Scope for industrial applications of production scheduling models and solution methods, Computers & Chemical Engineering, 62, 161-193

B. Kim and S. Kim, 2001, Extended model for a hybrid production planning approach, International Journal of Production Economics, 73(2): 165-173.

S. H. Kim and Y. H. Lee, 2016, Synchronized production planning and scheduling in semiconductor fabrication, Computers & Industrial Engineering, 96:72-85.

S. Martins, P. Amorim, G. Figueira and B. Almada-Lobo, 2017, An optimization-simulation approach to the network redesign problem of pharmaceutical wholesalers. Computers & Industrial Engineering, 106, 315-328.

M. Vieira, S. Moniz, T. Pinto-Varela and A.P. Barbosa-Póvoa, 2018, Simulation-optimization approach for the decision-support on the planning and scheduling of automated assembly lines. IEEE 13th APCA International Conference on Control and Soft Computing, 265-269.

Anton A. Kiss, Edwin Zondervan, Richard Lakerveld, Leyla Özkan (Eds.)
Proceedings of the 29th European Symposium on Computer Aided Process Engineering
June 16th to 19th, 2019, Eindhoven, The Netherlands. © 2019 Elsevier B.V. All rights reserved.
http://dx.doi.org/10.1016/B978-0-128-18634-3.50242-3

A Neural Network-Based Framework to Predict Process-Specific Environmental Impacts

Johanna Kleinekorte[a], Leif Kröger[a], Kai Leonhard[a] and André Bardow[a,b,*]

[a]*Institute for Technical Thermodynamics, RWTH Aachen University, Schinkelstraße 8, 52062 Aachen, Germany*
[b]*Institute of Energy and Climate Research (IEK-10), Forschungszentrum Jülich, Wilhelm-Johnen-Straße, 52425 Jülich, Germany*
andre.bardow@ltt.rwth-aachen.de

Abstract

Growing environmental concern and strict regulations led to an increasing effort of the chemical industry to develop greener production pathways. To ensure that this development indeed improves environmental aspects requires an early-stage estimation of the environmental impact in early process design. An accepted method to evaluate the environmental impact is Life Cycle Assessment (LCA). However, LCA requires detailed data on mass and energy balances, which is usually limited in early process design. Therefore, predictive LCA approaches are required. Current predictive LCA approaches estimate the environmental impacts of chemicals only based on molecular descriptors. Thus, the predicted impacts are independent from the chosen production process. A potentially greener process cannot be distinguished from the conventional route. In this work, we propose a fully predictive, neural network-based framework, which utilizes both molecular and process descriptors to distinguish between production pathways. The framework is fully automated and includes feature selection, setup of the network architecture, and predicts 17 environmental impact categories. The pathway-specific prediction is illustrated for two examples, comparing the CO_2-based production of methanol and formic acid to their respective fossil production pathway. The presented framework is competitive to LCA predictions from literature but can now also distinguish between process alternatives. Thus, our framework can serve as initial screening tool to identify environmentally beneficial process alternatives.

Keywords: predictive life cycle assessment, artificial neural network, early process development stage, comparison of pathway alternatives

1. Introduction

Today, the chemical industry is mainly based on fossil fuels leading to greenhouse gas (GHG) emissions. To reduce these GHG emissions and the dependency on fossil resources, process alternatives are developed using renewable resources such as CO_2 as feedstock. To quantify environmental reduction potentials resulting from technological change and to support the investment decision process, these process alternatives have to be assessed environmentally in an early design stage. An accepted method for environmental assessment is Life Cycle Assessment (LCA). Life Cycle Assessment is a holistic approach, which avoids emission shifting between life cycle phases or types of environmental impacts. However, LCA requires detailed data, which is limited in early process design stages. Therefore, so-called streamlined LCA approaches approaches are required.

Streamlined LCA approaches reduce the scope of the assessment or simplify the required data

(Hunt et al., 1998; Casamayor and Su, 2012). One possibility of streamlining is the use of machine learning. Current machine learning approaches for predictive LCA of chemicals are based on artificial neural networks (ANN). To transform the characteristics of a molecule into an input vector for the ANN, these models use molecular descriptors, e.g., molecular weight or functional groups (Song et al., 2017; Calvo-Serrano et al., 2017; Wernet et al., 2009). The main drawback of molecular descriptors is that the ANN cannot distinguish between production pathways leading to the same chemical. Therefore, molecular descriptors cannot resolve environmental reduction potentials due to process alternatives.

In this work, we propose an ANN-based framework that considers both, molecular and process descriptors. Thereby, the model can obtain process-dependent impacts and capture trade-offs of different production pathways.

2. Fully Automated Algorithm for Environmental Impact Prediction

Artificial neural networks are non-linear function approximators (Hornik et al., 1989), which usually have a greater generalization ability than linear regression models. The network consist of input, output and hidden layers, whereby each layer consists of neurons (see Figure 1, upper part). The neurons of each layer are connected by edges to the following layer. In a neuron, the inputs x_j are weighted according to their edge weight $w_{k,j}$, summed and the result is projected into a non-linear space by the activation function φ (see Figure 1, lower part). The edge weights are determined during the training process.

In this work, we propose a fully automated framework including the selection of a suitable subsets of descriptors, also called feature selection, and optimization of the network architecture (see Figure 2): In a first step, suitable descriptors are selected using stepwise regression. Afterwards, a genetic algorithm (GA) determines the optimal network architecture to predict a given environmental impact. The prediction quality of the resulting network is quantified and compared to the prediction error of a former loop. If the tolerance ε is reached, the GA stops and the optimal architecture is stored. In the last step, the generated ANN is used to predict the environmental impact for a given chemical.

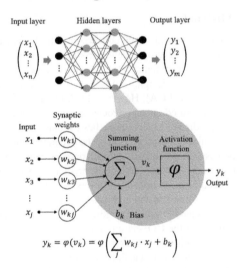

$$y_k = \varphi(v_k) = \varphi\left(\sum_j w_{kj} \cdot x_j + b_k\right)$$

Figure 1: General structure of an ANN, including a zoom of a single neuron. The input values x_j of a general neuron k are weighted by $w_{k,j}$ and summed subsequently. The bias b_k allows for shifting the output by constant value. The resulting output v_k is passed to an activation function φ, which is usually a sigmoid function. As a results of the sigmoid function, the output y_k lies in a range of $[-1; 1]$. The output is then passed to the neurons of the subsequent layer.

To distinguish between several production pathways leading to the same chemical, we use a combination of molecular descriptors and process descriptors. The process descriptors are based on the environmental indicators proposed by Patel et al. (2012).

2.1. Feature Selection

Feature selection reduce the input dimension by selecting a subset of suitable descriptors from all available descriptors to . We use stepwise regression as feature selection algorithm under the assumption that a sufficient linear correlation indicats also a non-linear correlation. Due to limited LCA data available for the training and to avoid overfitting, the final number of descriptors used in the ANN is limited to 10 % of the number of training samples.

We consider 185 possible descriptors in total, consisting of 178 molecular descriptors and 7 process descriptors. Molecular descriptors describe the main product of a process based on physical and chemical properties as well as classical functional groups. Process descriptors are based on the impact indicators proposed by Patel et al. (2012). E.g., the concentration of each component at the reactor outlet is calculated assuming ideal phase and chemical equilibrium. This process descriptor is assumed to correlate to the required separation effort. The temperature-dependent enthalpies and entropies are obtained from quantum mechanics calculations for each component. An additional process descriptor is given by the environmental impact EI of the reactants. The environmental impacts are summed according to the stoichiometric reaction equation

$$EI_{\text{stoichio}} = \sum_i \frac{v_i M_i}{v_{\text{mp}} M_{\text{mp}}} \cdot EI_i, i \in [1,...n], \text{ wherein } v_i$$

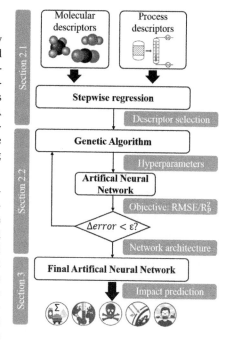

Figure 2: Algorithm to set up a neural network to predict environmental impacts. $\Delta error$ is the difference between the objective value of the former loop and the objective value of the current loop for a given architecture. ε is the chosen tolerance for convergence.

is the stoichiometric coefficient of reactant i out of n reactants, M_i is the molar weight of reactant i, v_{mp} is the stoichiometric coefficient and M_{mp} the molar weight of the main product. Considering the sum of environmental impacts of the reactants transforms the reaction equation into a scalar input for the ANN and thus delivers a lower bound of the expected environmental impact of the product. If co-products occur in the reaction, mass allocation is applied to divide the impact of the upstream on all products. After feature selection, the descriptors are normalized to the range of $[-1;1]$.

2.2. Network Setup

Our framework generates automatically a feedforward neural network for a given training set. The set up procedure of the neural networks requires the definition of so-called hyperparameters. These hyperparameters include the number of hidden layers, the number of neurons in each hidden layer, the regularization parameter, the number of maximum epochs and the kind of training algorithm. Typically, hyperparameters are chosen based on expert knowledge or using genetic algorithms. In this work, we use the GA of the MATLAB toolbox (The MathWorks, Release 2016a). The objective of the GA is to minimize the ratio between the root mean squared error (RMSE) and the coefficient of determination (R_P^2)). The coefficient of determination is given as

$$R_P^2 = \left(\frac{\sum_i (t_i - \bar{t})(y_i - \bar{y})}{\sqrt{\sum_i (t_i - \bar{t})^2 \sum_i (y_i - \bar{y})^2}} \right)^2$$

, wherein t_i is the target value of sample i, $\bar{t}$ the mean of all target values, y_i the estimation of sample i and $\bar{y}$ the mean of all estimations. A value of 1 represents perfect correlation, while a value of 0 corresponds to no prediction ability of the ANN. The objective is defined as ratio of two error measures to minimize the absolute error (RMSE) as well as possible trends (R_P^2). Minimizing only the RMSE tends to lead to constant predictions of the average impact value, while minimizing only the coefficient of determination improves predictions of trends but increases absolute prediction errors.

Initially, the available training data is divided into 3 sets: (1) a training set (containing 85% of the overall data), which is used to train the ANN in each loop of the GA, (2) a validation set (10%), which is used to qualify the generalization ability of the regarded network architecture in each GA loop, (3) and a final test set (5%), which is used to qualify the results generated by the GA. The validation set is used to calculate the objective value of the GA in each loop. The test set is used to calculate the final objective value and to validate the final optimized network architecture.

The GA suggests possible hyperparameters in each loop. Using the suggested hyperparameters, an ANN is trained on the training set. Afterwards, the generalization ability is tested on the validation set. This loop is iterated until the GA has found a local optimum. Due to the local nature of the GA, we use a multistart approach, running several GA instances parallel. The optimal architecture is used afterwards for estimating the regarded environmental impact category.

3. Case study: Component vs. Process-Specific Networks

The proposed framework is employed to generate 17 neural networks, each predicting one of 17 Recipe v1.08 (H) Midpoint categories (Goedkoop et al., 2009). Here, we exemplified the framework for predicting the impact on Climate Change of methanol and formic acid. For training, we use LCA data of 63 organic chemicals obtained from the Gabi Database (PE International, 2012). The training data includes two alternative pathways for methanol and formic acid, respectively. Since the number of data samples is limited, the number of descriptors selected during feature selection is set to 6. For the validation set, 6 samples are chosen randomly in the beginning of each GA run, corresponding to 10 % of the available training data. The samples are kept constant during the optimization of one GA run to guarantee convergence of the GA. For the test set, the two process alternatives for methanol and formic acid, respectively, are chosen manually to test the assessment of competing process alternatives. This procedure allows for validating the process-specific prediction ability.

3.1. Prediction Performance

To validate the prediction performance of the process-specific network, we compare the process-specific ANN to an ANN solely based on molecular descriptors, called component-specific ANN in the following (Figure 3). The ANN are compared based on the coefficient of determination R_P^2. The coefficient of determination is calculated for the training set of each network using leave-one-out cross validation. Each sample in the training set is left out once, the ANN is trained based on the remaining training set and the left out sample is predicted. Thus, the coefficient of determination is calculated using the prediction of each sample.

The process-specific ANN increase the prediction performance for 10 of 17 Recipe v1.08 (H) Midpoint indicators (Figure 3). The maximum improvement in terms of the coefficient of determination is achieved for Marine Ecotoxicity (MET) with an increase of 0.64 using the process-specific network in contrast to the component-specific network. The maximum decline is obtained for Photochemical Oxidant Formation (POF) with an reduction of 0.29. Although the absolute coefficient of determinations are still small with values lower than 0.4 for 16 of 17 impact categories, the use of process-descriptors has improved clearly prediction performance.

To compare our results to literature, we apply an literature-based ANN on our training data using the architecture and hyperparameters proposed by Wernet et al. (2008, 2009). The literature ANN achieves worse coefficients of determination for 14 of 17 impact categories compared to the process-specific ANN (Figure 3). Since Wernet et al. (2009) obtained coefficients of determina-

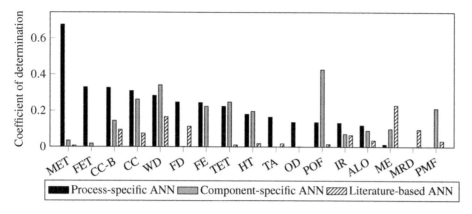

Figure 3: Comparison of the coefficient of determination for the process-specific ANN (black) and the component-specific network (grey), respectively, given for 17 Recipe 1.06 Midpoint categories: Marine Ecotoxicity (MET), Freshwater Ecotoxicity (FET), Climate Change exclusive biogenic carbon (CC-B), Climate Change (CC), Water Depletion (WD), Fossil Resource Depletion (FD), Freshwater Eutrophication (FE), Terrestrial Ecotoxicity (TET), Human Toxicity (HT), Terrestrial Acidification (TA), Ozone Depletion (OD), Photochemical Oxidant Formation (POF), Ionising Radiation (IR), Agricultural Land Occupation (ALO), Marine Eutrophication (ME), Mineral Resource Depletion (MRD), and Particulate Matter Formation (PMF). The shaded bars are the results from an ANN based on the descriptors and architecture given in Wernet et al. (2008, 2009).

tion of 0.41 up to 0.69 using similar hyperparameters, but other training data, the lower coefficient of determination shown in Figure 3 results from small training data set and its poor data quality. The coefficient of determination obtained for Climate Change by leave-one-out on our training set is comparable to the $R_P^2 = 0.31$ presented in Song et al. (2017).

3.2. Process-Specific Prediction

The proposed process-specific network is able to distinguish between alternative production pathways. To show this potential, the impact on Climate Change is predicted for two production alternatives for both methanol and formic acid (Figure 4). For this purpose, a network is trained without data for the production pathways of methanol and formic acid. Two pathway alternatives are then predicted: a fossil-based pathway, utilizing synthesis gas from steam methane reforming as carbon source, and a CO_2-based pathway utilizing CO_2 from an ammonia plant as carbon source. For both pathways, hydrogen is supplied by steam methane reforming and electricity is assumed as the current European grid mix. The impacts on Climate Change of both pathways are predicted using both the component-specific and the process-specific network.

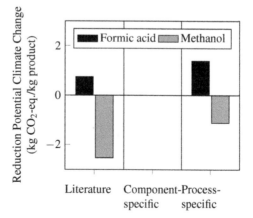

Figure 4: Comparison of the reduction potential in Climate Change for methanol and formic acid, calculated as the difference between fossil-based pathway and CO_2-based pathway alternative.

The predictions are compared to literature LCA values from Gabi (PE International, 2012) for the fossil-based pathways and from Artz et al. (2018) for the CO_2-based pathways.

In literature, the CO_2-based production of formic acid causes a smaller impact on Climate Change than the fossil-based production pathway, resulting in a positive reduction potential (Figure 4). In contrast, the CO_2-based production of methanol is less environmentally friendly than the fossil-based pathway, resulting in a negative reduction potential. The component-specific ANN is not able to distinguish between different production pathways. Therefore, no reduction potential can be identified using the component-specific ANN. In comparison, the pathway-specific ANN can distinguish between process alternatives, resulting in a positive predicted reduction potential of the CO_2-based pathway for formic acid and a negative predicted reduction potential of the CO_2-based pathway for methanol. Although the potentials are overestimated by the process-specific ANN, the trends in reduction potentials are predicted correctly. Thus, the process-specific neural network can be used for comparison of pathway alternatives in early design stages.

4. Conclusion

In this work, we present a neural network-based framework to predict process-specific environmental impacts. The framework is fully automated including feature selection and architecture optimization of the ANN. In addition to molecular descriptors, process descriptors are used to enable process-specific predictions. The process descriptors increase the coefficient of determination up to 0.64. Importantly, process descriptors enables the network to distinguish between production pathways. The process-specific prediction ability is exemplified for methanol and formic acid: for each chemical, two production scenarios are predicted, one fossil-based and one CO_2-based. The trends in the environmental impact are predicted correctly, showing that our framework can serve as initial screening tool for identifying environmentally beneficial process alternatives. More training data would be desirable to increase the quantitative accuracy.

5. Acknowledgement

The authors thank the German Federal Ministry of Education and Research (BMBF) for funding within the project consortium "Carbon2Chem" under Contract 03EK3042C.

References

J. Artz, T. E. Muller, K. Thenert, J. Kleinekorte, R. Meys, A. Sternberg, A. Bardow, W. Leitner, 2018. Sustainable conversion of carbon dioxide: an integrated review of catalysis and life cycle assessment. Chemical reviews 118 (2), 434–504.

R. Calvo-Serrano, M. González-Miquel, S. Papadokonstantakis, G. Guillén-Gosálbez, 2017. Predicting the cradle-to-gate environmental impact of chemicals from molecular descriptors and thermodynamic properties via mixed-integer programming. Computers & Chemical Engineering 108, 179–193.

J. L. Casamayor, D. Su, 2012. Integration of detailed/screening lca software-based tools into design processes. In: Design for Innovative Value Towards a Sustainable Society. Springer, pp. 609–614.

M. Goedkoop, R. Heijungs, M. Huijbregts, A. de Schryver, J. Struijs, R. van Zelm, 2009. Recipe 2008 - a life cycle impact assessment method which comprises harmonised category indicators at the midpoint and the endpoint level.

K. Hornik, M. Stinchcombe, H. White, 1989. Multilayer feedforward networks are universal approximators. Neural Network 2 (5), 359–366.

R. G. Hunt, T. K. Boguski, K. Weitz, A. Sharma, 1998. Case studies examining lca streamlining techniques. The International Journal of Life Cycle Assessment 3 (1), 36.

A. D. Patel, K. Meesters, H. den Uil, E. de Jong, K. Blok, M. K. Patel, 2012. Sustainability assessment of novel chemical processes at early stage: Application to biobased processes. Energy & Environmental Science 5 (9), 8430–8444.

PE International, 2012. Gabi lca software and lca database.
    URL http://www.gabi-software.com/databases/gabi-databases/

R. Song, A. A. Keller, S. Suh, 2017. Rapid life-cycle impact screening using artificial neural networks. Environmental science & technology 51 (18), 10777–10785.

I. The MathWorks, Release 2016a. Matlab neural network toolbox.

G. Wernet, S. Hellweg, U. Fischer, S. Papadokonstantakis, K. Hungerbühler, 2008. Molecular-structure-based models of chemical inventories using neural networks. Environmental science & technology 42 (17), 6717–6722.

G. Wernet, S. Papadokonstantakis, S. Hellweg, K. Hungerbühler, 2009. Bridging data gaps in environmental assessments: Modeling impacts of fine and basic chemical production. Green Chemistry 11 (11), 1826–1831.

Anton A. Kiss, Edwin Zondervan, Richard Lakerveld, Leyla Özkan (Eds.)
Proceedings of the 29th European Symposium on Computer Aided Process Engineering
June 16th to 19th, 2019, Eindhoven, The Netherlands. © 2019 Elsevier B.V. All rights reserved.
http://dx.doi.org/10.1016/B978-0-128-18634-3.50243-5

Multi-objective spatio-temporal optimisation for simultaneous planning, design and operation of sustainable and efficient value chains for rice crop

Stephen S. Doliente,[a,b] Sheila Samsatli[a,*]

[a] *Department of Chemical Engineering, University of Bath, Claverton Down, Bath BA2 7AY, United Kingdom*

[b]*Department of Chemical Engineering, College of Engineering and Agro-Industrial Technology, University of the Philippines Los Baños, Batong Malake, Los Baños, Laguna 4031, Philippines*

[*]S.M.C.Samsatli@bath.ac.uk

Abstract

The rice value chain, especially in the Philippines, can potentially progress in terms of efficiency and sustainability, as well as create more value by effectively utilising the crop residues of rice production. However, generating multiple products will lead to a complex decision-making process in supply chains, especially in the case of rice crop, which is vital for the Philippines' food security. Thus, systematic planning techniques are needed to address important decisions along the stages in rice value chains. This study presents a mixed integer linear programming (MILP) model developed for the planning, design and operation of multi-product rice value chains. In order to capture the spatial-dependencies of the problem, such as the candidate locations for rice farms and processing facilities, and location of demands, 81 cells represented the Philippines. The temporal aspect of the model considers a long planning horizon, out to 2050, and the time varying demands for white rice and the crop residues of farming and milling. The model determines design and operating decisions such as where to locate the farms and processing facilities, what products to produce, how to transport resources, among others. Different objectives are considered such as maximisation of the net present value and minimisation of CO_2 emissions. The Pareto set generated represents the optimal solutions representing the different trade-offs between economic gain and environmental protection.

Keywords: rice value chain optimisation, multi-product value chain, multi-objective optimisation, spatio-temporal modelling, mixed-integer linear programming

1. Introduction

Rice consumption in many developing countries of the tropics is rapidly growing due to rising population and increasing urbanisation. Compelled to secure sufficient rice supply, these nations had been enhancing farm productivity and expanding agricultural land. Despite resource scarcity, rice production in these nations has remained a highly land-use-, energy-, and water-intensive activity (Sims et al., 2015). Recently, extreme-weather events related to climate change have caused damage and losses to these nations' agricultural sectors such as rice. All of these drivers instil more pressure and inevitably disrupt the food-energy-water nexus that support rice value chains in the tropics (Keairns et al., 2016). Therefore, understanding and formulating better strategies in managing and operating rice value chains are crucial. In the case of the Philippines, rice sufficiency improvements have yet to result in economic benefits due to mismatched policies and lack of infrastructure (Briones, 2014). Harvest losses and inefficient logistics interrupt the country's rice supply chain (Mopera, 2016). Rice cultivation is a major greenhouse gas emitter in the Philippines because the majority of the carbon-rich crop residues, rice straw and rice husk, are burnt and untapped. If these wasted resources are managed and valorised, the rice value chain can contribute to the country's economic and environmental sustainability. Motivated by the desire for rice value chains to be more secure, efficient, profitable and sustainable, the aim of this work is to develop a mixed integer linear programming (MILP) model for planning, design and operation of multi-product rice value chains. The multi-objective optimisation model is used to perform a number of case studies to determine the Pareto set of solutions that represent a compromise between economic and environmental criteria.

2. Problem statement

The problem involves simultaneous determination of design and operating decisions, such as: where to farm rice crop; how much farm land to allocate; what and how much resources to utilise and what products to generate (e.g. food, energy, fuels and chemicals); what conversion technologies to invest in and where to locate them; whether or not to densify/pre-process crop residues before conversion and/or transportation; and how to transport raw materials and distribute products to customers; all of these in order to maximise of the net present value of the value chain while at the same time minimising CO_2 emissions.

3. Mathematical model

The mathematical model was based on the Value Web Model (Samsatli & Samsatli, 2018; Samsatli et al., 2015), a spatio-temporal MILP model configured for the United Kingdoms' energy value chains for natural gas and wind power. For the first time, this model was applied to rice value chains with a vastly different configuration for the Philippines. A planning horizon of 32 years, from 2018 to 2050, was considered. The model comprised multiple time scales: planning periods $y \in \mathbb{Y}$, seasonal time steps $t \in \mathbb{T}$, day type of the week $d \in \mathbb{D}$ and hourly intervals $h \in \mathbb{H}$. For the spatial representation, Philippines was divided into 81 cells corresponding to its provinces (Fig. 1). Each cell, $z \in \mathbb{Z}$, has information on the available farm land, the paddy rice yield, the cost of paddy rice production, existing technologies, if any, and existing transport infrastructures, if any. In addition, each cell has coordinates (x_z, y_z) calculated from the population density, in order to represent the demand centres and to calculate the transport distance $d_{zz'}$ (km) from cell z to cell z'.

Figure 1. Spatial representation of the Philippines in the model

3.1. Objective function

Eq. (1) shows the objective function to be minimised:

$$Z = \sum_{my} w_m \left(\mathcal{I}_{my}^{P} + \mathcal{I}_{my}^{Q} + \mathcal{I}_{my}^{fp} + \mathcal{I}_{my}^{fq} + \mathcal{I}_{my}^{vp} + \mathcal{I}_{my}^{vq} + \mathcal{I}_{my}^{i} + \mathcal{I}_{my}^{e} + \mathcal{I}_{my}^{U} - \mathcal{I}_{my}^{Rev} \right) \tag{1}$$

Each term inside the parenthesis represents an impact in year y on performance metrics m. These include the following: the total net present capital impact of technologies for conversion $\mathcal{I}_{my}^{P}$ and transport $\mathcal{I}_{my}^{Q}$; the total net present fixed operating impact of technologies for conversion $\mathcal{I}_{my}^{fp}$ and transport $\mathcal{I}_{my}^{fq}$; the total net present variable operating impact of technologies for conversion, $\mathcal{I}_{my}^{vp}$ and transport $\mathcal{I}_{my}^{vq}$; the total net present impact of importing resources $\mathcal{I}_{my}^{i}$; the total net present impact of exporting resources $\mathcal{I}_{my}^{e}$; the total net present impact of utilising resources $\mathcal{I}_{my}^{U}$; and the total net present revenue from the sale of resources (e.g. white rice, bio-electricity, biofuels, etc.) $\mathcal{I}_{my}^{Rev}$.

3.2. Constraints

Eq. (2) expresses the resource balance that considers various input and output terms for every resource $r \in \mathbb{R}$ in cell z during hour h of day type d in season t of planning period y. Eq. (3) states that the rate of utilisation, U_{rzhdty}, cannot exceed the maximum availability u_{rzhdty}^{max} (t/h or MW). Eq. (4) states that the total amount of biomass, represented by index c, utilised in each season cannot exceed the seasonal availability. Eqs. (5) and (6) are the local and national land footprint constraints, which respectively ensure that the allocated area A_{czy}^{Bio} (ha) for biomass c in cell z during planting period y, cannot be more than the maximum portion of suitable area in each cell z; and total allocated area for biomass cannot be more than the maximum portion of the total suitable area at the national level. The net rate of resource production (or consumption) P_{rzhdty} (t/h or MW) is expressed by Eq. (7); where α_{rpy} is the conversion factor of a resource r in technology $p \in \mathbb{P}$ in planning period y. The rate of operation of a single conversion technology $\mathcal{P}_{pzhdty}$ is limited by its minimum and maximum capacities, p_p^{min} and p_p^{max}, respectively, as given by Eq. (8). Furthermore, Eq. (9) states that the number of commercial conversion technologies invested N_{pzy}^{PC} cannot exceed the maximum number BR_{py} that can be built in planning period y. The net rate of transport Q_{rzhdty} (t/h or MW) of resource r between cells z and z' is given by Eq. (10) where $\bar{\tau}_{lr,dst,y}$ and $\bar{\tau}_{lr,dst,y}$ are the conversion factors for distance-dependent and distance-independent transport technology $l \in \mathbb{L}$ of resource r during planning period y, respectively. Eqs. (11) and (12) limit the operation of a transport technology $q_{lzz'hdty}$ up to its maximum capacity and the total operation rate of all transport technology up to the capacity of infrastructure $b \in \mathbb{B}$, respectively. The model categorises between two demands: those that must always be complied D_{rzhdty}^{comp} (t/h or MW) and those that may be optionally complied D_{rzhdty}^{opt}. Since the optimisation determines what and how much product to generate from a resource and what demands to satisfy, Eq. (13) determines the optional demand satisfied D_{rzhdty}^{sat}. Lastly, the rate of import I_{rzhdty} in Eqs. (14) and rate of export E_{rzhdty} in Eq. (15) cannot exceed the maximum import and export capacities I_{rzhdty}^{max} and E_{rzhdty}^{max}, respectively (all in t/h or MW).

$$U_{rzhdty} + I_{rzhdty} + P_{rzhdty} + Q_{rzhdty} \geq D_{rzhdty}^{comp} + D_{rzhdty}^{sat} + E_{rzhdty}$$
$$\forall\, r \in \mathbb{R}, z \in \mathbb{Z}, h \in \mathbb{H}, d \in \mathbb{D}, t \in \mathbb{T}, y \in \mathbb{Y} \tag{2}$$

$$U_{rzhdty} \leq u_{rzhdty}^{max} \qquad \forall\, r \in \mathbb{R} - \mathbb{C}, z \in \mathbb{Z}, h \in \mathbb{H}, d \in \mathbb{D}, t \in \mathbb{T}, y \in \mathbb{Y} \tag{3}$$

$$\sum_{hd} U_{czhdty} n_h^{hd} n_d^{dw} n_t^{wt} \leq A_{czy}^{Bio} Y_{czty}^{Bio} \qquad \forall\, c \in \mathbb{C}, z \in \mathbb{Z}, t \in \mathbb{T}, y \in \mathbb{Y} \tag{4}$$

$$\sum_{c} A_{czy}^{Bio} \leq f_{zy}^{loc} A_{zy}^{Bio,max} \qquad \forall\, z \in \mathbb{Z}, y \in \mathbb{Y} \tag{5}$$

$$\sum_{cz} A_{czy}^{Bio} \leq f_{y}^{nat} \sum_{z} A_{zy}^{Bio,max} \qquad \forall\, y \in \mathbb{Y} \tag{6}$$

$$P_{rzhdty} = \sum_p \mathcal{P}_{pzhdty}\alpha_{rpy} \qquad \forall r \in \mathbb{R} - \mathbb{C}, z \in \mathbb{Z}, h \in \mathbb{H}, d \in \mathbb{D}, t \in \mathbb{T}, y \in \mathbb{Y} \tag{7}$$

$$N^{PC}_{pzy}p^{min}_p \leq \mathcal{P}_{pzhdty} \leq N^{PC}_{pzy}p^{max}_p \qquad \forall p \in \mathbb{P}^C, z \in \mathbb{Z}, h \in \mathbb{H}, d \in \mathbb{D}, t \in \mathbb{T}, y \in \mathbb{Y} \tag{8}$$

$$\sum_c N^{PC}_{pzy} \leq BR_{py} \qquad \forall p \in \mathbb{P}^C, y \in \mathbb{Y} \tag{9}$$

$$Q_{rzhdty} = \sum_{z'|v_{z'z}=1} \sum_{l \in \mathbb{L}} [(\bar{\tau}_{lr,dst,y} + \bar{\tau}_{lr,dst,y}d_{zz'})q_{lz'zhdty}] + \sum_{z'|v_{zz'}=1} \sum_{l \in \mathbb{L}} [(\bar{\tau}_{lr,dst,y} + \bar{\tau}_{lr,dst,y}d_{zz'})q_{lzz'hdty}] \tag{10}$$
$$\forall r \in \mathbb{R}, z \in \mathbb{Z}, h \in \mathbb{H}, d \in \mathbb{D}, t \in \mathbb{T}, y \in \mathbb{Y}$$

$$q_{lzz'hdty} \leq \sum_{b \in \mathbb{L}} q^{max}_l N^B_{bzz'y|LB_{lb}=1 \wedge v_{zz'}=1} \qquad \forall l \in \mathbb{L}; z, z' \in \mathbb{Z}; h \in \mathbb{H}, d \in \mathbb{D}, t \in \mathbb{T}, y \in \mathbb{Y} \tag{11}$$

$$\sum_{l \in \mathbb{L}} q_{lzz'hdty}LB_{lb} \leq b^{max}_b N^B_{bzz'y} \qquad \forall b \in \mathbb{B}; z, z' \in \mathbb{Z}; h \in \mathbb{H}, d \in \mathbb{D}, t \in \mathbb{T}, y \in \mathbb{Y} \tag{12}$$

$$D^{sat}_{rzhdty} \leq D^{opt}_{rzhdty} \qquad \forall r \in \mathbb{R}, z \in \mathbb{Z}, h \in \mathbb{H}, d \in \mathbb{D}, t \in \mathbb{T}, y \in \mathbb{Y} \tag{13}$$

$$I_{rzhdty} \leq I^{max}_{rzhdty} \qquad \forall r \in \mathbb{R}, z \in \mathbb{Z}, h \in \mathbb{H}, d \in \mathbb{D}, t \in \mathbb{T}, y \in \mathbb{Y} \tag{14}$$

$$E_{rzhdty} \leq E^{max}_{rzhdty} \qquad \forall r \in \mathbb{R}, z \in \mathbb{Z}, h \in \mathbb{H}, d \in \mathbb{D}, t \in \mathbb{T}, y \in \mathbb{Y} \tag{15}$$

3.3. Network superstructure

Whereas food only (i.e. white rice) is produced in conventional rice value chains, many non-food products can also be generated as supplementary sources of income. The value chain superstructure in Fig. 2 shows potential pathways for multi-product generation from rice crop and its residues, e.g. the rice straw from farming and the rice husk from milling. The residues can be processed via thermo- and bio-chemical conversion technologies such as boiler-steam turbine, anaerobic digestion and combine cycle gas turbine (CCGT) for energy (i.e. bio-electricity); lignocellulosic (LC) fermentation and pyrolysis for fuels (i.e. bio-ethanol, pyrolysis oil); and gasification and Fischer Tropsch (FT) synthesis for chemicals (i.e. bio-naphtha). The value chain (Fig. 2) also includes a pelletiser for the densification of the residues. The inputs to the optimisation model include the spatial and temporal availability of paddy rice and product demands (white rice, bio-electricity, bio-ethanol and bio-naphtha). The model also requires the characteristics of each conversion and transport technologies (capital expenditure, operation and maintenance costs, conversion factors etc). The key constraints are given in Section 3.2. For a given objective (maximise NPV or minimise CO$_2$ emissions), the model determines the optimal location for rice crop farming and the sites of the conversion technologies. The demands for food has to be satisfied all the time but the model can choose whether to produce non-food products from rice straw and/or rice husk, up to a specified maximum demand for a non-food product. Furthermore, the model also decides between barge and/or truck transport of white rice and/or paddy rice and the operation of these transport networks.

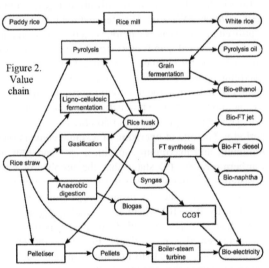

Figure 2. Value chain superstructure for rice crop, showing some of the potential conversion pathways

3.4. Multi-objective optimisation

The model was formulated as a multi-objective optimisation problem to allow the evaluation of trade-offs between conflicting objectives (Samsatli & Samsatli, 2018). The weighting method was used to transform the multi-objective optimisation problem into a single objective optimisation problem using different combinations of the weighting factors for the economic and environmental impacts. The weighting method can be expressed as $\min Z = \sum_1^N w_m Z_m$, where Z_m represents the individual impacts; w_m are the weighting factors, which can assume values between 0 and 1, with $\sum_{1=1}^N w_m = 1$. The optimal solutions obtained at different combinations of the weighting factors form a Pareto set, which represents the compromise between the different objectives (Ang et al., 2010). In this study, the performance metrics are $m \equiv \{Cost, CO_2\}$. Therefore, when $w_m \equiv \{1, 0\}$, the problem is the maximisation of net present value; and when $m \equiv \{0, 1\}$ the problem is the minimisation of CO_2 emissions.

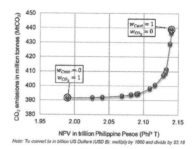

Figure 3. Pareto set for the multi-objective rice value chain optimisation problem

4. Case studies

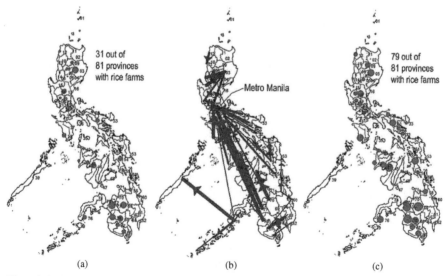

(a) (b) (c)

Figure 4. Optimisation results: (a) allocated farm lands in case 1; (b) transport of white rice by barges in case 1; and (c) allocated farm lands in case 2.

Two cases have been considered for this paper. Case 1 examines the optimal solution for maximising NPV and minimising CO_2 emissions for the Philippine rice value chains considering food production only. The Pareto set generated in Fig. 3 shows the compromise between these two objectives for a planning horizon of 32 years with white rice demand of 11 Mt/yr. The highest point on the right represents the optimal solution for maximising NPV. For this solution, the NPV is PhP 2.14 T (USD 40.26 bn) and the CO_2 emissions are 439 MtCO_2. The lowest point in the left represents the optimal solution for minimising CO_2 emissions, with NPV of PhP 1.99 T (USD 37.43 bn) and CO_2

emissions of 392 MtCO₂. Moving from the bottom-left of the curve to the top-right represents solutions with increasing profit at the expense of higher CO_2 emissions. Depending on the policies, this Pareto set presents an opportunity for the Philippines to decarbonise its current rice value chain. The model has also determined barges as the main mode of transport mode in delivering white rice to provinces with the highest demand (for example Metro Manila the capital city in cell 23), as shown in Fig. 4(b). Compared to transport by truck, the cost and CO_2 emissions of transport by barge is significantly lower, which is more beneficial for an archipelagic country like the Philippines. Furthermore, compared to the existing trend of farming rice, despite suitability, in all provinces of the country (e.g. using >50% of the total farm land area), the model recommends rice crop farming in selected provinces of the country with high paddy rice yield and low farming cost as shown in Fig. 4(a). The Case 2 study determined the profitability of a multi-product rice value chain wherein demand for white rice at 11 Mt/yr was always satisfied while annual demands for bio-electricity, bio-ethanol and bio-naptha were *optionally* satisfied at up to 908 GWh/yr, 2.1 TWh/yr, and 17 GWh/yr, respectively. The model revealed that this multi-product rice value chain is profitable at NPV of PhP 2.41 T (or USD 45 bn) with 1.3 GtCO₂ emissions. In contrast to Case 1, the model suggested rice crop farming throughout the country (rice farms are now present in 79 out of 81 provinces – see Fig. 4(c)). When transformed into a multi-product rice value chain, the currently food-only rice value chain in the Philippines can become more profitable at PhP 0.6 m/ha (USD 11 k/ha), a 12% increase in NPV, since the non-food products can become additional revenue sources. Furthermore, the model recommended an optimal integrated rice value chain with *full* utilisation of rice husk and rice straw processed by LC fermentation, gasification, FT synthesis, boiler-steam turbine and CCGT. The model sited these technologies locally where the resources are available. Finally, the model also recommended transport by barge for both white rice and paddy rice. Paddy rice transport facilitated movement of rice husk to comply with demands of non-food products.

5. Conclusions

A multi-objective, spatio-temporal MILP model was developed for the simultaneous planning, design and operation of multi-product rice value chains. For the food-only scenario in the Philippines, the model generated a Pareto set representing the trade-offs between economic gain and environmental protection. The model recommended rice crop farming in provinces with high yields and low farming cost. To meet the demand for white rice, model showed that it is more cost-effective and environmentally sound to use barge over truck. For the multi-product rice value chain for the Philippines, the model was able to demonstrate that it was more profitable over a food-only rice value chain. Furthermore, advanced biomass conversion technologies to process rice straw and rice husk are going to be important for the feasibility of a multiproduct rice value chain.

References

Ang, S. M. C., Brett, D. J. L., & Fraga, E. S., 2010, Journal of Power Sources, 195, 2754-2763.

Briones, R. M., 2014, Philippine Institute for Development Studies. Discussion Paper Series No. 2015-04.

Keairns, D. L., Darton, R. C., & Irabien, A., 2016, Annual Review of Chemical and Biomolecular Engineering, 7, 239-262.

Mopera, L. E., 2016, Journal of Developments in Sustainable Agriculture, 11, 8-16.

Samsatli, S., & Samsatli, N. J., 2018, Applied Energy, 220, 893-920.

Samsatli, S., Samsatli, N.J., & Shah, N., 2015, Applied Energy, 147, 131-160.

Sims, R., Flammini, A., Puri, S., & Bracco, S., 2015, In: Food and Agriculture Organization United States Agency for International Development. ISBN 978-92-5-108959-0.

Anton A. Kiss, Edwin Zondervan, Richard Lakerveld, Leyla Özkan (Eds.)
Proceedings of the 29th European Symposium on Computer Aided Process Engineering
June 16th to 19th, 2019, Eindhoven, The Netherlands. © 2019 Elsevier B.V. All rights reserved.
http://dx.doi.org/10.1016/B978-0-128-18634-3.50244-7

Dynamic Optimisation and Visualisation of Industrial Beer Fermentation with Explicit Heat Transfer Dynamics

Alistair D. Rodman,[a] Megan Weaser,[b] Lee Griffiths,[b] Dimitrios I. Gerogiorgis[a*]

[a] *School of Engineering (IMP), University of Edinburgh, Edinburgh, EH9 3FB, UK*

[b] *Carling House, Molson Coors Brewing Co, Burton-On-Trent DE14 1JZ, UK*

D.Gerogiorgis@ed.ac.uk

Abstract

Demand for beer products fluctuates throughout the year, with a substantial increase occurring for the duration of global sporting events. Beer fermentation is frequently the production bottleneck due to its significant duration (often exceeding 1 week). Therein lies a strong incentive for fermentation optimisation and batch duration reduction such that maximum production capacity can be elevated during peak demand periods. Beer fermentation optimisation has received considerable attention, however a limitation of prior work is the universal assumption of direct and instantaneous control of fermentor temperature. With the addition of only two more ODEs to the system model, heat transfer dynamics is approximated in this work. A novel, comprehensive visualisation of the attainable performance maps for key process variables is presented, obtained via a large-scale dynamic simulation campaign of viable cooling policies. These attainable performance maps are compared to equivalent results produced previously, to elucidate how fermentor performance varies once production scale increases beyond the point of the previous simplifying assumption. Utilising orthogonal polynomials on finite elements allows a finite dimensional optimisation NLP problem to be formulated for ethanol yield maximisation, which has been solved with IPOPT. Optimal operation involves a novel cooling policy to effectively manage the active yeast population in the fermentor for improved performance versus previous approaches for beer fermentation.

Keywords: Dynamic optimisation, Dynamic simulation, Beer fermentation

1. Introduction

Fermentation is a key step in the production of consumer alcohol products, frequently presenting a production bottleneck due to its significant duration (often >1 week). Demand for beer products fluctuates throughout the year, with a substantial increase occurring for the duration of global sporting events (e.g., the World Cup). Therein lies a strong incentive for process optimisation and batch duration reduction such that maximum production capacity can be elevated during peak demand periods without the need to increase plant equipment sizes, which would incur significant capital investment for vessels which would be underutilised for much of the year. Beer fermentation is an extremely complex biochemical process (Vanderhaegen et al., 2006), whose mathematical representation and parameter estimation are possible only via reduced-order models (Dochain, 2003) considering only the most essential subset of reactions via selective aggregation. The system's complexity (>600 species) renders comprehensive parameter estimation infeasible and dynamic optimisation cumbersome.

A number of studies towards improved beer fermentation via temperature profile optimisation have been performed on the basis of simplified models, exploring the inherent trade-off between attainable ethanol concentration and required batch time. These have addressed both the single objective case for batch time minimisation or ethanol yield maximisation (de Andrés-Toro et al., 1998; Carrillo-Ureta et al., 2001) or both (Rodman and Gerogiorgis, 2016). A rigorous dynamic optimisation has been performed via a NLP formulation employing orthogonal collocation on finite elements, computing the optimal temperature trajectory for the process duration (Rodman and Gerogiorgis, 2017). It is highly uncommon for brewers to monitor batch progression online during the fermentation process. Rather a prescribed temperature manipulation is followed, after which select terminal state concentrations may be sampled to confirm an adequate product quality. Herein, the design of a model predictive controller would be of little value to industry, who seek off-line computation of optimal dynamic manipulations to be implemented with a generic temperature feedback control loop.

A limitation of prior work (in the interest of simplification towards computational manageability) is the assumption of direct and instantaneous control of fermentor temperature. This is less realistic as the scale increases; industrial fermentation tanks can exceed 4,000 hectolitres, so the characteristic lag time for heat transfer from the reaction broth through the fermentor cooling jacket cannot be reliably considered as negligible. To this end, we build upon our published contributions by explicit modelling of the fermentor cooling jacket heat transfer dynamics. This requires only two addition model ODEs, where the rate of change of the reactor contents' temperature becomes a function of the jacket power supply and thermophysical properties of the system.

2. Beer Fermentation Modelling

The de Andrés-Toro (1998) fermentation model describes five components in the wort: ethanol (C_E), sugar (C_S) biomass (X_i) and two flavour-tarnishing compounds, diacetyl (C_{DY}) and ethyl acetate (C_{EA}). The single sugar species represents the sum of all sugars in the wort. Biomass is distinguished into three forms: active (X_A), latent (X_L) and dead (X_D) cells. Active cells can promote fermentation and duplicate and grow over time; however, a portion will die and no longer contribute to fermentation, settling at the bottom of the vessel. Latent (lag) cells are unable to promote fermentation, but over time they develop into active cells, responsible for consumption of the fermentable material in the wort. A diagram of the model scheme, and the corresponding kinetic rate equations are in Fig. 1. Arrhenius growth rates (μ_i) describe species progression with a stoichiometric yield factor (Y) and an inhibition factor, f, on ethanol production. A fundamental simplifying assumption of the fermentation model is that spatial variation of concentrations and temperature may be neglected. While some degree of variation is inevitable in practice, CO_2 generation from the primary reaction pathway and the cylindroconical vessel leads to considerable mixing. The lumped parameter model has been extensively validated as suitable for representing the process on an industrial scale.

The model is extended to consider heat transfer between the exothermic fermentor contents and the surrounding cooling jacket. Eq. 13 defines the bulk temperature inside the vessel, where energy generated by sugar fermentation at a rate of (ΔH) heats the wort according to its mean physio-thermal properties (ρ_R, C_{PR}).

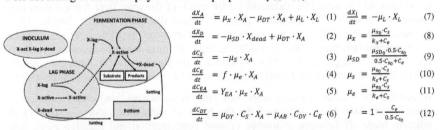

$$\frac{dX_A}{dt} = \mu_x \cdot X_A - \mu_{DT} \cdot X_A + \mu_L \cdot X_L \quad (1)$$

$$\frac{dX_D}{dt} = -\mu_{SD} \cdot X_{dead} + \mu_{DT} \cdot X_A \quad (2)$$

$$\frac{dC_S}{dt} = -\mu_S \cdot X_A \quad (3)$$

$$\frac{dC_E}{dt} = f \cdot \mu_e \cdot X_A \quad (4)$$

$$\frac{dC_{EA}}{dt} = Y_{EA} \cdot \mu_x \cdot X_A \quad (5)$$

$$\frac{dC_{DY}}{dt} = \mu_{DY} \cdot C_S \cdot X_A - \mu_{AB} \cdot C_{DY} \cdot C_E \quad (6)$$

$$\frac{dX_l}{dt} = -\mu_L \cdot X_L \quad (7)$$

$$\mu_x = \frac{\mu_{x0} \cdot C_s}{k_x + C_e} \quad (8)$$

$$\mu_{SD} = \frac{\mu_{SD0} \cdot 0.5 \cdot C_{S0}}{0.5 \cdot C_{S0} + C_e} \quad (9)$$

$$\mu_s = \frac{\mu_{s0} \cdot C_s}{k_s + C_s} \quad (10)$$

$$\mu_e = \frac{\mu_{e0} \cdot C_s}{k_e + C_s} \quad (11)$$

$$f = 1 - \frac{C_e}{0.5 \cdot C_{S0}} \quad (12)$$

Figure 1: Model for dynamic simulation of beer fermentation (de Andrés-Toro, 1998).

Table 1: Heat transfer model parameters.

Symbol	Definition	Units	Value
T	Reaction Temperature	°C	–
T_C	Jacket Temperature	°C	–
T_{C0}	Coolant Feed Temperature	°C	4
F	Coolant Feed Rate	m^3 hr^{-1}	–
ΔH	Enthalpy of Reaction	kJ kg^{-1}	587
ρ_R	Mean Density of Wort	kg m^{-3}	1,030
ρ_C	Density of Coolant	kg m^{-3}	1,042
C_{PR}	Wort Heat Capacity	J kg^{-1} K^{-1}	4,065
C_{PC}	Coolant Heat Capacity	J kg^{-1} K^{-1}	3,914
A_h	Heat Transfer Area	m^2	221.4
U	Overall Heat Transfer Coefficient	W m^{-2} K^{-1}	200
V_R	Wort Volume	m^3	400
V_C	Jacket Volume	m^3	3.8

The cooling rate is a function of the jacket temperature (T_C), the jacket/fermentor heat transfer area A_h, overall heat transfer coefficient (OHTC, U) and wort volume (V_h). The temperature of the jacket (volume = V_C) is described by Eq. 14, cooled with fresh coolant at ($T = T_{C0}$) at a volumetric rate (F). Heat losses to the surroundings are negated as the vessel jacket is close to ambient conditions. A typical large capacity industrial specification vessel is modelled. The 400,000 L vessel has a 3,800 L jacket, cooled by a propylene glycol:water mixture. A 4 °C coolant temperature is first assumed, with $U = 200$ W m^{-2} K^{-1}. Table 1 details the additional parameter values used in Eq. 13–14.

$$\frac{dT}{dt} = \frac{\frac{dC_s}{dt}\Delta H}{\rho_R C_{PR}} + \frac{A_h U}{V_R \rho_R C_{PR}}(T - T_C) \tag{13}$$

$$\frac{dT_C}{dt} = \frac{F}{V_C}(T_{C0} - T_C) + \frac{A_h U}{V_C \rho_C C_{PC}}(T_C - T) \tag{14}$$

3. Dynamic Simulation of Attainable Performance

Prior work investigating the potential for process improvement versus current plant fermentation operation involved an algorithm that rapidly generated and simulated plausible temperature manipulations adhering to realistic operability heuristics, while relying on the simplifying assumption that vessel temperature may be directly manipulated (Rodman and Gerogiorgis, 2016). Attainable performance maps were generated, allowing rapid visualisation of the viable output space permitted by operational modifications. To elucidate how fermentor performance varies once production scale increases, we perform a similar simulation campaign using the extended model. To exhaustively enumerate comparable performance maps for the most simple cooling policies five discrete coolant rates ($F = [0:5:20]$) are permitted, which may be only be switched at the end of a 20 hr interval across the 160 hr span considered. Thus $5^9 = 1,953,125$ cooling policies are considered for implementation.

Fig. 2 depicts the performance maps from the new cooling policy campaign (grey markers) alongside those from Rodman and Gerogiorgis (2016) which assumed direct temperature control (black markers). An immediate observation across all metrics is that the addition of the temperature model considerably reduces the broad range of viable terminal product concentrations, particularly for both by-product species. This is despite the solution set for coolant control being significantly larger (~1.9×10^6 policies) than that for direct temperature control (~1.75×10^5). The observation can be attributed to the fact that in the absence of any external heat source the vessel temperature can only rise from its initial temperature at a rate governed by the exothermic fermentation reaction: there is considerably less scope to vary the vessel temperature during primary fermentation, where the temperature is capped by how quickly it can rise from the reaction enthalpy. This contrasts with prior work where the vessel temperature could instantaneously rise as high as 16 °C, which is not feasible in practice, leading to a subset of simulations corresponding to very short batch times and high by-product concentrations, highlighted by the black markers lying to the upper left in Fig. 2a.

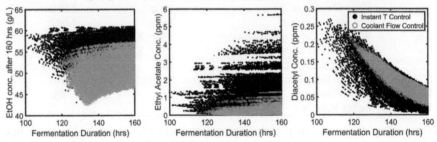

Figure 2: Comparison between attainable envelopes: direct $T(t)$ control vs. $F(t)$ control.

Of all cooling policies considered, the most favorable for alcohol (EtOH) maximisation is presented in Fig. 3 with the corresponding state trajectories. Here, the preferable approach is to begin cooling the jacket once the temperature exceeds 15 °C after 60 hr, increasing the cooling duty at 80 hr and reducing at 100 hr before turning the coolant flow off at 120 hr to allow the temperature to rise. This manages the active yeast population, preventing substantial cell death if the temperature were to rise further.

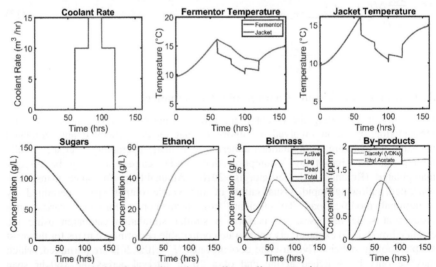

Figure 3: Best solution from exhaustive cooling policy campaign.

4. Dynamic Optimisation

Here, we describe the formulation of the nonlinear dynamic optimisation problem to maximise EtOH concentration. To formulate a finite dimensional optimisation problem, orthogonal polynomials on finite elements are used to approximate the control (coolant flowrate) and state (system species concentrations) trajectories to produce a large scale nonlinear programme (NLP). The differential algebraic equation (DAE) system is converted to a system of algebraic equations (AEs), where decision variables of the derived NLP problem include the coefficients of the linear combinations of these AEs.

The optimisation is performed using the DynOpt package in MATLAB (Cizniar et al., 2006). Fixing 16 elements across the time horizon, with 3 collocation points per element for the nine state ODEs and one collocation point (piecewise constant) per element for the control (F) leads to a $16 \times (3+1) \times 9 + 16 \times 1 \times 1 = 592$ variable NLP, able to accurately approximate the continuous dynamics of the system. For a single objective ethanol maximisation problem the NLP is solved using IPOPT (Wächter and Biegler, 2006). A multi-start initialisation procedure is used to best approximate global optimality.

The optimal control profile is presented in Fig. 4 with the corresponding jacket and vessel temperature trajectories. Similarities are observed between Figs. 3 and 4, with intermediate cooling to manage the active cell population evident as being the optimal policy to maximise the ethanol yield. The optimal policy computed here is to rapidly begin cooling after a favorable temperature is reached (~15 °C) and to slowly lower the coolant rate over the following 70 hr to facilitate gradual, sustained cooling of the wort.

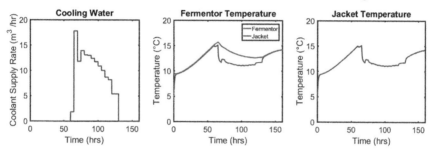

Figure 4: Optimal solution for maximum EtOH in 160 hrs.

5. Optimal Solution Sensitivity with respect to Jacket Design Variables

We investigate the impact of coolant temperature and U on the optimal cooling policy. The same dynamic optimisation problem is solved for a range of values for these process parameters. Fig. 5 shows the solutions for a range of values of U, while Fig. 6 shows the same for coolant temperatures. It is shown in Fig. 5 that with decreasing U, the same cooling policy profile form is utilised with increasing required coolant rate. Beyond a critical point ($U < 100$ W m^{-2} K^{-1}) the solution is at the maximum coolant rate. The third panel highlights how the jacket temperature differs in these cases with the increasing coolant rate, which lead to near identical temperature trajectories in the fermentor (centre panel), with the exception of the dark blue case ($U = 50$ W m^{-2} K^{-1}), which takes a less favourable path due to being limited by the maximum coolant rate. A wide range of values of U have been considered here; clarification of practiced values will further elucidate promising dynamic coolant rate trajectories for implementation.

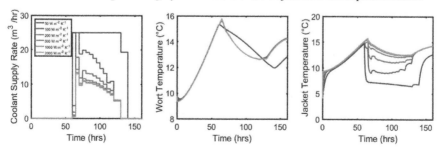

Figure 5: Overall Heat Transfer Coefficient (U) effect on optimal coolant supply rate.

In Fig. 6, it is similarly demonstrated how the same cooling policy form is computed regardless of the coolant temperature. Across all cases considered, it is possible to maintain essentially the same jacket-side temperature by adjusting the coolant rate such that the wort temperature and corresponding fermentation progression are unchanged. This highlights that in the common scenario where coolant feed temperature varies over time (ambient condition variability), adjustments can be made by varying the volumetric supply rate to adhere to the same temperature progression within the batch itself.

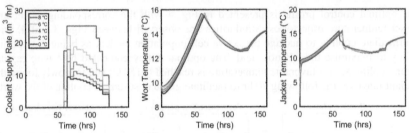

Figure 6: Coolant temperature effect on optimal coolant supply rate.

8. Conclusions

Consideration of explicit fermentor jacket heat transfer marks a significant improvement over the fidelity of prior work, which assumed that temperature may freely manipulated. Visualisation of attainable performance reveals that a vast portion of operation cases considered previously are unobtainable on an industrial scale, highlighting the importance of the improved approach. Utilising orthogonal polynomials on finite elements allows a finite dimensional NLP optimisation problem to be formulated for ethanol yield maximisation, which has been solved with the IPOPT solver. Optimal operation involves a novel cooling policy to effectively manage the active yeast population in the reactor, capable of improved performance versus established approaches. Sensitivity analysis of the solution dependence on heat transfer process and design parameters shows how adjustments in the coolant rate can be made to maintain the preferred optimal temperature trajectory of the fermenting wort in the vessel.

References

B. de Andrés-Toro, J.M. Giron-Sierra, J.A. Lopez-Orozco, C. Fernandez-Conde, J.M. Peinado and F. García-Ochoa., 1998, A kinetic model for beer production under industrial operational conditions, *Math. Comput. Simulat.*, 48, 1, 65–74.

C. Boulton and D. Quain. 2008, *Brewing yeast and fermentation*, Wiley.

G. Carrillo-Ureta, P. Roberts and V. Becerra, 2001, Genetic algorithms for optimal control of beer fermentation, *Proc. IEEE Int. Symp. Intell. Control*, 391–396.

M. Cizniar, M. Fikar and M.A. Lati, 2006, MATLAB dynamic optimisation code DYNOPT, User's guide, Bratislava, Slovak Republic.

D. Dochain, 2003, State and parameter estimation in chemical and biochemical processes: a tutorial, *J. Proc. Control*, 13, 8, 801–818.

A. Rodman and D.I. Gerogiorgis, 2016, Multi-objective process optimisation of beer fermentation via dynamic simulation, *Food Bioprod. Proc.*, 100, A, 255–274.

A. Rodman and D.I. Gerogiorgis, 2017, Dynamic optimization of beer fermentation: sensitivity analysis of attainable performance vs. product flavor constraints, *Comput. Chem. Eng.*, 106, C, 582–595.

B. Vanderhaegen, H. Neven, H. Verachtert, and G. Derdelinckx, 2006, The chemistry of beer aging – a critical review, *Food Chem.*, 95, 3, 357–381.

A. Wächter and L.T. Biegler, 2006, On the implementation of an interior-point filter line-search algorithm for large-scale nonlinear programming, *Math. Programming*, 106, 1, 25–57.

Anton A. Kiss, Edwin Zondervan, Richard Lakerveld, Leyla Özkan (Eds.)
Proceedings of the 29th European Symposium on Computer Aided Process Engineering
June 16th to 19th, 2019, Eindhoven, The Netherlands. © 2019 Elsevier B.V. All rights reserved.
http://dx.doi.org/10.1016/B978-0-128-18634-3.50245-9

Statistical Modelling for Optimisation of Mash Separation Efficiency in Industrial Beer Production

Qifan (Frank) Shen,[a] Megan Weaser,[b] Lee Griffiths,[b] Dimitrios I. Gerogiorgis[a*]

[a] *School of Engineering (IMP), University of Edinburgh, Edinburgh, EH9 3FB, UK*

[b] *Carling House, Molson Coors Brewing Co, Burton-On-Trent DE14 1JZ, UK*

D.Gerogiorgis@ed.ac.uk

Abstract

Mash separation is a critical pre-processing step in beer production, ensuring that a high-quality stream of solubilised grain carbohydrates and nutrients (wort) is fed to the fermentors, in which sugars are then biochemically converted to ethanol. This essential pre-fermentation step is performed via either of two key units (lauter tuns or mash filters); the output quality of the clarified liquid stream (wort) depends on numerous critical variables (grain composition and size distribution, mash mixture physicochem. properties, brewing recipe, separation conditions). While first-principles mathematical descriptions may remain elusive, a multitude of available (input-output) industrial data can be used to improve understanding. This paper explores causality via statistical (Partial Least Squares) models for two types of beer, and performs a sensitivity analysis using the proposed input-output correlations towards mash separation improvements. Strong wort volume and incoming feed quality to the mash filter emerge as having the strongest effect on filtration time, a key industrial performance metric for optimisation.

Keywords: Statistical modelling, mashing, separation, beer, Partial Least Squares (PLS)

1. Introduction

Mashing is an enzymatic conversion and extraction process, during which the insoluble high molecular-weight (HMW) substances in the malt are converted into soluble sugars by physical and enzymatic action (Bamforth, 2006), to ensure that a purified stream of dissolved grain carbohydrates is reliably fed to the fermentor (Meussdoerffer, 2009). The ground malt is mixed with hot water in the mash vessel to enhance extraction rates. This process involves the participation of several enzymes, notably α- and β-amylases, whose activity is crucial for high efficiency. Thus, mashing temperature greatly affects the efficiency of mashing. In practice, temperatures close to the upper limit of enzyme activity are used, as they increase the filtration run-off rate. Wort concentration, pH and mashing time are key parameters for high mashing efficiency. Three common variants employed are infusion, decoction, and temperature-controlled mashing (Bühler, 1996).

Mash separation is a solid-liquid separation process, during which malt compounds dissolved during mashing are separated from grain components, thus increasing extract yield and removing substances adversely affecting beer flavour (Bühler, 1996). Modern breweries typically produce 10-12 batches per day, including maintenance downtime. Mash separation is a rate-limiting process, therefore the strategic goal of increasing its efficiency can boost productivity and reduce total production cost (Bamforth, 2006).

The two leading technologies for mash separation employ either a mash filter or a lauter tun, exploiting wort permeability through beds in very different ways (Bamforth, 2006). A *mash filter* is operated under external pressure, which drives a filtrate flow through the bed. The filtrate remaining in the filter cake is extracted by squeezing the latter (spent grains) using flexible membranes incorporated into the unit. A *lauter tun* maintains permeability through large mean particle size and employ mechanical force to loosen the bed (the filtrate is obtained by gravity). Although mash unit design varies, wort separation equipment normally use solid substances in mash as filter medium, and therefore the properties of solid particles play an important role in mash separation. Mash filtering achieves higher filtration performance in terms of wort clarity and filtration efficiency, at a higher capital investment and maintenance cost (Bühler, 1996).

Mash filters are made of rectangular polypropylene plates, which are pressed together by a hydraulic ram to create a water-tight pocket between them. Alternates are equipped with elastic membranes which can be inflated with compressed air to squeeze the mash against permeable polypropylene cloth. The space between uninflated cloths is 40 mm and is reduced significantly to about 25-30 mm when compressed (Bamforth, 2006).

Figure 1 illustrates mash filter operation. Mash is pumped into it via the bottom inlets and chambers are vented until the filter is filled; then point vents are closed and wort outlets are opened (*filling*). Withdrawing strong wort, the mash is continuously pumped into the mash filter, and solid particles deposited in the chambers accumulate, forming a filter cake (*filtration*). When mash transfer is complete, inlets are closed and elastic membranes are inflated by compressed air to squeeze the filter (*pre-compression*). Once membranes are inflated, sparge water is fed to the chamber (*sparging*), and the membrane is inflated again to recover filtrate remaining in the filter (*final compression*). The cycle is repeated until wort concentration is acceptably low: then, plate separation cause spent grain cake drop (*discharge*) to a hopper/screw conveyor (O'Rourke, 2003).

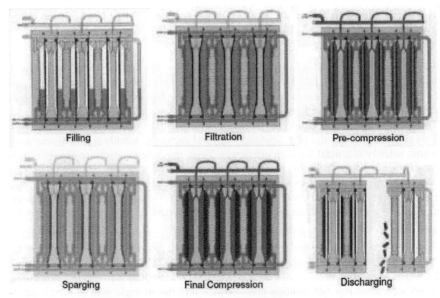

Figure 1. Mash filter operation stages (O'Rourke, 2003).

2. Methodology

The importance of process modelling, simulation and optimisation for the beer industry is well documented (Muster-Slawitch et al., 2014), although biochemical complexity of unit operations, mashing, fermenting and aging is immense (Vanderhaegen et al., 2006). The optimisation of mash separation is crucial, due to the substantial time requirement, and its influence on final product quality (Schneider et al., 2005; Kühbeck et al., 2006). First-principles models of filtration operations are useful (Wakeman & Tarleton, 1999), but their parameterisation may be laborious and elusive. Conversely, a multitude of statistical (data-driven) techniques can be used to improve our understanding of this industrial unit operation, on condition of production data availability and reliability.

This statistical modelling paper aims to explore causality in mash separation variability, develop input-output correlations for beer brands (here, type A) and propose viable improvements to the process by means of the Partial Least Squares (PLS) method, which is an established approach (MacGregor et al., 1994; Esposito Vinzi et al., 2010). A literature survey reveals that statistical (esp. PLS) models have been employed to improve product quality in the food (Poveda et al., 2004) and drink (Aznar et al., 2003) industry, and more specifically in regard to bottled lager beer aging (Liu et al., 2008). Nevertheless, to the best of our knowledge, such a statistical (esp. a PLS) approach has hitherto not been pursued to improve mash separation efficiency, despite its industrial importance (Schneider et al., 2005; Montanari et al., 2005; Kühbeck et al., 2006).

The parameters related to mash separation are not only numerous but also highly correlated, inducing significant modelling challenges. Conventional modelling methods, e.g. multiple linear regression (MLR) and ordinary least squares (OLS) cannot meet the set goals. Therefore, PLS regression is employed to capture mash separation variability. The algorithm is presented in Fig. 2 (Esposito Vinzi et al., 2010; Mehmood et al., 2012).

Step 1. Transform X and Y to normalized matrices F_0 , E_0 (m: indep., n: dep. variables)

Step 2. Extract the first component t_1 as the eigenvector corresponding to the largest eigenvalue of matrix $E_0^T F_0 F_0^T E_0$ (score vectors: $\hat{t}_1 = E_0 t_1$).

Step 3. Calculate the residual matrix E_0 as: $E_0 = t_1 p_1' + E_1$, where: $p_1 = \dfrac{E_0' t_1}{\|t_1\|^2}$

Step 4. Replace E_0 with the residual matrix E_1; extract the second component, t_2.

Repeat above steps to get all components.

Step 5. Determine the number of components k by cross-validation.

Step 6. Perform ordinary least squares on F_0 using extracted components ($t_1, \ldots, t_r$).

$$F_0 = \hat{t}_1 \beta_1^T + \cdots + \hat{t}_k \beta_k^T + F_k$$
$$t_i = u_{i1}^* x_1 + \cdots + u_{im}^* x_m \,{\scriptstyle(i=1,\ldots,k)}$$
$$Y = t_1 \beta_1 + \cdots + t_k \beta_k$$

Step 7. Perform regression modelling to obtain the input (x)-output (y) correlation as:

$$y_j = a_{j1} x_1 + \cdots + a_{jm} x_m \ (j = 1, \ldots, n)$$

Figure 2: Partial Least Squares (PLS) algorithm used for input (X) and output (Y) data.

3. Results and Discussion

A statistical (PLS) model is derived for the mash separation of a beer brand (type A) this PLS regression model can accurately predict the filtration time of the mash separation process, as per Table 1 (for $r=4$ PLS variables selected). The 14 variables (x_i) in order, are: QM1, QM2, QM3: incoming feed quality parameters; MT: mashing time; CTT: compression total time; WTV: wort to copper volume; PVV: presparging volume; SWV: strong wort volume; CT1: compression1 time; SF3: sparge3 flowrate; PCV: pre-compress volume; SF1: sparge1 flowrate; SF2: sparge2 flowrate; SV3: sparge3 volume.

Table 1: Correlation coefficients of the 14 key variables considered (4 PLS variables).

	QM1	QM2	QM3	MT	CTT	WTV	PVV	SWV	C1T	SF3	PCV	SF1	SF2	SV3
QM1	1.00													
QM2	0.82	1.00												
QM3	**0.90**	0.86	1.00											
MT	0.44	0.47	0.47	1.00										
CTT	-0.50	-0.61	-0.51	-0.47	1.00									
WTV	0.83	0.59	0.71	0.37	-0.32	1.00								
PVV	0.04	0.13	0.13	0.26	-0.33	-0.16	1.00							
SWV	0.06	-0.07	0.13	0.19	0.10	0.09	-0.12	1.00						
C1T	-0.40	-0.58	-0.42	-0.46	**0.94**	-0.20	-0.32	0.09	1.00					
SF3	0.29	0.43	0.42	0.54	-0.64	0.09	0.27	0.15	-0.59	1.00				
PCV	-0.80	-0.71	-0.77	-0.52	0.47	-0.65	-0.07	-0.25	0.36	-0.55	1.00			
SF1	0.51	0.51	0.58	0.55	-0.51	0.31	0.19	0.25	-0.40	0.86	-0.72	1.00		
SF2	0.47	0.53	0.57	0.56	-0.57	0.25	0.25	0.16	-0.47	**0.92**	-0.70	**0.97**	1.00	
SV3	0.98	0.82	0.90	0.45	-0.50	0.86	-0.04	0.07	-0.38	0.36	-0.85	0.57	0.53	1.00

The variable importance projection (VIP) value is used to determine whether a variable should be retained in the PLS model: it is a weighted sum of squares of PLS weights, calculated from the amount of dependent variable variance of each PLS component, and it is an indicator showing the contribution of each independent variable to the model:

$$VIP = \sqrt{\frac{p}{Rd(Y;t_1,\ldots,t_m)} \sum_{h=1}^{m} Rd(Y;t_h) w_{hj}^2} \tag{1}$$

Here, w_{hj} is the weight of the jth predictor variable; $Rd(Y;t_h)$ is the Y-variance fraction explained by component t_h; $Rd(Y; t_1,\ldots t_m)$ is the cumulative fraction of Y-variance explained by components $t_1,\ldots t_m$, and p are the components retained (this work: $p = 4$). The VIP values that have been computed for the 14 independent variables of our PLS model are presented in Table 2 (all others with VIP < 0.8 scores have been discarded).

Table 2: VIP values of 14 retained independent variables (retention criterion: VIP >0.8)

	QM1	QM2	QM3	MT	CTT	WTV	PVV	SWV	C1T	SF3	PCV	SF1	SF2	SV3
VIP	0.83	0.83	0.84	0.81	1.23	0.81	1.03	1.11	1.10	1.52	0.88	1.32	1.46	0.84

3.1. PLS model from odd-numbered batches validated vs. even-numbered batches

The input-output correlation obtained for filtration time (the critical output of industrial significance) vs. 14 independent variables from odd-batch data (in the order defined):

Filtration time (s) $= y = 626.9886 + 0.2404x_1 + 0.9265x_2 + 0.5015x_3 + 32.2516x_4 +$
$$+ 48.5300x_5 - 0.3995x_6 - 62.6566x_7 + 11.1106x_8 + 23.0814x_9 -$$
$$- 2.2046x_{10} - 2.0042x_{11} - 2.9928x_{12} - 3.3516x_{13} + 2.0214x_{14} \quad (2)$$

The agreement of PLS model with plant data is great (error ca. 4% in outlier batch 24).

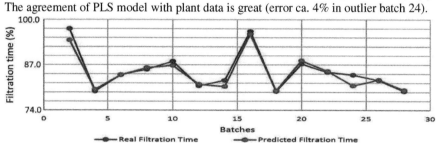

Figure 3. Filtration time prediction of odd (model) vs. even (data) separation batches.

3.2. PLS model from even-numbered batches validated vs. odd-numbered batches

Similarly, the input-output correlation obtained for filtration time (the critical output) vs. the same 14 independent variables from even-batch data (in the order defined):

Filtration time (s) $= y = 1879.0100 + 0.2234x_1 - 0.3054 x_2 + 0.2676 x_3 + 18.119 x_4 +$
$$+26.3076x_5 + 3.8418x_6 - 60.2435x_7 + 12.3015x_8 - 3.8380x_9 -$$
$$-2.3208_{10} - 0.8541x_{11} - 3.9576x_{12} - 3.0716x_{13} + 2.8110x_{14} \quad (3)$$

Great agreement of PLS model with plant data, again (error 3%, 4% in batches 1, 19).

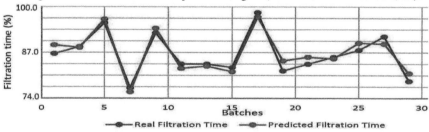

Figure 4. Filtration time prediction of even (model) vs. odd (data) separation batches.

3.3. Sensitivity analysis

Clearly, SWV, QM1, QM3, SF1, SF2 (in decr. order) are the strongest filtration drivers.

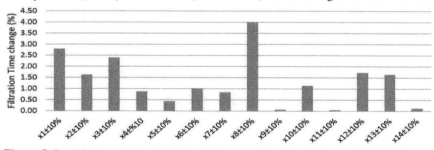

Figure 5: Sensitivity analysis of relative influence of input variables on filtration time.

4. Conclusions

This paper develops and discusses statistical (Partial Least Squares/PLS) models of mash separation for two beer types, and presents a sensitivity analysis in order to better understand which operational modifications are feasible and most promising in order to improve mash separation efficiency and benefit beer production and its supply chain. The agreement obtained between predicted and real filtration time is very good, and the analysis is repeated for different (odd/even) data subsets, to explore trends and outliers. A sensitivity plot shows that strong wort volume (SWV) and feed quality (QM1, QM3) have the strongest effect on filtration time (y), the key industrial performance metric. This concurs with industrial experience, but none of these are convenient manipulation variables: however, sparging flowrates (SF1, SF2) are, and thus can be effectively used. Mashing time (MT), pre-sparging and (pre-)compression metrics are not key factors. PLS models may require (re-)calibration to avoid drift, but clearly have business value.

Acknowledgements

The authors acknowledge instrumental help and helpful discussions with Molson Coors staff engineers and technicians, and EPSRC/IAA and MSc in Adv. Chem. Eng. project funding (School of Engineering, University of Edinburgh). Dr. Dimitrios I. Gerogiorgis also acknowledges a Royal Academy of Engineering (RAEng) Industrial Fellowship.

References

M. Aznar et al., 2003, Prediction of aged red wine aroma properties from aroma chemical composition: Partial least squares regression models, *J. Agr. Food. Chem.* 51, 2700–2707.

C. Bamforth, 2006, *Brewing: New Technologies*, Woodhead Publishing, Cambridge, UK.

T. Bühler, 1996, *Effects Of Physical Parameters in Mashing on Lautering Performance*, Doctoral Thesis, Loughborough University, UK.

V. Esposito Vinzi et al. (eds.), 2010, *Handbook of Partial Least Squares: Concepts, Methods and Applications*, Springer, Heidelberg.

F. Kühbeck et al., 2006, Influence of lauter turbidity on wort composition, fermentation performance and beer quality in large-scale trials, *J. Inst. Brewing* 112, 222–231.

J. Liu et al., 2008, Multivariate modeling of aging in bottled lager beer by principal component analysis and multiple regression methods, *J. Agr. Food Chem.* 56, 7106–7112.

J.F. MacGregor et al., 1994, Process monitoring and diagnosis by multiblock PLS methods, *AIChE J.* 40, 826–838.

T. Mehmood et al., 2012, A review of variable selection methods in Partial Least Squares regression, *Chemometr. Intell. Lab.* 118, 62–69.

F.G. Meussdoerffer, 2009, A comprehensive history of beer brewing, in *Handbook of Brewing: Processes, Technology, Markets*, Wiley-VCH, Weinheim, Germany.

L. Montanari et al., 2005, Effect of mashing procedures on brewing, *Eur. Food Res. Tech.*, 221, 175–179.

B. Muster-Slawitsch et al., 2014, Process modelling and technology evaluation in brewing, *Chem. Eng. Process.* 84, 98–108.

J.M. Poveda et al., 2004, Application of partial least squares (PLS) regression to predict the ripening time of Manchego cheese, *Food Chem.* 84, 29–33.

J. Schneider et al., 2005, Study on the membrane filtration of mash with particular respect to the quality of wort and beer, *J. Inst. Brewing* 111, 380–387.

R. Wakeman, S. Tarleton, 1999, *Filtration: Equipment Selection, Modelling, Process Simulation*, Elsevier, Amsterdam.

Anton A. Kiss, Edwin Zondervan, Richard Lakerveld, Leyla Özkan (Eds.)
Proceedings of the 29th European Symposium on Computer Aided Process Engineering
June 16th to 19th, 2019, Eindhoven, The Netherlands. © 2019 Elsevier B.V. All rights reserved.
http://dx.doi.org/10.1016/B978-0-128-18634-3.50246-0

End-of-Pipe Zero Liquid Discharge Networks for different brine water qualities

Fatima Mansour, Sabla Y. Alnouri*

Department of Chemical and Petroleum Engineering, American University of Beirut, P.O.Box 11-0236, Riyad El-Solh, Beirut, Lebanon
sa233@aub.edu.lb

Abstract

More often than not, wastewater generated from any sort of industrial process poses a problem in terms of disposal cost and hazardous environmental impact. Systems that seek to eliminate wastewater discharge from any given process are known as Zero Liquid Discharge (ZLD) systems. Such systems, often composed of a combination of different technologies, aim to minimize wastewater volume and maximize the amount of water that can be recovered. Economically speaking, it is challenging to alter existing processes in a way that can achieve zero discharge. End-of-pipe ZLD systems provide an intermediate solution, whereby additional units can be added, forgoing the need for structural change in the existing process. In addition, environmental discharge standards can be satisfied. This paper presents an overarching ZLD network structure and model that can be customized to yield a cost-effective end-of-pipe ZLD discharge system based on user specified data and conditions. In particular, this paper focuses on the design of different End-of-pipe ZLD systems using brine wastewater data from two different industries: i) wastewater from a coal seam gas plant, and ii) wastewater from a steel plating industry.

Keywords: Brine, Zero-Liquid-Discharge, Wastewater

1. Introduction

Increasing water scarcity and a growing water crisis worldwide have heightened the search for alternative water sources (Subramani and Jacangelo, 2014). Although desalination technologies are highly sought after to amend water scarcity, the brine they generate has also evolved to be a weighty problem (Morillo et al., 2014). Brine is the main by-product of many desalination process, and is always characterized by a high content of dissolved salts and a high salinity. As such, ZLD systems are gaining more attention, as they possess the ability to reduce wastewater discharge, generate recovered product water, and allow for compliance with strict environmental and wastewater disposal regulations. The optimal option for brine discharge depends on a number of factors, mainly brine quantity, quality, and cost (Giwa et al., 2017). Because of this, ZLD systems tend to be customized for each system based on individual system characteristics (Tillberg, 2004). As such, long term brine management has become of utmost importance, particularly with increasing environmental scrutiny and decreasing water supplies. Existing brine disposal methods, such as surface water discharge, deep well injection, and land application, do not eliminate the problem, but attenuate its impact. That and increasingly stringent environmental regulations have propelled the search for more suitable alternatives, mainly ZLD methods. Zero liquid discharge (ZLD) technologies have been gaining more attention as a means for brine minimization. The introduction of these options has encouraged their integration as end-of-pipe solutions to brine disposal (Subramani and Jacangelo, 2014). Moreover, the water quality generally plays a vital role in the design of any wastewater handling system (Alnouri et al. 2016). To provide a more generic methodology through which users can easily determine the best ZLD scheme for a certain system, this paper proposes an MINLP optimization model in which an economic objective, has been utilized to minimize system design costs, whilst selecting an optimal brine treatment system based on user input.

2. Methodology

The proposed model has been devised based on a ZLD superstructure that is illustrated in Figure 1 below. The ZLD network proposed consists of four different technology sets: chemical processing (set P), thermal (set T), membrane (set M), and brine processing (set B). The chemical processing technologies generally requires a chemical feed stream and results in the production of some salt product, aside from the concentrated brine produced. On the other hand, thermal and membrane technologies take in concentrated brine, and produce product water and further concentrated brine. Brine processing technologies use its feed of concentrated brine to produce salt and product water. The system is designed such that the entering feed water first undergoes chemical processing. Placing the chemical processing blocks first provides a starting point for treatment by removing salts and "softening" the brine for further treatment. The treated water stream can then move to either the thermal or membrane technology blocks, or to both of them. A portion of the streams leaving these two bins is recycled to the start of the system. A recycle stream is included in the design for enhanced system performance. The recycle is split from the concentrated brine leaving the thermal technology block, and another from the concentrated brine leaving the membrane technology block. These two streams then converge to join the original feed stream. The remaining brine enters the brine processing block. This block comes last because it essentially separates the salt from the water, thus resulting in water and salt production.

In accordance with the requirements for a specific system, a selection of technologies must be assessed in each block. The proposed approach selects the optimal technology in each block, or forgoes certain blocks altogether (mainly thermal or membrane), depending on mass balance consistency and other user specified requirements. Each technology comes with its own set of specifications, such as capacity limits and feed water characteristics (e.g. permissible inlet TDS and hardness content). Furthermore, each technology must have a specified water recovery, in addition to component rejection values that have been accounted for and included in system calculations. The system is designed such that different requirements can be met in terms of required product water yield and allowed purge quantities. Constraints can be placed on these values, and the overall ZLD scheme may be assessed based on those input parameters. The splits between technologies within a block, between different blocks, and for recycle and purge purposes can all be controlled by the user and specified to meet any particular standards. Hence, the constraints placed on the system can be specified according to one of the following forms: i) capacity constraints (the flowrate a particular technology can withstand), ii) composition constraints (the inlet or outlet composition limit to a particular technology), iii) and product water production (where the user specifies a certain requirement in terms of how much product water must be generated, as a percentage of the feed flow).

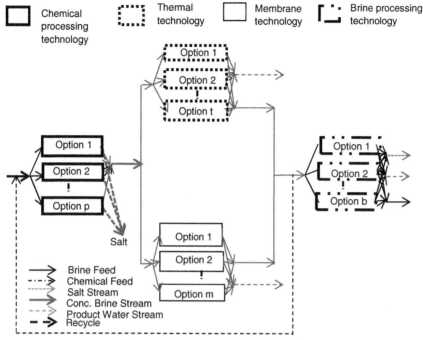

Figure 1. General ZLD superstructure scheme

3. Mathematical Formulation

The mathematical formulation of the proposed optimization model falls under the general form of a Mixed Integer Non-Linear problem, where a cost objective function (f(x)) is subjected to various equality and inequality constraints:

$$\min f(x) \tag{1}$$

$$h_i(x) = 0, \forall i \in I = \{1,,p\} \tag{2}$$

$$g_j(x) \leq 0, \forall j \in J = \{1,,m\} \tag{3}$$

Eq. (1) represents the objective function, f(x), which in the case of this work, is to minimize the overall cost of the ZLD system. The overall cost covers both capital and operating costs. nit. This is where the nonlinearity arises; the cost equations are derived from data sets collected from different literature sources (Mickley, 2008); each technology unit has two cost functions: one for operating costs) and another for capital costs. It should be noted that energy requirements are embedded in the operating cost models. Moreover, the total system cost is determined by summing the operating and capital costs of the technologies selected. The central variable in this model (x) is the split fraction vector, and it is of a continuous type. Hence, the split ratios for the flows between technology blocks, for individual feed flows between technologies within each block, and for the amount to be recycled, are all variables. The objective is subjected two sets of constraints. The first set, h(x), consists of equality constraints (Eq. (2)); these constraints cover the mass and component balances of the different flowrates and material throughout the system. The mass balance equations take into account technology specifications, particularly salt rejection and water recovery. The second set, g(x), pertains to inequality constraints (Eq. (3)). These constraints can be of several types: i) production and performance requirements, which refer to imposing a minimum or maximum amount of product (product water or salt) or waste (concentrated brine to be purged) to be generated, ii) capacity constraints, which place a limit of the feed flow allowed into a certain limit and/or a limit on the recycle flow, and iii) composition constraints, which place a restriction on the amount of a specific component allowed into (or out of) a technology (such as salinity). Aside from the constraints imposed for the mathematical consistency and practicality, the number and nature of constraints can be tailored to represent any reasonable user imposed constraints.

4. Case Study Illustration

This case study features two types of industrial wastewater: (1) wastewater generated from a steel plating mill and (2) wastewater from a coal seam gas plant. Each of these feed water types has different characteristics and would be expected to require different treatment methods. The main difference showcased for the purpose of this case study is TDS content, as provided in Table 1. In terms of technology appropriation, a handful of the commonly used ZLD technologies are selected for each block, as summarized in Table 2 below.

Table 1. Feed water Parameters, adapted from

Feed water Type	Flow (m³/h)	TDS (ppm)	Reference
Coal seam gas wastewater	100	62,450	(Pramanik et al., 2017)
Steel Plating Mill Wastewater	100	2,400	(Chimeng, 2014)

Table 2. Technology Selection

Chemical Processing	Thermal	Membrane	Brine Processing
Lime Softening	Multi-Effect Distillation	Brackish Water Reverse Osmosis (BWRO)	Brine Crystallizer
	Brine Concentrator	Seawater Reverse Osmosis (SWRO)	Evaporation Pond

As noted above, each technology comes with its own set of characteristics and constraints. An overall constraint is imposed such that 95% of the feed must be converted to product water. Aside from this constraint, there are capacity constraints (maximum inlet feed flow) on all of the selected technologies. The BWRO technology has a feed composition constraint in that the maximum allowed Total Dissolved Solids (TDS) into the unit is 6000 mg/L, whereas the brine concentrator can concentrate its feed stream up to 300,000 mg/L, placing a maximum on the allowed exit TDS. All of these values, in addition to each technology's water recovery (γ) and salt rejection (θ) values, are obtained from the

Optimization-based approach of End-of-Pipe ZLD systems for different
brine water qualities
1475

literature and used to characterize the system, as summarized in Table 3. The "what'sBest9.0.5.0" LINDO Global Solver for Microsoft Excel 2010, on a laptop with Intel® Core ™ i7-2620M, 2.7 GHz, 8.00 GB RAM, 64-bit Operating System, was used.

Table 3. Summary of system parameters and constraints, (Masnoon and Glucina, 2011)[a], (Mickley, 2008)[b], (Burbano and Brandhuber, 2012)[c], (Tillberg, 2004)[d]

Sets	Technology	Flow capacity (L/h)	γ	θ	TDS
Set T	Multi-Effect Distillation (MED)	625,000	0.3[a]	0.99[b]	NA
	Brine concentrator	158,987	0.95[c]	0.9[b]	300,000[d] (exit)
Set M	SWRO	5,333,333	0.5[b]	0.99[b]	NA
	BWRO	4,083,333	0.7[b]	0.99[b]	6000[b] (inlet)
Set P	Lime softening	NA	NA	0.5[b]	NA
Set B	Evaporation pond	157,725	NA	NA	NA
	Brine crystallizer	11,354	0.95[c]	0.85[b]	NA

The ZLD schemes illustrating the results generated from the solver simulations for each scenario are shown in Figures 2 and 3, respectively. In the first scenario, after the chemical processing unit, 5% of the concentrated brine is purged from the system. The solver only takes 95% of the feed into the thermal technology block, because that is all that is needed to satisfy the product water requirement (95% of the feed), purging the remaining 5% to reduce costs (as the cost functions depend on flow). In the thermal technology block, the concentrator technology is selected as the optimal technology, whereas in the brine processing block, the crystallizer is selected. This scheme train is of significance because the brine concentrator and brine crystallizer are the typical technologies most commonly implemented to achieve ZLD (Mickley, 2008).

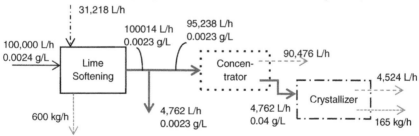

Figure 2. ZLD scheme for steel plating mill wastewater feed

The MED option is selected as a result of the exit TDS constraint being placed on the brine concentrator. Because of the higher inlet TDS in this scenario (62,450 versus 2,400) mixed with the recycled concentrated brine, the brine entering the thermal technology block has high salinity. Concentrating it further renders the exit TDS over the 0.3 composition limit placed on the concentrator, yielding MED as the only feasible choice. A recycle stream appears, where over 90% of the concentrated brine is exiting the MED unit is recycled back and mixed with the original feed. This option is feasible because the product water constraint (yielding 95% of the feed as product water) is already satisfied; and with the objective of minimizing cost, recycling becomes economical, instead of moving the entire flow to the brine processing block. In the brine processing block, both technologies are included in the scheme, due to the capacity constraint placed on the brine crystallizer (allowing a maximum of 11,354 L/h inlet flow). Consequently, the rest of the concentrated brine is re-diverted to the solar pond.

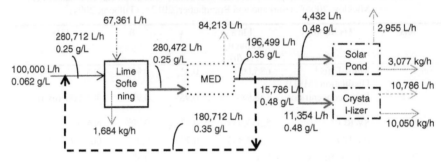

Figure 3. ZLD scheme for coal seam gas wastewater feed

5. Conclusions

The proposed assessment approach provides a simple and effective means of optimally designing ZLD systems, based on cost effectiveness. ZLD systems were found to depend on numerous factors, including but not limited to: the quality and quantity of the brine feed stream, the technologies that have been incorporated into the different technology blocks and their respective design parameters and constraints. In addition, any user-specified product water production or waste limit requirements can be incorporated.

Acknowledgment

The authors would like to acknowledge the financial support received from the University Research Board (Award# 103187; Project# 23308) at the American University of Beirut.

References

A. Burbano, P. Brandhuber, 2012. Demonstration of Membrane Zero Liquid Discharge for Drinking Water Systems. Water Environment Research Foundation.

A. Giwa,, V. Dufour, F. Al Marzooqi, M. Al Kaabi, M., S. W. Hasan, 2017. Brine management methods: Recent innovations and current status. Desalination, 407, 1-23.

A. Subramani, J. G. Jacangelo, 2014. Treatment technologies for reverse osmosis concentrate volume minimization: A review. Separation and Purification Technology, 122, 472-489.

F. Tillberg, 2004. ZLD-systems An Overview. Stockholm: Department of Energy Technology, Royal Institute of Technology.

J. Morillo, J. Usero, D. Rosado, H. El Bakouri, A. Riaza, A. and F. J. Bernaola, 2014. Comparative study of brine management technologies for desalination plants. Desalination, 336, 32-49.

M. Mickley, 2008. Survey of High-Recovery and Zero Liquid Discharge Technologies for Water Utilities. Alexandria, VA: WateReuse Foundation.

S. Masnoon, K. Glucina. 2011. Desalination: Brine and Residual Management. Global Water Research Coalition.

S. Y. Alnouri, P. Linke, S. Bishnu, M.M El-Halwagi, 2016, Synthesis and Design Strategies of Interplant Water Networks using Water Mains with Quality Specifications. Comput. Aided Chem. Eng., 38, 655-660

Anton A. Kiss, Edwin Zondervan, Richard Lakerveld, Leyla Özkan (Eds.)
Proceedings of the 29th European Symposium on Computer Aided Process Engineering
June 16th to 19th, 2019, Eindhoven, The Netherlands. © 2019 Elsevier B.V. All rights reserved.
http://dx.doi.org/10.1016/B978-0-128-18634-3.50247-2

A PLSR Model for Consumer Preference Prediction of Yoghurt from Sensory Attributes Profiles

Kexin Bi[a], Dong Zhang[b], Zifeng Song[a], Tong Qiu[a*], Yizhen Huang[b*]

aDepartment of Chemical Engineering, Tsinghua University, 100084 Beijing, China

bCOFCO Nutrition Health Research Institute, 102209 Beijing, China

* Corresponding author's E-mail: qiutong@tsinghua.edu.cn and 1289803467@qq.com

Abstract

Consumer preference investigations are extremely emphasized in marketing management strategy for food enterprises. In this communication, a partial least squares regression (PLSR) model for consumer preference prediction from sensory attributes profiles is developed and three brands of yoghurt are exampled to demonstrate the accuracy and practicability of the method. The model also provides the importance ranking of the analysed sensory variables, which may give a guidance for industrial production and sensory experiment design. Drawing on the results of PCA and hierarchical clustering analysis, the variable filtering process of PLSR is proved to be theoretically correct.

Keywords: Sensory attributes, Consumer preference, PLSR, Yoghurt

1. Introduction

Yoghurt is one of the most important commodities in people's daily life. Odour compounds, such as proteins, amino acids, fats, sugars, organic acids,[1] make a solid contribution in the sensory attributes of dairy. The consumer preferences for yoghurt are directly affected by their sensory quality. For dairy companies, consumer preferences for yoghurt data are essential to market competitiveness. Dairy products should be designed and improved according to consumer demand.[2] However, the data related to consumer is usually hard to be obtained, because the large-scale consumer preferences survey is unfeasible to carry out for new products[3] and the consumer preferences in different regions have significant differences in the flavor of dairy products.

Thus, a data-driven consumer preference prediction model is urgently needed by dairy companies. And in recent years, several researchers have focused on the sensory profiling of dairy products and proposed various correlation models to forecast the consumer overall likings. Bouteille et al[4] applied principal component analysis (PCA) to construct internal mapping between Temporal Dominance of Sensations (TDS) curves and liking scores or freshness sensation scores to evaluate plain yoghurts and yoghurt-like product. Oliveira et al[5] analyzed the relevancy between check-all-that-apply (CATA) frequency and overall linking by threshold statistics and advised the dairy companies to reduce the sugar addition in milk. Farah et al[6] utilized cluster analysis and multiple factor analysis to associate data from CATA test and acceptance test for overall impression and identified two significantly different clusters of yoghurts', whey-based beverages' and fermented milks' consumers. However, these models were lack of generalization ability and hard to deal with high dimensional data, making the model feeble in practicality and versatility.

To improve the overall performance of the model, a novel correlation model with PLSR was proposed to predict the consumers' overall liking of dairy products from sensory attributes profiles. The PLSR model could not only obtain the prediction value from multicollinear variables, but also filter out the key sensory attributes and classify the samples with its principal components. In consideration of the characteristics of data as well as the need in enterprise production, reducing the number of variables without affecting the accuracy of result fitting was also realized to reduce the cost of subsequent experiments.[7]

In summary, this communication intended to use the analytical sensory evaluation team to perform sensory evaluation on yoghurt samples and use multivariate statistical analysis methods to profile the sensory evaluation of different samples' flavors, which provided guidance for the improvement of sensory quality.

2. Sensory evaluation and data profiling method

2.1. Sensory data collection

The sensory attributes data was obtained by evaluators' rating of each sample. Through the training[8] of sensory sensitivity and expressive ability, a total of ten sensory experts were invited to complete the sensory evaluation. After tasting the series of flavored yoghurt samples, these experts were asked to describe all the sensations from the aspects of sight, smell and taste. The evaluation process is carried out in an independent evaluation room. Then the descriptors were screened through preliminary collation and multivariate statistical analysis methods, and the sensory attributes descriptors of the flavored yoghurt were finally determined.[9]

For the selected sensory attributes descriptors, each sensory expert would give the flavor intensities for a specific sample using a 10-point linear scale. The left end of the linear scale is extremely weak, and the right end is extremely strong. The evaluator can select any value from 0 to 10 as the attributes intensity evaluation value.

Consumer preference rating is an output variable in this study. Fifty heavy consumers of yoghurt products in Beijing were invited to the sensory lab to fill the overall liking questionnaires. The consumer preference test also adopted the 0 (extremely weak) to 10 (extremely strong) scoring mechanism.

2.2. Data characterization analysis

After the data collection process, we obtained a set of data including thirty-three kinds of yoghurt with different storage conditions evaluated by ten sensory experts on twenty-one sensory attributes and fifty consumers on overall likings. Due to the complexity of the input and output data, it's better to find out the data characterization before fitting.

For sensory attributes data, there were three brands of yoghurt stored under eleven different conditions that were evaluated by experts. So the input data could be summarized with a 33*21 matrix, where 33 (3*11) was the quantity of samples and 21 was the quantity of sensory attributes. To extract the key information of the data, Principal component analysis (PCA)[10] was applied to search for the relationship between the high-dimensional sensory attributes profiles. To profile the variable multicollinearity of yoghurt odour, the scoring plot of two PCs was generated and displayed in Figure 1.

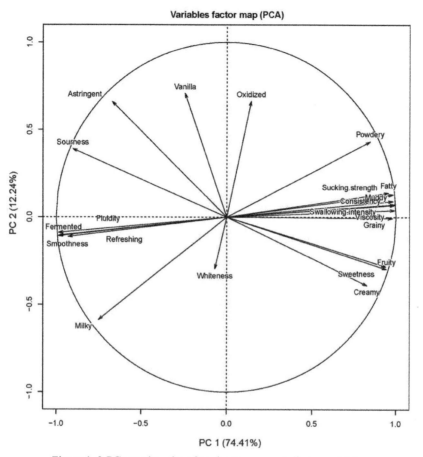

Figure 1. 2-PCs scoring plot of yoghurt sensory attributes variables

As shown in Figure 1, there were several overlapped variables in the scoring plot, which might contain the almost same information for result fitting and interfere the accuracy of regression analysis. However, only utilizing several PCs to predict the consumers overall liking might lose plenty of information. So a regression model with the function of variable filtering was needed.

2.3. PLSR model for preference prediction

The partial least squares regression (PLSR) was proposed by Wold[11] in 1979, and it was feasible to solve the multivariate regression problem with independent multivariate correlation. This model projected the predicted variables to a new space, so that the difference between the dependent variables was minimized. Then compared to PCA regression method, PLSR also considered the contribution of the input data for the output data, so the correlation during the regression process could be guaranteed.[12] What's more, an importance ranking of sensory attributes could be obtained from the regression process, which could tell the enterprise how to improve the products.

Thus, through the overall liking assignment and PLSR model building with training data, a complete consumer preference prediction of yoghurt from sensory attributes profiles could be applied for dairy companies or sensory labs.

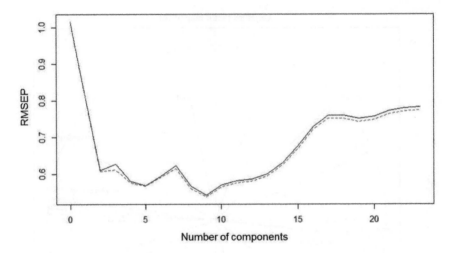

Figure 2. The fluctuation trend of RMSEP value affected by the number of components

3. Result and discussion

To execute the PLSR process, the quantity of variables that were employed to complete regression modeling should be determined. As shown in Figure 2, by analyze the root-mean-square error of prediction influenced by the number of components, a total of nine variables were finally picked up. The cumulative contribution rate of the nine selected variables reached the value of 94.52%, which gave a relatively comprehensive explanation of the origin input data.

Then the PLSR model was built on the R-3.4.1 and simulated on an Inter Core i7 processor of 2.80 GHz to predict the average liking scores of thirty-three collected samples. Figure 3 displayed the calculation results compared to actual data. Obviously, the calculation process demonstrated low relative errors that were below 10%.

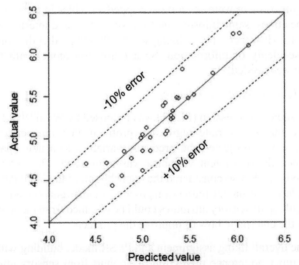

Figure 3. Comparison of predicted and actual value of consumers' preference rating.

Table 1. Properties and corresponding objective functions in simulated annealing algorithm

Variables	+/- Correlation	Relevant significance
Sweetness	+	>95%
Grainy	-	>90%
Sourness	+	>85%
Oxidized	-	>85%
Smoothness	+	>85%
Swallowing intensity	-	>80%

In addition to the overall liking information, the PLSR simulation process also provided the correlation properties of each variables. Table 1 listed top six sensory attributes with high relevant significance of the selected nine variables, and all these relevant significances were larger than 80%, showing a strong relationship between sensory attributes and consumers preference rating value.

From Table 1, a diary enterprise could obtain the product improvement plan for a typical brand of yoghurt, such as properly increasing the sweetness, sourness, smoothness and decline the grainy, oxidized, swallow intensity as much as possible. What's more, the PLSR process could give the sensory labs a reasonable streamlined solution to design the sensory attributes test. As shown in Figure 4, though PCA showed that there were several variables expressing the similar information in the 2-PC space, the PLSR only picked up two key attributes (Smoothness, Swallowing intensity) for regression process. And PLSR could found out some attributes with high relevant significance, such as oxidized, which didn't own a good variance explanation in PCA result.

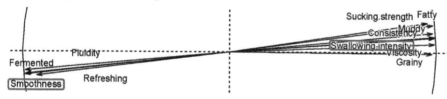

Figure 4. The sensory attributes with similar information in 2D PCA scoring plot

Finally, to demonstrate the description ability of the selected sensory attributes, a hierarchical clustering analysis was carried out. As shown in Figure 5, utilizing the selected sensory attributes as independent variables could clearly classify the three brands (670, 395, 481) of yoghurt, though the storage condition were different (1 to 5 for the storage month, R for Room temperature, H for high temperature, L for low temperature). This result demonstrated the correction of the sensory attributes selection from PLSR because these variables owned the ability to describe a specific b rand of yoghurt.

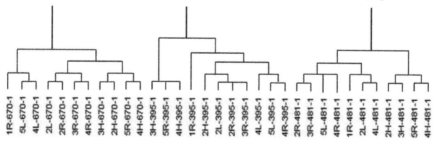

Figure 5. The hierarchical clustering analysis result using selected sensory attributes from PLSR

4. Conclusion

A PLSR model was developed to make connection between the collected sensory attributes profiles from trained sensory experts and the consumer preference rating. The accuracy and practicability of the method was demonstrated by the regression process of three brands of yoghurt. Besides, nine key sensory variables were selected by PLSR process and the correlation parameters between sensory variables and overall liking could give a reasonable advice for dairy enterprise to improve the yoghurt products. And the sensory labs could also update the experimental scheme according to the importance ranking of the sensory descriptors to save the overall consumption of the sensory evaluation.

Accurate and fast prediction of consumer preference can bring huge profits for food enterprise. The theory reliability of the PLSR method proposed in this work provides feasibility to applied the prediction model in other types of food. Hopefully, the utility of R based programs make the model easier to be applied as plug-in for Industrial Internet and promote the benefits for food industry.

Acknowledgements

The authors gratefully acknowledge the National Natural Science Foundation of China for its financial support (Grant No. U1462206).

References

[1]   Cheng, H. (2010). Volatile flavor compounds in yogurt: a review. Critical reviews in food science and nutrition, 50(10), 938-950.

[2]   Grunert, K. G., Bech-Larsen, T., & Bredahl, L. (2000). Three issues in consumer quality perception and acceptance of dairy products. International Dairy Journal, 10(8), 575-584.

[3]   Krishna, A. (2012). An integrative review of sensory marketing: Engaging the senses to affect perception, judgment and behavior. Journal of consumer psychology, 22(3), 332-351.

[4]   Bouteille, R., Cordelle, S., Laval, C., Tournier, C., Lecanu, B., This, H., & Schlich, P. (2013). Sensory exploration of the freshness sensation in plain yoghurts and yoghurt-like products. Food Quality and Preference, 30(2), 282-292.

[5]   Oliveira, D., Reis, F., Deliza, R., Rosenthal, A., Giménez, A., & Ares, G. (2016). Difference thresholds for added sugar in chocolate-flavoured milk: Recommendations for gradual sugar reduction. Food Research International, 89, 448-453.

[6]   Farah, J. S., Araujo, C. B., & Melo, L. (2017). Analysis of yoghurts', whey-based beverages' and fermented milks' labels and differences on their sensory profiles and acceptance. International Dairy Journal, 68, 17-22.

[7]   Lawless, H. T., & Heymann, H. (2010). Sensory evaluation of food: principles and practices. Springer Science & Business Media.

[8]   Murray, J. M., Delahunty, C. M., & Baxter, I. A. (2001). Descriptive sensory analysis: past, present and future. Food research international, 34(6), 461-471.

[9]   Giboreau, A., Dacremont, C., Egoroff, C., Guerrand, S., Urdapilleta, I., Candel, D., & Dubois, D. (2007). Defining sensory descriptors: Towards writing guidelines based on terminology. Food quality and preference, 18(2), 265-274.

[10] Mackiewicz, A., & Ratajczak, W. (1993). Principal components analysis (PCA). Computers and Geosciences, 19, 303-342.

[11] Gerlach, R. W., Kowalski, B. R., & Wold, H. O. (1979). Partial least-squares path modelling with latent variables. Analytica Chimica Acta, 112(4), 417-421.

[12] Helland, I. S. (2001). Some theoretical aspects of partial least squares regression. Chemometrics and Intelligent Laboratory Systems, 58(2), 97-107.

Anton A. Kiss, Edwin Zondervan, Richard Lakerveld, Leyla Özkan (Eds.)
Proceedings of the 29th European Symposium on Computer Aided Process Engineering
June 16th to 19th, 2019, Eindhoven, The Netherlands. © 2019 Elsevier B.V. All rights reserved.
http://dx.doi.org/10.1016/B978-0-128-18634-3.50248-4

Fouling and Cleaning of Plate Heat Exchangers for Milk Pasteurisation: A Moving Boundary Model

Abhishek Sharma and Sandro Macchietto

Department of Chemical Engineering, Imperial College London

South Kensington Campus, London SW7 2AZ, UK

s.macchietto@imperial.ac.uk

Abstract

Plate heat exchangers (PHEs), widely used in the food industries, use large amounts of energy. Fouling reduces their thermal and hydrodynamic performance, and requires periodic cleaning (often after few hours), with productivity losses. Cleaning uses large amounts of water and produces wastes. Improving productivity, minimising energy, water and wastes, while ensuring effective pasteurization is of high interest and can be supported by simulation models. Current models describing the dynamic behaviour of PHEs with fouling typically rely on simplified assumptions, with limited predictive ability. More detailed models based on Computational Fluid Dynamics (CFD) are very computationally expensive for complete PHEs, not always better at fitting experimental data, and practically infeasible to use for operations optimisation and control.

A 2D distributed, dynamic model described by Guan and Macchietto (2018) for a single PHE channel, is extended to enable the modelling of full PHEs with multiple plates arranged in any design configuration (e.g. with parallel, countercurrent, mixed parallel-countercurrent flow). The model is validated here for two such arrangements (A1 and A2) against experimental results. Four milk fouling models are assessed: two deposition mechanism (due to aggregate or denatured proteins), each with/without deposit re-entrainment. No single model was found capable of describing both arrangements. A1 was predicted best by models with aggregate proteins deposition, A2 with fouling by denatured proteins.

A dynamic model for cleaning of fouled surfaces was integrated with the moving boundary model and validated against experimental data. This enabled for the first time in the literature the simulation of Cleaning-in-Place (CIP) procedures, establishing the duration of cleanings, and the seamless simulation of entire single and multiple heating-CIP cycles, with either pre-set or condition-based termination conditions for each phase. It is shown there is large scope for optimising the performance and productivity of overall heating-CIP cycles, and an optimal operation is demonstrated for arrangement A1.

Keywords: energy recovery, plate heat exchanger, fouling, cleaning, dairy processes, dynamic model

1. Introduction

PHEs, widely used in milk production, suffer from rapid fouling, requiring frequent cleaning (every 5-10 hr) of the equipment to recover thermal and hydraulic performance. A PHE productivity (kg milk processed/day) clearly depends on the time out for cleaning. Modelling provides a way to predict the extent of fouling and optimise the operation of exchangers individually and in networks, aiming at more efficient processes. Many studies have dealt with modelling of PHEs for milk pasteurisation. Some used a set of differential and algebraic equation models developed from mechanistic mass, energy and reaction balances to capture the fouling dynamics (e.g. Georgiadis et al., 1998a, Jun and Puri, 2006), others used a Computational Fluid Dynamics (CFD) approach. The former is faster but typically needs simplifying assumptions associated with lack of detailed knowledge of underlying phenomena, leading to poor predictive ability. CFD incorporates better the effects of detailed geometry on fluid dynamics, but is computationally very expensive (often days of CPU), even when (other) simplifying assumptions are made, e.g. by focusing on a single channel, fixing temperatures (thus decoupling thermal and hydraulic aspects), or decoupling hot and cold sides (e.g. Bouvier et al. 2014). Cleaning has been much less investigated, and complete detailed models for simulation, monitoring, control and optimisation of full heating and cleaning cycles for PHEs are still not available.

Guan and Macchietto (2018) proposed an advanced dynamic, moving boundary deposition model for a single PHE channel, building on Shell and Tube exchangers work (e.g. Georgiadis et al., 1998b, Diaz Bejarano et al., 2016). The model includes spatially varying, temperature dependent fluid properties, a detailed description of deposition rates over time, and thermal and hydraulic behaviour coupled to deposit layer formation. The result is a more predictive model which is computationally rather inexpensive. However, the Guam and Macchietto (2018) model only describes a single channel, not a complete PHE. A complete PHE model is developed here, with multiple models for aggregation and deposition of milk, and is validated against experimental data for the heating phase, for two quite different PHE design and flow arrangements (Fig.1). A dynamic model of cleaning of a fouled PHE was also integrated and validated. The overall model is then used to simulate and optimise for the first time full Heating and Cleaning in Place (CIP) cycles. Details of the work in this paper are given in Sharma (2018).

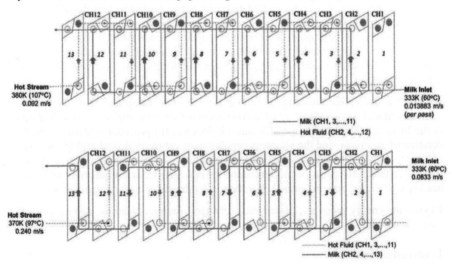

Figure 1. PHE arrangements A1 (top) and A2 (bottom). (Guan & Macchietto, 2018)

2. Models of PHE, Fouling and Cleaning

The thermo-hydraulic model of a single PHE channel (dynamic, 2D-distributed, characterising deposit growth at any point on plates using a moving boundary) detailed in Guan & Macchietto (2018) was used, with a few changes. Temperature dependent physical properties were modified for both heating and milk fluids (De-Jong, 1997, Gut and Pinto, 2003). A correlation between Nusselt, Reynolds and Prandtl numbers, Eq. 1, that better fits the conditions in the PHEs studied (Cavero, 2013) was used.

$$Nu(x) = 0.0902 \, Re(x)^{0.663} Pr(x)^{0.333} \qquad \forall \, i = L, R \qquad (1)$$

Milk fouling depends on denaturation and/or aggregation of the β-lactoglobulin protein but the exact kinetics is uncertain. Four different literature fouling models, differing in respect of source (aggregated or denatured proteins) and re-entrainment, were used here: i) Fouling with aggregate proteins (Model A); ii) with aggregate proteins and entrainment (Model AE); iii) with denatured proteins (Model D) and iv) with denatured proteins and entrainment (Model DE). All are represented by Eq. (2), but with different parameters:

$$\frac{d\delta}{dt} = \beta \frac{\lambda_d}{h_f^0} k_d C^n - k_\tau \delta \qquad (2)$$

δ is the deposit thickness, λ_d its thermal conductivity, h_f^0 the heat transfer coefficient in clean conditions, k_d the deposition rate constant, k_τ the re-entrainment rate constant, and β an adjustable constant. k_d is represented in the Arrhenius form. C is the aggregate protein concentration in models A, AE, and the denatured protein concentration in models D, DE. Models A and D parameters are taken from Georgiadis & Macchietto (2000) and De-Jong (1997), respectively. $k_\tau = 0$ in models A, D and $k_\tau = 1.3 \ 10^{-3}$ /s in models AE, DE (Fryer & Slater, 1985). The proportionality constant β is fitted to data in each case, as it depends on geometry (e.g. plate corrugations) and operating conditions.

For extension to a whole PHE, a hierarchical model structure was used (Fig. 2). Each channel inherits the single channel fouling model. Channels are linked in a parent PHE model according to its specific arrangement configuration. Connectivity conditions ensure proper connection of temperature, heat flux and flow between adjacent channels. Adiabatic heat flux boundary condition were used for the end plates. A Heating phase is considered, followed by a CIP comprising three phases (Washing with water, Cleaning with hot detergent solution, and Rinsing with water). Heating (the fouling producing phase) is modelled as described above. Mechanical displacement from Washing and Rinsing is very difficult to capture, so a fixed time duration was assumed, with no effects on the deposit. Cleaning was implemented by integrating the model of Bird & Fryer (1992) into the moving boundary model (with deposit removal at each point of a plate as in Eq.2) and validated (Sharma, 2018). We can then simulate a single Heating-CIP cycle, multiple cycles or other operation sequences. The termination of each phase may be based on fixed elapsed time, or some other conditions being achieved (condition-based).

The final DAE system (comprising 16, 15 and 7 equations for the thermal, fouling and CIP models, respectively) was implemented in gPROMS (2018). Differential equations were discretised with a second order backward difference for flow in positive x-direction, second order forward difference for flow in negative x-direction, and a central second order scheme in the y-direction (across deposit), with 20 and 8 differential elements in the x- and y-directions, based on mesh independency tests. For a full Heating-CIP cycle this results after discretisation in 43,485 equations (33,523 for the model + 9,962 initial conditions), with computing times for simulation of ~ 5 minutes.

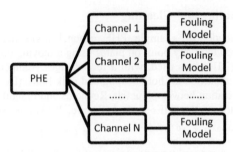

Figure 2. Hierarchical decomposition of solution models. Equipment topology is set in "PHE"

3. Results, Validation and Discussion

Arrangements A1 and A2 were simulated with geometry and operating conditions in Georgiadis & Macchietto (2000), starting from a clean PHE. Figure 3 shows the deposit distribution for A1 after 6.7 hrs of heating. The deposit depth increases along the channel length and is asymmetrical for plates at either side of a channel, due to different plate temperatures. The deposition profiles reflect the calculated bulk protein concentration profiles (not shown). Model responses were validated against experimental fouling data (collected deposit mass in each channel) reported in Georgiadis & Macchietto (2000), with good results. Errors are mostly under 10% for A1 (Fig. 4) and 20% for A2 (not shown). Errors in the initial channels are larger as there is little deposit there due to the low temperatures. There is a very small drop in milk exit temperature by the end of the heating phase (0.1 °C). For A1 (parallel channels), ΔP in a single channel (11) rises from 3.98 kPa when clean to 4.11 kPa. For A2 (parallel-countercurrent channels), ΔP across the PHE goes from 26.6 kPa to 29 kPa. The deposited mass for A1 are better predicted by fouling models A, AE, those for A2 by fouling models D, DE, maybe due to different dependence of deposition on protein aggregation and denaturation with geometry, flow and temperature. The large range in β, used as a fitting parameter by all authors in Fig. 4, confirms that the milk fouling kinetics in the literature warrants further investigation.

A complete Heating-CIP procedure is defined, with a Heating phase with termination condition based on reaching an overall critical deposit mass of 16 g/m² (Georgiadis and Macchietto, 2000), followed by Washing for 3 minutes, then Cleaning until < 0.01 g deposit is on each plate, with a final Rinse phase of 3 minutes. Fig. 5 shows the evolution of deposited mass over two cycles operated according to this procedure, for A2 with fouling model D. After Heating, the deposit mass of 3-6 g on each plate is effectively reduced by CIP to the set low level, the bottleneck being the plate with highest deposit. With a looser termination criterion of 0.1 g of deposit per plate, cleaning times are almost halved (from 0.31h to 0.15h for A1 and 0.30h to 0.15h for A2), decreasing the cycle time and increasing productivity (cycles/day). Care must be taken to ensure this is still safe and acceptable in terms of the milk product quality. With the Cleaning termination condition at the lower, safer level, the Heating time was varied, while maintaining the deposited mass under the critical mass specified. A plot (Fig. 6) of cleaning costs and throughput vs cycle time for arrangement A1 (here with fouling model A, to demonstrate the framework flexibility) shows that the largest productivity is obtained with the longest possible cycle time (7.2 h), as more fouling in the Heating phase is not offset by the longer time required for Cleaning dirtier plates.

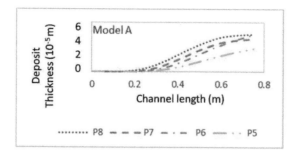

Figure 3. Deposit thickness distribution on A1 plates along channel length after 6.7h of heating

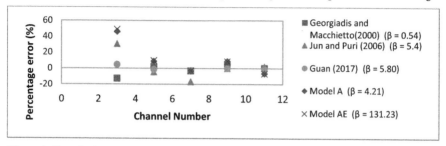

Figure 4. Error in deposited mass in each channel for four milk fouling models (arrangement A1)

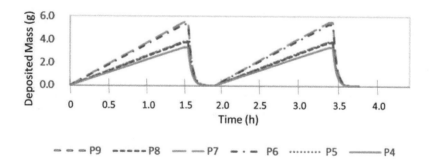

Figure 5. Evolution of deposited mass over two Heating-CIP cycles (A2, fouling model D)

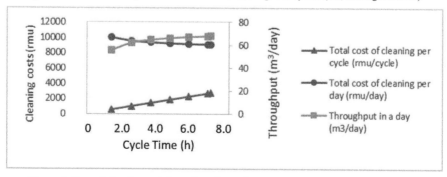

Figure 6. Cost of cleaning and throughput for various cycle times (A1, fouling model A)

4. Conclusions

The novel method presented allows us to easily and flexibly model PHEs of different configurations (e.g. with parallel, countercurrent, mixed parallel-countercurrent flow) and with various fouling, re-entrainement and cleaning models. This was demonstrated for two configurations which were implemented and validated. As far as we are aware, this is the first model describing in detail both fouling deposition and removal in PHEs. Results show that a good match to experimental data is achieved for two PHEs used for milk pasteurisation, in low computation time. It was demonstrated that the model may be used to study and optimise the dynamic operation of a complete PHE, including typical heating and cleaning policies. Its ability to predict deposit distribution on the plates will be also useful for monitoring and control.

References

Bird, M. R. and Fryer, P. J., 1992. An analytical model for cleaning of food process plant.. Food Engineering in a computer climate, ICHEME Symposium Series , Issue 126, pp. 325-330.

Bouvier, L, A. Moreau, G. Ronse, T. Six, J. Petit, G. Delaplace, 2014. A CFD model as a tool to simulate β-lactoglobulin heat-induced denaturation and aggregation in a plate heat exchanger, *Journal of Food Engineering*, 136, 56-63.

Cavero, C., 2013. Análise dinâmica de um processo contínuo de pasteurização em trocadores de calor a placas, PhD Thesis: University of São Paulo.

De Jong, P., 1997. Impact and control of fouling in milk processing. *Trends in Food Science & Technology*. 8 (12), 4011-405.

Diaz-Bejarano, E., Coletti, F. and Macchietto, S., 2016. A new dynamic model of crude oil fouling deposits and its application to the simulation of fouling-cleaning cycles. *AIChE J*. 62 (1), 90-107.

Fryer, P. J. and Slater, N. K. H., 1985. A direct simulation procedure for chemical reaction fouling in heat exchangers. *The Chemical Engineering Journal*, 31 (2) 97-107.

Georgiadis, M. C. and Macchietto, S., 2000. Dynamic modelling and simulation of plate heat exchangers under milk fouling. *Chemical Engineering Science*. 55 (9), 1605-1619.

Georgiadis, M. C., Rotstein, G. E. and Macchietto, S., 1998a. Modelling and simulation of complex plate heat exchanger arrangements under milk fouling. *Comp. Chem. Engnrg*, 22., S331-S338.

Georgiadis, M. C., Rotstein, G. E. and Macchietto, S., 1998b. Optimal design and operation of heat exchanger arrangements under milk fouling. *AIChE J*, 44 (9), 2099-2111.

gPROMS, 2018. Process Systems Enterprise Limited, www.psenterprise.com,

Guan, S. and Macchietto, S., 2018. A Novel Dynamic Model of Plate Heat Exchangers Subject to Fouling. *Computer Aided Chemical Engineering*, Vol 43 Part B, 1679-1684.

Gut, J. A. and Pinto, J. M., 2003. Modeling of plate heat exchangers with generalized configurations. *International Journal of Heat and Mass Transfer*, Volume 46, p. 2571–2585.

Jun, S. and Puri, V. M., 2006. A 2D dynamic model for fouling performance of plate heat exchangers. *Journal of Food Engineering*. 75 (3), 364-374.

Sharma, A., 2018. MSc Thesis, Imperial College London.

Anton A. Kiss, Edwin Zondervan, Richard Lakerveld, Leyla Özkan (Eds.)
Proceedings of the 29th European Symposium on Computer Aided Process Engineering
June 16th to 19th, 2019, Eindhoven, The Netherlands. © 2019 Elsevier B.V. All rights reserved.
http://dx.doi.org/10.1016/B978-0-128-18634-3.50249-6

Satellite based Vegetation Indices variables for Crop Water Footprint Assessment

Haile Woldesellasse[a], Rajesh Govindan[a], Tareq Al-Ansari[a]*

Division of Sustainable Devlopment, College of Science and Engineering, Hamad Bin Khalifa University, Qatar Foundation, Doha, Qatar

*talansari@hbku.edu.qa

Abstract

The global population has quadrupled over the last century. This has increased the global food demand and water sector. In the state of Qatar, the annual freshwater extraction from aquifers is four times the rate of natural recharge, and the depletion is driven by agriculture which represents only 1.6% of the total land area of Qatar, providing approximately 8-10% of domestic food consumption and contributing 0.1% to the domestic GDP. Considering the need for the sustainable intensification of food production systems, satellite technology has the ability to provide a frequent monitoring mechanism enabling the availability of physically-based spatial information useful for reliable environmental monitoring studies. The objective of this research paper is to assess the demand side water footprint of crops using satellite-driven technology in order to optimise the supply side irrigation requirements. The key vegetation indices, such as Normalized Difference Vegetation Index (NDVI) and Normalized Difference Water Index (NDWI) estimated from the remotely acquired information from the Landsat satellite is used for the water footprint assessment. Finally, the water resources demand in agriculture is met by optimising the water supplied from the various decentralized treated sewage effluent plants (TSE). A mixed integer non- linear programming (MINLP) formulation was used to model the spatially-dependent demands, and the TSE plants allocated 80% and 20% of their capacity to fields 1 and 2 respectively. The findings of this paper are promising and a high correlation of -0.93 is found between NDVI and crop water demand, demonstrating that satellite images can be used to monitor the crop vegetation development. vegetation stress can be differentiated using NDVI, thereby demonstrating its applicability for agricultural and water sectors.

Keywords: Crop water footprint, NDVI, NDWI, Satellite remote sensing.

1. Introduction

It is expected that global food production will need to increase to meet the demands of a growing population. Going forward, it is important increase global food produciton within a sustainable development framework which optimises the use of resources and in consideration of the relationship between energy, water and food (EWF) resources (Al-Ansari et al. 2016). Considering food security, and the need to enhance agriculture production, many countries within the Middle East are facing challenges related to the sustainable provision of water required for enhanced agriculture products. Agricultural production is heavily reliant on available water resources, of which excessive exploitation has contributed towards the depletion of groundwater and salt intrusion. The majority of farms in the Arabian Peninsula which is characterised by a hyper arid climate depend on irrigation rather than rain fed Haddadin (2002). Considering water scarcity, there is a need to develop methods like water footprint to adequately monitor water consumption. The water footprint (WF) of a crop is defined as the amount of water consumed for crop production, where blue and green water footprint represents irrigation and rain water usage respectively Moreover, the WF concept can be used to understand international trading of water known as virtual water (Romaguera et al. 2010). Remote sensing has been used within agriculture and hydrology studies to monitor environment conditions. In this study, remote sensing data is used for the assessment of water footprint of crops in a Qatar case study, as an extension of the work previously conducted by the authors (Woldesellasse et al., 2018). It focuses on developing a neural network model which predicts water demand of alfalfa crops over a period of one year. The novelty is demonstrated using macro-scale water demand estimations from satellite imagery as inputs for optimising water supply in the field, linking irrigation systems and decentralized treated sewage effluent (TSE) plants. The study details the location of the plants, the crop water demand (m^3/ha) and the total amount of water is calculated by multiplying the area of the fields with the crop water demand.

2. Vegetation indices for water footprint estimation

The WF is defined as the sum of the volume of fresh water (blue, green and grey) used to produce the product along its supply chain (Hoekstra et al. 2009). A global study conducted between 1995 – 1999 determined the WF of crops based on crop yields and crop water requirement per country (Allen et al. 1998), Hoekstra and Hung (2005). The model assumes an ideal crop condition such as, optimal soil moisture conditions, disease free, single cropping pattern, well-fertilized and grown in large fields with 100% coverage. In a similar study, Liu et al., 2009 provided the blue and green components of water use for different agricultural crops during the period between 1998-2002 and at a spatial resolution of 30 arc minutes. The input datasets for their model includes, maps of harvested area, climatic data from CRU, soil parameters, and crop fertilizer application obtained from statistical reports. The limitations of the study include coarse data resolution, and an optimum ideal soil water conditions is considered in the study.

Remote sensing techniques can be a useful tool to overcome these limitations by providing high spatial, temporal and spectral resolution data using advanced image classification methods. A global assessment of crop WF using remote sensing techniques was conducted by (Romaguera et al. 2010). The study proposed an approach driven by remote sensing techniques for retrieving the global actual irrigation using evapotranspiration, precipitation, and surface runoff as an input dataset. The Satellite instruments used for estimating these parameters include Meteosat and Climatic Prediction Center Morphing Technique (CMORPH) simultaneously. In agriculture, spectral information is used to characterise the biophysical features of plants using vegetation indices. Vegetation indices are a unitless radiometric measures which are calculated by using two or more bands in the wavelength range of VIS, NIR and SWIR. (Rouse et al., 1974) proposed the most commonly used index known as Normalized Difference Vegetation Index (NDVI). It is calculated as the ratio of the reflectance difference and sum of the NIR and Red bands. Green plants strongly absorb the blue and red bands because of the chlorophyll and simultaneously exhibited a strong reflectance in the NIR region. The value of NDVI ranges between -1 to 1, where below zero values indicate clouds and water, the positive values close to zero indicate bare soil, and higher positive values indicate vegetation. Sparse vegetation ranges from 0.1 to 0.5, and dense vegetation is illustrated by the value starting 0.6 and above (Stancalie et al. 2014).

The NDVI index is used for the application of many vegetative studies, including crop yield estimation, drought monitoring, monitoring of crop conditions, their developmental stage and biomass estimation (Stancalie et al. 2014). Thus, the information provided by different satellite instruments contribute to effective monitoring and management of agriculture, grasslands, forestry and above all help to understand the way the ecosystem works and its interactions with the atmosphere and human activities. Similarly, Normalized Difference Water Index (NDWI) can be used to provide an indication of the water content in leaves, through monitoring the humidity of the vegetation cover. The value of NDWI ranges from -1 to 1, with the green vegetation lies in the range of -0.1 to 0.4 (Gu et al. 2007). In retrieving the vegetation water content information, the combination of the two spectral bands eliminates the variation in information due to the leaf internal structure and dry matter content.

3. Methodology

The first part of the methodology consists of analysing satellite images acquired from Landsat 7 for the year 2002. NDVI and NDWI are calculated to assess the crop vegetation and water stress of crop over the given time period. NDVI is calculated using the Red and Near Infrared channels, which are Band 3 and 4 respectively using the GIS tool. The vegetation phase displays a seasonal pattern in terms of WF which depends on its water intake and temperature. Under ideal condition, where there is an abundance of water available within minimal evapotranspiration (ET), crops display green vegetation characteristics. Therefore, NDVI and NDWI are high because the corresponding water content of the crop is high. Alternatively, high temperature increases the ET and crop water stress, thereby lowering the NDVI and NDWI. The correlation of NDVI with crop water demand (m^3/ha) and temperature is also assessed to check the interaction between them. The total water demand of the fields, D, is then calculated by multiplying the crop water demand by the area of the fields. This study considers the optimisation of water supplied from the treated sewage effluent (TSE) plants to meet the water demand of the alfalfa fields, D, considered in (Woldesellasse, Govindan and Al-Ansari 2018). The water is pumped from the TSE plants and is stored at optimum locations (x,y) prior to its distribution to the alfalfa fields. The objective of the nonlinear optimisation is to minimize the cost of water distribution. The Non-Linear Programming sequence used to allocate the water irrigated to the fields is detailed below:

Constraint:

$q_{11} + q_{12} = 1$

$q_{21} + q_{22} = 1$

$C_1.q_{11} + C_2.q_{21} \geq D_1$

$C_1.q_{12} + C_2.q_{22} \geq D_2$

$q_{11}, q_{12}, q_{21}, q_{22} \geq 0$

Objectives - Minimise:

$(C_1.q_{11}.r_{11}.Cost_{11} + C_2.q_{21}.r_{21}.Cost_{21} + C_1.q_{12}.r_{12}.Cost_{12} + C_2.q_{22}.r_{22}.Cost_{22})$

Where , C_i is the capacity of the two TSE, D_i is the demand of the two fields, q_{ij} indicates the fraction of water transported from source plants to alfalfa fields, $Cost_{ij}$ is the cost of water distribution, and r_{ij} is the length of transportation from the TSE to the alfalfa fields.

4. Results and Discussion

The state of a crop is monitored using a timeseries measurement of NDVI. Figure 2 illustrates the change in colour of the circular fields throughout the year, whilst surrounding areas remain almost constant. The values range from 0 to 1, indicated by the colour bar ranging from dark blue to red. Winter is the period where the temperature is cool in arid regions such as Qatar, and harvesting is most of the time during this period. Generally, the winter months demonstrate a high NDVI compared to the summer months. The highest NDVI recorded is 0.646 in January and lowest 0.194 in June.

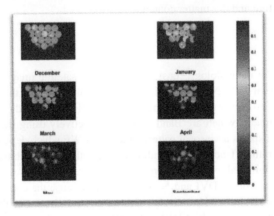

Figure 2. Normalized Difference Vegetation Index (NDVI).

Furthermore, the field images illustrated in Figure 2 can be better interpreted through a comparison with Figure 3, where the NDVI is plotted against temperature and crop water demand. The objective is to ascertain the correlation between vegetation indices and the crop water demand, and to ultimately utilise satellite information to allocate water to agricultural fields. In Figure's 3a and 2b, NDVI displays an inverse relationship with temperature and crop water demand. This is due to the fact that the greenness of a crop is affected by its water stress. Since precipitation is sporadic in arid regions, agricultural fields are irrigated using water from underground or TSE. The analysis clearly illustrates the effect of hot temperature and low precipitation on agriculture, in which the NDVI fluctuates between high to low scores during the low and high temperature respectively. However, it should be noted that a low NDVI does not always imply water stress, as the field can also be a just harvested bare ground. NDWI is an indicator of the water content in the leaves of crops. During dry season, the vegetation state is affected due to water stress, and vegetation index, i.e. NDWI contributes towards monitoring of droughts as the two bands used for calculation are responsive to changes in water content. In this study the NDWI is calculated for the same area using the data acquired from Landsat satellite. The result is illustrated in Figure 4, where NDWI behaves similar to NDVI as it is responsive to the crop water demand and temperature, and thereby can be used as a good indicator of water footprint.

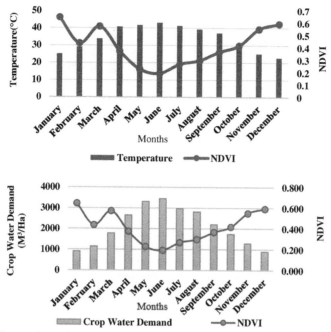

Figure 3. Comparison of (a) NDVI and temperature, b) NDVI and Crop Water Demand.

In general NDWI is high throughout the year except, during the winter-summer transition months or summer months where temperature is very high, and where ET exceeds the normal range. Therefore, during these periods, NDWI falls below 0.2 as illustrated in Figure 4. Overall, the results demonstrate consistency with the previous study done. A high correlation of -0.93 is determined between NDVI and the crop water demand, from which it can be concluded that remote sensing instruments can be used for agricultural applications, including for the assessment of the crop water demand. Considering the optimisation framework applied in this study, the TSE plants (Doha West and Doha South) with a capacity of 7.37×10^6 m^3 and 7.20×10^6 m^3 respectively. Outcomes of the study conclude that 80% of the water production in each plant is distributed to field 1, the remaining 20% of the water capacity in each plant is distributed to field 2.

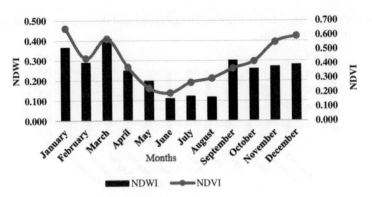

Figure 4. Comparison of NDVI and NDWI.

5. Conclusions

In this study, the correlation between vegetation indices and crop WF is assessed and an inverse relationship is found. This is a good indicator of vegetation phase. The results are further confirmed by considering both the NDVI and NDWI, and their interactions with temperature and ultimately the WF. Finally, the study concludes that satellite images can be used as input data to allocate water irrigated to agricultural fields.

References

Al-Ansari, T., A. Korre, Z. Nie & N. Shah. 2016. Integration of biomass gasification and CO2 capture in the LCA model for the energy, water and food nexus. In *Computer Aided Chemical Engineering*, 2085-2090. Elsevier.

Allen, R. G., L. S. Pereira, D. Raes & M. Smith (1998) FAO Irrigation and drainage paper No. 56. *Rome: Food and Agriculture Organization of the United Nations,* 56, e156.

Gu, Y., J. F. Brown, J. P. Verdin & B. Wardlow (2007) A five-year analysis of MODIS NDVI and NDWI for grassland drought assessment over the central Great Plains of the United States. *Geophysical Research Letters,* 34.

Haddadin, M. J. (2002) Water issues in the Middle East challenges and opportunities. *Water Policy,* 4, 205-222.

Hoekstra, A. Y., A. K. Chapagain, M. M. Aldaya & M. M. Mekonnen (2009) Water footprint manual. *State of the Art,* 1-131.

Hoekstra, A. Y. & P. Q. Hung (2005) Globalisation of water resources: international virtual water flows in relation to crop trade. *Global environmental change,* 15, 45-56.

Romaguera, M., A. Y. Hoekstra, Z. Su, M. S. Krol & M. S. Salama (2010) Potential of using remote sensing techniques for global assessment of water footprint of crops. *Remote Sensing,* 2, 1177-1196.

Stancalie, G., A. Nertan, L. Toulios & M. Spiliotopoulos. 2014. Potential of using satellite based vegetation indices and biophysical variables for the assessment of the water footprint of crops. In *Second International Conference on Remote Sensing and Geoinformation of the Environment (RSCy2014),* 92290K. International Society for Optics and Photonics.

Woldesellasse, H., R. Govindan & T. Al-Ansari. 2018. Role of analytics within the energy, water and food nexus–An Alfalfa case study. In *Computer Aided Chemical Engineering*, 997-1002. Elsevier.

Anton A. Kiss, Edwin Zondervan, Richard Lakerveld, Leyla Özkan (Eds.)
Proceedings of the 29th European Symposium on Computer Aided Process Engineering
June 16th to 19th, 2019, Eindhoven, The Netherlands. © 2019 Elsevier B.V. All rights reserved.
http://dx.doi.org/10.1016/B978-0-128-18634-3.50250-2

Life Cycle Assessment of Petroleum Coke Gasification to Fischer-Tropsch Diesel

Ikenna J. Okeke, Thomas A. Adams II*

Department of Chemical Engineering, McMaster University, 1280 Main St. W, Hamilton, ON, Canada, L8S 4L8.
tadams@mcmaster.ca

Abstract

This work presents a novel study on the life cycle assessment of Fischer-Tropsch diesel (FTD) production via petroleum coke (petcoke) gasification with and without carbon capture and sequestration (CCS). A detailed analysis which focuses on evaluating and quantifying the well-to-wheel (WTW) environmental impacts of converting petcoke to FTD and its subsequent combustion is discussed. The overall process inventory includes mass and energy balances data from our Aspen Plus model, the US Life Cycle Inventory Database, and the GREET model. Life cycle impact assessment (LCIA) categories of the petcoke-derived diesel (PDD) were calculated using TRACI 2.1 v1.04 in SimaPro 8.5.0.0 which is compared against both conventional petroleum and oil-sands derived diesel. In addition, two different plant locations in Canada (Alberta and Ontario) which respectively have very high and low electricity grid emissions were considered as they are representative of grid emissions across the world. Results of the analysis showed an overall reduction in fossil fuel depletion of up to 80% and 83% for Alberta and Ontario respectively when compared to the fossil derived diesel. When operated with CCS, the WTW GHG emissions were 7% lower than conventional petroleum diesel and 18% lower than oil-sands diesel for the plant located in Ontario. However, when located in Alberta, the WTW GHG emissions were 49% higher than conventional petroleum diesel and 30% higher than oil-sands diesel. This is due to the significant electricity requirement for the PDD process and the vast differences in carbon intensities of the Ontario and Alberta grids. Overall, the other impact categories for the PDD process showed to be significantly lower than the petroleum and oil-sand derived diesel.

Keywords: Petroleum coke, Fischer-Tropsch, Life cycle assessment, CO_2 Capture.

1. Introduction

Petroleum coke ("petcoke") is regularly produced as a by-product of crude oil refining, especially from heavy crudes commonly produced in Canada. While some forms of petcoke are useful for steel making or other speciality purposes, it is often stockpiled in large quantities as a waste product. Although the stockpiled petcoke can be combusted for heat or electricity generation, it is more carbon intensive than coal, and this along with other environmental concerns makes combustion undesirable or even prohibited by government regulation. Hence, instead of the traditional stockpiling practice, petcoke can also be converted to liquid fuels both as a means of disposal and to help meet increasing fuels demand.

Recent studies have shown the technical and economic feasibility and the high fuel, energy, and carbon efficiencies of converting petcoke to chemicals and fuels (Salkuyeh et al. 2015, Okeke et al. 2018). Although several studies have presented the life cycle assessment (LCA) of other synthetic fuels processes such as gas-to-liquids (Forman et al. 2011), coal-to-liquids (Jaramillo et al. 2009), and biomass-to-liquids (Xie et al. 2011) processes, to the knowledge of the authors, no work has reported the LCA study of petcoke conversion to liquid fuels. Therefore, this study presents the first-of-a-kind detailed cradle-to-grave LCA of petcoke gasification to FTD so as to ascertain the environmental feasibility of this petcoke disposal approach.

2. Life cycle methodology and process description

2.1. Goal, scope, and boundaries

The goal of the LCA study is to evaluate the environmental impacts associated with the production of FTD from petcoke gasification operating with or without CCS. The scope covered in this analysis includes all the direct and indirect material and energy inputs with its associated output and emissions in the entire life cycle of the PDD process. The energy allocation method was employed for the distribution of emissions between diesel and gasoline fuels produced. Overall, the WTW system boundaries comprise of the cradle to plant exit gate emissions, the diesel fuel transportation and distribution (T&D), and the diesel combustion in a compression ignition direct injection (CIDI) vehicle as shown in Figure 1. The WTW inventory was normalized to 1 MJ$_{PDD}$ combusted which is compared against conventional petroleum and oil-sands derived diesel (GREET, 2017). The impact categories considered were: ozone depletion potential (ODP), global warming potential (GWP), smog formation (SF), acidification potential (AP), eutrophication potential (EP), carcinogenic potential (CP), non-carcinogenic potential (NCP), respiratory effects (RE), ecotoxicity potential (ETP), and fossil fuel depletion (FFD).

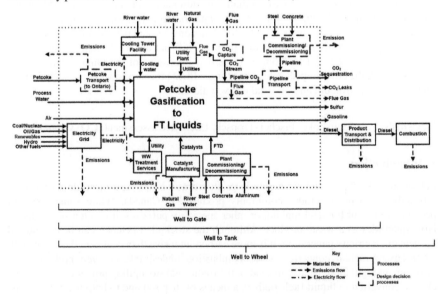

Figure 1: Complete WTW system boundary of the PSG designs studied

2.2. Process description

Figure 1 shows the system boundary for the WTW LCA of the PDD process which is based on our prior design (Okeke et al. 2018). Depending on the case, stockpiled petcoke at the refinery yard in Fort McMurry, Alberta is either directly sent to a petcoke standalone gasification (PSG) plant nearby, or, transported by train 3114 km to Sarnia, Ontario. The petcoke undergoes pre-processing prior to being fed to the petcoke E-gas gasifier operated with 99.5% pure oxygen from an air separation unit. Syngas is produced at 1426°C and 56 bar with the syngas cooling provided by boiler feed water sent through the radiant tubes, thereby, producing high-pressure steam (HPS).

The gasifier is designed for a carbon conversion of 99% (Amick 2000) while slag is collected which consists of unconverted carbon and trace metals such as nickel and vanadium (Basu 2006). An adiabatic water-quench saturation temperature of 200°C is employed to remove the entrained slag in the syngas while ammonia is removed from the syngas based on its high solubility in water to tolerable limits in order to avoid downstream catalyst deactivation (Pendyala et al. 2017). Water-gas shift and COS hydrolysis are employed to raise the H_2/CO ratio and convert COS to H_2S respectively.

Acid gases such as H_2S and CO_2 are removed from the syngas using the pre-combustion based MDEA solvent. The two-stage removal process is employed: H_2S removal in an H_2S absorber and CO_2 capture (for the CCS design) in a CO_2 absorber enhanced with piperazine. The absorbed H_2S is sent to the Claus unit for sulfur production while the captured CO_2 is compressed to 153 bar and sent via CO_2 pipeline to sequestration site. The minimum energy requirements for the overall system was determined by performing a heat exchanger network design (Okeke et al. 2018) of which all the required utilities and its associated emissions were considered.

FTD is produced by converting the syngas in a cobalt aluminium-based catalyst filled slurry reactor operating at 240°C and 30 bar. The slurry reactor design configuration consists of a once-through conversion of 80% CO and a chain growth probability, α, of 0.92. Up to 90% of the unconverted syngas was recycled to the FT reactor while the other portion is combusted primarily to produce HPS needed in the plant and the remaining HPS is sold. The boiler flue gas is emitted to the atmosphere which accounts for the direct process emission. The mass and energy balance results for the petcoke-to-liquids portion of the life cycle were determined and described in the prior work (Okeke et al. 2018) and incorporated within the present work. The system electricity demand is supplied from the local electricity grid, with the environmental impacts of the Alberta and Ontario grids considered (Environment and Climate Change Canada 2018).

Transportation of the PDD from the plant location to a bulk terminal for subsequent distribution to refueling stations was carried out using a combination of barge, pipeline, and rail for the transportation and truck for distribution based on the GREET model data (GREET, 2017). The subsequent combustion emissions of 1 MJ of PDD is evaluated in a CIDI vehicle with an assumed fuel consumption of 27.5 miles per gallon. In this analysis, we considered four cases, which were PSG plants with and without CCS, located either in Alberta or Ontario.

3. Results and discussion

Table 1 shows the plant gate-to-gate (GTG) process life cycle inventory of the PSG design with and without CCS. Both PSG processes were designed to consume the same amount of petcoke per FTD produced. However, a 27% increase in electrical consumption for the plant with CCS is observed due to the parasitic energy load of the CCS technology compared to the design without CCS.

Table 1. The total GTG life cycle inventory for the petcoke to diesel process.

Inventory	PSG-CCS	PSG
Input flows (kg)		
Petcoke	0.091	0.091
Petcoke transport, by train (t-km)	0.283	0.283
Catalyst, aluminium-based	6.11×10^{-5}	6.11×10^{-5}
Air	0.483	0.483
Process water	0.148	0.143
Cooling water, cooling tower	35.677	28.518
Waste water treatment (m^3)	1.81×10^{-4}	1.77×10^{-4}
LP steam generation (GJ)	5.67×10^{-4}	2.86×10^{-4}
MP steam generation (GJ)	-2.83×10^{-6}	-2.83×10^{-6}
HP steam generation (GJ)	-3.65×10^{-4}	-3.65×10^{-4}
Fired heat (GJ)	9.55×10^{-6}	9.55×10^{-6}
Electricity consumed (MW)	9.98×10^{-4}	7.88×10^{-5}
Output flows (kg)		
Products flow (MJ)		
Diesel	1	1
Gasoline	0.371	0.371
Sulfur (kg)	5.19×10^{-3}	5.2×10^{-3}
Sequestered CO_2 (kg)	0.177	0
Emissions flow (kg)		
Carbon monoxide	0	8.06×10^{-5}
Carbon dioxide	7.09×10^{-3}	0.186
Water (g)	1.57×10^{-2}	1.65×10^{-2}
Argon	1.51×10^{-3}	1.52×10^{-3}
Nitrogen	6.32×10^{-2}	6.33×10^{-2}
Nitrogen oxides	8.92×10^{-6}	8.92×10^{-6}

Similarly, there is an approximately 50% and 20% reduction in heating and cooling water requirements when the plant is operated with no CCS. Comparing the direct emissions of both plants, the GHG emissions for the design with CCS were 71 tCO_2eq/MJ_{PPD} while the design without CCS emits up to 186 tCO_2eq/MJ_{PPD}. This amounts to a 162% increase in GWP and defeats the overall goal of alternative sources of fuel with reduced environmental impacts. Therefore, it can be concluded that it is not environmentally desirable to dispose petcoke by converting it to diesel without using CCS. The complete WTW environmental impacts of the PDD plants operating with or without CCS for the 2 locations studied compared against the conventional petroleum and oil-sands derived diesel with the oil-sands diesel as the reference process is shown in Figure 2.

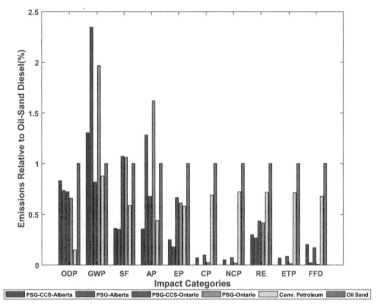

Figure 2:WTW life cycle impact categories relative to oil-sands derived diesel. The "Alberta" cases are cases where the PSG plant is located at the source of the petcoke stockpile. The "Ontario" cases indicate that the stockpiled petcoke is transported to Ontario for conversion. "Conv. Petroleum" is for diesel fuel from conventional oil.

Analyzing the performance of the plants when located in Alberta, it can be observed from Figure 2 that the GWP for the PSG plant with CCS is 49% and 30% higher than the petroleum and oil-sands derived diesel processes respectively. Such a huge GHG emission above the reference processes is due to the high carbon-intensive electricity grid in Alberta. Without CCS, the GHG emissions rises as high as 167% and 135% above the petroleum and oil-sands derived diesel respectively. Comparing the GWP of the plants when located in Alberta and Ontario, there is a significant 37% and 16% reduction in GHG emissions for the Ontario plants with and without CCS respectively. Furthermore, this Ontario plant showed up to 7% and 18% reductions in GHG emissions when operated with CCS compared to the conventional and oil-sands derived diesel plants respectively. When the economics (Okeke et al. 2018) and GHG emissions reductions of the PSG design operated with CCS are considered, there is an incentive in constructing the plant. Although the GWP is higher in Alberta, its SF, AP, EP, RE, and ETP are 66%, 48%, 63%, 32%, and 20% lower than that of the Ontario plant respectively. These increased impacts are linked to the train transit emissions during petcoke transportation to Ontario. It is noteworthy to mention that the impacts of ODP, CP, and NCP are not significant but were presented for completeness purposes and thus, not discussed. Finally, FFD at Alberta is 19% higher than in Ontario because of Alberta's grid high dependence on fossil energy. However, its FFD is still 70% and 80% lower than that of the conventional petroleum and oil-sands derived diesel processes respectively.

4. Conclusions

A novel WTW study on the LCA of PDD operating with or without carbon capture at different locations in Canada was presented and compared against conventional petroleum and oil-sands derived diesel. It was found that the GHG emissions of the PSG with CCS plant at Alberta and Ontario were 137 and 86 tCO$_2$eq/MJ$_{PPD}$ respectively. This implies that the plant can compete favorably with the status quo refinery GHG emission if its sited in locations with grid emissions like that of Ontario. Similarly, it is generally preferable to transport petcoke from Alberta to Ontario to take advantage of its lower grid carbon intensity than to gasify petcoke at the source in Alberta. Also, the extent of FFD for the design at Alberta and Ontario are between 70-80% and 75-83% lower compared to the conventional and oil-sands derived diesel respectively. Overall, it can be concluded that disposing petcoke via diesel production is an avenue to be considered when GHG emissions and FFD are a concern.

Although this work focused on the WTW LCA of petcoke only gasification operated with or without CCS, in our future work, we will explore the WTW environmental impacts of other designs such as combining petcoke gasification and natural gas reforming to FTD so as to ascertain the technology which has the most reduced impact on the eco-system.

5. Acknowledgments

Support for this research is made possible through funding from an NSERC Discovery grant (RGPIN-2016-06310).

References

P. Amick, (2000), Gasification of petcoke using the e-gas technology at wabash river. 2000 Gasification Conference, Houston, Texas.

P. Basu, (2006), Combustion and gasification in fluidized beds, CRC press.

Environment and Climate Change Canada, (2018). "E-Tables-Electricity-Canada-Provinces-Territories." Retrieved 30-06-2018, from http://data.ec.gc.ca/data/substances/monitor/national-and-provincial-territorial-greenhouse-gas-emission-tables/E-Tables-Electricity-Canada-Provinces-Territories/?lang=en.

G. S. Forman, T. E. Hahn, and S. D. Jensen, (2011), "Greenhouse gas emission evaluation of the GTL pathway." Environmental science & technology **45**(20): 9084-9092.

Jaramillo, P., et al. (2009). "Greenhouse gas implications of using coal for transportation: Life cycle assessment of coal-to-liquids, plug-in hybrids, and hydrogen pathways." Energy Policy **37**(7): 2689-2695.

I. J. Okeke and T. A. Adams II (2018), "Combining petroleum coke and natural gas for efficient liquid fuels production." Energy, 163, 426-442.

V. R. R.. Pendyala, W. D. Shafer, G. Jacobs, M. Martinelli, D. E. Sparks, and B. H. Davis, (2017), "Fischer–Tropsch synthesis: effect of ammonia on product selectivities for a Pt promoted Co/alumina catalyst." RSC Advances **7**(13): 7793-7800.

Salkuyeh, Y. K. and T. A. Adams II (2015), "Integrated petroleum coke and natural gas polygeneration process with zero carbon emissions." Energy **91**: 479-490.

X. Xie, M. Wang, and J. Han, (2011), "Assessment of fuel-cycle energy use and greenhouse gas emissions for Fischer– Tropsch diesel from coal and cellulosic biomass." Environmental science & technology **45**(7): 3047-3053.

GREET 2017, The Greenhouse Gases, Regulated Emissions, and Energy Use in Transportation Model. (available at: http://greet.es.anl.gov/)

Anton A. Kiss, Edwin Zondervan, Richard Lakerveld, Leyla Özkan (Eds.)
Proceedings of the 29th European Symposium on Computer Aided Process Engineering
June 16th to 19th, 2019, Eindhoven, The Netherlands. © 2019 Elsevier B.V. All rights reserved.
http://dx.doi.org/10.1016/B978-0-128-18634-3.50251-4

A Systematic Parameter Study on Film Freeze Concentration

Jan-Eise Vuist[a], Maarten Schutyser[a,*] and Remko Boom[a]

[a]*Food Process Engineering Wageningen University, P.O. Box 17, 6700 AA, Wageningen, The Netherlands*
maarten.schutyser@wur.nl

Abstract

Film freeze concentration is an alternative method to concentrate aqueous streams compared to suspension freeze concentration. Major advantage is that the equipment is less complex and thus capital costs are in principle lower. In our research we investigated especially how hydrodynamics, applied freezing temperatures and solution properties influence inclusion of solutes in ice and ice yield during film freeze concentration. For this we carried out both lab-scale experiments and CFD simulations. Model solutions of sucrose and maltodextrin were concentrated in a stirred vessel by growth of an ice layer at the bottom freezing plate. For varying stirring speeds, feed concentrations and freezing plate temperature profiles we determined the solute inclusion in the grown ice and the ice yield. When increasing stirrer speeds a decreasing amount of solute included in ice was found at constant freezing plate temperature. This can be explained because the transport of the solute molecules in the boundary layer is diffusion limited. An increase in shear above the surface reduces the thickness of this layer and therefore less solute is included in ice at high shear rates. CFD simulations were carried out to describe the hydrodynamics near the surface and to relate the shear rate to the impeller Reynolds number. Moreover, the CFD simulations could explain the increased solute inclusions for higher concentrations of sucrose as higher viscosities lead to significant reduction of shear rates close to the ice layer. The CFD simulations will facilitate easier translation of the obtained results for a differently designed film freeze concentration system. Sucrose and maltodextrin appeared to behave very similar with respect to inclusion behaviour, which may be explained from their similar diffusivities. Ice growth rate is found another important factor and is very much influence by applied freezing temperatures. Our experiments showed that there is a critical ice growth rate. If this ice growth rate is exceeded more solutes will be included in the ice layer. In this case the solute molecules will not have the chance to move away from the ice boundary. The next step in our research is modelling of the ice growth rate as function of the freezing plate temperature to optimise both ice yield and solute inclusion.

Keywords: freeze concentration, solute inclusion, modelling

1. Introduction

Concentration of aqueous food streams is a routine operation in food industry to increase their shelf-life and to reduce their transportation volume. To concentrate these streams multiple techniques are available based on evaporation, physical separation or freeze concentration. While thermal evaporation is the most common industrial technique and is readily applicable to many streams, such as dairy or sugar, it has drawbacks for streams containing heat-sensitive components as reviewed by Sánchez et al. (2009).

Heat-sensitive components such as proteins and flavours either become damaged by the heat-treatment or may be stripped away with the water from the product stream. Both phenomena result in a loss of product quality. During freeze concentration, these components are exposed only to low temperature and therefore minimally damaged. Because freeze concentration produces minimal thermal damage, it is considered a mild concentration technique.

Freeze concentration can be performed in two main modes: suspension freeze concentration and film freeze concentration. During suspension freeze concentration the ice is formed in suspension. The ice growth and the degree of supercooling are maintained by continuously seeding the suspension with fresh crystals. The ice crystals are then washed and melted in a continuous washing filter. This technique is commercially available and has applications in the fruit juice and beer industry. During film freeze concentration the ice crystal(s) are grown on the wall of a heat exchanger. Growing the ice crystals on the wall eliminates the need for a washing filter and reduces the complexity of the unit operation. This article will focus on the film freeze concentration process.

The objective of this study is to study the relationship between inclusion behaviour of solutes and the hydrodynamic conditions near the growing ice layer during film freeze concentration. Sucrose and maltodextrin are selected as the model solute components. To study their inclusion behaviour, a custom build freeze cell is used. Previous studies by Miyawaki et al. (1998) have often used progressive freeze concentration set-ups, which are optimized for studying solute partitioning behaviour at the ice water interface. In this study we chose for a different set-up that provides more realistic hydrodynamic conditions. This is especially relevant for scaling-up the film freeze concentration process. Previous studies by Jusoh et al. (2009) already indicated that hydrodynamic conditions in the liquid are relevant to inclusion behaviour. To gain more insight into this, we used a computational fluid dynamics (CFD) model to describe the hydrodynamic conditions near the ice-liquid boundary. The outcomes of this model are linked to the observed inclusion behaviour during the freeze concentration experiments.

2. Material and Methods

2.1. Experimental

The film freeze concentration experiments were conducted in a small-scale test apparatus designed after Genceli et al. (2007). The test apparatus was constructed from an acrylic cylinder with 90 mm internal diameter. The bottom of the cylinder was constructed out of a stainless steel chamber, which was cooled using a cooling fluid. The vessel was equipped with a pitched two-bladed stirrer. To monitor the temperature during the experiments, we placed thermocouples in the vessel at 1, 2, and 5 mm from the bottom surface, one thermocouple at 10 mm below the liquid surface and two thermocouples were placed in the in- and outgoing liquid flow. All thermocouples (type T) were connected to 24-bit data logger (National Instruments, cDAQ-9214, ±0.01 °C relative accuracy, USA).

Sucrose and maltodextrin solutions with various concentrations were used as feed solutions. The sucrose (342.30 g mol^{-1}) was obtained from Sigma-Aldrich (BioXtra, >99.5 %) and the maltodextrin DE12 ($\approx$ 3423.0 g mol^{-1}) from Roquette Freres (Glucidex 12). The feed solutions were pre-cooled to 4 °C before use. To determine the sucrose concentration in the sample, we used a refractometer (Anton Paar, Abbemat 500, Austria). To determine the maltodextrin concentration, we heated the pre-weighed solutions overnight in an oven at 105 °C and weighed the samples afterwards. The solutions were reused for a maximum of 5 days and separate solutions were used for experiments on the same day.

At the start of the freeze concentration experiments, a droplet (100 μL) of pure water was deposited on the freeze plate. This ice droplet prevented the initial super cooling. Initial supercooling will cause fast initial freezing and very strong inclusion of solutes at the beginning of the freezing process and can even lead to ice crystals forming in suspension rather than on the freeze plate. When the droplet was completely frozen, the feed solution was introduced through a filling port at the top of the cylinder. Only when the liquid submerged the stirrer blades, the stirring was started to prevent major inclusions of air in the solution. The experiments were stopped after a half hour. At the end of the experiment the liquid was removed from the cylinder and the ice was melted. Both fractions were weighted and their contents analysed. All experiments were carried out in duplicate.

2.2. Modelling

A COMSOL Multiphysics 5.3a model was used to estimate the fluid flow induced by the stirrer near the bottom of the cylinder. The fluid flow is described by the Navier-Stokes equation (Batchelor (2000)). The model used a geometry similar to the experimental set-up. For the outer walls and the bottom walls of the cylinder a no-slip boundary condition was applied. A symmetry boundary condition was applied to the fluid surface. The behaviour at the fluid surface is assumed to have no influence on the ice formation. To simulate the stirring a rotating material coordinate system was used. The wall condition for the stirrer is equal to the rotational speed of the material coordinate system and the wall condition for the outside and bottom wall the rotational speed is equal to 0.

The system was simulated for stirring rates of 50 and 150 rpm (Re = 703 to 9845) with laminar flow model and a turbulent $\kappa - \varepsilon$ model was used for the simulation of 500 rpm (Re = 7030, 20674, 32818 respectively for 36, 18, 6 wt% solutions). For the wall treatment the built-in wall functions of COMSOL Multiphysics 5.3a were used. The mesh is based on the physics induced sequence with the fine settings.

Three sugar solutions were simulated at their freezing points. The density of the sucrose solutions was estimated by calculating mass fraction average. The viscosity of the solution was measured with a rheometer (Anton Paar MCR-301, geometry: DG26.7/200/AL, Austria) using a shear rate sweep in the range of 1-100 s^{-1}. The freezing point of the solutions was determined using the melting curve from differential scanning calorimetry (TA Instruments DSC-250, USA). The variables are summarised in table 1. These values are similar to the ones observed by Bubnick and Werner (1995).

3. Results

3.1. Ice formation

Table 1: Used constants for CFD simulations

Solution [wt% Sucrose]	ρ [kg m^{-3}]	η [mPa s]	T_{fp} [°C]
6	1021	2.1	-0.3
18	1072	3.6	-1.9
36	1156	11.1	-4.5

The ice yield of film freeze concentration was compared for different stirrer rates and cold wall temperatures. The ice yield showed little dependence on the stirrer rate. A clear dependence was found on the cold wall temperature. The lower temperatures yielded more ice because the driving force for heat transfer is higher. There was a slight dependence on the solution concentration due to the freezing point depression at higher sucrose concentrations (figure 1).

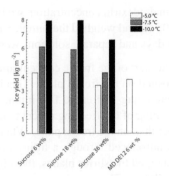

3.2. Sucrose and Maltodextrin inclusion

The effective solute inclusion (equation 1) versus the Reynolds numbers were compared for the same conditions as in figure 1 (figure 2). The solute inclusion decreased at increasing Reynolds numbers for all different solutions and cold wall temperatures. No major differences between the solutions could be found at Reynolds number below 10,000, the laminar regime. At Reynolds numbers above 10,000, differences were clear between the different solutions and the different cold wall temperatures. The 18 wt% sucrose solutions showed more inclusion than the 6 wt% sucrose solutions. The 6 wt% maltodextrin solution showed a similar behaviour to the 6 wt% sucrose solution. A lower cold wall temperature increased the amount of inclusions due to an increased freezing rate.

Figure 1: Effect of applied freeze plate temperature and solute concentration on the ice formation after freezing for a half hour. Ice yield for sucrose and maltodextrin DE12 solution averaged over the stirrer speeds.

$$K[-] = \frac{C_S[wt\%]}{C_L[wt\%]} \tag{1}$$

An increase of the stirrer speed increases the shear above the ice surface which improves the transport of the solute away from the ice-liquid boundary and therefore reduces the diffusion limitation of the freeze concentration. The solute inclusion will increase at lower cold wall temperatures. For a constant cold wall temperature, the freezing initially proceeds much faster than later. This implies that the inclusion rate in the initial stages is larger than later. Thus, by adjusting the cold wall temperature such, that it starts not too cold and gradually cools further, we may avoid the initial large amount of inclusions.

If one compares the freezing of solutions with different solutes, one observes that solutions with maltodextrin show less inclusion. We expect that this is because of the larger

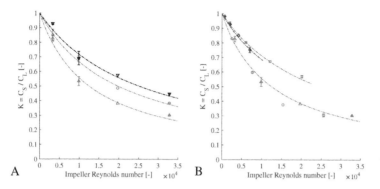

Figure 2: Effect of impeller Reynolds number on the effective solute inclusion.
A.) 6 wt% sucrose at △ -5.0 °C; ◯ -7.5°C; ▽ -10.0°C below the freezing point of the solution.
B.) △, 6 wt% sucrose; ☐, 18 wt% sucrose; ◇, 36 wt% sucrose;
◯ 6 wt% maltodextrin DE12 at -5.0 °C.
The error bars indicate standard deviation. The lines fitted to guide the eye.

molecular weight, giving rise to less melting point depression, hence a larger equilibrium concentration at a given freezing temperature, a stronger concentration gradient towards the solution and therefore faster diffusion from the freezing surface.

3.3. Fluid flow modelling

The shear rates present at the bottom of the cylindrical device was derived from the CFD simulations. The bottom surface, was taken from the result (figure 3). From this the average shear rate was calculated and compared to the Reynolds number (Re = $\frac{\rho N D^2}{\mu}$), with ρ in $kg \cdot m^{-3}$, N in s^{-1}, D in m, and μ in $Pa \cdot s$. The shear rate increased with increasing Reynolds number. Higher shear rates near the surface leads to a reduction of the thickness of the boundary layer. A thinner boundary layer will aid the transport from the solute away from the ice-water boundary and thus will lead to a solute inclusion close to the equilibrium between the ice and the solution. The diffusion through the boundary layer is the rate-limiting step in the freeze concentration process. The simulations show that there is a correlation between the surface shear rate and the impeller Reynolds number found for the vessel. This correlation is specific for our geometry. This

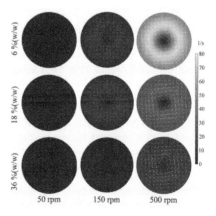

Figure 3: Shear rate magnitude in the freeze cell, from simulation. The arrows indicate the direction of the shear.

correlation is a useful tool for translation of the found results to a pilot scale set up.

Higher viscosities of the sucrose solutions will diminish the intensity of the fluid flow near the ice surface. The lower fluid flow will lead to higher rate of inclusions due to the aforementioned effects. To reach similar inclusion rates the stirrer speed needs to be increased proportionally. The increase in viscosity during a longer and larger freeze concentration process leads to a decrease in performance of the process.

4. Conclusion

In this study the inclusion behaviour of sucrose and maltodextrin during film freeze concentration was found to be closely linked with the stirrer speed and the ice growth rate. Fluid flow simulations showed that there is a direct link to the hydrodynamic conditions above the ice surface. The transport of the solute molecules in the boundary layer is diffusion limited and therefore an increase in shear above the surface reduces the thickness of this layer. The ice growth rate is the other important factor. Our experiments showed that there is a critical ice growth rate. Lewis et al. (2015) showed if the ice growth rate is exceeded more solute will be included in the ice layer . In this case the solute molecules will not have the chance to move away from the ice boundary. At a high enough stirrer speed and a sub critical ice growth rate the inclusions will reach a minimum dictated by a phase partitioning equilibrium as found by Jusoh et al. (2009). To improve the process and decrease the inclusions, experiments with a decreasing cold wall temperature will be carried out.

5. Acknowledgements

This work is an Institute for Sustainable Process Technology (ISPT) project. Partners in this project are TNO, Royal Cosun and Akzo Nobel.

References

G. K. Batchelor, 2000. An Introduction to Fluid Dynamics. Cambridge University Press, Cambridge, United Kingdom.
URL http://ebooks.cambridge.org/ref/id/CBO9780511800955

Z. Bubnick, E. Werner, 1995. Sugar technologists manual : chemical and physical data for sugar manufacturers and users TT -, 8th Edition. Bartens, Berlin, Germany.

F. E. Genceli, M. Lutz, A. L. Spek, G.-J. Witkamp, dec 2007. Crystallization and Characterization of a New Magnesium Sulfate Hydrate $MgSO4 \cdot 11H2O$. Crystal Growth & Design 7 (12), 2460–2466.
URL https://doi.org/10.1021/cg060794e

M. Jusoh, R. Yunus, M. Hassan, dec 2009. Performance Investigation on a New Design for Progressive Freeze Concentration System. Journal of Applied Sciences 9 (17), 3171–3175.
URL http://www.scialert.net/abstract/?doi=jas.2009.3171.3175

A. Lewis, M. Seckler, H. Kramer, G. Van Rosmalen, 2015. Industrial crystallization: Fundamentals and applications. Cambridge University Press, Cambridge, United Kingdom.

O. Miyawaki, L. Liu, K. Nakamura, sep 1998. Effective partition constant of solute between ice and liquid phases in progressive freeze-concentration. Journal of Food Science 63 (5), 756–758.
URL http://doi.wiley.com/10.1111/j.1365-2621.1998.tb17893.x

O. Miyawaki, L. Liu, Y. Shirai, S. Sakashita, K. Kagitani, jul 2005. Tubular ice system for scale-up of progressive freeze-concentration. Journal of Food Engineering 69 (1), 107–113.
URL http://linkinghub.elsevier.com/retrieve/pii/S0260877404003528

J. Sánchez, Y. Ruiz, J. M. Auleda, E. Hernández, M. Raventós, aug 2009. Review. Freeze concentration in the fruit juices industry. Food Science and Technology International 15 (4), 303–315.
URL http://journals.sagepub.com/doi/abs/10.1177/1082013209344267

Anton A. Kiss, Edwin Zondervan, Richard Lakerveld, Leyla Özkan (Eds.)
Proceedings of the 29th European Symposium on Computer Aided Process Engineering
June 16th to 19th, 2019, Eindhoven, The Netherlands. © 2019 Elsevier B.V. All rights reserved.
http://dx.doi.org/10.1016/B978-0-128-18634-3.50252-6

Simulation-based reinforcement learning for delivery fleet optimisation in CO_2 fertilisation networks to enhance food production systems

Rajesh Govindan, Tareq Al-Ansari[*]

Division of Sustainable Devlopment, College of Science and Engineering, Hamad Bin Khalifa University, Qatar Foundation, Doha, Qatar

*talansari@hbku.edu.qa

Abstract

As part of the drive for global food security, all nations will need to intensify food production, including those situated in hyper arid climates. The State of Qatar is one such example of a national system that whilst it is presented with environmental challenges, seeks to enhance food security. There is a consensus that CO_2 fertilisation of agricultural systems has the potential to enhance their productivity. In this paper, the authors present a novel study that involves the development of a simulation model of a GIS-based CO_2 fertilisation network comprising of power plants equipped with CO_2 capture systems, transportation network, including pipeline and roadways, and agricultural sinks, such as greenhouses. The simulation model is used to specifically train the CO_2 distribution agent in order to optimise the logistical performance objectives of the network, namely delivery fulfilment and network utilisation rates. The Pareto non-dominating solutions correspond to an optimal CO_2 delivery fleet size of around 1-2 trucks for an average year in the simulation example considered.

Keywords: CO_2 fertilisation, Simulation, Logistics, Reinforcement Learning

1. Introduction

With the changing climate and growing population, there exists a global drive towards the intensification of food production. Those countries particularly situated in hyper arid regions that seek to boost their agricultural productivity are faced with the challenges pertaining to the sustainable usage of their land, energy and water resources. As such, the importance of understanding the relationship between food security and the natural environment is recognised, as agricultural productivity essentially depends on the condition of the latter. This introduces the notion of 'sustainable intensification', which is related to the need to meet the nutritional demands of a growing population without furthering the depletion of resources and degradation of the environment (Godfray and Garnett, 2014). To this effect, there has been much discussion on the effect of elevated levels of CO_2 in the atmosphere on agricultural productivity. The consensus is that it is possible to obtain higher crop yields because CO_2 is an essential compound in the photosynthesis process where its carbon content is utilised to form carbohydrates. Furthermore, increasing the concentrations of CO_2 inhibits the loss of water, also known as transpiration, due to increased resistance from leaf stomata. The natural fertilisation of crops using CO_2 offsets the effects of increasing temperature and decreasing soil moisture (Lawlor and Mitchell, 1991). Moreover, CO_2 fertilisation of plants not only results in increases in dry matter production but is also accompanied by improvements of water utilisation efficiency (WUE) (Kimball and Idso, 1983).

Previously, the authors developed a GIS-based EWF nexus model to explore the potential of enhancing agricultural productivity through CO_2 fertilisation at the farm level (Al-Ansari et al., 2018). It was noted that, the approach is promising from an environmental perspective, as the emission reduction that can be potentially achieved by the CO_2 fertilization network is between 0.5 - 1 Mt/year considering the integration of capture technology for both CCGT and BIGCC, with the potential benefit of saving precious water resource at the agricultural fields. However, the overall viability must also depend on the economic feasibility of the transport network. As such, there is a requirement of using a combination of pipelines and roadways (using trucks) for the transportation of CO_2 from the source to the sink. It is likely that a dedicated CO_2 fertilisation network for enhancing agricultural productivity could either remain underutilised or sometimes unable to meet the demands of new greenhouses that are expected to be built, and hence, extended studies are necessary to address such logistical issues. Furthermore, increasing utilisation reduces the amount of net traffic (in vehicle kilometres) needed to transport a given quantity of freight (in tonne-kilometres), which in turn corresponds to decreased energy consumption and reduced CO_2 emissions from transportation itself (McKinnon, 2010).

The objective of the research presented in this paper is to develop a network simulation model comprising of spatially-distributed and autonomous learning agents representing sub-systems participating in a national-scale CO_2 supply chain logistics problem. For the purposes of illustration in this study, the business logic to enable agent simulation representing CO_2 distribution centres and greenhouses, was implemented. The authors posed the logistics problem as non-linear and multi-objective optimisation model using Markov Decision Processes (MDP) framework. This allows to determine the optimal solution representing the size and routing behaviour of a fleet of delivery trucks that transport CO_2 from the distribution centres in order to simultaneously fulfil the demand at various greenhouse locations (dependent on seasonal weather conditions) and improve the network utilisation rate.

2. Available data and assumptions

The information was derived from multimodal datasets for Qatar, each of which provides spatial information for: (a) distribution of land use for farming activities; (b) distribution of solar irradiance; (c) the locations of the major sources of CO_2 emissions; and (d) possible modes of CO_2 transport, namely pipelines and road transport. A meta-heuristic formalism was previously introduced by the authors which helped in determining the promising locations for new greenhouse developments as illustrated in Fig 1 (Al-Ansari et al., 2018). This served as the basis for setting up the CO_2 fertilisation network simulation model in the current study. It should, however, be noted that the locations of CO_2 distribution centres were assumed to be at the intersections of pipelines and roadways for simplification purposes and are not necessarily optimised.

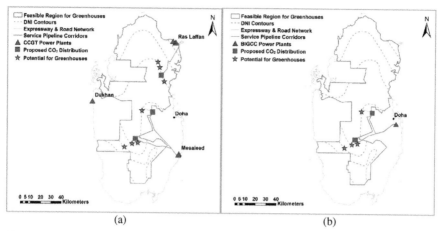

Fig 1. Optimised locations of greenhouses in the State of Qatar for the case of CO_2 captured from: (a) CCGT plants; and (b) BIGCC plants (After Al-Ansari *et al.* (2018)).

3. Methodology

3.1. CO₂ fertilisation network simulation

The network simulation model was setup using the Java-based AnyLogic® platform. The necessary spatial information was used in order to accurately capture the road networks in Qatar, which is used by the fleet in the model to transport CO_2 between the distribution centre and greenhouses, as illustrated in Fig 2.

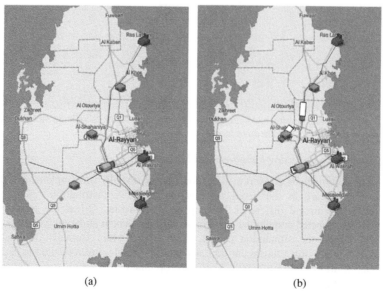

Fig 2. Spatial simulation of the CO_2 fertilisation network indicating power plants; CO_2 distribution centre; and greenhouses for: (a) initial state of the fleet; (b) moving state of the fleet during simulation.

The simulation period corresponds to one year capturing the essential variability in the average seasonal temperatures, *i.e.* minimum, maximum and mean temperatures, at different locations in Qatar. The business logic for the agents representing greenhouses was implemented using the modified form of Hargreaves original equation for evapotranspiration (ET), which uses local weather data, such as the temperatures and solar irradiance distribution data, given by (Hargreaves and Samani, 1985):

$$ET = k \times 2.3 \times 10^{-4} \times \left(T_{max} - T_{min}\right)^{0.5} \times \left(T_{avg} + 17.8\right) \times R_a \qquad (1)$$

Where, k is a factor which varies from 0.79 – 1.02. Increasing ET values (in mm) is an indicator of increasing CO_2 demand, and thus a maximum allowable water loss was assumed as the trigger condition for greenhouses to place an 'order' for more CO_2 from the distribution centre. Currently, this amount of CO_2 is assumed to be fixed, and the stock is fully utilised before a new order can be placed; no monetary transaction is associated with the placement of an order.

Meanwhile, at the distribution centre, the business logic for fleet management is illustrated by the process diagram in Fig 3. The 'seizing' of a truck from the fleet resource pool occurs whenever a demand trigger (order) is received. On the other hand, the 'loading' and 'unloading' times are assumed to be stochastic.

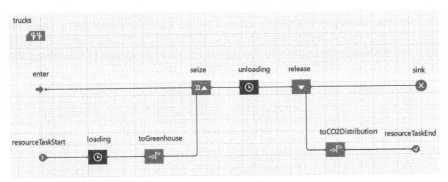

Fig 3. The process model setup in AnyLogic® representing the business logic for fleet resource pool management at the CO_2 distribution centre.

3.2. Fleet optimisation using Reinforcement Learning

The classical formalism of Markov Decision Process (MDP) was implemented to aid the learning feature in the agent representing the CO_2 distribution centre. More specifically, a temporal difference learning approach called Q-learning was used to maximise the expected cumulative value of an action (a) taken under a given state (s) of the agent during the simulation period of one year. More formally, the network objectives, namely order fulfilment time and utilisation rate, are modelled as Bellman Eq 2 and 3 (Sutton and Barto, 1998) to give the optimal policy for all states and actions in Eq 4 (Sutton and Barto, 1998):

$$Q\left(S_t = s, A_t = a\right) = E\left[r + \gamma V\left(S_{t+1} = s'\right) | S_t = s, A_t = a\right] \tag{2}$$

$$V\left(S_t = s\right) = E\left[r + \gamma V\left(S_{t+1} = s'\right) | S_t = s\right] \tag{3}$$

$$Q_*\left(S_t = s, A_t = a\right) = \max\left\{Q_\pi\left(S_t = s, A_t = a\right)\right\}, \forall s \in S, a \in A \tag{4}$$

Where 'r' is the immediate reward signal for an action 'a' in state 's'; γ is the discount factor whose value lies between 0 (short-term minded or greedy agent) and 1 (long-term minded agent), which indicates the importance the optimisation gives towards future rewards. Finally, the update of the action value function (Q) from a given time step to the next is given by (Sutton and Barto, 1998):

$$Q\left(S_{t+1} = s, A_{t+1} = a\right) \leftarrow Q\left(S_t = s, A_t = a\right) + \dots$$
$$\alpha\left[r + \gamma \max_{a'}\left\{Q\left(S_{t+1} = s', A_{t+1} = a'\right)\right\} - Q\left(S_t = s, A_t = a\right)\right] \tag{5}$$

4. Results and discussion

One of the performance metrics used by the agent (and receiving the corresponding reward signal) is the time it takes for the distribution centre to fulfil an order depends on both the speed of the truck (assumed constant) and the route it takes (assumed initially stochastic). The other performance metric used by the agent is the fleet utilisation rate. Q-learning was performed with respect to these performance metrics over the simulation period of one year.

The results obtained from the multi-objective optimisation using the Q-learning approach listed in Table 1 demonstrates that there exist Pareto non-dominating solutions for the optimal fleet size for the CO_2 distribution centre. It is therefore expected that the optimal fleet size is around 1-2 trucks for the simulation example considered. However, it is important to note that the study carried out thus far is a small network for the purposes of illustrating the technique, and does not consider the entire CO_2 fertilisation network, such as the power stations (see Fig 2).

Table 1. Objective function values using the Q-learning approach for various fleet sizes.

Size of the fleet	Order fulfilment time of the network (orders per day)	Utilisation rate in the network (%)
1	2.1	65
2	2.34	36
3	2.33	23
4	2.47	19
5	2.39	15

5. Conclusions

The study discussed in this paper provides an illustrative example of how simulation-based optimisation can aid in making large-scale logistical decisions for network-based applications having economic implications. With the recent advent of artificial intelligence, non-linear, multi-objective and sequential decision-making techniques such as reinforcement learning is useful in solving hard problems whose objective functions do not have closed form analytical representations, such as the CO_2 fertilisation network for the enhancement of agricultural productivity.

It is envisaged that the results obtained have important implications for planning the expansion of greenhouse networks, particularly for the State of Qatar, where self-sustenance and food security has become a priority. In this regard, it has been considered that both the environmental and economic dimensions are equally important. As such, the utilisation of waste streams such as CO_2 emissions for growing crops not only provides for food security, but also enhances the efficiencies of energy and water utilisation as highlighted in the research carried out by the authors.

With regards to the limitations of the current work, only a single agent has been considered thus far and hence there is scope to extend the work to develop multiple learning agents in large-scale network optimisation. The business logic considered are also likely to be simplified for the current illustration, and hence the simulator requires incorporating agent functions for real world scenarios, *e.g.* in multi-echelon supply chains where dynamic distribution of resources, inventory and supply risk management and long-term supply chain planning and development are some of the core problems to be generally solved.

References

Al-Ansari, T., Govindan, R., Korre, A., Nie, Z. and Shah, N., 2018. An energy, water and food nexus approach aiming to enhance food production systems through CO_2 fertilization. In Computer Aided Chemical Engineering, 43, 1487-1492, Elsevier.

Godfray, H.C.J. and Garnett, T., 2014. Food security and sustainable intensification. Phil. Trans. R. Soc. B, 369(1639), 20120273.

Hargreaves, G.H. and Samani, Z.A., 1985. Reference crop evapotranspiration from temperature. Applied engineering in agriculture, 1(2), 96-99.

Lawlor, D. W. and Mitchell, R. A. C., 1991, The effects of increasing CO_2 on crop photosynthesis and productivity: a review of field studies, Plant, Cell and Environment, 14, 807-818.

Kimball, B., and Idso, S., 1983, Increasing atmospheric CO_2 effects on crop yield, water use and climate, Agricultural Water Management, 7, 55-72.

McKinnon, A., 2010. Green logistics: the carbon agenda. Electronic Scientific Journal of Logistics, 6(1).

Sutton, R.S. and Barto, A.G., 1998. Reinforcement learning. MIT Press Cambridge, MA.

Anton A. Kiss, Edwin Zondervan, Richard Lakerveld, Leyla Özkan (Eds.)
Proceedings of the 29th European Symposium on Computer Aided Process Engineering
June 16th to 19th, 2019, Eindhoven, The Netherlands. © 2019 Elsevier B.V. All rights reserved.
http://dx.doi.org/10.1016/B978-0-128-18634-3.50253-8

Systematic decision-support methodology for identifying promising platform technologies towards circular economy

Dominic Silk, Beatrice Mazzali, Isuru A. Udugama, Krist V. Gernaey, Manuel Pinelo, John Woodley, and Seyed Soheil Mansouri

Process and Systems Engineering Centre, Department of Chemical and Biochemical Engineering, Technical University of Denmark, DK-2800 Kgs. Lyngby, Denmark
seso@kt.dtu.dk

Abstract

This work presents a systematic computer-aided decision-support framework to tackle challenges faced when attempting to identify and develop resource recovery projects. The framework is hierarchical and has been developed to facilitate the industries' transition towards more sustainable production, providing users with the opportunity to evaluate and support decisions on not only a financial basis, but also from a comprehensive techno-economic and sustainability angle. To this end, the framework collects relevant aspects within the sustainability field and addresses concepts such as circular economy, waste management and 'legacy' compounds to name some of the more pertinent topics. The proposed methodology is generic and can be applied to a wide variety of potential resource recovery projects. The application of the framework is highlighted through two case studies, phosphorus recovery from a phosphate-based wastewater stream; and monosaccharide recovery from wheat straw liquor.

Keywords: Sustainability, Decision-support framework, Technology readiness level (TRL), Resource recovery, Circular Economy

1. Introduction

For over 200 years, the world economy has been built upon a linear economic model, leading for never before seen economic and societal transformation (Hayward et al., 2013). However, this human induced change has led to unprecedent global environmental degradation and rapid resource depletion, together with an unprecedented climate impact. The proposed countermeasure to this impending catastrophe is the circular economy model. In this model, resources are no longer extracted and discarded but instead products and revenues are generated from what in a linear economy would be considered waste. In this context, resource recovery, being the facilitating process to enhance attainable resource value and avoid resource deployment, perfectly fits the proposed paradigm shift. Today, the use of resource recovery is widely expressed in various industrial processes, including applications related to the dairy industry (Udugama et al., 2017) and biodiesel refining (Mansouri et al., 2019). In petrochemical operations, waste reintegration is also practiced widely (Mansouri et al., 2017). Consequently, resource recovery can be observed as a key missing element required to close an imperfect production cycle. Resource recovery project implementation enables companies to achieve several sustainable development goals, such as responsible production and consumption, whilst improving their product portfolio and reducing dependency on external players in the supply chain.. Historically, little importance has been placed upon resource recovery, namely due to resource abundance, absence of reliable opportunity identification methods, and ease of waste treatment regulation compliance. In the domain of process systems engineering, this multi-dimensional and multi-scale problem has been an area of active interest with works on techno-economic sustainability analysis looking to address parts of these requirements (Gargalo et al., 2014, Gargalo et al., 2016, Hoseinzade et al., 2019). However, while these frameworks and methodologies can be adopted to perform a somewhat detailed analysis they do not consider the overall role of techno-economics and customer readiness feasibility in the pursuit of resource recovery technology deployment in an industrial context.

Based on the above considerations, the authors have identified the need to develop a methodology to manage the multi-faceted and multi-disciplinary intricacies of deploying resource recovery concepts, with focus placed on employing separation technology for recovering resources. The proposed methodology not only explicitly takes into consideration the techno-economics aspects but also implications of technology development and customer readiness in implementing resource recovery projects as well as taking implicitly taking into account the environmental impact of recovering a resources in both early and late stage of the decision making process. The proposed a decision-support methodology is intended to employed in introducing the resource recovery paradigm within traditional organisations and provide them the necessary tools to transition to a circular economy.

2. Methodology

The framework is decomposed into five segments, each defined by a characteristic question. Figure 1. illustrates the overall work flow of the methodology.

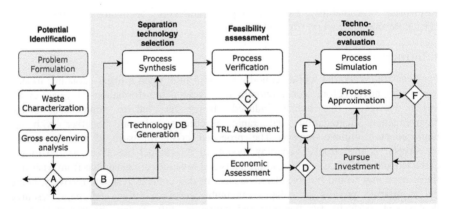

Figure 1. Schematic representation of the workflow of the methodology. Grey framework steps indicate input/output.

Systematic decision-support methodology for identifying promising
platform technologies towards circular economy
1515

2. 1 Step 1: Problem formulation. The first and most instrumental task of framework construction is the problem definition. It is here that the aim, objectives, and scope of the design problem are outlined mathematically. The objective function is formulated to fit the user's agenda, and demonstrates the desired outcome of the design problem. The subsequent algorithms are structured in a broad to narrow manner intended to thin the field of candidates that do not satisfy the objective function early on. This approach facilitates swift progression through the framework, especially in instances that manual input is required.

2.2 Step 2: Potential identification. The first screening phase, potential identification, preliminarily assesses whether resource recovery is relevant in the context of the stream studied and objective function set by the user. To account for all intentions, the gross sustainability potential is applied, providing a quantitative assessment of financial, environmental, and societal aspects. The information required in this section is easily attainable via the waste characterization algorithm, facilitating the rapid and efficient evaluation of many cases. The objective of this section is to identify valuable resources that can potentially justify the economic, environmental and societal aspects that are vital in implementing a resource recovery project and acts as a preliminary screening mechanism. Note: The gross sustainability potential in this step implicitly takes into account the overall environmental impact of recovering given resource, as such in a situation a pollutant or a "legacy" compound is captured, this will be reflected in the gross sustainability potential calculation. If a full circular economy model is desired, the potential identification step can be bypassed as all raw-materials are utilized irrespective of their value, generating no waste.

2.3 Step 3: Separation technology selection. In the next step of the framework, viable separation technologies for the recovery of the stream constituents are identified. Such technologies can then be sequenced to provide a physical feasible process synthesis for the recovery of the afore-determined target components. Separation task flowsheets should be furnished by process synthesis techniques where applicable, with the remainder subjected to technology database generation and manual pairing. A feedback loop is integrated with the feasibility assessment step to facilitate revisions of tasks that are not technically viable.

2.4 Step 4: Feasibility assessment. In the case of process synthesis, physically feasible flowsheets are automatically generated. This can be problematic as some tasks may be unsuitable or impractical, thus a feasibility assessment is carried out in the next step. This ensures that only technologically feasible flowsheets progress through the framework, each task is verified against openly available industrial implantations as well as academic literature where the concept of technology Readiness Level is applied. Preliminary financial viability is also assessed at this stage, by comparing the streams process revenue (Gross sustainability potential) to its estimated capital cost (Tsagkari et al., 2016). The payback period is then calculated and must surpass the user specified threshold to progress through the framework.

2.5 Step 5: Techno-economic evaluation. Now that basic technological and financial feasibility is proven, further granularity must be obtained to form a foundation for investment. If the user intends only on an order of magnitude estimate of the Net

Present Value, process approximation methods should also be applied to determine the associated operating costs. For more detailed assessments, process simulation is suggested. Once simulated, the model can be applied to determine recoveries, equipment costs, and operational expenses, ultimately resulting in improved NPV accuracy. If financially satisfactory, the framework is concluded with application of the WAR algorithm application to ensure environmental compliance. At this point the user should have a clear picture as to whether or not to pursue investment.

Note: the application of WAR algorithm as a part of this step implicitly takes into account the environmental impacts of a given resource recovery solution. The detailed economic analysis conducted will also account for the economic impact (marginal cost of abatement) which takes into account the environmental impact of reducing/eliminating a polluting compound.

3. Case studies

The application of the above framework is demonstrated through two case studies: phosphorus recovery from a phosphate-based wastewater stream; and monosaccharide recovery from wheat straw liquor. Besides the Biofuel and waste water treatment operations, the framework applicability is far reaching, providing potential in various other industries, such as petrochemicals and pharmaceuticals,

3.1 Monosaccharide recovery

In the case study the intention is to recover economically viable components from a wheat straw process stream (30,000 ton/yr), thereby improving economic potential and reducing waste, this was identified by applying step 1. Applying the potential identification, it was found that xylose ($241/m^3$) and arabinose ($\$236/m^3$) are the two most lucrative components in the waste stream. Applying step 2, a feasible process was synthesized for the case study where thermodynamic insights for process synthesis were used (Jaksland et al., 1995). The feasibility assessment step required five passes for an adequate flowsheet to be produced, as depicted in figure 2.

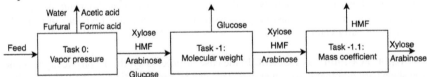

Figure 2. Output of step 2, final separation task flowsheet for monosaccharide extraction case study, generated by TaskGen software

Separation technologies were then paired to the successful separation tasks, resulting in a flowsheet consisting of detoxification via evaporation, glucose removal with nanofiltration, and monosaccharide purification with an ion exchange resin. Applying step 4 it was confirmed that the process sequence is a feasible one. In step 5, the recovery rates of the target components were then determined via process simulation, as seen in Table 1. Finally, the financial feasibility of the case was evaluated subject to uncertainty analysis, providing the results seen in Table 2.

Systematic decision-support methodology for identifying promising
platform technologies towards circular economy
1517

Table 1: Results from process simulation in the techno-economic evaluation step

Target	Flow	Recovery
Xylose	16.36 kg/hr	56.5%
Arabinose	86.79 kg/hr	43.6%

Table 1: Results in million USD from the techno-economic evaluation step (10% DR).

Case	Present value	Capital cost	Development cost	Net Present value
Mono	39.1	-8.2	-11.0	19.6

Based on the results of the demonstration of the proposed methodology, the conventional bio-ethanol production process could be feasibly diversified with the recovery of supplemental products, prior to the fermentation process.

3.2 Phosphorus recovery

In this case, the framework was applied to a wastewater stream containing phosphate. Applying step 1, the problem is formulated as follows: "Given a pre-treated waste water stream of 14 mg/l of phosphate in water, 550000 m^3/day, return a feasible separation technique to recover the desired resource, maximizing the resource recovery net present value meanwhile minimising the environmental impact".

For step 2, potential identification, the user obtains an overview of the compound details to evaluate the most valuable resource on which the analysis can be carried forward. In this step, the gross sustainability potential and environmental impact was computed for both phosphorus and water. For phosphorous, the environmental influence and geopolitical considerations were significant factors behind the decision to target the resource for recovery. In this case, the geopolitical influence draws from the location of the case study, New Zealand, where Phosphorous-shortage has been increasing problem over the years (Cordell et al., 2014). Based on this consideration and on the higher phosphorus environmental impact value, phosphorus has been selected as the resource to be further analysed in the framework. The subsequent application of step 3, via technology database generation, illustrated that there are many potential phosphorus recovery pathways, but the application of step 4 illustrated that Crystallisation of phosphorus to struvite, using magnesium hydroxide was the only technologically mature and hence feasible pathway. The pathway was subjected to preliminary financial feasibility assessment by applying the capital cost estimation technique (Tsagkari et al., 2016). This yielded an estimated capital cost of USD 576 Million and a yearly process revenue of USD 0.62 Million. Based on this finding the financial infeasibility of the project is evident, and that the recovery of phosphorus in the form of struvite crystal is not a process to consider for implementation. Resultantly, the framework iterates back to step 2, to pick the next most feasible separation pathway. However, as no other technology meets the maturity requirement, the case exits the framework unsuccessfully.

5. Conclusions

In this work, a systematic computer aided framework has been developed to systematically evaluate the techno-economic and environmental potential of a resource recovery initiative. The merit of the developed framework was then demonstrated on two practical case studies where the framework was able to systematically identify resource recovery potential, and synthesize first pass separation process designs. The diverse nature of the case study waste stream composition illustrates the generic structure, thus versatility of the framework. Analysis of the first case study demonstrated the framework's ability to identify and develop solutions in the presences of multiple target resources, and facilitate generation of lucrative technically feasible separation pathways. This established the frameworks applicability in the context of handling realistic waste streams, often containing numerous significant constituents. In the second case study, the framework facilitated the "weeding out" of numerous process technologies that illustrated phosphorus recovery potential, but were not currently at a sufficient technology readiness for industrial implementation. This illustrated the value of the multi-faceted approach undertaken by the framework, as feasible yet technologically underdeveloped solutions are quickly discarded in favour of the development of solutions that are of a sufficient technology development streamlining the long and resource intensive development phase.

Acknowledgments

This work partially recived financial support from the Carlsberg Foundation of Denmark (Grant Number CF17-0403)

References

Cordell, D., Mikhailovich, N., Mohr, S., Jacobs, B., White, S. (2014) 'Austalian sustainable phosphorus futures'. Rural industries research and development. 14, pp 1-59.

Gargalo C. L., Carvalho A., Gernaey K.V., Sin G., (2016), 'A framework for techno-economic & environmental sustainability analysis by risk assessment for conceptual process evaluation', Biochemical Engineering Journal, Volume 116, pp. 146-156.

Gargalo C. L., Carvalho A., Matos H.A., Gani R., (2014), 'Techno-Economic, Sustainability & Environmental Impact Diagnosis (TESED) Framework', Computer Aided Chemical Engineering, Volume 33, pp. 1021-1026.

Hayward, R., Lee, J., Keeble, J., McNamara, R., Hall, C., Cruse, S., Gupta, P., Robinson, E., (2013) 'The UN Global Compact-Accenture CEO Study on Sustainability 2013', UN Global Compact Reports. 5(3), pp. 1–60.

Hoseinzade L., Adams T.A., (2019), 'Techno-economic and environmental analyses of a novel, sustainable process for production of liquid fuels using helium heat transfer', Applied Energy, Volume 236, pp. 850-866.

Mansouri S.S. et al. (2019) Economic Risk Analysis and Critical Comparison of Biodiesel Production Systems. In: Tabatabaei M., Aghbashlo M. (eds) Biodiesel. Biofuel and Biorefinery Technologies, vol 8. Springer, Cham

Mansouri, S. S., Udugama, I. A., Mitic A., Flores-Alsina X., Garney K. V., (2017) 'Resource recovery from bio-based production processes: a future necessity?', Current Opinion in Chemical Engineering. 18, pp. 1–9.

Udugama, I. A., Mansouri S.S., Mitic A., Flores-Alsina X., Garney K. V., (2017) 'Perspectives on Resource Recovery from Bio-Based Production Processes: From Concept to Implementation', Processes, 5(3), p. 48.

Jaksland C. A, Gani,R., and Lien K.M. (1995), "Separation process design and synthesis based on thermodynamic insights," Chem. Eng. Sci., vol. 50, no. 3, pp. 511–530.

M. Tsagkari, J. L. Couturier, A. Kokossis, and J. L. Dubois, "Early-Stage Capital Cost Estimation of Biorefinery Processes: A Comparative Study of Heuristic Techniques," ChemSusChem, vol. 9, no. 17, pp. 2284–2297, 2016.

Anton A. Kiss, Edwin Zondervan, Richard Lakerveld, Leyla Özkan (Eds.)
Proceedings of the 29th European Symposium on Computer Aided Process Engineering
June 16th to 19th, 2019, Eindhoven, The Netherlands. © 2019 Elsevier B.V. All rights reserved.
http://dx.doi.org/10.1016/B978-0-128-18634-3.50254-X

A systemic approach for agile biorefineries

Michelle Houngbé[a], Anne-Marie Barthe-Delanoë[a], Stéphane Négny[a]

[a]*Laboratoire de Génie Chimique, Université de Toulouse, CNRS, INPT, UPS, Toulouse, France*

annemarie.barthe@ensiacet.fr

Abstract

Biorefineries is one of the main pathways towards energy transition. However, they are challenged by demand and supply variability. Moreover, the standalone and highly specialized biomass processing system has to face several internal and external constraints. To tackle this unstable ecosystem, it is necessary to provide agility in terms of both physical structure and organization. In this context, the ARBRE project proposes to redesign the biorefinery, at local scale, by introducing a collaborative network of stakeholders and the servitization of the biomass processing steps, and by integrating digital concepts from Industry 4.0. This paper proposes a framework to design agile biorefinery with a systemic approach by modelling the knowledge about the system and its environmental characteristics through a metamodel.

Keywords: Agility, Biorefinery, Collaboration, Industry 4.0, Virtual Organization, Meta-modelling

1. Introduction and motivations

Despite an increase of biomass processing projects these last years, the development of biorefineries is significantly slowed down because of several constraints.

First, the biorefinery concept is similar to the oil refinery principles [1] . Thus, the four major steps of the whole biomass processing are performed into a unique plant. Each step of the process requires specific devices, and specific operating conditions, according to the chosen biomass and the targeted bioproduct. In case of any changes regarding the biomass (type, quantity, quality, etc.) or the bioproduct, this physical, centralized, standalone and highly specialized plant cannot function at all. Adaptation to changes would require building a new plant, which is unrealistic considering investment and operational costs. Then biorefineries face high variability due to several constraints as: the external parameters influencing the availability of the biomass (seasonality, crop rotation), the natural biomass degradation or else the impurities inside the biomass feedstock (e.g. weeds, silica, minerals or soil). Therefore, this huge variability challenges the biomass supply leading to production irregularities. Additional challenging factors are the supply spreading and the geographic dispersion of both biomass feedstocks and biorefineries, increasing the nervousness of this supply chain.

All the aforementioned hurdles underline the need to design the biorefinery with its entire ecosystem to take into account uncertainties. This disruptive way to design biorefineries should be done on both the physical structure (equipment and apparatus) and the organisation (supply chain) to provide the required agility to face uncertainties.

Thus, our goal is to answer to the following research question: "How can biorefineries development be fostered with required agility to cope this unstable environment?" In the literature, biorefineries are analysed at different scales: molecular, equipment, process and supply chain scale [2] . Nevertheless, designing the biorefinery with a systemic approach, from the harvest to the final customer, as an agile collaborative network, is a new approach. This is the goal of our proposal.

2. State of art of agility in chemical process industry

Agility is used as well as a concept, a paradigm, a tool or else a method to qualify a professional sector or a means of production, in a context of changing, volatile and uncertain environment, characterized by the globalization and a production customization for customers. This concept was defined by several authors in the last years and leans on major development axis: responsiveness, flexibility, information and communications technology, and adaptability [3]. A relevant definition of agility that could be applied to the biorefinery system has been proposed by [4] as the ability of a system to "lead as quickly as possible, on the one hand, to the detection of its mismatch to a given context, on the other hand, to the set up of the required adaptation".

In the manufacturing industry literature, the expression "agile manufacturing" has been the subject of extensive researches for the last two decades. Whereas, in the process and the chemical industry areas, a few articles address about the agility issue. They often deal with subjects partially related to agility (as defined above) as flexible process, reconfigurable manufacturing, responsiveness system, modular production, with a technological or a chemical approach, at microscopic or mesoscopic scales. Considering, the whole biomass processing system, at the macroscopic scale, including both the chemical process and the environment, the entire biomass supply chain is then taken into account for the biorefinery design. Existing researches on this topic consider agility as the optimization criteria to design a standalone plant. but this standalone centralized plant cannot cope with the adaptation requirements to be considered as agile.

To switch from this centralized biomass processing paradigm, a few research works focus on decentralization by creating a collaborative network. [2] restrains its network to two stakeholders and assesses the decentralization based only on the cost of the transportation. Regarding [5], the concept of virtual biorefinery deals with a "loose coupling of decentralized stakeholders" where the biomass processing is dispatched according to the optimal biomass supply chain. In this work, the pre-treatment step is separated from the chemical process treatment. Within a service economy approach, the servitization aims to make the transition from material products (tangible) to global services (intangible). It enables to develop full services leaning on existing resources (equipment, devices, workforce...) and a know-how. A type of servitization, Product-Service System, has been studied by [6] in a manufacturing company in an organizational and a collaborative way. This type of work does not yet address the biomass processing. However, because of the high number of farmers involved in cooperatives, it could be economically attractive to enhance this collaborative behaviour with the share of the investment into apparatus and skilled workforce.

Plus, Information and Communication Technologies have a predominant position into agricultural and industrial sectors. They can provide a huge amount of data about weather forecast, crop state, device monitoring (e-farming) or market (economics) that are relevant to support decision making about biomass processing. Among new technologies, Complex Event Processing could support data collection and analysis by detecting patterns over a stream of data [7]. It could be used to study the variability of the entire ecosystem: territory, weather, stakeholders and complex biomass chemical treatment.

3. ARBRE: A collaborative software platform to design and monitor agile biorefineries

The part aims to present the methodological framework enabling to design and monitor agile biorefineries as a collaborative network, using stakeholders' services and reusing existing apparatus. Moreover, it can enhance the territorial anchorage of the biomass processing by choosing a local scale (biomass supply radius of fifty kilometres). This framework will be implemented as a decision support system (software platform) available for the biomass processing stakeholders, from farmers to customer.

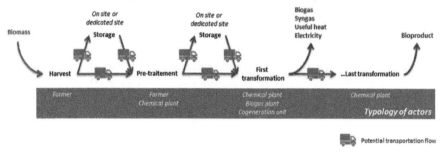

Figure 1: Agile biorefinery concept

The agile biorefinery concept leans on two main axes. Within a systemic approach, the first axis considers that the chemical process could be performed not into a unique standalone plant but among a network of several existing plants and locations. Thus, the chemical process is decentralized. Each processing step is servitized. When a change occurs into the system (context, stakeholder, etc.), each actor of the biomass supply chain could propose a service matching the consequent need for adaptation. This service could address one or several steps of the biomass processing. The process is then easily reconfigured. Then, the collaboration between the stakeholders creates the biorefinery as a virtual enterprise, relevant to the biomass to process, the targeted bioproduct and the context. The role of each stakeholder and/or the choice of the process (steps, apparatus) can evolve according to the occurring changes: biomass, bioproducts (type, purity, quantity), context (weather, market, etc.). This agile biorefinery will use existing equipment, devices and workforce, which is highlighting the concept of "virtual biorefinery" as an agile collaborative process. The second axis focuses on the biomass processing itself. To design this process according to the available input, the targeted output and the system constraints, it is possible to combine knowledge management with a technological approach. The objective is to match a typology of services proposed by the stakeholders (farmers, chemical industry, cogeneration unit, local authority) with a typology of biomass processing steps, while taking into account constraints such as the location, the price, etc.

4. The agile biorefinery metamodel

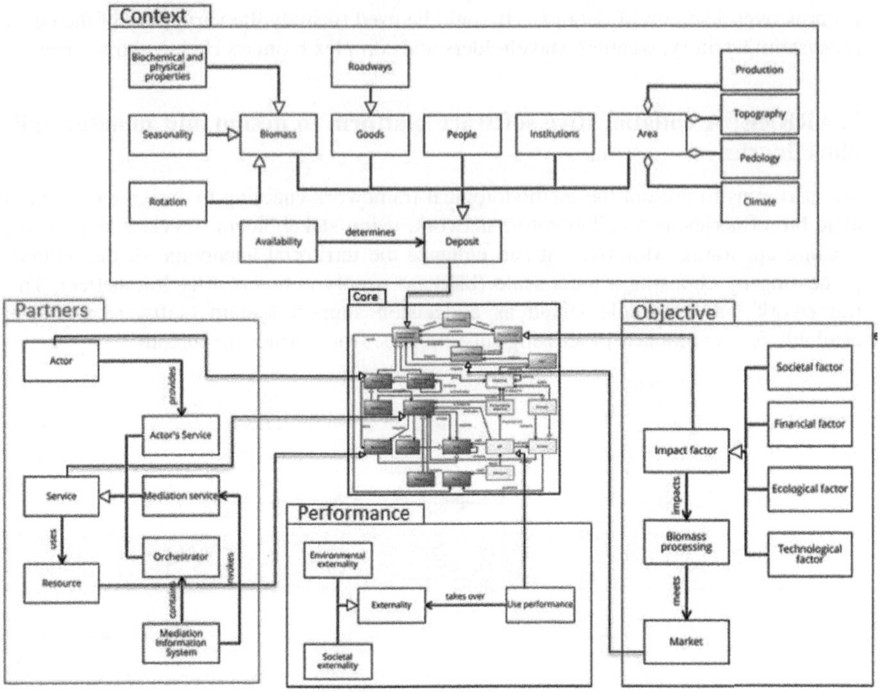

Figure 2: Agile biorefinery metamodel

4.1. The CORE metamodel as support

The Model Driven Engineering (MDE) approach provides a better understanding of complex systems, as biorefineries, thanks to a high level of abstraction. This engineering practice is focused on the model, which is at the heart of the process. It defines the modelling language. In this way, it is adapted to define the required and relevant data about the agile biorefinery system. To this end, we used the CORE metamodel as defined into [8]. Organized in layers, the core of this metamodel represents any collaborative situations, by including general concepts categorized into four main packages: the context (what is the environment?), the partners (who are the actors and their services?), the collaboration objectives and the performance. Then, each successive layer defines a business domain, which inherits concepts from the core. Besides, in this way, it has been successfully used in various application domains as crisis or business organization. So, based on this CORE metamodel, a specific layer has been extended to model the agile biorefinery. Each concept and link have been defined in this application area. They are generic enough to be adapted to any type of agile biorefinery and specialized enough to be relevant to the biomass processing domain. This agile biorefinery metamodel has to be instantiated with real data, in order to ensure the validity of the obtained model and so the relevancy of the concepts.

4.2. The "Context" package

The context enables to determine the characteristics of the collaborative environment. The deposit concept is the entry point. It inherits from the environment component concept from the CORE package. Its availability (related to the quantity) is crucial to start a collaboration. Therefore, the deposit is about the organic deposit usable which does not include: the organic matter needing to return to the ground, the unusable organic matter and the organic matter already involved into a chemical transformation process. It contains biomass, which is specified by biochemical and physical properties (chemical composition, humidity, shape, size), a seasonality (biomass capacity to be productive and harvested over one or several seasons) and the crop rotation (crop diversification including intermediate crops). The deposit is set on an area owing topographic, soil and climate features, with a given kind of production (intensive, organic, reasoned). The good concept is related to the required equipment ensuring the proper management of the deposit. This one is specified with the roadways concept, enabling to determine the accessibility to the deposit (fields path, paved road). The deposit concept embodied the people concept, essential for the public acceptability issues. Institutions concept, as political and decision-making bodies, are key players to provide public aids and subsidies. The context is crucial to characterize the environment where the collaboration takes place.

4.3. The "Actor" package

Partners constitute the collaborative network. They participate actively in the biomass processing. In the servitization context, actors from the collaborative network bring information about their ability to provide a service, according to their existing resources (material, human, special skills) and their constraints. They inherit from the CORE partner package. The proposed services ensure one or several biomass processing and transformation steps. Moreover, Mediation Information System (MIS) coordinates the collaborative process, from the data collection until the conversion to information, as an orchestra conductor. It makes connections and shares information among all the actor services. The whole collaboration partners aim to reach a common goal.

4.4. The "Objective" package

The objective package defines a target for the agile biorefinery. This concept inherits from the objective concept from the CORE package. The main purpose is to process the biomass, by meeting the market. The market determines the supply, the demand and the competitive requirements between stakeholders. However, impact factor concept can foster or disadvantage the biomass processing. This concept highlights issues about the agile biorefinery operation. It encompasses criteria about sustainability (social, economic and environmental) and the system's evolution capabilities. It is specified into four types of impact factors: the societal factor concept relates to the impacts on the extended community organization (annoyances); the financial factor concept relates to process, equipment and maintenance work cost evolution (trend in price for biomass); the ecologic factor concept concerns changes about living beings and their environment which can affect the biomass processing (biodiversity trend); the technological factor concept gathers innovations and evolutions about devices, processes and methods that could affect also the biomass processing. Achievements enable to assess the performance.

4.5. The "Performance" package

Agile biorefinery performances are measured thanks to Key Performance Indicators (KPI) about the overall result of the collaboration and also each service used. To this end, the use performance concept inherits from KPI concept, in the CORE metamodel. KPIs ensure positive or negative impact of the agile biorefinery on a societal level as economic drive or job development (measurable with meetings and surveys) and on an environmental level as pollution (measurable with sampling program and measurement campaign in the environment).

5. Conclusion and future work

To increase the development of biorefineries, it is crucial to take into account its whole ecosystem, which is marked by a high variability. Agility represents an innovative way in the chemical and process engineering to reconfigure easily the biomass transformation process, according to the constraints and the stakeholders. Being aware of environmental, economic and human factors enables to keep the system functional (operational) and relevant. Thus, the ARBRE project proposes to transform and decentralize the biomass transformation process, leading to a collaboration-based biorefinery. In this paper, the metamodel supporting the characterization of the agile biorefinery has been presented. It defines the concepts and the links between them, enabling to model the environment of the collaboration according to: the context, the actors, the objectives and the performances. The next step of the ARBRE project will focus on the instantiation of the proposed biorefinery metamodel into a model, prior to its use for collaborative process deduction. This will be achieved with real data provided by local authorities from Southern France (interviews, feasibility study, etc.).

Acknowledgements

This research work is funded by the French Research Agency regarding the research project ARBRE (Agility foR BioRefinerieEs)[Grant ANR-17-CE10-0006], 2017-2021. The authors would like to thank the project partners for their advice and comments regarding this work.

References

[1]   B. Kamm, P. R. Gruber, et M. Kamm, « Biorefineries–Industrial Processes and Products », in *Ullmann's Encyclopedia of Industrial Chemistry*, American Cancer Society, 2016, p. 1-38.
[2]   D. Yue, F. You, et S. W. Snyder, « Biomass-to-bioenergy and biofuel supply chain optimization: Overview, key issues and challenges », *Computers & Chemical Engineering*, vol. 66, p. 36-56, juill. 2014.
[3]   A. Soepardi, P. Pratikto, P. B. Santoso, et I. P. Tama, « An updated literature review of agile manufacturing: classification and trends », *International Journal of Industrial and Systems Engineering*, vol. 29, n° 1, p. 95-126, janv. 2018.
[4]   A.-M. Barthe, « Prise en charge de l'agilité de workflows collaboratifs par une approche dirigée par les événements », Univerité de Toulouse, École Nationale Supérieure des Mines d'Albi-Carmaux conjointement avec l'INSA de Toulouse, 2014.
[5]   B. Rapp et J. Bremer, « Paving the Way towards Virtual Biorefineries », *Green Technologies: Concepts, Methodologies, Tools and Applications*, p. 1901-1921, 2011.
[6]   S. Dahmani, X. Boucher, et S. Peillon, « Industrial Transition through Product-Service Systems: Proposal of a Decision-Process Modeling Framework », in *Collaborative Systems for Reindustrialization*, 2013, p. 31-39.
[7]   O. Etzion et P. Niblett, *Event Processing in Action*, 1st éd. Greenwich, CT, USA: Manning Publications Co., 2010.
[8]   F. Bénaben, M. Lauras, S. Truptil, et N. Salatgé, « A Metamodel for Knowledge Management in Crisis Management », in *2016 49th Hawaii International Conference on System Sciences (HICSS)*, 2016, p. 126-135.

Anton A. Kiss, Edwin Zondervan, Richard Lakerveld, Leyla Özkan (Eds.)
Proceedings of the 29th European Symposium on Computer Aided Process Engineering
June 16th to 19th, 2019, Eindhoven, The Netherlands.
http://dx.doi.org/10.1016/B978-0-128-18634-3.50255-1

Targeting material exchanges in industrial symbiosis networks

Ana Somoza-Tornos, Valeria Giraldo-Carvajal, Antonio Espuña, Moisès Graells[*]

Chemical Engineering Department, Universitat Politècnica de Catalunya, EEBE, C/ Eduard Maristany 16, 08019 Barcelona, Spain

moises.graells@upc.edu

Abstract

Industrial symbiosis plays an important role in the process industry, where an effective management of resources can bring both economic and environmental benefits. Traditionally, the design and extension of eco-industrial parks has been mainly based on intuition and experience; yet, the number of actors involved, the number of flows to manage and the different nature of the materials and energy exchanged lead to complex problems that can benefit from the use of systematic decision-making methods. The aim of this contribution is to produce useful tools to identify the most favourable synergies and transformations for a process network. The proposed model has been tested on a case study that consists of an ethylene and chlorine based eco-industrial park. Results confirm the capabilities of the proposed targeting methodology to match sources and sinks of resources.

Keywords: Industrial symbiosis, process networks, targeting, eco-industrial parks, optimization, sustainability.

1. Introduction

In the last years, there has been a growing awareness of the importance of applying circular economy approaches to close material, energy and water cycles (Merli et al., 2018). With their focus on closing loops in industrial processes, Industrial Symbiosis (IS) principles have been widely applied in many specific sectors (van Ewijk et al., 2018; Deschamps et al., 2018). A shared concern is engaging industries to join: the more participants are involved, the better environmental performance is achieved.

However, current eco-industrial parks (EIP) and resource exchange designs are mainly ad-hoc Industrial Symbiosis approaches, based on identifying opportunities through expert analysis. These strategies, even after a systematic local search, usually lead to sub-optimal solutions. In light of this, there is a need of systematic methods aimed at coping with the complexity of the problems by exploring only feasible and promising alternatives. Previous works have focused on the development of tools for transformation companies that might make profit of connecting sources and sinks of resources, and thus reducing the final waste involved (Somoza-Tornos et al., 2017 and 2018).

The aim of this work is to develop efficient targeting methods that identify the most promising synergies for industrial symbiosis while discarding infeasible links. A material network is designed to model the exchange of materials that become profitable for the involved actors (sources, transformers and sinks). Conservation laws and thermodynamic constraints are used to discern between the resulting alternatives.

2. Problem statement

The system under study is illustrated in Figure 1.

The targeting problem can be stated as follows: Given a set of waste streams j that could be potentially treated to satisfy the raw materials demand of streams k; a set of chemical reactions that may take place between the i products composing the mentioned streams: and other available data, including complete economic data, technical constraints and thermodynamic parameters. The decisions to be made comprise the amount of waste processed by the system, whether or not it is transformed, the requirements of external feeds or demands, how the products are distributed to satisfy the needs of customers and which side products have to be disposed.

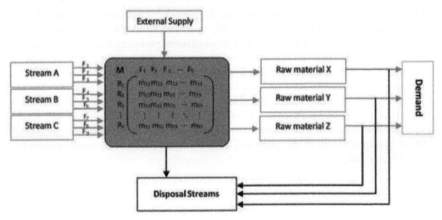

Figure 1. Material network scheme.

3. Mathematical formulation

The problem is formulated as a MILP that finds the optimal synergies between waste producers and raw materials consumers.

The total inlet to the system includes waste streams W_{ji} and potential supply of products required to complete the transformation ES_i (Eq. (1)).

$$\sum_j W_{ji} + ES_i = F_i^{in} \quad \forall i \tag{1}$$

Eq. (2) defines the mass balance of the system considering the inlet, outlet and generation terms, the last one calculated through stoichiometric coefficients R_{mi} and the extent of the reaction F_m^{gen}.

$$F_i^{in} + \left(\sum_m R_{mi} \cdot F_m^{gen}\right) = F_i^{out} \quad \forall i \tag{2}$$

The result of the transformation F_i^{out} is then divided in two, the amount sent to customers F_{ki}^{rm} and the side products that are unassigned F_{li}^d (Eq. (3)). This balance is completed with the introduction of the term F_i^{ed} to represent the external demand that new partners may have.

$$F_i^{out} = \sum_k F_{ki}^{rm} + \sum_l F_{li}^{d} + F_i^{ed} \quad \forall i \tag{3}$$

z_k is defined in Eqs. (4),(5) as a binary variable that takes a value of 1 if the amount sent to the customers, F_{ki}^{rm}, is greater than the demand.

$$D_{ki} \cdot z_k \leq F_{ki}^{rm} \quad \forall k, i \tag{4}$$

$$F_{ki}^{rm} - D_{ki} \leq M \cdot z_k \quad \forall k, i \tag{5}$$

Hence, when the demand is surpassed, the profit of selling C_k it is penalized with a cost for the excess of delivery C_k^d.

$$f_k^1 \leq M \cdot z_k \quad \forall k \tag{6}$$

$$f_k^1 \leq \left(\sum_i D_{ki} \cdot C_k \right) - C_k^d \cdot \sum_i (F_{ki}^{rm} - D_{ki}) \quad \forall k \tag{7}$$

On the contrary, when demand is not covered, only the amount sent to the customer must be taken into account for the profit calculation.

$$f_k^2 \leq M \cdot (1 - z_k) \quad \forall k \tag{8}$$

$$f_k^2 \leq \left(\sum_i F_{ki}^{rm} \cdot C_k \right) \quad \forall k \tag{9}$$

The energy balance of the system is calculated as in Eq. (10), where Q_m^{exc} denotes the amount of energy added or extracted from the system.

$$\left(\sum_i R_{m,i} \cdot F_m^{gen} \cdot H_i \right) = Q_m^{exc} \quad \forall m \tag{10}$$

Binary variable y_m is defined in Eqs. (11),(12) to differentiate processes that require heating or cooling and apply costs accordingly.

$$Q_m^{exc} \leq M \cdot y_m \quad \forall m \tag{11}$$

$$-Q_m^{exc} \leq M \cdot (1 - y_m) \quad \forall m \tag{12}$$

Eqs. (13),(14) apply when heat is extracted from the system, and cost parameter CQ_{out} is considered.

$$f_m^3 \leq M \cdot y_m \quad \forall m \tag{13}$$

$$-f_m^3 \leq Q_m^{exc} \cdot CQ_{out} \quad \forall m \tag{14}$$

Conversely, when heat is added to the system, the cost is calculated through Eqs. (15),(16).

$$f_m^4 \leq M \cdot (1 - y_m) \quad \forall m \tag{15}$$

$$-f_m^4 \leq -Q_m^{exc} \cdot CQ_{in} \tag{16}$$

The objective function to be maximized is the economic balance shown in Eq. (17). It considers the profit obtained from satisfying the demand of current companies and potential new partners. Aggregated cost parameters associated with the different transformation routes are considered at this step. These aggregated costs, including capital and operational costs plus indirect costs like transportation and management, must be estimated according to the specific circumstances, and the sensibility of the results to these estimations must be adequately assessed.

$$OF = -\left(\sum_i \sum_l F_{li}^d \cdot C_l\right) - \left(\sum_i \sum_j W_{j,i} \cdot C_j\right) - \left(\sum_m F_m^{gen} \cdot C_m^R\right) - \left(\sum_i Es_i \cdot C_i^{es}\right) + \left(\sum_i F_i^{ed} \cdot C_i^{ed}\right) + \sum_k f_k^1 + \sum_k f_k^2 - \left(\sum_m f_m^3 + \sum_m f_m^4\right)$$

(17)

The resulting model for the targeting can be posed as follows:

TSym   min [OF]

s.t. Eqs. (1)-(17)

4. Case study

The capabilities of the model are illustrated in a case study consisting of an eco-industrial park based on ethylene and chlorine, with 10 available waste streams and 7 demands of raw material have been defined. The considered compounds include acetic acid, benzene, chlorine, vinyl chloride, ethanol, ethylbenzene, ethylene, ethylene dichloride, ethylene oxide, hydrochloric acid, oxygen, tetrachloroethylene, trichloroethylene, vinyl acetate and water. Eqs. (18)-(26) show the reactions that the park would consider can take place between the components by a transformation company.

$$C_2H_4 + Cl_2 \rightarrow C_2H_4Cl_2 \tag{18}$$

$$3Cl_2 + C_2H_4Cl_2 \rightarrow 4HCl + C_2Cl_4 \tag{19}$$

$$2C_2H_4 + 4HCl + O_2 \rightarrow 2C_2H_4Cl_2 + 2H_2O \tag{20}$$

$$C_2H_4Cl_2 \rightarrow C_2H_3Cl + HCl \tag{21}$$

$$C_2H_4Cl_2 + Cl_2 \rightarrow C_2HCl_3 + 3HCl \tag{22}$$

$$C_2H_4 + H_2O \rightarrow CH_3CH_2OH \tag{23}$$

$$C_2H_4 + \frac{1}{2}O_2 \rightarrow CH_2OCH_2 \tag{24}$$

$$C_2H_4 + CH_3COOH + \frac{1}{2}O_2 \rightarrow CH_3CO_2CHCH_2 + H_2O \tag{25}$$

$$C_6H_6 + C_2H_4 \rightarrow C_6H_5C_2H_5 + H_2O \tag{26}$$

5. Results

The resulting MILP problem, featuring 1209 equations, 1064 continuous variables and 159 binary variables, has been modeled in GAMS 23.8.2 and solved with CPLEX 12.4.

Four different scenarios have been defined to examine the chances of incorporating new participants in the symbiotic network. These new participants can either be a source of waste or raw materials consumers, all presenting their own capacity limitations.

a) Base case of the existing eco-industrial park (EIP)
b) New companies could join the EIP and offer new sources of waste
c) New partners could join the EIP and take advantage of generated waste
d) New companies could both as a source and sink of resources

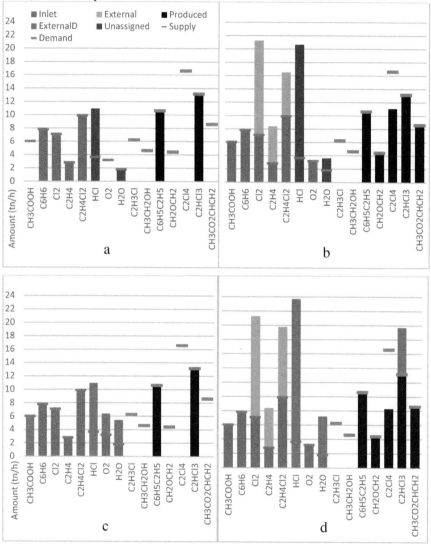

Figure 2. Waste usage and raw materials satisfaction for scenarios a, b, c, d.

Figure 2.a depicts the waste usage and the raw materials demand satisfaction for the existing EIP. The first case, where no external supply is available, is constrained by the limit in the waste supply. Reactions (22) and (26) are active to produce ethylbenzene and trichloroethane. The lack of ethylene does not allow acidic acid to be used in reaction (25) and there are sources of an excess of HCl and water that is not reused. Figure 2.b shows the effect of finding new partners that may be a source of waste. By adding new producers of chlorine, ethylene and ethylene dichloride to the park, more of the demands are internally covered and thus the external requirements of raw materials are reduced. Transformations (19), (24) and (25) would have to be activated to produce ethylene oxide, tetrachloroethylene and vinyl acetate, thus increasing the amount of waste processed and the profit of the entire complex. This would increase even more the excess of side products. In Figure 2.c the opposite case is represented, where new partners would only be interested in raw materials production. As the waste supply was limiting the base case, only the side products in excess can be used, resulting in a reduced grow of the EIP. When these limitations are overcome in Figure 2.d, the most promising ways of making the EIP grow are identified, and so are the transformations that the policy-makers should foster.

6. Conclusions

This work has addressed the development of a tool to identify the most promising routes to match sources and sinks of resources, even when a transformation step is required. This will help to reduce the complexity of the analysis required during the synthesis and design of industrial processing networks. Hence, the model offers policy-makers a method to systematically identify and assess opportunities for increasing the integration of process networks in industrial complexes. Thus, Administrations may use their resources to incentive partners that will ensure economically feasible synergies with the ultimate goal of reducing waste. An adequate reformulation of the objective function may also allow these companies to identify their opportunities, and even the different members of the industrial network the best cooperation opportunities (multi-objective approach). Future work will also focus on the application of combined targeting-synthesis methodologies to systematically analyse in detail the resulting proposals.

Acknowledgements

Financial support received from the Spanish "Ministerio de Economía, Industria y Competitividad" and the European Regional Development Fund, both funding the research Projects AIMS (DPI2017-87435-R) and PCIN-2015-001/ELAC2014/ESE0034 is fully acknowledged.

Ana Somoza-Tornos thankfully acknowledges financial support received from the Spanish "Ministerio de Educación, Cultura y Deporte" (Ayuda para la Formación de Profesorado Universitario - FPU15/02932).

References

J. Deschamps, B. Simon, A.Tagnit-Hamou, B. Amor (2018). Is open-loop recycling the lowest preference in a circular economy? Answering through LCA of glass powder in concrete. J. Clean. Prod. 185 14-22.

R. Merli, M. Preziosi, A. Acampora (2018). How do scholars approach the circular economy? A systematic literature review. J. Clean. Prod. 178, 703-722.

A. Somoza-Tornos, M. Graells, A. Espuña (2018). Evaluating the effect of separation and reaction systems in industrial symbiosisComp. Aided Chem. Engng 43, pp. 749-754.

A. Somoza-Tornos, M. Graells, A. Espuña (2017). Systematic Approach to the Extension of Material Exchange in Industrial Symbiosis. Comp. Aided Chem. Engng 40-B, pp.1927-1932.

S. van Ewijk, J. Y. Park, M. R. Chertow (2018). Quantifying the system-wide recovery potential of waste in the global paper life cycle. Resour. Conserv. Recycl. 134, 48-60.

Anton A. Kiss, Edwin Zondervan, Richard Lakerveld, Leyla Özkan (Eds.)
Proceedings of the 29th European Symposium on Computer Aided Process Engineering
June 16th to 19th, 2019, Eindhoven, The Netherlands. © 2019 Elsevier B.V. All rights reserved.
http://dx.doi.org/10.1016/B978-0-128-18634-3.50256-3

Modeling and Multi-Objective Optimization of Syngas Fermentation in a Bubble Column Reactor

Elisa M. de Medeiros,[a,b,*] John A. Posada,[a] Henk Noorman,[a,c] Rubens Maciel Filho[b]

[a]*Department of Biotechnology, Delft University of Technology, van der Maasweg 9, Delft 2629 HZ, The Netherlands*

[b]*School of Chemical Engineering, University of Campinas, Av. Albert Einstein 500, Campinas 13084-852, Brazil*

[c]*DSM Biotechnology Center, PO Box 1, Delft 2600 MA, The Netherlands*

E.MagalhaesdeMedeiros@tudelft.nl

Abstract

Ethanol may be produced from waste materials via a thermochemical-biochemical route employing gasification and syngas fermentation by acetogenic bacteria. This process is considered promising, but commercialization might be hindered by sub-optimal choices of design and operating conditions. In the present work, process systems engineering (PSE) techniques were applied for the optimization of a large-scale syngas fermentation bioreactor. Starting with the development of a dynamic model for a bubble column reactor with gas recycle, the multiple system outputs were studied with Principal Component Analysis to assist in the definition of relevant objective functions, and artificial neural networks were used to approximate the steady-state responses with fast and accurate functions. This framework was then used to conduct a multi-objective optimization aiming at maximizing the ethanol production rate, lower heating value efficiency and ethanol titer, while also minimizing acetic acid titer and reactor volume.

Keywords: bioethanol, syngas fermentation, multi-objective optimization, neural networks, principal component analysis

1. Introduction

Syngas fermentation is a biological process in which facultative autotrophic bacteria called acetogens convert CO, H_2 and CO_2 into cell mass, acids and alcohols. Most wild-type strains produce acetic acid and ethanol −and this is the focus of this work−, but other carboxylic acids and higher alcohols may also be obtained (Wainaina et al., 2018). Since syngas may be produced from the gasification of a diversity of feedstocks, including various waste materials, and syngas fermentation is conducted under mild conditions, this process has been gaining growing attention alongside other conversion pathways for waste to fuels and chemicals. Nonetheless, bench-scale experiments reported in the literature are often marked by difficulties with low productivity and unsatisfactory product selectivity. Moreover, modeling and optimization studies on this process are scarce in the literature, which makes it difficult to conduct appropriate economic and environmental assessments. This work therefore aims to fulfill part of this demand by presenting a bubble column model and a multi-objective optimization (MOO) framework using neural network surrogate models. The ultimate goal is to provide insights into the process conditions and reactor design that enhance reactor performance and demonstrate how the proposed methods may be used for this purpose.

2. Syngas bioreactor model

2.1. Dynamic bubble column

The bubble column reactor (BCR) hydrodynamic model is based on the Axial Dispersion Model described by Deckwer (1992) and it considers the following main assumptions: (i) isothermal operation; (ii) axial dispersion is neglected in the gas phase; (iii) in the liquid phase substances are carried differentially by convective and dispersive flows; (iv) volumetric mass transfer coefficient ($k_L a$), gas hold-up (ε_G) and liquid dispersion coefficient (D_L) are constant throughout the column; (v) atmospheric pressure at the top of the column. The system consists of 13 types of state variables distributed in space: the concentrations [mol/m^3] of 6 chemical species (i = CO, H$_2$, CO$_2$, ethanol or EtOH, acetic acid or HAc, and H$_2$O) distributed in the gas ($C_{G,i}$) and liquid ($C_{L,i}$) phases, and cell biomass ($C_{L,X}$) which is present in the liquid phase. It is therefore described by $13 \cdot N$ ordinary differential equations, where N is the number of discretization points considered. The general governing equations are formulated in Eqs. (1)-(3), where x is the spatial variable and u_G is the superficial gas velocity calculated at each point from a mole balance, as the gas expands due to reduced hydrostatic head or shrinks due to consumption by the cells. The hydrodynamic parameters $k_L a$ and ε_G are calculated using correlations by Heijnen and van't Riet (1984) and Deckwer et al. (1974), and a indicates the column operation mode: -1 if counter-current (present case), 1 otherwise. The saturation concentrations $C_{L,i}^*$ and $C_{G,i}^*$ are calculated using Henry's law constants and vapor pressures at 36 °C. In Eqs. (1)-(3), the boundary conditions ($x = 0$ and $x = L$) follow the model by Deckwer (1992), and the spatial derivatives are calculated using 1st and 2nd order finite differences.

$$\frac{\partial C_{G,i}}{\partial t} = -\frac{u_G}{\varepsilon_G} \cdot \frac{\partial C_{G,i}}{\partial x} - \frac{C_{G,i}}{\varepsilon_G} \cdot \frac{\partial u_G}{\partial x} - \frac{\dot{n}_{MT}}{\varepsilon_G} \tag{1}$$

$$\frac{\partial C_{L,i}}{\partial t} = -\frac{a u_L}{(1-\varepsilon_G)} \cdot \frac{\partial C_{L,i}}{\partial x} + D_L \frac{\partial^2 C_{L,i}}{\partial x^2} + \frac{\dot{n}_{MT}}{(1-\varepsilon_G)} + v_i C_{L,X} \tag{2}$$

$$\frac{\partial C_{L,X}}{\partial t} = -\frac{a u_L}{(1-\varepsilon_G)} \cdot \frac{\partial C_{L,i}}{\partial x} + D_L \frac{\partial^2 C_{L,i}}{\partial x^2} + \frac{\dot{n}_{MT}}{(1-\varepsilon_G)} + (\mu - k_d) C_{L,X} \tag{3}$$

$$\dot{n}_{MT} = k_L a_i (C_{L,i}^* - C_{L,i}), \; i = CO, H_2, CO_2 \tag{4a}$$

$$\dot{n}_{MT} = -k_L a_i (C_{G,i}^* - C_{G,i}), \; i = EtOH, HAc, H_2O \tag{4b}$$

The biokinetic variables v_i, μ and k_d refer to the specific secretion or production rate [mmol.g^{-1}.h^{-1}] of each component i, the specific biomass growth rate [h^{-1}] and the biomass decay rate [h^{-1}]. This kinetic model considers ethanol and acetic acid formation as well as cell growth, using CO/H$_2$ as energy sources and CO/CO$_2$ as carbon sources. The parameters that are used in the calculation of these variables were adjusted using experimental data obtained in a continuous stirred-tank reactor with high ethanol concentration, and the detailed modeling and statistical analysis are at the moment unpublished but under submission to a scientific journal (de Medeiros et al., Forthcoming).

2.2. Steady-state surrogate models

Steady-state solutions were generated by integrating the BCR ODE system over a sufficiently long time range using the MATLAB stiff solver *ode15s*. However, this procedure was proved very computationally expensive for some choices of process parameters, specifically when very low dilution rates were used in combination with high recycle ratios, with some case scenarios taking as long as three minutes to be solved. Since the multi-objective evolutionary algorithm requires a large number of objective function evaluations, the objectives were approximated with fast surrogate models. Similar strategies have been adopted before for a variety of chemical processes employing different types of models, such as artificial neural networks (ANNs) (Chambers and Mount-Campbell, 2002) and Kriging models (Quirante and Caballero, 2016). In the present work, multi-layer feedforward ANNs were trained with Bayesian regularization backpropagation to approximate responses that would be relevant for the optimization, namely: the concentrations of ethanol and acetic acid at the column outlet and the lower heating value (LHV) efficiency η_{fuel}, the latter being defined as the ratio between the total ethanol LHV and the total energy input considering syngas LHV and heat intake associated with the power consumed in the gas compressor. ANN architectures were selected after several tries, with the final ANNs containing two hidden layers with 10 to 20 neurons per hidden layer depending on the case. The data set used for ANN training and validation consisted of 2,780 points comprising random combinations of the operating and design parameters considered as model inputs (and optimization decision variables): dilution rate (D [h^{-1}]), fresh gas residence time (GRT [min]), gas recycle ratio (RR), column height (L [m]), and column aspect ratio (L/d_C). For the gas compressor, the power consumption was calculated in Aspen plus as function of the gas composition and flow rate at the top of the column, and the pressure at the bottom; Aspen Simulation Workbook was employed to facilitate the automatic execution of all considered scenarios. The output of the ANNs is depicted in Fig. 1 against the rigorous solutions (i.e. obtained with integration of the ODE system), where for each response the results are normalized with relation to the minimum and maximum values of the corresponding rigorous response.

3. Formulation of MOO problem

3.1. Decision variables and fixed conditions

The MOO problem aims at finding optimal combinations of the decision variables described in Sec. 2.2 within the following pre-defined ranges: *0.001 ≤ D ≤ 0.1* ; *5 ≤ GRT ≤ 50* ; *0 ≤ RR ≤ 0.8* ; *10 ≤ L ≤ 30* ; *3 ≤ L/d_C ≤ 6*. Some variables are fixed in this study but should be analyzed in connection with other unit operations in future studies of the global process. These are: the fresh syngas composition (fixed at 60% CO, 35% H_2, 5% CO_2), which is affected by the choices of design and operating conditions in the gasifier; and liquid inlet composition (considered pure water), which could contain small amounts of acetic acid if distillation waste streams are recirculated within the process. Cell recycle is also considered with a fixed purge fraction of 10% of the cells.

3.2. Selection of objectives using Principal Component Analysis

Process performance often needs to be measured by multiple criteria which are conflicting, such that a decision must be made based on Pareto-optimal solutions. Nonetheless, a too large number of objectives might bring unnecessary complexity to the problem, both in terms of computational effort and interpretation of the results. With that in mind, the multivariate technique of Principal Component Analysis (PCA) can be employed to uncover relationships among objectives and reduce the dimensionality of the problem (Sabio et al., 2012). In a nutshell, this is achieved by finding a new coordinate system that explains the variance in the original data set using a smaller dimensionality. In this study, PCA was applied to a set of 8 potential objectives calculated from the BCR model results at various conditions. Fig. 2 shows the coefficients (loadings) of the two first principal components (PC), which together explain nearly 87% of the data variance. These coefficients represent the contribution of the original variables to the PCs, thereby pointing out aligned or opposing tendencies. From this analysis, the gas conversions X_{CO} and X_{H2} were removed from the MOO since their tendencies are well captured by the LHV efficiency η_{fuel}. The other efficiency, η_{total}, considers the LHV of both ethanol and unconverted syngas. Since ethanol production is accompanied by CO_2 formation, this efficiency has an opposite trend to the conversions, eventually reaching 1 if no ethanol is formed; hence it was also removed from the analysis. The minimization objective functions were then defined with the 5 remaining metrics: (i) $OF_1 = -Prod$ (ethanol production rate); (ii) $OF_2 = -\eta_{fuel}$; (iii) $OF_3 = -C_{L,EtOH}$; (iv) $OF_4 = C_{L,HAc}$; (v) $OF_5 = V_R$ (reactor volume). Apart from the production rate and LHV efficiency that ideally would be maximized, the definition of objectives OF_3 and OF_4 takes into account possible energy expenses in downstream operations, and OF_5 is aimed at reducing CAPEX.

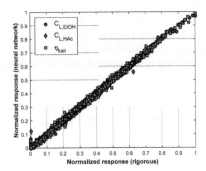

Figure 1. Validation of neural networks

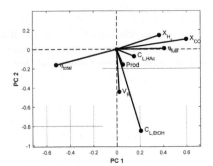

Figure 2. Principal Component Coefficients

4. Generation and analysis of the Pareto front

The Pareto front was generated in MATLAB using a hybrid algorithm that starts with multi-objective genetic algorithm (MOGA) and shifts to a local goal attainment method once the solution is near the optimum front. The short evaluation times due to the adoption of neural networks made it possible to test the hybrid MOGA with different population sizes, with and without the hybrid search. Fig. 3 presents the pairwise projections of the Pareto front, specifically for the objective OF_1 (maximization of ethanol production rate) against the other four objectives. These plots can be used to delineate a range of desirable and feasible operation; for example, it is not convenient to aim at production rates higher than 2,700 kg/h, since this would require reactor vessels of at least 1,500 m^3 which might be unpractical in real operation. It can also be noted that, for the current system, the highest LHV efficiencies are achieved at lower EtOH production rates, which makes sense since higher gas conversions are obtained when the gas residence time is longer (lower gas velocity), thus leading to smaller mass transfer coefficients and reaction rates. The ethanol titer is bounded to a maximum 30 g/L for a large range of production rates, with increasing formation of undesirable acetic acid, while low production rates can lead to over 50 g/L ethanol with almost zero by-product. The normalized values of the parameters are shown in Fig. 4 for sub-sets selected from the boundary dashed lines depicted in Fig. 3, where it can be seen how the optimal parameters adapt to the objective functions. An ultimate decision would need to consider the relevance of these objectives on the outcome of the global process; for example, a lower production rate that is compensated by a more concentrated broth might in some cases be more advantageous than the other way around. Either way, the current results may be used to define an optimal region where the LHV efficiency is higher than 30%, the ethanol titer is higher than 20 g/L, the acetic acid titer is lower than 10 g/L, and the reactor volume is less than 1,500 m^3; the decision variables in this region are in the range: $0.02 \leq D \leq 0.03$; $32 \leq GRT \leq 50$; $0.56 \leq RR \leq 0.8$; $27.8 \leq L \leq 29.4$; $3.8 \leq L/dC \leq 4.7$.

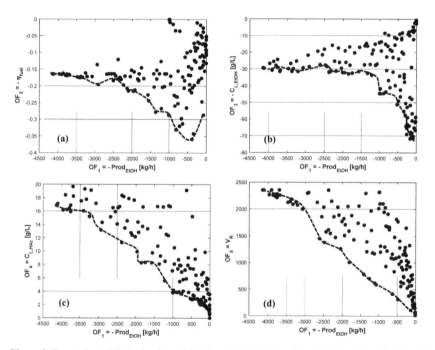

Figure 3. Pareto front 2-D Projections: (a)OF_1 x OF_2; (b)OF_1 x OF_3; (c)OF_1 x OF_4; (d)OF_1 x OF_5

5. Conclusions

In this study a modeling and optimization framework was built for syngas fermentation in bubble column reactors. First a spatial dynamic model was created to predict the concentrations of the chemical species and cells along the reactor in the gas and liquid phases; and the most relevant responses were approximated by artificial neural networks to reduce computational effort, so that objective functions could be evaluated efficiently during the multi-objective optimization search. We demonstrated how Principal Component Analysis can be used to reduce the complexity of the MOO problem without losing significant information. Although further studies are needed connecting the bioreactor to other unit operations, the five-objective Pareto front generated in this study can be used to pre-define an optimal range of process conditions and reactor size.

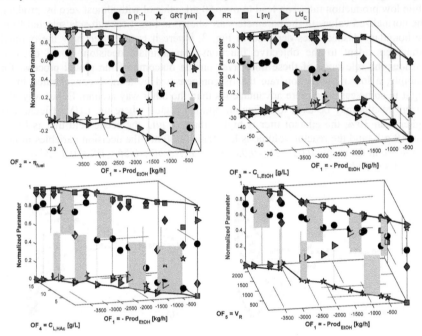

Figure 4. Selected Points from the Pareto Front Projected in 2-D vs Normalized Parameters

References

Chambers, M., Mount-Campbell, C. A., 2002. Process optimization via neural network metamodeling. Int. J. Prod. Econ. 79, 93–100.

Deckwer, W. -D., Burckhart, R., Zoll, G., 1974. Mixing and mass transfer in tall bubble columns. Chem. Eng. Sci. 29, 2177–88.

Deckwer, W. -D., 1992. Bubble Column Reactors. Chichester, New York: Wiley.

Heijnen, J. J., van't Riet, K., 1984. Mass transfer, mixing and heat transfer phenomena in low viscosity bubble column reactors. The Chemical Engineering Journal 28, B21–B42.

de Medeiros, E., Posada, J. A., Noorman, H., Maciel Filho, R. Dynamic modeling of syngas fermentation in a continuous stirred tank reactor: multi-response parameter estimation with statistical analysis. (Forthcoming).

Quirante, N., Caballero, J. A., 2016. Optimization of a sour water stripping plant using surrogate models. Computer Aided Chemical Engineering 38, 31–36.

Sabio, N., Kostin, G., Guillén-Gosálbez, G., Jiménez, L., 2012. Holistic minimization of the life cycle environmental impact of hydrogen infrastructures using multi-objective optimization and principal component analysis. Int. J. Hydrogen Energ. 37, 5385–5405.

Wainaina, S., Horváth, I. S., Taherzadeh, J., 2018. Biochemicals from food waste and recalcitrant biomass via syngas fermentation: A review. Bioresource Technol. 248, 113–121.

Anton A. Kiss, Edwin Zondervan, Richard Lakerveld, Leyla Özkan (Eds.)
Proceedings of the 29th European Symposium on Computer Aided Process Engineering
June 16th to 19th, 2019, Eindhoven, The Netherlands. © 2019 Elsevier B.V. All rights reserved.
http://dx.doi.org/10.1016/B978-0-128-18634-3.50257-5

Modeling of multi-effect desalination process operated with thermosolar energy applied to the northeastern Brazil

Diego P.S Cunha[a], Vanessa V. Gomes[a], Karen V. Pontes[a]

[a]Universidade Federal da Bahia, Rua Prof. Aristides Novis 02, Salvador-BA, 40210-630, Brazil
diego.cunha@ufba.br

Abstract

Desalination is one of the possible solutions to the world's potable water scarcity. In order to make the process feasible and sustainable, use of renewable energy is a very useful option. The multi-effect desalination (MED) process linked with solar energy is one of the most thermodynamically efficient and the one with lower cost by produced potable water. In this context, the present work focuses on modelling a MED process with thermosolar energy applied to the local conditions of northeastern Brazil. This region suffers from water scarcity but, on the other hand, shows the highest levels of solar radiation in Brazil. Despite that, the literature still lacks a study concerning the technical viability of operating a desalination process with solar energy in this semi-arid region. The steady-state model is based on mass and energy balances and empirical correlations from literature to calculate the enthalpy of water and brine as well as boiling point elevation and latent heat. The model is firstly validated against simulated results from the literature regarding the Plataforma Solar de Almeria. The results have shown that the implementation of a solar desalination plant in Northeast Brazil is technically feasible. The obtained model is useful for further process optimization towards minimization of investment and operational costs and to facilitate access to potable water in arid regions with high solar incidence.

Keywords: Desalination, MED, Solar, Brazil

1. Introduction

The need of potable water increases proportionally to the world's population growth; however, the planet's water sources are mostly salty water and the fresh water sources are scarce. According to Al-Rawajfeh et al. (2017), more than two billion people currently suffer from lack of water and until 2025 this number will rise to more than four billion people. One possible solution to increase the fresh water availability is the salty water desalination, since salty water is abundant through oceans and seas. Desalination is a process that aims to remove the excess of salt, microorganisms and some particles present on seawater or brackish water. According to Byrne et al (2015), this process may be accomplished in different ways, the most commercially used are: reverse osmosis (RO), multi stage flash desalination (MSF) and multi-effect desalination (MED).

Energy cost is one of the factors which has strong impact on desalination process total cost (Ghobeity and Mitsos, 2014; Shatat et al, 2013). The energy used by the main desalination industries comes predominantly from fossil fuels, which are responsible for increasing the release of carbon dioxide to the atmosphere, contributing to the greenhouse effect (Siirola, 2012). Due to this energy cost and carbon dioxide emission, a renewable and abundant energy source, like solar energy, would offer some benefits. Solar energy can be used as electric or thermal energy source.

MED and RO are the cheapest desalination processes and their costs in large scale commercial plants are around 1 USD/m³ (Shatat et al, 2013). On the other hand, when comparing desalination costs using renewable energies, Alkaisi et al. (2017) report that MED process with solar thermal energy has the lowest cost by m³ (2.5-3 USD/m³) while RO process with solar electric energy has a cost of 12.5-16.8 USD/m³. RO process requires mainly electrical energy and MED mainly thermal energy. El-Dessouky et al. (1998) also affirm that MED process has more operation flexibility, being able to operate from 0 to 100% of plant capacity, and presents shorter start time and less specific investment cost.

Considerable efforts have been made for the research of MED desalination plants coupled with solar energy. Desalination plant of Plataforma Solar de Almeria (PSA), in Spain, for example is one of those plants. It is a desalination plant composed of 14 effects with a design capacity of 72 m³/day. This plant has been modeled by several authors. Hatzikioseyian et al. (2003) develop a stationary model aiming to evaluate plant performance regarding energy consumption. Roca et al. (2012) develop a dynamic model to optimize distillate production. De La Calle et al. (2014) focus on thermal transient behavior of a MED plant's first cell, which is main responsible for plant energy consumption. In Turkey, Yilmaz and Söylemez (2012) simulate a stationary MED plant in Turkey regions using wind and solar energy. It has 12 effects and capacity of 1m³/day. Simulation results show that, although both energies are intermittent, solar energy is more stable than wind. In Jordan, Bataineh (2016) analyzes a stationary MED desalination plant combined with thermo compressor coupled with solar energy. It has 12 effects, capacity of 43,000 m³/day, 1,080,000 m² of solar collectors and a gas boiler back up to continuos operation. In this scenario, results show that solar energy can suply almost 70% of energy demanded for this plant. Bataineh (2016) also affirms that MED desalination plants powered by solar energies are able to produce more than 35 L/m²/day in regions with solar radiation above 4.8 kWh/m²/day.

The Northeast of Brazil is a semi-arid region where 11 million people suffers from fresh water scarcity (Rocha and Soares, 2015), although brackish water on underground wells is available. According to Fichter et al. (2017), the Northeast of Brazil has great potential to use solar energy because it presents solar radiation above 6 kWh/m²/day. Despite the investments and studies on desalination process in Middle East and North Africa, Brazil is still giving it first steps. Up to author's knowledge, there is no technical feasibility study of MED processes coupled with solar energy applied to the irradiation and water demand of the Brazilian semi-arid region. This paper, therefore, analyzes the technical feasibility of a MED desalination plant coupled with solar energy in Bom Jesus da Lapa (BA), a city in Northeast of Brazil. The study is based on a stationary model, validated with data from literature, considering different scenarios of solar radiation during the year.

2. Methodology

The desalination process model is based on Hatzikioseyian et al. (2003), which represents the PSA. Figure 1 represents the process layout which comprises the solar collectors and the MED process. In order to maintain plant's operation during night hours, the

water heated by collectors feeds two tanks which are perfectly insulated and store the
thermal energy. The hot water tank supplies the MED plant and the cold water returns to
the cold-water tank.

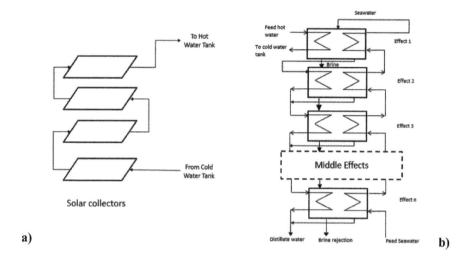

a)   b)

Figure 1 – Schematic of process layout: a) Solar collectors b) MED desalination plant.

Additionally to the mass and energy balance presented by Hatzikioseyian et al. (2003),
the following assumptions and equilibrium relations are considered in the present model:

- The temperature difference between saturated vapor and non-evaporated brine
  in equilibrium at an effect is obtained by the boiling point elevation (BPE),
  following Aly and El-Fiqi (2003);
- The mass production of saturated vapor at an effect is estimated by the latent
  heat of vaporization λ according to Roca et al. (2012);
- The enthalpy of brine and liquid water follows Sharqawy et al. (2010) model,
  which was validated here with experimental data from Chou (1968), presenting
  deviations up to 4 %;
- The vapor enthalpy follows Khademi et al. (2009), which was validated here
  with experimental data from Smith et al. (1996), presenting deviation up to 1 %;

Solar collectors' model was developed accordingly to Al-Ajlan et al. (2003), which focus
on flat plate collectors to heat water. The water outlet temperature is computed based on
the geometry of the collectors and on the following input variables: solar radiation,
ambient temperature, water mass flow rate and inlet temperature. The stationary model
was implemented in MOSAICmodeling (MOSAIC, 2018) and solved in MATLAB ®.

The base case for simulation has the same number of effects (14) and distillate production
capacity (72 m^3) as the PSA. The brackish water is fed at 30 °C and 0.3 bar with 7 g/L of
salt, considering the average brackish water salinity in Brazilian semi-arid region
(Menezes et al, 2011). According to Palenzuela et al. (2011), the hot water flow rate to
the first stage (Figure 1b) is 12 L/s, whose inlet and outlet temperatures are 74 °C and 70
°C, respectively. Considering the average water consumption recommended by the World
Health Organization guideline, that is 50-100 L per person per day (United Nations, 2018),

and the population from Bom Jesus da Lapa, which is around 60,000 people, the plant nominal capacity could supply approximately 2% of the population's fresh water need. The daily average solar radiation data from Bom Jesus da Lapa (BA) was collected by the RETScreen software (RETScreen, 2018) and is about 5.7 kWh/m²/day, while at Almeria this value is about 4.9 kWh/m²/day. The daily average solar radiation data considers a daily cycle with 24h but in Bom Jesus da Lapa (BA) the sun shines during 12h. Therefore, the average solar radiation considered for simulation is the double of the collected data.

3. Results and discussion

Figure 2a compares the model prediction with simulated data from Hatzikioseyian et al. (2003). Results show good fit of data as deviations up to 1% were obtained. Furthermore, the predicted total condensate water production is 5017.4 L and the predicted total brine production is 2982.6 L presenting deviations of 0.3% and 0.6%, respectively. Figure 2b shows the temperature and brine salinity profile along the desalination effects. The temperature decreases due to decreasing pressure in order to enable an efficient flash evaporation and brine salinity increases due to evaporation and production of fresh water.

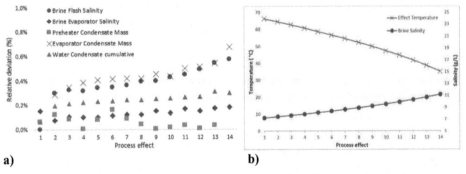

a)   b)

Figure 2: a) Validation of MED model: deviations of the model predictions against the data reported by Hatzikioseyian et al. (2003); b) Profile of temperature and brine salinity along process effects.

The solar collectors' model also presents good fit with Al-Ajlan et al. (2003), with deviations up to 0.7 %. The collectors' area needed to supply thermal energy to the MED base case is about 2929 m², which is 9.6% higher than the PSA. According to Bataineh (2016), though, the Almeria's plant has a gas boiler back-up system which can supply up to 30% of desalination thermal energy needs, what can explain the lower collectors' area in the PSA. Table 1 shows the required area for the solar collectors for MED plants with different capacities. It further shows the percentual of the population which would be benefited by the desalination plant.

The daily solar radiation at Bom Jesus da Lapa (BA) varies every month of the year according to Figure 3. Considering the solar area of the base case 2,929 m², the collectors' hot water temperature and the distillate production for the varying solar radiation are shown in Figure 3. The average distillate production is 24.5 L/m²/day for an average solar radiation of 5.7 kWh/m²/day Although Bataineh (2016) affirms that MED plants coupled with solar energies are able to produce more than 35 L/m²/day in regions with solar radiation above 4.8 kWh/m²/day, another source of energy, besides solar energy, is used. Furthermore, the plant studied by Bataineh (2016) is a large scale plant combined with thermo compressor system which may increase desalination energy efficiency.

Therefore, the distillate production obtained for Bom Jesus da Lapa is acceptable and cannot be directly compared with the value reported by Bataineh (2016).

Table 1 : Sensibility analysis of water production and solar collectors' area.

Water Production capacity(m³/day)	Solar Area(m²)	City's population supplied (%)
72	2,929	2 %
144	6,182	4 %
288	12,365	8 %
576	23,628	16 %

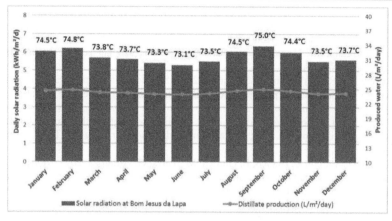

Figure 3 – Model Results and solar radiation data

4. Conclusions

This paper developed a model for a MED plant operated with thermosolar energy, which was successfully validated with data from literature. The model was used to evaluate the technical feasibility of the process in the northeastern Brazil, where the population suffers from water scarcity. The results show that is possible to produce fresh water using solar energy due to the high irradiation rates. Further studies might focus on the economic feasibility analysis to implement this MED desalination plant in Bom Jesus da Lapa (BA).

Acknowledgements

The authors would like to thank CAPES (Coordenação de Aperfeiçoamento de Pessoal de Nível Superior) and CNPq (Conselho Nacional de Desenvolvimento Científico e Tecnológico) for the financial support.

References

S.A. Al-Ajlan, H. Al Faris, H. Khonkar, 2003. A simulation modeling for optimization of flat plate collector design in Riyadh, Saudi Arabia. Renewable Energy, v. 28, n. 9, p. 1325–1339.

A. Al-Rawajfeh, S. Jaber, H.Etawi, 2017. Desalination by renewable energy: A mini review of the recent patents. Hemijska industrija, v. 71, n. 05, p. 451–460.

A. Alkaisi, R. Mossad, A. Sharifian-Barforoush, 2017. A Review of the Water Desalination Systems Integrated with Renewable Energy. Energy Procedia, v. 110, n. December 2016, p. 268–274.

N.H. Aly, A.K. El-Fiqi, 2003. Thermal performance of seawater desalination systems. Desalination, v. 158, n. 1–3, p. 127–142.

K.M. Bataineh, 2016. Multi-effect desalination plant combined with thermal compressor driven by steam generated by solar energy. Desalination, v. 385, p. 39–52.

J.C. Chou, 1968. Thermodynamic properties of aqueous sodium chloride solutions from 32 to 350°F,PhD thesis, Oklahoma State University.

A. De la Calle, J. Bonilla, L. Roca, P. Palenzuela, 2014. Dynamic modeling and performance of the first cell of a multi-effect distillation plant. Applied Thermal Engineering, v. 70, n. 1, p. 410–420.

H. El-Dessouky, H.M. Ettouney, A.L Imad, 1998. Steady-state analysis of the multiple effect evaporation desalination process. Chemical Engineering Technology, v. 21, p. 15–29.

A. Ghobeity, A. Mitsos, 2014. Optimal design and operation of desalination systems: New challenges and recent advances. Current Opinion in Chemical Engineering, v. 6, p. 61–68.

A. Hatzikioseyian, R. Vidali, P. Kousi, 2003. Modeling and thermodynamic analysis of a Multi-Effect Distillation (MED) plant for seawater desalination. National Technical University of Athens journal (NTUA).

M.H. Khademi, M.R. Rahimpour, A. Jahanmiri, 2009. Simulation and optimization of a six-effect evaporator in a desalination process. Chemical Engineering and Processing: Process Intensification, v. 48, n. 1, p. 339–347.

J.S. Menezes, V.P. Campos, T.A.C. Costa, 2011. Desalination of brackish water for household drinking water consumption using typical plant seeds of semi arid regions. Desalination, v. 281, p. 271–277.

P. Palenzuela, D. Alarcón, J. Blanco, E. Guillén, M. Ibarra, G. Zaragoza, 2011. Modeling of the heat transfer of a solar multi-effect distillation plant at the Plataforma Solar de Almería. Desalination and Water Treatment, v. 31, n. 1–3, p. 257–268.

RETScreen. "Natural Resources Canada.", 2018, http://www.nrcan.gc.ca/energy/software-tools/7465, Accessed September 2018.

L. Roca, L.J. Yebra, M. Berenguel, A. de la Calle, 2012. Dynamic modeling and simulation of a multi-effect distillation plant. p. 883–888.

R. Rocha, R.R. Soares, 2015. Water scarcity and birth outcomes in the Brazilian semiarid. Journal of Development Economics, v. 112, p. 72–91

M.H. Sharqawy, J.H. Lienhard, S.M. Zubair, 2010. Thermophysical properties of seawater : a review of existing correlations and data. Desalination and Water Treatment, v. 16, n. 1–3, p. 354–380.

M. Shatat, M. Worall, S. Riffat, 2013. Opportunities for solar water desalination worldwide: Review. Sustainable Cities and Society, v. 9, p. 67–80.

J.J. Siirola, 2012. A Perspective on Energy and Sustainability. In: I.A. Karimi, R.B.T. Srinivasan-Computer Aided Chemical Engineering (Eds.). 11 International Symposium on Process Systems Engineering. [s.l.] Elsevier. v. 31p. 1–7

J.M. Smith, H.C. Van Ness, M.M. Abbott. Introduction to Chemical Engineering Thermodynamics. 5. ed.

United Nations. "The human rigth to water and sanitation", 2018, http://www.un.org/waterforlifedecade/pdf/human_right_to_water_and_sanitation_media_brief.pdf, Accessed October 2018.

I.H. Yilmaz, M.S. SÖYLEMEZ, 2012. Design and computer simulation on multi-effect evaporation seawater desalination system using hybrid renewable energy sources in Turkey. Desalination, v. 291, p. 23–40.

Anton A. Kiss, Edwin Zondervan, Richard Lakerveld, Leyla Özkan (Eds.)
Proceedings of the 29th European Symposium on Computer Aided Process Engineering
June 16th to 19th, 2019, Eindhoven, The Netherlands. © 2019 Elsevier B.V. All rights reserved.
http://dx.doi.org/10.1016/B978-0-128-18634-3.50258-7

A stochastic environmental model to deal with uncertainty in life cycle impact assessment

Andreia Santos,[a,*] Ana Barbosa-Póvoa,[a] Ana Carvalho[a]

[a]*Centro de Estudos de Gestão do IST (CEG-IST), Av. Rovisco Pais, 1049-001 Lisbon, Portugal*

andreia.d.santos@tecnico.ulisboa.pt

Abstract

Life cycle assessment (LCA) is the most applied methodology to compare the environmental impacts of different products and processes. As uncertainty may be present in the first three steps of an LCA an uncertainty analysis should be considered allowing higher confidence in the results. However, few LCA studies include this analysis and when it is included is mostly applied to the LCI step of an LCA. Thus, the goal of this study is to develop a methodology to apply uncertainty analysis on the characterization factors used at the midpoint level in the LCIA step. The methodology employs Monte Carlo simulation as the stochastic modelling technique. A case study comparing the use of softwood and hardwood in the production of unbleached Kraft pulp is utilised to better explain the application of the developed methodology. As main results it can be concluded that the use of softwood as a raw material is overall worse for the environment than the use of hardwood when considering the production of unbleached Kraft pulp.

Keywords: Life Cycle Assessment, Uncertainty, Monte Carlo, ReCiPe, Pulp Production

1. Introduction

Life cycle assessment (LCA) is a four-step methodology for calculating the environmental impact of a product or service. The first step of this methodology, goal and scope definition, consists of defining the main objectives of the study, the functional unit (a representative element of the system being study), and system boundary. In the next step, life cycle inventory (LCI), all the flows that go in (e.g. raw materials and electricity) and out (e.g. emissions and solid waste) of the system are collected. The third step, life cycle impact assessment, consists in categorizing and characterizing the life cycle environmental impact of the system under study by converting the flows collected in the previous step into environmental impact using different parameters such as characterization factors. There are several LCIA methods that consider different categories and parameters. In the last step, results interpretation, the LCA study results are analysed and interpreted to identify the hotspots (areas to be prioritized for action) of the system and suggest possible improvements (ISO 14000 2006).

The aforementioned methodology is the most used to support environmental decision making and for this reason, the reliability of LCA results is of extreme importance. As uncertainty is always present in LCA (Björklund 2002), an adequate uncertainty analysis should accompany LCA studies to help decision makers having more confidence in the results. A literature review was conducted to understand how uncertainty in LCA has been addressed and it was concluded that uncertainty analysis has been increasingly included in LCA studies usually in the form of sensitivity analysis but also through Monte Carlo simulation and scenario analysis. Furthermore, uncertainty in the LCIA step is rarely treated and there is no clear guidance on how to take uncertainty into account. For these reasons, the main goal of this work is to develop a methodology capable of addressing the uncertainty associated with the characterization factors applied in the LCIA step by exploring Monte Carlo simulation. The developed methodology is applied to a case-study to show its usability in the decision-making process.

2. Methodology

The methodology followed in this study comprises three main steps:

Step 1 – Case Study Description

The first step of the methodology consists in applying the first two steps of an LCA to the case study in analysis.

Step 2 – Environmental Impact Assessment

In this step, the third step of an LCA, Life Cycle Impact Assessment, is deterministically and stochastically applied to the case study under analysis. To deterministically apply the LCIA, a deterministic environmental assessment model was developed in Excel considering the LCIA method ReCiPe 2008 with Hierarchist perspective (Goedkoop, et al. 2013), which considers 18 midpoint categories that can be further aggregated into three endpoint categories. After being normalized and weighted, these endpoint categories are added to calculate a single score (SS) that reflects the overall environmental impact of the system under study. To stochastically apply the LCIA, Monte Carlo simulation is applied to the deterministic model using the @Risk software (Palisade Corporation 2016). Three main tasks were followed for the stochastic analysis:

- Uncertain Parameters Identification – The characterization factors of 14 out of the 18 midpoint categories have uncertainty associated. The other four midpoint categories, agricultural land occupation, urban land occupation, natural land occupation, and water depletion do not have uncertainty associated.

- Probability Distributions Assignment – From the 14 midpoint categories consider as having uncertain characterization factors, probability distributions should be assigned to the characterization factors of the most impactful midpoint categories (identified using a pareto analysis). The @Risk software was used to fit several probability distributions to the characterization factors available for each impactful midpoint categories. The Akaike Information Criteria statistics was used as a goodness-of-fit statistics to help determine the best fit.

- Simulation Settings Definition – Instead of defining the number of iterations at the beginning of the simulation, the @Risk software should be run in Auto-Stop mode until all results reach an established convergence tolerance (3% and a confidence level of 95% are the default settings). This means that the simulation will stop when the estimated mean of the results simulated is within 3% of its actual value and is accurate 95% of the time.

Step 3 – Results Analysis

The last step of the methodology followed in this study coincides with the last step of the LCA methodology and consists in analysing and comparing the deterministic and stochastic environmental impacts obtained in the prior step.

3. Case-Study

The first stage in wood pulp production (i.e. pulping) consists in growing and harvesting the trees. Then, the trunk of each tree is bucked into logs, which are delivered to mills where they are debarked (bark contains few cellulose fibres, the main component of paper products) and chipped. The next stage consists in separating the cellulose fibres from the other wood components such as lignin (the "glue" that cements the fibres together), extractives (e.g. fats, waxes and alcohols), minerals and other inorganics. This can be done through different processes but the most common is the Kraft (sulphate) process. In this process, the wood chips are pressure-cooked with water and chemicals, sodium hydroxide (NaOH) and sodium sulphide (Na2S), in a digester. The resulting pulp is screened to remove uncooked wood and washed to remove the spent cooking mixture. Depending on the final product, the pulp can be bleached after being produced due to the brown coloration of the raw pulp caused by the presence of lignin that was not removed during cooking (Bajpai 2012) (Teschke 2011). In Figure 1, the pulp life cycle is represented.

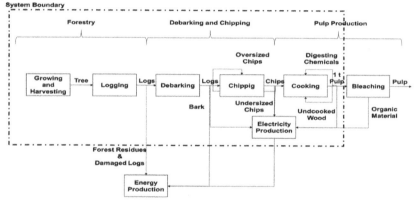

Figure 1 - Pulp life cycle including system boundary and functional unit

The species of the wood used as raw material is one of the characteristics that mostly influences the final properties of the pulp produced. Each tree species can be classified into one of two main families, the hardwoods and the softwoods.

In this paper, two different processes are compared – Case A (production of unbleached Kraft softwood pulp) and Case B (production of unbleached Kraft hardwood pulp). The system boundary considered is cradle-to-gate including the processes within the dashed rectangle in Figure 1 and the functional unit selected is 1 t of unbleached Kraft pulp produced in Europe.

The LCI of each system analysed was retrieved from the ecoinvent v3.3 database (ecoinvent 2018) with consequential system model which was assessed through the software SimaPro 8.4.0 (PRé Consultants 2018).

The results obtained through the deterministic and stochastic application of the LCIA to the case study presented are presented and discussed in the next section.

4. Results Analysis

4.1. Deterministic Results

After applying the deterministic environmental assessment model to the case study, it can be concluded that producing 1 t of unbleached Kraft softwood pulp (Case A) results in an

overall environmental impact of about 300 Pt while the same process using hardwood (Case B) results in an overall environmental impact of 238 Pt, which implies that the use of softwood as a raw material is worse for the environment than the use of hardwood.
A pareto analysis was conducted considering the normalised results at the midpoint level of using softwood as a raw material to produce 1 t of unbleached pulp (Figure 2).

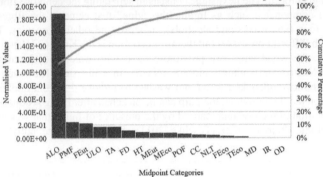

Agricultural Land Occupation (ALO), Particulate Matter Formation (PMF), Freshwater Eutrophication (FEut), Urban Land Occupation (ULO), Terrestrial Acidification (TA), Fossil Depletion (FD), Human Toxicity (HT), Marine Eutrophication (MEut), Marine Ecotoxicity (MEco), Photochemical Oxidant Formation (POF), Climate Change (CC), Natural Land Transformation (NLT), Freshwater Ecotoxicity (FEco), Terrestrial Ecotoxicity (TEco), Metal Depletion (MD), Ionising Radiation (IR), and Ozone Depletion (OD)

Figure 2 - Pareto analysis conducted considering the results at the midpoint level of Case A

By analyzing Figure 2, five midpoint categories (ALO, PMF, FEut, ULO, and TA) are responsible for about 80% of the overall environmental impact of producing 1 t of softwood unbleached Kraft pulp. Agricultural land occupation is the most contributing midpoint category because this process requires a continuous use of forestry land area. When comparing the results at the midpoint level of Case A and Case B, it can be concluded that the use of softwood as a raw material to produce 1 t of unbleached Kraft pulp has a higher result (worst for the environment) in 14 of the 18 midpoint categories considered in the ReCiPe 2008 method.

4.2. Stochastic Results

From the five midpoint categories identified in Figure 2 as being the most impactful in terms of environmental impact, three have uncertainty in the characterization factors (PMF, FEut, and TA) has previously explained. The simulation ran until the convergence of results previously stipulated was obtained which occurred after 2000 iterations.
Several stochastic results were obtained, and Figure 3 presents a summary of these results through box and whisker plots regarding the difference between Case A and Case B. Looking at the right side of Figure 3, it can be concluded that the difference between the single scores will be at least 61.90 Pt and at most 70.53 Pt. There is a 25 % chance that this difference will be lower than 62.05 Pt and higher than 62.35 Pt and a 50 % chance that it will be between these values. The inter-quartile range could be used as a measure of data variability and is calculated through the difference between the third quartile and the first quartile presented in Figure 3. Using this measure and looking at the left side of Figure 3, it can be concluded that the difference between terrestrial acidification shows the lowest variability while the difference between freshwater eutrophication shows the greatest variability.

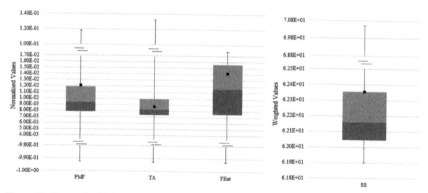

Figure 3 - Box and whisker plots of the difference between the three most impactful midpoint categories (on the left) and single score (on the right) of Case A and Case B

The characterization factor assigned to nitrogen oxides in the PMF midpoint category is the one that most contributes to the variability of the difference between the single scores of Case A and Case B (right side of Figure 3). In the other two most impactful midpoint categories, the characterization factor assigned to the emission of phosphate to the water in the FEut midpoint category and ammonia to the air in the TA midpoint category are the ones that most contributes to the referred variability.

4.3. Results Comparison

The difference between the deterministic single score for Case A and Case B (black square in the right side of Figure 3) and the difference between the value of the three most impactful midpoint categories for Case A and Case B (black squares in the left side of Figure 3) are represented in Figure 3. The comparison of the deterministic and stochastic results allows the following conclusions to be drawn:

1. The deterministic difference between the single score of Case A and Case B is about 62.36 Pt (see square in the right side Figure 3). A positive difference means that Case A is overall worst for the environment than Case B.
2. The difference between the single score of Case A and Case B, in the stochastic scenario, is always positive and ranges from 61.90 Pt and 70.53 Pt (see right side Figure 3), which means that even with uncertainty Case B has always lower environmental impact. Therefore, the confidence in this conclusion is high.
3. The deterministic difference between the single score of Case A and Case B is close to the third quartile of the stochastic results (see right side of Figure 3). This reveals that the deterministic difference between the single score of both cases is overestimated – there is 75 % chance that this difference is actually lower and therefore the overall improvement from Case A to Case B is lower than the initial estimated (62.36 Pt).
4. The deterministic difference between Case A and Case B considering the two of the most impactful uncertain midpoint categories (PMF and FEut) is also close to the third quartile of the stochastic results (see left side of Figure 3), which also points out to an overestimation of the improvements obtained by switching from softwood to hardwood. In the case of terrestrial acidification the deterministic difference between Case A and Case B is closer to the median – the deterministic results obtained for this midpoint category are more robust.
5. These overestimations of results lead to the conclusion that there is a high chance that the environmental impact improvements obtained by switching from the use

of softwood to hardwood to produce unbleached Kraft pulp are smaller than expected.

5. Conclusions

A methodology to address uncertainty in the characterization factors used in the LCIA step of an LCA was proposed and applied to a case study where the use of softwood and hardwood in the production of 1 t unbleached Kraft pulp were compared. From this application it was concluded that the use of hardwood is, overall, better for the environment even when considering the uncertainty associated with the characterization factors at the midpoint level. However, the improvements obtained by switching from softwood to hardwood to produce unbleached Kraft pulp are expected to be smaller than the deterministic assessment revealed. This reinforces the problem with deterministic LCA – it leads to misleading perception regarding the differences between the environmental profile of different products or processes.

Future research should include the application of uncertainty analysis to other LCIA methods besides ReCiPe 2008 along with the study of other parameters such as the normalization and weighting factors.

Acknowledgements

The authors grateful acknowledge the project funding (POCI-01-0145-FEDER-016733) and PhD grant SFRH/BD/134479/2017.

References

Z. Allen, Year, Article or Chapter Title, Journal or Book Title, Volume, Issue, Pages

P. Bajpai, 2012, Brief Description of the Pulp and Paper Making Process, Biotechnology for Pulp and Paper Processing, 7-14, Springer, Boston, United States of America

A. Björklund, 2002, Survey of approaches to improve reliability in lca, The International Journal of Life Cycle Assessment, 7, 64–72

Ecoinvent, 2018, Introduction to ecoinvent Version 3, https://www.ecoinvent.org/database/introduction-to-ecoinvent-3/introduction-to ecoinvent-version-3.html (accessed September 06, 2018)

M. Goedkoop, R. Heijungs, M. Huijbregts, A. De Schryver, J. Struijs, and R. van Zelm, 2013, ReCiPe 2008 - A life cycle impact assessment method which comprises harmonised category indicators at the midpoint and the endpoint level, Ministerie van Volkshuisvesting, Ruimtelijke Ordening en Milieubehee, Netherlands

ISO 14000, 2006, International Organization for Standardization, www.iso.org (accessed December 04, 2017)

Palisade Corporation, 2016, "@Risk User's Guide.", Ithaca, United States of America

PRé Consultants, 2018, SimaPro, https://simapro.com/ (accessed September 06, 2018)

K. Teschke, A. Keefe, G. Astrakianakis, J. Anderson, D. Heederik, S. Kennedy, K. Torén, 2011, Chapter 72 - Pulp and Paper Industry, Encyclopaedia of Occupational Health and Safety, Fourth Edition

Anton A. Kiss, Edwin Zondervan, Richard Lakerveld, Leyla Özkan (Eds.)
Proceedings of the 29th European Symposium on Computer Aided Process Engineering
June 16th to 19th, 2019, Eindhoven, The Netherlands. © 2019 Elsevier B.V. All rights reserved.
http://dx.doi.org/10.1016/B978-0-128-18634-3.50259-9

Green Supply Chain: Integrating Financial Risk Measures while Monetizing Environmental Impacts

Cátia da Silva[a*], Ana Paula Barbosa-Póvoa[a], Ana Carvalho[a]

[a]*CEG-IST, University of Lisbon, Av. Rovisco Pais, 1049-001 Lisboa, Portugal*

catia.silva@tecnico.ulisboa.pt

Abstract

Nowadays the growing concerns about the environment and society in general led companies to invest in a sustainable development. In this way, companies' design and planning decisions need to encompass simultaneously economic, environmental and social concerns. In addition, considering the complexity associated with market uncertainties, there is the need to consider risk management. This work is developed along this line and proposes a mixed integer linear programming model (MILP) that accounts for the economic and environmental performances in the same objective function by monetizing environmental impacts and simultaneously considering the most popular risk measures in the literature, CVaR. The goal is to maximize the difference between the expected net present value and the environmental impact while minimizing the associated risk. The augmented ε-constraint method is used to generate a Pareto-optimal curve in order to determine the trade-off between the objective functions. Conclusions can be drawn based on decision makers' risk profile as well as how monetization can support the decision maker's decision. A European supply chain case study is explored.

Keywords: supply chain, sustainability, risk management, monetization, uncertainty

1. Introduction

The importance of supply chain (SC) has been well-known for years in both scientific and industrial communities. As supply chain is a complex logistic system that covers the set of all activities from raw materials to final products' sales, it involves a careful and efficient management so that it is possible to obtain satisfactory results for the company. In fact, supply chain management intends to optimize customer value and to achieve a competitive advantage in the market. However, in the past, this optimization only included the economic performance of the supply chain. Taking into account the increasing companies' competitiveness and governmental pressures, SC has extended its goals to consider environmental and social concerns as well (Barbosa-Póvoa et al., 2017). The management of those economic, environmental and social concerns resulted in the appearance of sustainable supply chain (SSC) concept, which is a recognized area by the World Commission on Environment and Development. If managing a supply chain was a difficult task due to the higher number of variables, entities and products that can be included, then including environmental and social concerns makes this difficulty and complexity even greater. Thus, managing sustainable supply chains towards efficient and sustainable objectives is a challenge, especially if this encompasses the design and planning of the chain. Besides sustainability and responsiveness that supply chain management has to obey, risk management is also a current reality that needs to be accounted when considering the supply chain design and planning problems (Barbosa-Póvoa, 2017). In this way, the typical design and planning models for minimization of supply chain costs need to become more holistic. In fact, the majority of real-world supply chain problems are dynamic, because there are several uncertain factors, like raw material prices, demand, labour costs, among others (Cristonal et al., 2009). For this reason, there is the need to develop risk management tools that efficiently address the uncertainty involved. In the literature, there are several examples of the application of risk measures in supply chain design and/or planning. However, this application is mainly done in the financial area, which means that only the economic performance of the supply chain is assessed. Nevertheless, risk can have impact in multiple dimensions, such as financial, environmental and social aspects and, therefore, that should be accounted when analysing risk in supply chains, particularly in sustainable supply chains. This paper presents a mixed integer linear programming model (MILP) that accounts for the economic and environmental concerns in the same objective function by monetizing environmental impacts and considering a risk measure. The goal is to maximize the difference between economic and environmental performances while minimizing the associated risk.

2. Environmental monetization methodologies

There are several methods to assess environmental impacts. The most used methodology is life cycle assessment (LCA). One of the most controversial phases of LCA is the life cycle impact assessment (LCIA), as it encompasses the assignment of weighting factors by decision-makers and it may be difficult to choose the best method to use. On the LCIA phase, there are methods that assess environmental impacts with scores and others that monetize. According to the European Commission, EPS 2000 is a quite complete method and it has the uncertainties fully specified when compared to other LCIA methods. For this reason, in this study, monetization is performed by using EPS. The environmental impacts are quantified in a monetary unit through EPS, which is mainly derived from future costs (raw material depletion), direct losses (production) and willingness-to-pay WTP (health, biodiversity, and aesthetic values). The WTP values resulted from academic knowledge to proactively reduce environmental impacts (Steen, 1999).

3. Problem description and model characterization

The problem intends to study the design of a generic SC, where raw materials flow from suppliers to factories and final products are obtained. These final products move to warehouses or directly to markets. In addition, at the markets, end-of-life products can be recovered and sent to warehouses or directly to factories to be remanufactured. Taking into account the possible set of locations of SC entities, production and remanufacturing technologies, possible transportation modes between entities, and products within the SC, the main objective is to obtain the SC network structure, supply and purchase levels, entities' capacities, transportation network, production, remanufacturing and storage levels, supply flow amounts, and product recovery levels, in order to maximize profit, minimize environmental impacts, and minimize financial risk. The mixed integer linear programming (MILP) model used to solve this problem is based on da Silva et al. (2018). This model was extended to consider simultaneously two objective functions through the augmented ε-constraint method (Mavrotas, 2009), namely the maximization of the difference between economic and environmental performances, and the minimization of the financial risk, considering demand uncertainty. The first objective function is represented by Eq. (1), which is the maximization of the difference between the expected net present value (eNPV) and expected environmental impact (eEnvImpact). The economic performance is assessed through the eNPV (represented by Eq (2)), which is obtained through the sum of each node's probability multiplied by the discounted cash flows (CF_{Nt}) in each period t and for each node N at a given interest rate (ir). These CF_{Nt} are obtained from de net earnings (difference between incomes and costs). The costs include raw material costs, product recovery costs, production/remanufacturing operating costs, transportation costs, contracted costs with airline or freight, handling costs at the hub terminal, inventory costs, and labour costs. In addition, for the last time period, it is considered the salvage values of the SC ($FCI\gamma$). The environmental performance is assessed through the eEnvImpact (represented by Eq. (3)), LCA is performed on the transportation modes, in the technology involved, and on entities installed in the SC boundaries, using EPS 2015 (updated version of EPS 2000). The Life Cycle Inventory is retrieved from the Ecoinvent database (assessed through SimaPro 8.4.0 software). The LCA results are expressed in Environmental Load Units (ELU) and used as input data (ei) in Eq. (3), particularly in the environmental impact of transportation (first term), in the environmental impact of entity (second term), and in the environmental impact of technology (third term). The second objective function is represented by Eq. (4), where it is intended to minimize the financial risk, which is modelled through the adoption of conditional value at risk (CVaR). CVaR evaluates the likelihood of obtaining the difference between eNPV and eEnvImpact lower than value-at-risk (VaR). VaR is the minimum difference between NPV and EnvImpact for a given confidence level (α).

$$\max (e\mathrm{NPV}_N - e\mathrm{EnvImpact}_N) \tag{1}$$

$$\mathrm{eNPV}_N = \sum_N pb_N \left(\sum_{t \in T} \frac{CF_{Nt}}{(1+ir)^t} - \sum_\gamma FCI_\gamma \right) \tag{2}$$

$$\mathrm{eEnvImpact}_N = \sum_N pb_N \left(\sum_{\substack{t \in T \\ (a,m,i,j) \in NetP}} ei_{ac} pw_m d_{ij} X_{maijtN} + \sum_{i \in I_f \cup I_w} ei_{ic} YC_i \right.$$
$$\left. + \sum_{\substack{t \in T, i \in I_f \\ (m,g) \in H}} ei_{mgc} pw_m (P_{mgitN} + R_{mgitN}) \right) \tag{3}$$

$$\min Risk = \max CVaR_\alpha = \max \left(VaR - \frac{1}{1-\alpha} \sum_N (pb_N \cdot \delta_N) \right) ; \ \delta_N \geq VaR - NPV_N \tag{4}$$

4. Case study

The model is applied to an electronic components' producer located in Italy (Silva et.al., 2018). The company's suppliers are located in Verona. Currently, this company owns a factory and a warehouse with enough capacity to meet the demand of their existing clients. These clients are located in three main European markets, namely in Spain, Italy, and Germany. The SC capacity cannot meet a demand increase and four new potential clients induce the company's decision-makers to study different SC possibilities of expansion. Given the locations of the clients, Leeds and Hannover are possible locations for installing new factories, which may lead to significant changes in the transportation modes, particularly for clients that are located outside Europe. Also, other possible warehouse locations close to the markets have to be considered. There are two types of products (fp1 and fp2) that are currently being produced through two technology types.

There are two new technologies available to produce those products. Furthermore, end-of-life products can be recovered and remanufactured into final products. In addition, transportation can be performed only by road or through a combination of road, air and sea transportation modes. There are two types of road transportation available (Truck1 and Truck2) that vary in terms of capacity, investment costs, depreciation rate, variable costs, and vehicle consumption. This work accounts for product's demand uncertainty through a scenario tree approach, due to the fact that this is a method that allows the discretization of stochastic data over the time horizon and can be adjusted during the planning horizon. In this scenario tree (Figure 2), a node N characterizes a possible state and the arcs represent the evolutions it may have. Each node has a specific probability and a path from the root to a leaf node represents a scenario.

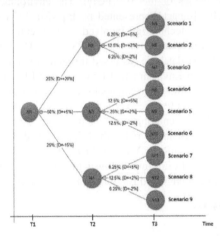

Figure 1 . Scenario tree - values for probability and demand (D) variation are represented

5. Results

Based on the SC reality, two cases are studied to understand the different decisions that can be made, considering the design and planning of the SC. Case A considers a stochastic approach since products demand is uncertain and does not considers a risk measure (risk neutral); while case B considers uncertainty in products' demand and studies the trade-offs between the expected value of (eNPV-eEnvImpact) and the associated financial risk, assessed through CVaR. Case B is analysed for the extreme values, namely the highest and lowest associated risk. Table 1 shows the eNPV and the eEnvImpact obtained in both cases (A and B). For case B, assuming a confidence level of 95%, it is necessary to compute all the scenarios and assess the cumulative distribution function (Figure 2). Thus, VaR is equal to -40043x10^4€., meaning that, at the end of the time horizon, the difference between eNPV and eEnvImpact for the design and planning of this supply chain is going to be, at least, this value. Thus, it is possible to obtain the Pareto curve assessed through CVaR (Figure 3).

Comparing the different cases, it can be seen that the risk neutral case has an eNPV of 97729x10^4 € and a correspondent eEnvImpact of 218570x10^4 €. Regarding case B, the results show that when the difference between eNPV and eEnvImpact increases the related risk also increases. Analysing in detail the extreme points of the Pareto curve, it is visible that the lowest risk decision is associated with the lowest eNPV and the highest eEnvImpact. In contrast, the highest risk decision offers better results for eNPV comparing with the risk neutral case, showing the relevance of considering risk measures.

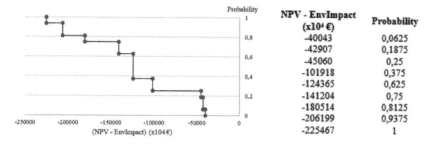

NPV - EnvImpact (x10⁴ €)	Probability
-40043	0,0625
-42907	0,1875
-45060	0,25
-101918	0,375
-124365	0,625
-141204	0,75
-180514	0,8125
-206199	0,9375
-225467	1

Figure 2. Cumulative distribution function to determine VaR

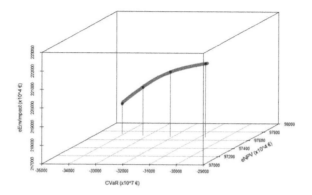

Figure 3. Pareto curve using CVaR risk measure

Table 1. Results for the economic (eNPV) and environmental (eEnvImpact) performances.

	Case A	Case B	
	Risk-neutral	High-risk	Low-risk
	Stochastic	*CVaR*	
eNPV (x10⁴)	97729 €	97745 €	97675 €
eEnvironmental impact (x10⁴)	218570 €	218600 €	220940 €

Table 2 shows SC structure decisions considering the results depicted in Table 1. It can be seen that there are some changes across the cases A and B. Additionally to the already existent factory, all cases are going to have the same new factories installed. The results also show that in all cases there is the need to expand the existing capacity by opening new warehouses. Regarding suppliers' allocation, it can be seen that in risk neutral and high-risk cases, where environmental impacts are lower, the results show that most factories are supplied by the closest suppliers. This can be explained by the lower environmental impact of transportation. On the other hand, in the low-risk case, all factories are supplied by all suppliers, which results from the balance between the lower costs of raw materials and fewer transportation costs. Regarding transportation, the truck with more capacity (Truck2) is preferred in all cases, since this has a lower environmental impact. In terms of intermodal transportation, sea option is preferred in all scenarios, while air transportation is not used. Going deeper into the results, it can be seen that the network structure of the risk-neutral case is similar to the high-risk case, due to the fact that both SC entities and transportation modes are the same, with some differences in capacities involved and in the product flows between entities.

Table 2. Summary of the SC decisions considering the different cases

| | | Case A | Case B | |
		Risk-neutral	High-risk	Low-risk
Factories		The same in all cases		
Warehouses		The same in all cases		
Suppliers allocation		Factories are supplied in totally by the closest supplier		All factories are supplied by all suppliers
Production		Most production of fp1 is in the existing warehouse		
		Most production of fp2 is divided between two new warehouses		
Inventory		fp1 and fp2 divided between Verona and Sofia; and most inventory of fp2 kept in Verona		fp1 and fp2 divided between Verona and Sofia
Transportation	Road	4 Truck1 and 17 Truck2	5 Truck1 and 17 Truck2	5 Truck1 and 18 Truck2
	Air	Not used		
	Sea	Used in all cases		

6. Conclusion

This study aimed to develop an optimization model for the design and planning of a generic SC, where the maximization of the difference between the expected economic and environmental performances was considered against the minimization of financial risk under uncertainty on the products demand. Considering the assessment of risk, CVaR was used. To show the trade-offs between (eNPV – eEnvImpact) and risk, a Pareto optimal curve was developed through the application of the augmented ε-constraint method. It is important to notice that this analysis was only possible, due to the monetization, which was able to quantify the environmental impacts in a monetary unit. This allowed to include in the same objective function both economic and environmental impacts. Furthermore, from the analysis made, it was clear that results are influenced by differences in risk strategies and this proves the importance of risk management in solving real-life problems. For future work, further research should be done on this topic to evolve monetization approaches in order to be a reliable alternative to evaluate environmental impacts. Also, an extension of this work should consider different risk measures and a more comprehensive study of uncertainty so as to even better conclude on its adequacy.

7. Acknowledgements

This work was supported by Portugal 2020, POCI-01-0145-FEDER-016418 by UE/FEDER through the program COMPETE2020 and FCT and Portugal 2020 under the project PTDC/EGE-OGE/28071/2017 and Lisboa - 01-0145-FEDER-28071

References

A.P. Barbosa-Póvoa, C. da Silva, A. Carvalho, 2017, Opportunities and Challenges in Sustainable Supply Chain: An Operations Research Perspective, European Journal of Operational Research, http://doi.org/10.1016/j.ejor.2017.10.036

B. Steen, 1999, A systematic approach to environmental priority strategies in product development (EPS), Version 2000 – General system characteristics, CPM Report 1999:4 CPM, Chalmers University of Technology, Goteborg, Sweden

M. Cristonal, L. Escudero, J. Monge, 2009, On stochastic dynamic programming for solving large-scale planning problems under uncertainty, Computers Operation Research, 36, 2418-28

C. da Silva, A. P. Barbosa-Póvoa, A. Carvalho, 2018, Sustainable Supply Chain: Monetization of Environmental Impacts, Computer Aided Chemical Engineering, 43, 773-778

G. Mavrotas, 2009, Effective implementation of the ε-constraint method in multi-objective mathematical programming problems, Ap Math Comp, 213(2), 455-465

Anton A. Kiss, Edwin Zondervan, Richard Lakerveld, Leyla Özkan (Eds.)
Proceedings of the 29th European Symposium on Computer Aided Process Engineering
June 16th to 19th, 2019, Eindhoven, The Netherlands. © 2019 Elsevier B.V. All rights reserved.
http://dx.doi.org/10.1016/B978-0-128-18634-3.50260-5

Life cycle assessment of Jatropha jet biodiesel production in China conditions

Haoyu Liu, Tong Qiu*

Department of Chemical Engineering, Tsinghua University, 100084 Beijing, China
Corresponding author's E-mail: qiutong@tsinghua.edu.cn

Abstract

Jet biodiesel derived from energy crop has been promoted by the Chinese government to decrease oil dependence of the aviation industry. This study carried out a life cycle assessment on jatropha-based jet biodiesel in China condition to evaluate the energy balances and GHG emissions. This life cycle process covered data from Southwest China with the help of CNPC. Detailed information with practical data in the production process and by-products recycling ways were provided. 18 cases were utilized to compare the influence of different planting output, oil extraction model and recycling pathway. Centralized extraction model and by-product re-used as fertilizer and biogas can be developed in the future. The formation of a multi-scene analysis results is useful for policy makers

Keywords: Life-cycle analysis, jet biodiesel, Jatropha, By-product, Scenario analysis

1. Introduction

China had become the world's largest oil consumer and GHGs emitter (EIA. 2014). The aviation industry, which had since long been pointed out as fastest growth contributors to global warming in the transportation sector, emited 2% of global GHGs emissions (Jong, S. D et al. 2017). A huge increasing demand of jet fuels not only led to an increase crude oil imports and price but also an potential risk for energy safety (Han J. et al. 2013).

Jet biodiesel derived from energy crop had been promoted by the Chinese government to decrease oil dependence of the aviation industry, according to the Medium and Long-term Development Plan for Renewable Energy issued by the National Development and Reform Commission, 2007. However, the controversy surrounding the social impacts such as food safety and environmental impacts promoted the government to impose a ban on expanding grain-based jet biofuel production(Hong, Yuan et al. 2009). In order to find a truly sustainable jet biodiesel, the feedstock should be an effective alternative to mitigate GHGs emissions, should be economically feasible to plant, and should not threat food security.

Jatropha curcas appeared to be an idealized feedstock to produce jet fuel in China conditions (Hou J et al. 2011). These inedible crop grows widely in the Southwest China due to its stable price, drought tolerance, and little agricultural inputs (Blanco-Marigorta et al.2013). While some environmental impact studies were carried out from India (Kumar S, 2012), Mexico (Fuentes A, 2018), US (Han J. et al. 2013; Stratton R W, 2011), Nigeria (Onabanjo T, 2017), a few studies had been carried out in China. Ou X et al. (2009) examined the jatropha curcas as one of six biofuel pathways in China, considering the fossil energy consumption and greenhouse gas emissions. Other impact categories were taken into account to evaluate the sustainability of jatropha curcas comprehensively by Hou J (2011). Both of them considered LCA on the same scenario. Wei et al. (2018) carried out a LCA of corn-based jet fuel production using Fischer-Tropsch synthesis. Except this, there are few practical data regarding jatropha production process and few case about the use of the by-products from jatropha oil extraction. In addition, studies have little knowledge about whether distributed production or centralized production is better.

Besides, jet biodiesel are considered as a low-carbon alternative to fossil fuels. At a country level, the China National Petroleum Corporation (CNPC) had promoted Jatropha jet biodiesel production under the policies and financial supports from the government since 2011. The documents by government and companies both target mitigation of GHG emissions as central sustainability criteria. However, there are a few studies on the GHG balance of jet biodiesel from jatropha to give advice to policymaking.

This study carried out a life cycle assessment on jet biodiesel made from jatropha in China condition to evaluate the energy balances and GHG emissions in order to provide more detailed information and data in the production process and by-products recycling. Different allocation methods were combined to deal with various by-products. Scenario analysis were utilized to predict different plant output and oil extraction model. Three by-products from oil extraction recycling pathways were investigated. The formation of a multi-scene analysis results is useful for policy makers.

2. Methods

2.1. LCA framework

A cradle-to-gate LCA was performed with GaBi ts Education software. The analysis focus on determine the energy consumption and GHGs emissions of jet biodiesel produced by jatropha in Southwest of China, to compare the results with different oil extraction model and to find an eco-friendly by-products recycling pathways.

2.2. System boundary and functional unit

The stages considered and the calculation logic of the energy consumption and the GHGs of the entire process is present in the Figure 1, including the plant cultivation, production and transport process. Effects of fertilizer emissions and land use change will not be considered for the moment due to the data given in the IPCC Guidelines have large deviations. The assessments were compared on the basis of their GHG emissions in g

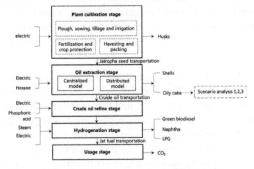

Figure 1. Product system boundary of this study.

Table 1 Allocation method for by-products

By-product	Method	By-product	Method
Husk	Energy method	Jatropah oily cake recycling as fertilizer	Displacement method
glycerin	Displacement method	Jatropah oily cake recycling as forage	Economic method
Naphtha	Energy method	Jatropah oily cake recycling as biogas	Displacement method
LPG	Energy method	Green biodiesel	Energy method

CO_2 per MJ and energy consumption in MJ per Kg of jet biodiesel produced from jatropha.

2.3. Product system description

In the plant stage, three planting output conditions were investigated: low output, basic output and high output which depends on the management level and effect of fertilizer. There were two kinds of oil extraction models: distributed extraction model where famers extracted jatropha on local plant transported oil to the refine factory, and centralized extraction model where here the jatropha was transported to the centralized extraction factory. The hydrogenation process was based on technology from the UOP company.

2.4. Method to deal with of by-products

In this case, there were various by-products in the production process. There are four allocation method: displacement method, and the mass, energy, market method. Different allocation methods showed in the Table 1 were combined for different by-product, considering the practical data and usage of the by-product.

2.5. Scenarios for reusing the by-products

For the further applications, the jatropha oily cake that can be used as forage, fertilizer or biogas will be comprehensive utilized. The data were gained from the factory and the farmers. The possible systems for recycle the jatropha oily cake are divided into 3 scenarios: Scenario 1: All the jatropha oily cake is used to fertilizer; Scenario 2: After detoxification, all the jatropha oily cake is used as forage to replace soybean; Scenario 3: All dried distiller produce biogas, where need extra coal.

2.6. Life cycle inventory

Based on the investigation with the technology developer and farmers, open LCI databases and references, the part of key LCA parameter is showed in the Table 2.

Table 2 Parameter about jet biodiesel production (To produce 1 kg jet biodiesel)

Item Parameter	Value						Item Parameter	Value					
Jatropha Output	Low	Basic	High	Low	Basic	High	Jatropha Output	Low	Basic	High	Low	Basic	High
Oil Extraction Model	Distributed			Centralized			Oil Extraction Model	Distributed			Centralized		
Planting							**Oil Refine**						
Land (ha)	0.0081	0.0026	0.0015	0.0070	0.0023	0.0013	Electric (Kwh)	0.0498	0.0498	0.0498	0.0498	0.0498	0.0498
N fertilizers (kg)	2.0268	0.6563	0.2369	1.7579	0.5692	0.2055	Phosphoric acid (kg)	0.0418	0.0418	0.0418	0.0418	0.0418	0.0418
P fertilizers (kg)	1.0539	0.3413	0.0939	0.9141	0.2960	0.0814	Refined oil (kg)	2.0180	2.0180	2.0180	2.0180	2.0180	2.0180
K fertilizer (kg)	1.8241	0.5907	0.1848	1.5821	0.5123	0.1602	**Hydrogenation**						
Electric (Kwh)	0	0.5011	0.2011	0	0.4388	0.1745	Hydrogen (kg)	0.0710	0.0710	0.0710	0.0710	0.0710	0.0710
Gasoline (kg)	0	0.1555	0.0883	0	0.1349	0.0766	Steam (kg)	0.8350	0.8350	0.8350	0.8350	0.8350	0.8350
Husks (kg)	4.7613	4.2407	3.8378	4.1297	3.6781	3.3287	Gas (kg)	0.0795	0.0795	0.0795	0.0795	0.0795	0.0795
Fruit (kg)	12.8685	12.1162	11.2676	11.1615	10.5089	9.7903	Electric (Kwh)	0.3120	0.3120	0.3120	0.3120	0.3120	0.3120
Seed (kg)	8.1072	7.8755	7.4498	7.0317	6.8308	6.4616	Green biodiesel (kg)	0.0170	0.0170	0.0170	0.0170	0.0170	0.0170
Oily Extration							Naphtha (kg)	0.5330	0.5330	0.5330	0.5330	0.5330	0.5330
Electric (Kwh)	0.3506	0.3406	0.3222	0.2488	0.2417	0.2286	LPG	0.2590	0.2590	0.2590	0.2590	0.2590	0.2590
Hexane (kg)	0	0	0	0.0144	0.0140	0.0132	**Transportation**						
Shells (kg)	2.3878	2.3195	2.1942	2.0501	1.9915	1.8839	Electric (Kwh)	0.0056	0.0056	0.0056	0.0056	0.0056	0.0056
Oily cake	3.3188	3.1577	2.8616	2.5214	2.3831	2.1289	Diesel (g)	0.3029	0.3029	0.3029	0.3029	0.3029	0.3029
Crude oil	2.3195	2.3195	2.3195	2.3195	2.3195	2.3195							
Transport gasoline (kg)	0.0040	0.0040	0.0039	0.0105	0.0102	0.0096							
Transport diesel (kg)	0.0096	0.0095	0.0094	0.0253	0.0245	0.0232							

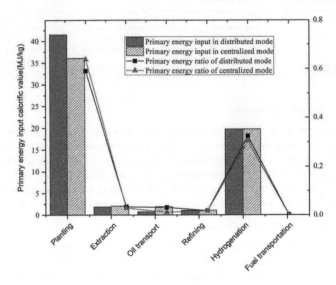

Figure 2. Energy consumption of the basic case in each production process.

3. Result and discussion

3.1. Energy consumption in the basic condition

The contribution of different processes and different oil extraction model to the energy consumption are showed in the Figure 2. The total energy consumption is 61.04 MJ primary energy/kg jet biodiesel in the basic output of centralized extraction model, while the energy product occupy 78.24 MJ including all the by-product. The total energy efficiency is 0.78 MJ primary energy/MJ jet biodiesel. Because of a lot of fertilizer input, the planting unit exhibits high contribution to energy consumption. Besides, the centralized extraction model is benefit for mitigation energy consumptions, it may be because of its advanced technology and high rate of extraction, but it also need time to develop the famers' market. To reduce the GHGs and enhance energy efficiency, enhancing the separation efficiency or using by-product to supply energy can be considered.

3.2. Scenario between output, extraction model and recycling of by-products

To illustrate the impact of output, oily cake extraction model and recycling of by-products, Figure 3 shows that the emissions for the Jatropha range between -36.28 and 466.69 g CO_2 /MJ for 18 cases. Scenario 2 increased a lot of GHGs emissions and the detailed reason will be discussed in the next part. For different planting output, it's obvious that the harvest yield is the key factor. The high output cases even cause negative emissions in the scenario 1 and 3, this may be because the by-products replaced a high emission substitution and the professional plantation management increased the efficient of fertilizers. From the Figure 3 it can be concluded that scenario 1 and 3 have positive environment impact, comparing to the fossil fuel, whose baseline is 91.8 g CO_2 /MJ in the US renewable fuel standards (WTW).

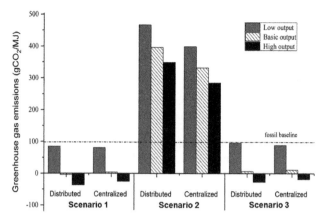

Figure 3. 18 cases with different planting output, recycling scenarios and extraction models.

3.3. Influence of recycling by-product

To examine the high emission in the scenario 2, Figure 4 shows the GHGs emission in each process where the jatropha oily cake from the oily extraction process will be used as forage. The detoxification for recycling need huge fuel and chemical substances input such as methyl alcohol. But the high economic value of forage, such as soybean, is an considerable reason to recycle by-product. From the scenario analysis, the conclusion that comprehensive utilization about the by-product especially like jatropha oily cake will have huge influence to the environment impact can be drawed.

4. Conclusion

The life cycle analysis about the jatropha jet biodiesel in the southwest China was conducted in this study. 18 cases with different oil extraction model, planting output and recycling pathway for by-products was utilized to compare the influence to energy consumption and GHGs emissions in the production process. The result shows that centralized extraction model is better than distributed extraction model especially in the low planting output condition. Scenario 1 and 3 showed the systematic and suitable utilization of by-product will improve the environment impact, with the emission of 80.1-88.28 g CO_2 /MJ even in the low output case of centralized model. The practical data regarding plant process and cases about the use of the by-products is valuable to the researchers. Besides, the formation of a multi-scene analysis results is useful for producer and policy makers.

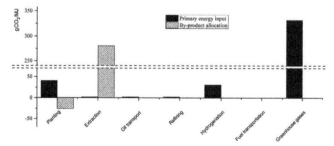

Figure 4. The GHGs emission of Scenario 2 in each process.

Acknowledgements

The authors gratefully acknowledge the National Natural Science Foundation of China for its financial support (Grant No. U1462206).

References

Blancomarigorta A M , SuárezMedina, J, Veracastellano A . Exergetic analysis of a biodiesel production process from Jatropha curcas[J]. Applied Energy, 2013, 101(1):218-225.

EIA, The U.S. Energy Information Administration, 2014. China is Now the World's Largest Net Importer of Petroleum and Other Liquid Fuels. http://www.eia.gov/todayinenergy/detail.cfm?id¼415531.

Fuentes A , Carlos García, Hennecke A , et al. Life cycle assessment of Jatropha curcas biodiesel production: a case study in Mexico[J]. Clean Technologies & Environmental Policy, 2018(1):1-13.

Han J , Elgowainy A , Cai H , et al. Life-cycle analysis of bio-based aviation fuels[J]. Bioresour Technol, 2013, 150(4):447-456.

Hong, Y., Z. Yuan and J. Liu (2009). "Land and water requirements of biofuel and implications for food supply and the environment in China." Energy Policy 37(5): 1876-1885.

Hou J , Zhang P , Yuan X , et al. Life cycle assessment of biodiesel from soybean, jatropha and microalgae in China conditions[J]. Renewable & Sustainable Energy Reviews, 2011, 15(9):5081-5091.

Kumar S . A comprehensive life cycle assessment (LCA) of Jatropha biodiesel production in India[J]. Bioresource Technology, 2012, 110(32):723-729.

Onabanjo T, Lorenzo G D, Kolios A. Life-cycle assessment of self-generated electricity in Nigeria and Jatropha biodiesel as an alternative power fuel[J]. Renewable Energy, 2017, 113.

Ou X, Zhang X, Chang S, Guo Q (2009) Energy consumption and GHG emissions of six biofuel pathways by LCA in (the) People's Republic of China. Appl Energy 86:S197-S208.

PRC State Development and Reform Commission. Medium and long-term development plan for renewable energy. Beijing: PRC State Development and Reform Commission; 2007.

S.C. Mao, Z.Y. Li, C. Li Application of biodiesel produced from Jatropha curcas L. seed oil Chin Oils Fats, 32 (2007), pp. 40-42

Stratton R W , Wong H M , Hileman J I . Quantifying variability in life cycle greenhouse gas inventories of alternative middle distillate transportation fuels[J]. Environmental Science & Technology, 2011, 45(10):4637-44.

Wei T , Jun X , Kai Y . Life cycle assessment of jet fuel from biomass gasification and Fischer-Tropsch synthesis[J]. China Environmental Science, 2018.

Consulting report of China Engineering Academy, 2014, Development, Demonstration and Application of Key Technology of Biofuels Industry in China.

Anton A. Kiss, Edwin Zondervan, Richard Lakerveld, Leyla Özkan (Eds.)
Proceedings of the 29th European Symposium on Computer Aided Process Engineering
June 16th to 19th, 2019, Eindhoven, The Netherlands. © 2019 Elsevier B.V. All rights reserved.
http://dx.doi.org/10.1016/B978-0-128-18634-3.50261-7

Life-Cycle Environmental Impact Assessment of the Alternate Subsurface Intake Designs for Seawater Reverse Osmosis Desalination

Abdulrahman H. Al-Kaabi[a], Hamish R. Mackey[a,*]

aDivision of Sustainable Development, College of Science and Engineering, Hamad bin Khalifa University, Qatar Foundation, Doha, Qatar

hmackey@hbku.edu.qa

Abstract

The study carried out a life-cycle environmental impact assessment of two reverse osmosis plants located within the Arabian Gulf using three different electric submersible pump (ESP) designs for subsurface intake of seawater and compared these with an existing open intake design. The study used life cycle assessment to quantify impacts for construction and operation of the plant for the various intake options. All values were compared to a functional unit of 1.0 m^3 of produced desalinated water based on individual well capacities of $15,000 \text{ m}^3/\text{d}$ and a design life of 50 years. Results showed that the subsurface system performed better across the various impact categories than the open intake. Construction phase impacts of the beach well were insignificant in comparison to operational phase with the exception of abiotic depletion potential, which was still minor. Nevertheless, of the three subsurface beach wells evaluated the slickline ESP design was consistently better than the other two subsurface options, albeit very similar to coil tubing ESP.

Keywords: Desalination, Subsurface Intake, Slickline Electric Submersible Pump, Beach Well, Life-Cycle Assessment, Construction Phase Impacts, Environmental Impact

1. Introduction

Global freshwater scarcity affects at least 2.8 billion people globally. Of this population, at least 43% (1.2 billion people) have no access to clean drinking water (United Nations, 2014; Mekonnen and Hoekstra, 2016). In response, use of desalination as a reliable alternative water source is increasing (Hoekstra and Mekonnen, 2012). Reverse osmosis (RO), which employs semipermeable membrane barriers, is the current desalination method of choice due to its energy efficiency compared to thermal techniques (Raluy et al., 2006).

Subsurface intakes, particularly if abstracting from brackish ground water, are a superior RO feed source to open seawater intakes due to better chemistry, microbiology and physical properties (Stein et al., 2016). In contrast to open seawater intakes which have high fouling potential requiring significant pre-treatment, subsurface intakes allow natural filtration of sediments, salt precipitates, and active biological treatment, which eliminates or reduces the need of pre-treatment (Missimer et al., 2015; Stein et al., 2016).

There are six types of subsurface seawater intakes, which are classified according to onshore location (vertical wells and beach infiltration gallery) and offshore location in the seabed (slant wells, horizontal wells, radial collector wells, and seabed infiltration galleries). Vertical wells, also commonly called beach wells, are the simplest subsurface intake systems, constructed according to designs and processes that are similar to conventional groundwater production wells. They are typically located along the beach, where they are screened within unconsolidated beach sand and alluvial materials. They are fitted with pumps, installed vertically above the wells, which induce radial inflow of seawater and inland groundwater into the wells. The rate and sustainability of seawater recharge of the beach wells is limited by local hydrogeology, especially the transmissivity of the aquifers (Missimer et al., 2015). Thus, beach wells have a limited production capacity. A recent modeling study, demonstrated that a beach well in Qatar drilled to a depth of 100 m can produce a maximum of only 16,000 m^3 per km^2 in the coastal zone, which can be considered suitable for small-capacity seawater desalination plants (Baalousha, 2016).

The conventional beach well layout includes an electric submersible pump (ESP) design, which was utilized as early as 1928 (Takacs, 2017). Newer ESP designs, such as coil tubing and slickline, were developed to boost production capacity and efficiency in the oil industry and could find useful application for subsurface seawater RO intake (Dieuzeide et al., 2014). The proposed alternative beach well layout designs using coil tubing ESP and slickline ESP designs are shown in Figure 1. Both designs have been proven worldwide in the oil industry (Dieuzeide et al., 2014) and as summarized in Table 1, provide operational and installation advantages, with slickline ESP systems being optimal. Given these benefits this paper aims to evaluate the environmental impact of ESP systems for subsurface seawater RO intakes and assess their performance over conventional vertical wells and open intakes. This is conducted through the use of life cycle assessment (LCA) of environmental loads associated with the construction and operation of the RO plant and intake systems.

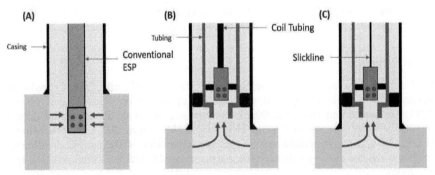

Figure 1: Proposed alternative beach well layouts; A. Current conventional well design; B. Proposed Coil tubing ESP subsurface intake and; C. Proposed Slickline ESP subsurface intake.

Table 1: Comparison between typical beach well design and ESP coil tubing and slickline
subsurface intake designs

Design aspects	Typical beach well	Coil tubing beach well	Slickline beach well
Maximum Run Deviation	90°	90°	60°
Installation Mean	Specialized truck or rig	Coil tubing unit	Slickline unit
OPEX	Higher	Lower	Lowest
Production Downtime	3-6 months	1 month	10-15 days

2. Materials and Methods

2.1. Life Cycle Assessment Goal and Scope

LCA is a widely-used process to measure the environmental impacts of different life cycle stages of a system, or compare system alternatives. A cradle to gate boundary LCA was undertaken to investigate the environmental impacts associated with the construction of the various intake systems and the plantwide process operation. The assessment scenarios included an open intake system and three alternative subsurface intake systems (conventional, coil tubing and slickline ESP). The LCA covered materials and diesel associated with machinery for the intake construction, as well as power, material and chemical inputs for the operation phase of the entire plant. The pretreatment and RO unit construction were excluded due to lack of suitable data, but would be identical for all three subsurface systems. All values were compared to a functional unit of 1.0 m³ produced desalinated water based on individual well capacities of 15,000 m³/d and a design life of 50 years. LCA was carried out with GaBi software.

2.2. Life cycle impacts

The LCA considered the following impact categories: global warming potential (GWP), acidification potential (AP), eutrophication potential (EP), ozone layer depletion potential (ODP), abiotic depletion potential of elements (ADP-e), abiotic depletion potential of fossil reserves (ADP-f), freshwater aquatic ecotoxicity potential (FAETP inf.), photochemical ozone creation potential (POCP), human toxicity potential (HTP inf.), marine aquatic ecotoxicity (MAETP inf.), and terrestrial ecotoxicity potential (TETP inf.).

2.3. Life cycle inventory

Data from two existing RO plants located across the Arabian Gulf were utilized for this comparison, both using similar pre-treatment processes of DAF and UF, and operating currently with open intakes. Plant 1 had a capacity close to 175,000 m³/d while Plant 2 had a capacity of close to 275,000 m³/d. Plants with subsurface intakes were modelled with the DAF pre-treatment units removed based on previous literature that indicates even the UF could potentially be dropped (Shahabi et al., 2015). Construction requirements associated with subsurface wells and open intake were taken from an oil and gas company operating within Qatar. Pumping heads were calculated based on 40 m drawdown for subsurface wells in the local hydrogeology (Baalousha et al., 2015). Distance from the different subsurface wells and RO plant were assumed to be similar.

3. Results and discussion

3.1. Intake construction impacts

Open intakes exhibited the highest overall environmental impacts related to construction in the areas of GWP, AP, EP, ODP, ADP-f, FAETP inf., MAETP inf. and POCP. The conventional ESP intake in contrast showed maximum impacts with respect to ADP-e, HTP inf., MAETP inf. and TETP inf. Coil tubing and slickline ESP intakes generally exhibited lower impacts than conventional ESP, with slickline ESP consistently exhibiting the lowest impact of the ESP systems with the exception of ADP-e, which was 100% for all three ESP systems (Figure 2).

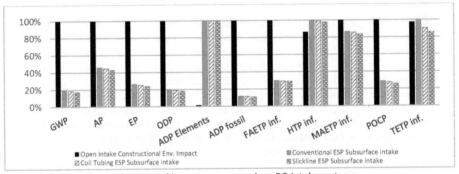

Figure 2: Construction environmental impacts across various RO intake systems.

3.2. Operational impacts of ESP intakes versus conventional vertical wells

The impacts from plant-wide operation at Plant 1 and Plant 2 were significantly higher than construction impacts for all environmental impact categories assessed, as shown in Figure 3. The low construction impacts is due to the minimal excavations and due to the long lifetime of the two plants of 50 years. The most significant contribution was to ADP-e which accounted for around 6.5% of total ADP-e impacts due to the high use of metals in the intake system.

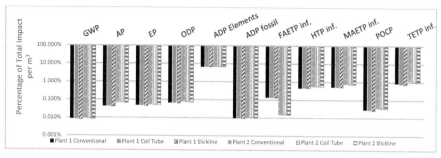

Figure 3: The percentage of total impacts associated with construction of the intake itself at Plant 1 and Plant 2 for the three different types of subsurface intakes.

3.3. Benefits of using subsurface intakes

Figure 4 demonstrates the relative environmental impacts at the two modelled plants using open and subsurface intakes for the operation phase only, given the minor impact from the intake construction phase. It is clear that subsurface intakes provide significant improvements across all selected environmental impact categories, although significant differences exist between the two plants based on chemical usage for the pretreatment.

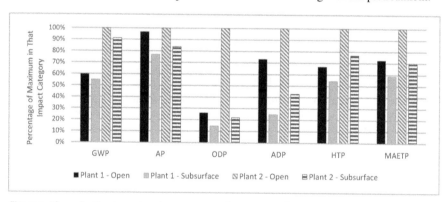

Figure 4: The reduction in various impact categories associated with moving from open intake to subsurface intake (modelled as conventional vertical well) with natural gas as energy source.

4. Conclusions

SWRO has the potential to alleviate water scarcity in coastal regions. While subsurface intakes are suitable due to their low environmental impacts, vertical beach wells have bottle neck effects on their potential production capacity. The current study evaluated the environmental impacts of conventional ESP layout design for beach wells and optimized ESP beach well designs, slickline and coil tubing. The conclusions deduced from the LCA are as follows:

- Construction phase impacts of subsurface beach well intakes are minimal compared to the operational phase requirements.
- Operational phase reductions in impacts are significant for subsurface intakes compare to conventional open-water intakes, particularly for ADP and ODP.

- While only small differences exist between the impacts associated with various beach well designs, slickline ESP consistently had the lowest construction related environmental impacts and is therefore the most suitable candidate given its proven economical value also,.

References

A. Dehwah & T. Missimer, 2016, Subsurface intake systems: Green choice for improving feed water quality at SWRO desalination plants, Water Res, 88, 216-224

A. Hoekstra & M. Mekonnen, 2012, The water footprint of humanity, Proceedings of the National Academy of Sciences of the United States of America, 109(9), 3232-3237

D. Williams, 2015, Slant Well Intake Systems: Design and Construction. Intakes and Outfalls for Seawater Reverse-Osmosis Desalination Facilities. In: Missimer T.M., Jones B., Maliva R.G., Environmental Science and Engineering, 275-320

G. Raluy, L. Serra, & J. Uche, 2006, Life cycle assessment of MSF, MED and RO desalination technologies. Energy, 31(13), 2361-2372.

G. Takács, 2018, Electrical submersible pumps manual: Design, operations, and maintenance, Oxford, United Kingdom: Elsevier, Gulf Professional Publishing, an imprint of Elsevier.

H. Baalousha, 2016, The potential of using beach wells for reverse osmosis desalination in Qatar, Modeling Earth Systems and Environment, 2(2), 97

I. Muralikrishna & V. Manickam, 2017, Principles and Design of Water Treatment. In: Environmental Management, 209-248

M. Mekonnen & A. Hoekstra, 2016, Four billion people facing severe water scarcity, Science Advances, 2(2), e1500323-e1500323

M. Shahabi, A. McHugh & G. Ho, 2015, Environmental and economic assessment of beach well intake versus open intake for seawater reverse osmosis desalination, Desalination 357, 259-266

S. Al-Mashharawi, A. Dehwah, K. Bandar & T. Missimer, 2014, Feasibility of using a subsurface intake for SWRO facility, south of Jeddah, Saudi Arabia, Desalination and Water Treatment, 55(13), 3527-3537

N. Hilal, A. Ismail, T. Matsuura & D. Oatley-Radcliffe, 2017, Membrane characterization, Amsterdam, Netherlands, Elsevier

S. Stein, A. Russak, O. Sivan, Y. Yechieli, E. Rahav, Y. Oren & R. Kasher, 2016, Saline groundwater from coastal aquifers as a source for desalination, Environmental Science & Technology, 50(4), 1955-1963

T. Dieuzeide, A. Bemba, D. Kusuma & G. Anderson, 2014, Wireline Retrievable Electric Submersible Pump: Innovative and Valuable Completion for Offshore Fields. International Petroleum Technology Conference

T. Missimer, B. Jones & R. Maliva, 2015, Intakes and outfalls for seawater reverse-osmosis desalination facilities: Innovations and environmental impacts, Springer

T. Missimer & R. Maliva, 2018, Environmental issues in seawater reverse osmosis desalination: Intakes and outfalls, Desalination 434, 198-215

United Nations, 2014, International decade for action 'WATER FOR LIFE'

Anton A. Kiss, Edwin Zondervan, Richard Lakerveld, Leyla Özkan (Eds.)
Proceedings of the 29th European Symposium on Computer Aided Process Engineering
June 16th to 19th, 2019, Eindhoven, The Netherlands. © 2019 Elsevier B.V. All rights reserved.
http://dx.doi.org/10.1016/B978-0-128-18634-3.50262-9

Optimization of biofuel supply chain design via a water-energy-food nexus framework

Dulce Celeste López-Díaz,[a*] Luis Fernando Lira-Barragán,[a] José Maria Ponce-Ortega,[a] Mahmoud M. El-Halwagi[b,c]

[a]*Chemical Engineering Deparment, Universidad Michoacana de San Nicolás de Hidalgo, Morelia 58060, México.*

[b]*Chemical Engineering Department, Texas A&T University, College Station TX 77843, USA.*

[c]Adjunt Faculty at the Chemical and Materials Engineering Department, King Abdulaziz University, Jeddah 21589, Saudi Arabia.

Sanasa_486@hotmail.com

Abstract

Biofuels are attractive energy renewable sources to satisfy the current growing energy demand, but large-scale production requires the installation of complex biorefinery systems that involve strategic decisions for design and operation such as to determine the location, feedstocks, production capacities, as well as the impact on the surrounding environment. This work proposes an optimization model for the design of a biorefinery system through an efficient supply chain considering economic (minimization of the total annual cost) and environmental objectives (minimization of the overall CO_2 emission and the water consumption along the nexus). This approach considers the inherent uncertainty on prices of feedstocks and products, biofuel demands, biomass requirements, weather conditions, agricultural production, etc. Finally, a case study located in Mexico is solved for several scenarios to illustrate the capacity of this approach. The solution shows the trade-off among the considered objectives affected by uncertainty in the biorefinery system.

Keywords: Biorefineries, Supply chain, Nexus, Uncertainty, Material flow analysis.

1. Introduction

Nowadays, the high energy demand has allowed the development of other energy sources such as of biofuels. This energy alternative represents an attractive and renewable option to partially reduce the consumption of fossil fuels. Nevertheless, large-scale production requires the installation of several biorefineries. A biorefinery is an industrial plant that from biomass produces high value-added chemicals and biofuels (bioethanol and biodiesel) [1]. Currently, the design of sustainable processes is essential [2]. Recent advances in processing technologies enable the conversion of non-food biomass resources for the controversy that exists against the food sector through second generation biofuels where the use of residues involves higher environmental benefits. With the variety of feedstocks, processing technologies, portfolio or products, geographic and market conditions, involves complex decisions to determine the optimal location, equipment, and operation of these industrial plants [3]. The objective is to synthetize a supply chain to satisfy the market demand while the production of feedstock is determined, as well the capacity and location of the processing facilities to convert this biomass to biofuels. Also, to design a robust biorefining system and to have a close estimation of the environmental impact that this energy sector can produce, the uncertain effects should be considered as weather conditions, agriculture parameters, conversion factors, etc. We propose an MFA (Material Flow Analysis) [4] approach for the watershed segmentation in reaches, where in each of them exists the possibility to install a biorefinery. Within each reach, the flow and main characteristics are assumed constant, even considering that several tributaries can feed to each reach. In this regard, this approach accounts for water extractions and discharges associated with domestic, agricultural and industrial uses, in addition to the natural phenomena such as filtration, evaporation and precipitation. Thus, mass balances, property tracking, and physical, chemical, and biochemical phenomena are included for each reach. Also, it estimates the overall CO_2 emissions by the industrial activity related to the biofuels production, and the balance considers the CO_2 fixation by the cultivation lands where the feedstocks are cultivated. The optimization approach is driven by economic (maximization of the total annual costs for the whole supply chain) and environmental (Minimization of the overall greenhouse emission) goals considering simultaneously the effects on the watershed. Because water is the resource of interaction between the sectors that make up the nexus.

2. Model Formulation

The problem is focused on satisfying the biofuel demand in a specific region where several feedstocks are candidate to be processed with existing and new potential cultivation areas, processing technologies, potential location for installing biorefineries, and large portfolio of bioproducts. Given the installation and operation of a biorefinery system (set of biorefineries to supply great demands) the impact on the environment must be considered in a sustainable design, in this case according with numerous studies the water reservoir is the most affected by this industrial activity. Surrounding watershed provides water for different facilities, as well as pollutant effluents are discharged. Implementing a Material Flow Analysis technique, the watershed is analysed to determine the environmental impact through water balances accounting for the availability of resources and environmental constraints along the watershed to avoid affectations for existing tasks as agricultural, residential or industrial activities. The proposed complex industrial system studied as a water-energy-food nexus is design through synthetizing the supply chain obtaining the production and operation specification. Also, the uncertainty that affects several parameters is considered.

The optimization approach is described and proposed as follows:

Watershed at initial conditions before the implementation of the biorefining system

$$Q_{r,s}^{Initial-Out} = Q_r^{In} + P_{r,s} + D_{r,s} + H_{r,s} + \sum_{t=1}^{N_{(t)}} FT_{r,t,s} + \sum_m w_{m,r,s}^{Discharge-feedstocks\_Initial} \tag{1}$$
$$- \sum_m w_{m,r,s}^{Used-feedstocks\_Initial} - L_{r,s} - U_{r,s}, \qquad \forall r, \forall s$$

Watershed at final condition (once the biorefining system is establish)

$$Q_{r,s}^{Out} = Q_r^{In} + P_{r,s} + D_{r,s} + H_{r,s} + \sum_{t=1}^{N_{(t)}} FT_{r,t,s} + \sum_m w_{m,r,s}^{Discharge-feedstocks} + \sum_k w_{k,r',s}^{Discharge-biorefinery} - \tag{2}$$
$$\sum_m w_{m,r,s}^{Used-feedstocks} - \sum_k w_{k,r',s}^{Used-biorefinery} - L_{r,s} - U_{r,s}, \qquad \forall r, \forall s$$

Wastewater discharges from new biorefineries

$$w_{k,r',s}^{Discharge-biorefinery} = \sum_m \alpha_{k,m,s}^{ww-biorefinery} \, P_{k,m,r',s}^{Bio}, \qquad \forall k, \forall r', \forall s \tag{3}$$

Production at the cultivating fields

$$f_{m,r,s}^{Prod-feedstock} = \alpha_{m,r,s}^{fedstock} \left[a_{m,r}^{exist} + a_{m,r}^{new} \right], \qquad \forall m, \forall r, \forall s \tag{4}$$

Activation of binary variables for biorefineries

$$F_{m,r',s} \leq F_{m,r'}^{max} \, y_{r',s}^{Biorefinery}, \qquad \forall r', \forall s \tag{5}$$

Production of biofuels

$$P_{k,m,r',s}^{Bio} = \alpha_{k,m,s}^{Bioref} \, F_{m,r',s}, \qquad \forall k, \forall m, \forall r', \forall s \tag{6}$$

Distribution of products to the markets

$$P_{k,r',s} = \sum_j P_{k,r',j,s}^{Prod-market}, \qquad \forall k, \forall r', \forall s \tag{7}$$

Economic equations
Capital cost for biorefineries
The capital cost for biorefineries has associated fixed and variable charges as follows:

$$CapCost_s^{Biorefineries} = k_F \left[\sum_m \sum_{r'} FC_{m,r'}^{Biorefinery} \, y_{r',s}^{Biorefinery} + \sum_m \sum_{r'} VC_{m,r'}^{Biorefinery} \left(F_{m,r',s} \right)^{\sigma_{m,r'}^{Biorefinery}} \right], \quad \forall s \tag{8}$$

Objective function
Total annual profit
The total annual profit accounts for all the sale gains, minus the cost of installation and operation of biorefineries, the agriculture cost to cultivate the biomass, water costs, and product and raw material transportation costs.

$$Profit_s = G_s^{Products} + G_s^{Grains} - CapCost_s^{Biorefineries} - OpCost_s^{Biorefineries} - OpCost_s^{Feedstocks} -$$
$$OpCost_s^{Water-biorefineries} - OpCost_s^{Water-feedstocks} - TCost_s^{Feedstocks} - TCost_s^{Grains} - TCost_s^{Products} \tag{9}$$

For this proposed system, the profit has a different value for each scenario, the involved uncertainty drastically impacts the economic target.

Expected profit
Considering the strong fluctuation that affect the profit, the model formulation defines the expected profit, which corresponds to the distribution profile for all the scenarios:

$$MProfit = \overline{Profit_s} = \frac{1}{scenario(s)} \sum_s Profit_s(s), \qquad \forall s \tag{10}$$

Worst Profit
It is particularly important to maximize the worst scenario for the economic performance. Thus, the decision makers can decide among risky configurations for the uncertainty management in the project.

$$MProfit \geq WProfit \tag{11}$$

Total water requirements for the proposed biorefining system

$$Water_s^{required} = Q_{s\,final}^{Initial} - Q_{s\,final}, \qquad \forall s \tag{12}$$

Expected fresh water consumption
The available sites to install industrial plants will define the requirements of water for this location, also it is important to include the expected value for the fresh water consumption:

$$MWater = \overline{Water_s} = \frac{1}{scenario(s)} \sum_s Water_s^{required}(s) \tag{13}$$

Worst fresh water consumption

In the same way, it is essential to know the worst scenarios for the performance of the fresh water consumption in the biorefining system:

$$WWater \geq Water_s^{required}, \qquad \forall s \tag{14}$$

CO2 emissions

The overall CO_2 emissions are estimated by the industrial and transportation operation where the capture of emission by cultivating areas is considered:

$$CO2_s^{total} = CO2_s^{prod} P_s^{biototal} - CO2_s^{abs} A_s^{total}, \qquad \forall s \tag{15}$$

3. Results

The proposed optimization model corresponds to a mixed integer non-lineal programming (MINLP) model. The model was coded in GAMS (General Algebraic Modelling System), and the data of the case study were charged. The uncertainty was considered for several parameters, where N-scenarios are generated through the Latin Hypercube Sampling method under normal distributions. Monte Carlo simulation is implemented to evaluate the risk in quantitative analysis for the parameters analyzed.

A case study was solved to evaluate the capacities of the proposed model. This case study is located in the central part of Mexico that is considered as a high-potential region due to the availability of resources and the predominant agricultural activity. The balsas river is considered as the watershed of analysis to estimate the environmental impact for the water consumption and the effluent discharges using the MFA approach. The watershed was divided in 23 fixed reaches based on hydrologic and geographic studies considering some parameters of the watershed system such as precipitation, filtration, evaporation or discharges due to industrial or residential activities to estimate the biomass and biofuel production. The total demand of biofuels is required by 10 markets, where each one is represented by a city along the studied region. The feedstocks are second generation biomass, where residues of corn, wheat, sorghum, and sugar are used to produce bioethanol, and for biodiesel production are cultivated jatropha and palm oil. The biofuel demands for each market were estimated using projections by the Energy Council of Mexico (SENER), and SENER-BID-GTZ (2016) projected for the next decade that the 10 % of the total energy demand will be satisfied by biofuels. The model was solved using the solver CPLEX. The optimization model consists of 309,650 continuous variables, 84,947 constraints, and 23 binary variables. The CPU time and the average solution for the objectives in expected and the worst scenarios are presented in Table 1. The time horizon is a year with an investment time of ten years. 50 scenarios were generated using Monte Carlo simulations.

The optimization problem was solved for the economic targets. At first, it is obtaining the maximization of profit being the greatest value that represents the optimistic economic condition, and the lowest represents the worst condition. Figure 1 shows the profit and water demands, where 3 interested scenarios were chosen. The solution for each scenario represents a specific configuration for the supply chain, as well the composition and operation of the biorefining system. For example, in scenario 3 with 94 % probability of occurrence, the expected profit is \$US 7.58×10^8/y and 3.24×10^7 m^3/y of demand of fresh water. where 7 biorefineries are required to satisfy the biofuel demands in the considered markets. The total biofuel production is 6.71×10^7 L/y of bioethanol and 8.66×10^7 L/y of biodiesel, the distribution between the type of biomass used and produced biofuels are presented in the Figure 2. But for the greatest economic risk that we have the worst profit that can be obtained of US\$ 7.32×10^8/y and 3.25×10^7 m^3/y of fresh water consumption.

Also, the satisfied demand of feedstocks and products are estimated. The availability of land to cultivate crops in the region was considered but also the possibility of increment, in the presented solution 19, 377 ha of new area should be cultivated to satisfy the biomass demand. For scenario 2, with 70 % of occurrence, it is possible to obtain approximately at least a profit of US$ 673 million for the worst profit, when the profit can reach up US$ 720 million (achieving an increase of 6.98 %). For the optimistic environmental conditions, the system has the lowest consumption of water, where the system with the greatest probability of occurrence yields a profit of US$ 1.68 x10^8/y and the demand of water is 2.07x10^7 m^3/y with a biofuel production of 4.4859 x10^7 L/y for bioethanol and 1.701459x10^8 L/y for biodiesel, for this scenario the area requires to be increased in 34,224 Ha to cultivate corn, 55 Ha for wheat, and 3,268 ha for oil palm.

Table 1. Case study solutions.

Case	CPU TIME (s)	AVERAGE SOLUTION
Maximizing Mprofit	2426	US$6.79 x$10^8$/y
Maximizing Wprofit	90.4	US$6.44 x$10^8$/y
Minimizing Mwater	244.7	2.01 x10^7 m^3/y
Minimizing Wwater	27.05	5.04 x10^5 m^3/y

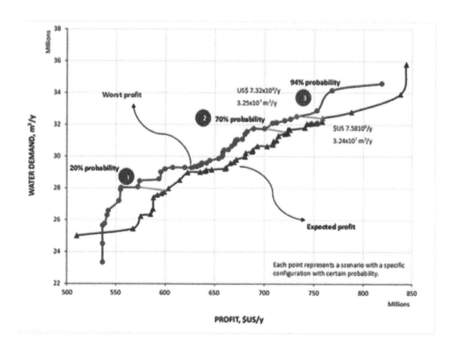

Figure 1. Pareto curve for the profit vs water demand.

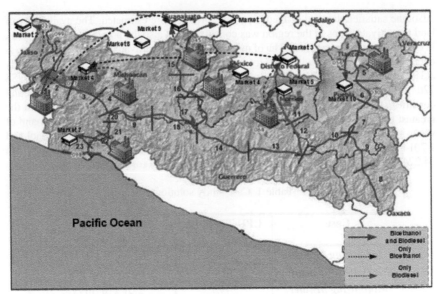

Figure 2. Optimal configuration for the biorefinery system in scenario 3 for expected profit.

4. Conclusions

The proposed mathematical approach incorporates the use of agricultural residues to reduce the environmental impact, as well as an analysis in the watershed to understand the interconnection between sector that works as a nexus, the sale of grains as economic retribution in the system, and the uncertainty to have a robust model.

The optimization approach is an efficient tool that determines the location and operation logistics of biorefineries, new cultivation areas and associated food profits and biomass wastes, fresh water demands, production of feedstock and biofuels, production and capture of greenhouse gas emission, routes of distribution considering environmental and economic objectives. The tradeoffs between these objectives were analyzed and discussed with focus on the strong impact of some environmental constraints on the selection of feedstock growth and location as well as production capacities and impact on the surrounding watershed.

References

[1] J.E. Satibañez-Aguilar, R. Morales-Rodriguez, J.B. Gonzales-Campos, J.M. Ponce-Ortega, 2016, Stochastic design of biorefinery supply chains considering economic and environmental objectives, Journal of Cleaner Production, 136, 224-245.

[2] P. Staurt, M.M. El-Halwagi, 2013, Integrated biorefineries: Design, analysis, and optimization, Taylor and Francis/CRC, Avingdon, Virginia, USA.

[3] B. Kamm, M. Kamm, 2004, Principles of biorefineries, Applied Microbiology and Biotechnology, 64, 2, 137-145.

[4] F. López-Villareal, L.F. Lira-Barragán, V. Rico-Martinez. J.M. Ponce-Ortega, M.M. El-Halwagi, 2014, An MFA optimization approach for pollution trading considering the sustainability of the surrounded watersheds, Computers and Chemical Engineering, 63, 1, 140-151.

Anton A. Kiss, Edwin Zondervan, Richard Lakerveld, Leyla Özkan (Eds.)
Proceedings of the 29th European Symposium on Computer Aided Process Engineering
June 16th to 19th, 2019, Eindhoven, The Netherlands. © 2019 Elsevier B.V. All rights reserved.
http://dx.doi.org/10.1016/B978-0-128-18634-3.50263-0

A multiscale model approach for cell growth for lipids and pigments production by *Haematococcus pluvialis* under different environmental conditions.

Alessandro Usai [a,b], Jon Pittman[c] , and Constantinos Theodoropoulos[a,b*]

[a]*School of Chemical Engineering and Analytical Science, University of Manchester, M13 9PL, UK*
[b]*Biochemical and Bioprocess Engineering Group, University of Manchester, M13 9PL, UK*
[c]*School of Earth and Environmental Sciences, University of Manchester, M13 9PL, UK*

*k.theodoropoulos@manchester.ac.uk

Abstract

The development of new technologies is essential to contrast the increasing global CO_2 emissions. Microalgae-based biorefineries present a promising tool to reduce the presence of CO_2 in our emissions (Bekirogullari et al 2018, Bekirogullari et al 2017). The development of processes for the simultaneous production of more than one product are essential to increase the possibility of making such biorefineries profitable (García Prieto, Ramos, Estrada, Villar, & Diaz, 2017). Microalgae eukaryotic cells are able to produce pigments and lipids. The production of lipids and pigments could vary depending on the environmental stress conditions such as light intensity, nutrients concentration, medium salinity, and temperature (D'Alessandro & Antoniosi Filho, 2016). *Haematococcus pluvialis* produces lutein, β-carotene, and lipids during growth, which makes it a potential candidate for the simultaneous production of high value products. *H.pluvialis'* division pattern ranges from the formation of two daughters up to 32 cells by multiple-fission (Shah, Liang, Cheng, & Daroch, 2016). New born cells achieve after a dimensional growth step a critical size point, where after that, they are able to keep growing and reproduce without any further input in the system, this point is called commitment point (CP). The achievement of an higher number of CP lead to a higher number of cells in the day/night cycle (Bišová & Zachleder, 2014). The above phenomena reaching one commitment point are described using a population balance model (PBM), in order to evaluate the influence of some environmental conditions on the cell growth. The model was fitted against synchronized experimental data, and then tested on a long duration experiment.
Keywords: microalgae; population balance models, synchronisation.

1. Introduction

The capability to produce high-value products such as neutral lipids, pigments, carbohydrates, proteins, and other products makes the development of the microalgal biorefinery processes interesting. The development of microalgae biotechnology is a possible candidate to solve significant issues such as the depletion of fossil fuels through the production of biofuels, wastewater removal through the consumption of NH_4^+, NO_3^-, and PO_4^{3-} and land utilization reduction utilising less land per litre of products produced. Moreover, they have a variety of metabolites produced which can be useful in many fields (Mata, Martins, & Caetano, 2010). Furthermore, CO_2 emissions from fossil-fuel shows an increase during the last decades, especially in developing countries (Olivier, Schure, & Peters, 2017), and microalgae are considered within a biotechnology perspective for CO_2 sequestration (Singh & Ahluwalia, 2013).

Microalgae, during cultivation, cover a series of essential steps leading to multiplication. Starting from the new born cell, the first phase is represented by the cell volume growth up to reach the commitment point (CP) (Bišová & Zachleder, 2014). This first cycle phase could be observed through a synchronisation of the cell population, and then following the cell population volume growth through the light cycle (Hlavová, Vítová, & Bišová, 2016). The given procedure could be improved and applied to other cell constituents, such as lipids and/or pigment. In this work the basis for a volume-structured PBM model was constructed, and the parameters related with the volume growth rate were estimated from synchronous experiments, followed by a prediction of a long-term experiment. Such models can give accurate predictions of individual cell growth and can be subsequently efficiently coupled with kinetic expressions and/or metabolic-level models to predict the production of lipids pigments and of metabolites within.

2. Materials and Methods

Haematococcus pluvialis strain FLOTOW (1844) CCAP 34/6 was purchased from Culture Collection of Algae and Protozoa (CCAP) Scotland, United Kingdom. The medium is carbon-free NIES-C. Two experiments were carried out and the starting point for both experiments is a cell culture inoculated at 5.5×10^3 cell mL^{-1} in 215mL medium in a glass bottles (500mL) with a sponge cap. Both of them were incubated at 27°C, stirred at 130 rpm, and illuminated at 60 µmol $m^{-2}s^{-1}$ by using white fluorescent neon, and considering 16/8 light dark cycle. The first experiment examines microalgae synchronisation, following an adapted procedure suggested by (Hlavová *et al* 2016), and the second experiment is a long-term photoautotrophic growth. The synchronisation experiment reaches around 2.0×10^5 cells mL^{-1} after 16 days. Starting from this point cells were kept at this concentration for 72 hours diluting the culture with NIES-C medium at the beginning of every light cycle. Starting from the 72th hour regular sampling was performed until the end of the light cycle, in order to follow the dimension, and the number of the cell population by using Cellometer Auto T4 purchased from Nexcelom Bioscience LLC. In the second experiment samples were taken starting from time zero every seven days until the 3tth day, and the volume and number population properties weremeasured. All of the experimental results presented are the mean of duplicate experiments.

3. Model development

The proposed model is based on the PBM model equations firstly proposed by Fredrickson (1967). The model is represented by an integro-partial differential equation (Eq.1) including a vector of initial conditions (Eq.2), and boundary conditions (Eq.3). The particle state is assumed to depend only on the volume which is an internal coordinate. The choice of the volume is related to the considerations given in many works where the volume seems to be the variable related to the CP achievement in microalgae (Bišová & Zachleder, 2014; Šetlík & Zachleder, 1984; Zachleder & Vítová, 2016), which aims to give a solid basis for future works in this direction. The term Ψ represents the number concentration density function, r_v and Γ^v are the rate of change for the particle state vector, and the transition rate for the birth term, respectively. They depend on the volume v, and on the attenuated light I_{att}.

$$\frac{\partial \Psi(v,t)}{\partial t} + \frac{\partial [r_v \Psi(v,t)]}{\partial v} = 2 \int_v^\infty \Psi(v',t)\Gamma^v(v',I_{att})p(v,v')dv' - \Psi(v,t)\,\Gamma^v(v,I_{att}) \tag{1}$$

$$\Psi(v,t) = \Psi^0(v) \text{ for } t = 0 \text{ and } \forall v \tag{2}$$

$$\Psi(v,t) = 0 \text{ for } t > 0 \text{ and } v = 0 \tag{3}$$

The model was reduced in order to perform an ideal fitting procedure. The birth term is set to zero, and only the part regarding the volume growth considered (Eq.4).

$$\frac{\partial \Psi}{\partial t} + \frac{\partial (r_v \Psi)}{\partial v} = 0 \tag{4}$$

$p(v,v')$ represents the partitioning distribution function, which considers how mother cells internal properties are redistributed in daughter cells. The one used here is able to get unequal partitioning (Hatzis, Srienc, & Fredrickson, 1995), q being a coefficient.

$$p(v,v') = \frac{1}{\beta(q,q)} \frac{1}{v'} \left(\frac{v}{v'}\right)^{q-1} \left(1 - \frac{v}{v'}\right)^{q-1} \tag{5}$$

Light attenuation functionalities related to volume-structured PBM models are not present in literature. In this model a functionality is given (Eq.6), which takes into account the ratio between the total cells volume V_{cell}^T, and the total volume of the culture V^T, α_{att} being a fitting parameter. The total volume of the cells is given by the first order moment of the distribution multiplied by the total cell volume (Eq.7).

$$I_{att} = I_0 \left[1 - \frac{V_{cell}^T}{V^T}\right]^{\alpha_{att}} \tag{6}$$

$$V_{cell}^T = V^T \int_0^\infty v\Psi(v)dv \tag{7}$$

The rate of change for the particle state vector considers the light attenuated by using a Monod functionality (Monod, 1949), where $K_{I_{att}}$ indicates the half saturation light term. Moreover, the volume is considered to influence the volume growth, in particular the raising volume is considered to make the volume growth rate lower, it is represented by an inhibition functionality with K_v inhibition constant. The latter right-hand term represents the decay rate, and μ_C is the decay rate constant (Eq.8).

$$r_v(v,I_{av}) = \mu_v \frac{I_{att}}{I_{att}+K_{I_{att}}} \frac{K_v}{v+K_v} - \mu_C v \tag{8}$$

The transition rate between mother cells and daughters is expressed as a function of the rate of change for the particle state vector multiplied by the gamma function γ^v (Eq.9). The functionality is like the one used in previous a previous work, considering the mass as internal variable, for microalgae (Concas, Pisu, & Cao, 2016). The probability to undergo division increases when the cells are approaching the critical volume v_c. σ_c^2 is the variance of the Gaussian functionality considered in equation 11.

$$\Gamma^v(v, I_{av}) = r_v(v, I_{att})\gamma^v(v) \tag{9}$$

$$\gamma^v(v) = \frac{f(v)}{1 - \int_0^v f(v')dv'} \tag{10}$$

$$f(v) = \frac{1}{\sqrt{2\pi\sigma_c^2}} \exp\frac{-(v-v_c)^2}{2\sigma_c^2} \tag{11}$$

4. Results and discussion

Parameter estimation has been carried out for the parameters involved in the cell volume growth description by using the data obtained from the synchronisation experiment, following a possible ideal procedure (Fadda, Cincotti, & Cao, 2012). The partial model showed in equation 4 was utilized. The initial volume distribution for the parameter estimation was obtained by fitting the initial number histogram percentage for the synchronisation experiment (Fig.1a) and converting it in terms of the number concentration density function. The parameters are obtained by fitting the mean volume during the synchronisation experiment (Fig.1b). The experimental volume increases in agreement with the model predictions. The normalized density function during the volume growth process (Fig.1c) in the light period agrees qualitatively with the experimental distributions (data not showed). The q parameter was specifically used for binary fission in microalgae (Concas et al., 2016).

The set of parameters obtained was used to predict the long-term experiment. The initial volume distribution showed in figure 1d was obtained by fitting the number histogram percentage of a duplicate experiment at time zero of the long-term experiment, and then it was converted to a number concentration density function. N_0, σ_0, and μ_0 were obtained by using this fitting. The critical volume was considered to be the double of the initial volume at time zero of the synchronisation, where the cells should be averagely at the minimum volume related to daughter cells (Bišová & Zachleder, 2014). The σ_c used was the same as the one of the initial distribution. The results of the complete model prediction are shown in figure 1e, where the total cell number shows a good agreement with the experimental data.

Table 1. Definition of model parameters and their respective values.

Parameter	Definition	Value	Units
N_0	Initial total cell number	9.11×10^5	Cells
σ_0	Initial distribution parameter	2.30×10^3	μm^3
μ_0	Mean initial volume	6.94×10^3	μm^3
α_{att}	Light attenuation shape parameter	8.84×10^3	-
q	Coefficient symmetric beta function	4.00×10^0	-
K_v	Volume inhibition constant	5.43×10^3	μm^3
$K_{I_{att}}$	Light half saturation constant	2.24×10^2	$\mu mol\ m^{-2}s^{-1}$

μ_c	Volume decay rate	5.38×10^{-5}	h^{-1}
μ_v	Volume growth rate	1.22×10^3	$\mu m^3 h^{-1}$
V_T	Total culture volume	2.15×10^2	mL
I_0	Incident light intensity	6.0×10^1	$\mu mol\ m^{-2}s^{-1}$
v_c	Mean critical volume	9.86×10^3	μm^3
σ_c	Critical volume distribution parameter	2.30×10^3	μm^3

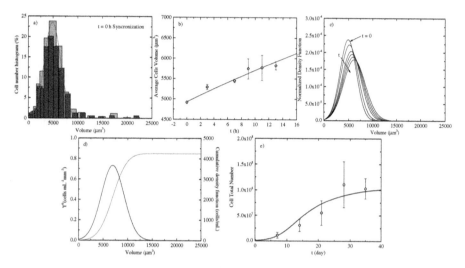

Figure 1. a) Cell number histogram (percentage) at time zero of synchronisation b) Experimental average cell volume (blue diamonds) and model fitting (black line) in synchronisation experiment. c) Normalized density function evolution during cell volume growth in synchronisation model fitting. d) Initial density function (black line), and cumulative density function (red line) at time zero of long-term experiment e) Experimental cell total number (black circles), and model prediction (black line) for the long-term experiment.

5. Conclusions

A population balance model was proposed to describe the growth of microalgae cells. Biomass growth is limited by the light attenuation, and the model takes into account the observed phenomena affecting growth rate decreasing in correspondence of the mean volume growth. The model parameters were estimated observing the volume growth phase during the light cycle in a short duration experiment trying to exclude other phenomena. Model predictions for the long-term experiment show the potential ability of the model to predict cell populations in term of total cell numbers for long-term experiments from a fitting data from short-term experiments. The fitting procedure could be adopted extended to other variables, making the model able to work in a wide range of operative conditions, and including the production of individual metabolites in order to take into account added-value products.

References

Bekirogullari, M, Pittman J & Theodoropoulos, C (2018). Multi-factor kinetic modelling of microalgal biomass cultivation for optimised lipid production. Bioresource Technology.

Bekirogullari, M, Fragkopoulos I, Pittman J & Theodoropoulos, C (2017). Production of Lipid-Based Fuels and Chemicals from Microalgae: An Integrated Experimental and Model-based Optimization Study. Algal Research, 23, 78-87.

Bišová, K., & Zachleder, V. (2014). Cell-cycle regulation in green algae dividing by multiple fission. *Journal of Experimental Botany, 65*(10), 2585–2602.

Concas, A., Pisu, M., & Cao, G. (2016). A novel mathematical model to simulate the size-structured growth of microalgae strains dividing by multiple fission. *Chemical Engineering Journal, 287*, 252–268.

D'Alessandro, E. B., & Antoniosi Filho, N. R. (2016). Concepts and studies on lipid and pigments of microalgae: A review. *Renewable and Sustainable Energy Reviews, 58*, 832–841.

Fadda, S., Cincotti, A., & Cao, G. (2012). A novel population balance model to investigate the kinetics of in vitro cell proliferation: Part II mumerical solution, parameters' determination, and model outcomes. *Biotechnology and Bioengineering, 109*(3), 782–796.

Fredrickson, A. G., Ramkrishna, D., & Tsuchiya, H. M. (1967). Statistics and dynamics of procaryotic cell populations. *Mathematical Biosciences, 1*(3), 327–374.

García Prieto, C. V., Ramos, F. D., Estrada, V., Villar, M. A., & Diaz, M. S. (2017). Optimization of an integrated algae-based biorefinery for the production of biodiesel, astaxanthin and PHB. *Energy, 139*, 1159–1172.

Hatzis, C., Srienc, F., & Fredrickson, A. G. (1995). Multistaged corpuscular models of microbial growth: Monte Carlo simulations. *BioSystems, 36*(1), 19–35.

Hlavová, M., Vítová, M., & Bišová, K. (2016). Synchronization of green algae by light and dark regimes for cell cycle and cell division studies. In *Methods in Molecular Biology* (Vol. 1370, pp. 3–16). https://doi.org/10.1007/978-1-4939-3142-2_1

Mata, T. M., Martins, A. A., & Caetano, N. S. (2010). Microalgae for biodiesel production and other applications: A review. *Renewable and Sustainable Energy Reviews, 14*(1), 217–232.

Monod, J. (1949). a Certain Number. *Annual Reviews in M, 3*(XI), 371–394.

Olivier, J. G. J., Schure, K. M., & Peters, J. A. H. W. (2017). TRENDS IN GLOBAL CO 2 AND TOTAL GREENHOUSE GAS EMISSIONS 2017 Report Trends in global CO2 and total greenhouse gas emissions: 2017 Report, (December). Retrieved from http://www.pbl.nl/sites/default/files/cms/publicaties/pbl-2017-trends-in-global-co2-and-total-greenhouse-gas-emissons-2017-report_2674.pdf

Šetlík, I., & Zachleder, V. (1984). The multiple fission cell reproductive patterns in algae. In *The Microbial Cell Cycle* (pp. 253–279).

Shah, M. M. R., Liang, Y., Cheng, J. J., & Daroch, M. (2016). Astaxanthin-Producing Green Microalga Haematococcus pluvialis: From Single Cell to High Value Commercial Products. *Frontiers in Plant Science, 7*(April). https://doi.org/10.3389/fpls.2016.00531

Singh, U. B., & Ahluwalia, A. S. (2013). Microalgae: A promising tool for carbon sequestration. *Mitigation and Adaptation Strategies for Global Change, 18*, 73–95.

Zachleder, V., & Vítová, M. (2016). The cell cycle of microalgae. https://doi.org/10.1007/978-3-319-24945-2

Anton A. Kiss, Edwin Zondervan, Richard Lakerveld, Leyla Özkan (Eds.)
Proceedings of the 29th European Symposium on Computer Aided Process Engineering
June 16th to 19th, 2019, Eindhoven, The Netherlands. © 2019 Elsevier B.V. All rights reserved.
http://dx.doi.org/10.1016/B978-0-128-18634-3.50264-2

Ecosystem Services Valuation and Ecohydrological Management in Salt Lakes with Advanced Dynamic Optimisation Strategies

A.G. Siniscalchi[(a,c)], C. Garcia Prieto[(a,c)], E.A. Gomez [(b,e)], A. Raniolo [(b,d)], R.J. Lara[(b)], M.S. Diaz[(a,c)]*

[a] *Planta Piloto de Ingeniería Química (Universidad Nacional del Sur-CONICET), Camino la Carrindanga Km 7, Bahía Blanca 8000, Argentina*
[b] *Instituto Argentino de Oceanografía (IADO-CONICET), Camino La Carrrindanga Km7, Bahía Blanca, Argentina*
[c] *Departamento de Ingeniería Química, Universidad Nacional del Sur, Bahía Blanca*
[d] *Departamento de Ingeniería, Universidad Nacional del Sur, Bahía Blanca*
[e] *Universidad Tecnológica Nacional Facultad Regional Bahía Blanca, Bahía Blanca*

Abstract

In this work, we propose a mathematical model within a dynamic optimization framework to address management and economic valuation of ecosystem services in a salt lake basin, located in a semiarid region in Argentina (Chasicó Lake). Ecosystem services are valuated by considering water, land and food provision, water flows and disturbance regulation, as well as cultural aspects. Numerical results provide useful insights on optimal management strategies effects and how they can improve services provided by the ecosystem under study.

Keywords: Dynamic Optimisation, Hydrological Mathematical Model, Ecosystem Services, Drought and Flood Mitigation

1. Introduction

The first step to develop useful tools for decision-making in the management of salt lakes is the identification of meteorological, hydrological and ecological components and driving forces of floods, drought and hypersalinisation. Additionally, ecosystem services valuation is essential to assess the relative contribution of the natural capital stock and to effectively manage sustainable human well-being (Costanza et al., 2014). In this work, we extend our previous ecohydrological models formulated as optimal control problems to plan management strategies in salt lakes (Siniscalchi et al., 2018). We formulate optimisation models for dampening extreme hydrological events and related lake salinity oscillations and to irrigate a drought-resistant crop in the grassland basin of a Pampean salt lake, also including equations for the calculation of crop water requirement dynamics. Further, we identify and valuate the ecosystem services offered by the studied endorheic basin and relate them to the management strategies proposed in this work. Numerical results provide control variables profiles to plan actions and their ecological and hydrologic effects.

2. 2. Mathematical Model

2.1 Hydrological Model

We formulate dynamic water mass balances for a salt lake and an artificial reservoir located nearby. Eqn. (1) and (2) show water and salt mass balances for the salt lake.

$$\frac{dm}{dt} = \left[Q_{pp}(t) \left(\frac{V}{h} \right) + Q_{river}(t) + Q_{gw} - Evap(t) \left(\frac{V}{h} \right) + Q_{res}(t) \right] \delta_w / 1000 \tag{1}$$

where m is total water mass in the lake (kg); δ_w, is water density (kg. m^{-3}), which is assumed constant; V corresponds to salt lake volume and h is average depth. Q_{pp} (L.day^{-1}.m^{-2}) corresponds to precipitations, Q_{river} (L.day^{-1}) is the tributary flowrate, Q_{gw} (L.day^{-1}) is groundwater flowrate, (Q_{res}) is the flowrate of a water stream that can be diverted from an artificial reservoir to the salt lake for salinity and volume control in drought periods. $Evap$ (L.day^{-1}.m^{-2}) corresponds to evaporation per unit area in the lake. As salt concentration in both groundwater and the tributary is negligible, we assume that salt mass is constant within the lake ($d(CsV)/dt=0$) and salt concentration (Cs) is calculated as:

$$\frac{dCs}{dt} = - \frac{Cs}{V} \left[Q_{pp}(t) \left(\frac{V}{h} \right) + Q_{river}(t) + Q_{gw} - Evap(t) \left(\frac{V}{h} \right) + Q_{res}(t) \right] \delta_w / (\delta_w 1000) \tag{2}$$

Equation (3) is the water mass balance for the artificial reservoir of fixed area, located upstream of the tributary, assuming constant water density.

$$\frac{dVres}{dt} = Q_{pp}(t)A\, \delta H_2O - Evap(t)A\delta H_2O - Q_{res}(t) - Q_r(t) \tag{3}$$

where $Q_r(t)$ is the flowrate of a water stream used for crop irrigation, calculated based on crop evapotranspiration ($ETc(t)$) as:

$$Q_r(t) = ETc(t) - Q_{pp}(t) \tag{4}$$

$ETc(t)$ is calculated for a hypothetic *Chenopodium quinoa* crop after Penman-Monteith, considering the evapotranspiration of a hypothetic crop ($ET0(t)$) and a cultivation coefficient for quinoa (Kc, Garcia et al., 2017) as follows:

$$ET0(t) = \frac{k*\Delta*Rn+\gamma*\frac{900}{Tair+273}*Fwind*(Prvap-Tvap)}{\Delta+\gamma*(1+0.34*Fwind)} \tag{5}$$

$$ETc(t) = Kc(t) * ET0(t) \tag{6}$$

Irrigation crop water requirement taken from the artificial lake and diverted flowrate for irrigation are calculated as:

$$CropWR(t) = ETc(t) - Q_{PP}(t) \tag{7}$$

$$Q_r(t) = CropWR(t) * A_{cult} \tag{8}$$

where A_{cult} corresponds to total cultivated area.

Evaporation is calculated taking into account energy and momentum balances (Penman 1948).

$$Evap = \frac{10}{\lambda v}\left[W\,Rn\,(t) + (1-W)F_{wind}\left(T_{vap0} - T_{vap}\right)\right] \tag{9}$$

where W is a weighting factor of the radiation effects on evaporation; $Rn(t)$ is net radiation (ly.d^{-1}), F_{wind} expresses wind effects as available energy to evaporate water (km.d^{-1}) and (T_{vap0} - T_{vap}) is a vapour saturation deficit. Additional algebraic equations include latent heat of vaporization for water and forcing functions, represented by Fourier series, including tributary and groundwater flowrate, air temperature, relative humidity, solar radiation, mean wind rate and related meteorological variables.

The model has three differential and twenty five algebraic equations and has been implemented within a control vector parameterisation environment in gPROMS (2017).

2.3 Dynamic Optimisation Model

We formulate management problems for water bodies as optimal control problems subject to the differential algebraic equation (DAE) systems described in the preceding sections. The goal is to keep salt concentration in the lake (Cs) around a desired value of 23 kg.m^{-3} within a drought climate scenario, considering the flowrate of a diverted stream from the artificial reservoir (Q_{diver}) as the control variable. The objetive function is:

$$\min Z = \int_0^{tf} (Cs(t) - 23)^2 \, dt$$

$Qres$

st

DAE system (1)-(9)

$Cs(0) = Cs^0$, $V(0) = V^0$, $V_{res} = V_{res}^0$, $Q_{diver}^L \le Q_{diver} \le Q_{diver}^U$ $Cs^L \le Cs \le Cs^U$ $V^L \le V \le V^U$

2.4 Ecosystem Services Valuation

Ecosystem Services for the salt lake basin were identified in previous work (Siniscalchi et al., 2018) and global valuation for each service was carried out following de Groot et al. (2012), in USD.ha^{-1}.yr^{-1}. These values were multiplied by the corresponding areas (basin, lake and natural reserve).

3. Case Study

Chasicó Lake is located in a depression, 20 m below sea level in the southwest of the Chaco-Pampean plain (38°38″ S 63°03″W), Argentina. The maximum and average depths of this water body are 13.2 and 8 m, respectively. This salt lake (6,500 ha) is part of a natural Reserve (7,800 ha) situated in an endorheic basin (376,400 ha). The hydrologic regime of this basin is closely related to climatic conditions, with dry and wet cycles (D'ambrosio et al., 2013). During the last century, there were flood events in 1918, 1923, 1924, 1976, 1978, 1983, 1993, 2001and 2002, alternating with dry periods during 1930–1970 and 2006–2014 (Lara, 2006). Salinity values in the lake during the last fifty years ranged between 100 g.L^{-1} (1963) and 16 g.L^{-1} (1993). A wide variation in lake extension and salinity is the result of natural periods of drought and flooding, and is reflected on changes in water quality and biota (Kopprio et al., 2014). It also affects the economy of the region through losses in cropland area and investment uncertainty in relation to tourism and sport fishing of silverside. The severe impact that droughts and floods cause to inland water resources and to the associated socioeconomic activities requires close monitoring of this lake.

4. Results and Discussion

4.1 Salinity and flood control in Chasicó Lake

In previous work, we proposed a hydrological model for Chasicó Lake calibrated and validated with collected salinity data throughout the last ten years (Siniscalchi et al., 2017, 2018). In this frame, we proposed the construction of an artificial reservoir for water storage during wet periods. In this work, we extend the model to drought scenarios, in which the reservoir is used to damp salinity oscillations in Chasicó Lake and to irrigate a drought-resistant crop. Dynamic optimization problems are formulated within a control vector parameterization framework in gPROMS (PSEnterprise, 2017). Annual rainfall and evaporation during the dry period were 400 and 1100 mm, respectively. Figure 1 shows rainfall and evaporation profiles during a 1.5-year dry period. The basis for hydric calculations for cultivation of *Chenopodium quinoa* in Chasicó basin is 300 ha, which corresponds to 23% of the entire area of the natural reserve that may be used for crop cultivation. Figure 2 shows rainfall and CropWR profiles for quinoa crops, from sowing to harvesting, throughout a drought period in which total CropWR is 1254.2 m^3.ha^{-1}. The maximum CropWR, 2.8 L.d^{-1}·m^{-2}, corresponds to the onset of grain maturation.

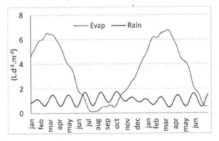

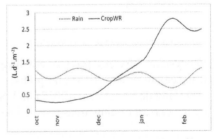

Figure 1. Rainfall and evaporation for 1.5-year drought period in Chasicó Lake

Figure 2. Rainfall and CropWR for *Chenopodium quinoa* crop in Chasicó basin (Oct.15th-Feb.15th)

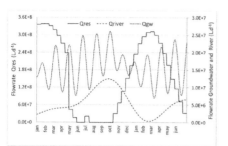

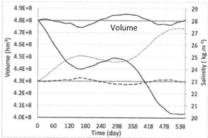

Figure 3. Inflow profiles into Chasicó Lake from the river (Q$_{river}$), groundwater (Q$_{GW}$) and reservoir (Q$_{res}$), for a 555-day dry period

Figure 4. Volume (continuous lines) and salinity profiles (dotted lines) with and without optimal management

Figure 3 shows inflow profiles into Chasicó Lake from the river (Qriver), groundwater (QGW) and reservoir (Q_{res}), being the latter the control variable. Q_{res} peaks correspond to Qriver minima and evaporation maxima (see Fig.1). Chasicó Lake volume and salinity profiles with and without optimal management through water input from the artificial reservoir (Q_{res}) are shown in Fig. 4. It is important to remark that the water volume stored during four wet years in an artificial reservoir is enough for keeping salinity at an optimum (23 g.L^{-1}) and simultaneously irrigating a 300 ha-quinoa crop during ca. 1.5 years, thus increasing the socio-ecologic system resilience to extreme drought periods.

3.2. Ecosystem Services Valuation

In this work ecosystem services were valuated for the Chasicó region (Gomez, 2015), including regulating (climate, water flows, disturbance dampening), provision (water, land and food) and cultural services (tourism and recreation). Following de Groot et al. (2012), valuation was carried out by multiplying a reference value (MMUS$/(ha.y) for each service by the corresponding area, as follows. For water flows, disturbance dampening, water and land supply, a basin surface of 376,400 ha was taken. A lake surface of 6,500 ha was used in the case of food supply (silverside). To calculate cultural services, the nature reserve surface (7,800 ha) was taken. Ecosystem Services with the highest economic value (in MMUS$/y) were water flows (2100) and disturbance dampening (1120), further supporting the importance of optimal salinity and lake volume regulation, as well as the construction of a buffering reservoir. Provisioning services valuation include land (450), water (154) and food (4) supply. Additionally to lake food supply services, optimisation results show the possibility of irrigating a 300 ha-quinoa crop. This would produce an improvement of the basin economic value due to quinoa sales of 1.5 MMUS$/y, considering a 2 t/ha yield. Cultural services were valuated in 33 MMUS$/y.

4. Conclusions

In this work, we propose detailed mechanistic models of salt lakes as constraints for optimal control problems to provide tools for planning water body management and exploring the associated effects. Numerical results show the optimal profile for the water stream flowrate that should be diverted from an artificial reservoir to mitigate drought effects by keeping lake salt concentration at the physiological optimum for a high-commercial value fish and irrigation of 300 ha of drought resistant crops (quinoa). Water flows and disturbance regulation correspond to ecosystem services with highest economic value for the basin where Chasicó Lake is located. Dynamic models provide useful insights for water flow regulation and food provision.

References

Costanza R., de Groot R., Sutton P., van der Ploeg S., Anderson S.J., Kubiszewski I., Farber S., Turner K. 2014. Changes in the global value of ecosystem services. Global Environmental Change 26:152–158

D'ambrosio, G.T., Bohn, V.Y., Piccolo, M.C., 2013. Evaluación de la sequía 2008–2009 en el oeste de la Región Pampeana (Argentina). Cuadernos Geográficos 52 (1), 1–17.

de Groot R.,Brander L., van der Ploeg S., Costanza R., Bernard F., Braat L., Christie M., Crossman N., Ghermandi A., HeinL., Hussain S., Kumar k P., McVittie A., Portela R., Rodriguez L.C., ten Brinkm P., van Beukering P. 2012. Global estimates of the value of ecosystems and their services in monetary units. Ecosystem Services 1: 50–61

FAO 1998. Evapotranspiración del cultivo Guías para la determinación de los requerimientos de agua de los cultivos. ISSN 0254-5293. p.322

García Villanueva J, Huahuachampi J; SotoL 2017. Determinación de la demanda hídrica del cultivo de quinua QML01 (Chenopodium Quinoa Willd) en la Molina. Anales Científicos, 78 (2): 200-209. ISSN 2519-7398 (Versión electrónica)

Gómez C. 2015 Revalorización y reutilización turística del Sector Norte del Partido de Villarino. Caso: Laguna Chasicó. http://sedici.unlp.edu.ar/handle/10915/60544

Kopprio G.A, R. Freije, M. Arias-Schreiber, R.J. Lara, 2014. An ecohydrological adaptive approach for a salt lake in semiarid grasslands of Argentina, Sustain. Sci, 9: 229-238.

Lara R.J. 2006. Climate change, sea-level rise and the dynamics of South American coastal wetlands: case studies and the global frame, In Wissenschaftliche Zusammenarbeit mit Argentinien: Begrenzung und Zuversicht Arbeits und Diskussionspapier (Frühwald, ed.), Bonn: Von Humboldt Foundation 40-50.

Millennium Ecosystem Assessment (MEA) 2005. Ecosystem and human well-being: Current state and trends Washington. C., USA: Island Press

Process Systems Enterprise, 2017, gPROMS, www.psenterprise.com/gproms.

Siniscalchi A.G., Kopprio G., Raniolo L.A., Gomez E.A., Diaz M.S., Lara R.J.2018. Mathematical modelling for ecohydrological management of an endangered endorheic salt lake in the semiarid Pampean region, Argentina. Journal of Hydrology 563: 778–789.

Siniscalchi A.G., García Prieto, VC., Raniolo A., Gomez E., Lara R.J., Diaz M.S.2018 Ecohydrological management and valuation of ecosystem services in salt lakes with advanced dynamic optimisation strategies. AIChE Annual Meeting 2018. Oct 28 - Nov 2. Pittsburgh, PA.

Viglizzo E.F., Franka F.C. 2006. Ecological interactions, feedbacks, thresholds and collapses in the Argentine Pampas in response to climate and farming during the last century. Quaternary International 158:122–126.

Anton A. Kiss, Edwin Zondervan, Richard Lakerveld, Leyla Özkan (Eds.)
Proceedings of the 29th European Symposium on Computer Aided Process Engineering
June 16th to 19th, 2019, Eindhoven, The Netherlands. © 2019 Elsevier B.V. All rights reserved.
http://dx.doi.org/10.1016/B978-0-128-18634-3.50265-4

Life cycle design of indoor hydroponic horticulture considering energy-water-food nexus

Yasunori Kikuchi,[a,b,c,d]* Yuichiro Kanematsu,[b] Tatsuya Okubo[b,c]

a Integrated Research System for Sustainability Science, The University of Tokyo Institutes for Advanced Study, 7-3-1 Hongo, Bunkyo-ku, Tokyo 113-8654, Japan

b Presidential Endowed Chair for "Platinum Society," The University of Tokyo, 7-3-1 Hongo, Bunkyo-ku, Tokyo 113-8656, Japan

c Department of Chemical System Engineering, The University of Tokyo, 7-3-1 Hongo, Bunkyo-ku, Tokyo 113-8656, Japan

d Center for Environment, Health and Field Sciences, Chiba University, 6-2-1 Kashiwa-no-ha, Kashiwa, 277-0882 Chiba, Japan

kikuchi@platinum.u-tokyo.ac.jp

Abstract

Agriculture has become one of the largest concerns in the field of sustainability science, while it is a major consumer of resources such as nitrogen, phosphorus, water, and land. Plant factories, indoor hydroponic farming systems, can be divided into two types according to the light source: plant factories with sunlight (PFSLs) and plant factories with artificial light (PFALs) also called as vertical farming. Through the previous studies, it was revealed that these plant factories reduced the use of irreplaceable resources for food production, i.e., phosphorus, water, and land area, at the expense of additional energy consumption compared with conventional Japanese horticulture systems. In this study, we are tackling with the sustainable agriculture mixing conventional and advanced farming technologies, especially plant factories. Based on the comprehensive assessment, an analysis was conducted on the current energy usage and applicability of technology options in plant factories. The results demonstrated that the regional characteristics should be taken into account in the design of plant factories, especially for the insulation and lighting sources in PFAL. The market fluctuation has also large impacts on the profitability of plant factories.

Keywords: vertical farming, power grid, variable renewable energy

1. Introduction

Agriculture has become one of the largest concerns in the field of sustainability science (Kajikawa et al., 2007), while it is a major consumer of resources such as nitrogen, phosphorus, water, and land (Schnoor, 2014). Although phosphorus is essential for food production (UNEP, 2011), it has not been effectively recycled (Matsubae et al., 2011) and there is a strong need to close the phosphorus loop (Christen, 2007). Many countries face food insecurity and starvation induced by a water deficit (Yang et al., 2003), and water resource management is crucial to food management. Land use is an important index of food environmental impacts (Heller et al., 2013).

Plant factories, indoor hydroponic farming systems, can be divided into two types according to the light source: plant factories with sunlight (PFSLs) and plant factories with artificial light (PFALs) (Kozai et al., 2015). Because PFSLs and PFALs feature semiclosed and closed cultivation rooms, respectively, plants cultivated in these factories can be protected from pathogens and harmful insects, leading to high stability of food production. As a proof of such concept of plant factories, a comprehensive analysis was conducted for the multiple aspects and technology options of plant factories: the use of nitrogen, phosphorus, and potassium (NPK) fertilizers, the consumption of water, the occupation or transformation of land, and greenhouse gas (GHG) emissions for the demonstration PFSL and PFAL located at Chiba University in Kashiwa, Japan (Kikuchi et al., 2018a). It was revealed that these plant factories reduced the use of irreplaceable resources for food production, i.e., phosphorus, water, and land area, at the expense of additional energy consumption compared with conventional Japanese horticulture systems.

Decentralized energy systems have been recognized as a solution to increase the resource utilization including locally available renewable resources (Kikuchi et al., 2018b; 2016). control systems for voltage and frequency should be equipped to accept the massive implementation of variable renewable energy (VRE) such as photovoltaic and wind turbine power generation, e.g., virtual power plant or demand response systems (Kikuchi, 2017) considering the future energy supply and demand structure (Kikuchi et al., 2014; 2016).

In this study, we are tackling with the sustainable agriculture mixing conventional and advanced farming technologies, especially plant factories. An analysis was conducted on the current energy usage and applicability of technology options in plant factories for designing the life cycles of plant factories considering their harmonization with future energy systems.

2. Materials and methods

2.1. Overview of plant factories in this study

Figure 1 shows the plant factories examined in this study (Kikuchi et al., 2018a). The PFSL and PFAL examined in this paper were demonstration factories located at Chiba University in Kashiwa, Japan, where long-stage, high-density cultivation and a 10-stage vertical horticulture system were adopted to produce fresh tomatoes and lettuce in hydroponic culture using rockwool and polyurethane, respectively, as the culture media. The life cycle of plant factories shown in Figure 1(a) includes the market survey and process design before farming of vegetables. As summarized in Figure 1 (b) which was obtained from the comprehensive analysis based on the same methods by Kikuchi et al., (2018), the use of irreplaceable resources for food production can be mitigated at the expense of additional fossil energy consumption indicated as the GHG emission.

2.2. Analysis on current energy utilization and technology options

The suppression of VRE has been conducted in Kyushu area in Japan shown in Figure 2. Especially in remote islands in the area, the capacity of suppression is increasing due to the depopulation within such area and additional installation of photovoltaic power generation.

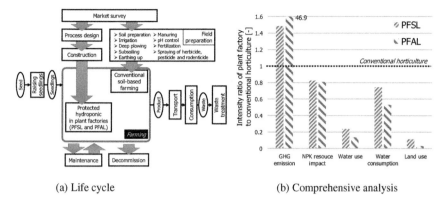

| (a) Life cycle | (b) Comprehensive analysis |

Figure 1 Overview of plant factories examined in this study (Kikuchi et al., 2018a)

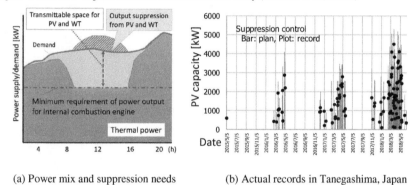

| (a) Power mix and suppression needs | (b) Actual records in Tanegashima, Japan |

Figure 2 Suppression of VRE in Japan (Kyushu Electric Power Company, 2018)

Considering the harmonization of plant factories with such decentralized energy systems implementing massive amount of VRE, the properties of power demands in plant factories are analysed by utilizing the automatically logged power consumption in the actual PFSL and PFAL in this study for every 1 minute from May 2011 to September 2014. The daily power consumption [Wh/day], maximum power demand [W], and peak demand time are selected as the properties of power demand. These properties can be changed by applying technology options in plant factories. The options to be considered for the plant factories in this study are the utilization of unused heat, a solid-oxide fuel cell, PV power, improved electric devices such as heat pumps and lighting, insulation, and the installation of all options. Their applicability and consequences are analysed and discussed in this paper.

3. Results and discussion

3.1. Current power consumption

Figure 3 shows the power demand properties in PFSL and PFAL in this study. The maximum daily power consumptions were 674 kWh/day and 3.32 MWh/day for PFSL and PFAL, respectively. The maximum power demand were 52.4 kW and 174 kW for PFSL and PFAL, respectively. The largest power demand in PFSL was air conditioning

(Kikuchi et al., 2018a), which gets higher in the morning, evening, and night during winter season. Due to the semiclosed environment of PFSL, the power demand is strongly affected by the seasonal changes, which means not easily controllable. The lighting in the PFAL, the largest power demand, is based on the production amount of vegetable (Kikuchi et al., 2018a). The peak demand time was fluctuated even though the power demand was at the maximum in 2014, which means that the lighting plans are controlled not by the time, but the operation considering the characteristics of plant growth.

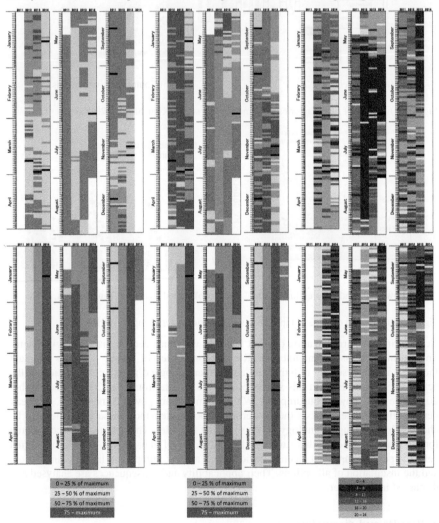

(a) Daily power use [Wh/day]   (b) Maximum power demand [W]   (c) Peak demand time

Figure 3 Power demand properties indicated as the ratio to the maximum during all periods (a,b) and ranges of time (c); upper and lower diagrams show the results of PFSL and PFAL, respectively

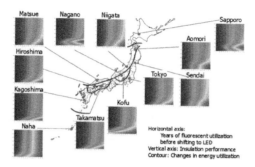

Figure 4 Regional differences of effects of LED implementation in PFALs considering their insulation

3.2. Applicability of technology options

The power demand properties shown in Figure 3 should be taken into account for the harmonization of plant factories with decentralized energy systems in the future. The application of technology options should also be addressed for such harmonization. For example, the implementation of fuel cell into PFSL can achieve the net-zero GHG emission (Kikuchi et al., 2018a), which means that the PFSL can supply power derived from fuel cell. This is partly because the power demand for air conditioning can be supplied by fuel cell as electricity and hot/cold heat with absorption chiller. As for the PFALs, their lighting needs electric power, which cannot be supplied by implementing small scale fuel cell and photovoltaic. At this time, the suppressed power due to the limitation of load following capacity in power grid shown in Figure 2 can be utilized in PFALs as a kind of virtual power plant or demand response, if the power demand can be controlled along with the condition of power grid.

The implementation of LED can also be considered for mitigating power demand in PFALs, where the relationship with insulation of PFALs should be carefully examined. Figure 4 shows the regional differences of effects of LED implementation in PFALs considering their insulation. Because the air conditioning in PFAL is always cooling mode, the lower insulation can reduce power demand in some cold area. However, if the fluorescent is applied at the design of PFALs, the LED implementation can disturb the design rationales and increase the power demand for air conditioning.

4. Conclusions

Food supply security can be also addressed by implementing plant factories (Kikuchi et al., 2018a). They can become a new energy demand to support the regional systems to supply foods. The radical changes in future energy systems should be taken into account for designing the life cycle of plant factories. For this purpose, computer-aided approaches are strongly needed for designing process and operation of plant factories.

Acknowledgement

The authors are grateful to Ms. Hanako Tominaga for her efforts in the modelling and simulation of plant factories. Part of this study is financially supported by JSPS KAKENHI under grant numbers 16H06126 (Grant-in-Aid for Young Scientists A).

Presidential Endowed Chair for "Platinum Society" in the University of Tokyo are supported by the KAITEKI Institute Incorporated, Mitsui Fudosan Corporation, Shin-Etsu Chemical Co., ORIX Corporation, Sekisui House, Ltd., and the East Japan Railway Company.

References

K. Christen, 2007, Closing the phosphorus loop. Environ. Sci. Technol. 41, 2078.

M.C. Heller, G.A. Keoleian, W.C. Willett, 2013, Toward a life cycle-based, diet-level framework for food environmental impact and nutritional quality assessment: A critical review. Environ. Sci. Technol. 47, 12632–12647.

Y. Kajikawa, J. Ohno, Y. Takeda, K. Matsushima, H. Komiyama, 2007, Creating an academic landscape of sustainability science: An analysis of the citation network. Sustain. Sci. 2, 221–231.

Y. Kikuchi, Y. Kanematsu, M. Ugo, Y. Hamada, T. Okubo, 2016, Industrial symbiosis centered on a regional cogeneration power plant utilizing available local resources: A case study of Tanegashima. J. Ind, Ecol, 20, 276–288.

Y. Kikuchi, S. Kimura, Y. Okamoto, M. Koyama, 2014, A scenario analysis of future energy systems based on an energy flow model represented as functionals of technology options. Appl. Energy. 132, 586–601.

Y. Kikuchi, Y. Kanematsu, R. Sato, T. Nakagaki, 2016, Distributed cogeneration of power and heat within an energy management strategy for mitigating fossil fuel consumption. J. Ind. Ecol. 20, 289–303.

Y. Kikuchi, 2017, Simulation-Based Approaches for Design of Smart Energy System: A Review Applying Bibliometric Analysis, J. Chem. Eng. Jpn., 50(6), 385-396.

Y. Kikuchi, Y. Kanematsu, N. Yoshikawa, T. Okubo, M. Takagaki, 2018a, Environmental and resource use analysis of plant factories with energy technology options: a case study in Japan, Journal of Cleaner Production, 186(10) 703-717.

Y. Kikuchi, Y. Oshita, M. Nakai, A. Heiho, Y. Fukushima, 2018b, A computer-aided analysis on regional power and heat energy systems considering socio-economic aspects: A case study on an isolated island in Japan, Comput. Aided Chem. Eng., 43, 1347-1352.

T. Kozai, G. Niu, M. Takagaki (ed.), 2015, Plant factory: An indoor vertical farming system for efficient quality food production. Elsevier, Tokyo.

Kyushu Electric Power Company, 2018, Report on the suppression control of renewable energy in Tanegashima, http://www.kyuden.co.jp/9437tanegashima5628

K. Matsubae, J. Kajiyama, T. Hiraki, T. Nagasaka, 2011, Virtual phosphorus ore requirement of Japanese economy. Chemosphere. 84, 767–772.

J.L. Schnoor, 2014, Agriculture: The last unregulated source. Environ. Sci. Technol. 48, 4635–4636.

United Nations Environment Program (UNEP), 2011. Phosphorus and food production. UNEP, Paris, pp. 34–45.

H. Yang, P. Reichert, K.C. Abbaspour, A.J.B. Zehnder, 2003, A water resources threshold and its implications for food security. Environ. Sci. Technol. 37, 3048–3054.

Anton A. Kiss, Edwin Zondervan, Richard Lakerveld, Leyla Özkan (Eds.)
Proceedings of the 29th European Symposium on Computer Aided Process Engineering
June 16th to 19th, 2019, Eindhoven, The Netherlands. © 2019 Elsevier B.V. All rights reserved.
http://dx.doi.org/10.1016/B978-0-128-18634-3.50266-6

Giving added value to products from biomass: the role of mathematical programming in the product-driven process synthesis framework

Aleksandra Zderic,[a] Alexandra Kiskini,[b] Elias Tsakas,[c] Cristhian Almeida Rivera,[d] Edwin Zondervan,[e]

[a] *Cosun, the Netherlands*

[b] *Wageningen University, the Netherlands*

[c] *Maastricht university, the Netherlands*

[d] *OPCW, the Netherlands*

[e] *Bremen University, Germany*

* *edwin.zondervan@uni-bremen.de*

Abstract

In the first years of the 2000's the late professor Peter Bongers introduced together with his co-workers at Unilever a design methodology that could be applied in the development of new products and processes for structured food products; the product-driven process synthesis method (PDPS). The method was successfully employed in the following years, designing new products from different bio based sources. Although researchers used the method and even made improvements; the structural incorporation of mathematical programming tools has been lacking and this seems to be a crucial component for decision-making processes. In this contribution we will discuss the possibilities to extend the PDPS framework with several of these optimization tools.

Keywords: Product-driven process synthesis, mathematical programming, agro-food products.

1. Introduction

We can agree that the procedures and methods for the development of (petro) chemical processes and products have matured significantly over the last 50 years. However, in new areas such as the bio-based-, food-, pharmaceutical- and water sectors, where feedstock, processing and product are dynamic and complex these developments lag and right now such methods are not at hand. A more structured approach for the synthesis of process and product was developed in the last decade by the late professor Peter Bongers, the so called product-driven process synthesis (PDPS) approach. This method makes use of the synergy of combining product and process work streams. The PDPS delivers a hierarchy of design levels of increasing details, where complex and level-interacting decisions are made from one level to another.

The application of the PDPS approach has resulted in a significant financial benefits in the fast-moving consumer goods industry and lead to novel and differentiating alternatives to current manufacture practices. Although the PDPS is a systematic approach, many of the decisions made at the different levels of the PDPS are based on expertise and knowledge brought in via experienced engineers and scientists (so called heuristics). It is clear that model-based optimization and mathematical programming could further enhance this decision process. In this work we will outline our vision for the integration of systematic optimization methods into the PDPS.

2. Product driven process synthesis (PDPS)

The product driven process synthesis methodology has been described in detail in (Almeida Rivera et al., 2016) and consists of nine levels of increasing detail, where complex and emerging decisions are made. These levels include the: framing level, consumer wants, product function, input–output level, task network, mechanism and operational window, multiproduct integration, equipment selection and design, and multiproduct equipment integration. Each level follows the general design paradigm cycle (Siirola, 1996): scope and knowledge, generate alternatives, analyse performance of alternatives, evaluate and select, report.

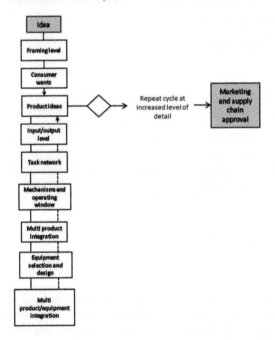

Figure 1: The product-driven process synthesis framework of 9-levels.

At all levels estimates of economic potential, turndown ratio, changeover time, sustainability, process complexity, scalability, hygiene design, among others, and all of them at various degrees of detail are made. The aim is also to include a multi criteria decision-making approach to assist in the "evaluate and select" activity (De Ridder et al. 2008). This approach, termed BROC (benefits, risks, opportunities, and costs), is used in industrial R&D settings and combines quality function deployment (QFD) and analytic network process (ANP). QFD is a method for structured product planning and development that enables a development team to specify clearly the customer wants and needs, and then to evaluate each proposed product or service capability systematically in terms of its impact on meeting those needs (Cohen, 1995).

The backbone of QFD is the so-called "house of quality", which displays the customer wants and needs, and the development team technical response to meeting those wants and needs. ANP is a commonly used benchmarking technique to compare alternatives. QFD has been applied to several PDPS cases but only in a qualitative fashion.

3. Examples of successful application of the PDPS to agro-food products

In the last years the PDPS has been successfully employed in several projects in the agro-food industry. These projects focused on the mild disclosure of added value chemicals from biological raw materials. In (Zondervan et al, 2015) a new process is designed for recovering vitality ingredients such as antioxidants from liquid tea-streams via solvent swing adsorption. In (Zderic et al, 2016) the scope was extended and novel methods for disclosing the antioxidants directly from fresh tealeaves using pulsed electric fields and ultrasound techniques were developed. In both studies the PDPS plays an important role for decision-making. (Zderic et al, 2017) also used the PDPS to extract oilbodies from soy.

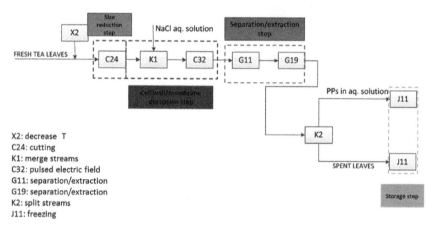

X2: decrease T
C24: cutting
K1: merge streams
C32: pulsed electric field
G11: separation/extraction
G19: separation/extraction
K2: split streams
J11: freezing

Figure 2: Example of a task-network for the isolation of antioxidants from fresh tealeaves.

In (Jankowiak et al, 2013) the PDPS was used as tool to find new pathways for isolating isoflavones from Okara and in work by (Kiskini et al., 2016) an extensive portfolio of different products that can be obtained from sugar beet leaves was set out, including proteins, carbohydrates, lipids and phenolics.

Although all studies delivered valuable insights and pointed to a strategy for an advanced product portfolio; the main strategy in decision-making was based on brainstorm sessions, experimental work and expert knowledge available in the project teams. The use of mathematical optimization techniques to facilitate decision-making would have certainly been of value, knowing the combinatorial complexity resulting from many different choices that can be made at each of the levels of the PDPS.

4. The use of mathematical optimization in the PDPS framework

4.1. The House of Quality in the PDPS
In the "framing" and the "consumer wants" levels of the PDPS one important question that needs answering is how to link the (qualitative) properties that a consumer wants in

a product to measurable indicators (such as physical properties or product formulations). For example: there might be a clear link between the "mouth feel" and the viscosity of a mayonnaise. The House of Quality (HoQ) is a graphic tool that establishes such links. The HoQ comes in different appearances, but an example is given in figure 3 below, which was taken from (Dawson et al., 1999). The figure shows the interactions between the physical properties of a pencil and how a user experiences the use of the pencil.

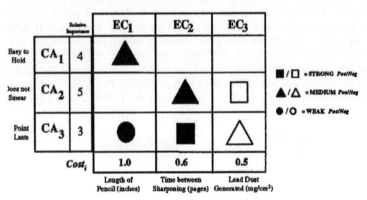

Figure 3: The Pencil design HoQ without roof.

The HoQ has been widely proposed as a method for capturing the voice of the customer when developing requirements for new products. But the methodology lacked a formal mechanism for trading off customer preferences with technical feasibility and economic reality. However, in the work of (Dawson et al., 1999) a non-linear mathematical program for determining the optimal engineering specifications during new product development as a function of elicited customer value functions, engineering development and production costs and development time constraints is proposed. Such models are shown to be computationally feasible for realistic-size problems and to outperform heuristic approaches.

4.2. Superstructure optimization of task networks in de PDPS

In the "Task network" level of the PDPS different definitions of the fundamental tasks that are needed to convert a raw material into product are taken from a cluster of tasks and its subgroup. Then, a network is made from the selected tasks and clusters. This work flow is often done on the basis of knowledge available within the project team. This leads to sound proposals for the task networks, but there are no guarantees that all possible network configurations have been evaluated. This is where superstructure optimization can be brought into the PDPS.

Figure 4 shows the general format of a superstructure, which consists of sources (raw materials), sinks (products), arcs (possible flows of mass/energy) and blocks (contained unit operation- or task models). In superstructure optimization the main question is, which route to take through the network in such way that a criterion is optimized (for example, costs, environmental impact, product yield).

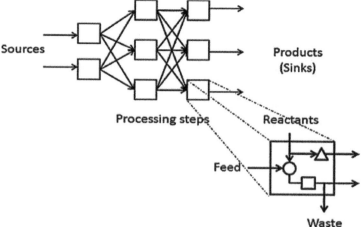

Figure 4: General format of a superstructure (Nawaz et al., 2011)

Of course, there are many possible routes to be evaluated. The number of alternatives increases factorial with the number of components that flow through the network, the number of technology options available and the number of processing stages allowed. The arcs and the blocks in the super structure can be translated into a mathematical program that can be used to evaluate many hundreds or thousands of possible pathways. The general structure of such model is:

$$\min f(x,y)$$
$$\text{s.t.}$$
$$g(x,y) = 0$$
$$h(x,y) > 0$$
$$x_L \le x \le x_U$$
$$x \in X; y \in Y$$

$$(1)$$

With f being the objective function (economic- or environmental criterion), x and y the decision variables, g the equality constraints (often mass- and energy balances), h the inequality constraints (often operating limits). There are also logical constraints (could be equalities as well as inequalities). The decision variables x and y are bounded and defined over the domains X and Y, respectively (this determines the nature of the problem; continuous or discrete). Often these models are captured as mixed integer linear (or nonlinear) programming problems, which can be handled well with appropriate software.

4.3. The use of propositional logic in the PDPS
As a complement to the use of superstructure models to generate and evaluate task networks a more intensive use of propositional logic is foreseen. Propositional logic is a tool that can be used to translate heuristics into mathematical constraints to reduce network complexity. The development of propositional logic in process design finds its roots in the work by (Raman & Grossmann, 1991). Propositional logic can be used to convert logical expressions into inequalities. As an example: "If the dryer or the hydrocyclone is selected, then do not use a crystallization step". This heuristic can be written as a logic expression:

$P_D \vee P_H \Rightarrow \neg P_C$ (2)

Removing the implication yields:

$$\neg(P_D \lor P_H) \lor \neg P_C \tag{3}$$

After applying de Morgan's theorem:

$$(\neg P_D \land \neg P_H) \lor \neg P_C \tag{4}$$

Distributing the OR over the AND:

$$(\neg P_D \lor P_C) \land (\neg P_H \lor \neg P_C) \tag{5}$$

After assigning the corresponding binary variables in the above conjunction two inequalities are obtained:

$$y_D + y_C \le 1 \tag{6}$$

$$y_H + y_C \le 1 \tag{7}$$

Which can be added to the superstructure model. The house of Quality, superstructure optimization and propositional logic are three examples of applying quantitative methods to the PDPS.

5. Conclusions

Product driven process synthesis has over the years proven to be an effective tool for process- and product design for novel agro-food based products. The method deservers further improvements by the inclusión of mathematical programming tolos that can be used to link sensory attributes with product properties (House of quality), the application of superstructure optimization for the evaluation of different task networks and the use of propositional logic in the development of constraints for the superstructure optimization.

References

C. Almeida Rivera, P. Bongers, E. Zondervan, A structured approach for product driven process synthesis in foods manufacture, Computer Aided Process Engineering, 2016, 39, pp. 417-441.

L. Cohen, Quality function deployment: How to make QfD work for you, 1995, Wesley Publishing Company Inc.

D. Dawson, R. Askin, Optimal new product design using quality function deployment with the empirical value function, Quality and Reliability Engineering International, 1999. 15, pp. 17-32.

D. De Ridder, C. Almeida Rivera, P. Bongers, S. Bruin, S. Flapper, Multi-criteria decision-making in product driven process synthesis, Computer Aided Process Engineering, 2008, 25, pp. 1021-1026.

L. Jankowiak, D. Mendez, R. Boom, M. Ottens, E. Zondervan, A. Van der Goot, A process synthesis approach for isolation of isoflavons from Okara, I&EC Research, 2015, 54(2), pp. 691-699

M. Nawaz, E. Zondervan, J. Woodley, R. Gani, Design of an optimal biorefinery, Computer Aided Process Engineering, 2011, 29, pp. 371-376

A. Kiskini, E. Zondervan, P. Wierenga, E. Poiesz, H. Gruppen, Using product driven process synthesis in the biorefinery, Computers and Chemical Engineering, 2016m 91, pp. 257-268

R. Raman, I. Grossmann, Relation between MILP modeling and logical inference for chemical process synthesis, Computers and Chemical Engineering, 1991, 15(2), pp. 73-84.

J. Siirola, Industrial application of chemical process synthesis, Advances in chemical engineering, Process synthesis, 1996, pp. 1-61.

A. Zderic, C. Almeida Rivera, P. Bongers, E. Zondervan, Product driven process synthesis for the extraction of oilbodies from soy beans, Journal of food engineering, 2016, 185, pp. 26-34.

A. Zderic, E. Zondervan, Product driven process synthesis: the extraction of polyphenols from tea, Journal of food engineering (2017), 196, pp. 113-122.

E. Zondervan, M. Monsanto, J. Meuldijk, Product driven process synthesis for the recovery of vitality ingredients from plant materials, Chemical Engineering Transactions, 2015, 43, pp. 61-66.

Anton A. Kiss, Edwin Zondervan, Richard Lakerveld, Leyla Özkan (Eds.)
Proceedings of the 29th European Symposium on Computer Aided Process Engineering
June 16th to 19th, 2019, Eindhoven, The Netherlands. © 2019 Elsevier B.V. All rights reserved.
http://dx.doi.org/10.1016/B978-0-128-18634-3.50267-8

Bottom-up method for potential estimation of energy saving measures

Anna S. Wallerand[a*], Ivan Kantor[a] and François Maréchal[a]

[a]*École Polytechnique Fédérale de Lausanne (EPFL) Valais Wallis, Switzerland*
anna.wallerand@epfl.ch, anna.wallerand@gmail.com

Abstract

Technology and energy saving potentials in the industrial sector are key data that policy makers rely on to drive decision making. A literature review reveals a distinct discrepancy between *general potential estimation* and *detailed design* studies. The former relies principally on top-down conceptual approaches based on temperature levels of the processes' thermal requirements, while the latter aims at mathematically optimized solutions for specific technologies in specific processes under specific economic conditions.

This work attempts to close this gap by proposing a bottom-up method for potential estimation of energy saving measures. To this end, a generalized optimization framework was developed which aims at generating a database of optimal solutions which are independent from most economic and environmental input data. By fixing these data, the optimal solution for the same process type in different countries and under various criteria can be identified in the database. The method also increases accessibility of optimization techniques by providing the solution database which then requires limited input parameters and background knowledge to provide expert guidance toward optimal utility integration. The method was applied to the dairy industry, highlighting that energy saving potentials can be achieved through heat recovery and integration of heat pumps, mechanical vapor re-compression, and co-generation units economically favourable especially in Japan and Switzerland.

Keywords: dairy industry, heat pump, co-generation, MINLP, mathematical programming, generalized optimization

1. Introduction

Constant efforts are undertaken to estimate energy saving and technological potentials in the industrial sector. Such analyses are used to formulate policy recommendations, determine target markets, and advance technological development. A state-of-the-art analysis is introduced concerning both *general potential estimation* and *detailed design* studies, before highlighting the contribution of this work and the outline of the study.

General potential estimation: This section concerns studies which aim at estimating energy saving potentials in the industrial sector. Most such studies are based on top-down approaches relying mainly on the temperature levels of the thermal requirements in certain industries and with specific technologies. Concerning heat pump integration, Wolf et al. (2014) presented a study which focused on the German low temperature sectors, estimating the overall energetic potential of heat pumps used as hot utilities. Similarly, Brückner et al. (2015) performed such an analysis for the European industrial sector with compression and absorption heat pumps. For other technologies, analogous studies have been conducted by Lauterbach et al. (2012) (solar thermal) and Campana et al. (2013) (organic Rankine cycles), as representative examples.

Though top-down approaches offer a clear and simple approach to estimate general potentials, conclusions drawn from such analysis may be misleading. Firstly, they account only for hot utility requirements neglecting the double impact of certain utilities (especially heat pumps) according to the 'more in, more out' principle of pinch analysis. Secondly, they do not consider possible process improvements and efficiency measures such as internal heat recovery.

Bottom-up approaches have also been presented by Wilk et al. (2017) who analyzed different process types and the heat pump integration potential for a set of energy price ratios, and Müller et al. (2014) who suggested an approach for solar thermal integration to the liquid food industry based on the available land surface area and process temperature levels. These approaches are often limited in their scope (in terms of economic and environmental boundary conditions) and lack rigorous methods.

Detailed design: Various detailed design studies are present in academic literature and typically focused on few technologies and processes. Studies by Kamalinejad et al. (2015), Yang et al. (2017), and Becker (2012) presented optimal integration approaches of heat pumps in different industrial processes relying on mathematical programming and superstructure-based design. Most recently Wallerand et al. (2018) presented a comprehensive heat pump integration method based on mathematical programming and expert insight. For organic Rankine cycles, superstructure-based approaches are also common, such as presented by Yu et al. (2017) and Kermani et al. (2018).

Though such approaches offer rigorous solutions to complex problems, the scope of treated problems is usually limited to specific processes, technologies, and economic and environmental boundary conditions.

Synthesis: The state-of-the art analysis can be summarized in two points.
1. Potential studies are either conducted coarsely by top-down approaches, or in a non-rigorous and limited in scope with bottom-up analyses.
2. Detailed design studies provide rigorous methods, however limited to specific conditions.

This work attempts to bridge this gap by suggesting a bottom-up method for potential estimation of energy saving measures through heat recovery and technology integration. The methodology is presented in section 2, followed by the results in section 3 presenting the solutions for various perspectives (from policy-makers to plant managers). The work is summarized and critically reviewed in section 4.

2. Methodology

The methodology presented in this work aims at reducing computational time while producing a large set of feasible and potentially 'good' solutions. The expensive optimization problem is therefore solved once, independent from most input parameters and its solutions are stored in a database from which a large number of scenarios with varying input conditions can be derived. An overview of this approach can be seen in Figure 1. The approach is clarified below in more detail.

Comprehensive solution space generation: The goal of this part is to formulate an op-

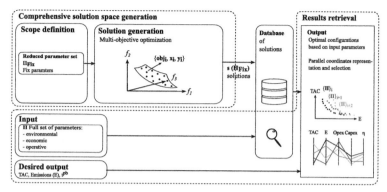

Figure 1: Bottom-up method for potential estimation.

timization problem independent of most typical input parameters. Usually the objectives of main interest for such analyses are the *total annual system cost (TAC)* and/or the *total emissions*. Both of these objectives require a wide range of input parameters ranging from economic (energy prices, investment cost parameters, interest rate, CO_2 tax, maintenance cost), operative (operating time), process specific (product mass flow rates, specific requirements), equipment specific (lifetime, efficiency, fuel consumption) to emissions associated with electricity and natural gas consumption (environmental parameters). Even though the final objectives are *total system cost* and/or the *total emissions*, a different set of objectives is selected at this stage, which requires fewer input parameters with the goal of generating a large set of potentially optimal solutions. The choice of objective depends on the exact goal of the study. In this case, the focus was placed on national differences, which have a drastic effect on energy prices, interest rates, and environmental parameters.

Therefore, the new set of objectives was defined as *consumption* of all resources and *investment cost (capex)*, which requires less input parameters to be fixed. The choice of the new set of objectives requires engineering common sense and needs to be carefully adapted to each problem. The new objectives are fed to a multi-objective optimization framework, similar to the one described by Wallerand et al. (2018) and Weber et al. (2007) to generate a set of solutions. The solution space generation needs to be carried out only once and requires a few days of computational effort.

Database: Depending on the number of objectives, the previous step generates a multi-dimensional surface of solutions. In this study, the resource consumption was distinguished between electricity and natural gas consumption, which led to a total of three objectives. Upon successful completion of the optimization, all solutions and the corresponding values of their decision variables (solution properties) can be stored in a database for later access and evaluation.

Results retrieval: The results retrieval procedure works in two steps: Initially, a solution together with its entire set of properties is retrieved. Subsequently, knowing the optimal values of each decision variable, a user can evaluate the original objective function as well as any other indicators given a new set of parameters (e.g. energy prices, operating time, CO_2 tax). This leads to a new set of objective values among which the user can identify their preferred choice. In this case, the lowest cost and lowest emission solutions were selected. This step is accessible without knowledge of optimization algorithms thus encouraging utilization of such methods by a variety of audiences which is briefly demonstrated in section 3.3. The results retrieval has been implemented in a tool, which can be found online.

3. Results

3.1. Case study

The underlying case study of this section is a dairy plant as presented by Kantor et al. (2018) with a two-stage refrigeration cycle, free cooling water (from a nearby river) and an existing boiler. The heat pump superstructure suggested by Wallerand et al. (2018) with various fluids and a co-generation engine (Becker (2012)) were studied as potential new technology options. Decision variables were the heat pump fluids, temperature levels, and number of stages, as well as the T_{min} of the heat recovery in the process. The heat exchanger network cost was graphically approximated as proposed by Townsend and Linnhoff (1984).

3.2. Comprehensive solution space generation

The three-dimensional surface of solutions resulting from the multi-objective optimization described in section 2 is shown in Figure 2. It can be seen that points closest to the reference case have lowest capital expenses (capex), while lower electricity and/or natural gas consumption necessitates an increase in capex.

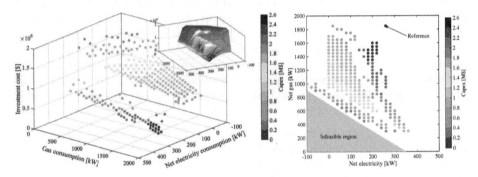

Figure 2: Solutions from multi-objective optimization considering consumption of natural gas and electricity with the corresponding investment required.

3.3. Results retrieval

The retrieval of results is illustrated in Figure 3 for the three exemplary countries, namely Switzerland (CH), Germany (DE), and the United States of America (US). The countries were selected due to their different economic and environmental conditions, as illustrated in the figure, ranging from a high natural gas price and low electricity/natural gas ratio in CH to a low natural gas price and high electricity/natural gas price ratio in the US. The total emissions and TAC are calculated for each solution in the database, but only the minimum cost solutions are selected (indicated with grey circles).

Plant operators: The minimum cost points from Figure 3 are depicted with more cost details in Figure 4, indicating the potential interest of a plant operator in reducing energy-related costs or emissions. CH is visibly the country with highest potential where up to 70% emission reductions could be achieved with heat recovery (firstly), with heat pumping providing options with payback times below 5y. This is in stark contrast to the US context where solutions are not economically attractive compared to the reference case. Here, and in DE, the highest emission reductions can only be achieved with heat pumps and co-generation units, due to the high emissions related to generation of electricity.

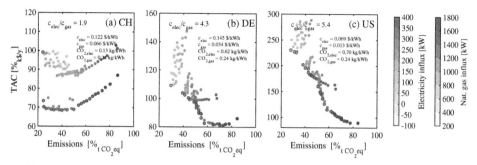

Figure 3: Results retrieval: minimum cost solutions, 8000 h/y operation, 0 $/t CO_2.

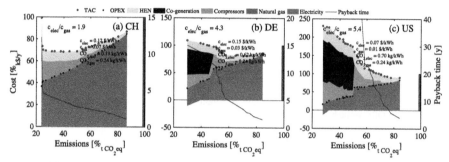

Figure 4: Minimum cost solutions for operating time 8000 h/y, 0 $/t CO_2.

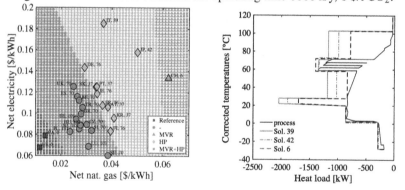

Figure 5: Utility map for operating time 2500 h/y (left), 0 $/ton CO_2; composite curves of solutions for Italy, Japan and Switzerland (right).

Policy-makers: Another type of analysis can be conducted for policy-makers as shown in Figure 5. With this analysis, the lowest emitting solution for each country (resource price) is selected with a payback time below 3 years. The countries with highest resource prices (Italy (IT), Japan (JP) and Switzerland (CH)), have the highest heat pump potential. The other solutions are principally based on improving process heat recovery to achieve the potential reduction in consumption.

4. Conclusions

This work proposes a bottom-up method for potential estimation of energy saving measures. To this end, a generalized optimization framework was developed based on multi-objective optimization which aims at generating a database of optimal solutions which are independent from most economic and environmental input data. By fixing this data, the optimal solution for the same process type in different geographical, economic and environmental contexts are retrieved from the database to address various objective criteria.

The method was applied to the dairy industry investigating energy saving and technological potentials through heat recovery and utility integration with heat pumps, mechanical vapor re-compression, and co-generation units. A set of countries was determined in which industrial heat pumping is particularly beneficial from environmental and economic perspectives, namely: Switzerland, Japan, and Sweden. Use cases for plant operators and policy-makers were briefly demonstrated and a variety of additional analyses could be conducted using the solution database; one such example was briefly presented.

References

H. C. Becker, 2012. Methodology and Thermo-Economic Optimization for Integration of Industrial Heat Pumps. Ph.D. thesis, École Polytechnique Fédérale de Lausanne, Lausanne.

S. Brückner, S. Liu, L. Miró, M. Radspieler, L. F. Cabeza, E. Lävemann, Aug. 2015. Industrial waste heat recovery technologies: An economic analysis of heat transformation technologies. Applied Energy 151, 157–167.

F. Campana, M. Bianchi, L. Branchini, A. De Pascale, A. Peretto, M. Baresi, A. Fermi, N. Rossetti, R. Vescovo, Dec. 2013. ORC waste heat recovery in European energy intensive industries: Energy and GHG savings. Energy Conversion and Management 76, 244–252.

M. Kamalinejad, M. Amidpour, S. M. M. Naeynian, Jun. 2015. Thermodynamic design of a cascade refrigeration system of liquefied natural gas by applying mixed integer non-linear programming. Chinese Journal of Chemical Engineering 23 (6), 998–1008.

I. Kantor, A. S. Wallerand, M. Kermani, H. Bütün, A. Santecchia, Raphaël Norbert, Sahar Salame, Hélène Cervo, Sebastian Arias, Franz Wolf, Greet van Eetvelde, François Maréchal, 2018. Thermal profile construction for energy-intensive industrial sectors. In: Proceedings of ECOS 2018. Portugal.

M. Kermani, A. S. Wallerand, I. D. Kantor, F. Maréchal, Feb. 2018. Generic superstructure synthesis of organic Rankine cycles for waste heat recovery in industrial processes. Applied Energy 212, 1203–1225.

C. Lauterbach, B. Schmitt, U. Jordan, K. Vajen, 2012. The potential of solar heat for industrial processes in Germany. Renewable and Sustainable Energy Reviews 16 (7), 5121 – 5130.

H. Müller, S. Brandmayr, W. Zörner, Jan. 2014. Development of an Evaluation Methodology for the Potential of Solar-thermal Energy Use in the Food Industry. Energy Procedia 48, 1194–1201.

D. Townsend, B. Linnhoff, 1984. Surface area targets for heat exchanger networks. In: IChemE Annual Research Meeting, Bath, UK.

A. S. Wallerand, M. Kermani, I. Kantor, F. Maréchal, Jun. 2018. Optimal heat pump integration in industrial processes. Applied Energy 219, 68–92.

C. Weber, F. Maréchal, D. Favrat, 2007. Design and optimization of district energy systems. In: V. P. a. P. S. Agachi (Ed.), Computer Aided Chemical Engineering. Vol. 24 of 17th European Symposium on Computer Aided Process Engineering. Elsevier, pp. 1127–1132.

V. Wilk, Bernd Windholz, Reinhard Jentsch, Thomas Fleckl, Jürgen Fluch, Anna Grubbauer, Christoph Brunner, Daniel Lange, Dietrich Wertz, Karl Ponweiser, 2017. Valorization of industrial waste heat by heat pumps based on case studies of the project EnPro. In: Proceedings of the 12th IEA Heat Pump Conference. iea Energy Technology Network, Rotterdam.

S. Wolf, U. Fah, M. Blesl, A. Voß, R. Jakobs, Dec. 2014. Analyse des Potenzials von Industrie - Wärmepumpen in Deutschland. Tech. rep., Universität Stuttgart, Institut für Energiewirtschaft und Rationelle Energieanwendung (IER), Stuttgart.

T. Yang, Y. Luo, Y. Ma, X. Yuan, Dec. 2017. Optimal synthesis of compression refrigeration system using a novel MINLP approach. Chinese Journal of Chemical Engineering.

H. Yu, J. Eason, L. T. Biegler, X. Feng, Dec. 2017. Process integration and superstructure optimization of Organic Rankine Cycles (ORCs) with heat exchanger network synthesis. Computers & Chemical Engineering 107, 257–270.

Anton A. Kiss, Edwin Zondervan, Richard Lakerveld, Leyla Özkan (Eds.)
Proceedings of the 29th European Symposium on Computer Aided Process Engineering
June 16th to 19th, 2019, Eindhoven, The Netherlands. © 2019 Elsevier B.V. All rights reserved.
http://dx.doi.org/10.1016/B978-0-128-18634-3.50268-X

Generating Efficient Wastewater Treatment Networks: *an integrated approach comprising of contaminant properties, technology suitability, plant design, and process optimization*

Kirti M. Yenkie,[a*] Sean Burnham,[a] James Dailey,[a] Heriberto Cabezas,[b] Ferenc Friedler[b]

[a]*Department of Chemical Engineering, Henry M. Rowan College of Engineering, Rowan University, Glassboro NJ – 08028, USA*

[b]*Pazmany Peter Catholic University, Institute for Process Systems Engineering and Sustainability, Budapest, Hungary*

yenkie@rowan.edu

Abstract

The rise in world population and industrialization in developing nations has tremendously increased the demand for water and has resulted in wastewater contaminated with several pollutants. Thus, wastewater treatment (WWT), reuse, and safe disposal have become crucial for sustainable existence. We believe that generation of a maximal structure (superstructure) comprising of all possible treatment methods and flow patterns using a systems approach, followed by optimization to decide the best treatment pathway, will enable efficient designing of WWT networks. In this work, the technologies/methods involved in WWT such as sedimentation, filtration, membranes, adsorption, activated sludge, etc. are modelled using material and energy balances, equipment design, costing and environmental impact. Utilizing systematic methods (*e.g.* mixed-integer non-linear programming, MINLP), we frame the WWT network selection as an optimization problem for cost and energy minimization along with sustainable goals. In our analysis, we demonstrate a case study of Municipal WWT and determine the best strategy in compliance with the 1972 US EPA's Clean Water Act to reuse the treated water for cropland irrigation. In the next step, we use the P-graph approach for solving the same problem and this tool provides a ranked list of candidate solutions.

Keywords: water, technology model, optimization, P-graph, sustainable process design

1. Introduction

Water is a necessary commodity, without its services in the industry, agriculture and domestic sectors the human species would cease to exist. The population rise has resulted in increased environmental pollution, majorly influencing urban water resources with issues such as low dissolved oxygen, bacterial contamination, and disruption in aquatic flora and fauna. Thus, strategies for judicious water consumption, pollution prevention, and efficient WWT are required for sustainable existence. Water pollution control boards and federal regulatory agencies (NIH, 2015; US EPA, 2013) have a well-defined list of pollutants and their safety limits, however, with advancements in industrial processing, new and unknown contaminants enter the waste streams ultimately leaching in groundwater sources. Hence, the need for improved technologies and stage-wise WWT methods (Melo-Guimarães et al., 2013; Ponce-Ortega et al., 2009). Additionally, treatment plants need policies to minimize costs and energy requirements while maintaining the necessary regulations. Some factors that affect technology selection in WWT network are the wastewater characteristics, purity requirements, reliability, sludge handling, and costs.

2. Materials and Methods

2.1. Treatment Stages and Technologies

WWT is usually considered as a four-stage process: pretreatment, primary, secondary, and tertiary. An overview of the treatment stages and technologies involved are described in Figure 1 (Inc et al., 2002; Lipták and Liu, 2000; Liu et al., 2015). Depending on the purity standards and the number of contaminants in the waste stream, WWT plants may utilize one or more technologies from each stage, for example, a treatment path might follow screening, sedimentation, adsorption, and bleaching.

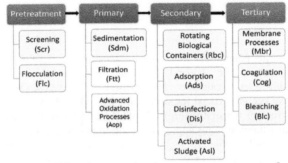

Figure 1 Traditional stage-wise wastewater treatment scheme

Pretreatment methods are used to enhance the operations in the following stages. Primary stage involves the removal of solid wastes. Sdm and Ftt are very successful at removing solid waste from wastewater. Aop can remove other harmful agents such as active pharmaceutical ingredients (API). Secondary stage methods may utilize microorganisms (Rbc, Asl) and disinfectants (Dis) to purify wastewater (Deng and Zhao, 2015; Shukla et al., 2006). Tertiary stage removes leftover contaminants and colour imparting impurities.

2.2. Heuristics, Technology Information and Engineering Judgement

Heuristics can guide in shortlisting the appropriate technology options. For example, characterizing the properties of contaminants present is a stream coming from a specific source of wastewater. Few identifiers for contaminant properties include physical state, size, and shape, density, relative amount, toxicity index, boiling point, biological affinity, chemical reactivity, etc. The second set of information should include the treatment technology options, the principle or driving force behind the purification and the process conditions or specifications that are essential while designing a treatment technology. Table 1 enlists a few technology options with their defining characteristics. However, heuristics can sometimes lead to the elimination of important non-intuitive solutions and thus a more systematic approach, which is more inclusive and unaffected by prior biases, is desired for efficient design.

Table 1. Wastewater treatment technologies and their important characteristics

Technology	Principle/driving force	Specifications/Process conditions
Sedimentation (Sdm)	Density gradient Particle settling velocity	Size, Particle density, Tank depth, Residence time
Filtration (Ftt)	Particle size	Average flux, Pressure gradient, Filtration rate
Advanced Oxidation Process (Aop)	Oxidation reactions for contaminant degradation	Ozone, peroxide, UV reactor
Disinfection (Dis)	Chemical, radiation	Chlorine dosage, Acid/Alkali treatment, UV radiation
Rotating Biological Contactors (Rbc)	Biological mechanism	Biological film on discs, speed of rotation, aeration rate
Membranes (Mbr)	Particle/molecular size Sorption/Diffusion Pressure	Pore size, Mol. wt. cut-off Flux, Pressure gradient, Type - MF, UF, NF and RO
Activated Sludge (Acs)	Microbial activity	Detention time, mixing efficiency, aeration rate

Sources: (Deng and Zhao, 2015; Ho and Sirkar, 1992; Lipták and Liu, 2000; Liu et al., 2015; Melo-Guimarães et al., 2013)

2.3. Wastewater Treatment Superstructure

Superstructure is a systematic representation of all possible treatment technologies available in the four stages of wastewater treatment. It also shows the flow from the initial wastewater stream towards the final purified water stream. Figure 2 shows the current WWT superstructure for treatment of municipal wastewater. The bypass (Byp) option is inlcuded in stages 2, 3 and 4. The current superstructure consists of 4 treatment stages, 10 technologies, 38 streams, 3 splitters, and 3 mixers.

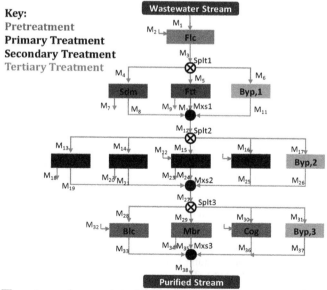

Figure 2 The stage-wise wastewater treatment superstructure for municipal wastewater treatment to reuse it for cropland irrigation

2.4. Framework for Evaluation

Step#1: Formulate the case study by defining the input stream composition and output stream purity specifications

Step#2: Construct an optimization model as a Mixed-Integer Non-Linear Programming (MINLP) problem that consists of:
- Mass and energy balances
- Design equations, cost calculations: capital (equipment) and operating (utility, materials, consumables, labor and overhead requirements) costs
- Binary integer variables to select technology in use by holding a (0,1) value

Step#3: Minimize treatment costs while still reaching target values for safe reuse/ disposal of wastewater written as a mathematical objective function

Step#4: Develop the optimization model in GAMS (General Algebraic Modeling System) and solve by global MINLP solvers such as BARON (Branch-and-Reduce Optimization Navigator) (Kılınç and Sahinidis, 2018)

Step#5: Implement the same case study in P-graph studio (Friedler et al., 1992; Heckl et al., 2010) that utilizes concepts from graph theory to obtain the maximal structure, list alternative paths (solutions) in the superstructure and rank them

3. Results and Discussions

The case study details with the list of inlet contaminant concentrations and the purity requirements for cropland irrigation are presented in Table 2 (Page et. al., 1996). Inlet wastewater flowrate is assumed at 100 m^3/h and plant operates for 330 days annually.

Table 2. Municipal case study contaminant concentrations and purity levels

Contaminant	Inlet Concentration (g/L)	Outlet Specification (g/L)
Solids (Settleable)	200	≤ 0.5
Metals (Pb, Cu, Zn, Ni)	0.1	≤ 0.05
Chemicals (Acids, chlorides, organics, and inorganics)	1	≤ 0.5

3.1. Solution from the GAMS-MINLP problem

The model in GAMS had a total of 267 equations, 199 variables, and 12 binaries. The solution time was less than a minute with a relative gap of 0.00001. The cost of the wastewater treatment per hour was obtained as 6252.27 USD/h. The technologies selected (Flocculation, Flc, Sedimentation, Sdm and Adsorption, Ads) and active streams are highlighted in Figure 3 and the cost distribution chart shows that the materials cost (flocculent and adsorbent) is the highest contributor. Utility costs are negligible as the selected technologies do not require significant amounts of electricity, cooling or heating.

3.2. Solution from the P-graph framework

The same case study was implemented in P-graph; the maximal structure and the comparable structure to the GAMS-MINLP solution is presented in Figure 4. The cost of wastewater treatment from P-graph is 6219.14 USD/h, which is comparable to the GAMS solution (6252.27 USD/h). This cost discrepancy is due to the linearized cost (capital and operating) equations required in P-graph. However, the cost deviation in values predicted by GAMS and P-graph is less than 1% and we gain additional insights which were not possible in the GAMS-MINLP framework.

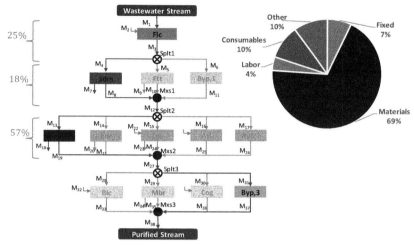

Figure 3. Solution for Municipal WWT case study from GAMS showing the active technologies and streams with cost distribution chart on the right

However, the structure (Flc-Sdm-Ads) is ranked#3 in the optimal feasible structure list. The rank#1 (2673.7 USD/h, consisting of Flc-Sdm-Rbc) and rank#2 (5033.33 USD/h, consisting of Flc-Ftt-Rbc) low-cost feasible structures (networks) are not predicted by the GAMS model. We tested these low-cost structures in GAMS and found that the purity levels of the outlet streams were lower than desired for cropland irrigation. We plan to investigate the differences in costing due to linearizations in the future analysis.

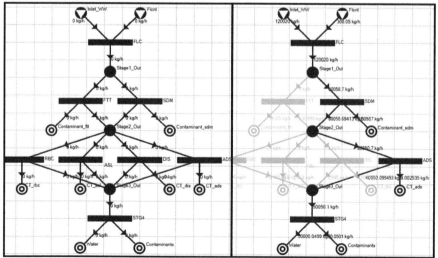

Figure 4. The Municipal WWT case study implemented in P-graph showing the maximal structure on the left and the feasible structure comprising of flocculation, sedimentation, and adsorption with the active streams on the right.

4. Conclusions

In this work, we have proposed a systematic framework for designing wastewater treatment networks. The heuristic-based superstructure is modelled as a MINLP problem in GAMS and the optimal network includes, Flc (stage-1) – Sdm (stage-2) – Ads (stage-3) – Byp (stage-4) with a cost of 6252.27 USD/h. The same case study is modelled in P-graph studio and the results show that the GAMS optimal network is ranked #3 with a comparable cost of 6219.14 USD/h. We believe that the comparative analysis provides a complete perspective into ranked solutions and identifies options, which may prove cost effective by comprising on certain purity constraints. In future, we plan to evaluate the networks with multiple outlet streams with different specifications. Furthermore, we will extend the analysis to include sustainability index for the ranked list of optimal networks predicted by both these approaches.

References

Deng, Y., Zhao, R., 2015. Advanced Oxidation Processes (AOPs) in Wastewater Treatment. Curr. Pollut. Rep. 1, 167–176.

Friedler, F., Tarján, K., Huang, Y.W., Fan, L.T., 1992. Graph-theoretic approach to process synthesis: axioms and theorems. Chem. Eng. Sci. 47, 1973–1988.

Heckl, I., Friedler, F., Fan, L.T., 2010. Solution of separation-network synthesis problems by the P-graph methodology. Comput. Chem. Eng., Selected Paper of Symposium ESCAPE 19, June 14-17, 2009, Krakow, Poland 34, 700–706.

Ho, W., Sirkar, K., 1992. Membrane Handbook. Springer, New York, USA.

Tchobanoglous, G., Burton, F.L., Stensel, H.D., 2002. Wastewater Engineering: Treatment and Reuse, 4th edition. ed. McGraw Hill Higher Education, Boston.

Kılınç, M.R., Sahinidis, N.V., 2018. Exploiting integrality in the global optimization of mixed-integer nonlinear programming problems with BARON. Optim. Methods Softw. 33, 540–562.

Lipták, B.G., Liu, D.H.F.,2000. Wastewater treatment. Boca Raton, Fl, London : Lewis.

Liu, Y., Yu, H.-Q., Ng, W.J., Stuckey, D.C., 2015. Wastewater-Energy Nexus. Chemosphere, Wastewater-Energy Nexus: Towards Sustainable Wastewater Reclamation 140, 1.

Melo-Guimarães, A., Torner-Morales, F.J., Durán-Álvarez, J.C., Jiménez-Cisneros, B.E., 2013. Removal and fate of emerging contaminants combining biological, flocculation and membrane treatments. Water Sci. Technol. 67, 877–885.

Page, 1996. Use of Reclaimed Water and Sludge in Food Crop Production.

Ponce-Ortega, J.M., Hortua, A.C., El-Halwagi, M., Jiménez-Gutiérrez, A., 2009. A property-based optimization of direct recycle networks and wastewater treatment processes. AIChE J. 55, 2329–2344.

Shukla, A.A., Etzel, M.R., Gadam, S., 2006. Process Scale Bioseparations for the Biopharmaceutical Industry. CRC Press.

US EPA, O., 2013. Summary of the Clean Water Act. US EPA. URL https://www.epa.gov/laws-regulations/summary-clean-water-act.

Anton A. Kiss, Edwin Zondervan, Richard Lakerveld, Leyla Özkan (Eds.)
Proceedings of the 29th European Symposium on Computer Aided Process Engineering
June 16th to 19th, 2019, Eindhoven, The Netherlands. © 2019 Elsevier B.V. All rights reserved.
http://dx.doi.org/10.1016/B978-0-128-18634-3.50269-1

A process systems engineering approach to designing a solar/biomass hybrid energy system for dairy farms in Argentina

Carolina Alvarez C. Blanchet[a], Antonio M. Pantaleo[b,c] and Koen H. van Dam[b,*]

[a]*Energy Futures Lab, Imperial College London, South Kensington Campus, London SW7 2AZ, UK*
[b]*Department of Chemical Engineering, Imperial College London, South Kensington Campus, London SW7 2AZ, UK*
[c]*Department of Agricultural and Environmental Sciences, University of Bari, Italy*
k.van-dam@imperial.ac.uk

Abstract

Argentina targets a 20% share of renewables in the energy mix by 2025 and a 15% emissions reduction by 2030, while at the same time removing subsidies for grid electricity. This paper aims to provide a feasible solution for small farm owners for sustainable and affordable energy: a grid-connected hybrid system able to fulfill the demand in a cheaper, reliable and sustainable way. The case study of an existing dairy farm in Carmen de Areco, Buenos Aires, is taken. A grid-connected hybrid system with solar photovoltaics, unheated anaerobic digestion (AD) coupled to an internal combustion engine and storage system, was selected. The size of the hybrid system was optimized via a mathematical model that compared different technologies. The interrelation between sources was economically optimized in order to match the demand on an hourly basis. The scenario comparison defined the optimal solution for this case study, which was the installation of an unheated AD plant with 2.4 kW capacity, and 16 solar panels with a capacity of 5.2 kW, added to a shift in the demand profile. The initial investment required is 17,042 USD, with a payback of 3.4 years and a GHG reduction of 275.9 tons of CO_2 eq per year.

Keywords: bioenergy, solar energy, dairy farm

1. Introduction

New targets defined by Argentina include an increase in the share of Renewable Energy (RE) to 20% in the energy mix by 2025 and a 15% emissions reduction by 2030 (Porcelli and Martínez, 2018). Together with this shift towards sustainability, the government implemented a subsidy scheme removal in 2016, generating a rise in electricity bills that affected end users in domestic and business segments, including rural sector (Observatorio Economico Social, 2017). The affordability and reliability concerns of the rural grid increased the problems for agricultural farmers. This project aims to provide a feasible solution for small farm owners: a grid-connected hybrid system able to fulfill the demand in a cheaper, reliable and sustainable way. The case study of an existing dairy farm in Carmen de Areco, Buenos Aires, is taken.

The power industry is facing new network challenges worldwide with the rise of the renewables. The supply/demand equilibrium can be compared to a RE curtailment in case of an excess of generation, or to a rapid power compensation from fossil fuel power plants in case of unavailability of renewable resources. RE intermittency introduced a new need to the grid operators: today's power plants need to be more flexible, operate within a large load range and minimize emissions in order to react and compensate for possible RE changes. The higher the RE share in the energy mix, the higher the flexibility needed from back up fossil fuel plants to ensure heat and power availability when demanded. Gas-fired power plants will support this balancing given that nuclear energy cannot offer a rapid answer. Available capacity and response times are the key factors for conventional plants, since RE are intermittent (Qadrdan et al., 2017; Welch and Pym, 2017).

In this case study, the combination of an intermittent renewable with a fast responding, flexible biogas plant can deliver secure, reliable, and low-carbon energy. The constant biomass input into an anaerobic digester can produce a secure amount of biogas, being able to balance the supply. The surplus RE generation, rather than being constrained, can be stored (in gas or electricity form) for future use ensuring maximum flexibility. Affordable and reliable electricity can be provided thanks to a fully integrated hybrid system as the one presented before, minimizing environmental impacts and providing dispatchability and flexibility.

2. Approach

The literature mention two approaches: sizing the system and resources needed regarding the demand (Pantaleo et al., 2013), or calculating the energy supply from the resources available (Vis et al., 2010; Elbersen et al., 2012). In this case study, both approaches are proposed, but the heat and power demand will be the decisive approach. First, biomass and solar energy potential are estimated (see Section 3.1) then a demand analysis will define the system sizing; finally, a cost minimization procedure will determine the hybrid system to be implemented, based on the heat and power demand (see Section 3.2). The interrelation between the different sources was economically optimized in order to match the demand on an hourly basis. A baseline scenario was run with available farm data, obtaining a benchmark grid-connected hybrid system configuration. An economic analysis evaluated the investment profitability. Afterwards, multiple scenarios are run, evaluating demand side response, possible regulations outcomes, and technological upgrades (mesophilic AD technologies).

2.1. Case study

The case study is a family business in which the owner takes strategic investment decisions and operates the farm. The company has 7 workers, divided between stable and cattle operation, milk extraction operation, milk refrigeration and raw milk transfer. Having 304 cows, of which 256 in actual milk production, producing approximately 6,000 L of milk per day, it is a small-scale company. Cattle confinement is limited to twice a day for the milking. The mixture of manure, urine and water is held in an open lagoon. This is a normal structure for small and medium farms but lacks in design can potentially generate contamination problems. The decrease in water quality can affect the animal nutrition, minimizing milk production and productivity. In addition, bad water quality can potentially contaminate milk. Moreover, open tanks generate odors, favor insect proliferation

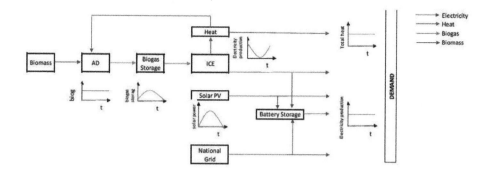

Figure 1: Hybrid System 2 (HS2) model (PV + AD).

(cow health and milk quality could be endangered), and CH4 emissions (Alessandretti, 2011). The farm's installations are common in small farming operations, including the milking room, the milk process and the waiting stable.

2.2. Hybrid PV + AD system

Eight different hybrid systems were compared, with a hybrid system based on biogas production from an AD digester, an ICE to generate electricity and heat and an array of PV and a storage system (see Figure 1) considered as most attractive option for this case study. The operation modes are the following:

- MODE 1: when the solar power is sufficient to meet the demand and charges the battery with the excess energy produced.

- MODE 2: when the solar power is sufficient to meet the demand, but not to charge the battery, the grid will charge it during off-peak hours if the grid price is lower than the bioenergy price.

- MODE 3: when solar power is insufficient to meet the demand and the ICE starts generating to meet the demand. Excess energy produced will charge the battery.

- MODE 4: when solar power is insufficient to meet the demand and the ICE starts generating to meet the demand, but not to charge the battery. The grid will charge it during off-peak hours.

- MODE 5: when solar power and ICE generation are insufficient to meet the demand and grid electricity is cheap (valley period), the grid starts providing the electricity needed.

- MODE 6: when solar power and ICE generation are insufficient to meet the demand and grid electricity is expensive (peak period), the battery starts providing the electricity needed.

- MODE 7: when solar power, ICE and battery are insufficient to meet the demand, the grid will provide the electricity needed, independent on its price.

The automatic selection of this modes by the energy management system (EMS) will define the instant operation of the system.

3. Methodology

3.1. Energy potential

Bioenergy production and potential depends on the number of cows in the farm, and on breeding and nutritional characteristics (Pantaleo et al., 2013). The total solid manure collected depends on the total number of cows producing milk (N^C), the number of hours in a closed stable (H^C) (manure collection is unfeasible in the fields) and a tabulated potential daily Manure Production per Cow (P^C). Total Manure Production per Day (P^d) becomes:

$$P^d = N^C * \frac{H^C}{24} * P^C \tag{1}$$

From (1) it is possible to obtain the volume of biogas generated per day.

The potential for solar energy varies depending on the power rating of the solar panels installed, the number of panels, the day of the year and the hour of the day. For these calculations, different literature was revised and used: McEvoy et al. (2003); Wenham et al. (2013); Mahachi (2016). Calculations take into account: Installation Relative Azimuth, Installation Tilt, Case Study's Latitude (φ) Sun radiation (P_{sun}), Day number (d), Solar Time max (ω), Temperature max avg, Sun radius (R_{sun}) and Relative distance Sun/Earth (r_o). Potential solar energy produced was calculated for winter and summer solstice, considering the available space for panels.

3.2. Optimisation model for the sizing of hybrid systems

An optimisation model was built in AIMMS with a modular structure. Demand and energy supply were divided into sections in order to organize and aggregate them in different ways. The different sections were demand per day, bioenergy per day, solar energy per day, storage per day, grid per day and finally the optimization and total emissions. The model represents the current Argentinian situation in which it is not possible to sell the energy surplus to the grid.

Demand parameters include hourly electricity demand, maximum electricity demand during peak, and minimum electricity demand during off-peak (in kW). Bio-energy parameters are those ones reported in eqn (1), together with investment and operational bioenergy plant costs. Solar energy parameters include peak power, area of panels, roof area, solar variation, capacity factor, CAPEX and OPEX. Further input parameters include storage investment and potential, grid connection operating costs (energy and power prices during peak and off-peak periods), and emissions.

The objective function for the model then is:

$$\text{minimise } Cost^{Tot} = Cost^{BG} + Cost^{SP} + Cost^{sto} + Cost^{NG} \tag{2}$$

where $Cost^{Tot}$ is the total cost, $Cost^{BG}$ is the cost for bioenergy, $Cost^{SP}$ is the cost for solar power, $Cost^{sto}$ storage costs and $Cost^{NG}$ the cost of supply from the national grid. This

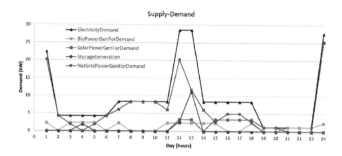

Figure 2: Demand and supply per energy source with the optimised hybrid PV and AD solution.

is subject to constraints on meeting demand, maximum available supply from biogas, available supply from the solar installation, space constraints, battery constraints, and national grid constraints, as well as mass and energy balances.

The objective function in Eqn 2 represents yearly costs considering an annualized CAPEX, as well as fixed and variable OPEX of each technology, so both investment and operational costs of each technology are considered. This economic optimization aims to find the most cost-effective combination of technologies, sizing a hybrid system capable of supplying the electricity demand. However, according to the Argentinian energy market regulation limitations, the model does not take into account selling surplus renewable energy to the grid.

4. Results

For the baseline scenario, the optimal solution was a bioenergy installation of 2.3 kW, a solar energy installation of 5.2 kW (supplied by 16 solar panels) and a storage capacity of 29.9 kW. Figure 2 shows the demand and supply profile with this installation.

The baseline scenario optimization results are very specific of the techno-economic and regulative input conditions assumed for the dairy farm under investigation, and which could be, at some extent, extended to other rural farms in Argentina. However, changes in regulations, demand profile or in the hybrid technology are possible. In order to assess the influences of these key factors, different scenarios have been evaluated: 1) national grid prices before subsidies scheme removal; 2) shifting load demand (mid-night peak) and keeping milk refrigeration on during peak period; 3) regulations evolution, selling electricity to the grid; 4) Mesophilic AD tank with electricity retrofit; 5) mesophilic AD tank with CHP (heat retrofit). Results from these scenarios are shown in Figure 3.

For this small-scale dairy farm the recommended investment is scenario 2: installing a bioenergy plant with 2.4 kW capacity (and no CHP), and 16 solar panels (345 Wp) with a total capacity of 5.2 kW. To avoid the purchase of a battery, a modification in the demand profile is endorsed: a switch of the milking process to after midnight will be sufficient, being able to keep the milk refrigerated after 6 pm. This solution requires an initial investment of 17k USD with 317 USD monthly savings for the first year, and with a payback of 3.4 years. This leads to a GHG reduction of 275.9 tons of CO_2 eq per year.

Description		Scenario (-1) Before subsidies scheme removal	Baseline-Scenario 1 Baseline	Scenario 2 Shifting demand	Scenario 3.1 FIT UK	Scenario 3.2 FIT UK with demand shift	Scenario 4.1 Mesophilic AD with ICE	Scenario 4.2 Mesophilic AD with ICE and demand shift	Scenario 5.1 Mesophilic AD with CHP	Scenario 5.2 Mesophilic AD with CHP and demand shift
Bio-Energy installation	kW	0.0	2.3	2.4	2.3	2.3	3.1	3.1	3.1	3.1
Solar Energy installation	kW	0.0	5.2	5.2	5.2	5.2	5.2	3.4	5.2	1.9
Number of solar panels		0	16	16	16	16	16	10	16	6
Storage Capacity	kW	0.0	29.9	0.0	29.9	0.0	30.1	0.0	29.0	0.0
Final Total Investment	USD	0.0	33684.2	17042.3	33684.2	16579.2	36469.4	16310.8	38071.0	16326.2
Monthly savings (Year 1)	USD	0.0	706.5	317.3	714.9	312.8	687.1	251.9	752.3	297.4
Emission Reduction	tons of CO_{2eq}	0.0	275.3	275.9	275.3	275.5	274.7	273.3	277.5	274.6
NPV savings	USD	0.0	62646.8	27134.2	63445.7	26729.1	57768.6	19304.5	65057.4	25410.7
Payback period	Years	0.0	3.1	3.4	3.0	3.3	3.4	4.0	3.2	3.4

Figure 3: Comparison of the different scenarios results

5. Discussion and future work

The results show that the proposed model can generate cost-optimal solutions to meet the demand profile of a small dairy farm, and that additional scenarios can be explored to derive optimal solutions in different regulatory framework or technical hybrid solutions implemented. In order to generalize the model, the scheduling of different RE could lead to a complete and more integrated approach. For example, a solar section could have subsections with solar PV (panels, HCPV) and solar thermal (flat plates, PT, HCSP). The model could define which RE and technology or combination of RE and technologies would be the most appropriate from a cost-effective point of view for a particular case study, with results likely to depend on demand profile, location, space constraints etc. Another approach for further work would be a more detailed temporal analysis, considering seasonal variations of resources and energy demand. Various political and economic risks can be faced while developing this project, which is highly uncertain and depends on Argentina's policy stability, but latest trends demonstrate a promising future market.

References

E. Alessandretti, 2011. Propuesta energética sostenible para establecimientos agropecuarios. Instituto Tecnologico de Buenos Aires (ITBA).

B. Elbersen, I. Staritsky, G. Hengeveld, L. Jeurissen, J. Lesschen, 2012. Outlook of spatial biomass value chains in eu28. D2. 3 Biomass Policies project.

T. Mahachi, 2016. Energy yield analysis and evaluation of solar irradiance models for a utility scale solar pv plant in south africa. Ph.D. thesis, Stellenbosch: Stellenbosch University.

A. McEvoy, T. Markvart, L. Castañer, T. Markvart, L. Castaner, 2003. Practical handbook of photovoltaics: fundamentals and applications. Elsevier.

Observatorio Economico Social, 2017. Energia electrica en Argentina y Santa Fe. Rosario. Available at: http://www.observatorio.unr.edu.ar/llego-la-factura-de-la-luz/.

A. Pantaleo, B. De Gennaro, N. Shah, 2013. Assessment of optimal size of anaerobic co-digestion plants: an application to cattle farms in the province of Bari (Italy). Renewable and Sustainable Energy Reviews 20, 57–70.

A. M. Porcelli, A. N. Martínez, 2018. Una inevitable transición energética: el prosumidor y la generación de energías renovables en forma distribuida en la legislación argentina nacional y provincial. Actualidad Jurídica Ambiental (75), 4–49.

M. Qadrdan, H. Ameli, G. Strbac, N. Jenkins, 2017. Efficacy of options to address balancing challenges: Integrated gas and electricity perspectives. Applied Energy 190, 181–190.

M. Vis, et al., 2010. Harmonization of biomass resource assessments Volume I: Best practices and methods handbook.

M. Welch, A. Pym, 2017. Flexible natural gas/intermittent renewable hybrid power plants. In: ASME 2017 11th International Conference on Energy Sustainability. American Society of Mechanical Engineers.

S. R. Wenham, M. A. Green, M. E. Watt, R. Corkish, A. Sproul, 2013. Applied photovoltaics. Routledge.

Anton A. Kiss, Edwin Zondervan, Richard Lakerveld, Leyla Özkan (Eds.)
Proceedings of the 29th European Symposium on Computer Aided Process Engineering
June 16th to 19th, 2019, Eindhoven, The Netherlands. © 2019 Elsevier B.V. All rights reserved.
http://dx.doi.org/10.1016/B978-0-128-18634-3.50270-8

Modelling and Simulation of Supercritical CO_2 Oil Extraction from Biomass

Rui M. Filipe,[a,b,*] José A. P. Coelho,[a,c] David Villanueva-Bermejo,[d] Roumiana P. Stateva[e]

[a]*Instituto Superior de Engenharia de Lisboa, Instituto Politécnico de Lisboa, 1959-007 Lisbon, Portugal*

[b]*Centro de Recursos Naturais e Ambiente, Instituto Superior Técnico, Universidade de Lisboa, 1049-001, Lisbon, Portugal*

[c]*Centro de Química Estrutural, Instituto Superior Técnico, Universidade de Lisboa, 1049-001 Lisbon, Portugal*

[d]*Instituto de Investigación en Ciencias de la Alimentación CIAL (CSIC-UAM), 28049 Madrid, España*

[e]*Institute of Chemical Engineering, Bulgarian Academy of Sciences, 1113 Sofia, Bulgaria*

rfilipe@isel.ipl.pt

Abstract

The aim of this work is to model the kinetics of the supercritical CO_2 extraction of oils from biomass, namely from industrial grape seeds and chia seeds. A model introduced by Sovova and Stateva (2015) is used and solved using gPROMS ModelBuider. The supercritical extraction experiments performed at different temperature, pressure and flow rate conditions, provide the data to the modelling studies and for model parameter estimation, using gPROMS parameter estimation.

The partition coefficient and the maximum oil content corresponding to monolayer adsorption was estimated and the absolute average relative deviation between the experimentally measured and simulated extraction curves for the cases examined is within the range (1-9) %, which may be considered as a good agreement taking into consideration the very complex nature of the systems under study.

Keywords: Grape seeds, chia seeds, supercritical CO_2 extraction, extraction kinetics, mathematical modelling.

1. Introduction

Increasing attention is being drawn to the effective use of waste biomass and vegetal material as renewable resources of high added value compounds with applications in food, cosmetics, pharmaceutical industries, biodiesel production, etc. Grape (*Vitis vinifera* L.) seed biomass contains typically (8-15) % (w/w) of oil rich in long chain polyunsaturated fatty acids (PUFAs) and antioxidants (Bail et al., 2008; Fernandes et al., 2013). Although representing about (20-25) % of the biomass generated by the wine industry, it is still considered a disposable material and rarely valorised. Chia (*Salvia hispanica* L.) seeds are attracting great attention due to its health beneficial properties and are being studied as a source of protein and dietary fiber. Furthermore, they have a high oil content (20-35 % w/w), rich in PUFAs, mainly the omega-3 α-linolenic acid which represents (60-65) % of total fatty acids (Porras-Loaiza, 2014).

Supercritical CO_2 extraction (SCE) is known to prevent or minimize the degradation of bioactive compounds due to the comparatively low temperatures applied and oxygen free atmospheres. It also allows obtaining solvent-free products. Unsurprisingly, it is currently establishing itself as the viable and eco-friendly alternative to the use of organic solvents. Yet, kinetic data are not abundant, and, particularly for chia seeds oil extraction they are scarce and superficial.

In view of the above, the aim of our work is to model the kinetics of the SCE of oils from industrial grape seeds, obtained directly from a Portuguese wine industry (Coelho et al., 2018), and from two sets of chia seeds - high oil content (HOCS), and underutilized low oil content (LOCS) seeds, by applying an efficient solution method to a recently introduced kinetics model (Sovova and Stateva, 2015).

2. Methodology

The seeds were prepared prior to extraction. Grape seeds, which were supplied crushed, have been dried for 48 h at 343 K while the chia seeds were cleaned from straws and other extraneous material prior to being grounded and sieved. No other pretreatment was applied. The average particle diameter (d_p) was calculated to be 0.62 mm for the grape seeds and 0.37 mm for the chia seeds.

The SCE of the grape seeds was performed at temperatures of (313 and 333) K, pressures of (20, 30 and 40) MPa and flow rates of (1.8, 2.3 and 2.8) x 10^{-3} kg·min^{-1} of CO_2. Regarding the chia seeds, SCEs from LOCS were carried out at (25 and 45) MPa, (313 and 333) K and a CO_2 flow rate of 40 x 10^{-3} kg·min^{-1}. The extractions of HOCS were performed at 45 MPa, 313 K and several CO_2 flow rates, namely (27, 40 and 54) x 10^{-3} kg·min^{-1}. The amount of sample used for the experiments was 17 g (grape seeds) and 130 g (chia seeds). A detailed description of the units employed for the extraction of grape and chia seeds can be found, respectively, in the studies reported by Coelho et al. (2018) and Villanueva-Bermejo et al. (2017).

The grape and chia seed oils are very complex mixtures of mainly triacylglycerols (TAGs) with minor amounts of other compounds. Due to this complexity, and in order to reduce the size of the kinetics modelling task, the oils composition is usually being represented by model TAGs. Following the results of the composition analyses, the grape seed oil was represented by triolein (Coelho et al., 2018), while chia seed oil should be represented by trilinolenin. However, considering the lack of any experimental information on the VLE of trilinolenin+scCO$_2$ and the total lack of data (both experimental and estimated) on the thermophysical properties of trilinolenin, triolein was selected as the model TAG to exemplify the chia seed oil.

To calculate the solubility of the representative TAG in the scCO$_2$, the VLE of the two-component system (triolein+CO$_2$) was modelled applying the predictive Soave-Redlich-Kwong (PSRK) cubic EoS (Holderbaum and Gmehling, 1991).

To simulate the extraction curves in terms of yield and composition of the oil extracts for both the grape and chia seeds, the novel model of Sovova and Stateva (2015), which reflects the complex relationship between kinetics and phase equilibria (solubility), was applied. Although the model was originally developed for multicomponent systems extraction, it remains valid for single component systems too.

This model considers homogeneous concentration in the extractor at both solid and fluid phases and assumes that the extracts are located on the surface of the solid particles. This assumption allows neglecting internal diffusion, which is compatible with finely ground substrates where the diffusion path in the particles is short and the extract is easily accessible, resulting in negligible internal mass transfer resistance. Further details on the model can be found elsewhere (Sovova and Stateva, 2015)

The model was deployed, validated and executed using gPROMS ModelBuilder (Process Systems Enterprise, Ltd.), an equation-oriented modelling and optimisation platform for steady-state and dynamic systems. gPROMS offers an integrated framework where all the tasks can be performed within the same environment, with embedded solvers that can usually complete the simulations without further user input provided that a complete and well formulated model, with the adequate initial conditions, is available.

gPROMS Modelbuilder parameter estimation was used to estimate the partition coefficient (K) for all the cases studied. The maximum oil content corresponding to monolayer adsorption (w_t) was also estimated for the chia seeds cases. gPROMS uses a maximum likelihood parameter estimation problem and attempts to determine values for the uncertain physical and variance model parameters that maximize the probability that the mathematical model will adequately predict the values obtained from the experiments.

In order to compare the fitting accuracy, a standard deviation measure, the absolute average relative deviation, *AARD*, was also calculated after the parameter estimation using Eq. (1), where N is the total number of experimental points, and e_i^{exp} and e_i^{est} - the *i*-th experimental and estimated point, respectively.

$$AARD = \frac{100}{N} \sum_{j=1}^{N} \frac{\left| e_i^{exp} - e_i^{est} \right|}{e_i^{exp}} \qquad (1)$$

3. Results and discussion

Figure 1 shows the experimental and simulated cumulative extraction curves and Figure 2 - the simulated composition profiles for the solid and fluid phases of the SCE of grape seeds oil. The positive effect of the pressure increase on the extraction rate is clearly demonstrated on the yield, while the effect of temperature is not so well-defined. The simulated profiles for solid and fluid phases depicted in Figure 2, where the increase in pressure results in higher extraction rates, are compatible with the behaviour depicted in Figure 1.

For LOCS (Figure 3) analogous effect of the pressure is observed, while the temperature effect on the rate of extraction and on the total amount of extracted oil is more significant. Figure 4 depicts the profiles for HOCS at different scCO$_2$ flows, with the yield increasing faster for higher flows.

From the analysis of Figure 1, 3 and 4, a steepest ascent in cumulative mass fraction may be observed for chia seeds, indicating that the oil is more readily available than in the in the grape seeds.

The parameter K was estimated for all systems studied, while w_t - only for the chia seeds cases. The values obtained are presented in Table 1. The AARD values obtained ranged approximately from (2 to 9) % for the grape seeds and LOCS, and from (1 to 7) % for HOCS, which should be considered as a good agreement taking into consideration the very complex nature of the systems under study.

The new modelling approach applied incorporates in a rigorous way the interplay between phase equilibria (solubility) and kinetics, and the results obtained demonstrate that albeit the simplifications introduced in representing the grape and chia seed oils, there is a good qualitative and quantitative agreement between the experimental and calculated extraction yields at the SCEs operating conditions examined. It should be emphasized that although the internal diffusion is neglected, the model can still deliver adequate results, providing that the particle size is small, as is the case for the matrix used in this work.

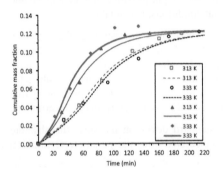

Figure 1. Experimental (symbols) and simulated (lines) extraction curves for grape seeds at different temperatures, p = 30 MPa (dashed lines) and p = 40 MPa (solid lines) and scCO$_2$ flow rate of 1.8 x 10^{-3} kg·min^{-1}.

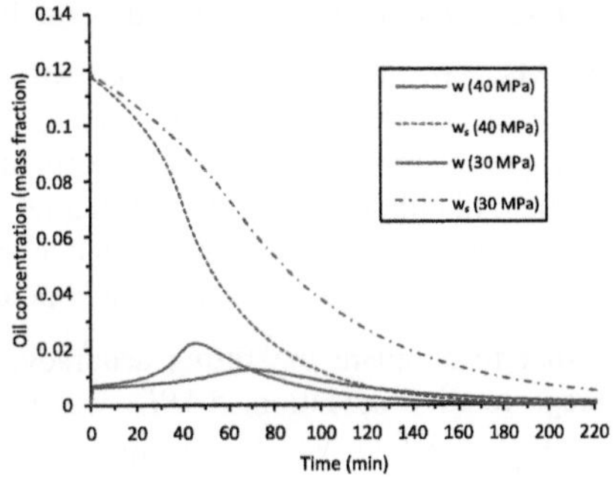

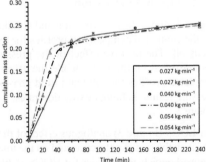

Figure 2. Simulated profile of oil concentration in the grape seeds (w_s) and scCO$_2$ (w). T = 333 K and scCO$_2$ flow rate of 1.8 x 10^{-3} kg·min^{-1}.

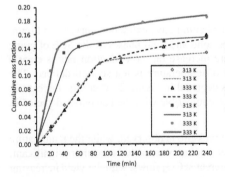

Figure 3. Experimental (symbols) and simulated (lines) extraction curves for LOCS at different temperatures, p = 25 MPa (dashed lines) and p = 45 MPa (solid lines) and scCO$_2$ flow rate of 1.8 x 10^{-3} kg·min^{-1}

Figure 4. Experimental (symbols) and simulated (lines) extraction curves for HOCS at different at different flow rates. T = 333 K and p = 45 MPa.

The values of K are directly related with the extraction rate and the values presented in Table 1 follow generically the same trend observed in Figures 1, 3 and 4, namely K increases with the increase of pressure and temperature. Regarding the $scCO_2$ flow, the trend variation of K is not clear and shows to be system dependent. This behaviour is compatible with the results previously reported by other authors (Honarvar et al., 2013).

Table 1. Estimated values for parameters K and wt.

System	P (MPa)	T (K)	F (10^{-3} kg·min^{-1})	K (kg plant ·kg^{-1} CO_2)	w_t	AARD (%)
Grape	30	313	1.8	2.24E-1	-	4.51
Grape	30	333	1.8	2.22E-1	-	8.84
Grape	40	313	1.8	3.65E-1	-	6.73
Grape	40	333	1.8	4.33E-1	-	8.64
Grape	40	333	2.3	5.58E-1	-	6.68
Grape	40	333	2.8	3.46E-1	-	2.16
LOCS	25	313	40	3.22E-3	8.04E-2	4.82
LOCS	25	333	40	1.36E-2	8.04E-2	9.49
LOCS	45	313	40	4.27E-3	5.76E-2	5.64
LOCS	45	333	40	2.47E-2	5.20E-2	2.13
HOCS	45	313	27	2.00E-2	6.01E-2	1.22
HOCS	45	313	40	1.69E-2	8.25E-2	2.47
HOCS	45	313	54	9.28E-3	8.07E-2	6.65

4. Conclusions

This work presents the results of the SCE of biomass from industrial grape seeds, obtained directly from a Portuguese industry without preliminary treatment, and from two sets of chia seeds. The influence of the operating conditions on the extraction yield were analysed in detail and reported.

To simulate the extraction process, a kinetic model assuming that the extracts are located on the surface of the solid particles, thus neglecting internal diffusion, was applied. The solubility in $scCO_2$ of the oil, represented as triolein, at the temperatures and pressures of interest to both experiments were obtained applying a rigorous thermodynamic framework, which allowed to interweave the respective solubility functions in the mass balance equations.

The model equations were then integrated using gPROMS ModelBuilder. The qualitative and quantitative agreement between the experimental and simulated extraction curves in terms of yields for both the grape and chia seeds oil was good taking into consideration the complex nature of the systems examined. We can also conclude that the model applied can be successfully used in finely ground matrix where the internal diffusion contribution to the extraction phenomena is very small and, thus, negligible.

Acknowledgements

The authors acknowledge the funding received from the European Union's Horizon 2020 research and innovation programme under the Marie Sklodowska-Curie grant agreement No 778168. J. Coelho and R. M. Filipe are thankful for the financial support from Fundação para a Ciência e a Tecnologia, Portugal, under projects UID/QUI/00100/2013 and UID/ECI/04028/2013.

References

S. Bail, G. Stuebiger, S. Krist, H. Unterweger, G. Buchbauer, 2008, Characterisation of various grape seed oils by volatile compounds, triacylglycerol composition, total phenols and antioxidant capacity Food Chemistry, 108, 1122–1132.

J. Coelho, R. Filipe, M. Robalo, R. Stateva, 2018, Recovering Value from Organic Waste Materials: Supercritical Fluid Extraction of Oil from Industrial Grape Seeds, Journal of Supercritical Fluids, 141, 68-77.

L. Fernandes, S. Casal, R. Cruz, J.A. Pereira, E. Ramalhosa, 2013, Seed oils of ten traditional Portuguese grape varieties with interesting chemical and antioxidant properties, Food Research International, 50, 161-166.

T. Holderbaum, J. Gmehling, 1991, PSRK: A Group Contribution Equation of State Based on UNIFAC, Fluid Phase Equilibria. 70,251-265.

B. Honarvar, S. Sajadian, M. Khorram, A. Samimi, 2013, Mathematical modeling of supercritical fluid extraction of oil from canola and sesame seeds, Brazilian Journal of Chemical Engineering, 30, 159–166.

P. Porras-Loaiza, M. T. Jiménez-Munguía, M. E., Sosa-Morales, E., Palou, A. Lopez-Malo, 2014, Physical properties, chemical characterization and fatty acid composition of Mexican chia (Salvia hispanica L.) seeds, International Journal of Food Science and Technology, 49, 571-577.

H. Sovova, R. P. Stateva, 2015, New Approach to Modeling Supercritical CO_2 Extraction of Cuticular Waxes: Interplay between Solubility and Kinetics, Industrial and Engineering Chemistry Research, 54, 17, 4861-4870.

D. Villanueva-Bermejo, F. Zahran, M.R. García-Risco, G. Reglero, T. Fornari, 2017, Supercritical fluid extraction of Bulgarian Achillea millefolium, Journal of Supercritical Fluids, 119, 283-288.

Anton A. Kiss, Edwin Zondervan, Richard Lakerveld, Leyla Özkan (Eds.)
Proceedings of the 29th European Symposium on Computer Aided Process Engineering
June 16th to 19th, 2019, Eindhoven, The Netherlands. © 2019 Elsevier B.V. All rights reserved.
http://dx.doi.org/10.1016/B978-0-128-18634-3.50271-X

Optimisation of the integrated water – energy systems: a review with a focus in Process Systems Engineering

Christiana M. Papapostolou[a], Emilia M. Kondili[a*], Georgios T. Tzanes[b]

[a] Optimisation of Production Systems Laboratory,
[b] Soft Energy Applications and Environmental Protection Laboratory,

Dept. of Mechanical Engineering, University of West Attica, Campus 2, P. Ralli & Thivon Street, 12244, Egaleo, Greece,

*ekondili@uniwa.gr

Abstract

The analysis of issues within the 'water-energy nexus' has become a topic of increasing attention for the scientific and policy communities. Lately various water, energy and food nexus systems are developed to identify, analyse and better understand the interactions between water and energy systems with the purpose of managing these critical resources in the context of various conflicting requirements on top of the imperative need of their optimal use. The objectives of the present work are at first to identify the wide spectrum of problems being encountered under the term Water – Energy Nexus and to review the existing and recent advancements in the Process Systems Engineering (PSE) field that exploit optimisation models, methods and tools for the analysis and the operation of various integrated water energy systems. In addition, the work aims to provide a perspective to PSE and CAPE experts thus expanding their scope and interests and, at the same time, to provide the WEN with the knowledge, experience, tools and excellent quality researchers able to assist to the solution of these significant global problems.

Keywords: PSE and CAPE fields, Water-Energy Nexus, Optimisation models

1. Introduction – Objectives and Scope of the Work

Water – energy nexus (WEN) deals with the efficient integration of water and energy resources in order to tackle a wide range of design, operation and optimisation issues for systems including both of these two resources. Many problems such as the supply and the continuous availability of the resources, their quality and safety, the environmental and social impacts, along with the cost and/or profit emerged from their exploitation, are problems encountered in a very wide range of real situations. Food is in most cases involved due to its strong interaction with the other two utilities.

However, the present paper focuses mainly to the Water – Energy Nexus, mainly due to the types of problems that are analysed. The water and energy interactions are complex and they affect one another directly and indirectly. The dynamic behavior and the presence of multiple conflicting energy/water demands and the complex interactions between various sources and sinks of water and energy greatly impact decision-making within the water-energy nexus, taking also into account that related decisions depend strongly on the interests of the different stakeholders involved. Therefore, a systematic and holistic analysis of the nexus is critical and there is a very wide scope for optimisation of the whole system, supporting unbiased decision-making.

2. Background of the Work - WEN representing a wide range of problems

By their definition PSE and CAPE fields investigate solutions for integrated systems and have made a tremendous progress all these years in the understanding, modeling and solving a wide range of problems met initially in process systems and now in a very wide spectrum of fields. Therefore, this integrated approach for problem-solving is very familiar to PSE and CAPE communities and many research works deal with this type of problems. The models, methods and tools of PSE may really be the most added value approaches for the WEN problems due to their integrated approach and their ability to address the complexity of these systems. PSE 2018 conference had a dedicated session in the WEN and in the last 2-3 ESCAPE events there are papers that explicitly deal with this issue. The basic parameters and dimensions of these problems are:

- The type of the problem under consideration.
- The geographic area, i.e. the spatial dimension of the problem
- The scope of the work, i.e. is it for households, for cities, for small isolated communities, for a whole country.
- The temporal dimension, i.e. for which time horizon is the problem being solved.
- The stakeholders and interested parties.
- The objective of the optimisation. For the same problem it will be different for a private investor that might seek for maximum profit, the entire society might seek for minimum environmental impacts, maximum security of water and energy supply, a local authority that might seek for the optimal policy / pricing / management of the resources, for the government that might seek for the best water and energy infrastructure investments.
- The importance for the optimal solution or for a set of feasible solutions.

There are very interesting and very recent review papers that deal with various aspects of the research work relevant to Water Energy (and possibly Food) Nexus. Garcia and You (2016) analyse relevant research works concerning Water – Energy and Food Nexus (WEFN) highlighting challenges and areas where PSE may find interesting research opportunities.

Dai et al. (2018) made an extensive survey of recent scientific literature on the water-energy nexus. The authors identified 70 studies and 35 were selected as comprehensive case studies for review. In their work the reviewed studies have been classified and assessed according to groupings based on both geographic scale and their 'nexus scope'. The review has also concluded that, while many studies aim to develop new methods and frameworks to comprehensively assess interactions between water, energy and other elements, none can or do provide a singular framework for performing a "nexus study".

In the context of these excellent review works the current study mainly aims to highlight different types of problems encountered within the WEN scope as well as some modern methods and tools employed to optimise the nexus. It should be stressed that the list of problem types is indicative and various other very significant works have been carried out. The selection has been based on the innovative character of the problem and/or the method being developed for the purposes of the specific work.

3. Types of problems tackled with the Water Energy Nexus

3.1. Water Supply

Although many papers have dealt with the production and the use of energy, much less attention has been paid to the water side of the problems.

Sea Water Desalination

The production of clean water through membrane processes requires large amounts of energy. On the other hand, in remote and isolated from central networks communities, there are is a significant technical potential for the simultaneous production of water and energy. A methodology to design desalinated water supply systems including solar energy as an energy source option is proposed by Herrera – Leon et al. (2018). The methodology consists of two stages and uses a MINLP model. The objective of the first stage is to minimize the total cost of the whole system, while the objective of the second stage is to minimize the greenhouse gases (GHG) emissions related to desalinated water transport by considering the opportunity of installing photovoltaic panels. The MINLP model considers the installation and operation costs of the desalination plants, pumping stations, pipes, and photovoltaic panels.

Optimisation of Water Supply systems

A very interesting review has also been carried out by Vakilifard et al. (2018), acknowledging that the utilization of WEN in optimising water supply systems not only ensures the sustainability of the water supply for increasing water demand but also diminishes water-related energy and environmental concerns. The work also presents a review highlighting knowledge gaps in optimisation models related to the water energy nexus in water supply systems or "water supply side of the nexus". Several major gaps are identified, including the lack of optimisation models capturing spatial aspects as well as environmental impacts of the nexus problems. The main gap is the absence of models for optimising long-term planning of water supply system considering renewable energy within an urban context.

3.2. Water Energy Nexus Optimisation in Remote Communities

Remote communities render a very special issue of investigation within the WEN research umbrella: WE supply problems in these communities are quite intense and require immediate solutions that can only be provided by optimisation approaches. Following that, various research works have been carried out by pioneering in the field teams, indicatively focused in the Greek, Aegean islands and/or in the Mediterranean Region and/or in Maldives islands.

Aiming at providing water and energy sustainable supply solutions for the isolated Aegean islands, Bertsiou et al. (2018), investigated the development of a Hybrid Renewable Energy System (HRES) in Fournoi, a small island in northern Aegean. The proposed methodological framework is designed to produce hydropower for the coverage of electricity needs and for the desalination of the amount of water that is needed for domestic and agricultural use. Data about the island's population and data of water and electricity consumption are collected, as well as, meteorological data from the nearby station. The authors make a Scenario Analysis with three different scenarios.

In the same field Papapostolou et al. (2017, 2018) implemented an optimisation framework for the optimal design and operation of water and energy supply chains (WESCs) accounting both economic and environment considerations. The framework consisting of a representation following the Resource-Task-Network approach along with a developed MINLP model and its variations, was implemented and tested in different island topographies (a medium-size and a very small island in Cyclades Complex), so as to prove the most profitable WESC accounting: on the side of energy supply the options of diesel, PV, wind and battery storage, whilst on the side of water supply, the options of the desalination plant vs the traditional water transfer via ships, both acting operating complementary to the island's existing water tank. The results of the work provide a water supply strategy allowing in the case of RES excess the desalination plant to operate "free of charge" so as to allow the water intake to the water tank, a water to be supplied at the island network at a time whereas both peaks loads in energy and water demand occur (thus more expensive). Additionally, results evidence that energy autonomy and water demand fulfilment is possible for small-scale islands as well in a cost-effective way, that being achieved through RES, allowing a reduction of CO_2 emissions more than 90%.

3.3. Water Energy Nexus Optimisation in Urban – Residential Sector

Cities are concentrations of demand to water and energy systems that rely on resources under increasing pressure from scarcity and climate change mitigation targets. However, the effect of the end-use water and energy interdependence on urban dynamics had not been studied. In the work of De Stercke et al. (2018), a novel system dynamics model is developed with an explicit representation of the water-energy interactions at the residential end use and their influence on the demand for resources. The model includes an endogenous carbon tax based climate change mitigation policy which aims to meet carbon targets by reducing consumer demand through price. It also encompasses water resources planning with respect to system capacity and supply augmentation. Using London as a case study, the authors show that the inclusion of end-use interactions has a major impact on the projections of water sector requirements.

4. Modern Methods and Tools in the Optimisation of WEN

Certainly the well-known optimisation methods and tools (Mathematical Programming as MILP, MINLP, global optimisation, etc.) have extensively been used for the solution of the WEN problems. However, in the latest advancements the authors have used modern approaches concerning optimisation modelling and solution of the corresponding problems. Game Theory and Graph theory are two of the modern methods that have been identified in WEN optimisation works.

4.1. Game theory

Land use optimization can have a profound influence on the provisions of interconnected elements that strongly rely on the same land resources, such as food, energy, and water. Avraamidou et al., (2018b) have developed a hierarchical FEW-N approach to tackle the problem of land use optimization and facilitate decision making to decrease the competition for resources. The authors have formulated the problem as a Stackelberg duopoly game, a sequential game with two players – a leader and a follower. The government agents are treated as the leader (with the objective to minimize the competition between the FEW-N), and the agricultural producers and land developers as the followers (with the objective to maximize their profit).

Nie at al (2018) in the context of their work for land use optimisation analyse the FEW-N for crop-livestock systems with the development and implementation of a global optimisation land allocation framework, providing an adaptive data-driven modeling method based on limited realistic data to predict yields for production components.

For the development of a feasibility model, Namany et al. (2018) developed a bottom-up approach in which the self-interested stakeholders, i.e. the players, are represented based on their own decision variables and objective functions at various levels, i.e. resources and environment systems, engineering systems, and their integration and operations. For this purpose, a combination of optimization models based on linear programming, stochastic programming, and game theoretic approach, represented by a Stackelberg competition, were developed. A case study is set in the State of Qatar with the objective of enhancing food security using hypothetical scenarios. The results obtained demonstrates that interesting interactions between systems can potentially result in the achievement of desired objectives under properly regulated markets.

4.2. Graph theory

Some very interesting modern tools seem to emerge that are exploited in the optimisation of WEN. For example, Tsolas et al. (2018) present a scalable and systematic method for the design and optimization of complex water-energy nexus using graph theory-based network representation and a novel (WEN) diagram. The authors show that for specified external grid demands, the optimal nexus configuration with minimum water and energy generation is the one without any redundant subgraphs. They then propose a systematic method to identify and eliminate redundant cycles, flows and entities within a nexus leading to (i) minimum generation/extraction of water and energy resources from the environment, or (ii) maximum yield of water and energy to meet external demands. Their approach results in optimal nexus configurations that also satisfy operational constraints, restrictions and water quality specifications. The approach is demonstrated using case studies on water-energy nexus systems focusing on power generation, seawater desalination, groundwater and surface water at regional and national scales.

5. Conclusions and Future Prospects

The interest and research challenge in the modelling and optimisation of WEN in various scales, expressions, communities is continuously increasing. There is a very important optimisation scope in these problems since there are a lot of conflicting demands in the resources involved. The present work made an effort to highlight the wide spectrum of real problems that may be encountered under the term Water – Energy Network. For so different problems it is clear that there is no unique modelling framework that could describe all of them. On the other hand it is obvious that there is a serious optimisation scope in these problems due to the different stakeholders, objectives and characteristics. Advanced model development that will describe the details and characteristics of the problems is still required since the interactions between the energy and water resources are complex and not yet clarified. The other important issue in this discussion is the implementation of the above in real problems and how the developed knowledge and knowhow will be useful and exploitable in real operating water – energy systems. Furthermore, it is certain that many investors (on a local or global perspective) will emerge finding business opportunities. Legislation, regulations and infrastructure expansion will definitely be needed in order to make this knowledge exploitable for the social and investor's benefit.

Acknowledgements

"The Post-doctoral Research of Dr C. Papapostolou was undertaken with a scholarship fund by IKY, Act "Supporting Post-Academic Researchers" with resources from the Operational Program "Human Education Development and Lifelong Learning", priority axes 6,8,9, co-funded by the European Social Fund - ESF and the Greek State ".

References

S. Avraamidou, A. Milhorn, O. Sarwar, .E.N. Pistikopoulos, 2018, Towards a Quantitative Food-Energy-Water Nexus Metric to facilitate Decision Making in Process Systems: A Case Study on a Diary Production Plant, Proc. of ESCAPE28, Computer Aided Chemical Engineering, 43

S. Avraamidou, B. Beykal, I.P.E. Pistikopoulos, E.N. Pistikopoulos, 2018, A hierarchical Food-Energy-Water Nexus (FEW-N) decision-making approach for Land Use Optimization, Computer Aided Chemical Engineering, 44

M. Bertsiou, E. Feloni, D. Karpouzos, E. Baltas, 2018, Water management and electricity output of a Hybrid (HRES) in Fournoi Island in Aegean Sea, Ren. Energy, 118, 790–798

J. Dai, S. Wu, G. Han, J. Weinberg, X. Xie, X. Wu, X. Song, B. Jia, W. Xue, Q. Yang, 2018, WEN: A review of methods and tools for macro-assessment, Appl.Energy, 210, 393–408

S. De Stercke, A. Mijic, W. Buytaert, V. Chaturvedi, 2018, Modelling the dynamic interactions between London's water and energy systems from an end-use perspective, Applied Energy, 230, 615–626

J. Garcia, F. You, 2016, The water-energy-food nexus and process systems engineering: A new focus, Computers and Chemical Engineering, 91, 49–67

S. Herrera-León, A. Kraslawski, L.A. Cisternas, 2018, A MINLP model to design desalinated water supply systems including solar energy as an energy source, CACE, 44, 1687–1692

S. Namany, A-A Tareq, G. Rajesh, 2018, Integrated techno-economic optimization for the design and operations of energy, water and food nexus systems constrained as non-cooperative games, Computer Aided Chemical Engineering, 44, 2018, 1003–1008

Y. Nie, S. Avraamidou, J. Li, E.N. Pistikopoulos, 2018, Land use modeling and optimisation based on Food-Energy-Water Nexus: A case study on crop-livestock systems, Computer Aided Chemical Engineering, 44

C. Papapostolou, E. Kondili, J.K. Kaldellis, 2017, Optimising the total benefit of water resources management in combination with the local energy systems in remote communities taking into account sustainability, Computer Aided Chemical Engineering, 40, 2689–2694

C.M. Papapostolou, E.M. Kondili, G. Tzanes, 2018, Optimisation of water supply systems in the water – energy nexus: Model development and implementation to support decision making in investment planning, Computer Aided Chemical Engineering, 43, 1213–1218.

R. Segurado, M. Costa, N. Duić, M.G. Carvalho, 2014, Integrated analysis of energy and water supply in islands. Case study of S. Vicente, Cape Verde, Energy, 92, 639–648

S.D. Tsolas, M.N. Karim, M.M.F. Hasan, 2018, Optimization of water-energy nexus: A network representation-based graphical approach, Applied Energy, 224, 230–250

G.T. Tzanes, D. Zafirakis, C. Papapostolou, K. Kavadias, J.K. Kaldellis, 2017, PHAROS: An Integrated Planning Tool for Meeting the Energy and Water Needs of Remote Islands using RES-based Hybrid Solutions, Energy Procedia, 142, 2586–2591

N. Vakilifard, M.A. Anda, P. Bahri, G. Ho, 2018, The role of water-energy nexus in optimising water supply systems – Review of techniques and approaches, Renewable and Sustainable Energy Reviews, 82, Part 1, 1424–1432

Anton A. Kiss, Edwin Zondervan, Richard Lakerveld, Leyla Özkan (Eds.)
Proceedings of the 29th European Symposium on Computer Aided Process Engineering
June 16th to 19th, 2019, Eindhoven, The Netherlands. © 2019 Elsevier B.V. All rights reserved.
http://dx.doi.org/10.1016/B978-0-128-18634-3.50272-1

On the role of H_2 storage and conversion for wind power production in the Netherlands

Lukas Weimann[a], Paolo Gabrielli[b], Annika Boldrini[a], Gert Jan Kramer[a] and Matteo Gazzani[a*]

[a]*Utrecht University, Princetonlaan 8a, 3584 CB Utrecht, The Netherlands*
[b]*ETH Zurich, Sonneggstrasse 3, 8092 Zurich, Switzerland*
m.gazzani@uu.nl

Abstract

Mixed integer linear programming (MILP) is the state-of-the-art mathematical framework for optimization of energy systems. The capability of solving rather large problems that include time and space discretization is particularly relevant for planning the transition to a system where non-dispatchable energy sources are key. Here, one of the main challenges is to realistically describe the technologies and the system boundaries: on the one hand the linear modeling, and on the other the number of variables that can be handled by the system call for a trade-off between level of details and computational time. With this work, we investigate how modeling wind turbines, H_2 generation via electrolysis, and storage in salt cavern affect the system description and findings. We do this by implementing methodological developments to an existing MILP tool, and by testing them in an exemplary case study, i.e. decarbonization of the Dutch energy system. It is found that modeling of wind turbines curtailment and of existing turbines are key. The deployment of H_2 generation and storage is driven by the interplay between area availability, system costs, and desired level of autarky.

Keywords: MILP, wind turbines, energy storage, technology modeling, energy transition

1. Introduction

A high penetration of non dispatchable renewable energy sources (NDRES) comes with the necessity for energy storage capacity. While batteries show a very high round-trip efficiency and suitability for intra-day storage, they are not suited for long-term storage, especially seasonal, due to their energy losses over time. (Gabrielli et al. (2018a)) The production of hydrogen via power-to-gas (PtG) is a more promising candidate for such long-term storage. Nevertheless, the large volumes required on a national or even international scale require alternatives to conventional gas tanks. Hydrogen storage in salt caverns is a proven technology (Lord et al. (2014)) that features large point storage capacity, especially in the Netherlands and in Germany. Previous studies focusing on Germany (Welder et al. (2018)) have already shown the potential of the described system. Understanding the trade-offs between offshore vs. onshore wind farms, and planning the replacement of old wind installations is a challenging task that benefits from the adaption of rigorous MILP frameworks. Therefore, this work aims at grasping the aforementioned trade-offs by analyzing a hypothetical Dutch energy system consisting of onshore and offshore wind turbines, and PtG systems with hydrogen storage in salt caverns. Furthermore, the extent of autarky achievable with such a system will be quantified.

2. Methodology

The model discussed in this work builds upon the MILP modeling framework reported by Gabrielli et al. (2018b). Following, we focus on new developments within this framework.

2.1. Wind turbine modeling

Various approaches to model wind turbines showing different levels of detail can be found in literature, ranging from detailed wind power plant modeling as reported by Gebraad et al. (2017) to simplified models for the use in MILP energy system optimization (Weber and Shah (2011)). In this work, we model the wind turbines using power curves as a function of cut-in, rated, and cut-out windspeed (v_{in}, v_r and v_{out}, respectively) as described by Jerez et al. (2015).

Implementation into the MILP framework: Using wind profiles for the full analyzed time horizon allows to tackle the non linearity arising from the power curve in a pre-processing step. The maximum power output P_{max} for a wind turbine is calculated for every hour of the year and passed on to the optimization as a constant vector. Note that P_{max}, being the uncurtailed output for a given windspeed, is different from P_r, the rated power correlating to v_r. The actual output P_{out} is then calculated as

$$P_{out,i,t} \leq P_{max,i,t} \cdot S_i$$
$$0 \leq S_i \text{ for all } i \in \{1,I\} \text{ and } t \in \{1,T\} \tag{1}$$

where the integer decision variable S is the number of turbines, I the number of types of turbines and T the length of the time horizon. The optimization can choose to build new turbines from a discrete set (offshore: 3.5 MW and 6 MW, onshore: 0.9 MW, 2.5 MW and 4.5 MW). The treatment of existing turbines is more complicated owing to the limited amount of information open databases provide.

Existing turbines: Detailed modeling of existing turbines is required to decide upon replacement and continuous operation. While it is easy to model a certain turbine in detail, the vast variety of existing turbines calls for a more generalized approach, which is described in this section. A data set consisting of 43 turbines, accounting for about 78 % of Dutch turbines, was used. The only clear correlation was found between the P_r and the total integral of the power curve. Based on this observation and the aforementioned set of turbines, an algorithm to calculate estimates of v_{in}, v_r and v_{out} for arbitrary turbines was developed. The algorithm takes P_r and the manufacturer of the wind turbine to be analyzed as an input as well as the 43 turbines data set. Knowing that the maximum hourly windspeeds are usually around 10-20 m s^{-1}, v_{out} usually ranging from 20-25 m s^{-1} is of minor importance and hence neglected here. If a match for the manufacturer is found, the set is reduced to that manufacturer. Otherwise, the full set is used. The resulting data set is compared for P_r. Depending on how many matches are found, 3 cases are distinguished.

- *Case 1: 1 match for P_r found:* The algorithm uses the match from the dataset to simulate the turbine of interest.

- *Case 2: >1 match for P$_r$ found:* A target power curve integral I_t is determined as

$$I_t = \frac{\sum_{i=1}^{N_m} \int_0^{v_{r,max}} P_i(v_{in}, v_r, v) dv}{N_m} \tag{2}$$

where $v_{r,max}$ is the maximum rated windspeed found in the data set and N_m is the total number of P_r-matches found. v_{in} and v_r are then calculated solving eq. (3)

$$\min_{v_{in}, v_r} g = \left| \int_0^{v_{r,max}} P(v_{in}, v_r, v) dv - I_t \right|$$
$$\text{s.t. } v_{in,min} \leq v_{in} \leq v_{in,max} \tag{3}$$
$$v_{r,min} \leq v_r \leq v_{r,max}$$

v_{out} is returned as the average of the values in the considered data subset.

- *Case 3: No turbines found* In this particular case, the turbines with the next higher and next lower P_r are chosen. If more than one turbine each is found, their integrals are averaged as seen in eq. (2) to end up with one upper and one lower value. I_t is then obtained by interpolation. If P_r of the turbine of interest is higher or lower than all turbines in the considered subset, the whole subset forms the basis for a linear fit that allows to calculate I_t by means of extrapolation. Once I_t is obtained, the remaining procedure is identical to Case 2, i.e. solving eq. (3) for v_{in} and v_r and averaging v_{out}.

Curtailment: While the inequality in eq. (1) already allows for curtailment, this formulation is imprecise from a physical point of view since wind turbines are curtailed in a discrete manner. They are either curtailed, i.e. turned off, or operated following their power curves. To account for this effect, the curtailment C was introduced as an additional integer decision variable, describing how many turbines are turned off. The power output can then be formulated as

$$P_{out,i,t} = P_{max,i,t} \cdot (S_i - C_{i,t})$$
$$0 \leq C_{i,t} \leq S_i$$
$$0 \leq S_i \tag{4}$$
$$r^2 \pi \cdot \sum_i S_i \leq A_{max}$$

Note that the strict equality in eq. (4), as compared to eq. (1), is reducing the flexibility of the system. This formulation allows for a physical-based description of curtailment, but it also increases significantly the complexity of the problem as discussed in the results. The total number of turbines is constraint by the maximum available land A_{max}. It is assumed that the distance between two wind turbines has to be at least 500 m, giving each turbine a radius r of 0.25 km.

2.2. *Power-to-gas and hydrogen storage*

The power-to-gas system consists of a polymer electrolyte membrane (PEM) electrolyzer and fuel cell, operating with water and air, respectively. Their modeling is reported in detail by Gabrielli et al. (2018b). Hydrogen is considered to be stored in cylindrical salt caverns as described in Gabrielli et al. (2018c).

3. Results

The problem was formulated in MATLAB R2018b using YALMIP (Löfberg (2004)) and solved with Gurobi v8.1 on an Intel Xeon E5-1620 3.60 GHz machine with 16 GB RAM.

Case study 1:   To quantify the effect of discrete compared to continuous curtailment, a hypothetical offshore-island - aimed at providing the whole Dutch electricity demand - was analyzed. The set of technologies encompasses a 3.5 MW wind turbine ($v_{in} = 3\,\text{m s}^{-1}$, $v_r = 14\,\text{m s}^{-1}$), a 6 MW wind turbine ($v_{in} = 4\,\text{m s}^{-1}$, $v_r = 13\,\text{m s}^{-1}$) and a PtG system. Due to the absence of salt caverns off-shore, traditional tank storage was assumed. In addition, the currently existing Dutch offshore turbines are considered but can be replaced through new turbines. The analysis was conducted with and without PtG system, both with discrete and continuous curtailment, ultimately resulting in 4 scenarios whose results are summarized in Table 1. As opposed to case study 2, no spatial constraints are assumed. As expected, dissipation of electricity occurs with discrete curtailment only and results from the fact that the costs for dissipation and import are equal at 10^5 euro per kW h. Hence, building an additional turbine and dissipating energy can be more favorable than importing. The application of PtG systems reduces the dissipation by an order of magnitude and the import by a quarter. Surprisingly, the PtG system is not utilized with continuous curtailment. Given the assumptions used, it is most likely that this is a result of the missing cost benefit of avoiding dissipation by using storage. Using discrete curtailment, the computational time is increased by about 80 % without PtG and by 1000 % with PtG. This indicates that the computational effects of discrete curtailment scale with the complexity of the system as a whole. This can easily lead to unsolvable systems due to a lack of memory, as experienced in case study 2. The preference of the 3.5 MW turbines over the 6.0 MW turbines can be explained by comparing their v_{in} with the wind profile, showing that the 3.5 MW turbine operates for 6366 hours per year while the 6.0 MW turbine only operates for 4674 hours per year. This significant downtime for both also explains the high amount of imported electricity despite the high amount of installed turbines. Furthermore, this also shows the need of large scale energy storage; the storage installed in scenario 4 is at its maximum of $4.9290 \cdot 10^7$ kW h.

Case study 2:   The system under investigation consists of 3 nodes, hereafter called onshore-node (ONN), offshore-node (OFFN), and cavern-node (CN). All data refers to the Netherlands and is hourly resolved for 2017. Wind turbines can be installed in ONN and OFF, and PtG systems with storage in CN and without storage in ONN. The country's electricity demand is an input profile for ONN and wind profiles are considered as input for ONN and OFFN. Full connectivity between the nodes is assumed for the electricity network while the hydrogen network needs to be built if required, i.e. investment costs occur. The optimization decides upon selection, sizing, and scheduling of the technologies as well as the flows between the nodes with the levelized costs of the total system as objective. While there is no export of electricity allowed, electricity can be imported at a

Table 1: Effect of discrete curtailment on technology selection. Scenarios: (1) no PtG, continuous curtailment (2) PtG, continuous curtailment (3) no PtG, discrete curtailment (4) PtG, discrete curtailment. For simplicity, only the storage of the PtG system is shown. Import, Dissipation and Storage are given in kWh.

Scenario	WTex	WT-3.5	WT-6.0	Import	Dissipation	Storage
1	79	13356752	0	1.08 E10	0	-
2	79	13356753	0	1.08 E10	0	0
3	79	13356753	0	1.08 E10	1.07 E06	-
4	79	13356752	0	8.27 E09	1.04 E05	4.93 E07

Table 2: Sensitivity of technology implementation towards level of autarky. The sizes for electrolyzer (PEMEC) and fuel cell (PEMFC) are in kW capacity, and for the wind-turbines (WT) in built units. The number next to *WT* refers to its maximum capacity in MW

Autarky	PEMEC	PEMFC	WT-0.9	WT-2.5	WT-4.5	WT-3.5	WT-6.0
0.625	42586	423.8	0	26944	4792	1077	24
0.475	0	0	0	15223	0	0	0
0.325	0	0	0	7520	0	0	0
0.175	0	0	0	3687	0	0	0
0.025	0	0	0	531	0	0	0
0	0	0	0	0	0	0	0

price of 0.032 euro per kWh and an emission factor of 0.676 kg_{CO_2} per kWh. A CO_2-tax of 20 euro per tonne was applied, corresponding to the current ETS prices. The available land for ONN was assumed to be 15 % of the total land (WorldBank (2018),McKenna et al. (2015)). For OFFN, a total available area of 2900 km^2 (Dutch Government (2014)) was assumed. Although being spatially reduced to only 3 nodes, the study is based on realistic assumptions and constraints in terms of electricity demand, wind speed, available area and existing turbines. Putting the focus on achievable autarky using PtG, and considering the issues found in case study 1, continuous curtailment is used and existing turbines are neglected. Since imported electricity is the only source of emissions, the maximum achievable autarky was determined by applying a CO_2-tax of 10^6 euro per tonne giving an autarky of 62.5 % (defined as the fraction of the produced electricity over the total demand). Following, the autarky was implemented as a constraint and the sensitivity of the technology selection towards this constraint was investigated using the real CO_2-tax (see Table 2). It can be observed that in order to achieve the upper limit of autarky, oversizing of production technologies and installation of PtG is necessary. As soon as the constraint on autarky is relaxed, undersizing of production technologies and compensation with import is the preferred combination. Note that this result is also sensitive to the import price, the CO_2-tax, and the emission factor of imported electricity. The difference in size between PEMEC and PEMFC indicates that the amplitudes in overproduction are greater than in overdemand. Nevertheless, it does not necessarily follow that the total production for a year is higher than the demand.

4. Conclusions

In this work, the implementation of wind turbines with its various aspects into an MILP framework, and in particular the consideration of existing turbines, was discussed. A first analysis of the Dutch case study shows how the level of targeted autarky affects the selection of technologies. For the considered assumptions and constraints in this study, an upper limit to autarky is given by the available land, and PtG is required to approach this limit. From a computational point of view, the description of the wind turbine curtailment strongly affects the problem complexity and can cause an up to 10-fold increase of computation time for discrete curtailment. The exact impact depends on the overall system complexity. This becomes especially crucial when existing turbines are considered, because each type of turbine is treated as separate technology. Hence, existing turbines intrinsically lead to a vast increase in system complexity.

5. Acknowledgement

ACT ELEGANCY, Project No 271498, has received funding from DETEC (CH), BMWi (DE), RVO (NL), Gassnova (NO), BEIS (UK), Gassco, Equinor and Total, and is co-funded by the European Commission under the Horizon 2020 programme, ACT Grant Agreement No 691712.

References

P. Gabrielli, M. Gazzani, E. Martelli, M. Mazzotti, 2018a. Optimal design of multi-energy systems with seasonal storage. Applied Energy 219, 408–424.

P. Gabrielli, M. Gazzani, M. Mazzotti, 2018b. Electrochemical conversion technologies for optimal design of decentralized multi-energy systems: Modeling framework and technology assessment. Applied Energy 221, 557–575.

P. Gabrielli, A. Poluzzi, C. Spiers, M. Mazzotti, M. Gazzani, 2018c. Seasonal energy storage through hydrogen salt caverns: Technology modeling for system optimization and assessment of emissions reduction potential. in preparation.

P. Gebraad, J. Thomas, A. Ning, P. Fleming, K. Dykes, 2017. Maximization of tha annual energy production of wind power plants by optimization of layout and yaw-based wake control. Wind Energy 20, 97–107.

S. Jerez, F. Thairs, I. Tobin, M. Wild, A. Colette, P. Yiou, R. Vautard, 2015. The CLIMIX model: A tool to create and evaluate spatially-resolved scenarios of photovoltaic and wind power development. Renewable and Sustainable Energy Reviews 42, 1–15.

J. Löfberg, 2004. Yalmip : A toolbox for modeling and optimization in matlab. In: Proceedings of the CACSD Conference. Taipei, Taiwan.

A. Lord, P. Kobos, D. Borns, 2014. Geologic storage of hydrogen: Scaling up to meet city transportation demands. International Journal of Hydrogen Energy 39, 15570–15582.

R. McKenna, S. Hollnaicher, P. O. v. d. Leye, W. Fichtner, 2015. Cost-potentials for large onshore wind turbines in europe. Energy 83, 217–229.

Dutch Government, 2014. White paper on offshore wind energy. https://www.government.nl/documents/reports/2015/01/26/white-paper-on-offshore-wind-energy, (accessed 14 November 2018).

C. Weber, N. Shah, 2011. Optimisation based design of a district energy system for an eco-town in the United Kingdom. Energy 36, 1292–1308.

L. Welder, D. Yberg, L. Kotzur, T. Grube, M. Robinius, d. Stolten, 2018. Spatio-temporal optimization of a future energy system for power-to-hydrogen applications in Germany. Energy 158, 1130–1149.

WorldBank, 2018. Arable land. https://data.worldbank.org/indicator/AG.LND.ARBL.ZS?view=chart, (accessed 14 November 2018).

Anton A. Kiss, Edwin Zondervan, Richard Lakerveld, Leyla Özkan (Eds.)
Proceedings of the 29th European Symposium on Computer Aided Process Engineering
June 16th to 19th, 2019, Eindhoven, The Netherlands. © 2019 Elsevier B.V. All rights reserved.
http://dx.doi.org/10.1016/B978-0-128-18634-3.50273-3

An MPEC model for Strategic Offers in a Jointly Cleared Energy and Reserve Market under Stochastic Production

Evangelos G. Tsimopoulos, Michael C. Georgiadis[*]

Aristotle University of Thessaloniki, Department of Chemical Engineering, 54124 Thessaloniki, Greece

mgeorg@auth.gr

Abstract

This work, based on Stackelberg hypothesis, considers a conventional power producer exercising their dominant position in an electricity pool with high penetration of wind power production. A bi-level optimization model is used to provide optimal offer strategies for the aforementioned producer in a jointly cleared energy and reserve pool settled through an hourly auction process. The upper-level problem illustrates the expected profit optimization of the strategic producer while the lower-level problem represents the energy-only market clearing process through a two-stage stochastic program. The first stage clears the day ahead market, and the second stage presents the system operation in balancing time though a set of plausible wind power production realizations. The bi-level problem is recast into mathematical programming with equilibrium constraints (MPEC) which is then reformulated into an MILP. These transformations occur using the Karush-Kuhn-Tucker optimality conditions and the strong duality theory. The suggested model provides optimal strategic offers and local marginal prices under different levels of wind penetration and network line transmission capacities.

Keywords: MPEC, energy-only markets, LMPs, strategic offers, wind uncertainty

1. Introduction

In recent years the electricity generation industry has experienced a remarkable penetration of renewable energy resources. However, the inherently uncontrollable fluctuations of renewable generation have resulted in the change of operational framework, the development of new tools to handle the stochastic nature of non-dispatchable production, and the redesign of market clearing algorithms (Conejo et al. 2011). Concerning the strong penetration of renewable sources supported by a generous mechanism of subsidized production and priority dispatch, the role of conventional energy production is diminishing. Although the market recognises the critical role of the thermal plants as capacity providers, the latter are faced with unequal treatment and have to adopt specific strategic behaviour to ensure competitiveness. Within the above context, and considering the conventional energy production, this work investigates the reaction of an incumbent firm and examines its incentives to exert market power and ensure its dominant position to avoid energy profit losses.

2. Problem Statement

This work analyses the optimal offering strategies of a conventional (thermal) power producer which participates with other conventional as well as wind power producers in a jointly cleared energy and balancing auction under network constraints. A bi-level programming model is developed based on Stackelberg leader-follower hypothesis. The upper-level problem determines expected profit maximization of the considered strategic producer which depend on clearing local marginal prices (LMPs) of day ahead (DA) and real time (RT) $\lambda_{n\omega}^{RT}$ market obtained at the lower level problem. On the other hand, the lower-level problem represents the clearing price process ensuing the least cost of energy dispatch conducted by the system operator (SO). Thus, the lower-level problem is formulated in a linearized DC network as two-stage stochastic programming. The first stage facilitates the DA market and results in the optimal anticipated dispatch (DA scheduled energy production), and the LMPs received as dual variables. The second stage represents the balancing market under the realization of all the plausible wind production scenarios and derives real time dispatch (reserve deployments) and RT prices (Morales et al. 2012).For the following formulation the indices i, o and j indicate the strategic conventional units, the nonstrategic conventional units and the wind farms respectively while the index d indicates the demands of the system. Additionally, the indices b and f refer to power blocks offered by the conventional units i, o and the wind farms j, the index k refers to power blocks consumed by load d and the index ω refers to wind power production scenarios. The sets Ψ_n^i, Ψ_n^o, Ψ_n^j and Ψ_n^d map the generation units and loads on to the system and the set Θ_n defines the connection of the bus n with the other buses of the network. The parameters c_{ib}, c_{ob}, c_{jf}^{DA} and u_{dk} represent the marginal cost of offered power blocks of generation units and the marginal utility of load blocks. In addition, the parameters c_i^{up}, c_o^{up}, c_i^{down}, c_o^{down} and $c_{j\omega}^{RT}$ indicate the cost of offered regulations and wind power realization at balancing stage. In addition, the parameter B_{nm} denotes the susceptance of the line n-m. The *here and now* decision variables P_{ib}^{DA}, P_{ob}^{DA} and W_{jf}^{DA} represent the scheduled production of conventional units and wind farms respectively and the δ_n^o represents the voltage angle at DA stage. The *wait* and *see* variables $r_{i\omega}^{up}$, $r_{o\omega}^{up}$, $r_{i\omega}^{down}$ and $r_{o\omega}^{down}$ refer to upward and downward reserves offered by units i and o, the $W_{j\omega}^{sp}$ indicates the wind power production spillage of wind farm j under scenario ω, the $L_{d\omega}^{sh}$ indicates the load shedding of demand d under scenario ω and the $\delta_{n\omega}$ indicates the voltage angle at RT stage. Finally, O_{ib}^{DA} defines the price offer of generation block b of strategic unit i in DA market while O_i^{up} and O_i^{down} define the price offerσ of upward and downward reserveσ of strategic unit i in RT market respectively.

2.1 Bi-level Formulation

The bi-level stochastic optimization model is formulated to derive its optimal offers as follows:

Upper level problem

$$
\text{maximize} \quad \sum_{(i\in\Psi_n^i)b} \lambda_n^{DA} P_{ib}^{DA} - \sum_{ib} c_{ib} P_{ib}^{DA} + \sum_{(i\in\Psi_n^i)\omega} \lambda_{n\omega}^{RT}
$$

$$
- \sum_{i\omega} \pi_\omega c_i^{up} r_{i\omega}^{up} - \sum_{(i\in\Psi_n^i)\omega} \lambda_{n\omega}^{RT} r_{i\omega}^{down} + \sum_{i\omega} \pi_\omega c_i^{down} r_{i\omega}^{down} \tag{1}
$$

Lower level problem

minimize

$$\sum_{ib} O_{ib}^{DA} P_{ib}^{DA} + \sum_{i\omega} \pi_\omega O_i^{up} r_{i\omega}^{up} - \sum_{i\omega} \pi_\omega O_i^{down} r_{i\omega}^{down}$$

$$+ \sum_{ob} c_{ob} P_{ob}^{DA} + \sum_{o\omega} \pi_\omega c_o^{up} r_{o\omega}^{up} - \sum_{o\omega} \pi_\omega c_o^{down} r_{o\omega}^{down}$$

$$+ \sum_{jf} c_{jf}^{DA} W_{jf}^{DA} + \sum_{j\omega} \pi_\omega c_{j\omega}^{RT} \left(W_{j\omega}^{RT} - \sum_f W_{jf}^{DA} - W_{j\omega}^{sp} \right)$$

$$- \sum_{dk} u_{dk} L_{dk}^{DA} + \sum_{d\omega} \pi_\omega vLOL_d L_{d\omega}^{sh} \tag{2}$$

s.t.

$$- \sum_{(i\in\Psi_n^i)b} P_{ib}^{DA} - \sum_{(o\in\Psi_n^o)b} P_{ob}^{DA} - \sum_{(j\in\Psi_n^j)f} W_{jf}^{DA}$$

$$+ \sum_{(d\in\Psi_n^d)k} L_{dk}^{DA} + \sum_{m\in\Theta_n} B_{nm}(\delta_n^o - \delta_m^o) = 0 \quad : (\lambda_n^{DA}) \quad \forall n \tag{3}$$

$$- \sum_{i\in\Psi_n^i} r_{i\omega}^{up} + \sum_{i\in\Psi_n^i} r_{i\omega}^{down} - \sum_{o\in\Psi_n^o} r_{o\omega}^{up} + \sum_{o\in\Psi_n^o} r_{o\omega}^{down} - \sum_{d\in\Psi_n^d} L_{d\omega}^{sh}$$

$$- \sum_{j\in\Psi_n^j} W_{j\omega}^{RT} + \sum_{(j\in\Psi_n^j)f} W_{jf}^{DA} + \sum_{j\in\Psi_n^j} W_{j\omega}^{sp}$$

$$+ \sum_{m\in\Theta_n} B_{nm}(\delta_{n\omega} - \delta_n^o + \delta_m^o - \delta_{m\omega}) = 0 \quad : (\lambda_{n\omega}^{RT}) \quad \forall n, \forall \omega \tag{4}$$

$$\sum_b P_{ib}^{DA} + r_{i\omega}^{up} \le \sum_b P_{ib}^{MAX} \quad : (\mu_{i\omega}^{max}) \quad \forall i, \forall \omega \tag{5}$$

$$\sum_b P_{ob}^{DA} + r_{i\omega}^{up} \le \sum_b P_{ob}^{MAX} \quad : (\mu_{o\omega}^{max}) \quad \forall o, \forall \omega \tag{6}$$

$$r_{i\omega}^{down} - \sum_b P_{ib}^{DA} \le 0 \quad : (\mu_{i\omega}^{min}) \quad \forall i, \forall \omega \tag{7}$$

$$r_{o\omega}^{down} - \sum_b P_{ob}^{DA} \le 0 \quad : (\mu_{o\omega}^{min}) \quad \forall o, \forall \omega \tag{8}$$

$$0 \le W_{j\omega}^{sp} \le W_{j\omega}^{RT} \quad : (\kappa_{j\omega}^{min}, \kappa_{j\omega}^{max}) \quad \forall j, \forall \omega \tag{9}$$

$$0 \le L_{d\omega}^{sh} \le \sum_k L_{dk}^{DA} \quad : (v_{d\omega}^{min}, v_{d\omega}^{max}) \quad \forall d, \forall \omega \tag{10}$$

The objective function (1) optimizes the expected profit of the strategic producer, and it is defined by the revenues from the day DA and RT market minus the actual incurred cost. The objective function (2) optimizes the expected cost of the power system operation conducted by SO. It consists of the scheduled production cost and the scenario dependent reserve deployment, spilling wind power and shedding load cost in real time operation. Constraint (3) enforces the energy balance at each node and the transmission capacity limits between them at DA. Thus, the total power flowing into bus *n*, which is the algebraic sum of generation and load at the bus, should be equal to the power flowing away from the bus. Constraint (4) counterbalances the imbalance occurred in RT due to the uncertain wind production arranging the reserve deployment and the load curtailments. Constraints (5), (6), (7) and (8) ensure that unit generation is above zero and below maximum capacity P_{ib}^{MAX} and P_{ob}^{MAX} of units i and o capturing the coupling between anticipated dispatch and deployed reserve. Constraints (9) and (10) indicate that the wind energy spillage cannot exceed the scenario dependent actual wind energy production $W_{j\omega}^{RT}$ and the involuntary load curtailment cannot exceed the actual load consumption. Moreover, other constraints enforce transmission capacity limits between two buses, upper and lower bounds of voltage angle at each bus. Finally, the bus $n1$ is defined as slack bus thus δ_{n1}^{o} and $\delta_{(n1)\omega}^{o}$ equal to zero.

2.2 MPEC formulation and linearization

Considering the continuity and the convexity of the lower problem, the bi-level problem is reduced to an MPEC through first-order KKT optimality conditions. In this case, the Lagrange multipliers have the same meaning with the dual variables in linear programming. Using disjunctive constraints to linearize the KKT complementarity constraints (Fortuny-Amat and McCarl, 1981) as well as the strong duality theorem in combination with some of the KKT equality constraints to eliminate the non-linear terms within objective function (1) (Ruiz and Conejo, 2009), the MPEC is reformed in an MILP solvable by commercial solvers such as GAMS/CPLEX.

3. Illustrative example

The proposed clearing market formulation is applied in a six-node system sketched in Fig. 1. The conventional generating units i1, i2, i3 and i4 belong to the strategic producer and the o1, o2, o3 and o4 belong to non-strategic producers. The technical data of the units is taken from Ruiz and Conejo (2009). Two wind farms j1 and

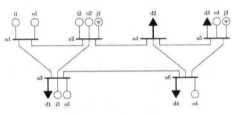

Figure 1. six-bus system

j2, located at bus n2 and n5, have installed capacity of 100 MW and 70 MW, and their scheduled power production is offered in one block with zero marginal cost. Wind farms' uncertain power production is realized through three scenarios, ω1 (high production) with 100 MWh and 70 MWh, ω2 (medium production) with 50 MWh and 35 MWh, and ω3 (low production) with 20 MWh and 15 MWh while occurrence probability of each scenario is 0.2, 0.5 and 0.3 respectively. A total demand of 1 GWh is allocated according to Figure 1. Load d1 accounts for 19 % and loads d2, d3 and d4 account for 27 % of the total demand each. Additionally, data about demand bids (energy and utility marginal cost) for each period of time comes from Ruiz and Conejo (2009). Finally, the value of the involuntary load reduction is 200 euro/MWh for all demands and all the connecting lines have a transmission capacity of 500 MW with susceptance equal to 9412 per unit.

4. Results

4.1. Uncongested network solution

Based on the above information the proposed MILP model is applied to the system and solved using GAMS/CPLEX. When the strategic producer offers at marginal cost, the DA clearing price is constant throughout the 24-period time at a level of 11.260 /MWh. However, when the strategic producer exerts their market power the DA

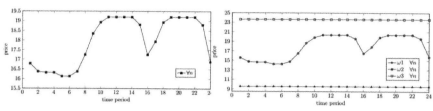

Figure 2. DA clearing prices [€/MWh]   Figure 3. RT clearing prices [€/MWh]

clearing price is raised, while fluctuating between 16.130 and 19.200 €/MWh as shown in Figure 2. Similarly, the RT clearing prices are raised too. More specifically, in high wind scenario

ω1 realization the RT price increases from 9.280 to 9.570 €/MWh, in medium wind scenario ω1 the RT clearing price leaves the level of 11.470 €/MWh and moves in a range between 14.254 and 20.394 €/MWh, and in low wind scenario ω3 the price rockets from 12.230 to 23.630 €/MWh as presented in Figure 3. In both cases, the prices are the same in all buses at each time period as line capacities are enough to facilitate the energy transaction. The proposed model results in an increase in the total expected profit of the strategic producer from 9,326 € to 67,556 € even though the producer's scheduled power production as well as its deployed reserve decreases in all wind scenarios.

4.2. Congested network solution

In the previous case the maximum power flow through line 3-6 is 227 MW. If the line capacity is reduced at the level of 240 MW, slightly above the maximum flow, the results

remain the same when the strategic producer acts as price taker. However, the proposed MILP formulation shows that the strategic producer can make offers in such a way that the system becomes congested resulting in different LMPs of DA and RT clearing prices at specific time periods as illustrated in Figures 4 and 5 respectively. It can be seen that bus n6 exhibits the highest price giving the strategic producer the opportunity

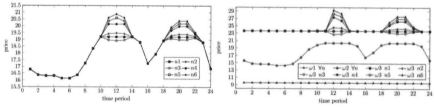

Figure 2. DA clearing prices [/MWh]   Figure 3. RT clearing prices [/MWh]

to increase the profit of unit i4. In that way the total expected profit of the strategic producer is slightly higher compared to that of uncongested network. Additionally, in the uncongested case the maximum power flow through line 4-6 is 24 MW. If the capacity of the line is reduced to 20 MW, slightly below the maximum flow, the network becomes congested under cost offer optimization, resulting in different LMPs and profit losses for the strategic producer. However, applying the proposed MILP formulation, the strategic producer changes the mixture of units' production keeping the line uncongested and the profit high.

4.3. Wind power production increment

In this case, the level of wind power penetration increases from 10 % to 14.16 % of the total installed capacity. More specifically, the power production of the wind farms j1 and j2 is 150 MW and 100 MW respectively in high wind scenario ω1, 75 MW and 50 MW in medium wind scenario ω2, and 30 MW and 20 MW in low wind scenario ω3. It can be seen in Fig.6 that the

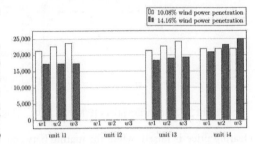

Figure 4. Profit of strategic units under different level of wind power penetration

expected profits of unitς i1 and i2 are reduced in all wind scenarios. Considering unit i4 the expected profits decrease in high wind scenario ω1; however, the expected profits increase in medium and low wind scenarios as the unit becomes more involved in reserve supply. Nevertheless, even if the total expected profits of unit i4 rise, the total expected profits of strategic producer decrease from 67,556 euro to 59,589 euro, indicating that wind power production can be used as a tool for market power mitigation.

5. Conclusions

In this work based on Stackelberg hypothesis, an MILP is developed to derive optimal offer strategies for a conventional power producer participating in a jointly cleared energy and reserve market under high penetration of wind power production. The model concerns energy-only markets.   Co-optimizing scheduled energy   and reserve deployment through two-stage stochastic programming, it gives insight information on LMPs and the way they are configured when the strategic producer exercises its dominant position in the market. Furthermore, the model provides information about how line capacities and network congestions can be used for the benefit of the strategic producer and indicates that wind power production can be used as an instrument of market power mitigation. Future work should involve a larger number of wind production scenarios and scenario reduction techniques. The model should also incorporate intertemporal constraints such as units' ramp-up and ramp-down limitations as well as non-convexities, e.g., unit minimum power production and unit start-up and shut-down

References

A. J. Conejo, J. M. Morales and J. A. Martinez, (2011). Tools for the analysis and design of distributed resources Part III: Market studies. IEEE Transactions on Power Delivery, 26(3), 1663-1670.

J. Fortuny-Amat and B. McCarl, (1981). A representation and economic interpretation of a two-level programming problem. Journal of the operational Research Society, 32(9), 783-792.

J. M. Morales, A. J. Conejo, K. Liu and J. Zhong, (2012), Pricing electricity in pools with wind producers. IEEE Transactions on Power Systems, 27(3), 1366-1376.

C. Ruiz and A. J. Conejo, (2009), Pool strategy of a producer with endogenous formation of locational marginal prices, IEEE Transactions on Power Systems, 24(4), 1855-1866.

Anton A. Kiss, Edwin Zondervan, Richard Lakerveld, Leyla Özkan (Eds.)
Proceedings of the 29th European Symposium on Computer Aided Process Engineering
June 16th to 19th, 2019, Eindhoven, The Netherlands.
http://dx.doi.org/10.1016/B978-0-128-18634-3.50274-5

Model-Based Bidding Strategies for Simultaneous Optimal Participation in Different Balancing Markets

Pascal Schäfer[a], Nils Hansmann[a], Svetlina Ilieva[b] and Alexander Mitsos[a,*]

[a]*Aachener Verfahrenstechnik – Process Systems Engineering, RWTH Aachen University, Aachen, Germany*
[b]*TRIMET Aluminium SE, Essen, Germany*
amitsos@alum.mit.edu

Abstract

Due to the increasing share of intermittent renewable energy sources, energy-intensive enterprises are nowadays motivated to actively participate in electricity markets. They can do so by marketing their operational flexibility either at spot markets or by offering ancillary services on demand, such as balancing power. We propose an optimization-based methodology to select which capacity to offer at which market. We extend our framework for optimizing the bidding strategy at the primary balancing market (Schäfer et al. (2019)) to a simultaneous participation in both the primary and the secondary balancing market. We solve the resulting optimization problem using a hybrid stochastic-deterministic approach. The framework applies a genetic algorithm with a subordinated mixed-integer linear solver that allows for marketing the remaining flexibility at spot markets through solving a scheduling problem. For an industrial aluminum electrolysis, we demonstrate the efficacy of the approach to identify profitable bidding strategies. We show for an exemplary week with price data from 2018 that by optimal participation in balancing markets, savings of more than 10 % in weekly costs are enabled compared to a non-flexible operation.

Keywords: Demand side management, Aluminum electrolysis, Ancillary service markets, Mixed-integer nonlinear programming, Model-based bidding strategies

1. Introduction

The increasing share of electricity generation from intermittent renewable sources and the involved increasing volatility at electricity markets represent a severe challenge for the energy-intense industry (Mitsos et al. (2018)). However, at the same time, flexible consumers can also benefit from these circumstances by adapting their consumption to fluctuating market signals (e.g., Ghobeity and Mitsos (2010); Mitra et al. (2012)). In addition to that, consumers also have the opportunity to offer parts of their flexibility range at ancillary service markets, which opens further promising revenue potentials (Paulus and Borggrefe (2011); Dowling et al. (2017)). It is thus crucial to have systematic approaches for selecting which capacity to offer at which market (e.g., Zhang et al. (2015); Otashu and Baldea (2018)), as usually a reserve capacity is not allowed to be marketed at the same time at multiple markets. Furthermore, in some important market environments like the European balancing market, the prices that market participators ask for the provision of the ancillary services are of high relevance as they not only determine the probability of acceptance of an offer but also define the potential revenues. In order to balance these trade-offs, we recently proposed to formulate the optimal bids at the market for primary balancing power (PRL) as a two-stage stochastic optimization problem where remaining capacities after acceptance or rejection of the individual bids can be used for optimization at spot electricity markets (Schäfer et al. (2019)).

In this work, we extend the formulation to cope with more than one balancing market simultaneously. More precisely, we also consider the market for secondary balancing power (SRL) that opens further revenue potentials, but also poses different rules for bidding than the PRL market. In order to cope with the increased complexity of the resulting optimization problem, we use a hybrid stochastic-deterministic solution approach. We apply the approach to an industrial aluminum electrolysis to calculate the optimal bidding strategy for one exemplary week from 2018. Finally, we discuss revenue potentials that are enabled via the optimal strategy in comparison to a pure marketing at the spot market and a single bidding at the PRL market only.

2. Primary and secondary balancing market

Both the German PRL and SRL market are auction markets where all participators are allowed to submit their individual bids. In case of PRL, one bid consists of a capacity [MW] reserved for balancing power for the entire horizon (one week) and a capacity price [EUR/MW] that the bidder is willing to accept as compensation. Once the auction is closed, those bids with the lowest capacity prices are accepted, until the total demand of PRL is satisfied. Note that if the bid is accepted, the capacity price is paid irrespectively of whether the held capacity is actually retrieved. The smallest capacity that can be offered at the PRL market is 1 MW. The increment is 1 MW. At the SRL market, there are three important differences: (i) the horizon is subdivided into periods that take into account the varying demand for balancing power in the course of the horizon, (ii) capacities for a load reduction and a load increase can be offered separately, and (iii) participators need to further submit an asked energy price [EUR/MWh] that is only paid for the actual retrieval of the balancing power. These energy prices determine how often the capacities are retrieved, i.e., capacities with low energy prices are retrieved more often. The smallest capacity that can be offered at the SRL market is 5 MW. The increment is again 1 MW.

3. Model-based bidding

3.1. Definitions and preliminary assumptions

For marketing the remaining flexibility, we herein only consider the day-ahead market. We assume that day-ahead prices for the entire week are known when the bidding takes place. Price data is retrieved from EPEX SPOT (www.epexspot.com). All data concerning balancing power is retrieved from the central auction platform REGELLEISTUNG.NET (www.regelleistung.net).

For each balancing power product j, we consider only one submitted bid. Placing more than one bid has been shown in our previous work to have only minor influences on potential revenues (Schäfer et al. (2019)). A bidding strategy thus comprises 14 variables: a capacity P_j and a capacity price CP_j for PRL as well as for each of the four considered SRL products (two load directions *POS* and *NEG* with each two periods denoted by *HT* and *NT*) plus an energy price EP_j for each of the SRL products. The decision variables are summarized in the vector $\mathbf{x}$. We assume that the individual bids are accepted once the submitted asked capacity prices are lower than the upcoming maximum accepted ask prices for the respective products. For methods to estimate the marginal price, we refer to the relevant literature (e.g., Box and Jenkins (1994)). To simplify the presentation and discussion, we herein assume that the maximum accepted ask price of product j is normally distributed around its real value μ_j with a standard deviation of $\sigma_j = 0.05\mu_j$. Thus, the probability of acceptance p_j for the offered capacity for product j reads

$$p_j = \frac{1}{2}\left(1 - \mathrm{erf}\left(\frac{CP_j - \mu_j}{\sqrt{2}\sigma_j}\right)\right), \quad j \in \{PRL, POSHT, NEGHT, POSNT, NEGNT\}. \quad (1)$$

Other probability distributions can be used as well, as long as the respective cumulative distribution functions can be computed. Note that when using the hybrid stochastic-deterministic solution approach presented, no further assumptions on the probability distribution are required, e.g., concerning availability of gradients or relaxations.

The compensation payments, i.e., the revenues from a balancing power product R_j, can be calculated using the following equations. In case of PRL, capacity and capacity price are just multiplied (Eq. (2)). In case of the SRL products, a second term is added for the compensation of the actual retrieved energy E_j over the entire horizon (Eq. (3)). We express the relation between E_j and OP_j via an exponential function. Parameters of the functions are fitted to historic retrieval data.

$$R_j = CP_j \cdot P_j, \quad j \in \{PRL\} \tag{2}$$

$$R_j = CP_j \cdot P_j + OP_j \cdot E_j, \quad j \in \{POSHT, NEGHT, POSNT, NEGNT\} \tag{3}$$

$$E_j = P_j \cdot \exp(a_j + b_j \cdot OP_j), \quad j \in \{POSHT, NEGHT, POSNT, NEGNT\} \tag{4}$$

3.2. Formulation of the optimization problem

We minimize the expected weekly costs ϕ, wherein the compensation payments from the balancing markets are considered as negative costs. ϕ itself can be written as a probability-weighted sum over mutually exclusive scenarios (Eq. (5)). The scenarios arise from the combination possibilities of acceptance and rejection of the individual bids. Considering one bid for each of the five balancing power products, there are $N = 2^5 = 32$ possible scenarios. Scenario i can be characterized by the set of accepted bids j_i^*. The probability for one scenario p_i as well as the revenues R_i can be calculated by summation or multiplication of the individual p_j's (Eq. (6)) and R_j's (Eq. (7)) respectively. Furthermore, in each of the scenarios, the remaining flexibility range, i.e., the part of the original flexibility range which has not been reserved for provision of balancing power, can be marketed at the day-ahead market in order to lower the weekly costs without the revenues (C_i's). This is done by solving an integrated scheduling model which is denoted as the function $\mathrm{C}^*(\cdot)$ in the following.

$$\min_{\mathbf{x}} \ \phi = \sum_{i=1}^{N} p_i (C_i - R_i) \tag{5}$$

$$\text{s.t.} \ p_i = \left(\prod_{j \in j_i^*} p_j \right) \cdot \left(\prod_{j \notin j_i^*} (1 - p_j) \right), \quad i = 1 \ldots N \tag{6}$$

$$R_i = \sum_{j \in j_i^*} R_j, \quad i = 1 \ldots N \tag{7}$$

$$\text{Eq. (1)-(4)} \tag{8}$$

$$C_i = \mathrm{C}^* \left(P_{j \in j_i^*}, E_{j \in j_i^*} \right), \quad i = 1 \ldots N \tag{9}$$

$$P_j \in \{1, \ldots, P_{j,max}\} \tag{10}$$

For setting up the function $\mathrm{C}^*(\cdot)$ for the process considered in the case study - an industrial aluminum process - we refer to Schäfer et al. (2019). Therein, we describe a mixed-inter linear programming (MILP) scheduling formulation for optimizing the electricity consumption of the process considering operating limits, off-design efficiency losses and opportunity costs due to the provision of balancing power. In this work, we only slightly adjust the model by accounting for surpluses or deficits in the energy balance of the cell due to the retrieval of SRL. The maximum capacity that can be offered in product j are set to 10 MW for PRL and 22 MW for SRL.

3.3. Hybrid stochastic-deterministic solution algorithm

As discussed in Schäfer et al. (2019), the mixed-inter nonlinear optimization program (MINLP) given in Eq. (5)-(9) shows a similarity with two-stage stochastic programs. That is, selecting a bidding strategy **x** represents the first stage, the subordinated MILP scheduling problems (Eq. (9)) represent the second stage. A direct solution of (5)-(9) using state-of-the-art deterministic MINLP solvers is considered prohibitive. One possible approach to a deterministic solution would be the use of tailored decomposition algorithms exploiting the structure.

As an alternative, the use of a hybrid stochastic-deterministic method appears promising here. More precisely, the stochastic solver that varies the first-stage variables **x** invokes an external MILP solver in a black-box manner. The single MILP problems for each scenario can be solved efficiently within short time using state-of-the-art MILP solvers. Furthermore, parallelization of the optimizations for single scenarios is easily possible. We herein use a genetic algorithm (GA) as stochastic solver. The framework (see, Figure 1) uses the GA supplied in MATLAB R2018a's Global Optimization Toolbox (Mathworks, Inc). The subordinated scheduling problems are implemented in GAMS v.25.1 (GAMS Development Corp.) and are solved using CPLEX v.12.8 (IBM Corp.).

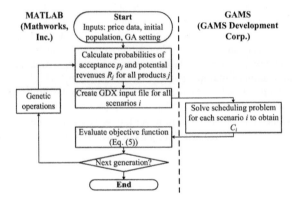

Figure 1: Flow diagram of the hybrid stochastic-deterministic solution algorithm.

4. Case study

As a case study, we calculate the optimal bidding strategy for the second calender week of 2018. We conduct all calculations on a server with two Intel® Xeon® CPUs X5690 with 3.47 GHz and 96 GB RAM. For solution of the subordinated MILPs in parallel, the GAMS Grid Facility is used. The GA is stopped after evaluation of 300 generations with a population size of 70. The respective convergence plots of the GA are given in Figure 2 (a). Six cases are considered: no balancing market participation, a pure bidding at the PRL market, and a simultaneous bidding at the PRL and the SRL market, each of them once with the possibility to optimally market the remaining flexibility at the spot market (day-ahead) and once without this possibility. The longest run (bidding at PRL and SRL market with optimization at the spot market) took approximately 72 h, all others finished in substantially less time. Note however that all runs converged to the final value within a relative tolerance of 10^{-3} in less than 100 generations. This runtime is not prohibitive considering a horizon of the bids of one week. Calculations have been repeated with

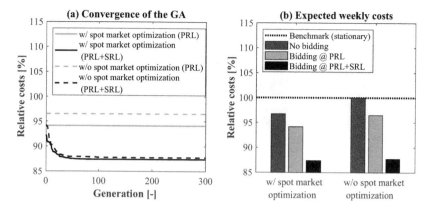

Figure 2: (a) Convergence plot of the GA. Lines indicate strongest individual of each generation. Solid lines: with optimal scheduling at spot market, dashed lines: without. Colors like in (b). (b) Economic evaluation of the optimal bidding strategy once with optimization at spot market and once without. The dotted black line is the benchmark case with constant production and spot prices. Blue bars: no balancing market participation, orange bars: bidding only at PRL market, purple bars: bidding at both PRL and SRL market.

different GA settings, however no significant differences neither in optimal variable values nor in the objective values have been found.

Figure 2 (b) further compares the expected costs following the optimal bidding strategy on both the SRL and the PRL market to a pure optimization at the spot market. We consider a stationary operation under day-ahead prices as a benchmark scenario. One sees strong economic benefits in actively marketing the operational flexibility in the considered week, which allows for overall savings of > 10 % compared to a non-flexible operation. Figure 3 gives additional information about the selected bidding strategies. Note that in both cases (with and without optimization at the spot market) the most probable scenario by far is that all bids are accepted and that therefore all capacities depicted likely have to be reserved for balancing power. One sees only a minor influence of the spot market optimization in the sense that slightly more conservative prices are asked (decreased probability of the scenario) and that only a little larger part of the entire flexibility remains that is not offered as balancing reserves. Together with Figure 2 (b), this indicates a strong advantage of the balancing over the spot markets for that week. That is, the optimal bidding strategy aims - when possible - at offering almost the entire flexibility range at balancing markets and ensures that all offered capacities are likely accepted by asking conservative prices. The ability to market remaining flexibility at the spot market does not significantly change this strategy and consequently, does not enable additional savings.

5. Conclusion

We present a framework for model-based bidding at the market for primary and secondary balancing power simultaneously. We solve the resulting optimization problem using a tailored hybrid stochastic-deterministic solution approach. Considering an exemplary week from 2018, we calculate the optimal bidding strategy for an industrial aluminum electrolysis. We demonstrate that the approach effectively identifies bidding strategies which allow for substantial savings. Furthermore, we show that a balancing market participation can show substantially greater economic potentials than a pure trading at spot markets. Future work should aim at the rigorous deterministic solution of the bidding problem, e.g., through application or tailoring of decomposition techniques

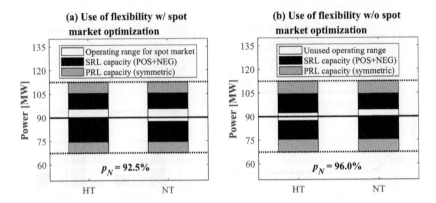

Figure 3: Capacity that is offered as balancing power (orange: PRL, purple: SRL) for the two time periods. (a) is with optimization at spot market and (b) without. Solid black line represents the nominal operating point, dotted black lines correspond to bounds of the operating range. p_N denotes the probability that all bids are accepted. Note that in (a), the remaining white part corresponds to the capacity available for spot market trading.

for MINLPs that would allow for a separation into nonlinear bidding and MILP scheduling sub-problems. Here, the reader is referred to, e.g., Li et al. (2012) for nonconvex generalized Benders decomposition or Mouret et al. (2011) for sophisticated Lagrangian methods.

6. Acknowledgment

The authors gratefully acknowledge the financial support of the Kopernikus project SynErgie by the Federal Ministry of Education and Research (BMBF) and the project supervision by the project management organization Projektträger Jülich. The authors further thank H. Hauck from TRIMET SE for valuable discussions.

References

G. E. P. Box, G. M. Jenkins, 1994. Time Series Analysis: Forecasting and Control, 3rd Edition. Prentice Hall PTR.

A. W. Dowling, R. Kumar, V. M. Zavala, 2017. A multi-scale optimization framework for electricity market participation. Applied Energy 190, 147–164.

A. Ghobeity, A. Mitsos, 2010. Optimal time-dependent operation of seawater reverse osmosis. Desalination 263 (1), 76–88.

X. Li, Y. Chen, P. I. Barton, 2012. Nonconvex generalized Benders decomposition with piecewise convex relaxations for global optimization of integrated process design and operation problems. Industrial & Engineering Chemistry Research 51 (21), 7287–7299.

S. Mitra, I. E. Grossmann, J. M. Pinto, N. Arora, 2012. Optimal production planning under time-sensitive electricity prices for continuous power-intensive processes. Computers & Chemical Engineering 38, 171–184.

A. Mitsos, N. Asprion, C. A. Floudas, M. Bortz, M. Baldea, D. Bonvin, A. Caspari, P. Schäfer, 2018. Challenges in process optimization for new feedstocks and energy sources. Computers & Chemical Engineering 113, 209–221.

S. Mouret, I. E. Grossmann, P. Pestiaux, 2011. A new Lagrangian decomposition approach applied to the integration of refinery planning and crude-oil scheduling. Computers & Chemical Engineering 35 (12), 2750–2766.

J. I. Otashu, M. Baldea, 2018. Grid-level "battery" operation of chemical processes and demand-side participation in short-term electricity markets. Applied Energy 220, 562–575.

M. Paulus, F. Borggrefe, 2011. The potential of demand-side management in energy-intensive industries for electricity markets in Germany. Applied Energy 88 (2), 432–441.

P. Schäfer, H. Westerholt, A. M. Schweidtmann, S. Ilieva, A. Mitsos, 2019. Model-based bidding strategies on the primary balancing market for energy-intense processes. Computers & Chemical Engineering 120, 4–14.

Q. Zhang, I. E. Grossmann, C. F. Heuberger, A. Sundaramoorthy, J. M. Pinto, 2015. Air separation with cryogenic energy storage: Optimal scheduling considering electric energy and reserve markets. AIChE Journal 61 (5), 1547–1558.

Anton A. Kiss, Edwin Zondervan, Richard Lakerveld, Leyla Özkan (Eds.)
Proceedings of the 29th European Symposium on Computer Aided Process Engineering
June 16th to 19th, 2019, Eindhoven, The Netherlands. © 2019 Elsevier B.V. All rights reserved.
http://dx.doi.org/10.1016/B978-0-128-18634-3.50275-7

Optimal CSP-waste based polygeneration coupling for constant power production

Ester de la Fuente, Mariano Martín[*]

Departamento de Ingeniería química- Universidad de Salamanca. Plz. Caídos 1-5, 37008, Salamanca, Spain

mariano.m3@usal.es

Abstract

This work addresses the integration of a biogas-based Brayton Cycle with a concentrated solar power facility. A heat exchanger network generates steam out of flue gas or molten salts to be used in a regenerative Rankine cycle to produce power. A multiperiod scheme is formulated for the design of the hybrid facility determining the waste required as back up to mitigate seasonal solar availability. The steam turbine is responsible for power production while the gas turbine works mainly as a combustion chamber. In the South of Spain, a 50% excess of biogas is produced during summer yielding a production cost of electricity of 0.17 €/kWh with an investment of 380 M€ for a production facility of 25MW. A scale-up study is also carried out.

Keywords: Circular economy, biogas turbine, CSP, renewable power, mathematical optimization

1. Introduction

The transition from fossil-based power to a renewable based economy faces an important challenge, the variability of the largest energy resource on Earth. Solar availability is highly volatile, it is subjected to seasonality in the longer term, but weather also results in hourly or subhourly fluctuations. Meeting power demand over time requires the use of integrated systems to help mitigate the variation in power production (Yaun and Chen, 2012). Hybrid concentrated solar power (CSP) facilities use a second energy source to mitigate the lack of Sun (Peterseim, et al 2013). Apart from the use of fossil resources, renewables have already been considered. Vidal and Martín (2015) optimized the operation of a facility that used lignocellulosic biomass gasification, providing hot streams to heat up the molten salts, and a gas turbine as buffer technology to operate a CSP facility. This integration suggested the use of a syngas powered gas turbine and the possibility of storing power in the form of hydrogen as energy vector if solar availability was enough to meet the power demand. The molten salts where not heated up using thermal energy from the hot gases nor a syngas furnace was used. However, the diversification of the use of lignocellulosic raw materials for the production of chemicals and the large and distributed availability of waste in isolated areas (Taifouris and Martín, 2018) suggest its use as back-up to mitigate the solar variability. In this work a mathematical optimization framework is developed to evaluate the integration of waste and CSP within a renewable-based power plant. The system considers waste anaerobic digestion and a gas turbine to produce power from residues. The hot flue gas and/or the heat transfer fluid, molten salts, can be used for the production of the steam required by the steam turbine of a regenerative Rankine cycle

2. Process description

The process consists of five major sections: Biogas production and purification, gas turbine, solar collection technologies, heat exchanger network and a regenerative Rankine cycle.

Waste and water are fed to a digestor. The mixture is anaerobically digested for 25 days to produce biogas and a digestate. The biogas is sent to purification to remove CO_2 and traces of NH_3, using a Pressure Swing Adsorption (PSA) system, and to eliminate the H_2S in a fixed-bed reactor. Once biomethane is obtained, a Brayton cycle is used to produce power. The flue gas produced is sent to a heat exchanger network (HEN) to produce steam.

The CSP facility consists of the heliostat field including the receiver and the molten salts storage tanks and a HEN. The salts are used to generate superheated steam to be used in a steam turbine following a regenerative Rankine cycle.

The Rankine cycle is common for both energy resources. In order for the molten salts and the flue gas to be used, a system of heat exchangers in parallel is considered. For heating up and evaporating water, the entire flow of the molten salts is used while only a fraction was used for reheating and overheating the steam. However, the higher temperature of the flue gas suggests a different lay out. The entire flow is used for overheating the steam before feeding the high-pressure turbine whereas it is split to reheat the steam before feeding it into the medium pressure turbine and to heat-up and evaporate the water coming from the deaerator. In the second stage of the turbine, a fraction of the steam is extracted at a medium pressure and it is used to heat up the condensate. The rest of the steam is finally expanded to an exhaust pressure, condensed and recycled. A cooling tower is used to cooldown the cooling water, see Figure 1.

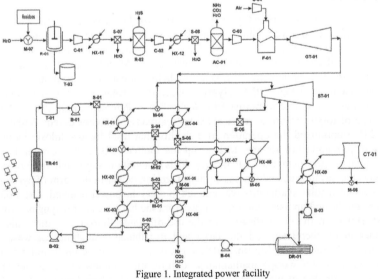

Figure 1. Integrated power facility

3. Problem formulation

The different technologies described in the flowsheet of Figure 1 are modelled using mass and energy balances, for the digester, the heat exchangers, mixers and splitters, thermodynamics for the gas compression and expansion, the steam turbine is modelled based on enthalpy and entropy balances where surrogate models are developed for the enthalpy and entropy of the steam as a function of the temperature and pressure (León and Martín, 2016), rules of thumb are used to estimate the compressor and turbine efficiencies (Walas, 1990), design equations are used to model the solar Receiver, and reduced order models are developed for the adsorption beds. The model is written in terms of the total mass flows, component mass flows, component mass fractions, and temperatures of the streams in the network. These are the main variables whose values have to be determined from the optimization. The components in the system correspond to the ones in the set $J = \{$ Wa, CO_2, O_2, N_2, H_2S, NH_3, $CH_4\}$

The problem is a multiperiod NLP of 9600 vars and 4800 eqs. written in GAMS, that is solved maximizing eq. (1) on a monthly basis for the selection of the best use of waste to mitigate the lack of sun, either the production of power in a biogas fuelled gas turbine and/or the use of the flue gas to produce steam together with the molten salts. The objective function includes the net power production and the annualized cost (K=1/3) of the mirrors, the digesters, each of $6000m^3$ at 365 €/m^3, and the linearized cost for the turbines as a function of the power

$$Z = C_{Electricity} \cdot (W_{(Turbgas)} + W_{Total} - W_{(Comp1)} - W_{(Comp2)} - W_{(Comp3)} - W_{(Comp4)})$$
$$-(120 \cdot AreaTotal + C_{Digestor} \cdot N_{digestor} + (335.27*(W_{(Turbgas)}) + 36.211) - (270.5* W_{Total} + 2 \cdot 10^6))*(K) \quad (1)$$

A case study is presented for a region in Spain, Badajoz, where hybrid plants are already installed but using fossil backup energy (http://es.csptoday.com). The large availability of poultry and solar incidence makes its integration an interesting alternative to the use of fossil resources. A net production of 25 MW is considered. The framework is flexible to study the use of other residues and evaluate the operation in other regions.

4. Results

4.1. Plant operation

The plant allocation in the northern hemisphere dictates its operation. Summer operation is based on the use of solar energy, reducing the use of waste and the usage of the digestors, see Figure 2. This is an important feature since the use of the digestors at full capacity can allow the production of chemicals such as methanol or methane (Hernández and Martín, 2016) for energy storage purposes. It turns out that, in summer, additional 12 MW of power are available. Furthermore, an interesting feature unveiled from the optimization is the fact that the gas turbine actually operates as a furnace providing little power but generating a high temperature stream to be used in the heat exchanger network. As a result, the two heat exchanger networks, the one that uses the molten salts and the one that uses the hot flue gas, operate the entire year. Figure 3 summarizes the operation of both turbines and shows the contribution of the solar energy and the poultry manure. In winter up to 40% of the energy is provided by the manure while in summer, the solar energy is enough to reach the desired power production of 25 MW.

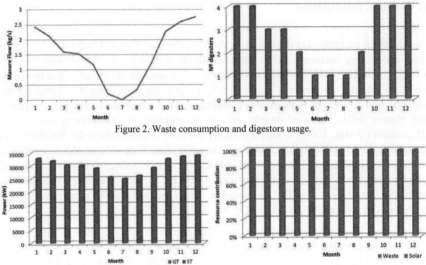

Figure 2. Waste consumption and digestors usage.

Figure 3. Contribution of the two sections of the process to power production

4.2. Economic evaluation

The production cost and investment costs of the integrated facility are estimated and compared to the stand alone CSP plant and other integrated facilities.

The investment cost is estimated using the factorial method (Sinnot 1999). It relies on the equipment cost, updated from Matche (Matche 2014). We consider the units described in the flowsheet given by Figure 1. Digesters cost is assumed to be 365 €/m^3 (Taifouris and Martín, 2018) and the heliostat costs are 120 €/m^2. We assume that each heliostat has an area of 120 m^2. The installed equipment is assumed to represent 1.5 times the equipment cost. Piping, isolation, instrumentation and utilities represent 20 %, 15 %, 20 % and 10 % of the equipment cost respectively. Land and buildings cost are assumed to be 8 M€, and the load of molten salts is priced at 0.665 €/kg. These items add up to the fix cost (319 M€). The fees represent 3 % of the fix cost, other administrative expenses and overheads and the plant layout represent 10 % of the direct costs (fees plus fix capital) and 5 % of the fix cost respectively. The plant start-up cost represents 15 % of the investment. The investment adds up to 380 M€. Figure 4 shows the contribution to the cost of the different sections of the facility, the solar field is the larger contribution, over a third of the total, followed by the turbine and the heat exchanger network. The digestors, a total of 4 in the facility, represent 7.5% of the units. An interesting comparison between this plant and the stand-alone plant presented in (Martín and Martín, 2013) reveals that the investment cost is around 50% larger, but the production capacity was variable over time. Thus, the investment to secure constant production capacity was of around 130 M€.

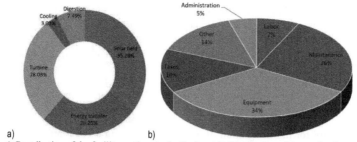

a)   b)

Figure 4.- a) Contribution of the facility sections to the final cost b) Distribution of the production costs

The production costs of the electricity generated, without considering the possibility of producing additional power, chemicals or credit from the digestate, can be seen in Figure 4b. The labour costs are assumed to be 0.5% of the investment, equipment maintenance around 2.5 % of fix costs, the amortization linear with time in 20 years, the taxes and the overheads 1 % of the investment each, and the administration around 5 % of labour, equipment maintenance, amortization, taxes and overheads. The production costs add up to 0.17 €/kWh, slightly higher than the values reported for the CSP facility of similar sized operating alone (Martín and Martín, 2013) that shows a production costs of around 0.15 €/kWh but that secures constant production using renewable resources. That additional 10% in the electricity cost is to secure the continuous production of power over time.

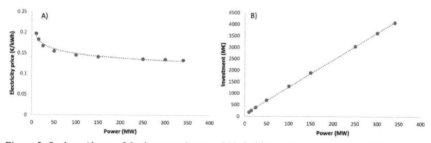

Figure 5.-Scale up/down of the integrated waste-CSP facility. A) Production cost; B) Investment

An additional comparison can be made with the integration of lignocellulosic biomass and CSP by Vidal and Martin (2015). Note that the collection of miscanthus is easier due to its higher density and easier transportation which allowed considering a larger facility of 340 MW benefiting from economies of scale. Furthermore, in that case, the integration considered the use of hot streams to heat up the molten salts, not to produce steam directly. The optimal integrated process consisted of the use of a syngas fueled gas turbine that produced power in parallel to the one obtained from the steam turbine from the CSP section. Finally, when solar power was enough to meet the power demand, the syngas was used to produce hydrogen using a WGSR, providing additional income. The larger scale allowed electricity production costs of 0.073 €/kWh with an investment cost of 3225 M€. In this work the integration is different since the hot streams were not intended to be used in heating up the molten salts but to directly produce steam. The gas turbine is merely used as a furnace and the excess of biogas can be stored or further processed to methane or methanol. Reaching that production capacity requires more than 50 digesters and processing around a peak of 40 kg/s of manure which is far larger facility than any available. For the proper comparison, a scale up study is required. Figure 5 shows the results of the scale up/down of the economics of the integrated facility analyzed in this work. For a 340 MW integrated facility, an investment of 4080M€ is required for an electricity production cost of 0.133 €/kWh. Note that no credit is assumed from the digestate or the additional biogas production capacity

5. Conclusions

A superstructure for the integration of waste-to-power and a concentrated solar facility has been proposed embedding anaerobic digestion, a Brayton Cycle, solar field, molten salts system and a regenerative Rankine cycle. The solution of the multiperiod optimization problem decides on the use of each energy source.

The low yield to power of the poultry manure and the allocation of the plant results in an excess of capacity during summer of about 50%, so that the demand of 25MW is met in winter. The steam turbine is the one responsible for producing the power, while the gas turbine operates mostly as a combustion chamber. The production cost to secure power production are 5% larger than the ones of the stand-alone plant, 0.17 €/kWh, but the investment cost increases by 50% up to 380 M€. However, the excess of capacity can be used to produce chemicals that will be an asset to the process in terms of additional income or providing flexibility to its operation. The digestate can also be further used as a fertilizer. The large number of units involved makes difficult that the future lower costs of heliostats result in large investment savings per kW installed. Integration of waste is suggested at smaller scales compared to lignocellulosic biomass. The formulation is useful to evaluate trade-offs related to the location of the plant as function of the availability of the raw materials, solar energy and waste.

Acknowledgments

The authors acknowledge PSEM3 group and JCYL under grant SA026G18

References

B. Hearnandez, M. Martín, 2016 Optimal Process operation for biogas reforming to methanol: effects of dry reforming and biogas compositionl IECR. 55 (23), 6677–6685,

L. Martín, M. Martín 2013. Optimal year-round operation of a Concentrated Solar Energy Plant in the South of Europe *App. Thermal Eng*. 59:627-633.

Matche (2014) http://www.matche.com/prod03.htm. (last accessed March 2018)

J.H. Peterseim, S. White, A Tadros, W Hellwig, 2013. Concnetrated solar power hybrd plants, which technolgies are best suited for hybridisation. Renew. Energ. 57, 520-532

R.K. Sinnot Coulson and Richardson, Chemical Engineering. 3ªEd. Butterworth Heinemann, Singapore. 1999.

M.R. Taifouris, M. Martín, 2018. Multiscale scheme for the optimal use of residues for the production of biogas across Castile and Leon, J. Clean. Prod., 185, 239-251

M. Vidal, M. Martín 2015. Optimal coupling of biomass and solar energy for the production of electricity and chemicals. Comp. Chem. Eng , 72, 273-283

S.M. Walas, 1990, Chemical Process Equipment. Selection and *Design* Butherworth Heinemann. Elsevier. Oxford UK.

Z. Yuan, B Chen, 2012, Process Synthesis for Addressing the Sustainable Energy Systems and Environmental Issues. *AIChE J.* ,58 (11), 3370-3389

http://es.csptoday.com/mercados/arranca-el-proyecto-de-la-planta-termosolar-astexol-2

Anton A. Kiss, Edwin Zondervan, Richard Lakerveld, Leyla Özkan (Eds.)
Proceedings of the 29th European Symposium on Computer Aided Process Engineering
June 16th to 19th, 2019, Eindhoven, The Netherlands. © 2019 Elsevier B.V. All rights reserved.
http://dx.doi.org/10.1016/B978-0-128-18634-3.50276-9

A Multiperiod Optimisation Approach to Enhance Oil Field Productivity during Secondary Petroleum Production

Emmanuel I. Epelle, Dimitrios I. Gerogiorgis[*]

School of Engineering (IMP), University of Edinburgh, Edinburgh, EH9 3FB, UK

D.Gerogiorgis@ed.ac.uk

Abstract

Water injection rate allocation across different injection wells in an oil and gas field undergoing secondary production is one of the most economical ways to increase hydrocarbon production. Exploiting a process systems engineering description of a rate allocation problem via mathematical optimisation can improve operational planning. We simultaneously address a production and injection optimisation problem with an economic objective function, the net present value (NPV), subject to practical constraints ensuring operational feasibility. A case study with a constant water injection rate is compared to another in which optimal dynamic injection rates and water sharing ratio among the respective injection wells are determined. With the formulated optimisation (NLP) problem, an optimal water injection strategy that reduces field water consumption and increases profitability is obtained. MATLAB's fmincon and IPOPT optimisation solvers are employed for solution, observing superiority of IPOPT compared to fmincon.

Keywords: Net present value (NPV), Optimal injection strategy, Production optimisation

1. Introduction

Secondary production via water flooding is one of the most implemented enhanced oil recovery methods in the oil and gas industry, involving the injection of water to increase the reservoir's pressure and sweep residual oil in order to consequently increase oil production. Cost reductions and improved productivity during secondary production are dependent on the dynamic rate allocation to different injection wells, the underlying reservoir permeability distribution and other rock and fluid properties. Applying reservoir and multiphase flow simulation (wellbore and pipeline hydraulics) in an integrated manner can aid practical operational decisions during early design stages (Kosmidis et al., 2004, Gerogiorgis et al., 2006; Van Essen et al., 2011). Current oil and gas production optimisation studies in the literature address problems such as well placement, production and drill rig scheduling, pipeline and surface facility routing, infrastructure installation and geological and economic uncertainty (Scheidt et al., 2010; Gupta and Grossmann, 2012; Gunnerud et al., 2013; Tavallali et al., 2013; Epelle and Gerogiorgis, 2017).

A common limitation of these studies that motivates our work is the lack of a detailed water flooding model that incorporates the intricacies of the reservoir behaviour and its dynamic fluid and well properties (gas oil ratio, water cut, productivity indices and well allocation factors). Here, we demonstrate that streamline simulation is a powerful tool that facilitates accurate and optimal determination of the injection rates to different wells. Compared to previous studies, in which the adopted wellbore geometries are vertical, the multiphase flow complexities in deviated well geometries are also incorporated in our approach. Furthermore, the inherent property of the petroleum production system that allows its components to be treated independently is significantly exploited. This is achieved by optimising an economic objective function that captures the overall system behaviour. The coupling of reservoir and pressure drop simulation via proxy-modelling enhances the adaptability of the implemented framework to real field operations via speedy computation.

2. Simulation Methodology and Optimisation Formulation

The model is first built in a numerical reservoir simulator using permeability, porosity and saturation data of a sandstone reservoir. The field contains four production wells and four injection wells (Fig. 1). Two of each class are horizontal wells and the others are vertical. In running the reservoir model, ECLIPSE 100 (Schlumberger, 2017), it is assumed that water injection commences in the third year. A production forecast for a 6-year duration is done to obtain the field pressure, productivity indices, water cut and gas oil ratios of the wells at each time step (Δt = 1 yr). These data are fed into a multiphase flow simulator, PIPESIM 2017 (Schlumberger, 2017), in which the wellbore and pipeline models are built.

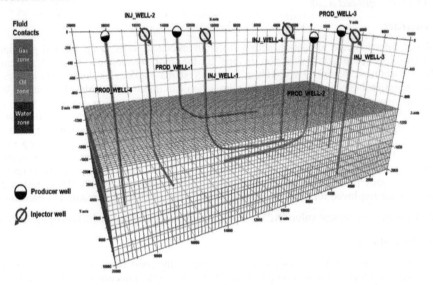

Figure 1: Reservoir structure with drilled wells. The surface network built in PIPESIM additionally consists of 2 main pipelines, 2 manifolds, 8 valves and 2 separators.

The objective function and operational constraints of the proposed formulation are outlined in Table 1. Compared to previous studies which maximise production (Kosmidis et al., 2004; Gunnerud et al., 2013), we maximise the net present value (NPV) by maximising oil and gas production (Eqs. 2–3) and limiting water production rates (Eq. 4) and injection costs (Eq. 5) (Guo and Reynolds, 2018). Here, TOP and TGP are the revenues from total oil and gas produced, respectively, TWP and TWI represent the cost incurred for water production and injection, respectively, r_{OP} and r_{GP} are oil and gas prices, respectively, r_{WP} and r_{WI} are the water production and injection unit costs, respectively, N_{prod} and N_{inj} are the number of producers and injectors, respectively, Δt is the timestep, b is the discount rate for reference time, t_{ref}; q_{OP}, q_{GP}, q_{WP} and q_{WI} are the flow rates of produced oil, produced gas, produced water and injected water at time t, respectively. Subscripts i, j and l represent the injection wells, production wells, and pipelines, respectively. The capacity constraint of the field is represented by Eq. 6, where $C_{p,sep}$ is the separator capacity.

The pressure-rate relationship of the production and injection wells and pipelines are described via proxy-modelling. Deciding which form of surrogate modelling technique to be used considered various factors, such as the feasibility of implementation for our particular case studies, the ability to capture the system's dynamic nonlinearities and accuracy, the solution time and the modelling effort and ease of automation. Of all candidate forms considered, an algebraic/polynomial proxy model was best suited for application, yielding an error of <2 % with respect to the simulation data obtained for wells and pipelines (Gunnerud and Foss, 2010). These proxy models developed for each time period in the entire time horizon constitute the constraints fed to the optimisation routine. The proxy models for description of component flowrates and pressure drops in pipelines are shown in Eqs. 7–9. Here, subscript l indicates the pipeline, superscript m the manifold and s the separator, so P_l, P_m and P_s represent the pipeline, manifold and separator pressures, respectively, while the subscript t denotes their values at the particular time interval. Variables q_{lo}, q_{lw} and q_{lg} are flowrates of oil, water and gas in pipeline l, respectively. Sensitivity analyses are performed over wells and pipelines at each Δt to obtain the proxy model coefficients a_n, b_n and c_n for $n = 1, 2, 3, 4$. Other constraints include the separator capacity and formation fracture pressure limitations. Forward flow of fluids towards the separator is described via Eqs. 10–11. Formation fracture due to high injection pressures (reducing the sweep efficiency) is avoided via Eq. 12, material balance constraints between wells and pipelines are denoted by Eq. 13 and the maximum allowable injection rate for the respective injection wells is limited by the available field water via Eq. 14. Here, P_l, P_m, P_s and P_j^w are the pipeline, manifold, separator and wellhead pressures, respectively, P_{inj} and P_{frac} are the water injection and formation fracture pressures and $Q_{inj,tot}$ is the total daily volumetric flowrate of water available for injection.

The number of variables, model nonlinearities and non-convexities and the adopted local optimisation solvers (fmincon and IPOPT) imply that solutions will be sensitive to the initial guess supplied. To address this challenge, streamline simulation is adopted for the determination of the dynamic well allocation factors (WAFs), which are used in the calculation of the dynamic injector efficiencies (Theile and Batycky, 2006; Azamipour et al., 2018). With these calculated efficiencies, the best performing injection well can be identified and consequently allocated the highest injection rate as an initial guess. Using these practical initial guesses obtained by applying streamline simulation, we enforce bounds on the decision variables, assuming a uniform distribution between the upper and lower bounds of the respective variables and perturb the initial guess 1,000 times between the bounds, thus facilitating convergence to a global solution. The final optimisation formulation consists of 198 variables and 252 constraints (nonlinear, linear, equality and inequality). These constraints and the objective function are written in MATLAB and solved using the fmincon and IPOPT solvers via the object-based OPTI toolbox platform.

Table 1: Objective function and constraints of the optimisation framework.

Objective Function		
$$\max NPV = \sum_{t=1}^{N_t} \frac{TOP^t + TGP^t - (TWP^t + TWI^t)}{(1+b)^{\frac{t}{t_{ref}}}} \quad (1)$$	$q_{p,j,t} = a_1 + a_2 P_{j,t}^w + a_3 (P_{j,t}^w)^2$	(7)
	$q_{p,i,t} = b_1 + b_2 P_{i,t}^w + b_3 (P_{i,t}^w)^2$	(8)
$$TOP^t = r_{OP} \sum_{j=1}^{N_{prod}} q_{OP,j,t} \quad (2)$$	$\Delta P_{l,t} = c_1 + c_2 q_{lg,t} + c_3 q_{lo,t} + c_4 q_{lw,t} + c_5(q_{lg,t})^2 + c_6(q_{lo,t})^2 + c_7(q_{lw,t})^2 + c_8 q_{lg,t} q_{lw,t} + c_9 q_{lg,t} q_{lo,t} + c_{10} q_{lo,t} q_{lw,t}$	(9)
$$TGP^t = r_{GP} \sum_{j=1}^{N_{prod}} q_{GP,j,t} \quad (3)$$	$P^s = P^m - P_{l,t}$	(10)
$$TWP^t = r_{WP} \sum_{j=1}^{N_{prod}} q_{WP,j,t} \quad (4)$$	$P^m < P_j^w$	(11)
$$TWI^t = r_{WI} \sum_{j=1}^{N_{prod}} q_{WI,j,t} \quad (5)$$	$P_{inj} < P_{frac}$	(12)
Constraints	$$\sum_{j=1}^{N_{prod}} q_{p,j,t} = q_{p,l,t} \quad (13)$$	
$$\sum_{j=1}^{N_{prod}} q_P \le C_{P,sep} \quad (6)$$	$$\sum_{j=1}^{N_{prod}} q_{w,inj} = Q_{inj,tot} \quad (14)$$	

3. Simulation and Optimisation Results

Two optimisation case scenarios are presented. In Scenario-1, the injection rates and production rates are optimised and in Scenario-2, the injection rates are maintained at the maximum allowable rate (constant) for each injection well and only the production rates optimised. Before discussing the optimisation results, the results of the simulation procedure are analysed. These results (Fig. 2) are the production and injection performances, based on a 6-year simulation forecast. It is observed that PROD_WELL-1 slightly outperforms PROD_WELL-2 (Fig. 2a); their respective equidistant locations from the INJ_WELL-1 (in the reservoir model – Fig. 1) and the similarity of the underlying local permeability distribution implies they experience similar levels of pressure support from the injection well. Also observed is the reduced productivity of the vertical wells (PROD_WELL-3 and 4) compared to the horizontal wells (PROD_WELL-1 and 2). This is inevitably due to the higher drainage area that the horizontal wells are exposed to in the reservoir. The large disparity observed in the well productivity profiles is not present in the streamline-derived injection efficiencies (Fig. 2b). Despite the favourable well geometry of INJ_WELL-2 (horizontal) compared to INJ_WELL-4 (vertical), they both exhibit similar efficiencies. The interaction of INJ_WELL-1 with the horizontal producers is the main factor contributing to its high efficiencies as determined from the WAFs obtained from streamline simulation. With the available daily field water injection rate and injection efficiencies, it was possible to propose initial injection rates for the optimisation algorithm.

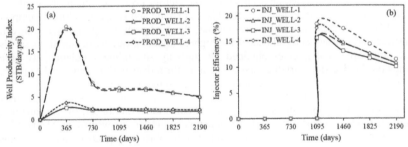

Figure 2: Production and injection well performances.

The production rate profiles for the oil, gas and water phases of the respective production wells for the two optimisation scenarios considered are compared in Fig. 3. The optimisation algorithm captures the increment in oil production at the start of injection (at 1,095 days) and the reduced decline rate after injection commences (Fig. 3a). Similar to the oil production rate profile, the gas production rates (Fig. 3b) of the respective wells also show an increment in production due to water injection and is more significant for the horizontal wells. The commencement of water injection is accompanied by a rapid increase in the water production rates, particularly for PROD_WELL-3 and 4. The horizontal wells (PROD_WELL-1 and 2) display a rather gradual increase in the water production rate compared to the vertical ones. It is also observed that the superior performance of the horizontal wells has been exploited by the optimisation algorithm for the improvement of the total oil production (Fig. 3a). This improvement mostly occurs in the plateau region (i.e. when constant oil production is observed). Compared to the oil production rates, the water and gas production rates of the horizontal wells only increase marginally; this happens towards the end of the production horizon (Fig. 3b–c). As for the other vertical producers, there is no difference between the observed production rates for the optimal and constant injection scenarios; this difference can be increased by incorporating well placement optimisation for the exploration of highly productive sites (with good permeability) in the reservoir. This would require the application of stochastic optimisation techniques and integer variables representing a grid block allocation method, introducing further complexity to the considered problem, which is beyond the scope of this paper.

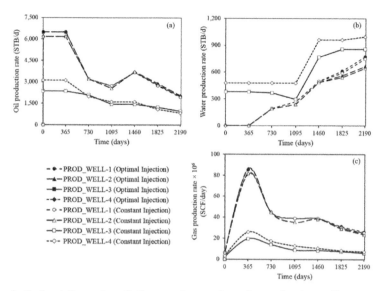

Figure 3: Optimal flow rates of oil gas and water from the production wells.

The optimal injection strategy and the corresponding cumulative injection rate for each well are shown in Figs. 4a–b. It is shown that a stepwise increase in the injection rate for INJ_WELL-1 and a constant injection for the rest of the wells is guaranteed to increase the NPV of the field by 6% (Fig. 4d) and reduce total water consumption by 11%. Furthermore, it is shown in Fig. 4c that the water production rates of PROD_WELL-3 and 4 (vertical wells) are higher than those of the horizontal wells, further demonstrating the desirable performance of the horizontal wells. Additionally, PROD_WELL-3 experiences a significantly lower water production compared to PROD_WELL-4 but with a similar oil production. Hence, PROD_WELL-3 can be considered the best performing vertical well.

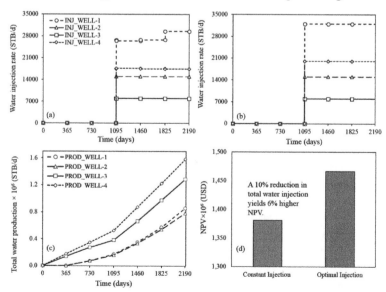

Figure 4: Water injection rates for (a) the optimal scenario and (b) the constant injection scenario; (c) cumulative water production rates of the optimal scenario and (d) the economic analysis of both scenarios.

A comparison of the solver performances showed that the MATLAB's fmincon was unable to yield satisfactory results that met convergence criteria. The IPOPT solver implementing an interior point line-search filter method was able to converge successfully.

4. Conclusion

A production and injection optimisation problem with an economic objective function subject to practical constraints that ensure operational feasibility has been thoroughly addressed. The application of streamline-based well allocation factors and injector efficiencies served as a reliable method of obtaining good initial guesses for the time-dependent injection rates, enabling faster performance of the optimisation algorithm. The implementation of algebraic proxy models was sufficient for accurate representation of the simulator output with a maximum error of <2 %. The optimisation framework has shown that a systematic variation of the injection rates, well head pressures and production rates yields increased field profitability. In the considered scenario, a 6% improvement in the NPV and a corresponding 11% decrease in the total field water consumption is attained.

5. References

V. Azamipour, M. Assareh, M.R. Dehghani and G.M. Mittermeir, 2017, An efficient workflow for production allocation during water flooding, *J. Energy Resour. Technol.*, 139, 3, 1–10.

E.I. Epelle and D.I. Gerogiorgis, 2017, A multiparametric CFD analysis of multiphase annular flows for oil and gas drilling applications, *Comput. Chem. Eng.*, 106, 645–661.

D.I. Gerogiorgis, M. Georgiadis, G. Bowen, C.C. Pantelides and E.N. Pistikopoulos, 2006, Dynamic oil and gas production optimization via explicit reservoir simulation, *Comput. Chem. Eng.*, 21, 179–184.

V. Gunnerud and B. Foss, 2010, Oil production optimization—A piecewise linear model solved with two decomposition strategies, *Comput. Chem. Eng.*, 34, 1803–1812.

Z. Guo and A.C. Reynolds, 2018, Robust life-cycle production optimization with a support-vector-regression proxy, *SPE Journal*, 23, 6, 2409–2427.

V. Gupta and I.E. Grossmann, 2012, An efficient multiperiod MINLP model for optimal planning of offshore oil and gas field infrastructure, *Ind. Eng. Chem. Res.*, 51, 19, 6823–6840.

V.D. Kosmidis, J.D. Perkins and E.N. Pistikopoulos, 2004, Optimization of well oil rate allocations in petroleum fields, *Ind. Eng. Chem. Res.*, 43, 14, 3513–3527.

Schlumberger, Eclipse User Manual, 2017, Technical Description, *Schlumberger Ltd.*

Schlumberger, Pipesim User Manual, 2017, *Schlumberger Ltd.*

C. Scheidt, J. Caers, Y. Chen and L. Durlofsky, 2010, Rapid construction of ensembles of high-resolution reservoir models constrained to production data. In *ECMOR XII-12th European Conference on the Mathematics of Oil Recovery*, Oxford, UK.

M.S. Tavallali, I.A. Karimi, K.M. Teo, D.Baxendale and S. Ayatollahi, 2013, Optimal producer well placement and production planning in an oil reservoir, *Comput. Chem. Eng.*, 55, 109–125.

M.R. Thiele and R.P. Batycky, 2006, Using streamline-derived injection efficiencies for improved waterflood management, *SPE Reservoir Eval. Eng.*, 9, 2, 187–196.

G. Van Essen, P. Van den Hof and J.D. Jansen, 2011, Hierarchical long-term and short-term production optimization, *SPE Journal*, 16, 1, 191–199.

Anton A. Kiss, Edwin Zondervan, Richard Lakerveld, Leyla Özkan (Eds.)
Proceedings of the 29th European Symposium on Computer Aided Process Engineering
June 16th to 19th, 2019, Eindhoven, The Netherlands. © 2019 Elsevier B.V. All rights reserved.
http://dx.doi.org/10.1016/B978-0-128-18634-3.50277-0

Fast Fourier Transforms for Microgrid Climate Computing

Paolo Fracas[a], Edwin Zondervan[a,b,*]

[a] Genport srl – Spinoff del Politecnico di Milano, Via Lecco 61, Vimercate, 20871, Italy

[b]University of Bremen, Leobener Str. 6, Bremen, 28359, Germany

paolo.fracas@genport.it

Abstract

Solar panels and wind turbines are key technologies for a sustainable low-carbon energy transition. The large diffusion of weather-dependent renewable energy generators face the challenge to fit the demand of uncertain loads with the most appropriate mix of distributed energy resources. The availability of accurate climate variables projections is essential to identify the best combination among distributed generators, energy storages and power loads in each geographical location. From the European Centre for Medium-Range Weather Forecasts (ECMWF) datasets of several climate variables are available for renewable energy resources yield forecasting. This paper presents different approaches to manipulate ECMWF datasets by combining Fast Fourier Transform with polynomial and forest tree regression models to predict climate variables over a one-year period, typical for microgrid simulations. An example of climate datasets forecasts related to the city of Bremen is presented with the evaluation of performances during test and training scenarios phases.

Keywords: FFT, IFFT, ECMW, Renewable Energy Systems, Climate Monthly Datasets.

1. Introduction

1.1. Scope of this work

The European Centre for Medium-Range Weather Forecasts (ECMWF) provides a large number of climate datasets for public use updated until three months of real time; these datasets are available in ERA5, the fifth generation ECMWF atmospheric reanalysis of climate variables, built by combining models from across the world into a globally consistent database. Our work has been focus on developing an approach to extrapolate multiple ERA5 datasets by manipulating just a single dataset. In order to reach this result, Fast Fourier Transform (FFT) has been combined with Inverse Fast Fourier Transform (IFFT) and polynomial regression. The novelty of work mainly consists in the application of this method to the most updated ERA5 gridded datasets. The result is a viable route for improving the accuracy in wind and solar energy simulations, by avoiding the errors introduced with real weather data (D. Quiggin et al. - 2012) and old average climate datasets (NASA/SSE - 2008).

1.2. State of Art

The spectral transform method has been successfully applied in climate datasets for more than thirty years. The Fourier Transform method was introduced to numerical weather prediction starting from the work of Eliasen et al. (1970) and Orszag (1970).

Later works show an extensive use of monthly climate data: Joly and Voldoire, (2009) have developed a method to manipulate gridded datasets with Fast Fourier Transform (FFT) to better understand the coupled ocean–atmosphere processes. Kentat et al. (2013) analysed climate measurements, satellite retrievals of monthly mean marine Wind Speed to validate the accuracy required in calculation of air–sea heat fluxes. Wang and Zeng (2015) have elaborated climate data to quantify the land surface air Temperature. Amendola et al. (2017) used Fourier Transform to recombine Gaussian distributions obtained via Neural Network for seasonal weather forecasts. Wang et al. (2018) using monthly mean, have constructed indices of boreal sea surface Temperature in equatorial Pacific. Andres and Agostaa (2014), elaborated the correlations among atmospheric, oceanic gridded data with lunar nodal by using filtered climate data and Fast Fourier Transform.

2. Methods to manage the climate datasets and detect the errors

Fourier analysis is a method for expressing a complex function as a sum of periodic components, and for recovering the function from those components. Cooley and Tukey (1965) and later Press et al. (2007) provided a computing approach to Fourier analysis for discretized counterparts: the Fast Fourier Transform (FFT).

2.1. Fast Fourier Transforms for monthly climate predictions

Manipulation of datasets arrays, computing optimizations, FFT algorithms and regression analysis has been conducted with Python programming language. The following ECMWF monthly means of daily means has been elaborated: solar net surface Radiation [kWh/m^2], 10 meter Wind Speed [m/s] Temperature at 2 meters from soil [K], Cloud Cover [%]. Two predicting curves of n-dimension respectively: n=72 and n=12 inherent the four climate datasets are generated. The first set of curves (n=72) was utilized to train the IFFT model; the second set (n=12), to test the model over one-year timeframe (June 2017 to June 2018). The developed FFT models (1) returned a transformed complex array ($y_d[k]$) and the corresponding frequency spectrum. A first complex array in the frequency domain (CAY) of k_1-dimension was generated with original n=72 datasets. A second complex array (CAA) of k_2-dimension, with n=12 array was created with the averages of monthly datasets.

$$y_d[k] = \sum_{n=0}^{N-1} e^{-2\pi j \frac{kn}{N}} x_d[n] \tag{1}$$

The two complex arrays have been used as input to return into the time domain the training and test predicting curves by using the inverse discrete transform (IFFT) as defined as:

$$x_f[n] = \frac{1}{M} \sum_{k=0}^{M-1} e^{2\pi j \frac{kn}{N}} y_f[k] \tag{2}$$

For testing purposes a typical test period of a simulation (n=12) was selected, suitable for conversion in hourly solar Radiation data (T. Khatib, and W. Elmenreich, 2015).

Three different routes have been pursued to return in the time domain, the predicting curves. The first and the second, the predicting curves were based on CAY and CAA complex arrays, where original spectrum have been considered. A third approach was based on a CAY complex array generated with a subset of the original frequencies, minimizing the Mean Squared Error (MSE). The MSE and the predicted array x_f [n] were minimized with an optimization algorithm that embeds a low-pass filter (LPF). LFP iterates the cut-off frequency of the complex array and attenuates signals with the frequencies (k_3) higher than the cut-off frequency as defined in Eq. 3 until the optimal MSE result were obtained. The best complex final array with k_3 spectrum, returned in the time domain thru the IFFT algorithm, the predicting curves (model validation).

$$\min f(k) = \frac{1}{N}\sum_{n=1}^{N}(x_d[n] - \frac{1}{M}\sum_{k=0}^{M-1}e^{2\pi j \frac{kn}{N}}y_f[k])^2$$

$$0 \le f_f[k] \le \frac{f_s}{2}$$

(3)

Finally the performances with R^2 index, which estimates the ratio between the square error and the variance. $R^2 = 1$ indicates the best performance:

$$R^2 = 1 - \frac{\sum_{n=1}^{N}(x_d[n] - x_f[n])^2}{\sum_{n=1}^{N}(x_d[d] - \mu_x)^2}$$

(4)

2.2. Detection of climate variable interrelations with different regression analysis

The scope was to define an alternative indirect method to build the predicting curves, starting from one of the variables. As proposed by S. Raschka (2015) an exploratory data analysis was executed to identify the presence of outliers, the distribution of the data, and the relationships between the variables, and then scatterplot was created from the matrix to visualize the pair-wise correlations. To quantify the linear relationship between the variable, a correlation matrix embedding the Pearson product-moment covariance coefficients was defined:

$$P_{x,y} = \frac{\sum_{n=1}^{N}\left[(x_d[n] - \mu_x)(y_d[n] - \mu_y)\right]}{\left[\sum_{n=1}^{N}(x_d[n] - \mu_x)^2\right]^{-2}\left[\sum_{n=1}^{N}(y_d[n] - \mu_y)^2\right]^{-2}}$$

(5)

where μ denotes the sample mean of the corresponding variable. The linear dependence between pairs of variables is strictly related to the value of Pearson coefficient within the range -1 and 1. A perfect positive linear correlation is expressed by $P_{x,y} = +1 / -1$, while no correlation if $P_{x,y} = 0$. The relationship among monthly climate variables with the Pearson's coefficient higher than 0.7 has been modeled by using: linear, quadratic and cubic polynomials. For variables with a weak Pearson's coefficient, as proposed by A. Liaw and M. Wienerthe (2002) the random forest method has been used. This method allows dividing a continuous regression curve into a sum of linear functions. The performance of the forest regression was again evaluated with the following parameters: R^2 and MSE.

3. Results

Three FFT-IFFT methods have been utilized to generate the predicting curves, reproducing the climate monthly data and compared the result with ERA5 datasets over the period June 2017 to June 2018. The reference location was: the city of Bremen, where the Latitude is 53.07 and the Longitude is 8.80; both in Decimal Degree. After detecting the frequency domain, the peak frequency of each dataset was identified. As showed in table 1, Radiance, Cloud Cover, Temperature were predictable with a limited spectrum (1,16-2,83 [Hz]), while Wind Speed reached the minimum MSE at a highest cut-off frequency (4,16 [Hz]). The Pearson's coefficient correlation matrix indicated a linear interrelation among the most predictable variables (Radiation, Temperature and Cloud Cover). After generating all the predicting training curves by filtering the frequencies at minimum MSE the R^2 index was measured. As reported in table 1 the best performing curves has been obtained with the filtered frequency dataset for Radiance while for Cloud Cover, Temperature and Wind Speed the average datasets has returned a better extrapolation.

Dataset	Radiance	Cloud Cover	Temp	Wind Speed
R2 test – average all spectrum	0,937	0,226	0,890	0,549
R^2 test - all spectrum	0,955	-0,014	0,868	0,172
R^2 test - filtered spectrum	0,956	0,196	0,877	0,114
Peak Frequency	1	1	1	1
Cutoff Frequency	1,16	1,16	2,83	4,16
MSE	0,058	0,003	4,025	0,223

Table 1 – Summary of the datasets performances.

An example of test results (Radiance) with a comparison among the three predicting curves is showed in fig.1. The profile of the curve of Temperature reflects a similar behavior. Although the predicting curves of Cloud Cover and Wind Speed did not match the original dataset, tendency of these curves were coherent to the original datasets.

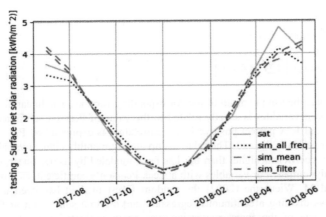

Fig. 1 - Comparison of test curves for Radiance.

After modeling the predicting curves, and visualizing a scattered matrix, the pair-wise correlations between the different variables and the Pearson Coefficients have been measured. The relationship between the Radiation and the other variables is characterized by a value higher than 0.6 (ssr-t2m = 0.81, ssr-tcc = 0.71, srr-si10 = 0.63), which reflects a stronger linear correlation. Looking at these Pearson's correlation coefficients, three different regression models have been implemented. The results delivered linear, quadratic and cubic polynomial regressions (fig. 2). The R^2 (0.65) was the same for all degree of regressions and confirmed the strong correlation among these variables.

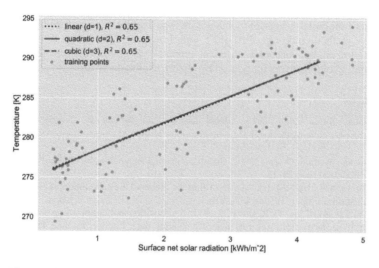

Fig. 2 - Linear, Polynomial regression and performance of two variables with a strong linear interrelation.

For the variables correlated with a lower Pearson coefficient (tcc-si10) the random forest regression obtained by subdividing the space into smaller regions and connecting these regions with piecewise linear functions gave a better result: in contrast to the polynomial regression models that returns $R^2=0.30$ with linear, and $R^2=0.34$ with quadratic and cubic regressions, forest tree regression has given $R^2=0.85$.

4. Conclusions

The world is facing massive energy and environment challenges due to global warming and the increase of energy demand. Renewable energy systems play an important role for a sustainable low-carbon energy transition thru microgrid, a major emerging technology concept that enables a tight integration of photovoltaic systems and wind turbines, with residential, commercial, industrial buildings. The design and management of a microgrid has to accomplish several challenging requirements while dealing with uncertainty of the renewable energy sources and the variability of load profiles. Model-based simulation tools are extensively used to identify the best-fit combination of distributed energy sources and the accuracy of input climate variables is pivotal for the quality of the simulations. The European Centre for Medium-Range Weather Forecasts (ECMWF) provides a large number of climate datasets but later works show that since now, their main use is in climate models and weather forecasts. In this paper, we presented a new application of these climate datasets for improving microgrid simulations. An innovative method based on Fast Fourier Transform combined with Inverse Fast Fourier Transform was used to convert original gridded ECMWF Solar Radiation, Wind Speed, Temperature, and Cloudiness datasets into the frequency domain and then the complex arrays have been transformed back into the time domain by filtering the frequency spectrum. The objective was minimizing the mean square error between the predicted and original climate variables. Three different routes have been utilized to train the algorithm. R^2 index applied to a further testing dataset indicated the resulting performances. Finally the Pearson's correlation coefficient among coupled datasets was used for the choice of the most suitable regression method to generate multiple strings of climate data, starting from a predicted dataset. This innovative method combining FFT-IFFT with linear regression proved to be easy and effective to predict accurate discrete climate variables suitable for microgrid simulations, avoiding errors introduced by historical climate records and observed data.

References

Aihui Wang, Xubin Zeng, 2015, Global hourly land surface air Temperature datasets: inter-comparison and climate change, International Journal of Climatology, 35, 13, pag.3959-3968.

Andy Liaw and Matthew Wiener, 2002, Classification and Regression by randomForest, Vol. 2/3, ISSN 1609-3631, pag.21-22.

Cooley, James, and Tukey, 1965, "An algorithm for the machine calculation of complex Fourier series," Math. Comput. 19, pag.297-301.

D. Quiggin, S. Cornell, M. T., R. Buswell, 2012, A simulation and optimisation study: Towards a decentralised microgrid, using real world fluctuation data, Energy, Volume 41, Issue 1, pag. 549-559.

E. Andres, Agostaa, 2014, The 18.6-year nodal tidal cycle and the bi-decadal precipitation oscillation over the plains to the east of subtropical Andes, South America. Int. J. Climatol. 34, pag.1606-1614.

Eliasen, E., Machenhauer, B., Rasmussen, E. ,1970, On a numerical method for integration of the hydrodynamical equations with a spectral representation of the horizontal fields. Report 2, Institut for Teoretisk Meteorologi, University of Copenhagen.

E. Kent, S. Fangohr, D. Berry, 2013, A comparative assessment of monthly mean Wind Speed products over the global ocean. Journal of Climatology 33, pag.2520–2541.

M. Wang, Z. Guan, D. Jin, 2018, Two new sea surface Temperature anomalies indices for capturing the eastern and central equatorial Pacific type El Niño-Southern Oscillation events during boreal summer, International Journal of Climatology, 38, 11, pag.4066-4076.

M. Joly, A. Voldoire, 2009, Role of the Gulf of Guinea in the inter-annual variability of the West African monsoon: what do we learn from CMIP3 coupled simulations?, Int. J. Climatol. 30, pag.1843–1856.

Orszag, S. ,1970, Transform method for calculation of vector coupled sums: application to the spectral form of the vorticity equation. J. Atmos. Sci., 27, pag.890–895.

Press, W., Teukolsky, S., Vetterline, W.T., and Flannery, B.P., 2007, Numerical Recipes: The Art of Scientific Computing, ch. 12-13. Cambridge Univ. Press, Cambridge, UK.

S. Amendola, F. Maimone, A. Pasini,F. Ciciulla and V. Pelinod, 2017, A neural network ensemble downscaling system (SIBILLA) for seasonal. Meteorol. Appl. 24, pag.157–166

S. Raschka, Python Machine Learning, 2015, pag.280-297.

T. Khatib, W. Elmenreich, 2015, A Model for Hourly Solar Radiation Data Generation from Daily Solar Radiation Data Using a Generalized Regression Artificial Neural Network, International Journal of Photoenergy, Vol. 2015, Article ID 968024, pag.13.

Anton A. Kiss, Edwin Zondervan, Richard Lakerveld, Leyla Özkan (Eds.)
Proceedings of the 29th European Symposium on Computer Aided Process Engineering
June 16th to 19th, 2019, Eindhoven, The Netherlands. © 2019 Elsevier B.V. All rights reserved.
http://dx.doi.org/10.1016/B978-0-128-18634-3.50278-2

A Financial Accounting based Model of Carbon Footprinting: Built Environment Example

Alex Veys[a], Virginia Acha[b], Sandro Macchietto[a,*]

[a] Department of Chemical Engineering, Imperial College London, South Kensington Campus, London SW7 2AZ, UK

[b] Imperial Business School, Imperial College London, South Kensington Campus, London SW7 2AZ, UK

* s.macchietto@imperial.ac.uk

Abstract

The built environment contributes up to 50% of the UK's carbon emissions. Most buildings' analyses focus on the "carbon operating cost" of buildings rather than the embodied "carbon capital cost" and do not take into account a building lifespan. In this paper a new carbon accounting framework is presented modelled on financial accounting principles, with a carbon balance sheet and profit and loss. It is argued that it is illogical to discount future emissions of carbon but that it is reasonable to depreciate carbon assets and liabilities. Like Value Added Tax (VAT) accounting in financial transactions, the proposed carbon accounting method enables a proper, explicit and transparent allocation of carbon assets and liabilities to multiple agents in complex supply chains over distinct temporal, spatial and organisational boundaries.

The approach is demonstrated by applying the model to two energy-efficient buildings of the UK National Energy Foundation, a new Light-Weight building and an older High Thermal Mass one, designed for a lifespan of 50 and 150 years, respectively. The analysis shows that cement based materials, metal and insulation are the major carbon liabilities, with concrete foundations the main structural component responsible. On an operating carbon basis, the newest building produces the least CO_2. In aggregate, the new building is more efficient with an annual carbon cost, including depreciated carbon capital costs, of ~14 tonnes of CO_2/year, vs. ~21 tonnes of CO_2/year for the longer lasting building. The best discounted annual capital carbon cost was for a (notional) building constructed by increasing the lifetime of the 50-year building to 150 years. However, in all buildings the discounted carbon capital cost was less than 11% of the total operating costs, showing that improvements in operating efficiency (carbon P&L) are the most important contribution to carbon emissions.

Keywords: CO_2 accounting model, built environment, building lifetime; construction techniques comparison, CO_2 efficiency assessment.

1. Introduction

We are regularly reminded of the need to radically cut emissions of CO_2 and other greenhouse gases (in this paper, collectively called "carbon" as a shortcut). Whether the cost of carbon externalities is better included through a carbon tax on products and services, or a regulatory framework with permits allocation and trading is hotly debated. Whatever the solution, the amount of carbon emitted will need to be counted and attributed, so that individuals and organisations may be held accountable for their contribution. This is more easily said than done, as most products and services involve multiple materials, parts, suppliers, distributors and end users in highly complex supply chains, across intersecting geographical, national, regulatory, business entity and legal boundaries, to name a few. To properly allocate carbon origination, use, transfer and cost requires agreed, transparent and easily applied carbon accounting mechanisms. While this seems like a daunting problem, it has been addressed and solved in other areas. In particular, most governments have found ways to tax whatever we, as individuals or corporates, produce, earn, trade, spend, or consume (with exclusions, allowances, progressive or regressive rates, etc.) in ways that are (broadly) socially accepted. One such mechanism taxes the "value added" in each step by each entity.

Conventional material and energy balances over a set system boundary and temporal timeframe define their accrual or depletion within the system (Fig. 1, left). The financial state of a system is similarly defined by a set of money balances, done according to conventional principles or rules, and incorporated in three key documents, a Balance Sheet (assets and liabilities at a given time), Profit and Loss (flows in/out from activities between balance sheets) and Cash Flow (covering the same time frame as the P&L but for cash only). Their relationship is schematically shown in Fig. 1, right. Three similar balances and documents were developed to account for carbon assets (value held) or liabilities (value due), flows and cost over defined physical, organisational and temporal boundaries. The value of carbon added (VCA) by a material flow or process, the overall total embodied carbon, EC (sum of all VCAs contributions up to a given time and carried forward) and the overall life cycle impact, LCC of a product or service (adding up all contributions from cradle to grave) are defined. The latter is similar to the values produced by a life cycle analysis, LCA. Entries from single activities or flows are aggregated into the carbon accounting reports (balance sheet, P&L).

Figure 1. Flows over a physical boundary (e.g. a building), left. Relationship between key financial accounting reports over organisation (e.g. a company) and temporal boundaries (e.g. 1 year), right.

1.1. Carbon accounting and buildings

Half of humanity lives in cities. In the UK, the built environment contributes up to 50% of carbon emissions. The UK construction industry consumes over 420 Mt of materials, 8Mt of oil and releases over 29 Mt of carbon dioxide annually, including a significant quantity of new materials disposed as waste (Hammond and Jones, 2008). Capital cost, not just economic but also in terms of the carbon embodied in materials and construction costs ("carbon capital") required to achieve such objectives is also important. Some estimate suggest that the embodied energy in domestic buildings might be ten times the annual operating energy requirements and in commercial buildings the ratio could be as high as 30:1 (Rawlinson and Weight, 2007). Most buildings' analyses and construction regulations focus on the "carbon running cost" of buildings rather than the "carbon capital cost" and do not take into account a building lifespan, which our method proposes to do. Finally, building design, construction, and use involve many entities, each of which will likely be required in the future to produce detailed, non-overlapping, mutually consistent carbon accounting reports. Counting and attributing carbon assets and liabilities is particularly challenging.

Traditional Life Cycle Analysis (LCA) does not account for temporal information and assumes a fixed horizon, often 100 years, with results sensitive to the chosen horizon. More recent advances addressed this aspect. Levasseur et al., 2010 proposed a Dynamic LCA (DLCA), of which there are versions related to the timing of processes, other to the proximity to the time horizon in the study scope. Collinge et al., 2018 applied two static and four time-resolved DLCA models to the analysis of two energy efficient buildings, however only to the buildings operating (use) phase. Pan et al., 2018 proposed a fundamental rethinking of system boundaries for the carbon emissions of buildings to include temporal, spatial, functional and methodological dimensions based on 12 variables. In spite of progress, a recent review of methodologies and applications for the carbon footprint of buildings (Fenner et al., 2018) concluded that there is no internationally accepted method for measuring, reporting, and verifying GHG emissions from existing buildings in a consistent and comparable way and indicated the need for a clear, accessible and consistent method.

Given the long lifetime of a building, financial concepts such as discounting (for future events) and depreciation (for past events) must be considered. A discount factor implicitly values carbon emitted in the future at lower price than emitted now. For example, the future cost of permits in a trading scheme, or the future cost of abatement technologies may decrease. As this is controversial and impossible to estimate on a long timescale (and arguably it is immoral to discount current liabilities to future generations), no discounting was applied. Depreciation is a measure of cost or value of an asset that has been consumed during a period. For carbon accounting purposes, assets should be written down at the faster of a model that estimates the decay of CO_2 in the atmosphere (such as the Bern model used in IPCC reports, Schimel et al., 1997) or the straight line that takes its value to zero at the end of its life, while liabilities should be depreciated over the lifetime of a product. Here, assets and liabilities were aggregated into net assets, depreciated at the shorter of the Bern rule or the lifetime of the building.

For all materials we used the ICE (Inventory of Carbon and Energy) database (Hammond and Jones, 2008). Together with relevant properties (e.g. thermal conductivity, density, specific heat) ICE gives the embodied energy and carbon of construction materials, by main groups, for Typical, Primary and Recycled material, with ranges and accuracy estimates. Values are qualified as being cradle to (factory) gate, to (building) site or to grave. At the project time, ICE1.5a contained records for 1400 construction materials, of which 22 (main materials only) were used (and cross checked against other sources). For each material the embodied carbon in ICE does not include an initial asset contribution, so an adjustment was made for biomass based materials. For example, manufacturing of softwood timber takes ~0.47KgCO$_2$/Kg, but assuming ~50%$_w$ of Primary wood is carbon drawn from the atmosphere, the initial carbon asset of softwood timber becomes -1.354KgCO$_2$/Kg. All timber products end up with a negative initial embodied carbon value.

2. Case study: Carbon accounting for buildings with different lifetime

The UK National Energy Foundation (NEF) was set up in 1990 to "provide help to improve energy efficiency in residential buildings". In addition to providing advice to companies and large corporations, they prepare inspectors for statutory home energy efficiency assessment and the UK National Home Energy Rating scheme. NEF is housed in two main buildings (Fig. 3), built to demonstrate energy efficiency practices. The Phase 1 building with ~1,000m^2 net floor area, finished in 1999, was designed based on a High Thermal Mass concept (to store heat and maintain the building's temperature) using conventional materials, with a lifespan of 150 years. It features amongst other devices a steel frame with brick outer skin, insulated cavity and poured concrete inner skin, a high thermal mass insulated roof, massive concrete highly insulated raft floor, internal light shelves to boost daylight and a 500W PV array. The Phase 2 building with ~450m^2 net floor area, finished in 2004, was designed based on a Lightweight Building concept using a timber frame and minimal use of concrete, with a lifespan of 50 years. It features amongst other devices a Ground Source Heat Pump (GSHP) driven by mains electricity feeding a 3 zone underfloor heating systems, solar hot water, eight Sunpipes to boost daylight and a 6.5kW rooftop PV array.

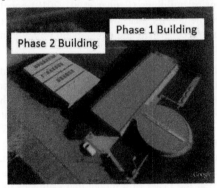

Figure 3. NEF Buildings Phase 1 and 2.

2.1. Carbon analysis

A carbon balance sheet was calculated for the two buildings, comprising carbon "capital" assets (carbon embedded in the building materials, positive or negative, discounted over the expected lifetime of the building) and annual carbon contributions from "operating" costs. As the two buildings are rather different, for ease of comparison calculations were also done for a notional building of the same dimension and plans as the Phase 2 building, but built with the more durable construction materials of Phase 1. This is denoted as Phase 2.1 building. The net carbon assets in each structure, based on detailed construction plans, drawings, records and material samples provided by NEF, are summarised in Fig. 4 by main type of materials used (left) and construction element (right). Some numerical values are shown in Table 1 (left). The embodied carbon in Phase1 Building is rather evenly distributed between foundations, floors, roofs and walls, while in the Phase 2 building over 95% is in foundations and floors, with the lightweight walls and roof hardly contributing. "Operating" carbon was calculated from a detailed analysis of energy suppliers and use in the buildings (Fig. 5). Table 1(right) summarises the buildings capital, operating and total discounted annual carbon.

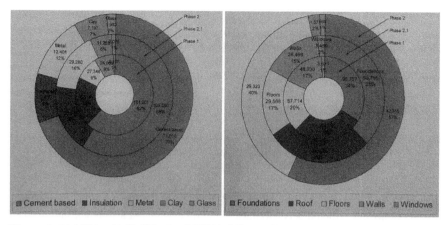

Figure 4. Liabilities in KgCO$_2$ for NEF Buildings Phase 1, Phase 2.1 and Phase 2 by main materials (left) and main building structure (right).

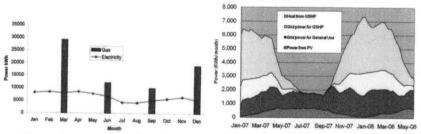

Figure 5. Energy used (kWh): Monthly average for Phase 1 building (left); usage profile for Phase 2 building over 17 months (right).

Table 1. Liabilities [KgCO$_2$] of Phase 1, Phase 2.1 and Phase 2 buildings by main materials and structure - errors based on estimated uncertainties of data sources (left); Summary of Carbon Opex, Capex and discounted overall P&L in annual terms (right).

Material	Phase 1 Footprint (kg CO2)	Phase 1 Std. Dev (kgCO2)	Phase 2.1 Footprint (kg CO2)	Phase 2.1 Std. Dev (kgCO2)	Phase 2 Footprint (kg CO2)	Phase 2 Std. Dev (kgCO2)
Timber Plywood	-10,971	5,485	-6,192	3,096	-25,427	12,714
Timber Chipboard	-11,776	5,888	-7,722	3,861	-1,069	535
Timber softwood	-1,226	613	-1,150	575	-6,362	3,181
Polyethylene Pipe 16	0	0	0	0	210	63
Plaster	497	248	274	137	0	0
Vapour Barrier	0	0	0	0	1,259	378
...						
Plasterboard	1,690	845	1,263	631	3,085	1,543
Aluminium	2,649	530	2,533	507	0	0
glass soda lime	2,198	835	2,075	789	1,983	753
Insulation ExPoly	0	0	0	0	3,984	1,195
Breeze Block	16,500	4,950	9,103	2,731	0	0
Steel Beams	24,697	7,409	16,409	4,923	0	0
Zinc	18,143	5,443	10,338	3,101	12,206	3,662
Insulation Celotex	57,786	8,668	33,165	4,975	0	0
Brick	24,000	7,200	13,241	3,972	7,197	2,159
Concrete	162,515	48,754	99,425	29,827	59,334	17,800
Total	290,000	52,000	170,000	32,000	71,000	23,000
rounded to nearest:	10,000	1,000	10,000	1,000	1,000	1,000

by STRUCTURE	Phase 1	Phase 2.1	Phase 2
Foundations	95,757	63,795	42,345
Roof	80,045	48,305	7,024
Floors	57,714	29,568	29,323
Walls	48,030	26,499	1,671
Windows	3,621	3,459	869
Supports	1,534	1,139	41
GSHP	0	0	728

Operating Carbon (annual P&L)		Phase 1	Phase 2.1	Phase 2	2/1 over 2
Energy (Heat)					
GSHP (Renewable)	kWh/ yr				
Grid GSHP (Carbon)	kWh/ yr				
Gas (Carbon)	kWh/ yr			19,869	
		70,400	40,800	8,089	
Energy (General)					
PV (Renewable)	kWh/ yr			5,053	
Grid (General)	kWh/ yr	52,900	23,100	15,994	
Total Heat	kWh/ yr	70,400	40,800	27,958	46%
Total General Elec	kWh/ yr	52,900	23,100	21,047	10%
Carbon Energy	kWh/ yr	123,300	63,900	24,083	165%
Renewable Energy	kWh/ yr			24,922	
Total Operating Energy	kWh/ yr	123,300	63,900	49,005	30%
Factor					
Gas	kgCO2/kWh	0.206	0.206	0.206	
Electricity	kgCO2/kWh	0.523	0.523	0.523	
Carbon from					
Gas	kgCO2/yr	14,502	8,405		
Electricity	kgCO2/yr	27,667	12,081	12,595	
Total Operating Carbon	kgCO2/yr	42,169	20,486	12,595	63%
Capital Expenditure (Completion B/S)					
Construction Carbon	kgCO2	290,000	170,000	68,000	
Waste at start	%	10%	10%	10%	
Waste at end	%	1%	1%	1%	
Lifetime	years	150	150	50	
Total Capital Carbon (annual depreciation)	kgCO2/yr	2,146	1,258	1,510	-17%
Carbon ratio: Capex / (Capex + Opex)	%	4.8%	5.8%	10.7%	
Annual footprint (P&L including depreciation)	kgCO2/yr	44,315	21,744	14,105	54%

3. Conclusions

The Phase 1 energy efficient building, built with traditional materials, embodied ~290 tonnes of "capital" CO_2 (with an estimated error of 52 tonnes), vs 71 tonnes (+/- 23 tonnes) for the Phase 2 building and 190 tonnes (+/- 32) for the notional 2.1 building. The lightweight, 50-year lifetime Phase 2 building was found to be the most carbon efficient even when carbon embodied in the longer lasting building was discounted over a much longer period. The carbon Capex of all buildings being low relative to their carbon operating costs, the impact of depreciation over longer or shorter building lifetimes was limited. However, it increased from 5% to 11% of discounted annual costs from Phase 1 to Phase 2. As buildings become more efficient, carbon capital cost becomes more important. Further details of method and case study are in Veys (2008).

The proposed (financial-inspired) carbon accounting method enables the assessment of carbon investments with distinct lifetimes. It also enables the attribution, holding and transfer of arising carbon assets/liabilities to multiple agents in a complex supply chain over distinct temporal, spatial and business boundaries. As for VAT tax accounting, it is sufficient for each organisation, product or service to include in its carbon balance sheet just the net liabilities arising between accounting periods from its carbon P&L flows.

Acknowledgements

The authors wish to thank the UK National Energy Foundation for providing the case study, data, information and many useful discussions, and Bath University for access to the ICE database.

References

Collinge W.O., H. J. Rickenbacker, A. E. Landis, C. L. Thiel and M. M. Bilec, 2018, Dynamic Life Cycle Assessments of a Conventional Green Building and a Net Zero Energy Building: Exploration of Static, Dynamic,Attributional, and Consequential Electricity Grid Models, *Environ. Sci. Technol.*, 52, 11429–11438

Fenner A. E., C. J. Kibert, J. Woo, S. Morque, M. Razkenari, H. Hakim and X. Lu, 2018, The carbon footprint of buildings: A review of methodologies and applications, *Renewable and Sustainable Energy Reviews,* 94, 1142–1152

Hammond, G. P. and C. I. Jones, 2008, Embodied energy and carbon in construction materials, Proceedings of the Institution of Civil Engineers - Energy, 161 (2) 87-98.

Levasseur A., P. Lesage, M. Margni, L. Descenes and R. Samson, 2010, Considering Time in LCA: Dynamic LCA and Its Application to Global Warming Impact Assessments, *Environ. Sci. Technol.*, 44, 3169–3174

Pan W., K. Li and Y. Teng, 2018, Rethinking system boundaries of the life cycle carbon emissions of buildings, *Renewable and Sustainable Energy Reviews*, 90, 379–390

Rawlinson S. and D. Weight, 2006, Sustainability: embodied carbon. *Building Magazine*, 12 October 2007, 88–91.

Veys A., 2008, A Financial Accounting based Model of Carbon Footprinting using an Example in the Built Environment, *MSc Thesis*, Imperial College London.

Schimel D., M. Grubb, F Joos, R. Kaufmann, R. Moss, W. Ogana, R. Richels, T. Wigley, 1997, Stabilization of Atmospheric Greenhouse Gases: Physical, biological and Socio-economic Impilcations, pp. 1-48 in *IPCC Technical Paper III*, IPCC, Geneva.

Anton A. Kiss, Edwin Zondervan, Richard Lakerveld, Leyla Özkan (Eds.)
Proceedings of the 29th European Symposium on Computer Aided Process Engineering
June 16th to 19th, 2019, Eindhoven, The Netherlands. © 2019 Elsevier B.V. All rights reserved.
http://dx.doi.org/10.1016/B978-0-128-18634-3.50279-4

Population Balance Equation Applied to Microalgae Filtration

Pui Ying Lee[a], Keat Ping Yeoh[a], Chi Wai Hui[a,*]

[a]*Department of Chemical and Biological Engineering, The Hong Kong University of Science and Technology, Clear Water Bay, Hong Kong SAR*

*kehui@ust.hk

Abstract

In this paper, a mathematical model is developed for microalgae filtration by applying Population Balance Equation. The main objective is to investigate the impacts of cell size distribution and filter average pore size on filtration efficiency and their interrelationships. The results of using standard filters with different pore sizes are compared and discussed. Experiments of microalgae filtration on Chlorella cells are performed to provide size distribution information of the cell cultures to validate the model. Behaviours of cells at different sizes during the harvesting process are examined and the obtained data is combined with the mathematical model to evaluate the harvesting efficiency along with operation time and filter depth. The model could be extended to predict the size specific filtration efficiency of microalgae under different operation conditions, which potentially helps determine the process parameters for optimization purposes.

Keywords: Microalgae harvesting, Filtration, Population Balance Equation

1. Introduction

Biofuel, as a renewable energy, is an alternative to the traditional fossil fuel. One of its bio-resources is microalgae, which is regarded to be promising with high lipid content and high photosynthetic efficiency. Microalgae harvesting has been the bottleneck of the whole microalgae production process since it accounts for 20-30 % of the cost (Danquah et al., 2009). Therefore, maximizing the harvesting efficiency as well as capacity and meanwhile minimizing the cost are of critical importance to the development of a scale-up process for industrial purposes.

Filtration is a commonly employed harvesting method for microalgae because of its high efficiency as well as stable operation conditions without using chemicals. Although research efforts have been dedicated to investigate the efficiency of microalgae filtration, few modelling works have been done to understand the relationships between different physical parameters. Previous studies usually consider the algal culture, which contains different cell sizes, as a whole for performance evaluations (Wicaksana et al., 2012; Zhao et al., 2017). However, not many concerns have been raised towards the interdependence of cell size distribution and the corresponding cell recovery although size distribution is a critical factor of the filtration process.

The main objective of this study is to apply Population Balance Equation (PBE) to simulate the microalgae filtration process. This equation has been commonly used in chemical engineering processes to describe evolutions of particle populations. There are few studies employing PBE for modelling microalgae growth (Pahija et al., 2017) and harvesting processes of sedimentation and centrifugation (Lee et al., 2018). So far there is no significant research contribution to the application of PBE to model the microalgae filtration process and to study the impact of cell size distributions or characteristics. In this study, a PBE model is proposed for investigating microalgae filtration efficiency, which derives additional size specific information of the cells during the process.

2. Methodology

Applying Population Balance Equation to microalgae filtration, a model is proposed with the aim to study size dependent recovery and filtration coefficient of the process. Experimental data is used to evaluate the process efficiency and to validate the mathematical model.

2.1. Population Balance Equation model
One of the simplified forms of PBE for studying particle size distributions commonly includes two terms as follow:

$$\frac{\partial n(t,d)}{\partial t} + \frac{\partial [G \cdot n(t,d)]}{\partial d} = 0 \tag{1}$$

The first term on the left side denotes changes in particle number density with time, while the second term represents the change with particle size, in which n is number density, t is time and G is particle growth rate. To model the microalgae filtration process, Eq. (1) is modified into the following form in this study:

$$\frac{\partial n(t,h,d)}{\partial t} + v\frac{\partial \big[[1 - \phi(h,d)] \cdot n(t,h,d)\big]}{\partial h} = 0 \tag{2}$$

The first term is similar to that in Eq. (1), while the second term is revised to be the cell number density change with h, which denotes traveling distance of cells inside the filter medium. ϕ is defined as filtration coefficient, so that the fluid velocity v times $[1 - \phi(h,d)]$ together gives the cell velocity inside the filter medium. The larger the coefficient, the smaller the cell velocity compared to fluid velocity, resulting in higher harvesting efficiency of microalgae. First order upwind discretization scheme is used to generate numerical solutions for Eq. (2).
Since the model parameters i.e. cell number density, cell size, time, filter depth and flux can be easily measured, the filtration coefficient can be back-calculated for evaluations of the process efficiency. In addition, the main purpose of the model application in this paper is to compare and study the impacts of different filter mediums on cell size distributions during microalgae filtration.

2.2. Experiments
Chlorella is a common green spherical microalgae species and the cell size ranges from 3 um to 9 um. In this study, an experimental setup of dead end vacuum filter with filter diameter of 50 mm is used to separate the cells from the cultivation medium. To

evaluate the impacts of filter pore size on the harvesting process, standard polypropylene filters with different average pore sizes of 6, 8, 10, 15 and 20 um are used. Each run of the experiment filters 10 mL microalgae solution for 1 min and the filtrate's optical density (OD) as well as size distribution are measured. In every sampling, 1.5 mL filtrate is transferred by an electronic pipette to a photospectrometer for OD measurement to obtain the solution concentration. Meanwhile, a micropipette is used to take 10 uL filtrate for imaging under a microscope and the images are processed and analysed by a computer software to give cell size distribution information. A calibration curve of the relationship between cell concentrations and OD is experimentally attained prior to conducting the filtration experiments.

3. Result

Performances of filtration is commonly evaluated by the change in liquid concentration after the process. In this study, the filtration efficiency is defined as the percentage change in optical density, which is proportional to the cell number density, from the original algal solution to the collected filtrate. The overall filtration efficiency represents the percentage change in cell number density of the solution as a whole without taking the specific change of cells at each size into consideration. After the experiments and data analysis, the overall filtration efficiencies of microalgae filtrations with different filter mediums are obtained. With the previously mentioned calibration curve, the cell number densities of the filtrate samples can be derived from the OD measurements. By fitting the cell size distribution results into the proposed PBE model, the filtration coefficient ϕ defined in the model can be derived.

As shown in Figure 1, the filtration efficiency is inversely proportional to pore size as expected and the relationship is observed to be almost linear. Although the pore size of from 15 to 50 um are obviously larger than the cell size range of 3 to 9 um, considerable amounts of cells are captured. Therefore, concerns are raised towards size selectivity of the microalgae filtration process and the mechanisms behind. With an original mean cell size of 5.381 um, Figure 2 shows the mean filtrate cell size, which drops with decreasing filter pore size. From the pore size of 6 um to 15 um, it is observed that smaller mean pore size filters can effectively capture smaller size cells since the cells cannot penetrate through the narrow pores. However, a drop of mean captured cell size is observed at 20 um pore size, implying that an increasing number of small cells are captured even though the pores become larger. This data briefly indicates the tendencies of the filters to capture cells at different sizes.

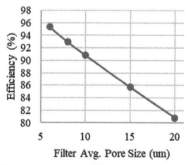

Figure 1. Impact of filter pore size on the overall cell harvesting efficiency

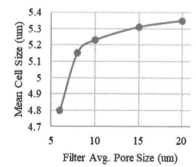

Figure 2. Impact of filter pore size on the mean filtrate cell size

Analysing size distributions of the captured cells can give additional information of microalgae filtration for studying the changes in mean cell size during the process. Figure 3 presents the size distributions of captured cells after the filtration experiments. It agrees with the previous findings that the number density of microalgae in the filtrate generally decreases when a filter with smaller pores is used. Applying Population Balance Equation to simulate the filtration process, the harvesting efficiencies of microalgae cells at each size changing along with time and travelling distance throughout the process are investigated. After numerical simulations of the developed PBE model with size distribution data from the experiments, the size specific filtration coefficients of microalgae filtered by medium of different pore sizes are obtained as shown in Figure 4.

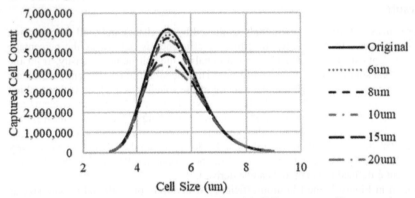

Figure 3. Size distributions of captured cells with different mean filter pore sizes

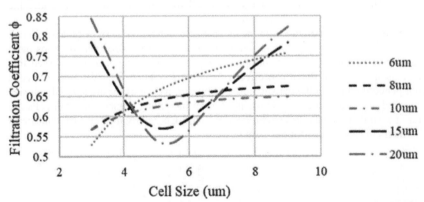

Figure 4. Derived size specific filtration coefficients of microalgae from the PBE model

Two different patterns are observed for smaller pore sizes (6, 8 and 10 um) and larger pore sizes (15 and 20 um). For small pore sizes of 6, 8 and 10 um, the coefficient increases with cell size, indicating that larger cells are more favourable to be captured. Among the three curves, the 6um pore size curve has the steepest slope, which means it is more sensitive to cell size and its selectivity for large cells is higher. For large pore sizes of 15 and 20 um, convex curves are observed, indicating that both large and small cells are more favourable to be captured. The 20 um pore size curve is comparatively steeper, such that the filter captures small and large cells better but oppositely it captures middle size cells worse than 15 um pore size filter.

4. Discussion

Comparing the overall filtration efficiencies of the tested standard filters, the filters with smaller pore size are generally more effective in capturing the microalgae cells. Studying the pattern of the derived size specific filtration coefficients gives additional insights of the process size selectivity as well as the filtration mechanisms behind. It is shown in Figure 4 that 6, 8 and 10 um filters demonstrate similar size selectivity patterns, which the direct interception mechanism is essentially dominant. 6 um filter gives the best overall filtration efficiency, but its performances for capturing 3 and 3.5 um cells are the worst. For 15 and 20 um filters, although their resulted changes in mean captured cell size are different, their derived filtration coefficients give similar patterns. The capture of more large cells can again be explained by the direct interception mechanism. Meanwhile, the tendency of capturing more small cells by larger pore filters shows that other mechanisms which are selective towards small cells are dominant. Particularly, the mechanism of electrostatic effect favours the capture of small cells since microalgae cell surface is known to be negatively charged and it is enhanced with increasing charge intensities of the cells. The convex pattern implies that in addition to direct interception, electrostatic effect is also affecting the process size selectivity. This is in alignment with the literature finding of a minimum filtration efficiency at a certain particle size in micron scale when the pore size to particle size ratio becomes large (O'Melia and Ali, 1978; Yao et al., 1971), confirming the results of this study.

With the PBE model, the actual size selectivity of the filters are derived. If only the overall filtration efficiency or mean cell size is considered, a very brief conclusion of the filters' capturing tendencies may be drawn. For example, although 15 um filter apparently tends to capture more large cells than small cells as shown in Figure 2, it actually tends to capture both small and large cells as shown in Figure 4. Its size selectivity pattern is also quite different from those of 6, 8 and 10 um filters, which this information is obtained by the model simulation. The new modelling approach to examine the size specific performance of microalgae harvesting derives additional information of the process size selectivity. This is useful for understanding the interrelationship between process parameters and cell characteristics, including cell size and surface charge.

5. Conclusion

A Population Balance Equation model is developed for modelling microalgae filtration, which is a common harvesting process in microalgae production. The model effectively describes changes in cell number densities with cell size, filter pore size, filtration time together with filter depth during the process. Filtration experiments are carried out on Chlorella cells by using filters different pore sizes. Experimental data, including the overall filtration efficiency, mean filtrate cell size and size distribution, are processed and analysed from a size specific approach. The results are combined with the developed mathematical model to study the impacts of cell size distribution and filter pore size on the filtration performance as well as their interrelationships. Furthermore, size specific capturing of the cells is investigated for a clearer picture of the process size selectivity.

The derived size specific filtration coefficient provides insights to anticipate the effectiveness of different filter mediums for the filtration process. Instead of analysing the cell culture as a whole, this model focuses on size selective properties of microalgae filtration. The PBE model can also be extended to study and predict microalgae size distributions throughout the filtration process with different parameters or set ups. Meanwhile, it is beneficial for the determination of process parameters, such as filter pore size, material, thickness and filtration time for specific size selective separation purposes as well as process optimization.

References

MK. Danquah, L. Ang, N. Uduman, N. Moheimani, GM. Forde, 2009, Dewatering of microalgal culture for biodiesel production: exploring polymer flocculation and tangential flow filtration, Journal of Chemical Technology and Biotechnology 84, 7, 1078-1083

PY. Lee, E. Pahija, Y. Liang, KP. Yeoh, CW. Hui, 2018, Population balance equation applied to microalgae harvesting, Proceedings of the 28th European Symposium on Computer Aided Process Engineering, Graz, Austria

C.R. O'Melia, W. Ali, 1978, The role of retained particles in deep bed filtration, Proceedings of the 9th International Conference, Stockholm, Sweden, 167-182

E. Pahija, Y. Liang, CW. Hui, 2017, Determination of microalgae growth in different temperature condition using a Population Balance Equation, Chemical Engineering Transactions 6, 721-726

F. Wicaksana, AG. Fane, P Pongpairoj, R. Field, 2012, Microfiltration of algae (Chlorella sorokiniana): Critical flux, fouling and transmission, Journal of Membrane Science 387-388, 83-92

K.M. Yao, M.T. Habibian, C.R. O'Melia, 1971, Water and wastewater filtration: Concepts and applications, Environmental Science and Technology 5, 11, 1105-1112

F. Zhao, H. Chu, Z. Yu, S. Jiang, X. Zhao, X. Zhou, Y. Zhang, 2017, The filtration and fouling performance of membranes with different pore sizes in algae harvesting, Science of the Total Environment 587-588, 87-93

Anton A. Kiss, Edwin Zondervan, Richard Lakerveld, Leyla Özkan (Eds.)
Proceedings of the 29th European Symposium on Computer Aided Process Engineering
June 16th to 19th, 2019, Eindhoven, The Netherlands. © 2019 Elsevier B.V. All rights reserved.
http://dx.doi.org/10.1016/B978-0-128-18634-3.50280-0

Contract Settlements for Exchanging Utilities through Automated Negotiations between Prosumers in Eco-Industrial Parks using Reinforcement Learning

Dan E. Kröhling, Ernesto C. Martínez[*]

INGAR (CONICET-UTN), Avellaneda 3657, Santa Fe, S3002 GJC, Argentina

ecmarti@santafe-conicet.gov.ar

Abstract

Peer-to-peer trading of utilities (heating, cooling and electric power) in an eco-industrial park (EIP) is key to realize significant economic and environmental benefits by exploiting synergistic co-generation and surplus trading. A crucial aspect of this symbiotic scheme is that each selfish company (prosumer) will participate in exchanging utilities to increase its own profits depending on its internal load. In this paper, an automated negotiation approach based on utility tokens is proposed to incentivize participation in a market of prosumer peers for trading surpluses in an EIP. During each negotiation episode, a pair of prosumers engage in a bilateral negotiation and resort to a learned policy to bid, concede and accept/reject using both private information and environmental variables. Contract negotiation revolves around agreeing (or not) on the price expressed in tokens of a utility profile. The time-varying value of the utility token accounts for contextual variables beyond the control of each prosumer. Simulation results demonstrate that reinforcement learning allows finding Nash equilibrium policies for smart contract negotiation in a blockchain environment.

Keywords: automated negotiation, blockchain, reinforcement learning, smart contracts.

1. Introduction

An EIP mainly consists of distributed co-generation (e.g. heating, cooling and electric power), shared storage systems and flexible/distributed demand/supply of intermediate products and waste materials (Chertow, 2007; Nair et al., 2016). A key aspect of this symbiotic scheme is that each potential partner company will only participate in exchanging utilities through loans or credits with the motivation of increasing its own profits. The selfish nature of each prosumer thus results in a conflict of interest which, if not resolved, may give rise to the failure of the EIP (Chew et al, 2009; Tan, et al., 2016). It is argued here that the Blockchain approach based on a distributed ledger (Sikorski et al., 2017) and smart contracts (Kristides and Devetsikiotis, 2016) can be used to foster decentralized, peer-to-peer automated negotiations between prosumer companies that want to trade their utility surplus with other peers in the park. Considering heterogeneous prosumer preferences, which are private information, each peer company (represented by a software agent) should learn to negotiate how to bid and what to accept in utility contract settlements by properly accounting for its context including weather, scheduled utility loads, season, oil prices, mood of the economy, etc. The strategic behavior of a prosumer (proactive consumer actively managing its consumption and production of different types of utilities) in EIPs depends significantly on the influence of these environmental conditions or contextual variables on its negotiation policy, since they affect not only a given agent preferences and decision-making policies, but also those of other agents. A novel context-driven approach for policy learning in automated peer-to-peer negotiations between prosumers of an EIP is proposed. Reinforcement learning (Sutton and Barto, 2018) is key to learn negotiate utility contracts via simulation by properly accounting for most relevant environmental variables, other peer negotiation policies, their models and contextual information, which provides each prosumer agent a competitive edge while incentivizing its participation in an EIP near to Nash equilibrium conditions with other prosumers.

2. Peer-to-peer negotiations of utility contracts

Utility prosumers are agents that both consume and may produce different types (heating, cooling and electric power) of utilities in an EIP. Each prosumer company is assumed to be equipped with one or more utility generation means (e.g. a boiler, solar panel, freezer, heating/quenching streams, etc), has a scheduled profile load and, possibly, a temporary surplus for some of its internal generated utilities and extra needs for other utilities. Peer-to-peer local trading of such surpluses by taking advantage of material and energy integration in an EIP gives rise to unique opportunities to lower emissions and reduce energy costs.

Each company in the EIP is modeled as a software agent, $i \in N$, where N is the set of engaged prosumer agents in the peer-to-peer utility market of the EIP at time t. It is assumed that, depending on the utility needs for their scheduled load and surplus, if any, each agent negotiates buying/selling utility profiles and their prices in tokens (τ_t). For example, in Fig. 1 agent A, at time t, seeks and agreement with agent O_3 to sell part of its heating power. Concurrently, agent A aims to buy renewable electric power from agent O_1 and some cooling from agent O_2. The context for these negotiations is private information (X_i) and environmental variables (Y_j). The specific type for an agent i is summed up in state variable v^i, known as "necessity", whereas its external context is given by a "risk" ρ^i. Each agent necessity accounts for the needs and surplus derived from capacity slacks and scheduled usage loads. In turn, risk accounts for the different factors that may influence the current and future values of the token $\tau_t, \tau_{t+1}, \dots, \tau_{t+H}$.

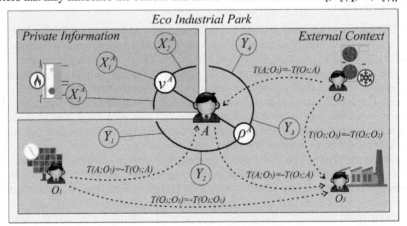

Fig. 1. Peer-to-peer negotiation setting at time t for utility surpluses in an EIP.

From the point of view of a central EIP authority, the exchange utility problem requires maximizing the overall value of surplus traded assuming perfect information about both public variables and private prosumer types, that is, their necessities and risks as follows

$$Max \ \sum_u \sum_t \sum_i T_t^u(i; -i) \tag{1}$$

subject to:

$$T_t^u(i) + T_t^u(-i; i) - T_t^u(i; -i) \geq NCT_t^u(i) \quad \forall i, u, t \tag{2}$$

$$T_t^u(i) + T_t^u(-i; i) - T_t^u(i; -i) - NCT_t^u(i) = MCT_t^u(i) - SCT_t^u(i) \quad \forall i, u, t \tag{3}$$

$$T_t^u(i; -i) \leq SCT_t^u(i) \quad \forall i, u, t \tag{4}$$

$$T_t^u(i), T_t^u(i; -i), NCT_t^u(i), MCT_t^u(i), SCT_t^u(i) \geq 0 \quad \forall i, u, t \tag{5}$$

where each variable represents:

- $T_t^u(i)$, the total amount of utility u in tokens traded by agent i at time t.
- $T_t^u(-i; i)$, the amount of u in tokens bought by i to all other agents $-i$ at time t.
- $T_t^u(i; -i)$, the amount of u in tokens sold by i to all other agents $-i$ at time t.
- $NCT_t^u(i)$, the overall power of u in tokens required by i in t to fulfil its necessity.
- $MCT_t^u(i)$, the max power of utility u in tokens that i could produce at time t.
- $SCT_t^u(i)$, the surplus capacity of u in tokens that i is ready to sell at time t.

The problem aims to maximize the number of tokens (utilities) traded between agents. This gives rise to a maximization of the EIP's social welfare while minimizing the surplus capacity loss over time, considering the maximum power per agent, its internal generation and scheduled usage. The assumption of a perfectly informed central authority is unrealistic in the selfish, competitive environment of any EIP. Prosumers are reluctant to provide their private information when behaving strategically.

3. Learning to negotiate

Bilateral negotiations between two agents using a discrete timeline will be used. The alternating offers protocol is used throughout. We will use a technique widely known, namely Reinforcement Learning (or RL) (Sutton and Barto, 2018). Prosumers in an EIP will learn to negotiate better by simulating negotiation episodes against each other using the Q-Learning algorithm and accumulate their experience rewards in a decision policy. A prior work on prosumers' negotiation can be found in (S. Chakraborty et. al., 2018).

The Q-Learning algorithm consists of a function that estimates the expected cumulative rewards for future time steps in negotiation episodes given the actual state s_t each agent is in, if it chooses for a certain action a. Which action to take is determined by the negotiation policy π derived from the Q-values. At the end of each episode, an agent receives the reward r which allows updating Q-values. To this aim, the learning rule is:

$$Q(s_t; a_t) = Q(s_t; a_t) + \alpha * [r_{t+1} + \gamma * max_a Q(s_{t+1}; a_t) - Q(s_t; a_t)] \tag{6}$$

Each state for the ith prosumer agents is defined by the tuple:

$$s = (v, \rho) \tag{7}$$

where v and ρ, as defined before, are the necessity and risk associated with the perceived state by an agent, including the token value and its utility needs/surplus.

Each action during the negotiation episode, on the other hand, will be defined by:

$$a = (RV, \beta) \tag{8}$$

where RV is the reserve value of the negotiation, that is, the worst agreement an agent is willing to accept, and β is the concession rate, or the way it concedes in negotiations.

The Q-learning algorithm hyper-parameters, as defined commonly in the literature, are set to $\alpha = 0,1$ and $\gamma = 0,9$. ε, that measures how much our agent will take random actions to explore different options, starts at 1 and slowly decreased towards 0. Next, we define the way our agents compute necessities and risks out of their context, considering time t as one entire trading day. Necessities will be given by:

$$v_t^b = 0.5 * g_t(e_t + p_t) + 0.3 * g_{t+1}(e_{t+1} + p_{t+1}) + 0.2 * g_{t+2}(e_{t+2} + p_{t+2}) \tag{9}$$

$$v_t^s = su_t \tag{10}$$

For the buyer, g_t is the probability of rescheduling production tasks at time t; e is the amount of energy needed for the period and p is the maximum utility power required. For the seller, su_t is the surplus utility capacity or power it has available to trade. The risk may be similar for both agents, but, as defined, when the risk is low, the buyer is in a stronger position, whereas when the risk is high, the seller is in a stronger position.

$$\rho_t = season_t * \tau_t \tag{11}$$

where $season_t$ represents the variability of demand and τ_t, the price of the token.

4. Simulation results

4.1. The nature of the negotiation learning process

In Fig. 2 is presented the learning experienced by prosumer agents via simulating negotiation episodes against each other. It is worth noting the random nature of the learning process in negotiation episodes as it is shown in the 10000 simulated outcomes and the rewards obtained by the agents. This is mainly due to a rather unknown context and private information our agents must deal with. Simulations with diverse environmental contexts were conducted, varying utility or maxim power needed, possibility of rescheduling tasks, seasonality influence and the token price. The average trend highlights how negotiating agents increasingly obtain better deals as simulations makes room for learning to concede and accept. From the sparsity of rewards obtained it seems contextual information influence less to the seller than to the buyer. In the latter, greater risks are at stake when a contract is not settled in due course.

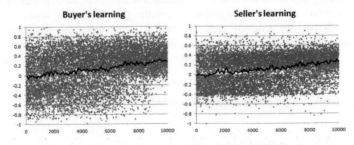

Fig. 2. Agents learning via simulation of negotiation episodes.

4.2. Reward functions

In Fig. 3, the functions used for rewarding negotiation outcomes of prosumer agents are shown. In the graph of the buyer's rewards when agreements are reached, it can be clearly perceived the importance of considering the risk ρ involved in the negotiation (the price τ of the token and the seasonality), given its reward function. On the other hand, it is very important for this agent to consider its necessity v when it decides not to agree, as rewards strongly depend on such private information and the possibility of rescheduling tasks and changing the utility load by levelling power peaks.

Similarly, for an agent in the role of a seller, its necessity and risk should be taken almost equally into account when deciding to agree the terms of an offer, but the agent deems better considering mainly its perceived risk when not agreeing. Such a reasoning helps illustrate the importance of correctly designing a reward function for guiding negotiation agents about what to bid, when to concede, what to accept or reject in contract settlements by considering what is at stakes in different scenarios.

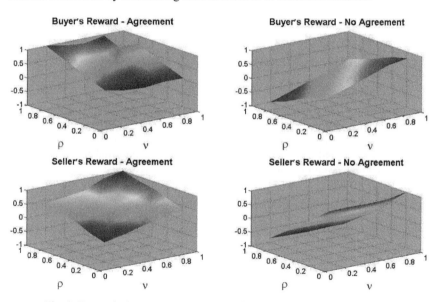

Fig. 3. Rewards for prosumer agents in different environmental contexts.

4.3. Agents' policies

In Fig. 4, the policies learned by prosumer agents are shown. Given v and ρ, agents behave differently depending on their role (buyer or seller). From a buyer's point of view, it will concede more slowly (β) and less (RV) as its perceived risk is higher, but it will concede faster and more as its necessity increases. From a seller's point of view, it will concede more slowly and to a lesser extent as its perceived risk is higher. Conversely, it will concede more quickly and to a lesser extent as its (internal) necessity increases. Having a near optimal policy for negotiating smart contracts is key for automating utility exchanges and incentivizing prosumer to participate in a peer-to-peer market. Social welfare in industrial parks is heavily dependent on negotiation policies that generate win-win deals that lower capital costs and carbon emissions.

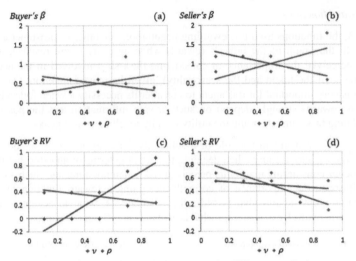

Fig. 4. Behavior of negotiation agents in different scenarios.

5. Conclusions

The importance of learning to negotiate utility surpluses in eco-industrial parks where prosumers do have private information has been addressed. Simulation results highlight that reinforcement learning is key to learn negotiation policies based on each agent internal necessity and the perceived risk in its environment. The proposed approach can be readily implemented using the Blockchain technology, industrial internet of things and smart contracts to enforce the agreed upon terms automatically.

References

S. Chakraborty et. al., 2018, Energy Contract Settlements through Automated Negotiation in Residential Cooperatives, IEEE International Conference on Communications, Control, and Computing Technologies for Smart Grids, DOI: 10.1109/SmartGridComm.2018.8587537.

M. Chertow, 2007, Uncovering Industrial Symbiosis, J. Industrial Ecology, 11, 1, 11-30.

I. M. L. Chew et al., 2009, Game theory approach to the analysis of inter-plant water integration in an eco-industrial park, J. Cleaner Production, 17, 18, 1611–1619.

K. Kristides and M. Devetsikiotis, 2016, Blockchains and Smart Contracts for the Internet of Things, IEEE Access, 4, 2292 – 2303.

S. K. Nair et al., 2016, Shared and practical approach to conserve utilities in eco-industrialparks, Computers and Chemical Engineering, 93, 221–233.

J. J. Sikorski et al., 2017, Blockchain technology in the chemical industry: Machine-to-machine electricity market, Applied Energy, 195, 234-246.

R. S. Sutton and A. G. Barto, 2018, Reinforcement Learning – An Introduction, The MIT Press, Cambridge, MA, USA.

R. R. Tan, et al., 2016, An optimization-based cooperative game approach for systematic allocation of costs and benefits in interplant process integration, Chemical Engineering Research and Design, 106, 43-58.

Anton A. Kiss, Edwin Zondervan, Richard Lakerveld, Leyla Özkan (Eds.)
Proceedings of the 29th European Symposium on Computer Aided Process Engineering
June 16th to 19th, 2019, Eindhoven, The Netherlands. © 2019 Elsevier B.V. All rights reserved.
http://dx.doi.org/10.1016/B978-0-128-18634-3.50281-2

Exergoeconomic Analysis for a Flexible Dry Reforming Power Plant with Carbon Capture for Improved Energy Efficiency

Szabolcs Szima*, Calin-Cristian Cormos

Babes-Bolyai University, Faculty of Chemistry and Chemical Engineering, Arany Janos 11, Cluj-Napoca, RO-400028, Romania

szima@chem.ubbcluj.ro

Abstract

Dry reforming of methane is an innovative alternative method for the production of hydrogen that uses CO_2 for the reforming process, thus lowering the carbon emissions of the plant. With the rise of renewable energy sources (solar, wind), the difference between the power supply and demand within the grid increased heavily, making flexibility of power plants more important for costs and emissions reduction. This paper presents the exergoeconomic analysis performed on the dry reforming technology designed for flexible power and hydrogen co-generation with carbon capture. The examined power plants' output is 450 MW net power with the possibility of hydrogen production of 200 MW_{th} (based on lower heating value). The CO_2 capture rate is 90 % (a part being used in the reforming process). The energy efficiency of the plant is 50 %, the exergy destruction rate is 370 MW and the specific capital cost of the plant is 1,295 M€/MW. The goal of this work is to offer an alternative to conventional steam reforming, improve the process of dry reforming by finding and improving the components of the system that contribute the most to the destruction of exergy and present a new flexible power plant that can produce both power and hydrogen. The evaluated design concepts were simulated using process flow modeling software and the mass & energy balances were used for the exergoeconomic analysis to improve the overall performance of the plant.

Keywords: dry methane reforming, flexible power plant, exergy analysis, exergoeconomic analysis,

1. Introduction

As atmospheric CO_2 concentrations are on the rise, a shift can be expected in global energy sources from conventional power sources to renewable ones, like solar and wind. This transition generates a stress on the electric grid as renewable energy sources power output depend heavily on the weather. The flexibility of a power system describes the ability of the system to balance out predictable and unpredictable changes on the supply or on the demand side. Cruz et al., (2018a) presents an extensive review on this issue and discusses potential options to ease the integration of renewable sources. Alemany et al., (2018) proposes the storing of electric energy during high production intervals. Several storing possibilities were proposed in the literature, the use of chemicals being heavily discussed (Chiuta et al., 2016).

With the rise of carbon capture technologies, the amount of available CO_2 is expected to increase in the future and as storage can be geographically limiting, the use of captured

CO_2 can be an alternative. Dry methane reforming uses CO_2 for the reforming of methane in the process of obtaining syngas. Several studies have been published in the past year on the subject (Salkuyeh et al., 2017). Aramouni et al. (2018) did a comprehensive review on catalysts and identified carbon deposition as the major issue in deploying the technology. As the carbon ratio is high in the reactor, carbon deposition on the catalyst is a barrier that needs solving. Developing new highly active catalysts is one of the approaches in resolving this issue.

In the current global context, as energy consumption is constantly on the rise while carbon mitigation is becoming more important a stalemate situation will be reached. In order to comply with the strict emissions reductions, set either large-scale carbon capture deployment is needed or a shift will be necessary towards energy carriers with a lower carbon footprint. Hydrogen is one promising alternative to fossil fuels, however this is conditioned to an adequate supply-chain. Hydrogen is obtained from natural gas reforming (48 %), petroleum refining (30 %) and coal gasification (18 %) (Cruz et al., 2018b). In order for hydrogen to be a viable option, either these technologies have to be redesigned with lower CO_2 emissions or a new, carbon-free way has to be implemented for its production. One option is water hydrolysis, however this technology is too expensive. Flexible power plants with carbon capture producing hydrogen can be an alternative to supply the necessary hydrogen.

This paper investigates the thermodynamic performance of dry methane reforming for the co-production of power and hydrogen through exergy analysis and offers a glance into the economics of the process. The plant is divided into four subunits, gas reforming, syngas processing, CO_2 capture and purification and the power island. A model of the plant is implemented in Aspen Plus process simulator to obtain the mass and energy balance of the process. Each division is evaluated separately in the analysis to better understand and identify the least efficient units.

2. Evaluation methodology

The plant was designed based on the general structure of a steam-methane reforming system, as presented in Figure 1. The gas reforming unit consists mainly of the reforming reactor, the syngas processing unit contains the water-gas shift reactors and the steam generator. The carbon capture unit is a gas-liquid absorption system using MDEA solution. Part of the captured CO_2 is recycled to the reforming unit, while the rest is sent to a storage site. The power island is designed using a Mitsubishi Hitachi Power Systems M701G2 series gas turbine combined cycle with the following specifications: 334 MW net power output, 39.5 % net power efficiency. The heat and energy balances from the process flow simulator are at the basis of the economic calculations of the plant.

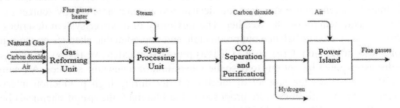

Figure 1. Simplified process flow diagram

The reactions describing the process of dry reforming are presented in Table 1. The reforming reaction is highly endothermic; hence 153 MW_{th} energy is necessary for the reforming reactor (this is provided by the burning of additional natural gas). The reverse-WGS reaction has a great impact on the final equilibrium and being endothermic, it is highly favored. Coke formation, the major issue in the case of dry reforming can be described by the last four reactions in the table, one of them being highly favored also because of its endothermic nature. In the process simulator, equilibrium reactors were considered, as equilibrium conditions can be easily achieved in these conditions.

Table 1. Governing reaction for dry reforming of methane

Reaction number	Name	Equation	ΔH_{298K} (kJ*mol^{-1})
Main reaction			
1	Dry methane reforming	$CH_4 + CO_2 \leftrightarrow 2CO + 2H_2$	+ 247
Side reaction			
2	Reverse water-gas-shift	$CO_2 + H_2 \leftrightarrow CO + H_2O$	+ 41.0
Coke formation			
3	Decomposition of methane	$CH_4 \leftrightarrow C + 2H_2$	+ 74.9
4	Disproportionation of CO	$2CO \leftrightarrow C + CO_2$	- 172.4
5	Hydrogenation of CO_2	$CO_2 + 2H_2 \leftrightarrow C + 2H_2O$	- 90.0
6	Hydrogenation of CO	$H_2 + CO \leftrightarrow H_2O + C$	- 131.3

Exergy analysis is a powerful tool for energy analysis of a system. By its definition, exergy is the maximum theoretical useful work obtained if a system S is brought into thermodynamic equilibrium with the environment by means of processes in which the S interacts only with this environment (Cruz et al., 2018b). In contrast to mass and energy laws, exergy is destroyed in a system due to irreversibility. A system is in equilibrium with its surroundings when there is no difference in temperature, pressure or chemical composition. In the present work, thermodynamic and chemical exergy was calculated with respect to the reference conditions (Table 2).

Table 2. Reference environment for exergy calculations

Thermo-mechanical exergy		Chemical exergy	
Parameter	Value	Component	Molar fraction
Temperature	18°C	N_2	0.7651
Pressure	1 atm	O_2	0.2062
		H_2O	0.0190
		Ar	0.0094
		CO_2	0.0003

The temperature and pressure data were used for the calculation of the thermodynamic exergy and the chemical composition for the chemical exergy. In both cases, atmospheric parameters were considered as a reference. For the calculation of the exergy flow and the exergy efficiency in the system, equations (1) and (2) were used.

Parameters were obtained from the exergy balance of the system. Simulation was performed using Aspen Plus process simulator.

$$E_f = E_p + E_d + E_l \tag{1}$$

$$\varepsilon = E_p / E_f \tag{1}$$

E_f- exergy flow; E_p - exergy of product; E_d - exergy destruction; E_l - exergy lost

The estimation of the capital costs for each unit were performed using capital cost correlations found in the literature (Cormos et al., 2017) based on the simulation results. On-site utility costs, owners' cost and land purchasing costs were estimated as 20 %, 15 % and 5 % of the total installed cost of the plant. Specific capital requirements were also calculated to ease comparison between technologies.

3. Results and discussion

Table 3. Exergy analysis results

	Exergy flow in (MW)	Exergy flow out (MW)	Exergy destruction (MW)	Exergy efficiency
Gas reforming unit	933.88	888.70	63.32	91.37 %
Syngas processing	888.70	863.20	33.47	93.72 %
CO_2 separation and purification	863.20	819.75	12.34	95.70 %
Power island	819.75	476.30	260.04	57.99 %

69 t/h of natural gas is required (900 MW_{th}) for the plant. The reforming, the water gas shift and the CO_2 capturing unit, all operate at 32 bar, whereas the gas turbine inlet pressure is at 19 bar. An expander is introduced to facilitate this pressure drop. The recovered energy is 2.3 MW. Combined with the gas turbine and the steam turbine the gross electric output is 476 MW_e. The ancillary energy for the system totals to 26 MW_e. This equals a net power output of 450 MW_e with a net electric efficiency of 50 %.

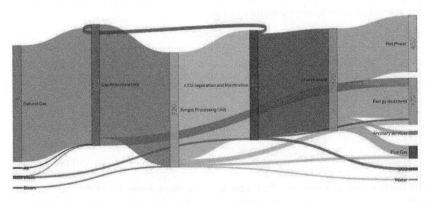

Figure 2. Exergy flow diagram for the process

The stream leaving the carbon capture unit is rich in hydrogen. With the addition of a pressure swing adsorption unit, the plant can produce high purity H_2 during low power demand offering a higher degree of flexibility to the system.

The results of the exergy analysis are presented in Figure 2 and Table 3. 934 MW exergy is introduced into the system with the natural gas stream. In the gas reforming unit with the compression of CO_2 another 38 MW of exergy is added. As a significant amount of heat is required for the endothermic reforming reaction, more than 60 MW of exergy is destroyed; the rest is fed into the Syngas processing unit (20 MW of exergy is transferred from the gas reforming unit to the syngas processing unit in the flue gasses stream leaving the heater). The syngas processing unit is significantly more efficient with an exergy efficiency of 93 %. The carbon capture unit is the most efficient, that is in part because 60 % of the captured CO_2 is recycled to the reforming unit. As expected, most of the exergy is destroyed in the power island; this is the subsystem with the highest grade of irreversibility, because of combustion. Energy and exergy analysis show similar results, the exergy analysis resulting lower efficiencies for the plant.

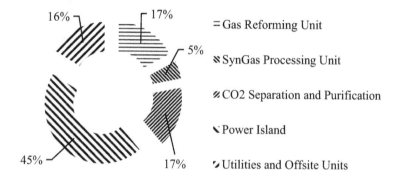

16% 17% 5% 45% 17%

= Gas Reforming Unit

≈ SynGas Processing Unit

≈ CO2 Separation and Purification

≈ Power Island

≈ Utilities and Offsite Units

Figure 3. Share of each subsystem in the total capital cost

Figure 3 presents the obtained costs share in the total capital cost requirement for each subsystem. The most expensive unit, the power islands' cost was estimated at 219 M€, the gas reforming unit at 83 M€, the CO_2 separation and purification unit at 80 M€. The syngas processing unit had the lowest cost, this was estimated at 24 M€. To obtain the total installed cost of the plant, utilities and offsite units capital costs were also added, these were estimated as 20 % of each units' cost resulting in 80 M€. The obtained costs for the proposed power plant are higher than in the case of a NGCC plant without carbon capture, however its comparable to the costs of a NGCC power plant with post-combustion capture.

Table 4. Capital cost data for the plant

Total Installed Cost	MM Euros	486.18
Total Investment Cost	MM Euros	583.41
Gross Power production	MWe	476.30
Net Power Production	MWe	450.30
Total Investment Cost per kWe (gross)	MM Euro / MWe	1,224.88
Total Invest ment Cost per kWe (equiv.)	MM Euro / MWe	1,295.60

Both exergy and economic analyses point to the power island as the unit that's improvement can have the highest positive impact on the plant. Exergy analysis shows, that the burning and boiling processes are responsible for this, however these are intrinsically low efficiency units. This calls for improvements to be made in other parts of the plant.

4. Conclusions

This paper presents the dry methane reforming technology as an alternative reforming process incorporated into a power plant for flexible operation. During low electricity demand, this can produce hydrogen for better efficiency. The system underwent a rigorous exergy analysis to better understand and to locate the least efficient units. Overall efficiencies for the two reforming systems are comparable, meaning dry reforming is a viable alternative for steam reforming if the issue of carbon deposition is addressed. The analysis also presented the effect of individual unit on the overall efficiency of the system simplifying the process of improvement. The highest grade of irreversibility was observed in the case of the power generation and reforming unit. Exergy analysis shows that most of the exergy is destroyed in the burner and in the boiler. The paper also presents an insight into the economics of the proposed flexible power plant, the power generation unit having the highest cost, the gas reforming and the CO_2 capturing unit having similar costs and the syngas processing unit having the lowest cost.

Acknowledgement

This work was supported by a grant of Ministry of Research and Innovation, CNCS – UEFISCDI, project ID PN-III-P4-ID-PCE-2016-0031: "*Developing innovative low carbon solutions for energy-intensive industrial applications by Carbon Capture, Utilization and Storage (CCUS) technologies*", within PNCDI III.

References

J.M. Alemany, B. Arendarski, P. Lombardi, P. Komarnicki, 2018, Accentuating the renewable energy exploitation: Evaluation of flexibility options, International Journal of Electrical Power & Energy Systems, 102, 131–151

N.A.K. Aramouni, J.G. Touma, B.A. Tarboush, J. Zeaiter, M.N. Ahmad, 2018, Catalyst design for dry reforming of methane: Analysis review, Renewable and Sustainable Energy Reviews 82, 2570–2585

S. Chiuta, N. Engelbrecht, G. Human, D.G. Bessarabov, 2016, Techno-economic assessment of power-to-methane and power-to-syngas business models for sustainable carbon dioxide utilization in coal-to-liquid facilities, Journal of CO2 Utilization, 16, 399–411

A.-M. Cormos, C.-C. Cormos, 2019, Techno-economic evaluations of post-combustion CO2 capture from sub- and super-critical circulated fluidised bed combustion (CFBC) power plants, Applied Thermal Engineering, 127, 106-115

M.R.M. Cruz, D.Z. Fitiwi, S.F. Santos, J.P.S. Catalão, 2018, A comprehensive survey of flexibility options for supporting the low-carbon energy future, Renewable and Sustainable Energy Reviews, 97, 338-353

P.L. Cruz, Z. Navas-Anguita, D. Iribarren, J. Dufour, 2018, Exergy analysis of hydrogen production via biogas dry reforming, International Journal of Hydrogen Energy, 43, 11688-11695

Y. Khojasteh Salkuyeh, B.A. Saville, H.L. MacLean, 2017, Techno-economic analysis and life cycle assessment of hydrogen production from natural gas using current and emerging technologies, International Journal of Hydrogen Energy, 42, 18894–18909.

Anton A. Kiss, Edwin Zondervan, Richard Lakerveld, Leyla Özkan (Eds.)
Proceedings of the 29th European Symposium on Computer Aided Process Engineering
June 16th to 19th, 2019, Eindhoven, The Netherlands.
http://dx.doi.org/10.1016/B978-0-128-18634-3.50282-4

Variations of the shrinking core model for effective kinetics modeling of the gas hydrate-based CO_2 capture process

Hossein Dashti,[a] Daniel Thomas,[b] Amirpiran Amiri,[b,*] Xia Lou[a]

[a]*Department of Chemical Engineering, WA School of Mines: Minerals, Energy and Chemical Engineering, Curtin University, Kent Street, Bentley WA 6102, Australia*

[b]*European Bioenergy Research Institute (EBRI), School of Engineering and Applied Science, Aston University, Birmingham, B4 7ET, United Kingdom*

a.p.amiri@aston.ac.uk

Abstract

The hydrate-based carbon dioxide (CO_2) capture (HBCC) process has been widely studied for CO_2 separation and sequestration. This paper aims to conduct a model-based investigation of the kinetics of the HBCC process. A variation of the shrinking core model (SCM) was developed for the analysis of this heterogeneous system under varying boundary conditions. The results revealed that while CO_2 diffusion through the hydrate layer is the dominant controlling mechanism, for a realistic scenario in which a time-dependent bulk gas concentration exists, the model results would better match the experimental data if the effects of the reaction rate were incorporated into the diffusion-based model. Sensitivity analysis showed that increasing the diffusivity through the hydrate layer significantly decreases the full conversion time of the water. Moreover, the effect of temperature change was investigated, and it was found that lower temperatures slow the hydrate growth rate. The model was demonstrated to be a computationally effective and time-efficient predictive tool that does not require high-speed computers for large-scale (reactor) applications.

Keywords: gas hydrate, CO_2 capture, shrinking core model

1. Introduction

Gas hydrates are solid clathrates comprising gas molecules (guests) such as CO_2, nitrogen (N_2), hydrogen (H_2), methane (CH_4) encased in a cage of water molecules (H_2O) connected to each other by hydrogen bonds. Hydrates form under specific thermodynamic conditions, including low temperature and high pressure. The small and nonpolar CO_2 gas forms hydrate structure I at temperatures lower than 283 K and pressures lower than 4.5 MPa (Sloan and Koh, 2008). CO_2 hydration under 203 K and 0.08 MPa has also been reported (Falabella, 1975). The potential of gas hydrate technology to capture CO_2 from flue gases has sparked research interest in investigating different aspects of this novel technology (Dashti and Lou, 2018; Dashti et al., 2015). Early research mostly focused on the experimental investigation of CO_2 hydrate formation and methods to improve the hydration kinetics and separation efficiency. In recent years, studies focused on understanding gas hydrate formation kinetics have increased (Yin et al., 2018). Many of these studies have been based on the methane hydrate formation process, and the kinetic model reported by Englezos et al. (1987) has been frequently used (Englezos et al., 1987).

This model follows the modeling approach for crystal growth from solution (Karpiński, 1980) and divides hydrate particle growth into two steps: 1) diffusion of the dissolved gas from the bulk solution to the hydrate-liquid interface and 2) the hydrate formation reaction at the interface. According to Lederols et al. (1996), gas hydrate formation is similar to the crystallization process, which is governed by nucleation followed by the rapid growth of hydrates; both are controlled by kinetics and mass and heat transfer phenomena (Lederhos et al., 1996).

The unreacted shrinking core model (SCM) has been widely used for solid reactions in which the intrinsic reaction rate is much higher than the mass transfer rate (Amiri et al., 2013; Amiri et al., 2015). This paper applies a modified SCM to model the CO_2 hydrate formation process assuming a single mechanism for a better understanding of gas hydrate formation kinetics.

2. Modeling methodology

2.1. Base model with a constant boundary condition

A variation of SCM with the steady boundary condition was applied to the CO_2 hydrate formation case study. In a batch reactor, however, the composition of the bulk gas surrounding the reactive particle is significantly dynamic, which conflicts with the boundary condition assumption in a mechanistic model.

In the typical SCM paradigm, the gas-solid reaction initiates at the outer surface of a solid particle and then moves towards the particle core, leaving a product layer behind. The unreacted core radius shrinks with time until full conversion of the solid particles to products, as shown in Figure 1. The model can be adopted to describe the CO_2 hydrate formation process, which includes three steps. Assuming effective agitation leading to the formation of water droplets that interact with CO_2 to form gas hydrates, the whole process involves (i) CO_2 diffusion through the gas film surrounding the hydrate particle to reach the particle's surface; (ii) CO_2 diffusion through the hydrate shell to reach the unreacted core (water) surface; and (iii) reaction of CO_2 with water (Eq. (1)) at the hydrate–liquid interface or unreacted core surface. The overall reaction can be expressed using Eq. (1):

$$CO_2 + nH_2O \leftrightarrow CO_2 \cdot nH_2O \qquad\qquad r_h = kC_{CO_2} \qquad\qquad (1)$$

where n is the hydrate number, 5.75 (Sun and Kang, 2016); r_h represents the overall hydrate formation rate; C_{CO_2} is the CO_2 concentration at the water-hydrate interface; and k is the reaction rate constant for reaction (1).

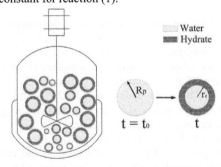

Figure 1: Schematic diagram of the CO_2 hydrate formation in a well-mixed batch reactor (Left) and a shrinking core hydrate particle (Right)

To build a comprehensive model, the system is first analyzed step by step, assuming that the analyzed step is rate limiting. The results are then compared with the experimental data. According to Eq. (1) and Figure 1 (right), for a liquid droplet with a spherical geometry, water conversion (X_{H_2O}) and the unreacted water core radius (r_c) can be correlated as follows:

$$X_{H_2O} = 1 - \left(\frac{r_c}{R_p}\right)^3 \tag{2}$$

where R_p is the particle radius. Assuming a rigid hydrate shell and constant R_p, step (ii) and step (iii) occur instantaneously. There is a linear relationship between the conversion rate and time (Eq. (3)).

$$t = \frac{\rho_{H_2O} R_p}{3 n k_g C_{CO_2}^b} X_{H_2O} \tag{3}$$

where ρ_{H_2O} is the molar density of water and k_g and $C_{CO_2}^b$ are the mass transfer coefficient of CO$_2$ through the gas layer surrounding the hydrate particle and the molar concentration of CO$_2$ in the gas bulk, respectively. Similarly, the conversion rate and rate relationship for step (ii) rate limiting and step (iii) rate limiting can be expressed by Eqs. (4) and (5), respectively.

$$t = \frac{\rho_{H_2O} R_p^2}{6 n D_e C_{CO_2}^b} [1 - 3(1 - X_{H_2O})^{\frac{2}{3}} + 2(1 - X_{H_2O})] \tag{4}$$

$$t = \frac{\rho_{H_2O} R_p}{n k C_{CO_2}^b} [1 - (1 - X_{H_2O})^{\frac{1}{3}}] \tag{5}$$

where D_e is the CO$_2$ diffusion coefficient in the hydrate layer.

2.2. Base model with a varying boundary condition

The base model presented earlier is based on the constant partial pressure of CO$_2$ in the bulk gas, $p_{CO_2}^b$. However in a closed system, when CO$_2$ capture in the hydrate cages is exceedingly fast, $p_{CO_2}^b$ declines with time. Given that the partial pressure of CO$_2$ in the bulk gas is the boundary condition in solving the mass transfer differential equations, the CO$_2$ concentration profile inside the particle, C_{CO_2}, might be influenced by the transient CO$_2$ partial pressure in the batch reactor. The base model for the hydrate layer diffusion-dominated case, step (ii), is extended. Eq. (6) defines the rate of CO$_2$ capture:

$$\frac{dN_{CO_2}}{dt} = -4\pi r^2 D_e \frac{dC_{CO_2}}{dr} \tag{6}$$

in which N_{CO_2} is the number of moles of CO$_2$. The steady-state assumption and integration yield Eq. (7) and Eq. (8):

$$\frac{dN_{CO_2}}{dt} \int_{r_c}^{R_p} \frac{dr}{r^2} = -4\pi D_e \int_0^{C_{CO_2}^b} dC_{CO_2} \tag{7}$$

$$\frac{dN_{CO_2}}{dt} \left(\frac{1}{r_c} - \frac{1}{r_p}\right) = -4\pi D_e C_{CO_2}^b(t) \tag{8}$$

Instant N_{CO_2} can be presented as a function of $r(t)$ according to Eq. (9):

$$\frac{1}{n} dN_{CO_2} = 4\pi \rho_{H_2O} r_c^2 dr_c \tag{9}$$

CO_2 concentration in bulk gas is the boundary condition for Eq. (6) and is related to the CO_2 partial pressure in the reactor by Eq. (10):

$$C_{CO_2}^b(t) = \frac{p_{CO_2}^b(t)}{RT} \tag{10}$$

Eq. (9) and (10) are substituted in Eq. (8) and integrated across the hydrate layer, resulting in Eq. (11):

$$\frac{R_p^2 \rho_{H_2O} RT}{6nD_e}\left[1 - 3\left(1 - X_{H_2O}\right)^{\frac{2}{3}} + 2\left(1 - X_{H_2O}\right)\right] = \int_0^t p_{CO_2}(t)\,dt \tag{11}$$

which describes the correlation between water conversion and CO_2 partial pressure changes and allows monitoring of the time-dependent mass diffusion resistance based on the reactor pressure readings. Similarly, the governing equations for the reaction control case with a varying boundary condition have been derived and are given below:

$$\frac{R_p \rho_{H_2O} RT}{nk}\left(1 - \left(1 - X_{H_2O}\right)^{\frac{1}{3}}\right) = \int_0^t p_{CO_2}(t)\,dt \tag{12}$$

Combining Eq. (11) with Eq. (12), a model considering both the diffusion step (ii) and the reaction step (iii) is produced according to Eq. (13):

$$\frac{R_p \rho_{H_2O} RT}{n}\left[\frac{R_p}{6D_e}\left[1 - 3\left(1 - X_{H_2O}\right)^{\frac{2}{3}} + 2\left(1 - X_{H_2O}\right)\right] + \frac{1 - \left(1 - X_{H_2O}\right)^{\frac{1}{3}}}{k}\right] = \int_0^t p(t)\,dt \tag{13}$$

The right-hand side of Eq. (13) presents the measured time-dependent CO_2 partial pressure inside the reactor. This must be provided practically, making the presented model a semi-empirical model.

3. Results and Discussion

3.1. Controlling mechanism identification

The models derived based on the individual mechanisms (Eqs. 3, 4, 5) were compared against a set of experimental data obtained from batch reaction vessel experiments at 203 K and 0.08 MPa (Falabella, 1975). As shown in Figure 2(A), for the model based on step (ii), diffusion through the hydrate layer best fits the data. The model developed based on gas film control (step (i)) predicts a linear relationship between conversion and time that departs greatly from the experimental data. This is not surprising and is consistent with the assumptions of Englezos et al. (1987) (Englezos et al., 1987). Judgments about the dominant mechanism become critically challenging when comparing the modeling results based on the intrinsic reaction with those based on the diffusion controlling model. While both models follow the practically observed non-linear clathrate conversion trend, the diffusion-dominated model seems more appropriate. This is more obvious at the early stages of the process, from zero to 55% conversion, because the reaction on the water-hydrate interface occurs after gas diffusion through the hydrate layer. At higher conversion ranges, above 55%, both models reasonably predict the experimental data. It can be concluded that while diffusion through the formed hydrate predominantly controls the hydration progress rate, the inherent reaction kinetic role is not negligible. The former becomes more significant when considering the depleting CO_2 partial pressure in the bulk gas surrounding the particle that occurs in a batch reactor.

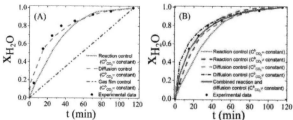

Figure 2: Comparisons of (A) the single-mechanism model prediction with the experimental data for constant bulk concentration and (B) the single- and combined-mechanism model predictions against the experimental data for constant ($C_{CO_2}^b$ = constant) and varying bulk concentration ($C_{CO_2}^b$ ≠ constant)

The assumptions, including the single mechanism control and constant CO_2 partial pressure in bulk gas, used in the base model derivation may be responsible for the observed deviation between the modeled and experimental results in Figure 2(A). The model was improved by accounting for combined mass transfer and reaction kinetics rates, both of which are influenced by the depleting CO_2 fraction in the surrounding gas. Figure 2(B) presents the prediction results for (i) the hydrate diffusion control model with varying CO_2 concentration in the bulk gas (Eq. (11)), (ii) the intrinsic reaction control model with varying CO_2 concentration in the bulk gas (Eq. (12)), and (iii) the combined model under a transient CO_2 concentration (Eq. (13)) and compares these results with the experimental data (Falabella, 1975). The results of the combined model fall between the hydrate layer diffusivity and reaction control results, thus improving the prediction results compared with the experimental data.

3.2. Model analysis
Due to the role of the diffusion rate demonstrated earlier, the accuracy of the diffusion coefficient estimation is of critical importance for model fidelity. The diffusivity of CO_2 in hydrates has been reported to be in the range of 1×10^{-16} m²/s to 2×10^{-14} m²/s (Liang et al., 2016). The effect of CO_2 diffusivity on the CO_2 hydrate conversion time was investigated as shown in Figure 3(A). The total conversion time increases dramatically with decreasing diffusivity because at lower diffusivity values, the diffusion of the gas into the inner layer of the gas hydrate shell is slower. As a result, it takes longer to form the gas hydrate particle and consequently reach the end of the conversion time.

In further analysis of the model, the effects of the gas hydrate formation temperature on the completion time were investigated. As shown in Figure 3(B), the effect of the temperature changes on the completion time is not as significant as the effects of the diffusivity on the completion time. At lower temperatures, CO_2 hydrate formation is faster. For example, at 250 K and 298 K, the completion times are approximately 107 min and 130 min, respectively.

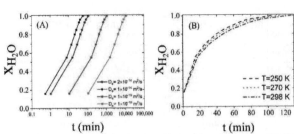

Figure 3: Effects of (A) CO_2 diffusivity and (B) hydrate formation temperature change on the conversion time

4. Conclusions

The purpose of the present study was to determine the dominant controlling mechanism during gas hydrate formation using a novel variation of SCM. The primary model was proposed based on a constant concentration of CO_2 in the bulk gas. The model was further improved by considering the transient CO_2 concentration in the bulk gas, which is a more realistic case. While the diffusion-based model reasonably predicts the CO_2 hydration rate under a constant bulk gas concentration condition, for the realistic scenario in which the bulk gas varies with time, the model must consider the reaction control role as well. The proposed model was successfully used to quantify the effects of the CO_2 diffusivity and the operating temperature on conversion.

References

A. Amiri, A.V. Bekker, G.D. Ingram, I. Livk, N.E. Maynard, 2013, A 1-D non-isothermal dynamic model for the thermal decomposition of a gibbsite particle, Chemical Engineering Research and Design, 91, 485-496.

A. Amiri, G.D. Ingram, N.E. Maynard, I. Livk, A.V. Bekker , 2015, An unreacted shrinking core model for calcination and similar solid-to-gas reactions, Chemical Engineering Communications, 202, 1161-1175.

H. Dashti, X. Lou, 2018, Gas hydrate-based CO_2 separation process: Quantitative assessment of the effectiveness of various chemical additives involved in the process, Springer International Publishing, Cham, pp. 3-16.

H. Dashti, , L. Zhehao Yew, X. Lou, , 2015, Recent advances in gas hydrate-based CO_2 capture, Journal of Natural Gas Science and Engineering, 23, 195-207.

P. Englezos, N. Kalogerakis, P.D. Dholabhai, P.R. Bishnoi, 1987, Kinetics of formation of methane and ethane gas hydrates, Chemical Engineering Science, 42, 2647-2658.

B.J. Falabella, 1975, A Study of Natural Gas Hydrates, University of Massachusetts Amherst, Ann Arbor, p. 180.

P.H. Karpiński, 1980, Crystallization as a mass transfer phenomenon, Chemical Engineering Science, 35, 2321-2324.

J.P. Lederhos, J.P. Long, A. Sum, R.L. Christiansen, E.D. Sloan, 1996, Effective kinetic inhibitors for natural gas hydrates, Chemical Engineering Science, 51, 1221-1229.

S. Liang, D. Liang, N. Wu, L. Yi, G. Hu, 2016, Molecular mechanisms of gas diffusion in CO_2 hydrates, The Journal of Physical Chemistry C, 120, 16298-16304.

E.D. Sloan, , C.A. Koh, 2008, Clathraye Hydrates of Natural Gases, third ed., Taylor & Francis Group, New York.

Q. Sun, Y.T. Kang, 2016, Review on CO_2 hydrate formation/dissociation and its cold energy application, Renewable and Sustainable Energy Reviews, 62, 478-494.

Z. Yin, M. Khurana, H.K. Tan, P. Linga, 2018, A review of gas hydrate growth kinetic models, Chemical Engineering Journal 342, 9-29.

Anton A. Kiss, Edwin Zondervan, Richard Lakerveld, Leyla Özkan (Eds.)
Proceedings of the 29th European Symposium on Computer Aided Process Engineering
June 16th to 19th, 2019, Eindhoven, The Netherlands. © 2019 Elsevier B.V. All rights reserved.
http://dx.doi.org/10.1016/B978-0-128-18634-3.50283-6

Optimization of semi-permeable membrane systems for biogas upgrading

Diego Filipetto[a], Federico Capra[a,b], Francesco Magli[a,b], Manuele Gatti[a,b], Emanuele Martelli[a,*]

[a]*Politecnico di Milano, Department of Energy, Via Lambruschini 4, 20156 Milano, Italy
Milano*

[b]*LEAP s.c.a r.l. – Laboratorio Energia e Ambiente Piacenza, Via N. Bixio 27/C, 29121, Piacenza, Italy*

[*] *emanuele.martelli@polimi.it*

Abstract

This work focuses on the optimization of a biogas upgrading process consisting of three CO_2-permeable membranes. A 1-D membrane model, capable of handling multi-component gas mixtures, is formulated as a set of differential algebraic equations and validated against experimental data. The model is used within a sequential algorithm to solve the simulation of the three-stage system. The key design variables (i.e., areas of each membrane module, stream pressures and recycle mass flow rates) are optimized using the recently developed MO-MCS algorithm, a derivative-free optimizer suitable for multi-objective problems. Results show that the Pareto-optimal solutions have a total plant cost in the range 720-770 €/(Nm^3/h of feed) and an energy efficiency of 87.7-89.2%.

Keywords: Biogas upgrading, membrane separation, black-box optimization, multi-objective optimization

1. Introduction

Biomethane is produced through the purification of biogas, a mixture of CH_4, CO_2 and other contaminants (e.g., H_2S, NH_3 and siloxanes) which is originated by the anaerobic digestion of biomass. Such biogas purification process is typically called "biogas upgrading". Among the available upgrading technologies (e.g. water scrubbing, amines, PSA), CO_2-permeable membranes appear as a promising option, especially for medium-low size plants (250-500 Nm^3_{CH4}/h), owing to their good separation efficiency, modularity, straightforward operation and maintenance, and low installation costs. However, since a single membrane stage cannot achieve high purity and recovery of the desired product (CH_4), it is advantageous to adopt integrated configurations which combine multiple modules placed in series and/or parallel, entailing recycle streams. Due to the presence of multiple design variables, optimization-based design approaches have been adopted by several authors. For example, Scholz et al. (2015) proposed a structural-optimization approach employing a flexible superstructure of possible process layouts. The modelling and optimization problems are solved together as a single MINLP, according to the equation oriented process optimization approach. Also, the choice of the optimal membrane material according to an economic criterion is tackled with the same approach. Gabrielli et al. (2017) adopt the black-box approach, combining a genetic algorithm with a 1-D model of the membrane modules, to minimize the compression energy and the required membrane area by selecting the optimal process layout.

This work investigates the numerical optimization of the most promising integrated membrane-based configuration found by Scholz et al. (2015) for biogas upgrading. First, the 1-D model proposed by Gabrielli et al. (2017) for co-current hollow-fiber membranes is extended to handle multi-component gas mixtures. Then, the model is validated with the experimental data available in Pan et al. (1986) and integrated into a modular Matlab algorithm to solve the overall process simulation. Finally, the key design variables (i.e., areas of each membrane module, stream pressures and recycle mass flow rates) are optimized using the recently developed MO-MCS algorithm (Capra et al., 2018), a model-based derivative-free optimizer for multi-objective problems.

2. Process description

The process configuration considered in this work is shown in Figure 1. A gas dryer and a bed of activated carbon, not shown in the figure, are placed upstream of the membrane system to remove H_2O and H_2S (and other contaminants) respectively. Then, sulphur-free dried biogas (stream 0) is mixed at atmospheric pressure with the recycle stream (stream 6), pressurized by the first compressor, and fed to the first membrane unit (stage A). The retentate flow (stream 2) can be further compressed (compressor 2, included in case $Pressure_{stage B} > Pressure_{stage A}$) and sent to the second membrane unit (stage B). Since the permeate flow of the first stage (flow 1P) may include a certain amount of CH_4, it is advantageous to perform a selective recycle including a third membrane unit (stage C) downstream of it. The permeate of stage C is expected to be nearly pure CO_2 with a very low content of CH_4, while the retentate stream (stream 5) can be mixed with stream 2P and recycled back to the process inlet (stream 6). According to Scholz et al. (2015), with the proper selection of the membrane material, this configuration combines high purity of the produced biomethane (96%$_{mol}$), high recovery of CH_4 (> 95%) and a good balance between energy consumption and investment cost.

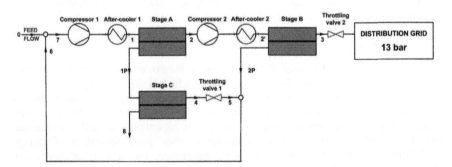

Figure 1. Scheme of the integrated membrane configuration considered in this study.

As far as the membrane material is concerned, we selected a dense polymer, specifically polyaramid (PA), on the basis of the results found by Scholz et al. (2015) and Havas in (2017). According to Scholz et al. (2015), this commercially available material can be assembled in hollow-fibers modules featuring a CO_2 permeance of 200 gpu and a CO_2/CH_4 selectivity of 20.

The main process data (dry biogas composition and pressure) and specifications (grid injection requirements and membrane material properties) are reported in Table 1 and can be considered representative of an anaerobic digestion plant fed with chicken-manure.

Table 1. Assumptions and specifications for the process.

Biogas flow rate	11.75	mol/s
Dry biogas CH_4 content	56%	$\%_{mol}$
Dry biogas CO_2 content	44%	$\%_{mol}$
Biogas feed pressure	1	bar
Grid injection pressure	13	bar
Max biomethane oxygen content	$< 0.6\%$	$\%_{mol}$
Max biomethane CO_2 content	$< 3\%$	$\%_{mol}$
Biomethane HHV range	$34.95 \div 45.28$	MJ/Sm^3
Wobbe index range	$47.31 \div 52.33$	MJ/Sm^3
Relative density range	$0.5548 \div 0.8$	-
Membrane CO_2 permeance (polyaramide, Havas and Lin, 2017)	$6.8 \cdot 10^{-8}$	$mol/(s \cdot Pa \cdot m^2)$
Membrane CO_2/CH_4 selectivity (polyaramide, Havas and Lin, 2017)	20	-

3. Problem statement

The objective of this work is to identify the set of Pareto-optimal solutions which exhibit the minimum total plant cost and maximum process efficiency. More in detail, the optimization problem can be stated as follows: "Given the biogas specifications and the membrane material properties, determine the set of Pareto-optimal design solutions which minimize the total upgrading plant cost (TPC) while maximizing the primary energy efficiency η_{PE} (chemical power of biomethane/total primary energy consumed by the process) and meet the specifications on the produced biomethane (minimum CH₄ fraction, allowed ranges of HHV, Wobbe index, etc)". The primary energy efficiency is computed as the ratio between the primary energy contained in biomethane (given by the product between mass flow rate of biomethane and its LHV) and the primary energy input of the process (i.e. the primary energy contained in biogas plus the equivalent primary energy consumed by the upgrading process to operate compressors and other auxiliaries, etc).

The independent optimization variables of the selected process configuration are the areas of the three modules and the pressure ratios of the two compressors and the pressure of the permeate side of stage A membrane. Given the limited amount of independent optimization variables, the black-box approach has been selected to tackle the multi-objective optimization problem. MO-MCS, a recently developed multi-objective model-based derivative-free algorithm (Capra et al., 2018), optimizes the independent design variables and, for each sampled solution, the process simulation is executed and the total plant cost is assessed.

4. Membrane module model

We propose an extension for multi-component gas mixtures of the 1-D model originally formulated by Gabrielli et al. (2017) for the membrane separation of binary gas mixtures. The peculiarity of such model is the use of dimensionless parameters and variables which attenuate the numerical problems arising in the solution of the set of differential equations.

$$\frac{d\Psi}{d\zeta} = \frac{m_1}{y_1'}(x_1 - \beta y_1') \tag{1}$$

$$\frac{dx_i}{d\zeta} = \frac{m_i}{\Psi y_i'}(x_i - y_i')(x_i - \beta y_i') \qquad \forall i = 1, \ldots, C-1 \tag{2}$$

$$0 = y_1'(x_j - \beta y_j') - \alpha_{1j}y_j'(x_1 - \beta y_1') \qquad \forall j = 2, \ldots, C \tag{3}$$

$$0 = \sum_{i=1}^{c} y_i' - 1 \tag{4}$$

$$\Psi|_{\zeta=0} = 1 \tag{5}$$

$$x_i|_{\zeta=0} = x_i^{FEED} \qquad \forall i = 1, \ldots, C \tag{6}$$

As described in Gabrielli et al. (2016), ζ is the non-dimensional variable representing the axial coordinate, $\Psi(\zeta)$ is the non-dimensional total molar flow rate on the retentate side (the ratio between the flow rate at point ζ and the one in the feed), $x_i(\zeta)$ is the molar fraction of the permeate bulk flow, and y_i' is the molar fraction at the membrane interface. The parameter β is the ratio between the permeate and retentate pressure, α_{1j} is the membrane selectivity of the j-th component with respect to the most permeable component (CO_2 in this application). The parameter m_i is defined as:

$$m_i = \frac{Q_i p^F A}{\delta U^F} \tag{8}$$

Where Q_i denotes the permeability of the i-th gas species, p^F is the feed pressure, A is the membrane area, δ is the membrane thickness and U^F is the total feed molar flow rate. Eq. (1) is the balance of total gas moles on the retentate side and it relates the variation of the total dimensionless flow on the retentate side Ψ with the total molar flux across the membrane. Eq. (2) relates the axial variation of concentration of species on the permeate side with the molar flux across the membrane. Eq. (3) expresses the membrane fluxes of the different gas species with respect with that of CO_2 ($i = 1$). Eq. (4) imposes that the sum of the molar fractions on the membrane interface is equal to one, while Eq. (5) and (6) are boundary conditions. Eq. (1-6) define a set of Differential Algebraic Equations (DAE) which is solved in Matlab® v. 2018-a with the *ode15s* algorithm.

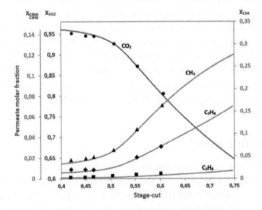

Figure 2. Validation of the module model with the experimental data of Pan et al. (1986).

The permeability of the various species have been validated against the experimental data provided by Pan et al. (1986) for stage cuts of the membrane comprised between 0.4 and 0.6. Figure 2 shows that the model fits well the experimental data, with maximum discrepancies in the permeate molar fractions limited to 1%.

5. Process simulation and optimization

The process flowsheet is solved using a sequential-modular approach with tearing. The optimal precedence ordering consists in calculating the units in this sequence: inlet gas mixer (0+6), first compressor, membrane stage A, second compressor, membrane stage B and stage C, valve and the mixer (5+2P). It is important to note that the system of DAE described in Section 4 is solved for each membrane module. The recycle stream 6 has been selected for tearing and the tear stream properties (i.e., molar flow rates of the gas species) is updated with the direct substitution. All the membrane stages operate at a constant temperature of 40 °C and any enthalpy change related to the Joule-Thomson effect or non-ideal mixing has been neglected with limited loss of accuracy, as suggested by the literature. Test cases have shown that convergence of the tear stream is reached in a few iterations (10-30, corresponding to 15-40 s of computational time) not justifying the adoption of more sophisticated algorithms (because the maximum eigenvalue of the Jacobian of the overall simulation function is in the range 0.2-0.3). After the process simulation has reached convergence, the total plant cost (overall capital expenditure) and the primary energy efficiency are computed considering literature data for the cost of the units (for the membranes we considered a lumped-cost of the equipment of 50 €/m^2) and the performance of small size reciprocating compressors.

It is worth noting that the output of the process simulation (i.e., TPC and η_{PE}) are in general noisy and non-smooth due to the round-off errors and convergence tolerances of the algorithms which are called to solve the DAE and the process simulation. Moreover, the problem may feature multiple minima. For these reasons, gradient-based optimization algorithms are not suitable and it is necessary to adopt global-search, black-box algorithms. Recently, Capra et al. (2018) have proposed a model-based, derivative-free, algorithm which uses a local model of the black-box function so as to save computationally expensive evaluations. The algorithm is an extension of the MCS algorithm (Huyer & Neumaier 1999) to the multi-objective case. It allows for well-spaced Pareto fronts thanks to the integration of the Normalized Normal Constraint method (Messac et al. 2003). In MO-MCS the search for the Pareto optimal points is assigned to the available workers (cores) so as to reach the maximum possible parallelization of computations.

6. Results

The process optimization has been performed considering three different limits on the CH$_4$ slip (fraction of CH$_4$ which is lost to the process and goes together with the vented CO$_2$): no slip limit, 2% and 1%. The three Pareto fronts returned by MO-MCS are shown in Figure 3. The required computational time to obtain a Pareto front is 3 hours on a workstation with Intel® Core™ i7-3770 CPU @ 3.40 GHz and 4 cores. Their optimality have been double-checked repeating the optimization with different setup of MO-MCS and performing sensitivity analyses. The biomethane specification constraints are active for all Pareto-optimal points. Figure 3 indicates that setting a limit of 2% on the CH$_4$ slip does not penalize the achievable efficiency and it does not lead to a considerable increase of TPC. Instead, 1% CH$_4$ slip calls for 40% higher recycle flow rates causing an appreciable increase of compressors costs and power consumption. For the case with slip < 2%, the total membrane area is approximately 1500 m² for the maximum efficiency design and 950 m² for the minimum total plant cost. Another interesting result is that no Pareto-optimal solution employs the intermediate compressor (compressor 2).

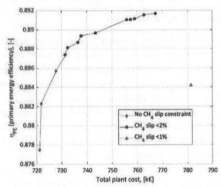

Figure 3. Pareto fronts returned by MO-MCS for three different limits of CH_4 slip.

7. Conclusions

The proposed 1-D model is suitable for multi-component gas mixtures and it can be efficiently adopted in sequential process simulation codes as well as derivative-free optimization algorithms, such as MO-MCS. The application to a biogas upgrading process with increasingly tight specifications has shown that the proposed approach is able to return very good approximations of the Pareto-front within a reasonable computational time. Optimization results indicate that the considered membrane process can achieve primary energy efficiency figures in the range 87.7%-89.2% with specific total plant costs ranging from 720-770 €/(Nm³/h of feed).

Acknowledgements

Politecnico di Milano acknowledges the Social Energy project (CUP E59J18000000009) funded by Regione Lombardia, while LEAP is grateful to the Gobiom project (CUP E82F16001020007, funded by Regione Emilia Romagna) which has set the basis for the membrane model development.

References

P. Gabrielli, M. Gazzani, M. Mazzotti, 2017, On the optimal design of membrane-based gas separation processes, Journal of Membrane Science, 526, 118–130

M. Scholz, M. Alders, T. Lohaus, M. Wessling, 2015, Structural optimization of membrane-based biogas upgrading processes, Journal of Membrane Science, 474, 1–10

C.Y. Pan, 1986, Gas separation by high-flux, asymmetric hollow-fiber membrane, AIChE Journal, 32(12), 2020–2027

D. Havas, H. Lin, 2017, Optimal membranes for biogas upgrade by removing CO2: High permeance or high selectivity?, Separation Science and Technology, 52(2), 186–196

F. Capra, M. Gazzani, L. Joss, M. Mazzotti, E. Martelli, 2018, MO-MCS a Derivative-Free Algorithm for the Multiobjective Optimization of Adsorption Processes, Industrial & Engineering Chemistry Research, 57, 9977-9993

Anton A. Kiss, Edwin Zondervan, Richard Lakerveld, Leyla Özkan (Eds.)
Proceedings of the 29th European Symposium on Computer Aided Process Engineering
June 16th to 19th, 2019, Eindhoven, The Netherlands. © 2019 Elsevier B.V. All rights reserved.
http://dx.doi.org/10.1016/B978-0-128-18634-3.50284-8

System-Level Optimisation of Combined Power and Desalting Plants

Houd Al-Obaidli,[a] Sarah Namany,[a] Rajesh Govindan,[a] Tareq Al-Ansari[a,*]

[a]*Division of Sustainable Development, College of Science and Engineering, Hamad Bin Khalifa University, Qatar Foundation, Doha, Qatar*

*talansari@hbku.edu.qa

Abstract

Nations in the Gulf Cooperation Council (GCC) utilise their vast oil and natural gas resources in order to satisfy the demand for water and power, which continues to increase due to population and economic growth. The abundance in fossil fuels and relatively low O&M costs make fuel-based technologies a de-facto choice when planning new power or desalting facilities which are proven to have the lowest levelized cost compared to other alternatives. However, energy intensive desalination systems have contributed to high greenhouse gas (GHG) emissions and increased costs. Recently, several facilities have been designed to include renewable energy in order to mitigate GHG emissions and improve the overall environmental welfare. This study evaluates different power and desalting configurations using both fossil-based and renewable energy sources to determine an optimal configuration that minimises CO_2 emissions at a relatively low levelized cost. Using the current utilities infrastructure in Qatar as a basis, six configurations were devised where three of them mimic current configurations and three were proposed. The technologies selected were limited to existing configurations: open cycle gas turbine (OCGT), combined cycle gas turbine (CCGT), multi-stage flash (MSF), and seawater reverse osmosis (SWRO), and three renewable technologies concentrated solar power (CSP), solar PV, and biomass integrated gasification combined cycle (BIGCC). At a system-level, optimisation of levelized costs vs global warming potential (GWP) was conducted using a stochastic programming framework. The findings conclude that current configurations are suboptimal and the infusion of renewable energy sources (RES) into the combined power and desalting plant (CPDP) infrastructure significantly improves CO_2 emissions with a reduction ranging between 52% and 67% and overall system levelized cost decrease between 8% and 32%.

Keywords: Portfolio optimisation, renewable energy, stochastic processes, levelized costs, environmental impact.

1. Introduction

Economic and population growth in the Middle-East since the hydrocarbon boom has increased the industrial and urban demand for water and energy resources. By 2030, water demand globally is expected to rise by 25%, with a 50% increase in energy (Madani et al., 2015). Considering the demand for both, water and power, the countries in the GCC bordering the Arabian Gulf, are reliant on co-generation power desalting plants (CPDP) which produce two useful utilities: electricity and freshwater. In fact, 44% of the global desalination capacity is located in the Middle-East and North Africa

(Voutchkov, 2016). Desalination is an energy intensive process in which associated emissions are a function of the input energy, i.e. natural gas, oil, etc.

The State of Qatar, a peninsula located in the heart of the Arabian Gulf, suffers from freshwater scarcity as it is located in a hyper arid region with renewable groundwater rate equivalent to 58 million m³ annually (Darwish, 2014). It exhibits a similar development profile to those witnessed by the rest of the region and in terms of its energy resources, Qatar possess approximately 25.4 trillion m³ of gas reserves and its annual output of liquefied natural gas is 77 million tons per year (Qatargas, 2018). Technology advancements in the provision and supply of natural gas, in addition to the fact that it is considered the cleanest of fossil fuels has enhanced the position of natural gas as an energy resource. In fact, natural gas, it is the sole driver of power generation in Qatar and has witnessed increasing demand in the global market, therefore offering further opportunities for the export of natural gas and associated products. Several utility plants based on affordable and clean natural gas have been constructed in Qatar in order to satisfy water and electricity increasing needs as illustrated in Table 1. The State of Qatar aims to diversify the energy mix and reduce GHG emissions through the integration of alternative and renewable energy and desalination technologies. This study serves as a guideline for decision makers offering a comparative tool for identifying an optimal mix of power and water systems minimizing environmental and economic impacts.

Table 1. List of CPDP facilities in Qatar (Kahramaa, 2017).

Plant	OCGT (MW)	CCGT (MW)	MSF (MIGD)	SWRO (MIGD)
RAF A	497			
RAF A1			45	
RAF A2			36	
RAF A3				36
RAF B	609		33	
RAF B1	376.5			
RAF B2	567		30	
RLPC		756	40	
Q Power		1,025	60	
RGPC		2,730	63	
M Power		2,007		
UHPC		2,520	76.5	60
Total	2,049.5	9,038	383.5	96

In terms of portfolio optimisation, Bhattacharya and Kojima (2012) examined the increase of renewable energy contribution in the energy supply portfolio for Japan utilizing mean-variance portfolio (MVP) optimisation. Fuss et al. (2012) examined the adoption of renewable energy and carbon-saving technologies across different socioeconomic scenarios. Cucchiella et al. (2012) analysed the electricity market in Italy using MVP, and reviewed a set of optimal portfolios with different energy mixes. The study concluded that it is necessary to include renewable energy into the portfolio to mitigate the high investment risk in this market. Lucheroni and Mari (2014; 2015; 2016) used stochastic levelized cost of energy (LCOE) theory to determine the optimal power generation portfolio for different scenarios. Govindan et al. (2018) estimated the LCOE for CCGT using stochastic modelling and applied it into a portfolio optimisation of different energy and water technology mixes as part of energy, water and food

(EWF) Nexus studies. Namany et al. (2018) expanded on the work by Govindan et al. (2018) and Al-Ansari et. al (2015; 2017) by applying a game theoretic approach using a Stackelberg duopoly model to the EWF Nexus. The study considered a CCGT competing against bio-energy with carbon capture and storage (BECCS). Darwish et al. (2014, 2015) provided a techno-economic analysis of different CPDP to improve energy and water security in Qatar in a cost-effective manner. The aim of this paper is to evaluate the integration of RES, consisting of PV, BIGCC and CSP, into the existing utilities infrastructure in terms of CO_2 emissions and levelized costs by applying stochastic modelling and multi-objective genetic algorithm optimisation.

2. Methodology

Using a stochastic optimisation framework that considers fluctuation in natural gas prices, this study models diverse power and desalting scenarios using a combination of renewable and non-renewable energy sources to identify the optimal configuration with the lowest levelized and environmental impact. The utilities infrastructure is segregated by typical configurations of CPDP as illustrated in Table 2. There are 3 existing configurations observed based on 2 power technologies OCGT and CCGT and 2 water desalination technologies MSF and SWRO. Typically, configurations tend to improve over time transitioning to a more reliable and efficient technology as was obvious from the transition of legacy OCGT systems in favour of CCGT and from expensive and energy intensive MSF to SWRO. Three new technologies were selected based on suitability for the environment in the Middle-East, CSP, PV and BIGCC.

Table 2 List of Current CPDP Configurations.

Configuration	OCGT	CCGT	MSF	SWRO
1	1.0		1.0	
2		1.0	1.0	
3		1.0	0.63	0.37

Figure 1 illustrates a comparison of the 7 technologies together with the 3 existing configurations. The aforementioned technologies form the baseline and will be used for comparison with 3 additional scenarios that are based on a portfolio optimisation using stochastic modelling of natural gas prices, thus providing a representation for natural gas price volatility. The model was based a mean gas price and a standard deviation as predicted by Govindan et al. (2018).

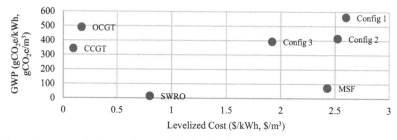

Figure 1. Current Water and Power Technologies and CPDP Configurations.

A multi-objective genetic-algorithm (GA) model was developed for the 7 technologies representing 7 decision variables with the aim of minimizing both the overall levelized costs of the system and GWP. Levelized costs and GWP factors per technology were selected from literature as described in Table 3. Subsequently, two optimisation objectives are formulated for the levelized costs and GWP respectively:

$$Minimize \sum_i c_i x_i \tag{1}$$

$$Minimize \sum_i e_i x_i \tag{2}$$

Where, c represents the levelized cost factors, and e represents the emission factors per technology.
The optimisation was subject to a number of constraints:

$$x_{OCGT} + x_{CCGT} + x_{CSP} + x_{PV} + x_{BIGCC} = 1 \tag{3}$$

$$x_{MSF} + x_{SWRO} = 1 \tag{4}$$

$$0 \le x_i \le 0.5 \tag{5}$$

Several iterations will be run to ensure a representative sample is obtained and the median pareto front will be selected for comparison with the existing configurations.

Table 3 Levelized Costs and GWP Factors per Technology.

Technology	Levelized Cost ($/kWh, $/m³)	Source	GWP (gCO₂e/kWh, gCO₂e/m³)	Source
OCGT	0.17	NREL OpenEI	490	IPCC (2014)
CCGT	0.093	Govindan (2018)	341.56	Al-Ansari (2017)
MSF	2.43	Darwish (2015)	73.35	Darwish (2015)
SWRO	0.8	Darwish (2015)	14.67	Darwish (2015)
CSP	0.16	NREL OpenEI	27	IPCC (2014)
PV	0.215	Al-Ansari (2017)	33.79	Al-Ansari (2017)
BIGCC	0.1	NREL OpenEI	230	IPCC (2014)

3. Results and Discussion

Thirty iterations were run using the GA model and a representative sample was obtained as illustrated in Figure 2. The median pareto front (Figure 3) is selected for comparison with the different configurations. Figure 4 illustrates the three manually selected points from the pareto front in comparison to the existing configurations. It is evident that current configurations are suboptimal although the more recent installations following configuration 3 display significant improvement and do approach the optimal solution in the right direction. However, without the integration of RES, the configurations will continue to be suboptimal and increasing share in the portfolio should be assigned to

CSP, BIGCC, and PV. Using configuration 5 as a representative proposed configuration and the comparison to exiting configurations reveal that configuration 5 displayed an improvement in GWP of 67%, 55%, and 52% in relation to configurations 1, 2, and 3 respectively. A similar comparison using levelized costs showed an improvement of 32%, 30%, and 8%. All 3 proposed configurations exhibited a significant share of renewables and a large reduction of natural gas-driven technologies from 100% to a range between 10% and 20%. Water desalination technologies on the other hand had an equal share in the portfolio due to the given constraints in the optimisation problem formulation.

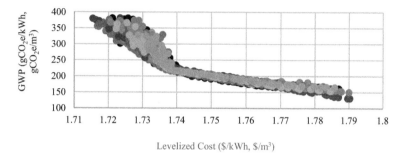

Figure 2. Pareto front for 30 iterations.

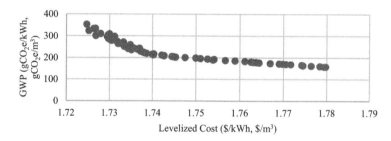

Figure 3. Median Pareto front.

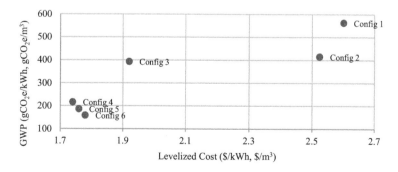

Figure 4. Comparison between 3 optimal configurations with existing ones.

4. Conclusions

In this paper, current CPDP configurations in Qatar are compared to a set of proposed configurations using non-conventional RES such as CSP, PV and BIGCC. Using stochastic modelling of natural gas prices and a multi-objective GA optimisation, it is concluded that the 3 current configurations evaluated are suboptimal. The optimal configurations are represented by a significant share of RES technologies which decrease both CO_2 emissions and levelized costs. Considering, water desalination, since there are only two technologies and both are constrained not to exceed a 50% share, the load is distributed to both of them equally. In the event that the constraint is removed, SWRO would dominate due to its significantly low levelized cost and GWP factors. However, such a scenario is not possible due to the large installed base of MSF units in Qatar. Similarly, gas-driven power generation units with a huge installed base, are also somewhat difficult to upgrade to RES. Meanwhile, RES integration can be considered for future network expansions or upgrade projects. Future research can consist of the integration of additional RES technologies such as nuclear, wind and geothermal and further evaluation of less-common water desalination technologies such as multiple-effect distillation (MED), membrane distillation (MD) and mechanical vapor-compression (MVC).

References

T. Al-Ansari, A. Korre, Z. Nie, and N. Shah, 2016, Integration of Biomass Gasification and CO2 Capture in the LCA Model for the Energy, Water and Food Nexus, Computer Aided Chemical Engineering, 38, 2085-2090.

―――, 2017, Integration of Greenhouse Gas Control Technologies within the Energy, Water and Food Nexus to Enhance the Environmental Performance of Food Production Systems, Journal of Cleaner Production, 162, 1592-1606.

M. Darwish , 2014, Qatar water problem and solar desalination, Desalination and Water Treatment, 52(7–9), 1250–1262.

M. Darwish, H. Abdulrahim, A. Hassan, 2015, Realistic Power and Desalted Water Production Costs in Qatar, Desalination and Water Treatment, 57, 4296-4302

R. Govindan, T. Al-Ansari, A. Korre, N. Shah, 2018, Assessment of Technology Portfolios with Enhanced Economic and Environmental Performance for the Energy, Water and Food Nexus, Computer Aided Chemical Engineering, 43, 537-542.

Kahramaa, 2017, Statistics Report 2016, Kahramaa Publications.

K.Madani, 2015, Using Game Theory to Address Modern Resource Management Problems, Grantham Briefing note 2, Imperial College London.

S. Namany, T. Al-Ansari, R. Govindan, 2018, Integrated Techno-economic Optimisation for the Design and Operations of Energy, Water and Food Nexus Systems Constrained as Non-cooperative Games, Computer Aided Chemical Engineering, 44, 1003–1008.

Qatargas, 2018, Global Energy Supplier, Retrieved from https://www.qatargas.com/english

N. Voutchkov, 2016, Desalination – Past, Present and Future, Retrieved from https://www.iwa-network.org

Anton A. Kiss, Edwin Zondervan, Richard Lakerveld, Leyla Özkan (Eds.)
Proceedings of the 29th European Symposium on Computer Aided Process Engineering
June 16th to 19th, 2019, Eindhoven, The Netherlands. © 2019 Elsevier B.V. All rights reserved.
http://dx.doi.org/10.1016/B978-0-128-18634-3.50285-X

Consequential Life Cycle Analysis for Food-Water-Energy-Waste Nexus

Yanqiu Tao, Fengqi You

Cornell University, Ithaca, New York, 14853, USA

Abstract

Animal manure, as a type of unavoidable organic wet biomass waste, contains abundant nutrients, moisture and pathogen, therefore how to properly handle them has become an arising problem. Opposite to the severe challenge, dairy manure can be viewed as stably supplied feedstocks for biorefinery, which is invariant to seasonal and climate change and is of huge quantities. In this study, to thoroughly explore the potential to treat manure as biorefinery feedstocks, we select three promising thermo-chemical technologies, namely slow pyrolysis, fast pyrolysis and hydrothermal liquefaction and quantify their corresponding environmental impacts. We adopt a consequential life cycle assessment (LCA) approach, and successfully capture the environmental impact both attributed to the target process and consequential to the affected processes and markets. Lastly, we compare the consequential environmental impact with the attributional environmental impact and find consequential LCA advantageous to capture environmental impact without overestimation.

Keywords: wet biomass, dairy manure, pyrolysis, hydrothermal liquefaction, consequential life cycle assessment

1. Introduction

Organic wastes treatment is recognized as one of the most challenging issues in the systems engineering of food-water-energy nexus (FWEN) (Garcia et al., 2016). For example, animal manure, as one main type of unavoidable organic wastes, is rich in nutrients, bacteria and aerobic organics, hence it causes severe degradation of water and soil quality and makes contribution to greenhouse gas emission if being composted or landfilled (Yue et al., 2014). However, if handled properly, manure wastes can be converted to value-added products, for example biochar, biogas and bio-oil, which may bring both environmental and economic benefits (Garcia and You, 2017).

To address this issue, we specifically focus on the treatment of dairy manure wastes in this study. The environmental impact of three promising technologies for organic waste conversion, namely slow pyrolysis, fast pyrolysis and hydrothermal liquefaction, are evaluated and compared. To assess the environmental impacts, we adopted a cradle-to-grave analysis technique named life cycle assessment (LCA). The methodology of LCA involves four steps: goal and scope definition, life cycle inventory analysis (LCI), life cycle impact assessment (LCIA) and interpretation. The first step, goal and scope definition, includes the definition of the goal of LCA, the functional unit, system boundary, allocation procedures, type of methodology, life cycle impact category and assumptions. The second step LCI involves compilation and quantification of LCI data such as mass and energy flow rates. Next, LCI data is assigned to life cycle impact categories such as global warming potential (GWP) and multiplied by the characterization factors of corresponding impact categories. Eventually the environmental impact indicators are obtained and interpreted during the final step. Notably, interpretation does not only happen in the end, but is conducted iteratively in every step of LCA.

LCA approach can be classified into two types: attributional LCA (ALCA) and consequential LCA (CLCA). The conventional ALCA takes a "snapshot" of existing conditions by only taking into account the environmental impacts of direct physical flows attributed to the life cycle of the products (Yang 2016). One underlying assumption of ALCA in which supply of inputs are unlimited implies the absence of both market effects and indirect environmental impacts. Therefore, ALCA is not suitable for situations where market effects and those indirect environmental impacts are not negligible, which accounts for the vast majority of the real world cases. Furthermore, another assumption, fixed input/output coefficients, eliminates possibilities for ALCA to capture any change in decision making to targeted processes.

In contrary to ALCA, by involving market constraints, CLCA is capable of capturing consequential environmental impacts in response to changes all through the targeted processes. CLCA systematically identifies the affected processes that change its production in response to the changes in both demand for inputs and supply for outputs, and sequentially expands its system boundary by including those indirectly affected processes. Several frequently-used methods to quantify the indirect environmental impacts are scenario analysis, simple partial equilibrium (PE) model, sophisticated PE model and computable general equilibrium (CGE) model (Garcia and You 2018). Considering model resolution, model maturity and the viability of the plan to incorporate CLCA into an optimization framework, it is the most appropriate to integrate the simple PE model in our CLCA study. Simple PE model replies on the microeconomic concepts of price elasticity of demand and supply, and it is confined to only one or a few economic sectors. While in reality, economic consequences can be passed on through multiple levels of markets, intuitively in this study, we aim to identify and analyze no more than two levels of markets and no more than one affected technology. Considering the inherently interrelated nature of FWEN, the CLCA approach is more reliable and may contribute to the improvement of systems engineering of more sustainable production, consumption, and distribution processes in the FWEN (Gong and You 2017).

2. System boundary, assumptions, and consequence identification

The goal of this consequential LCA is to assess the consequential life cycle environmental impacts of three thermo-chemical technologies: slow pyrolysis, fast pyrolysis and hydrothermal liquefaction (HTL), and make a comparison between the attributional and consequential environmental impacts. Pyrolysis is a thermo-chemical process that converts the waste biomass into a series of value-added products: product gas, bio-oil and biochar in the absence of oxygen (Zhang et al. 2014). Pyrolysis can be categorized as slow pyrolysis and fast pyrolysis, according to the heating rate and residence time (Gebreslassie et al., 2013). Particularly, the product yields vary with temperature, where the slow pyrolysis produces more product gas, biochar and less bio-oil, while the fast pyrolysis leads to more oil product (Goyal et al, 2008). Notably, by taking into account its capability of more than 100-year carbon sequestration and fertilizer use efficiency improvement, biochar is especially valuable from an environmental point of view (Roberts 2010). Hydrothermal liquefaction converts biomass into bio-oil, product gas, hydro-char and aqueous phase product at moderate temperature in water (Gong et al. 2014). Therefore, this process does not need the energy-intense drying process and may serve as an ideal approach to handle wet biomass (Gollakota et al, 2018).

The functional unit is defined as treatment of 1 t fresh dairy manure. System boundary for the slow and fast pyrolysis is shown in Figure 1. In both processes, we start from collection of 1 t fresh dairy manure, transport it to anaerobic digestion (AD) and solid-liquid separation (SLS) (Aguirre-Villegas and Larson 2017). The avoided emission resulted from the AD-derived biogas is not included in the system boundary. Then, the pre-treated feedstock is transported to the biorefinery and undergone thermal-drying prior to being pyrolyzed. The heat is assumed to be supplied by high pressure steam, which is vaporized using natural gas (Gong et al., 2014). The resulting wastewater is sent to a wastewater plant for treatment and recycling. The biochar from pyrolysis is used as a soil amendment, which is then transported and applied to soil. Product gas will be used to generate electricity, and bio-oil will be converted to gasoline and diesel after being hydrotreated and hydrocracked. System boundary for the HTL is shown in Figure 2. After AD and SLS, the digested dairy manure is transported to the biorefinery and sent to the HTL reactor directly. The resulting bio-oil will be converted to gasoline and diesel, gas will be emitted directly, hydro-char will be applied to soil as P, K fertilizer. Additionally, the aqueous phase is treated by anaerobic digestion and used for algae cultivation (Zhou et al. 2015).

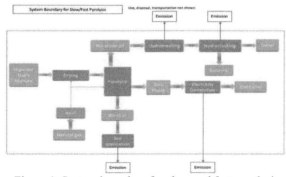

Figure 1. System boundary for slow and fast pyrolysis

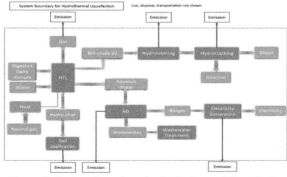

Figure 2. System boundary for hydrothermal liquefaction

While feedstocks are consumed, products are generated, the biorefinery interacts with the feedstock and product markets and breaks the current market equilibrium. The change in supply and demand in the feedstock and product market lead to change in the price and quantity, and eventually shifts to a new market equilibrium. This will also consequentially affect the market equilibrium in the downstream market. Therefore, it is important to identify the affected markets and capture the changes. There are five markets directly interacting with the target process: (1) Gasoline market. As an additional supply of gasoline is introduced into the market, the demand for fossil-based gasoline decreases, and the market reaches a new equilibrium. (2) Diesel market. The diesel market experiences the same trend as the gasoline market. (3) Electricity market. We assume that the price elasticity of demand is zero, and thus the increase in supply of electricity from the target process can be completely absorbed by other electricity suppliers. (4) Biochar market. The biochar market is very small and expanding, so we will not model the market behavior of biochar. Instead, we will only account for its improved fertilizer use efficiency and avoided soil emission during soil application. (5) Natural gas market. The natural gas will be consumed as a feedstock to generate heat for vaporization. The increased demand for natural gas will lead to increased price, and thus consequentially increased demand in other fossil-based fuels in the downstream market of heat generation. Two other modifications we made are: (1) Hydro-char. Because the hydro-char is sold as a substitute for P, K fertilizer, it will in fact avoid emissions during the use and disposal of the P and K fertilizers. (2) Dairy manure. Since the dairy manure market is absent, we will only calculate the avoided manure composting emissions.

3. Life cycle assessment results

Following the steps of consequential LCA, we obtain results as shown in Figure 3. The 100-year global warming potential (GWP) generated from consequential analysis for the treatment of fresh dairy manure using slow pyrolysis, fast pyrolysis and HTL are: 26.6, 34.7 and 15.7 kg CO_2 eq., respectively. HTL has the overall best performance with respect to the 100-year GWP. The energy-intensive process involved in the slow and fast pyrolysis process, thermal-drying, contributes to a large portion of the environmental burden. Moreover, the natural gas market accounts for a large part of the GWP, and this is mainly because the demand for natural gas consumed in our processes is fulfilled by the substitution of fossil-based fuel, which puts much consequential environmental burden on the target process. In contrary, the consequential GWP associated with the products are all negative, especially for the electricity. This result confirms that the products from these three technologies are truly environmental friendly. Attributional LCA is also conducted for comparison, the result is shown in Figure 4 and 5. The attributional LCA shows a similar trend that HTL has the best performance among the three technologies, except that the net GHG emissions are significantly higher. The 100-year GWP value of slow pyrolysis, fast pyrolysis and HTL are 85.9, 89.2 and 52.7 kg CO_2 eq., respectively. Since the attributional approach overlooks a series of consequences, it overestimates the overall environmental impacts of each target process.

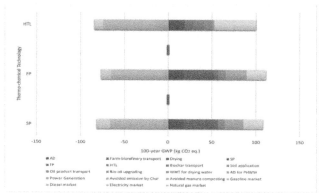

Figure 3. CLCA environmental impacts evaluated by 100-year GWP

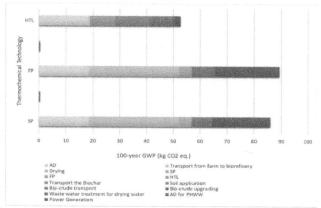

Figure 4. ALCA environmental impacts evaluated by 100-year GWP

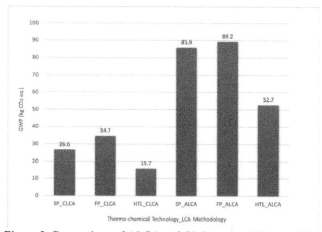

Figure 5. Comparison of ALCA and CLCA using 100-year GWP

4. Conclusion

We conducted attributional and consequential life cycle analysis for three types of thermo-chemical conversion of dairy manure. In both ALCA and CLCA, HTL showed the best environmental performance with respect to the 100-year GWP. Moreover, by including consequential environmental impacts associated with the affected processes and downstream markets, CLCA leads to a much lower GWP than ALCA does. In the future, we will further extend our research to consequential life cycle optimization and broaden the scope of our study by integrating spatial analysis tools and uncertainty analysis.

References

H.A. Aguirre-Villegas, R.A. Larson, 2017, Evaluating greenhouse gas emissions from dairy manure management practices using survey data and lifecycle tools. Journal of Cleaner Production, 143, 169-179.

D.J. Garcia, F. You, 2016, The water-energy-food nexus and process systems engineering: A new focus. Computers & Chemical Engineering, 91, 49-67.

D.J. Garcia, F. You, 2017, Systems engineering opportunities for agricultural and organic waste management in the food-water-energy nexus. Current Opinion in Chemical Engineering, 18, 23-31.

D.J. Garcia, F. You, 2018, Addressing global environmental impacts including land use change in life cycle optimization: Studies on biofuels. Journal of Cleaner Production, 182, 313-330.

B.H. Gebreslassie, M. Slivinsky, B. Wang, et al., 2013, Life cycle optimization for sustainable design and operations of hydrocarbon biorefinery via fast pyrolysis, hydrotreating and hydrocracking. Computers & Chemical Engineering, 50, 71-91.

A. Gollakota, N. Kishore, S. Gu, 2018, A review on hydrothermal liquefaction of biomass. Renewable & Sustainable Energy Reviews, 81, 1378-1392.

J. Gong, F. You, 2014, Global Optimization for Sustainable Design and Synthesis of Algae Processing Network for CO2 Mitigation and Biofuel Production Using Life Cycle Optimization. AIChE Journal, 60, 3195-3210.

J. Gong, F. You, 2014, Optimal Design and Synthesis of Algal Biorefinery Processes for Biological Carbon Sequestration and Utilization with Zero Direct Greenhouse Gas Emissions: MINLP Model and Global Optimization Algorithm. Industrial & Engineering Chemistry Research, 53, 1563-1579.

J. Gong, F. You, 2017, Consequential Life Cycle Optimization: General Conceptual Framework and Application to Algal Renewable Diesel Production. ACS Sustainable Chemistry & Engineering, 5, 5887-5911.

H. Goyal, D. Seal, R. Saxena, 2008, Bio-fuels from thermochemical conversion of renewable resources: a review. Renewable & Sustainable Energy Reviews, 12, 504-517.

K.G. Roberts, B.A. Gloy, S. Joseph, et al., 2009, Life cycle assessment of biochar systems: estimating the energetic, economic, and climate change potential. Environmental Science & Technology, 44, 827-833.

Y. Yang, 2016, Two sides of the same coin: consequential life cycle assessment based on the attributional framework. Journal of Cleaner Production, 127, 274-281.

D. Yue, F. You, S.W. Snyder, 2014, Biomass-to-bioenergy and biofuel supply chain optimization: Overview, key issues and challenges. Computers & Chemical Engineering, 66, 36-56.

Q. Zhang, J. Gong, M. Skwarczek, et al., 2014, Sustainable Process Design and Synthesis of Hydrocarbon Biorefinery through Fast Pyrolysis and Hydroprocessing. AIChE Journal, 60, 980-994.

Y. Zhou, L. Schideman, M. Zheng, et al., 2015, Anaerobic digestion of post-hydrothermal liquefaction wastewater for improved energy efficiency of hydrothermal bioenergy processes. Water Science and Technology, 2139.

Anton A. Kiss, Edwin Zondervan, Richard Lakerveld, Leyla Özkan (Eds.)
Proceedings of the 29th European Symposium on Computer Aided Process Engineering
June 16th to 19th, 2019, Eindhoven, The Netherlands. © 2019 Elsevier B.V. All rights reserved.
http://dx.doi.org/10.1016/B978-0-128-18634-3.50286-1

The Optimization of Heliostat Canting in a Solar Power Tower Plant

Nao Hu,[a] Yuhong Zhao,[a*] JieQing Feng[b]

[a]College of Control Science and Engineering, Zhejiang University, Hangzhou 310027, China

[b]State Key Laboratory of CAD&CG, Zhejiang University, Hangzhou 310058, China

yhzhao@zju.edu.cn

Abstract

In a solar power tower plant, solar radiation is reflected by the heliostat field and concentrated in the receiver at the top of the tower to provide energy for conventional thermodynamic cycle. For the heliostats composed of small facets, heliostat canting, which means applying specific rules for the orientation of single facets, can reduce the size of reflected energy spot and increase the concentration ratio and the optical efficiency of the heliostat filed. An optimization based on the paraboloid canting method is presented in this paper. The facets are aligned to fit the paraboloid to get a minimal spillage lost. A two-stage method combining Particle Swarm Optimization (PSO) and direct search method based on quadratic fitting model is designed to solve the optimization problem. The simulation results of two real plants demonstrate that the canting method with the optimal parameter can reduce the spillage lost significantly.

Keywords: Solar power tower plant, Heliostat canting, PSO, Direct searching optimization, Quadratic fitting model.

1. Introduction

Concentrated solar power (CSP) technologies offer promising options for high efficiency solar energy applications. Of all CSP technologies available, the solar power tower (SPT) can achieve higher temperature up to 1000ºC and hence higher efficiency (A.Boretti et al, 2017). In a SPT plant, the solar radiation is reflected and gathered by an array of mirrors called heliostat field to the aperture of receivers to form high-temperature steam.

A large heliostat is usually composed of several small flat facets and the power reflected by a flat mirror creates a large energy spot on the receiver plane due to the sun shape and various errors, such as optical, tracking, and positioning. This usually causes spillage lost, that part of the energy spot is out of the receiver area, and reduces the heliostat field efficiency. Therefore, the alignment of facets in a specific orientation on the heliostat frame, usually called heliostat canting, influences the focusing performance of the heliostat significantly. Applying specific rules for the orientation of single facets does not necessarily mean additional cost, since also standard canting methods require precise and controlled orientation of the facets (Buck and Teufel, 2009). On the other hand, the reduction of the spot size can decrease the spillage lost and provide heliostat aiming strategy more operating choices, thus a higher optical efficiency of the heliostat field and a better power distribution on the receiver can be achieved.

For the heliostat with azimuth-elevation tracking, traditional canting methods include on-axis canting and off-axis canting. Jones (1996) investigated the difference of the two strategies and concluded that off-axis canting can achieve slightly higher efficiency but it depends on the appropriate canting date and time. Moreover, dynamic off-axis canting, in which the facets are actively controlled such that the center of each facet is always perfectly focusing, is concerned in simulation and shows a remarkable reduction of spot size (Bonanos and Noone, 2012), but it's complex and costly in practical applications. Another canting method is aligning the facet normals parallel to the corresponding surface normals of a paraboloid. Buck and Teufel (2009) compared the three methods above and paraboloid canting gets the best result. However, no optimization method is employed to ensure the optimal alignment for the entire year in these methods.

An optimization of heliostat canting based on paraboloid pattern is presented to find the minimum annual spillage. In section 2, the Monte Carlo raytracing method is introduced and the optimization model is built. A two-stage method combining PSO and direct search method based on quadratic fitting model is designed to solve the optimization problem in section 3. Section 4 explores the practical applications of the proposed method in two heliostat fields, namely a test small heliostat in Hangzhou, China, and the large heliostat of PS10. Finally, conclusions are drawn and future research is indicated in Section 5.

2. Optimization model

2.1. Flux distribution on the receiver

The calculation of the flux distribution on the receiver surface is fundamental for the design and operation of a solar power tower plant. The sun is a disk rather than a point observed from any place on the earth. Therefore, the incident rays can't be considered parallel but a group of light described by the sun shape. The common way to calculate the flux distribution on the receiver can be divided into two classes, raytracing approaches and analytical approaches. Considering the high accuracy, a raytracing process using Monte Carlo method to address the sun shape is employed in this paper. An empirical sun shape model proposed by Walzel et al (1977) is adopted to evaluate the solar energy flux density $S(\alpha)$ for different values of the solar angle α:

$$S(\alpha) = \begin{cases} S_0 \left[1 - \lambda (\dfrac{\alpha}{\alpha_s})^4 \right], & \alpha \leq \alpha_s \\ \\ 0 & , \quad \alpha > \alpha_s \end{cases} \tag{1}$$

where S_0 is a parameter related with the direct normal irradiance, $\lambda=0.5138$ is a constant, $\alpha_s=4.6$ mrad is half the angle of the light cone and the solar angle α is the angle between the vector QP and QO as shown in Figure 1 and P is a point on the solar disk.

Consequently, a large number of rays are produced depending on the probability of the energy flux density by Roulette Wheel Selection Principle. Figure 2 illustrates the flux distribution of the solar disk by random points, where the closer the points appear to the center of the sun, the greater the number of the points is. In the raytracing process, a set of random points are generated on the heliostats and each point reflected a group of sun light illustrated in Figure 2.

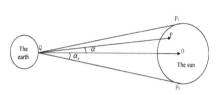

Figure 1 solar disk and solar angle

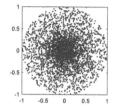

Figure 2 solar flux density

The raytracing method, however, needs huge amount of sun light be traced to get the accurate result and it is a time-costing process. Therefore, a parallel implementation of this process is utilized based on CUDA platform to speed up the simulation.

2.2. Paraboloid canting optimization

As mentioned above, the large reflected flux spot will cause spillage lost. A paraboloid canting method can be applied to improve the situation. The objective of the optimization problem is described as:

$$\min_{a,b} \quad f(a,b) = \sum_{t=1}^{n_s} \sum_{i=1}^{n_f} \left[1 - \frac{E_{t,i}^{in}(a,b)}{E_{t,i}(a,b)} \right] \tag{2}$$

where n_s and n_f are respectively the sample number in a year and the facet number, $E_{t,i}(a,b)$ is the total power reflected by the facet i at sample time t when aligned to fit the paraboloid determined by the parameters (a, b), and $E_{t,i}^{in}(a,b)$ is the part in the receiver area. The objection function represents the spillage lost of the specific heliostat for an entire year. The facets normal vectors are determined in the way showed in Figure 3. For the facet center at position (x, y, z) in the heliostat coordinate, x and y are predetermined by heliostat size and facet numbers, and z is calculated through the standard paraboloid equation and then the facet normal vector is set to the surface normal vector at this point. There are gaps between adjacent facets to ensure that no collision occurs when applying different canting angles.

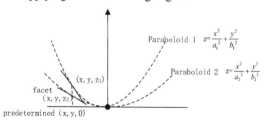

Figure 3 paraboloid canting

More than 500 samples are calculated to ensure the result behave well in the entire year. The sampling interval is three days and five or six uniformly spaced times are chosen in each sampled day according to the day length. The reflected spot at each sample time is obtained by Monte Carlo ray tracing method based on GPU. As the objective function is calculated through raytracing, the derivatives are not available. Also the high time cost of objective function evaluation and the noise of Monte Carlo method bring difficulties to the solving process.

3. Optimization algorithm

A two-stage method combining PSO and direct search method based on quadratic fitting model is designed to solve above optimization problem.

PSO is a metaheuristic as it makes few or no assumptions about the problem being optimized and can search very large spaces of candidate solutions. The output of this algorithm, with parameters selected appropriately can converge to the neighbourhood of the optimum after several iterations. More details about PSO can be found in Bonyadi and Michalewicz (2017). The fitness of swam is evaluated in each iteration and the rate of convergence falls down as the particles move to the neighbourhood of the optimum.

A direct search method with trust region is combined to find the optimal result in the neighbourhood of the preliminary result offered by PSO. The trust region method behaves well in the search of small scale so that the times of objective function evaluation to find the final optimal result can be reduced significantly.

During the iteration, a cheaper model is built around the current optimal point in place of the real expensive one. There are several research and optimization solver (Powell, 2007) using quadratic interpolation model. In this problem, however, due to the noise in the objective function evaluation, the interpolation may not represent the real model well, which leads to a slow convergence process or even to a failure. Thus, the quadratic approximation model, which is more robust to the noise, is used in the trust region. The algorithm procedure is as follows:

Step 1. Initialization: Choose an initial trust region radius $\Delta_0 > 0$ and an approximation set Y with L points derived from PSO. Determine $x_0 \in Y$ such that $f(x_0) = \min_{y \in Y} f(y)$.

Set values for parameters $\eta \in (0,1)$, $\theta_0 \in (0,1)$, $\theta_1 \in (1,2)$, maximal number of iterations k_{max}, error tolerance ε, maximal fitting error E_{max} and upper limit of the cardinality of the approximation set L_{max}. Set $k=0$.

Step 2. Using the current approximation set Y, build the quadratic fitting model $m_k(x)$ and compute the minimal point x_k^+ within the trust region $B_k = \{x_k + s \mid s \in R^n, \|s\| < \Delta_k\}$ such that $m_k(x_k^+) = \min_{x \in B_k} m_k(x)$. Compute $f(x_k^+)$ and the ratio:

$$\rho_k = \frac{f(x_k) - f(x_k^+)}{m_k(x_k) - m_k(x_k^+)} \tag{3}$$

Step 3. Update the approximation set and the trust region radius. For the successful step that $\rho_k > \eta$, include x_k^+ in Y and check the existing points to drop the ones far away from x_k^+. Set $\Delta_{k+1} = \theta_1 \Delta_k$. For the unsuccessful step that $\rho_k \leq \eta$, add more sample points to Y and set $\Delta_{k+1} = \theta_0 \Delta_k$. The point that maximizes the qualitative index $\delta(x)$ is added to Y until the fitting error is less than E_{max} or the cardinality of the approximation set $L > L_{max}$. $\delta(x)$ is described in Eq. (4).

$$\delta(x) = \sum_{j=1}^{L} \lambda \|x - x_j\| + (1 - \lambda)(S(x) - S_j(x))^2 \tag{4}$$

where x_j represents existing point in Y, and $\|x - x_j\|$ is the Euclidean norm, which is used to place new points in relatively unexplored region and as far away from the existing points as possible. $S(x)$ is the fitting model built using all points in set Y, $S_j(x)$ is the fitting model built using all points except point x_j. The impact of locating a sample point in the neighbourhood of x_j on $S(x)$ is measured in this way. With the weighting factor $\lambda \in (0,1)$, a new point can be placed to achieve space filling and high impact on the current model, without evaluation of the real objective function.

Step 4. Update the current iteration and check for convergence. For the successful step, set $x_{k+1} = x_k^+$, else $x_{k+1} = x_k$. Increase k by one and go to step 2 until the termination criterion, $k > k_{max}$ or $\|x_{k+1} - x_k\| < \varepsilon$ is satisfied.

The minimal annual spillage lost can be achieved using the optimization method above.

4. Case studies

The optimization is applied to get the canting paraboloid parameter for a test heliostat in Hangzhou, China and different heliostats in PS10, Spain. The two cases represent the small size and large size heliostats respectively, both of which are serving in the existing SPT plant and are two different technologies with their own advantages and disadvantages. Three typical heliostats positions in the PS10 field are considered, the two exactly in the northern direction of the receiver with different distance and the other one located in the northwest.

The optimization results are listed in Table1. The position of heliostat is represented in the coordinates system that x-coordinate points to the east, y-coordinate to north and z-coordinate to the zenith. In order to reveal the influence by the heliostat canting, a receiver size smaller than the actual receiver in the plant is used. It should be noted that the receiver size is also adjusted to get the optimal canting angles in the optimization process though it's fixed in an established plant. Since the spot size of heliostat far away from the receiver and the one in the northwest side is much larger than that of the nearer one and the one exactly in the north, the formers need larger receiver size to obtain acceptable spillage lost. The objective function calculation times in the trust region process and time consumption of the programme running on the computer equipped with NVIDIA GeForce GTX 1080 are also shown in Table 1. The optimal result can be obtained in less than an hour for various heliostats.

Table 1 heliostats parameters and optimization results

Heliostat No.	location	size (m²)	facets number	receiver size	position	optimal (a, b)	minimal lost	objective evaluation times	time consumption
1	Hangzhou	2	3×3	1.69 m²	(12, -107, 1.2)	(28, 32)	7.17 %	17	2461 s
2				16 m²	(0.84, 110, 11)	(25, 23)	14.63 %	18	3018 s
3	PS10	121	7×4	36 m²	(0.84, 698, 52)	(55, 52)	15.19 %	15	2846 s
4				28 m²	(139, 138, 15)	(30, 26)	14.22 %	20	2961 s

The reflected spots of the heliostats at 8:00, 12:00 and 16:00 in 26th, April under the circumstances that the facets are aligned in on-axis canting, dynamic off-axis canting and in the paraboloid determined by the optimization result separately are illustrated in Figure 4. Compared with the on-axis canting, the optimal paraboloid canting method reduces the spot size remarkably. The spot sizes of the optimal paraboloid canting are slightly larger than that of the dynamic off-axis canting which nearly represents an ideal situation. More significant reduction of the spot size can be achieved for the larger heliostat with more facets. Actually, the spillage lost in PS10's origin receiver (168 m²) can be decreased to zero using the optimal canting angle.

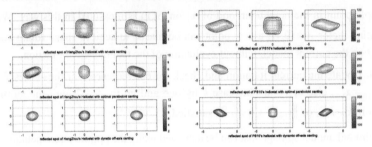

Figure 4 reflected spot of small heliostat (left) and large heliostat (right)

5. Conclusions

An optimization based on the paraboloid canting method is developed in this paper to reduce the reflected spot of the heliostats and improve the annual optical efficiency of the SPT plant. With the combined optimization method, the optimal result can be got at a low time cost, which is helpful when a large number of heliostats are designed in a SPT plant. The further research will be focused on the integrated optimization of the heliostat field and the receiver.

Acknowledgement

This work is supported by the National Natural Science Foundation of China (61772464).

References

A. Boretti, S.Castelletto, A.Z, Sarim, (2017), Concentrating solar power tower technology: Present status and outlook. Nonlinear Engineering. 10.1515/nleng-2017-0171.

A. M. Bonanos, C J. Noone, A. Mitsos, (2012), Reduction in Spot Size via Off-Axis Static and Dynamic Heliostat Canting, International Conference on Fuel Cell Science, Engineering and Technology. 2012: 567-571.

M. J. D. Powell, (2007), A view of algorithms for optimization without derivatives. Cambridge Na Reports Optimization Online Digest 2007, 5: 170-174.

M. R. Bonyadi, Z. Michalewicz, (2017), Particle Swarm Optimization for Single Objective Continuous Space Problems: A Review, Evolutionary Computation, 2017, 25(1): 1-54.

R. Buck., E. Teufel, (2009), Comparison and Optimization of Heliostat Canting Methods. Energy Sustainability, DLR. 131, 1015-1022.

S. A. Jones, (1996), A comparison of on-axis and off-axis heliostat alignment strategies. Office of Scientific & Technical Information Technical Reports.

Anton A. Kiss, Edwin Zondervan, Richard Lakerveld, Leyla Özkan (Eds.)
Proceedings of the 29th European Symposium on Computer Aided Process Engineering
June 16th to 19th, 2019, Eindhoven, The Netherlands. © 2019 Elsevier B.V. All rights reserved.
http://dx.doi.org/10.1016/B978-0-128-18634-3.50287-3

On-grid Hybrid Power System and Utility Network planning to supply an Eco-Industrial Park with dynamic data

Florent Mousqué[a], Marianne Boix[a*], Stéphane Négny[a], Ludovic Montastruc[a], Serge Domenech[a]

[a] *Laboratoire de Génie Chimique, Université de Toulouse, CNRS, INPT, UPS, Toulouse, France*

marianne.boix@ensiacet.fr

Abstract

Eco-Industrial Parks (EIP) aim to preserve environment while increasing competitiveness of companies. This paper presents a mathematical Mixed Integer Linear Programming (MILP) model to optimally grassroots design an EIP energy network comprising heat and electricity. In the utility system, heated steam is produced, network is designed by selecting boilers and turbines technologies and interconnection pipes between companies. Simultaneously in the on-grid Hybrid Power System (HPS) composed of wind turbines, solar Photovoltaic (PV) panels, steam turbines and external grid, the model can select which source to use to meet the power demand.

A case study of 10 industries from Yeosu real industrial park with seasonal data is provided to assess the model, first on economic comparison between stand-alone and EIP situation of companies, and secondly to determine which power source is profitable for HPS based on the sale price of electricity on the external power grid.

In conclusion, this model allows the optimal design of the energy network of an EIP with dynamic data and with the objective of minimizing the net present cost.

Keywords: Industrial Ecology, Eco-Industrial Parks, Utility Network, Hybrid Power System, Mathematical Optimization, Renewable Energies.

1. Introduction

Nowadays the risks of human-induced climate change are increasingly pointed out by scientific reports. These reports conclude that limiting global warming would require ambitious mitigation actions while achieving sustainable development which means a development leading to economic, environmental and social benefits (Brundtland et al., 1987). To these objectives, industry has an important role to play because classical linear model (extract, produce, consume, throw) is responsible for overwhelming extraction of natural resources and waste rejection.

Facing these environmental issues, Industrial Ecology (IE) provides an innovative way to produce goods and services based on a circular model (share, reuse, mutualize). Indeed, in these industrial ecosystems, energy and resources consumption are optimized, waste generation is minimized and effluents from one production process are used as raw materials for another (Frosch & Gallopoulos, 1989). An application to this concept are EIP, in which industries gather on a same site and thanks to their vicinity they share and exchange efficiently different flows (material, energy, utilities). Furthermore, while reducing their environmental impacts they can also achieve greater economic competitiveness. To implement EIP in a sustainable way, companies need methods and optimization tools to design appropriate inter-enterprises exchanges. While an increased number of research projects have been carried out in recent years, Boix et al. (2015) pointed out that only a few deal with energy flow management neither with coupled networks. However, a systemic approach studying interactions between the different flows and optimizing them simultaneously would allow a better synergy between companies.

In the great majority of industrial parks, the facilities to meet energy demand are utility systems producing utilities for processes (i.e. mainly heat, cold streams and compressed air) and HPS that produce electricity using several power sources.

In utility systems, main energy consumption concerns heat, due to its high calorific value and its ability to generate power by rotating turbine. Moreover, in these utility systems, Combined Heat and Power (CHP) is an energy-efficient way to produce simultaneously electricity and thermal energy. Indeed, compared to the separate generation of heat and power, cogeneration can improve energy efficiency between 10 and 40% (Madlener & Schmid, 2003) and thereby reduces CO_2 emissions. Through combined production raises other constraints such as operational planning of production. Yet a large number of publications are dedicated to optimizing the operational management of existing CHP, (Aguilar, 2007; Mitra, 2013) but only a few to design such facilities even less so adapted to EIP context with heat utility exchanges between companies. Afterwards, HPS, as a coupled system, is an efficient way to harness RE sources, relying on complementarity between sources. When it is connected to the external grid it is possible to compensate for the lack of production from other sources purchasing power, it is then called on-grid HPS. In addition, a decentralized production means to reduce reliance on external grid and reinforces the robustness of the power system.

In a previous publication, a study on on-grid HPS to supply electricity in an EIP was carried out using a mathematical LP model for the optimal design of this HPS with dynamic demand, on an hourly basis (Mousqué et al., 2018). This model has the possibility to select power sources to install between wind turbines, solar PV panels, biomass by combustion or also buy electricity from the grid.

To go further, this work extends the previous HPS model by presenting a generic optimization model to the grassroots design of both the utility network and the HPS of an EIP. The first purpose of this article is to assess economic interest that companies can encounter by sharing their heat and power facilities in comparison to their stand-alone situation. Another objective is to discuss which RE or energy-efficient electricity source can be profitable to supply EIP demand. Two optimal solutions are finally discussed depending on the sale price of electricity fixed by politic regulation. These results are obtained on a complex real problem using the proposed approach involving 10 companies in an EIP with real demands taken from Yeosu industrial park (Kim et al., 2010).

2. Problem statement

The aim is to design the exchanges of water, steam and electricity between processes inside a single company as well as exchanges between different companies, during several time steps. A generic superstructure (Figure 1) has been developed in order to show the different components taken into account in the model.

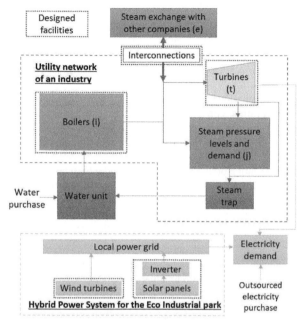

Fig 1. Superstructure of the utility system of an industry and HPS of the EIP.

In a company, steam flows through steam headers, different levels of steam pressure are taken into consideration. Pressure release valves are used to convert higher pressure steam into lower pressure steam by dropping its pressure. Steam is produced by boilers installed in companies. In the case studied in this publication, they are natural gas boilers. Different types of boilers can be installed depending on the maximum production capacity and the level of steam pressure produced. Boiler capacity range is between 50 and 100 % of its maximum power, because in this range boiler has a relatively constant efficiency (Aguilar et al., 2007) and also in seasonal operation it would be a big loss to operate the boiler at a low capacity, which means at a low efficiency level.

These power sources are selected to be RE technologies, i.e. wind turbines and photovoltaic panels, or energy-efficient such as cogeneration by mean of turbines. Data are dynamic with a seasonal variation (one-step for each season). In an EIP configuration, the power production is shared between companies and used to supply their demand. By connecting to external power grid, power can be bought and excess production can be sold. Finally, multi-stage extraction turbines can be selected between three technologies (i.e. 500 kW, 3 MW, and 15 MW). These turbines convert energy contained in high-pressure steam to a lower level of pressure or to condensed water. Higher is the pressure difference, higher the power production.

In this case, the production of renewable energy is considered as an average over a season. This estimate is only possible thanks to the external grid connection. Indeed, periods when production is higher or lower than the average production are supposed to be offset by the purchase or sale of electricity.

Water flow is a closed loop, after consumption from the process, water returns to the water unit, reducing water consumption. Each company has the possibility to buy water at every time step if necessary. Indeed, water losses by evaporating in water unit and steam losses by condensing trap or vent to evacuate overload are included.

3. Generic optimization model

Regarding the superstructure previously shown and the complexity of this model, due to the important numbers of continuous and binary design variables, a large-scale MILP model is developed. Due to its genericity, it is intended to be adaptable to the initial design of new EIP or to the retrofit of existing industrial parks, this last case by setting the initial structure.

The selected objective function is to minimize NPV of the designed utility network and HPS for a project duration considered of 20 years, with a discount factor of 7%. Constraints to model units are based on thermodynamic principles, mass and energy balances. Decisions variables are the different possible connexions between units and enterprises characterized by their type (VHS, HS, MS, LS, water…) and their flow and the decision to install or not a boiler and if it is installed which one is chosen? Capital and operational cost for installed boilers, turbines, pipes and the consumption of raw material are thus determined by the decision variables. Raw materials are fuel i.e. natural gas, treated water and electricity, which can be purchased or sold. Binary variables are used to select installed pipes between companies, installed technologies for boilers and turbines and operational state of boilers (i.e. switched on or off).

Given the high capacity of the superheated steam to be transported, losses in the interconnection pipes between the companies are considered negligible. Implemented data for boiler cost and steam pipes comes from economic evaluation manual (Chauvel et al., 2001) and are actualized using OECD Industrial Production Index. Renewable production is supposed to take place in France, with RTE report seasonal production (RTE, 2018) for these sources complete cost are those from ADEME (2016). Price for natural gas is set at 280 €/ton.

This model has 4101 constraints, 5103 variables including 540 binaries and it was solved with IBM ILOG CPLEX Optimizer.

4. Results and discussion

Two studies are conducted: an economic analysis of the case study to determine the interest of a company to take part of an EIP while the second one deals with the design of HPS power sources considering a different price for sale electricity.

This first study will compare Stand-alone and EIP solutions, both with a sale price of 0.05€/kWh which reflects the average price for electricity producer in France. While the second analyse will be a comparison between EIP (0.05) and EIP (0.10), latter case with a price of 0.10 €/kWh, considering politic aids to enhance creation of a local power grid. Detailed costs for these discussed optimal solutions are presented in Table 1.

Table 1
Overall economic comparison between Stand-alone and EIP situation.

Overall cost	Net Present Cost	Boilers cost	Power sources cost			Raw Materials cost				Pipes cost
			Turbines	Solar PV panels	Wind Turbines	Fuel	Water	Electricity purchased	Electricity sold	
Stand-alone	1,097,762 k€	56,294 k€	4,181 k€	- €	- €	1,040,671 k€	3,781 k€	4,311 k€	-11,478 k€	- €
EIP (0.05)	1,048,715 k€	36,186 k€	- €	- €	- €	1,001,113 k€	3,416 k€	4,430 k€	- €	3,569 k€
EIP (0.10)	982,324 k€	55,638 k€	56,647 k€	- €	38,594 k€	1,184,730 k€	3,923 k€	4,431 k€	-366,036 k€	4,396 k€

4.1. Stand alone and EIP (0.05) comparison

Observations show that while cost is slightly different for NPV of the EIP (0.05) (i.e. - 4.7 %), for water (i.e. -10.7 %) and for fuel (i.e. -4.0 %), main reduction observed is for the cost of installed boilers (i.e. a reduction of 55.6 %). Additionally in stand-alone situation, the model proposes to invest in turbines, produce electricity and sale it.

This is because single companies have oversized boilers during seasons with lower demand, and this overproduction cannot be valorised. To avoid dumping it, it is profitable to install turbines to value this overproduction. In EIP configuration, cost for installed boilers is significantly reduced. In fact, pipe cost being less expensive than gain resulting from creating interconnections, companies share their production and their demand. This share allows a better size and management of installed boilers. This is observable by an average production ratio of installed boiler at 77% of their maximal production in EIP against 60% in stand-alone. However, on project life span, fuel is the most important cost, and in comparison between cases, only a few difference is observed. More accurate fuel consumption results should be obtained by including boiler inertia on a daily basis.

4.2. Impact of the electricity sale price on the design

Figure 2 shows the optimal solution of the utility network for EIP (0.05) case, with boiler for each company and different pressure level exchanges, highlighting results that this model achieves and its complexity.

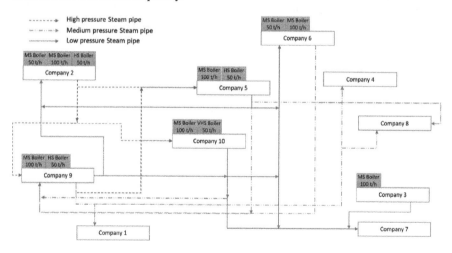

Fig 2. Designed boilers and utility steam exchange network for EIP (0.05).

With a sale price of 0.05 €/kWh neither turbines nor RE are selected, whereas when the sale price increases at 0.10 €/kWh both wind turbines and turbines are selected with three turbines of 15 MW. These choices can be justified with wind turbines and PV panels break-even point, which is calculated as respectively 0.094 €/kWh and 0.125 €/kWh, depending on discount factor and overall production and complete cost of each technology. For turbines, it is also dependent on utility system criteria, such as turbines and boilers technology (i.e. investment cost and efficiency), output pressure steam and if extra steam production is available to value it. However, over the lifespan of the project, the complete discounted cost for the installation of the turbines represents only a relatively small part. For example, for a 500 kW system operating at nominal power, the cost of the installation is calculated at 0.019 €/kWh. Only based on an economic criterion, external electricity price has been identified as a main parameter to decide which technology to use. Turbines can be considered as an efficient way to value overproduction or overheated steam. Currently, wind turbines are cheaper than PV panels, although the difference is small, the design will depend mainly on the renewable resource available on the site.

5. Conclusion

A model of grass-root design of an EIP utility system and HPS with an economic criterion has been developed to determine the facility to invest in and the exchanges between companies. In this model, the implementation of this EIP represents a slight NPV gain, but a significant reduction in the investment cost of boilers. For HPS, a too low price of outsourced electricity does not foster to create a local grid with cogeneration, wind turbines or PV panels. However, other criteria could be taken into account, such as public subsidies on investment or on sale price to incite such technologies. This mathematical model has validated a first frame to assess capability of MILP to be solved on a real case of 10 companies taken from Yeosu industrial park with dynamic seasonal data. Further improves could be achieved, using this model on a case study with dynamic data on a daily basis (day and night) allowing more accurate results for boiler management, more detailed analyse on RE sources variation production depending on weather conditions. Another interesting point is that due to fuel cost importance, other fuels and boilers technology could be added. Finally using a multi-objective optimization, including environmental objective is another way to explore.

References

ADEME. (2016). Coûts des énergies renouvelables en France. Edition 2016. France

Aguilar, O., Perry, S. J., Kim, J.-K., & Smith, R. (2007). Design and optimization of flexible utility systems subject to variable conditions. Chemical Engineering Research and Design.

Boix, M., Montastruc, L., Azzaro-Pantel, C., & Domenech, S. (2015). Optimization methods applied to the design of eco-industrial parks: A literature review. Journal of Cleaner Production.

Brundtland, G., Khalid, M., Agnelli, S., Al-Athel, S., Chidzero, B., Fadika, L.M et al., (1987). Our Common Future : world commission on environment and development. Oxford.

Chauvel, A., Fournier, G., Raimbault, C., & Pigeyre, A. (2001). Manuel d'évaluation économique des procédés. Editions Technip. France.

Frosch, R. A., & Gallopoulos, N. E. (1989). Strategies for Manufacturing, *261*(3), 144–153.

Kim, S. H., Yoon, S. G., Chae, S. H., & Park, S. (2010). Economic and environmental optimization of a multi-site utility network for an industrial complex. Journal of Environmental Management, 91(3), 690–705.

Madlener, R., & Schmid, C. (2003). Combined Heat and Power Generation in Liberalised Markets and a Carbon-Constrained World. *GAIA* - Ecological Perspectives for Science and Society, 12(2), 114–120.

Mitra, S., Sun, L., & Grossmann, I. E. (2013). Optimal scheduling of industrial combined heat and power plants under time-sensitive electricity prices. Energy, 54, 194–211.

Mousqué, F., Boix, M., Négny, S., Montastruc, L., Genty, L., & Domenech, S. (2018). Optimal on-grid hybrid power system for eco-industrial parks planning and influence of geographical position. Computer Aided Chemical Engineering (Vol. 43).

RTE - Réseau de transport d'électricité. (2018). Panorama de l 'électricité renouvelable en 2017.

Anton A. Kiss, Edwin Zondervan, Richard Lakerveld, Leyla Özkan (Eds.)
Proceedings of the 29th European Symposium on Computer Aided Process Engineering
June 16th to 19th, 2019, Eindhoven, The Netherlands. © 2019 Elsevier B.V. All rights reserved.
http://dx.doi.org/10.1016/B978-0-128-18634-3.50288-5

Evaluating the Benefits of LNG Procurement through Spot Market Purchase

Mohd Shahrukh,[a] Rajagopalan Srinivasan,[a] I.A.Karimi,[b]

[a]Department of Chemical Engineering, Indian Institute of Technology Madras, Chennai 600036, India

[b]Department of Chemical and Biomolecular Engineering , National University of Singapore, 119077, Singapore

raj@iitm.ac.in

Abstract

Natural gas (NG) is liquefied for shipping and storage purposes, this gas in liquid state is known as liquefied natural gas (LNG). The trade of LNG is usually regulated by contracts between suppliers and buyers. Historically long term contracts covering 20-25 years have been used; but more recently, due to emergence of new suppliers and consumers spot market purchases of LNG has also become possible. Nowadays, a consumer can opt to procure LNG via long term contracts or from spot market purchase or a combination of the two. We seek to evaluate the relative benefits of long term versus spot purchases. As a first step, in this paper, we report a mixed-integer linear programming (MILP) model to compare the cost of transportation through long term contracts as against spot market purchase. Several examples are solved; the results show that, in every case, spot market purchase is better compared to long term contracts.

Keywords: LNG, mixed integer linear programming, Long term contracts.

1. Introduction

Consumption of natural gas (NG) has grown steadily in recent years all over the world. Major factors that have contributed to this growth are environmental impact of conventional fossil fuels, abundant availability, and local market liberalization (Pospíšil et al., 2019). Despite NG being the most favored fossil fuel, its transportation to distant markets constitutes a challenge due to its gaseous state. Its liquefied form, liquefied natural gas (LNG), with its 600 time reduction in volume over NG, can be more easily transported using specially designed LNG carriers (Kumar et al., 2011).

Trade of LNG is usually regulated by contracts between suppliers and buyers. Historically, there were only a few buyers and suppliers of LNG. So contracts were generally 20-25 years long and thus provided security of demand and supply to sellers and buyers, respectively. In recent years, new suppliers have emerged, and LNG trade is increasing rapidly. Further, local NG market liberalization in many countries has also triggered many small companies to become LNG consumers. The emergence of these new players, especially small companies in the power sector and fertilizer manufacturers which use the regasified LNG as a feedstock has led to competitive LNG markets, but one characterized by high demand variability. Consequently, there is an increased need felt by regasification terminals to increasingly purchase LNG through short term (i.e., spot) contracts. In this paper, we seek to evaluate the relative benefits of long term versus spot purchases. Specifically, we develop a mathematical programming to quantify the transportation cost of LNG purchase for a given demand via spot market and long-term contracts.

1.1. Literature Review

In literature, procurement of LNG is considered as a contract (vendor) selection problem, mainly from a buyer's perspective. The problem is formulated from the point of a focal company that needs to procure a certain amount of LNG over a planning horizon. The company can buy LNG from suppliers distributed all over the world. A supplier may offer one or more contract options. Typically, a mathematical model can be used to choose a set of contracts (vendors), which allows the buyer to minimize the procurement cost. Till the 1970's, vendor selection was done using linear weighting model where each vendor was scored on multiple decision criteria and the vendor with maximum score selected. Later on, researchers started using linear programming (LP) for vendor selection (Khalilpour and Karimi, 2011).

Contract selection problems reported in literature are usually from single buyer's perspective. See for examples (Khalilpour and Karimi, 2011) and (Jang et al., 2017). No model has been proposed to make a comparative study between spot market purchase and long term contracts for a set of buyers. In this paper, we therefore develop separate models for the long term and spot purchase of LNG and compare the optimal transportation costs.

2. Problem Statement

Consider a set of buyers and producers of LNG distributed all over the world. Each consumer and producer has its own specific production and consumption profile. Let P (p=1, 2, ...P) be the total number of producer and C (c=1,2, ...C) be the total number of consumers such that I= P ∪ I. Each site has a storage tank. A heterogeneous fleet of V ships (v=1,2, ...V) is used to transport LNG and manage the inventory level at all sites. A site i has J_i identical jetties. Thus, at most J_i number of ships may load/unload at a site. For spot market purchase, ship moves among all sites to maintain inventory level at each site, while for long term contract, ships move only between sites that have contracts with each other.

Initial positions of ships, capacity, loading/unloading rate, material onboard are known. Site to site travel costs, number of sites, production, and consumption profile are also known along with duration of the planning horizon. Sites having long term contracts with each other are known. The aim is to administer the ship movements over the planning horizon so that inventory levels are maintained at all the sites for both spot market purchase as well as for long term contracts.

2.1. Example 1

This example consists of two ships; five ports in which two of them are production sites (Site 1 and Site 2) while three of them are consumption sites (Site 3, 4 and 5). Every port has one jetty at each site. Each port has storage tank with limited capacity. Planning horizon is of 10 days. At time zero both ships are empty. Initially, ship 1 is at Site 1 while ship 2 is at Site 2. Loading/Unloading rate of both the ships at every site is 12t/day. Complete data for this example is shown in Tables 1 and 2. Producer 1 has contract with Consumer 3 while Producer 2 have contract with Consumers 4 and 5.

Table 1: Supply and demand data for Example 1

Site	Type	Initial material (t)	Capacity (t)	Production / Consumption rate (t/day)
1	Producer	2	80	2
2	Producer	6	100	1.25
3	Consumer	3	25	0.5
4	Consumer	2	40	0.5
5	Consumer	5	60	1

Table 2: Travel time and travel cost for ships in Example 1

i→j	Travel time	Travel cost
i1→i2	1.62	38.90
i1→i3	1.75	42.07
i1→i4	2.05	49.21
i1→i5	1.99	47.72
i2→i3	0.16	3.95
i2→i4	0.82	19.76
i2→i5	0.73	17.50
i3→i4	0.95	22.89
i3→i5	0.86	20.64
i4→i5	0.33	7.80

3. Mathematical Model

The proposed model for spot market purchase is based on the model for bulk maritime logistics for the supply and delivery of multiple chemicals reported by (Li et al., 2010). They developed a model for a multinational company (MNC) which has sites distributed globally. The model considers multiple materials such that the product of one site is a raw material for another. The goal of the model was to develop routing and scheduling for a fleet of ships while ensuring the continuity of operation at all sites at minimum transportation costs. (Li et al., 2010)'s model has been simplified for the transportation of LNG by reducing the dimensions of all variables which explicitly considered the multiple materials to one (LNG). These variables denoted storage tank capacity, material unloading, material onboard etc. For long term contracts, this model is modified with additional constraints to model the stipulated contract terms.

3.1. Modeling long term contracts for procuring LNG

Long term contracts are between suppliers and buyers and specify the quality, quantity, payment terms, etc of the product to be delivered to the buyer over a specified period of time. For modelling long term contracts we consider a set of producers P (p=1,2,...P) with each producer having supply contracts with a set of customers L_{cp} ($L_{cp} \subseteq C$) . Since producer p has contracted only with customers L_{cp}, ships from p can only move to L_{cp}. Hence we write

$$\sum_{i=1}^{P} \sum_{j \notin L_{cp}} z_{vijk} = 0 \qquad 1 \le v \le V, 1 \le k \le K \qquad (1a)$$

$$\sum_{i \in Lcp} \sum_{j \notin L_{cp}, j \neq i} z_{vijk} = 0 \qquad 1 \le v \le V, 1 \le k \le K \qquad (1b)$$

where z_{vijk} is equal to 1 when ship v is at site i in slot k and at site j in slot $k+1$, otherwise it is zero.

3.2. Objective Function

The objective is to minimize total transportation cost over the planning horizon. Let Cv_{vij} be the cost of travelling from site i to site j via ship v. Then the objective function can be written as:

$$\sum_{v=1}^{V} \sum_{i=0}^{I+1} \sum_{j=0, j \neq i}^{I+1} \sum_{k=1}^{K} Cv_{vij} * z_{vijk} \qquad (2)$$

Eq.(1a) and Eq.(1b) are specific to the long term contract model; all other equations are common to both the models.

4. Case Studies

We have solved three examples to assess our model. These examples vary in planning horizon, distances between the sites and production and consumption rates. We used Cplex 12.8.0 on a Lenovo idea pad 510 workstation (core i5-7200U, 8GBmemory) for solving these examples running on windows 10. Results of these examples are summarized in Table 3.

4.1. Case Study 1

In this example, Site 2 has contracts with Sites 4 and 5 while Site 1 has a contract with Site 3. Results shows that Ship 1 loads 5.58 t of LNG during [0,0.465] from Site 2 then moves to Site 4 and Site 5 to deliver 4.0 and 1.58 t of LNG at time (in days) 1.28 and 3.53 respectively. Then, it moves to Site 2 again to load 5.48 t of LNG and delivers it to Site 5 from time [5.581, 6.038]. Ship 2 loads 0.78 t of LNG from Site 1 during [0, 0.065] and delivers it to Site 3. Then it moves back to Site 1 to load 2.21 t of LNG which is delivered to Site 3. Fig 1 shows the resulting inventory profile of the sites over the planning horizon.

When the same example is solved as a spot market purchase problem, there is a significant reduction in the transportation cost. Here Ship 1 loads 5 t of LNG from Site 2 and delivers it to Sites 4 and 5. Ship 2 loads 1.2 t of LNG from Site 1 during [0,0.1] and delivers it to Site 3 and then moves to Site 2 to load 5 t of LNG which is delivered at Site 5 in between [4.56,4.97]. It then returns to Site 2 to load 2.8 t which is delivered at Site 3 during [5.99, 6.22]. The total transportation cost is therefore reduced by 46.64%. Fig 2 shows the inventory profile of the sites over the planning horizon with spot purchases.

Table 3: Results for examples

Example	Planning Horizon	No. of Voyages Spot / Long Term	No of Iterations Spot / Long Term	Time (min: sec: ms) Spot / Long Term	Optimal Cost Spot / Long Term
1	10	7/9	5435k/211k	00:50:40/02:43:29	88.82/146.7
2	15	10/11	8264k/1632k	03:56:82/00:43:26	134.11/136.77
3	25	3/6	10k/13k	00:00:90/00:01:48	127.03/220.76

5. Discussion

We solved two other examples with different parameters (see Table 3). In these examples also, the transportation cost with spot market purchase is less than with the long-term contract. The difference in the transportation costs between the long term and spot purchases increases significantly when the producer and consumer sites are at large distances. These clearly indicate that spot purchases lead to a more optimal usage of the ships – i.e., the same demands can be more effectively met in the absence of long-term contracts. The current model relies on a number of simplifying assumptions. Here, the price of the LNG cargo has not been taken into account. In practice, purchases through long term contracts may have price advantage. The net effect of lower cargo price and a higher transportation cost therefore needs to be evaluated. Also, the current model assumes deterministic values for all parameters; in reality, these would vary based on demand. In future, we aim to develop a model that can incorporate effect of demand variability.

References

Jang, W., Hong, H.-U., Han, S.H., Baek, S.W., 2017. Optimal Supply Vendor Selection Model for LNG Plant Projects Using Fuzzy-TOPSIS Theory. J. Manag. Eng. 33, 04016035. https://doi.org/10.1061/(ASCE)ME.1943-5479.0000474

Khalilpour, R., Karimi, I.A., 2011. Selection of Liquefied Natural Gas (LNG) Contracts for Minimizing Procurement Cost. Ind. Eng. Chem. Res. 50, 10298–10312. https://doi.org/10.1021/ie200275m

Kumar, S., Kwon, H.-T., Choi, K.-H., Lim, W., Cho, J.H., Tak, K., Moon, I., 2011. LNG: An eco-friendly cryogenic fuel for sustainable development. Appl. Energy 88, 4264–4273. https://doi.org/10.1016/j.apenergy.2011.06.035

　　　　　　　　　　　　　　　　　　　　　　　　　　M. Shahrukh et al.

Li, J., Karimi, I.A., Srinivasan, R., 2010. Efficient bulk maritime logistics for the supply and delivery of multiple chemicals. Comput. Chem. Eng. 34, 2118–2128. https://doi.org/10.1016/j.compchemeng.2010.07.031

Pospíšil, J., Charvát, P., Arsenyeva, O., Klimeš, L., Špiláček, M., Klemeš, J.J., 2019. Energy demand of liquefaction and regasification of natural gas and the potential of LNG for operative thermal energy storage. Renew. Sustain. Energy Rev. 99, 1–15. https://doi.org/10.1016/j.rser.2018.09.027

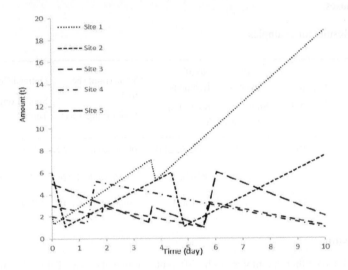

Fig 1: Inventory profile at sites while procuring LNG via Long term contracts

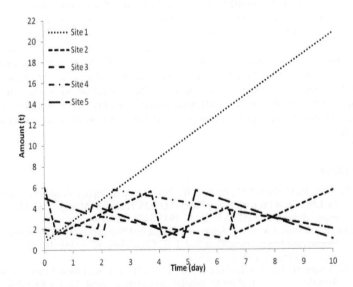

Fig 2: Inventory profile at sites while procuring via spot market purchase.

Anton A. Kiss, Edwin Zondervan, Richard Lakerveld, Leyla Özkan (Eds.)
Proceedings of the 29th European Symposium on Computer Aided Process Engineering
June 16th to 19th, 2019, Eindhoven, The Netherlands. © 2019 Elsevier B.V. All rights reserved.
http://dx.doi.org/10.1016/B978-0-128-18634-3.50289-7

Optimal design of biomass supply chains with integrated process design

Theodoros Damartzis*, François Maréchal

Industrial Process and Energy Systems Engineering (IPESE), Ecole Polytechnique Fedérale de Lausanne (EPFL) Valais, 1951, Sion, Switzerland

theodoros.damartzis@epfl.ch

Abstract

The current work focuses on the optimal design of biomass supply chains, addressing the conversion of raw materials into useful services. The latter include not only heat and power but also synthetic bio-based fuels (e.g. bio-SNG). Optimally pre-designed processing units representing the conversion facilities responsible for the transformation of biomass are employed. This permits the use of standardized units that act as building blocks within a MILP optimization problem, significantly simplifying the formulation and solution of the problem. These building blocks encompass a variety of transformation options and carry specific information regarding their capacity, total processing cost, overall efficiency and energetic performance. Then, the supply chain design optimization problem aims not only to identify the most suitable conversion plant type but also to determine their location by minimizing the overall cost, expressed as the capital and operating expenses of the processing units as well as the transport costs between the points of operation. The above formulation is applied for the conversion of animal manure to different end products in the district of Brig in southern Switzerland. Results reveal the predominance of large centralized bio-SNG producing units as a consequence of their higher conversion efficiency.

Keywords: supply chain design, bio-SNG, biomass conversion, animal manure.

1. Introduction

The increasing depletion of the fossil-based energy resources as well as the new mandates for the insertion of renewables in the global energy mix are continuously shaping the background for the energy transition. As renewable energy sources are presented to be promising alternatives to the traditional fossil resources, an intensification of the efforts to deploy sustainable biomass-based energy systems is ongoing not only from a strategic decision point of view, but also from an environmental perspective (d'Amore and Bezzo, 2016).

However, the introduction of bioenergy into the existing energy market necessitates not only the efficient design of the processing of raw materials but also the consideration of the spatial interactions of the conversion plants with the supply and demand points (Chomette et al., 2018). A number of recent studies address the issue of the biomass supply chain design. Costa Melo et al. (2018) present a data envelopment analysis approach (DEA) to identify the most suitable transport pathway of wood from Africa to Europe. How and Lam (2017) proposed a novel optimization approach that incorporating principal component analysis (PCA) and analytical hierarchy process

(AHP) to determine the optimal transportation design and processing hub location in an integrated biomass supply chain. The work of Ghaderi et al. (2016) provides a comprehensive review regarding the design of biomass supply chains. In this paper, a large number of published works are compared and a large diversity on modeling approaches is revealed. Focusing more on manure as a source material and the production of bio-SNG, Calderón et al. (2017) provide a framework for the design and optimization of bio-SNG supply chain using av explicit MILP multi-period scheme.

However, one should not forget that the role of process design within the definition of the supply chain is crucial. Thus, a bigger approach is needed that accounts for the integration of process design with the supply chain, highlighting the benefits and revealing potential synergies between the various steps of the conversion path from raw materials to services. Moreover, such a formulation will set the transition from the well-established process design level to the larger scale market level. To account for the role of the design of the processing units in the supply chain, expressed both as the type of conversion plant but also in relation to the output capacity, the current work presents a holistic approach based on the model presented by Chomette et al. (2018) for the design of wet biomass supply chains. To this end, the type and size of the conversion plants will play a decisive role in the formulation of the supply chain. The type of the plant refers to the various sub-processes that are used to convert the raw material to useful products as well as the formulation of the entire process flowsheet, while the size represents the output capacity in energy units. Using optimally designed conversion plants as predefined building blocks permits the synthesis of the entire supply chain through a MILP optimization problem that aims to minimize the total cost, expressed as the sum of the processing and transport expenses. The above formulation is then presented in an illustrative example for Brig, one of the highest manure producing districts of the canton of Valais in Switzerland.

2. Methodology

2.1. Manure conversion and design of the building blocks

The use of animal manure as the raw biomass material offers the possibility of using a plethora of conversion options of variable efficiencies depending on the desired end use as well as the conversion pathway that is followed. The common conversion process is anaerobic digestion, in which manure is converted to biogas through a series of complex (bio-)chemical reactions. The employment of pretreatment options of the raw biomass offers the chance to increase the biogas yield from the digester, inducing however an extra processing cost. Especially in cases where thermal pretreatment is used, extra degrees of freedom appear in the design process, related to the heat exchange between the pretreatment unit and the rest of the process. Biogas, being a mixture of CH_4 and CO_2 in a molar ratio of roughly 60:40 has to undergo through a cleaning and upgrading process to remove traces of impurities like H_2S, NH_3, siloxanes etc. and increase its methane content, respectively. The final stream of upgraded biogas consists mainly of CH_4 and can be used as bio-SNG for grid insertion or power production in engines or fuel cells. On the other hand, anaerobic digestion leaves an undigested stream that still contains a large portion of the initial carbon content. It is therefore of great interest to extract the latter in order to maximize the efficiency of the overall process. Hydrothermal gasification (HTG) offers the possibility of converting wet organic streams into methane using high pressure in a complex reactor scheme. The supercritical conditions that prevail within the HTG reactor ensure that the nitrous and phosphoric minerals are released unharmed in the residual output stream. Moreover, the use of catalytic methanation to convert the separated stream of CO_2 from the biogas is another option of further increasing the final bio-SNG yield. The methanation reactor can be fed with hydrogen produced through the electrolysis of water using renewables (e.g. solar power).

Using the MILP optimization framework described in Vigot et al. (2018), a number of possible configurations have been designed, each emphasizing in different process schemes and end products. Figure 1 shows the pareto front correlating the annualized process cost (in Swiss francs) with the overall energy efficiency of the process for a 100 kW unit.

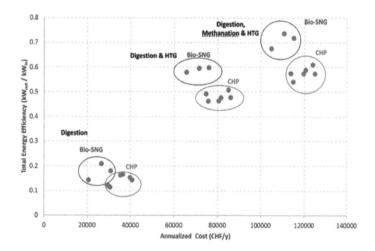

Figure 1. Pareto front of various process designs for manure conversion.

2.2. Employed flowsheets and typical sizes

Three different indicative flowsheets have been chosen to be used in the design of the supply chain, each focusing in a different end product. The employed flowsheets are depicted in Figure 2.

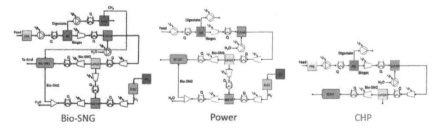

Figure 2. Employed flowsheets for the conversion of manure to different end products.

As already mentioned, the size of the conversion units plays an important part in the supply chain design. The specific costs of each flowsheet expressed as Swiss francs per kW produced is shown in Table 1. The typical sizes are selected with a reference on the raw material supply as well as the existing plants in the application area. The nonlinear dependence of the specific cost with the output capacity of each plant stems from the nonlinear dependence of the process efficiency with the plant dimensions.

Table 1. Specific costs (cost per kW) in Swiss Francs (CHF) of the employed flowsheets

	50 kW	100 kW	250 kW
Bio-SNG	2945.90	2292.54	1830.96
Power	2188.55	1821.40	1605.95
CHP	866.28	709.39	594.17

3. Application - Results

The three types of conversion plants presented in Section 2 together with the three typical sizes are used for the design of a supply chain for manure conversion in the commune of Brig in southern Switzerland. Six source sites have been identified and the corresponding manure potentials are taken from Thees et al. (2017). Using the design framework presented by Chomette et al. (2018), the results are illustrated in Figure 3.

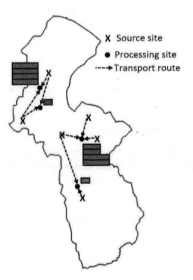

Figure 3. Supply chain results for manure conversion in the district of Brig, Switzerland

The horizontal length of the bars representing the processing units corresponds to the energetic output and specifically, 50 kW for small bars, 100 kW for medium bars and 250 kW for large bars. Similarly, the colors of the units correspond to the type of the plant, namely red for bio-SNG producing plants, blue for power production and green for CHP (also see Figure 2). It is seen that large centralized bio-SNG units are favored over the rest of the options due to the much larger energy efficiency of the process. In other words, these plants are able to convert more effectively a larger capacity of raw materials into bio-SNG. The centralized fashion in the placement is a direct consequence of the larger capital expenses that correspond to this type of conversion (Table 2). On the other hand, one can see that one biomass source is not limited to feeding a single conversion unit as shown in the northern part of the commune. According to the results, heat and power producing plants act as a supplement to the residual manure that is more expensive to be converted to bio-SNG. It has to be noted that this of course depends also on the initial choice of the building blocks. In other words, if an exhaustive investigation was made using all the available conversion options, it would be apparent that the highly efficient designs would be preferentially used over the other options. This of course, will increase the accuracy of the results but also drastically raise the computational effort and is the next step of this research.

4. Conclusions

A study for the design of manure supply chains was presented here. Based on a previous work, the approach involves the use of pre-designed processing plants representing the main steps of the biomass-to-services conversion chain. Such plants come equipped with information on their assigned capacity and type, the first being expressed as the energetic output of the conversion process while the second as the pathway from raw materials to products and are then used within a MILP design optimization framework. The latter is able to perform the spatial allocation by selecting the most suitable conversion type and size of each plant with regard to the supply of raw biomass by minimizing the total cost of the supply chain represented by the sum of the processing and transport costs. Three pre-defined process flowsheets, each accounting for the conversion of manure to different end products and with a different energetic efficiency, as well as three typical plant sizes selected by considering the domain of application are used for the design of the manure supply chain for the district of Brig in Switzerland. According to the model outputs, the conversion to bio-SNG in large plants of centralized fashion seems to be the most economically viable option.

Acknowledgements

This research project is financially supported by the Swiss Innovation Agency Innosuisse and is part of the Swiss Competence Center for Energy Research SCCER BIOSWEET.

References

A. J. Calderón, P.Agnolucci, L. G. Papageorgiou, 2017, An optimisation framework for the strategic design of synthetic natural gas (BioSNG) supply chains, Applied Energy, 187, 929-955.

G.A. Chomette, Th Damartzis, F. Maréchal, 2018, Optimal design of biogas supply chains, Computer Aided Chemical Engineering, 43(A), 669-674, 2018.

I. Costa Melo, A. Rentizelas, A. J. Paulo Nocera, J.S. Campoli, D.A. do Nascimento Rebelatto, 2018, An assessment of biomass supply chain : a DEA application. Athens Journal of Sciences, 5 (2). pp. 125-140.

F. d'Amore, F. Bezzo, 2016, Strategic optimisation of biomass-based energy supply chains for sustainable mobility, Computers and Chemical Engineering 87, 68–81.

H. Ghaderi, M. S. Pishvaee, A. Moini, 2016, Biomass supply chain network design: An optimization-oriented review and analysis, Industrial Crops and Products, 94 (30), 972-1000.

B.S. How, H. L. Lam, 2017, Novel Evaluation Approach for Biomass Supply Chain: An Extended Application of PCA, Chemical Engineering Transactions, 61, 1591-1596.

O. Thees, V. Burg, M. Erni, G., Bowman, R. Lemm, 2017, Biomass potenziale der Schweiz für die energetisce Nutzung, WSL Berichte, Heft 57.

M.A. Vigot, Th. Damartzis, F. Maréchal, 2018, Thermoeconomic design of biomass biochemical conversion technologies for advanced fuel, heat and power production, Computer Aided Chemical Engineering, 44(A), 1801-1806, 2018.

Anton A. Kiss, Edwin Zondervan, Richard Lakerveld, Leyla Özkan (Eds.)
Proceedings of the 29th European Symposium on Computer Aided Process Engineering
June 16th to 19th, 2019, Eindhoven, The Netherlands. © 2019 Elsevier B.V. All rights reserved.
http://dx.doi.org/10.1016/B978-0-128-18634-3.50290-3

Optimisation of the integrated water – energy systems in small group of islands communities

Christiana M. Papapostolou[a], Georgios T. Tzanes[b], Emilia M. Kondili[a*]

[a]*Optimisation of Production Systems Laboratory,*

[b]*Soft Energy Applications and Environmental Protection Laboratory,*

Dept. of Mechanical Engineering, University of West Attica, Campus 2, P. Ralli & Thivon Street, 12244, Egaleo, Greece

*ekondili@uniwa.gr

Abstract

Islands are very special geographical areas with distinct and unique characteristics, often presenting slow infrastructure advancement, rendering the 'Water-Energy Nexus' (WEN) subject of increasing attention at all levels of policy making. This works aims to contribute in the optimisation of infrastructure development for the integrated solutions of water-energy systems by devising suitable optimisation models that capture technical and economic aspects. The developed framework is applied to a set of two Aegean Sea islands, assessing alternative scenarios for energy and water planning, with vast RES penetration, either if each island operates in stand-alone mode or the two of them interact as an energy community, via an undersea cable. The results of the work evidence that the solution of interconnection between remote islands could be economic viable, thus leading to downsized plants' installations and decreased capital investment costs, as greater exploitation of RES is achieved.

Keywords: Water-Energy Nexus, Islands, Frameworks, Desalination, Interconnection.

1. Introduction

Water and energy are two fundamental resources with the supply of one highly depended on the availability of the other. The recognition that their scarcity underpins socio-economic development, has resulted a growing interest of policy makers in the so-called 'Water-Energy Nexus', a policy priority which is also reflected in the rising interest of the research community: the water-energy (WE) supply problem affects all types and sizes of contemporary communities around world due to the spatial characteristics of the water availability, also challenged by the availability and the security of energy supply.

With our special focus on small-scale and isolated economies and societies, like the ones of the Aegean Archipelagos that are more likely to be affected by WE problems, the present work seeks to introduce an integrated evaluation framework for the optimal design and operation of water and energy supply chains (WESCs) with special focus on the community-based perspective. To that end, two scenarios are currently elaborated: a) optimisation of the WESCs of two typical Aegean islands operating independently (stand-alone mode), resulting for each one the most economic and environmental viable energy and water supply solution for the upcoming 25 years and b) the two islands operating as a community, being interconnected via undersea cable, revealing possible WE synergies (energy interactions) for the community to exploit. The proposed WESC optimisation framework, apart from achieving the optimal design of the energy and water supply systems (considering RES-based hybrid energy configurations) of the two islands, also investigates the performance of their interconnection as a feasible and profitable / valuable solution.

2. Background of the work: The island communities' dimension in the WESC optimisation problem

Islands render a special field of study in the WESCs optimisation, focusing on the local specific characteristics of each community namely size, topography, population, energy and water demands. That is why, well-known island communities have been extensively studied by the researchers seeking to identify the optimum WESC, for now and for the future. Focusing on the Aegean Sea, aiming at establishing a sustainable electrification model, the "Green Island" model for (island) communities, Kaldellis et al. (2017) carried out a theoretical analysis estimating the contribution of wind energy curtailments (of both new and existing parks) on the island's energy supply. The results of the work, under the elaborated case study for the island of Kos strongly support the necessity of energy storage introduction, if one may compare the expected and the finally absorbed wind power of the installation: 9.74GWh (final absorbed wind power, according to the official data by the local operator), while the expected wind energy yield was 14.13GWh (on the basis of the available wind potential), thus almost curtailing 4GWh of clean electricity on an annual basis.

Acknowledging the necessity of sustainable energy supply at small-scale islands on the basis of RES, Bertsiou et al. (2018), investigated the implementation possibilities of a Hybrid Renewable Energy System (HRES) in Fournoi island, operating on the basis of hydropower for electricity generation aiming at meeting both drinking and agricultural water demands through desalination of sea water, with wind contribution used for supplying the local power network and for pumping and/or desalination operation depending on the scenario examined. The results of the work support the possibility of achievable high level of water and energy coverage and days of electricity autonomy, according to the water coverage consumption pattern.

Towards the same direction, Papapostolou et al. (2018), emphasising also on small-sized islands, thus energy and water vulnerable islands, studied the case of Irakleia, an island of 151 inhabitants during winter season. With the current energy supply both for the existing desalination plant and the electricity demands of the islanders being met by conventional fuels, which also demanding water imports via shipments to fulfil irrigation needs, the authors propose a RES-based desalination configuration (wind turbine, PV, energy storage battery) limiting the diesel operation to meet just the stability of the network. The results of the work also support high RES energy coverage minimising the CO_2 (diesel-based emissions by ~92%) balancing the calculated capex required for the specific solution (1.87MEUROS).

Tzanes et al. (2017), also dealing with the emerging issue of electrification of remote islandic Greek communities, developed an advanced software tool named Energy System Analysis (ESA), for evaluating different energy and water local scenarios, accounting also the available geographical information and the corresponding RES potential characteristics of the island examined. The study identified optimal solutions comprising HRES and desalination installations for the islands of Anafi and Agios Efstratios, with minimised LCA costs achieved for 85% annual coverage of electricity demand.

Consequently, one may conclude that using hybrid power systems results to better balances in energy supply while facilitates the matching of demand and production, with energy storage contributing also towards this direction. However, of great significance is also proven the consideration of larger spatial distribution of RES generation along with the aggregation of diversified load and water demands (community-based perspective), which, in the majority of the cases, leads to improved cost performances.

3. Case study characteristics

The proposed framework is implemented in two (medium-sized) remote islands of South Aegean Sea, namely Amorgos and Astypalaia. Both islands are characterised by a) high wind and solar potential, capable of achieving high RES capacity factors (Fig. 1a, 1b) and b) by significant water deficits, some of them met by ship transfers (Table 1). The load demand profile in these islands is characterised by high alteration, as tourism is the main economic activity. The most recent official available data (2013 for Amorgos and 2014 for Astypalaia, with no variations till today, as stated by the local authorities) are presented in Fig. 1c. In the case of water (in the absence of official recorded data) a typical (hourly-based) water demand profile is assumed, reflecting the islands' potable water and irrigation needs (~57,000 m³ for Amorgos and ~37,500 m³ for Astypalaia). The water and electricity needs have a similar pattern in time. However, in the case of water demand, the profiles present a quite low correlation. Therefore, the results of the work include an optimal WE solution for each island separately and in the second case as a community for a time-span of 25 years.

Table 1. Annual water deficits for Amorgos and Astypalaia (YPEKA, 2015)

Island	Irrigation (m³)	Potable (m³)	Ship transfers (m³)	Total (m³)	Peak demand (m³/h)
Amorgos	~7,500	14,000	~35,500	57,000	27,2
Astypalaia	12,500	25,000	-	37,500	24,8

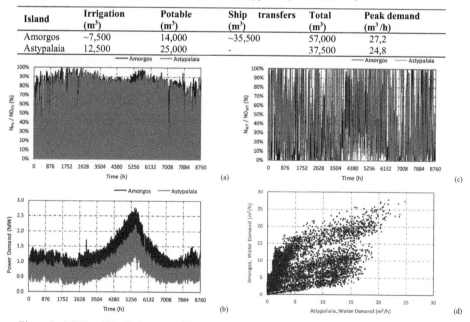

Figure 1. a) PV and b) wind power utilization factors, c) the load demand profiles and d) the islands hourly water demand correlation

4. Optimisation model

For the evaluation of the optimal power and water supply systems size in stand-alone mode, the optimisation model developed by Papapostolou et al. (2018) has been deployed. The key features of the model include two interlinked components, Eq.(1) and Eq.(2), the energy balance of the system (island or islands) and the island's water tank (WT) balance, for a time step of 1h, respectively:

$$\sum_{r=1}^{r=\max} Eg_r(t) + ESSg(t) - ESSs(t) - \sum_{l=1}^{l=\max} Ed_l(t) = 0 \tag{1}$$

$$WTank(t) = WTank(t\text{-}1) + \sum_{wr=1}^{wr=\max} F_{wr}(t) - Wd(t) \tag{2}$$

Where, in Eq.(1), 'Eg_r', represents the energy production of the resource, 'r' (wind, diesel and solar PV power installations), at time step, 't', '$ESSg$' and '$ESSs$', stands for energy storage either generated or stored accordingly, and finally the electricity needs are represented by 'Ed_l', with the index 'l' being equal to 1 for the present load demand and 2 for additional demand due to the desalination plant operation. Water balance described in Eq.(2), considers a typical water storage tank operation, whereas the level '$Wtank(t)$' at each time step 't' is a function of the water initially stored '$Wtank(t\text{-}1)$', as well as of the sum of the inflows '$Fwr(t)$' of water resources 'wr' minus the water demand '$Wd(t)$'.

In the case of integrated operation (two islands operating as a community in terms of exchanging energy), which becomes possible through an undersea cable installation, the main model modifications are applied in the system's electrical balance Eq.(3), the energy storage Eq.(4) and in the objective function Eq.(5):

$$\sum_{r=1}^{r=\max} Eg_r(t) + ESSg(t) - ESSs(t) - \sum_{l=1}^{l=\max} Ed_l(t) - \frac{Et_{load}^{isl}(t)}{\eta t} + Et_{load}^{isl+1}(t) * \eta t = 0 \tag{3}$$

$$ESSs(t) \leq dEres^+(t) - \frac{Et_{storage}^{isl}(t)}{\eta t} + Et_{storage}^{isl+1}(t) * \eta t, \tag{4}$$

$$f_{\to \min} = \frac{\sum_{r=1}^{r=\max} \left[N_r * IC_r * (1 + n * FC_r) \right]}{(1+i)^n} + \frac{n * \sum_{t=1}^{t=\max} Eg_{r=diesel} * OC_{diesel} + F_{ship}(t) * OC_{ship}}{(1+i)^n} \tag{5}$$

$$+ IC_{cable}(D, N_{cable})$$

More precisely, the 'Et' stands for energy transfers. The index '$load$' refers to the case of energy generated directly addressing (sent) to meet electricity needs, and the index '$storage$' indicates the energy transfers for charging the energy storage systems. The upper index indicates whether the energy is provided, at island 'isl', or drawn, from every next island '$isl+1$'. The term '$dEres^+$' represents the renewable energy surplus, which occurs when the renewable energy production potential is larger than actual utilisation of the installations. The losses of trading energy are represented by the factor 'ηt'. Concerning the objective function to be minimised Eq.(5), it represents the installation 'IC_r', maintenance 'FC_r' and operation costs 'OC_r' for 25-year time-period, of resource '*r*'. The last term 'cable' stands for the installation cost of the undersea cable, which is a function of the distance between the two islands and its power capacity (Fig. 2, Kokotsakis 2018). Both optimisation problems, were modelled as Linear Programming Problems (including 30 blocks of equations and 19 of respective variables) in GAMS (www.gams.coms) and solved with NEOS Server IBM ILOG CPLEX Optimizer (https://neos-server.org) with typical computational times starting from 5s. NEOS Server is hosted by the Wisconsin Institutes for Discovery at the University of Wisconsin in Madison with typical high-level hardware specs.

5. Results

5.1. Stand-alone, island operation

Model implementation resulted viable solutions (in terms of plant-sizes, see Table 2) for both examined scenarios. In stand-alone mode, the islands have a similar behavior, mostly utilising RES for the operation of the corresponding desalination plant (Fig. 2a), which is sized to 0.33 and 0.43 times peak hourly water demand, for Astypalaia and Amorgos island respectively (Fig. 2c). At the same time, the water tanks are mostly utilized until the electricity demand peaks (Fig. 2d).

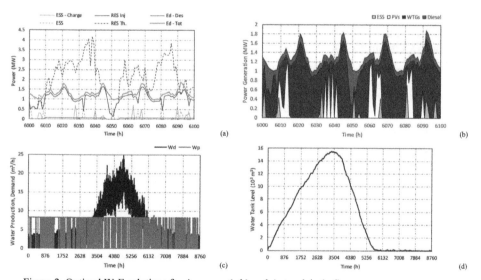

Figure 2. Optimal W-E solutions for Amorgos (a,b) and Astypalaia (c,d)

5.2. Community-based islands' operation

In the scenario of islands operating as a community, via an undersea cable of 1 MW capacity (40km in length), the flexibility in designing (allocation of resources) as well as exploiting individuals' RES surplus for covering the needs of the opposite island accordingly, results mainly to significant downsizing of thermal power and energy storage capacity as well as large increase in installed wind power (Table 2). In terms of cost, accordingly, the total expenditures get reduced from 50.2M€ (stand-alone scenario) to 41.36M€ (community-based scenario). The approx. 8.5M€ difference justifies a further research and application of the framework to islands with similar distance and features (different size, complementarity in energy demand), hence a large number of Aegean islands, heavily relying on diesel.

Table 2. Optimal Capacities- Modeling Results

Stand-alone, island-based operation optimal results				
Island	Diesel Plant (MW)	Wind Park (MW)	PV Park (MW)	ESS Cap. (MWh)	Desalination Plant (m³/d)
Amorgos	1.91	3.01	1.39	0.84	287
Astypalaia	2.04	1.74	0.73	0.23	199
Community-based (set of islands) operation optimal results				
Amorgos	1.83	2.88	1.26	0.46	235
Astypalaia	1.72	2.42	0.05	0.17	181

Conclusions

In this work an integrated WESCs optimisation framework has been introduced and tested comparing the single-island to the community-based (island) operation. The results evidenced that RES-based hybrid energy supply solutions including the desalination option are economically attractive for small-sized Aegean Sea islands while also serving at increasing the security of supply, especially under the community-based perspective and for reasonable cable sizes (length and capacity). The present framework will be further tested to other sets of islands (more than 2), with characteristics and complementarities regarding water and energy demand and RES potential (e.g. Santorini-Anafi (distanced ~24km)), for enabling strategic planning with regards to Aegean Sea islands water and energy future infrastructures.

Acknowledgements

"The Post-doctoral Research for Dr C. Papapostolou was undertaken with a scholarship fund by IKY, Act "Supporting Post-Academic Researchers" with resources from the Operational Program "Human Education Development and Lifelong Learning", priority axes 6,8,9, co-funded by the European Social Fund - ESF and the Greek State ".

References

M. Bertsiou, E. Feloni, D. Karpouzos, E. Baltas, 2018, Water management and electricity output of a Hybrid Renewable Energy System (HRES) in Fournoi Island in Aegean Sea, Renewable Energy, 118, 790–798

J.K. Kaldellis, G.T. Tzanes, C. Papapostolou, K. Kavadias, D. Zafirakis, 2017, Analyzing the limitations of vast wind energy contribution in remote island networks of the Aegean Sea Archipelagos, Energy Procedia, 142, Volume 142, 787-792

A. Kokotsakis, D. Zafirakis, 2018, Energy storage in offshore wind farm, in order to avoid over-dimensioning of energy transmission lines from wind farms to the core network, Master dissertation, Heriot-Watt University & Piraeus University of Applied Sciences

C.M. Papapostolou, E.M. Kondili, G. Tzanes, 2018, Optimisation of water supply systems in the water – energy nexus: Model development and implementation to support decision making in investment planning, Computer Aided Chemical Engineering, 43, 1213–1218

G.T. Tzanes, D. Zafirakis, C. Papapostolou, K. Kavadias, J.K. Kaldellis, 2017, PHAROS: An Integrated Planning Tool for Meeting the Energy and Water Needs of Remote Islands using RES-based Hybrid Solutions, Energy Procedia, 142, 2586–2591

[YPEKA] Ministry of Environmnet and Energy, 2015, River Basin Management Plan of the Aegean Islands (EL 14), Text only in Greek, URL: wfdver.ypeka.gr/el/management-plans-gr/, Accessed on September, 2018

Anton A. Kiss, Edwin Zondervan, Richard Lakerveld, Leyla Özkan (Eds.)
Proceedings of the 29th European Symposium on Computer Aided Process Engineering
June 16th to 19th, 2019, Eindhoven, The Netherlands. © 2019 Elsevier B.V. All rights reserved.
http://dx.doi.org/10.1016/B978-0-128-18634-3.50291-5

Novel Methodology for Cogeneration Targeting with Optimum Steam Level Placement

Julia Jimenez, [a,b,*] Adisa Azapagic, [b] Robin Smith[a]

[a] Centre for Process Integration, School of Chemical Engineering and Analytical Science, The University of Manchester, Manchester M13 9PL, UK.

[b] Sustainable Industrial Systems, School of Chemical Engineering and Analytical Science, The University of Manchester, Manchester M13 9PL, UK.

julia.jimenezromero@manchester.ac.uk

Abstract

This study aims to develop a novel method to synthesise site-wide heat recovery, distribution, cogeneration systems with optimum operating conditions of the steam mains. Previous approaches have simplified the problem to the extent that many important practical issues have been neglected and restricted the scope of the options included. The proposed methodology uses a combination of total site analysis and mathematical programming for a holistic approach to the steam system, which accounts for interactions between site utility system and processes. The optimisation problem involves the selection of more realistic operating conditions of the steam mains (superheating and pressure). The model will also account for water preheating, and superheating and desuperheating for process steam generation and use. Deaerators and let-down stations are also included in the analysis. The application of this methodology to a case study yielded a 7.6 % reduction in total energy requirement, compared to conventional utility system design method. The proposed approach addresses severe shortcomings in previous research on this topic and provides a foundation for future work to explore the next generation of sustainable utility systems.

Keywords: Cogeneration targeting, heat and power integration, total site integration, steam level optimization, mathematical programming.

1. Introduction

Sustainable development, as one of the most significant challenges faced by society, is closely related to the rational generation and use of energy. For this reason, in the process industries it is important to place a focus on the synthesis of energy-efficient sustainable utility systems and how this can shape energy use patterns. The energy transition of existing systems to meet future demands needs to be directed to be on a sustainable basis (Broberg Viklund, 2015). In the future, process utility systems will need to incorporate a much greater contribution from renewable energy sources. This will create a paradigm shift in the way such utility systems are designed and operated.

One of the key performance indicators for the synthesis of utility systems is cogeneration potential, which establishes objectives on heat and power generation as well as steam distribution and boiler fuel requirement (Ghannadzadeh et al., 2012). Steam mains pressures and superheating play an essential role in the performance of both heat and power generation at the site. Research has analyzed the influence of the steam levels on the energy targets in total site heat integration (Mavromatis and Kokossis, 1998; Shang and Kokossis, 2004; Beangstrom and Majozi, 2016). However, previous methods do not consider the degree of superheat and its effect on the potential shaft power generated by steam expansion.

In addition to these models, Kundra (2005) and Ghannadzadeh et al. (2012) presented targeting approaches, in which both sensible and latent heat of steam has been considered. However, previous methodologies have been based on the assumption of fixed steam mains pressure, usually based on heuristics. This neglects the close interrelation between the processes and the system and its subsequent implications. Moreover, the superheat for both the steam generation and use has been assumed to be the same. This premise might lead to thermodynamically infeasible solutions or further difficulties in practice, because of limitations associated with the materials used for construction, or due to the design complexity.

Sun et al. (2015) proposed a graphical approach to overcome the shortcomings in relation to assessing cogeneration potential and enhancing site-wide heat recovery methodically. Whilst the proposed methodology by Sun et al. (2015) allows for useful thermodynamic and physical insights for understanding some of the interactions within the system It does not provide a systematic decision-making approach to determine the optimum utility system performance since it does not allow the analysis of the trade-off between the cost of the additional steam and the profit from power generation.

Though there is extensive literature for cogeneration targeting in utility systems, the conventional concepts present a number of limitations and drawbacks for the selection of optimum site operating conditions. This study aims to offer the basis for a systematic approach to explore the next generation of sustainable utility systems. For the first time, a study in this area combines an extended transshipment method based on Shang and Kokossis (2004) with total site analysis and more realistic site conditions. In turn, this determines the total site heat and mass flow and ensures that the total operating cost is minimized.

The novelty of this work is derived from the requirement of increased practical and realistic conditions for both steam generation and usage. This is used in combination with an evaluation of the interactions between steam mains conditions (pressure and superheat) and system performance. The effect of process steam generation at a different temperature from the steam mains, as well as the efficiency and the exhaust temperature of the steam turbines based on steam conditions (superheating) and load, is also explored to provide a more realistic and accurate heat recovery. Ultimately, this methodology allows for power targeting of utility systems operating at optimum conditions. In essence, the study provides a framework for the analysis of sustainable steam systems.

2. Methodology

Cogeneration targeting with simultaneous steam mains selection requires making continues and discrete decisions, where non-linear energy terms are involved. Thus, to produce a linear model and avoid convergence and robustness issues, some properties are fixed during the optimization. Every optimization step is succeeded by a rigorous simulation, as shown in Figure 1. The following subsections provide a summary of these steps.

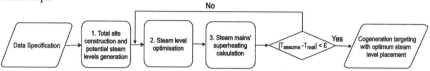

Figure 1. Schematic representation of optimization approach

Step 1 Total site construction and potential steam levels generation
The extended Total Site Profile (TSP) concept of Varbanov et al. (2012) is adopted to obtain the potential steam levels required for the heat recovery and power generation via the utility system. Apart from the stream data specification - i.e. number of processes involved, number of stream of each process, parameters including supply and target temperatures, as well as the heat capacity- the specific minimum approach temperature (ΔT_{min}^{PU}) is used to avoid misleading energy targets that can result from the use of inaccurate global minimum temperature difference (ΔT_{min}) values between process streams.

Saturated temperature (pressure) denotes the potential steam levels. The temperatures are obtained by partitioning the site-wide temperature range into intervals. Since the sink profile defines the quantity and quality of heat required, steam intervals are based on its temperature range. Operational constraints are also taken into account, such as minimum/maximum temperature for process and utility steam generation. Additionally, minimum temperature and pressure difference between each potential level are set by the designer, to guarantee a representative number of options and avoid unnecessary levels. Once generated the potential steam levels, they are classified based on the number of steam main required and the pressure ranges for each header.

Step 2 Steam level optimization
The MILP formulation is based on Total Site Heat and Mass Cascades (TSHMC) employing a transshipment model. An extension of Shang and Kokossis (2004) model enables recording energy and mass balances among process source/sink streams and potential steam levels. The TSHMC are formulated by the temperature intervals (defined by the steam levels) and comprise three cascades: source, steam and sink, as illustrated in Figure 2(a) Process streams act as steam sources or sinks, where the (residual) heat that cannot be used in the interval for either steam generation or use is going to the next lower temperature level, at the respective cascade. Heat flows from the process sources to sinks through steam. Utility steam is raised at boiler house at VHP conditions and distributed to the different headers by either passing through steam turbines or let down stations.

In the source cascade, the heat from process sources is used to raise steam at the steam level pressure, from Boiler Feed Water (BFW) conditions to superheating. The latter is a designer variable and is restricted by the source profile and the heat exchanger equipment. Regarding sink cascade, heat flows from steam level to process sinks via steam desuperheated. Steam is desuperheated prior its use by BFW injection. For the development of the TSHMC, let m^H, m^{ST}, m^{LD}, m^{Cmain}, m^{BFW} be the steam mass flowrates of the process steam generation, steam turbines, let-down stations, process sinks and BFW injected, respectively.

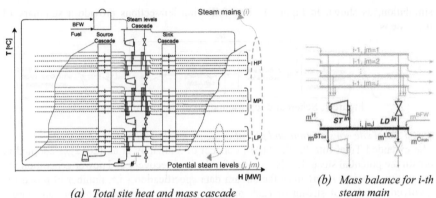

(a) *Total site heat and mass cascade*   (b) *Mass balance for i-th*
                                                  steam main

Figure 2. Schematic representation of optimization approach

The MILP formulation is based on the minimum annualised utility cost as the objective function, where the steam flows, heat loads, fuel consumption, electricity generation and cooling water requirement are continuous variables. Binary variables are associated with whether a steam main (defined as i) exists at a given condition, denoted by the set of potential steam levels j or jm. Binary variables also related to the steam turbines operation between steam mains (i,j) and $(i+1,jm)$ and the operating status of boilers at VHP conditions.

Energy and mass balance, as well as the electricity balance are the main equality constraints. While equipment size constraints -- that avoid equipment operating at lower efficiency -- are inequality constraints. Power generation and fuel consumption are estimated employing the linear models of Sun and Smith (2015) and Shang (2000), respectively. Both models accounts for full and part-load operating performance.

Step 3 Calculation of steam mains' superheating
Once the optimum steam level placement (saturated temperature/pressure) has been obtained, the actual superheating temperature of the header is calculated. The superheating is defined through a material and energy balance in each steam main. This is determined by the enthalpy and flow rate of the process steam generation, turbine exhausts, steam passing through let-down valves, and any BFW injected in the steam main. The calculations require top-down iterations that start with the utility steam and work down through the cascade from high to low pressure until superheating constraint is satisfied by all the steam mains. Turbine exhaust properties are obtained using the Willan's linear model presented in the Sun and Smith (2015) research, based on the inlet steam conditions and the steam main pressures. The steam is expanded through letdown valves at isenthalpic conditions. Finally, Step 2 and 3 are repeated until achieving convergence (usually 3-4 iterations).

3. Case Study

Site data was adapted from an example available in the literature (Sun et al., 2015). The number of streams and ΔT_{min}^{PU} for each process are detailed in Table 1. Power site demand is 40 MWh. The site energy requirement is satisfied using a steam system comprising a natural gas boiler, three distribution steam mains, a deaerator, expansion valves, steam turbines and a single cold utility (cooling water). The utility steam is generated in the boiler house at the very high pressure (VHP) main conditions (100 bar). The inlet temperature of the cooling water is 20 °C. Electricity is generated by single back-pressure steam turbines allocated between each level. The HP, MP and LP steam mains operating conditions are estimated to minimize the total operating cost.

Table 1. Summary of the stream data for case study

	Process A	Process B	Process C	Process D	Process E
No. of streams	4	8	9	6	9
ΔT_{min} (°C)	15	5	5	10	15

In order to assess the benefit of the methodology, the optimized system configuration is compared against a conventional design based on heuristics, where the HP, MP and LP mains pressure are 40, 20 and 5 bar, respectively. Figure 3 compares the two steam system configurations. Both systems were obtained based on the temperature specifications presented in Table 2. For the optimization, the present case study has considered 65 potential steam levels, based on the specifications and the sink profile temperature. The potential steam levels has been classified in 15 levels for HP ($\geq$ 30 bar), 42 levels for MP (6 – 30 bar) and 8 levels for LP (3 - 6 bar).

Table 2. Temperature specifications for the steam system

Constraints	Temperature [°C]
Maximum boiler steam temperature	570
Minimum process steam generation temperature (saturation)	134
Minimum steam main superheating	20
Degree of superheating for process steam generation	20
Degree of desuperheating for process heating	3

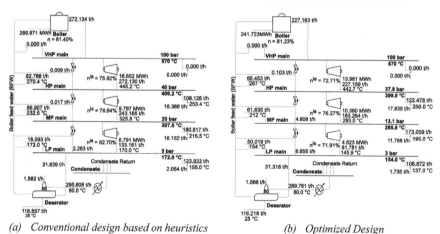

(a) *Conventional design based on heuristics*   (b) *Optimized Design*

Figure 3. Steam system configuration

The steam main pressures for the optimized system configuration are 37.8, 13.1 and 3 bar; respectively.The manipulation of steam main conditions affects site operational performance. Steam mains selection is affected by several factors i.e. process steam generation, turbine exhausts and let-down flows. Steam passing through let-downs is required to achieve steam balance and more important to maintain the minimum superheat of 20 °C in each steam main. However, this may result in a reduction in the power generation as it is observed in Table 3. Compared with conventional design cost, the proposed design diminishes the fuel and CW consumption by 16.35 % and 10.61 %, respectively. Even though the power generation is less than the traditional design, the total utility cost decreases 7.61 %.

Table 3. Comparison of steam system designs

Parameter	Conventional	Optimization	Difference in units
Fuel consumption [MWh]	288.97	241.72	- 47.25
Power Generation [MWh]	35.20	28.96	+ 6.24
Cooling Utility [MWh]	237.54	212.33	- 25.21
Fuel Cost [M£ y^{-1}]	51.19	42.82	- 8.37
Power Cost [M£ y^{-1}]	3.62	8.38	+ 4.76
Cooling Cost [M£ y^{-1}]	10.21	9.13	- 1.08
Operating Cost [M£ y^{-1}]	**65.02**	**60.33**	**- 4.69**

4. Conclusions

A new methodology has been developed to provide increased realism and accuracy in utility systems synthesis, operating at optimum conditions for future utility systems. The study shows the close relation between steam level selection and heat recovery and power generation enhancement. In an illustrative example, the new model presents a significant reduction of the total energy requirement at the site compared to a conventional design method (7.61 %). This proves that the energy requirement can be further reduced by holistically optimizing the steam mains operating conditions and the site heat recovery and cogeneration.

5. Acknowledgements

This publication has been funded by HighEFF-Centre for an Energy Efficient and Competitive Industry for the Future. The authors gratefully acknowledge the financial support from the Research Council of Norway and user partners of HighEFF. Also, a special gratitude to the National Secretariat for Higher Education Science, Technology and Innovation of Ecuador (SENESCYT) for its support.

References

Beangstrom, S. G. & Majozi, T. (2016). Steam system network synthesis with hot liquid reuse: II. Incorporating shaft work and optimum steam levels. *Computers & Chemical Engineering,* 85, 202-209.

Broberg Viklund, S. (2015). Energy efficiency through industrial excess heat recovery—policy impacts. *Energy Efficiency,* 8(1), 19-35.

Ghannadzadeh, A., Perry, S. & Smith, R. (2012). Cogeneration targeting for site utility systems. *Applied Thermal Engineering,* 43, 60-66.

Kundra, V. (2005). *To develop a systematic methodology for the implementation of R-curve analysis and its use in site utility design and retrofit.* Master of Science, The University of Manchester.

Mavromatis, S. P. & Kokossis, A. C. (1998). Conceptual optimisation of utility networks for operational variations— I. targets and level optimisation. *Chemical Engineering Science,* 53(8), 1585-1608.

Shang, Z. (2000). *Analysis and Optimisation of Total Site Utility Systems.* Ph.D. Thesis, The University of Manchester.

Shang, Z. & Kokossis, A. (2004). A transhipment model for the optimisation of steam levels of total site utility system for multiperiod operation. *Computers & Chemical Engineering,* 28(9), 1673-1688.

Sun, L., Doyle, S. & Smith, R. (2015). Heat recovery and power targeting in utility systems. *Energy,* 84, 196-206.

Sun, L. & Smith, R. (2015). Performance Modeling of New and Existing Steam Turbines. *Industrial & Engineering Chemistry Research,* 54(6), 1908-1915.

Varbanov, P. S., Fodor, Z. & Klemeš, J. J. (2012). Total Site targeting with process specific minimum temperature difference (ΔTmin). *Energy,* 44(1), 20-28.

Anton A. Kiss, Edwin Zondervan, Richard Lakerveld, Leyla Özkan (Eds.)
Proceedings of the 29th European Symposium on Computer Aided Process Engineering
June 16th to 19th, 2019, Eindhoven, The Netherlands. © 2019 Elsevier B.V. All rights reserved.
http://dx.doi.org/10.1016/B978-0-128-18634-3.50292-7

From renewable energy to ship fuel: ammonia as an energy vector and mean for energy storage

Francesco Baldi[a*], Alain Azzi[a] and François Maréchal[a]

[a]*Laboratory of Industrial processes and energy systems engineering, École Polytechnique Fédérale de Lausanne, EPFL Valais-Wallis, Rue de l'Industrie 17, 1950 Sion, Switzerland*
francesco.baldi@epfl.ch

Abstract

The stochastic and non-controllable nature of most of renewable energy sources makes it necessary to include extensive use of energy storage in national grids to overcome periods with low availability. Hence, effective energy storage technologies will be requiredfor achieving a 100% renewable energy system.

Currently used technologies are only partly suitable for this task. Batteries are efficient but expensive, and mostly suitable for daily storage. Hydrogen has a significant potential, but suffers from a low energy density and difficulties in handling and transportation. Among different potential solutions, ammonia was often pointed out as a high-density and low-cost hydrogen carrier.

In this paper, we analyze the efficiency of an ammonia-based pathway for the storage of excess energy from renewable energy sources, its transportation, and its final use. As a case study, we consider the use on board of a urban car transport vessel as the user of the stored energy. The energy efficiency and cost of the whole chain, from the production at the wind farm until its use on board of the ferry, is evaluated and compared with competitive alternatives, namely batteries and hydrogen storage.

The results show that, while not being a game-changer, the use of ammonia as mean for storing and transporting excess energy from renewable power plants is viable and, in combination with other storage systems, contributes to a relevant part of the share of total installed storage capacity in the most cost-effective solution.

Keywords: energy storage, renewable energy, ammonia, fuel cells

1. Introduction

The challenge of making human activities environmentally sustainable is one of the most important that humanity will have to face in the coming years. While the solution will be provided by a combination of different technological developments, it is widely recognized that renewable energy sources will take a major share of the task.

Given the stochastic and non-controllable nature of most of renewable energy sources, improving the performance of energy storage represents one of the most important challenges to face for achieving a 100% renewable energy system. Batteries, while being the most efficient storage technology, suffer from limitations in energy density and high cost. Hydrogen generated by electrolysis, while still providing a reasonably high round-trip efficiency, has limitations in energy density and in handling.

As a solution for long-term storage and for transportation, various authors have proposed the conversion of hydrogen to conventional fuels, such as syngas, methane, methanol, and Diesel, but none of these solutions would prove to be carbon neutral, unless the carbon is originated by biomass, or by carbon capture and storage.

The use of ammonia as fuel can provide a valid alternative to carbon-based fuels. Ammonia has a relatively energy density in its liquid phase, that can be achieved at conditions relatively close to ambient conditions. For this reason, we believe that there is an interest in proposing ammonia as an energy vector, and as a mean for energy storage.

In this paper, we aim at investigating the potential role of ammonia as a way to store renewable energy that is produced in excess during low-demand periods. We address this by looking at a specific case study, based on the hypothesis to use the extra energy to propel ferries. The study aims at comparing existing storage technologies (batteries and hydrogen) with the potential case of ammonia.

2. Method

In order to investigate the potential role of ammonia as means for storing renewable energy during periods of excessive availability, in this paper we refer to the case of an off-shore wind farm of a rated power of 200 MW. The excess energy can either be curtailed, and hence lost, or stored and then transferred to a harbor, where it is assumed that a car ferry will be the final user.

The process of storage and transportation of the excess energy is allowed to take different ways, as shown in Figure 1:

Electric way : Excess energy is stored in batteries, and then sent to shore via existing power lines. The power transmission is allowed whenever there is no excess energy generation.

Hydrogen way : Excess energy is used in an electrolyser to generate hydrogen, that is stored in either compressed or liquid form. It is here assumed that the hydrogen is stored in container-sized tanks, that can then be loaded on supply vessels that regularly serve the wind farm. A bi-weekly frequency is assumed for the purpose of this study, as reported by CIT.

Ammonia way : Excess energy is used to generate hydrogen from water in an electrolyser, and nitrogen from air in a cryogenic air separation unit. Hydrogen and nitrogen are then used in an ammonia synthesis plant (here assumed based on the Haber-Bosch technology) that converts them to ammonia, that is then stored in liquid form at low temperature. The transportation is assumed to happen in the same way as assumed for the hydrogen way.

Once delivered to the port, the different fuels can be loaded on board of the ferry that operates on daily round trips. Electrical energy can be stored onboard in batteries, and directly used to power electric motor. Hydrogen is assumed to be used in a proton exchange membrane fuel cell (PEMFC), a well-developed technology extensively tested for

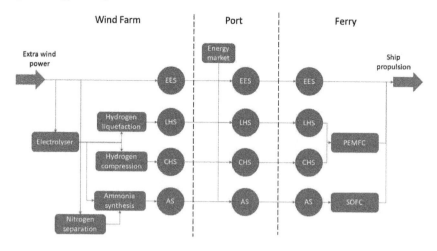

Figure 1: Representation of the superstructure for the optimization model

Name	Location	C_{inv}	Unit	η	Unit
Electrolyser	WF	400[1]	$\frac{EUR}{kW_{H_2}}$	0.73[1]	$\frac{kW_{el}}{kW_{H_2}}$
H_2 Compression plant	WF	180[2]	$\frac{EUR}{kW_{H_2}}$	0.91[3]	$\frac{kW_{el}}{kW_{H_2}}$
H_2 Liquefaction plant	WF	650[4]	$\frac{EUR}{kW_{H_2}}$	0.64[5]	$\frac{kW_{el}}{kW_{H_2}}$
N_2 Separation plant	WF	1450[1]	$\frac{EUR}{\frac{kg_{H_2}}{h}}$	0.11[1]	$\frac{kWh_{el}}{kg_{N_2}}$
NH_3 Synthesis plant	WF	3000[1]	$\frac{EUR}{kg_{NH_3}/h}$	0.64[1]	$\frac{kWh_{el}}{kg_{NH_3}}$

[1]Ikäheimo et al. (2018), [2]Law et al. (2013), [3]Tzimas et al. (2003), [4]Kelly (2007), [5]Gardiner (2009)

Table 1: Numerical assumptions employed in the study

use with hydrogen as fuel, also in maritime applications CIT. Ammonia is assumed to be used in a solid oxide fuel cell (SOFC); while the technology of SOFCs is less mature when compared to PEMFCs, particularly when powered by ammonia, recent tests have shown that there is no conceptual, or practical obstacle to a wider adoption of this technology CIT. The assumptions related to unit cost and efficiency are summarized in Table XXX. We assume that the maintenance cost does not significantly vary between different options and technologies and that, hence, can be excluded from the analysis.

The evolution of the excessive power generation over the year is represented using to the processed data provided by Zerrahn et al. (2018), originally retrieved in raw format from ?, and refer to the German electrical grid for the year 2014. This dataset provides hourly values for the curtailed energy from the combination of all renewable power plants. The implicit assumption of using this data in this paper is that the excess power is redistributed among different plants based on the installed power capacity. While this can be considered as a strong assumption, this approach provides a relatively accurate representation of the size and frequency of power curtailment in a European electrical grid. A representation

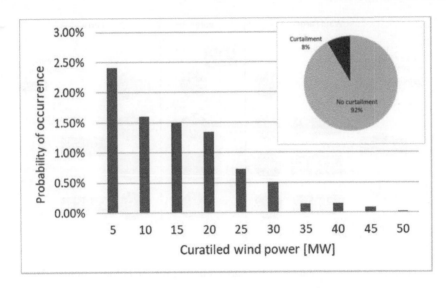

Figure 2: Wind curtailment power, probability of occurrence

of the resulting frequency of different curtailment powers is provided in Figure XXX.

Both the excess power and the power demand of the ferry are represented with a two-hour definition. As this would require a total of 4320 time steps for one year of operation, we assume that yearly operations can be summarized with a total number of four "reference periods" of the duration of two weeks, where the choice of the period length is based on the frequency available for transporting fuel from the wind farm to shore. The four reference periods where defined as follows:

- Maximum cumulated excess energy (for dimensioning the storage), occurring once per year

- Maximum peak excess power (for dimensioning the conversion units), occurring once per year

- Reference period at low cumulated excess energy, occurring 15 times per year

- Reference period at high cumulated excess energy, occurring 9 times per year

3. Results and discussion

The results for the optimal annualized costs in the different cases are provided in Figure 3. The option of using only electric energy storage in the form of batteries is clearly the least convenient case: as a consequence of the high specific investment cost, the optimizer chooses to reduce the total storage installed capacity, hence increasing the operational costs related to energy curtailment. The results also show that, for the selected case study, liquid hydrogen storage is more competitive in comparison to compressed hydrogen storage. Both the increased energy cost for liquefaction and the relatively high investment

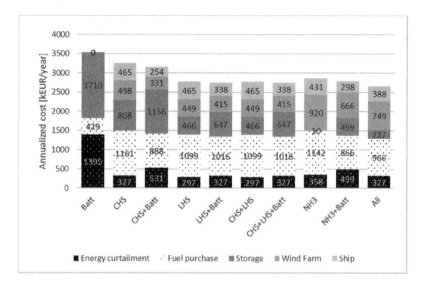

Figure 3: Comparison of operational, investment, and total costs for the different investigated cases

cost of the hydrogen liquefaction plant are offset by the lower investment cost for hydrogen storage.

When ammonia comes into play, a combination of effects can be observed. The use of ammonia alone as storage option is less profitable compared to the case of liquid hydrogen storage. This is mostly due to the high specific investment cost of the ammonia synthesis plant, which in this case is the largest contributor to the investment cost (32%), while the cost of the storage is only marginal (less than 1%). On the other hand, when other storage possibilities are included, the ammonia production plant is downsized (the installed capacity is reduced by 38%), and liquid hydrogen is used for intermediate storage. The detailed cost share in this case is shown in Figure 4: the electrolyser, the SOFC and the ammonia synthesis plant all contribute with 20-25% of the investment costs, while the use of a battery for part of the onboard power generation is favored because of the lower operational costs, particularly when energy is bought from the energy market.

4. Conclusion

References

M. Gardiner, 2009. Energy requirements for hydrogen gas compression and liquefaction as related to storage needs. Technical Report 9013, US Department of Energy, United States.

J. Ikäheimo, J. Kiviluoma, R. Weiss, H. Holttinen, 2018. Power-to-ammonia in future North European 100 % renewable power and heat system. International Journal of Hydrogen Energy 43 (36), 17295–17308.

B. Kelly, 2007. Liquefaction and pipeline costs.

K. Law, J. Rosenfeld, V. Han, M. Chan, H. Chiang, J. Leonard, 3 2013. U.s. department of energy hydrogen storage cost analysis. Tech. rep.

E. Tzimas, C. Filiou, S. Peteves, J.-B. Veyret, 2003. Hydrogen storage: state-of-the-art and future perspective. Tech. Rep. 20995, European Commission, Joint Research Centre (JRC), Directorate General (DG), Institute for Energy, Petten, The Netherlands.

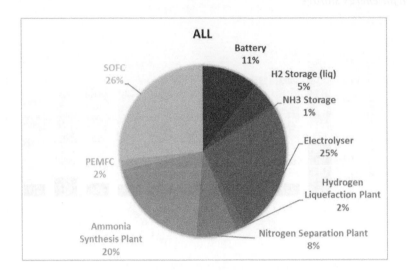

Figure 4: Investment cost breakdown, full case

A. Zerrahn, W.-P. Schill, C. Kemfert, Sep. 2018. On the economics of electrical storage for variable renewable energy sources. European Economic Review 108, 259–279.

Anton A. Kiss, Edwin Zondervan, Richard Lakerveld, Leyla Özkan (Eds.)
Proceedings of the 29th European Symposium on Computer Aided Process Engineering
June 16th to 19th, 2019, Eindhoven, The Netherlands. © 2019 Elsevier B.V. All rights reserved.
http://dx.doi.org/10.1016/B978-0-128-18634-3.50293-9

Are renewables really that expensive? The impact of uncertainty on the cost of the energy transition

Xiang Li[a,*], Stefano Moret[a], Francesco Baldi[a] and François Maréchal[a]

[a]*Industrial Process and Energy Systems Engineering, Ecole Polytechnique Fédérale de Lausanne, Sion 1951, Switzerland*
xiang.li@epfl.ch

Abstract

The dramatic evidence of climate change is making the transition to more renewable energy systems an urgent global priority. As energy planners normally look 20-50 years ahead, it is crucial to consider the key uncertainties stemming from inaccurate forecasts (e.g. of fuel prices, investment costs, etc.) in energy models to ensure making robust investment decisions. Nonetheless, uncertainty is to date seldom accounted for in energy planning models.

In this paper, we challenge the general perception that the transition to a more renewable energy system always comes at a higher price; to do this, we analyze the impact of uncertainty on the cost of this transition for a real-world national energy system. Concretely, we first generate a set of energy planning scenarios with increasing renewable energy penetration (REP); in a second stage, we perform an uncertainty analysis to compare the cost of these scenarios and thus to determine the significance of the difference in their total cost in presence of uncertainty. Our results show that increasing the amount of renewable energy in a national energy system is not necessarily associated to a higher cost, and can even lead to a cost reduction for some specific realizations of the uncertain parameters.

Keywords: Energy planning, Uncertainty, Energy transition, Mixed-integer linear programming

1. Introduction

The Earth is warming up at an alarming pace. Most countries have agreed on ambitious goals for reducing the human impact on the climate (Rogelj et al., 2016) and have started to make plans for a transition towards more sustainable systems (Kern and Rogge, 2016). A major part of this *energy transition* is related to an increased use of renewable energy sources (RESs). Many authors have thus used energy models to evaluate the impact of an increased share of RESs in energy systems planning. Jacobson et al. (2015) investigated the feasibility and cost of a fully renewable-powered system for the United States and concluded that, when externalities are accounted for, such system would be less expensive than a traditional energy system; Schill (2014) performed a similar analysis for the German energy system with the objective of achieving 86% REP by 2050, showing that this can be done with a relatively limited investment in energy storage capacity as long as a higher flexibility in the operation of fossil power plants is allowed.

Despite these results, and despite the fact that today there are several technological options for a cost-competitive generation of renewable energy, the public opinion often still perceives RESs as being far more expensive than fossil fuels. As an example, Ntanos et al. (2018) report the results of a recent survey in Greece, in which the respondents identified in the "high installation costs" the main reason for not using RESs. In this paper, we challenge this general perception. The main

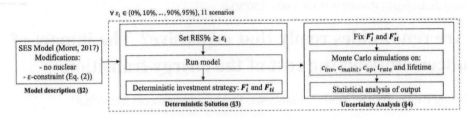

Figure 1: Modeling flowchart

contribution of our work compared to the literature is the consideration of multiple sources of uncertainty in the analysis. In fact, the main shortcoming of most works proposed in the literature lies in the use of "deterministic" models, which do not consider uncertainty and assume that long-term forecasts for the key parameters (e.g. fuel prices) are correct. However, long-term forecasts are inevitably inaccurate (Moret et al., 2017), and thus it is crucial to consider uncertainties in long-term energy modeling (DeCarolis et al., 2017; Mavromatidis et al., 2018).

In this paper, we propose an analysis of the cost and investment decisions for different degrees of REP in an energy system, with a particular focus on the impact of uncertainty in the process, taking the national energy system of Switzerland as a real-world case study. The methodological approach is shown in Figure 1. First, we model the Swiss energy system using the Swiss EnergyScope (SES) framework (Moret et al., 2014; Codina Gironès et al., 2015). SES, which we describe in Section 2, is a mixed-integer linear programming (MILP) model, which identifies the optimal investment and operation strategy to minimize the total annual cost of the energy system. In a second phase, the model is used deterministically (i.e. with all parameters at nominal values) to determine the optimal energy planning scenarios under increasing shares (ε_i) of RESs (Section 3). Third, the impact of uncertainty is evaluated: for each of the previously obtained scenarios, the investment strategy is fixed and the total cost of the system is evaluated with a Monte-Carlo based approach, considering as uncertain parameters all the costs, the discount rates and the lifetime of technologies. This allows to perform an uncertainty analysis of the solution (Section 4).

2. The energy model

The paper is based on the SES model, a framework for the strategic energy planning of the Swiss energy system. In particular, in this paper we use the open-source MILP version of the model by Moret (2017)[1]. It is a representative model of an energy system, including electricity, heating and mobility: given the end-use energy demand, the efficiency and cost of energy conversion technologies, the availability and cost of energy resources, the model identifies the optimal investment ($\mathbf{F}$) and operation ($\mathbf{F_t}$) strategies to meet the demand and minimize the total annual cost of the energy system. In comparison to other energy models, which often consider hourly timesteps and multi-stage investment plans, it has a lower level of detail, but it offers a reasonable trade-off between CPU time and accuracy; in particular, its multiperiod monthly formulation allows accounting for the main dynamics of concern in energy systems planning, such as seasonal variations, long-term energy storage management, and uncertainties in loads and renewable production.

The objective of the SES model is to minimize the annual total cost $\mathbf{C_{tot}}$ expressed by the sum of the annualized investment cost ($\mathbf{C_{inv}}$), of the annual operational costs ($\mathbf{C_{op}}$) and of the annual maintenance cost ($\mathbf{C_{maint}}$) (1).

$$\min \mathbf{C_{tot}} = \min \left(\sum_{j \in \mathscr{E}} \mathbf{C_{inv}}(j) + \sum_{j \in \mathscr{E}} \mathbf{C_{maint}}(j) + \sum_{r \in \mathscr{R}} \sum_{t \in \mathscr{T}} \mathbf{C_{op}}(r,t) t_{op}(t) \right) \tag{1}$$

[1]Model available at https://github.com/stefanomoret/SES_MILP and fully documented in (Moret, 2017)

where the sets $\mathscr{E}$, $\mathscr{R}$ and $\mathscr{T}$ represent the technologies, the resources (renewables and non-renewables as well as electricity import) and the time periods (months) respectively. $t_{op}(t)$ denotes the duration of the period t. The main constraints are the energy and resource balance in each period; the limited availability of resources; the capacity factor of different energy conversion technologies; the limits to the maximum installed capacity of different technologies; and the limits to grid capacity (accounting for the additional investments linked to a higher penetration of stochastic renewables).

In this paper, we modify the model presented by Moret (2017) as follows: *i*) we allow individual technologies to cover the full share of the demand in their sector, e.g. electric vehicles can satisfy all the private mobility demand ($f_{min,\%}(j) = 0$, $f_{max,\%}(j) = 1 \; \forall j \in \mathscr{E}$); *ii*) we allow freight transport to be fully satisfied by trains; *iii*) we impose the phase out of nuclear energy, in agreement with the Swiss energy strategy.

3. Deterministic energy planning

In addition, in order to quantify and to constrain the impact of RESs, Eq. (2) is added to the model formulation.

$$\sum_{r\in\mathscr{R}_{RES}}\sum_{t\in\mathscr{T}} \mathbf{F_t}(r,t)t_{op}(t) \geqslant \varepsilon \sum_{r\in\mathscr{R}}\sum_{t\in\mathscr{T}} \mathbf{F_t}(r,t)t_{op}(t) \tag{2}$$

where the parameter ε represents the minimal renewable penetration ratio in the energy system and the set $\mathscr{R}_{RES}$ is a subset of $\mathscr{R}$ composed of all renewable resources (solar, wind, wood, hydro, geothermal, synthetic natural gas, bioethanol and biodiesel). In Eq. (2), $\mathbf{F_t}(r,t)$ stands for the utilization of resource r at period t. In this paper we analyze 11 scenarios with different values of the ε parameter, ranging from 0 % to 90 % with 10 % steps, plus the "maximum REP" scenario with $\varepsilon = 95$ %.

The energy mixes of the resulting 11 scenarios are shown in Figure 2. It should be noted that scenarios 1, 2 and 3 are equivalent, due to the fact that the REP in the unconstrained (i.e. with 0 % minimum REP) case is approximately 20 %. Among the RESs, wind and (new) hydro-power are the most competitive from a cost perspective and reach their maximum potential at a REP level of

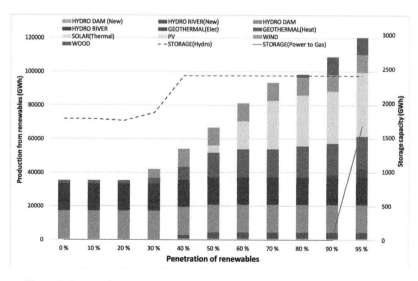

Figure 2: Renewable energy production and storage capacity in different REPs

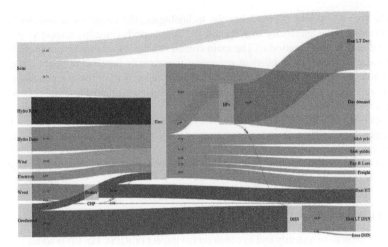

Figure 3: Sankey diagram for energy flows in 95 % REP scenario [TWh]

60 %. The use of solar energy emerges in the optimal solution only starting from a REP of 50 %, and it is only used extensively for REP ≥ 70 %. Wood is only used in very high REP scenarios, due to its high assumed cost in the model. The results of the deterministic optimization also show that the required storage capacity is relatively limited: hydro-electric storage reaches its peak in the 40 % REP scenario, at a value approximately 35 % higher than the base case, while power-to-gas is only activated in the 95 % REP scenario, allowing to reach a total seasonal storage capacity of approximately 4,000 GWh. Note that, due to the monthly resolution of the model, here dynamic storage technologies for balancing in-time supply-demand at a daily or more fine time resolution (such as batteries) are not considered. This choice is motivated by computational reasons.

How does an almost completely renewable-based energy system look like? This is shown in Figure 3, where the energy flows in Scenario 11 are represented. As expected, the demand is dominated by the use of electricity, with the exception of part of the high-temperature heat demand (provided by biomass boilers) and of the low-temperature heat demand (provided by thermal solar and by geothermal heat). The use of boilers and gas-driven heat pumps for low-temperature heat is completely discontinued, substituted by electric heat pumps.

4. Uncertainty analysis

As highlighted in the Introduction, the problem of strategic energy planning is subject to a high degree of uncertainty, which should not be neglected. In this article, we consider five main uncertainty categories: interest rate, investment cost, maintenance cost, resource prices, and technology lifetime. In the SES model, this translates to a total of 157 uncertain parameters. We assume that all these uncertain parameters are uniformly distributed between their lower and upper bounds, adopting the methodology and the uncertainty ranges defined by Moret et al. (2017).

In the uncertainty analysis, the decision variables related to the installed capacity of each technology, and the energy utilization for each mobility technology in each period are fixed to the optimal values calculated in the deterministic optimization for each scenario ($\mathbf{F}^*$ and $\mathbf{F_t}^*$, respectively). Then, for each of the 11 scenarios, the SES model is run for different realizations of the uncertain parameters with a Monte-Carlo approach. This corresponds to a situation where a decision for the strategic energy planning has been taken, and we aim to analyze the cost of the application of that specific strategy with different values for the uncertain parameters, corresponding to different

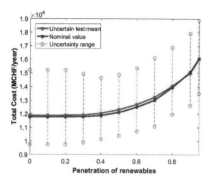

Figure 4: Nominal and uncertain mean total cost

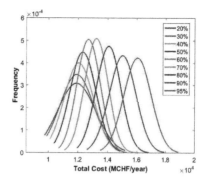

Figure 5: Nomalized total cost distribution

possible future scenarios. A total of 3,000 Monte-Carlo simulations (based on an observed convergence after 2,600 simulations) is run for each scenario.

In the Swiss energy system, a large share of the total annual cost (equal to 6,654 MCHF/y) derives from fixed costs, i.e. costs which are equal for all the solutions (such as the investment cost of existing hydroelectric power plants, of the electricity grid and of energy efficiency measures). In the simulation runs, we fix these costs to zero; this means that we do not consider them as uncertain in the simulations, but we add them to the simulation results at their nominal values.

Figure 4 shows the results of the uncertainty analysis. It can be osberved that the mean total cost values obtained by Monte-Carlo simulations are consistent with the deterministic results for each scenario, as in both cases the total cost increases with ε. However, the results also show that there is a high uncertainty in the total cost of the system for all REP levels, leading to a relatively high overlap between the different distributions (Figure 5). This can lead to the conclusion that, while on average increasing the share of renewable energy beyond 20 % involves a higher cost for the system, this is not necessarily true once uncertainty is accounted for.

A higher REP has also an impact on the variance of the total cost (Figure 6): given the high uncertainty on the price of different fossil fuels, excluding them from the energy mix generates a clear reduction in the uncertainty of operational costs. In contrast, the uncertainty on the investment increases, partially balancing the previous effect, as a consequence of both the larger uncertainty of the investment cost of technologies activated with higher REP, and of the overall higher investment cost of the system, which amplifies the uncertainty related to the interest rate. Overall, it can be concluded that cost uncertainty decreases when the share of RES increases.

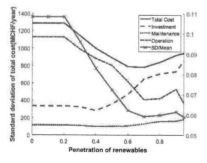

Figure 6: Total cost standard deviation breakdown

	Penetration of renewable energy							
	30%	40%	50%	60%	70%	80%	90%	95%
0%	0.44	0.33	0.23	0.17	0.12	0.06	0.00	0.00
10%	1.00	1.00	0.97	0.62	0.43	0.30	0.12	0.01
20%	1.00	1.00	1.00	0.97	0.76	0.51	0.35	0.18
30%	1.00	1.00	1.00	1.00	0.97	0.75	0.55	0.39
40%	1.00	1.00	1.00	1.00	1.00	0.95	0.77	0.57
50%	1.00	1.00	1.00	1.00	1.00	1.00	0.95	0.78
60%	1.00	1.00	1.00	1.00	1.00	1.00	1.00	0.94
70%	1.00	1.00	1.00	1.00	1.00	1.00	1.00	0.99
80%	1.00	1.00	1.00	1.00	1.00	1.00	1.00	1.00
90%	1.00	1.00	1.00	1.00	1.00	1.00	1.00	1.00
100%	1.00	1.00	1.00	1.00	1.00	1.00	1.00	1.00

(Cost increase wrt to base scenario)

Figure 7: Probability of total cost increase for different REP scenarios

Figure 7 shows the percentage of Monte-Carlo simulations where a given REP scenario was more expensive than the base case by a certain amount. An energy system with 70 % REP was less than 20 % more expensive than the base case in 76 % of the simulations, and in 12 % of the simulations it was actually less expensive than the base case. Even the 95 % REP case was, in most (78 %) cases, no more than 50 % more expensive than the baseline case, while the less ambitious 50 % REP scenario is at most 10 % more expensive than the baseline. These results are in accordance with what previously published in the academic literature (e.g. Jacobson et al. (2015)), and the cost difference is expected to be even lower if the external costs were included in the analysis.

5. Conclusions

Are renewable really that expensive? Based on the work that we present in this paper, the answer is that yes, a renewable-based energy system will most likely be more expensive than one where we apply no constraint on the share of renewable energy, but not always, and not by much. Furthermore, a renewable-based energy system will be, overall, inherently more robust against unexpected developments, particularly with respect to fluctuations in fuel prices.

The results of our uncertainty analysis, in fact, reveal that the variance in the total cost of the system is high, and of the same order of magnitude of the average cost difference between different scenarios. This suggests that, when considering uncertainty, increasing the amount of renewable energy in a national energy system is not necessarily associated to a higher cost, and can even lead to a cost reduction in some specific realizations of the uncertain parameters. These results are of particular value to energy planners, as they challenge the common perception of a very high cost related to the energy transition towards a renewable energy-based system. While these conclusions are drawn based on a specific case study (the Swiss energy system), we consider that they are sufficiently general to be applicable to other countries.

References

V. Codina Gironès, S. Moret, F. Maréchal, D. Favrat, Oct. 2015. Strategic energy planning for large-scale energy systems: A modelling framework to aid decision-making. Energy 90, Part 1, 173–186.

J. DeCarolis, H. Daly, P. Dodds, I. Keppo, F. Li, W. McDowall, S. Pye, N. Strachan, E. Trutnevyte, M. Usher, M. Winning, S. Yeh, M. Zeyringer, May 2017. Formalizing best practice for energy system optimization modelling. Applied Energy 194, 184–198.

M. Z. Jacobson, M. A. Delucchi, G. Bazouin, Z. A. Bauer, C. C. Heavey, E. Fisher, S. B. Morris, D. J. Piekutowski, T. A. Vencill, T. W. Yeskoo, 2015. 100% clean and renewable wind, water, and sunlight (wws) all-sector energy roadmaps for the 50 united states. Energy & Environmental Science 8 (7), 2093–2117.

F. Kern, K. S. Rogge, Dec. 2016. The pace of governed energy transitions: Agency, international dynamics and the global Paris agreement accelerating decarbonisation processes? Energy Research & Social Science 22, 13–17.

G. Mavromatidis, K. Orehounig, J. Carmeliet, Mar. 2018. Uncertainty and global sensitivity analysis for the optimal design of distributed energy systems. Applied Energy 214, 219–238.

S. Moret, 2017. Strategic energy planning under uncertainty. Ph.D. thesis, École Polytechnique Fédérale de Lausanne, Lausanne, Switzerland.

S. Moret, V. Codina Gironès, M. Bierlaire, F. Maréchal, Sep. 2017. Characterization of input uncertainties in strategic energy planning models. Applied Energy 202, 597–617.

S. Moret, V. Codina Gironès, F. Maréchal, D. Favrat, 2014. Swiss-EnergyScope.ch: A Platform to Widely Spread Energy Literacy and Aid Decision-Making. Chemical Engineering Transactions 39, 877–882.

S. Ntanos, G. Kyriakopoulos, M. Chalikias, G. Arabatzis, M. Skordoulis, Mar. 2018. Public Perceptions and Willingness to Pay for Renewable Energy: A Case Study from Greece. Sustainability 10 (3), 687.

J. Rogelj, M. den Elzen, N. Höhne, T. Fransen, H. Fekete, H. Winkler, R. Schaeffer, F. Sha, K. Riahi, M. Meinshausen, Jun. 2016. Paris Agreement climate proposals need a boost to keep warming well below 2 °C. Nature 534 (7609), 631–639.

W.-P. Schill, Oct. 2014. Residual load, renewable surplus generation and storage requirements in Germany. Energy Policy 73, 65–79.

Anton A. Kiss, Edwin Zondervan, Richard Lakerveld, Leyla Özkan (Eds.)
Proceedings of the 29th European Symposium on Computer Aided Process Engineering
June 16th to 19th, 2019, Eindhoven, The Netherlands. © 2019 Elsevier B.V. All rights reserved.
http://dx.doi.org/10.1016/B978-0-128-18634-3.50294-0

Comparative analysis of gasification and reforming technologies for the syngas production

Hussain A Alibrahim [a], Siddig SeedAhmed [a], Usama Ahmed [a], Umer Zahid [a] [*]

[a] *Department of Chemical Engineering, King Fahd University of Petroleum and Minerals, Dhahran, 34464, Saudi Arabia*

uzahid@kfupm.edu.sa

Abstract

This study is focused on the production of syngas using gasification and dry methane reforming (DMR) technologies. The two processes are compared as a standalone design for a syngas production capacity of 10,000 kmol/h with H_2 to CO ratio of 0.88. The two processes are then integrated to develop a novel design that can reduce the CO_2 emissions to a near-zero level with reasonable amount of energy requirement. In this study, Aspen Plus® has been employed as a process simulation tool. The results reveal that the DMR process consumes 3.4 times less energy compared to a gasification plant of same capacity. The result also shows that the integration of gasification and DMR design off-sets the negative aspects of gasification technology enabling it to utilize conventional solid fossil fuel with low emission levels.

Keywords: Gasification, dry reforming, syngas, energy.

1. Introduction

The production of syngas is one of the important process in the petrochemical industries from which many downstream products like ammonia, methanol and fertilizers are produced. In order to have a sustainable hydrocarbon fuel usage, the amount of carbon dioxide in the atmosphere must be reduced. As in the Paris agreement at COP21 conference in 2015, there are 195 countries who signed a ratification to reduce carbon emissions. The Paris accord can be realized by supplying clean fuels and renewables for the energy production, improving efficiency of the current energy consuming processes, and utilization of fossil fuels by eliminating the pollutants and GHG emissions. Syngas production technologies can play a role in bridging the gap for the use of fossil fuels while keeping the emissions within the limits. Syngas can be produced using gaseous fuels, liquid hydrocarbons and solid feedstocks. The gaseous and liquid feedstocks utilize technologies such as steam reforming, partial oxidation, dry reforming, and auto-thermal reforming. With solid feedstocks such as coal and biomass, gasification is the main technology available for the syngas production.

Recently, dry methane reforming is gaining a lot of attention as an environmentally friendly process for syngas production because it consumes two major GHGs, carbon dioxide and methane. It requires 363.6 kg of methane to consume 1 ton of CO_2 for the production of 90.9 kg of H_2 and 1.27 tons of CO. The main reaction taking place in the dry reforming is shown below:

$$CH_4 + CO_2 \leftrightarrow 2H_2 + 2CO \qquad \Delta H_{298} = 247 \text{ kJ/mol} \qquad (1)$$

Reverse water gas shift reaction also takes place in the DMR reaction, which combines the CO_2 and H_2 to produce CO and water. The reverse water gas shift reaction is shown below:

$$H_2 + CO_2 \leftrightarrow H_2O + CO \tag{2}$$

The reaction is favoured by low pressure and high temperature which means high energy requirement and high production cost. Gasification is a proven and commercialized technology to convert solid feedstocks into syngas along with other impurities such as H_2S and CO_2. Gasifiers are operated with limited amount of oxygen to conduct the pyrolysis reactions for the formation of syngas. Although there are many simultaneous complex reactions taking place in the gasifier, some of the main reactions are shown below:

$$C + \tfrac{1}{2} O_2 \leftrightarrow CO \qquad \Delta H_{298} = -110.5 \text{ kJ/mol} \tag{3}$$
$$C + H_2O \leftrightarrow CO + H_2 \qquad \Delta H_{298} = -131.4 \text{ kJ/mol} \tag{4}$$
$$C + CO_2 \leftrightarrow 2CO \qquad \Delta H_{298} = -172.0 \text{ kJ/mol} \tag{5}$$
$$C + H_2O \leftrightarrow CO + H_2 \qquad \Delta H_{298} = -41.0 \text{ kJ/mol} \tag{6}$$

Gasification process is usually following by a gas cleaning process where acid gases are removed from the syngas. Depending on the downstream requirement, another reaction known as water gas shift reaction is usually followed to convert CO to CO_2 and increase the syngas ratio as shown below:

$$CO + H_2O \quad \leftrightarrow \quad H_2 + CO_2 \quad \Delta H_{298} = -891 \text{ kJ/mol} \tag{7}$$

Many studies in the past have focussed on the syngas production technologies in order to improve the process performance by developing new catalyst materials, optimizing the process conditions, and integration with other processes. Mallick et al. (Mallick et al., 2018) analysed the effects of co-gasification of coal and biomass experimentally on the process performance. They concluded that the co-gasification of coal and biomass has synergetic effects with merits compared to the individual feedstock gasification. Li et al. (Li et al., 2018) studied the design of dry methane reforming with CO_2 utilization using nickel based catalyst. Their results showed that the catalyst development can be an important factor in the commercialization of DMR technology. Wu et al. (Wu et al., 2005) combined the coal gasification with steam reforming to evaluate the feasibility of process integration. Lim et al. (Lim et al., 2012) studied the combination of steam reforming with dry reforming. Their results showed that the combined process can reduce the net CO_2 emission by 67 % compared to the standalone steam reforming process. Ahmed et al. (Ahmed et al., 2017) studied the integration of gasification with steam reforming for power generation application with CO_2 capture unit.

Although, standalone syngas production technologies have been widely investigated in the literature, however, there is no significant research available that compared the DMR process with that of gasification process and considered their integration. The goal of this study is to compare the process performance of DMR and gasification technologies in terms of energy efficiency and economics. Aspen Plus® V9 is used to simulate the two processes. In order to have a fair comparative analysis, syngas production rate and ratio (H_2/CO) have been fixed for the two processes at 10,000 kmol/h and 0.88 respectively.

2. Process Design

2.1. Dry methane reforming

Dry methane reforming process has been simulated using the design outlined in the literature (Luyben, 2014). Peng-Robinson property method has been used for the simulation of DMR section while Amines package has been employed in the CO_2 removal section. The feed stream containing methane and excess CO_2 is compressed to the required reactor pressure of 4 bar. The temperature of feed stream after compression increases to about 130.6 °C which is further heated using a counter current process heat exchanger. The syngas formation reaction takes place at 4 bar and 1000 °C. The reactor effluent stream is used to cool down the incoming feed stream. The reactor effluent stream contains H_2, CO_2, unreacted CO_2 and CH_4, and small amount of water. The CO_2 is removed from the gas stream using a typical absorber-stripper system by employing MEA as a solvent. The syngas free of CO_2, contains 44 mol % of H_2 and 50.5 mol% of CO is compressed in multi-stage compressor train with intercooling to a product pressure of 30 bar. The capture CO_2 from the absorber is stripped from the solvent in the stripper and is recycled back to the feed compressor.

2.2. Gasification

Feedstock for the syngas production using gasification technology in this study is coal. The composition and heating value for the coal has been adapted from the literature (Ahmed et al., 2016). Multiple property packages have been used to precisely predict the properties of gasification simulation. Solids property method has been used for the feed preparation block, while Peng-Robinson has been used for the gasification, cleaning units and WGS blocks. Figure 1 shows the simplified block diagram of the gasification process for syngas production.

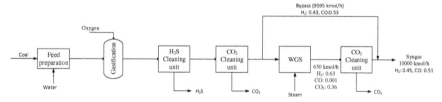

Figure 1: Simplified block diagram of gasification process for syngas production

Coal slurry is first prepared by crushing the coal to a size of 0.2 mm particle diameter along with water addition. The coal slurry is the pumped to the gasifier where the reactions take place in the controlled amount of oxygen. The gasifier effluent contains a mixture of gaseous products including CO, H_2, CO_2, H_2O, Ar, H_2S and minor quantities of other gases. The reactions in the gasifier takes place at 33 bar and 1200 °C. The hot effluent gas is cooled down in a series of heat exchangers to generate high pressure steam. The cooled gases at 152 °C are then sent to the cleaning unit where H_2S and CO_2 is removed from the syngas in separate absorbers. Methanol is employed as a solvent in the cleaning unit for the removal of acid gases from the syngas. The syngas coming out of the cleaning unit has a H_2/CO ratio of approximately 0.80. In order to meet the required H_2/CO ratio of 0.88, approximately 4.3 % of the total syngas is sent to the water gas shift reactor where CO is converted H_2 by reaction with steam. The additional

CO_2 produced from the water gas shift reaction is removed using methanol absorption to get the syngas product of required purity.

2.3. Integrated design

Integrated design combines the gasification and DMR process in series to enhance the process performance as shown in figure 2. The raw syngas produced from the gasification process is sent to the cleaning unit, where only H_2S is removed and the remaining gas is fed to the DMR section. Methane is also fed to the DMR unit along with the gases entering from the cleaning unit. In this way, the CO2 generation from the gasification process can be utilized as a feed material for the DMR process. As in the standalone designs, the integrated design is also set to produce 10,000 kmol/h of syngas with the syngas ratio of 0.88. The desired production rate and product purity has been maintained by varying the coal flowrate to the gasifier and methane flowrate to the DMR reactor respectively.

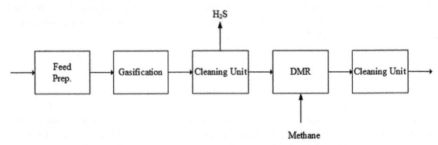

Figure 2: Simplified block diagram of integrated design for syngas production

3. Results

Energy analysis has been performed for the two systems to estimate the power requirement of various units in the process. Table 1 shows the power consumption of various units in the DMR and gasification designs.

Table 1: Energy consumption of various sections in DMR and gasification units

Dry methane reforming (MW)		Gasification (MW)		Integrated design (MW)	
Compression	33	Gasification unit	341	Gasification unit	271
Reactor	183	Cleaning unit	75	Cleaning unit	53
Stripper	7	WGS reactor unit	485	DMR	93
Cooling	42				
Total	265		901		417

The results shows that the DMR process consumes 3.4 times less energy compared to the same size of gasification unit for syngas production. The results also show that since the DMR process consumes CO_2 as a feed material, the amount of CO_2 consumed in DMR process is approximately 5 times compared to what the gasification process emits. This makes DMR process more environmentally sustainable for the syngas production. The gasification section in the integrated design produces around 500 kmol/h of CO_2 which is used a feed material in the DMR process. In order to maintain the required syngas ratio in the product, CH_4:CO_2 molar ratio in the DMR reactor is maintained at 2.3 in contrast to the standalone DMR design. Figure 3 shows the total energy consumption and CO_2 emissions for the three designs. The results show that the integrated design produce no CO_2 emissions and can reduce the energy requirement by approximately 53.6 % compared to the gasification process. Figure 4 shows the energy requirement per unit of syngas produced from the three processes. The results show that syngas can be produced with a lower energy requirement in case of an integrated design while mitigating the CO_2 emissions.

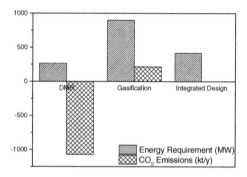

Figure 3: Energy consumption and CO_2 emissions comparison for the standalone and integrated designs

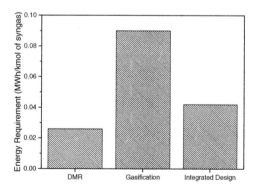

Figure 4: Specific energy requirement for the three designs

4. Conclusion

Dry reforming of methane is considered as an environmentally friendly process to produce syngas as it consumes two GHGs. The energy requirement for the DMR process is significantly lower than what is required in the gasification process. The O/C ratio is an important parameter which controls the syngas composition and its heating value. Adding WGS to gasification process can give a wide range of syngas ratios for downstream consideration. There is a need to further investigate the feasibility of the process by analyzing process economics.

Acknowledgment: The authors would like to acknowledge financial support from King Fahd University of Petroleum & Minerals (KFUPM) under the project no. USRG1803.

References

Ahmed, U., Zahid, U., Jeong, Y. S., Lee, C. J., & Han, C. (2016). IGCC process intensification for simultaneous power generation and CO_2 capture. *Chemical Engineering and Processing: Process Intensification, 101*, 72-86.

Ahmed, U., Kim, C., Zahid, U., Lee, C. J., & Han, C. (2017). Integration of IGCC and methane reforming process for power generation with CO_2 capture. *Chemical Engineering and Processing: Process Intensification, 111*, 14-24.

Li, B., Lin, X., Luo, Y., Yuan, X., & Wang, X. (2018). Design of active and stable bimodal nickel catalysts for methane reforming with CO 2. *Fuel Processing Technology, 176*, 153-166.

Lim, Y., Lee, C. J., Jeong, Y. S., Song, I. H., Lee, C. J., & Han, C. (2012). Optimal design and decision for combined steam reforming process with dry methane reforming to reuse CO_2 as a raw material. *Industrial & Engineering Chemistry Research, 51*(13), 4982-4989.

Luyben, W. L. (2014). Design and control of the dry methane reforming process. *Industrial & Engineering Chemistry Research, 53*(37), 14423-14439.

Mallick, D., Mahanta, P., & Moholkar, V. S. (2018). Synergistic Effects in Gasification of Coal/Biomass Blends: Analysis and Review. In *Coal and Biomass Gasification* (pp. 473-497). Springer, Singapore.

Wu, J., Fang, Y., Wang, Y., & Zhang, D. K. (2005). Combined coal gasification and methane reforming for production of syngas in a fluidized-bed reactor. *Energy & fuels, 19*(2), 512-516.

Anton A. Kiss, Edwin Zondervan, Richard Lakerveld, Leyla Özkan (Eds.)
Proceedings of the 29th European Symposium on Computer Aided Process Engineering
June 16th to 19th, 2019, Eindhoven, The Netherlands.
http://dx.doi.org/10.1016/B978-0-128-18634-3.50295-2

Model-based decision-support for waste-to-energy pathways in New South Wales, Australia

Koen H. van Dam[a*], Bowen Feng[b], Xiaonan Wang[b], Miao Guo[a], Nilay Shah[a] and Stephen Passmore[c*]

[a]Department of Chemical Engineering, Imperial College London, South Kensington Campus, London SW7 2AZ, UK
[b]Department of Chemical Engineering, National University of Singapore, 4 Engineering Drive 4, Singapore 117576
[c]Resilience Brokers Ltd, London, 10 Queen Street Place, London EC4R 1BE, UK
k.van-dam@imperial.ac.uk and stephen.passmore@resiliencebrokers.org

Abstract

Utility operators involved in energy, waste and water sectors are exploring technology options to generate energy in a sustainable way, while also dealing with waste streams including sewage and municipal solid waste from growing urban environments. The resilience.io platform was developed to study long term changes and development in an urban area of region under a range of policy, behaviour, economic and technological scenarios, to meet local resource demands in a cost-effective and sustainable way. A systems modelling approach studying waste-to-energy in New South Wales suggests initial findings are a suitable starting point for further exploration of the potential in the region using this approach as well as complementary tools and methodologies. Model results show 854 GWh renewable energy could be generated through waste-to-energy pathways in 2036.

Keywords: waste to energy, water, infrastructure modelling

1. Introduction

The Hunter region, in New South Wales, Australia, is undergoing an economic and sustainability transition. Hunter's Smart Specialisation Strategy (Regional Development Australia Hunter, 2016) is Australia's first policy aimed to identify the region's competitive strengths and formulate specific activities to make the most of the key growth areas. "In a post-mining boom economy", it encourages new ideas to "build on [the Hunter's] strong knowledge-base in mining, energy, agriculture and medical research" (ibid: p.14). It also highlights the wider context of climate related challenges, the call for a reduction in greenhouse gas emissions, an energy matrix with an increasing proportion of renewables, demographic shifts and urbanisation (Government of Australia, 2015).

The focus on energy security in a changing environment is reflected at the state level, where the government intends to ensure that NSW is well prepared for the current transformation of the energy markets. Therefore, key actions in the NSW State Infrastructure Policy 2018-2038 (NSW Government, 2018) ensure reliable and affordable energy, focus on supporting private investment in efficient energy generation and horizon-scanning for new approaches. Moreover, the NSW government has set an aspirational goal to achieve

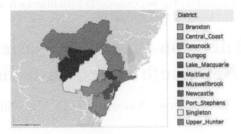

Figure 1: The study area in the Hunter Region, focusing on the area supplied by Hunter Water (highlighted) as well as neighbouring regions in Upper Hunter and Central Coast.

net-zero emissions by 2050 which can be clearly seen as a driver for innovation and development of new climate-compatible growth, jobs and livelihoods in the region.

Utility operators in urban environments, particularly those involved in energy, waste and water sectors, are exploring technology options to generate energy in a sustainable way, while also dealing with waste streams (e.g. sewage and municipal solid waste). Policies at different levels and from different angles appear to provide a window of opportunity to consider waste-to-energy (WtE) as a catalyst for green growth in the region. This paper explores this potential using an optimisation model representing the region and its resources, networks and technologies.

2. Methodology

The resilience.io platform (Triantafyllidis et al., 2018) was developed to study long term changes and development in an urban area of region under a range of policy, behaviour, economic and technological scenarios, to meet local resource demands in a cost-effective and sustainable way. It provides modules to allow forecasting of socio-demographic scenarios, simulating spatio-temporal activities, and planning investment and operational strategies. Agent-based modelling to simulate the population (with their activities and resource demands) is combined with resource-technology-network optimisation to find a suitable combination of technologies, their location and size, to meet these demands. Scenarios can be set up to vary the population, their activities, new technologies and economics properties. To study the nexus between water, energy and waste, this platform can then be used to identify opportunities based on modelled material flows and conversion technologies. Resources available and waste generated can be simulated or included from published data and regional forecasts, with specific technologies (with their economic and environmental performance) selected as input for the optimisation model.

3. Case study

In close cooperation with a local utility operator a model has been set up to simulate the current water and energy demand, supply and networks for a region in New South Wales, Australia, to explore potential future designs and interventions to take advantage of synergies at the water-waste-energy nexus. Figure 1 shows the main case study area. Key resources incorporated in the model include potable water, non-drinking water (industrial use), source water, electricity, gas, domestic waste (municipal solid waste, sewage),

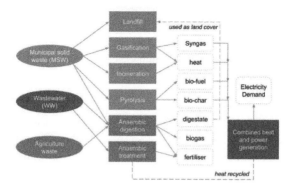

Figure 2: Possible pathways for processing and use of waste from households, waste water and the agriculture sector with technologies that can supply energy. Based on Wang et al. (2018)

Data type	Source
Sociodemographic data per region	Australian Bureau of Statistics (2018)
Water supply and demand	Hunter Water water consumption data
Recycle water supply	Hunter Water effluent reuse data
Wastewater generation and treatment	Hunter Water forward capital program data
Biosolids data	GHD Pty Ltd (2018)
Electricity supply and demand	Ausgrid (2016)
Gas supply and demand	Jemena (2011)
Municipal solid waste generation	WARR Survey provided by Hunter Water
Agricultural waste generation	MRA Consulting Group (2018)
Other organic waste data	MRA Consulting Group (2018)
Waste technology data	Jain et al. (2015); Lofrano (2012); Sun et al. (2015)
Water and wastewater technology data	Hunter Water data
Electricity generation technology data	Aemo (2018)
Existing water / wastewater network	Hunter Water network data
Existing electricity network	Jemena (2011)
Central Coast data	Central Coast Council (2018)

Table 1: Data sources used as input for the optimisation model

agricultural waste, sludge, and carbon dioxide. Potential conversion technologies are compared to alternatives for waste handling such as landfill (see Figure 2). The spatially explicit optimisation model selects which options can be attractive from an economical and environmental perspective, where to locate them to take advantage of local feedstock, and sizing to meet local demands as well as the creation of distribution networks.

The year 2016 is taken as base year, simulating 5 year intervals until 2036 which is the target year for the regional plan (NSW Government, 2018). The population is characterised by Sex, Age, Employment status, Income status, and annual population growth rate for this period is assumed to be the same level as in 2016. Demand data from 2016 was used as baseline, with the growth related to the population growth in the different regions from Figure 1. For modelling the current infrastructure networks, existing technologies for energy, waste and water, and for waste produced (municipal solid waste, agricultural waste, and waste water) a rich selection of local sources was used, see Table 1.

The agricultural and other organic waste data is from MRA's report to Hunter Water, which identifies a total of 3.3 million tpa of organic feedstock in Lower Hunter and Central

Coast. It is understood that the spatial distribution and amounts of accessible organic waste is unclear at this time. However, since the model investigates the development of Hunter region until 2036, all feedstocks are assumed to be accessible and their spatial distribution is assumed to follow the sociodemographic growth trends in different regions.

4. Scenarios

This section describes the waste to energy scenarios explored in this project using the systems model for the Hunter region described above, leading to insights on the potential contribution of waste to energy pathways in the region. Three scenarios related to waste to energy are explored:

1. Business-as-usual; a baseline scenario using the current infrastructure in the region for water and energy (electricity and gas)

2. Waste to energy scenario (WtE), in which biosolids, municipal solid wastes, agricultural wastes and other organic wastes are considered as the (co-)feedstock of renewable energy generation using anaerobic digestion (with thermal hydrolysis), incineration, gasification and pyrolysis.

3. Economic/policy scenarios, including changes feed-in-tariffs and carbon credit to study the sensitivities of waste to energy. Each scenario is considered at the current situation (based on 2016 data as a baseline) as well as forecasts at 5 year intervals until 2036 following the timeline of the 20-year regional blueprint published by the NSW government.

The Business as usual (BAU) scenario serves as a basis to analyse the changes in other scenarios. In terms of waste management, biosolids is modelled by using BAU options provided by GHD Pty Ltd (2018)'s thermal hydrolysis option comparison. The input waste streams are assumed to leave the system through recycle/recovery and landfill at different prices. The biosolids disposal cost via land application program is included in the system based on cost data provided by Hunter Water. Other than biosolids, MSW will have a landfill levy assumed to be AUD 78.20 dollar per tonne.

5. Results

The model simulation and optimization starts from year 2016 where no WtE facility is installed. With five year intervals, the model progressively suggests optimal WtE strategies up to 2036 based on both economic incentives and spatial-temporal constraints. The results are generated for biosolids, municipal solid waste, agricultural waste and other organic waste. Note that agricultural waste and other organic waste (post-consumer food waste, etc) are shown together as same technology data are used. Figure 3 shows the transportation of biosolids between the different sites in 2036 as a result of optimal allocation of WtE installations, and similar graphs can be generated specifically for e.g. municipal solid waste.

Four biosolids WtE options have been selected by the model at the timestamp of 2036. Plasma gasification with MSW as co-feedstocks (plasma), AD with thermal hydrolysis

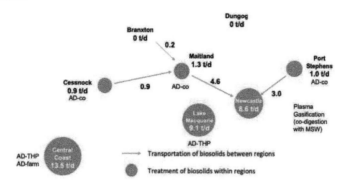

Figure 3: Treatment and transportation of biosolids between and in district (tonnes per day) after the optimisation of WtE technologies and their locations for 2036

(AD-THP) and AD with MSW and wastewater as co-feedstocks (AD-co) are favoured by the model using the objective function and defined economic and environmental performance indicators

Pyrolysis and incineration of biosolids are eliminated by the model, as they are not allocated for waste to energy conversion. The reason is presumed to be their inefficient renewable energy generation under the assumptions of the model. 16.5 GWh of energy can be generated from biosolids mainly in the form of natural gas. If all natural gas is converted to electricity, the value is 5.1 GWh, which can be used to reduce wastewater treatment energy consumption by 16%. Two centralized AD-THP are suggested by the model in Central Coast and Lake Macquarie to treat biosolids only, accounting for 70% of total biosolids in the study area. They are selected for major biosolids WtE pathway due to their relatively low costs. Plasma gasification are proposed to treat 8.6 tonnes per day biosolids together with municipal waste generated in or transported to Newcastle. The plasma plant is mainly used for municipal waste treatment, but the remaining capacity of the facility is identified by the model to be further utilised for biosolids WtE processing. AD co-digestion of MSW with biosolids are also proposed in Port Stephens, Maitland and Cessnock. The plants treat most of the MSW in the corresponding region and also handle a smaller potion of local biosolids. The remaining biosolids will be transported to Newcastle for centralized plasma treatment. The selection of an AD-co plant can be explained by the increased wastewater generation in 2036. With no wastewater reclamation facilities installed over 20 years, AD-co plants can effectively reduce the load of wastewater treatment plant and recover energy from biosolids. A farm AD plant is also allocated in Central Coast to recover energy from the remaining biosolids that will not be treated by the centralized AD-THP. This shows that the model only select decentralized facilities when centralized plants cannot treat all the waste in a region in an efficient way.

A carbon pricing scheme was introduced to Australian industrial sector from 2012 with a cost of 23 AUSD per tonne of emitted carbon dioxide, which was abolished later in 2014. The previous WtE scenario thus assumes zero carbon price in the optimization. In a further scenario, the carbon credit prices are set as alternatively 25 AUSD, 75 AUSD or 125 AUSD per tonne of CO_2 to study its effects on the three key indicators: CAPEX, OPEX and GHG emission.

6. Conclusions and final remarks

A total of 854 GWh renewable energy can be generated through WtE pathway in 2036. 94% of the generation from waste is in the form of biogas or syngas. By upgrading to natural gas and injecting into gas grid, the gas-to-electricity conversion losses can be avoided, and less energy transmission losses will incur. Biogas generation is thus considered to render more economic benefits than electricity. Decentralised technologies are optimal due to the large amount of waste distributed in each region. Organic feedstocks identified in the study region possess vast potential for WtE pathway, as the WtE energy contributed by agricultural waste in 2036 will cover about 80% of total generation.

The systems approach presented in this report can be seen as a starting point for the exploration of different visions and futures for the region, with local stakeholders at the centre of the decision making process. The data as well as simulated scenarios will enable well-informed decisions and further detailed examination of options deemed attractive at the systems level. Following a systems approach, the interdependencies between infrastructure systems and sectors will continue to be at the core of planning integrated solutions at the national, regional and local level.

Acknowledgements

The authors would like to express their thanks to Hunter Water Corporation, in particular David Derkenne, for their support and contribution to this study.

References

Aemo, 2018. Australian energy market operator website. https://www.aemo.com.au/.

Ausgrid, 2016. 2015-16 summary community electricity report. https://www.ausgrid.com.au/-/media/Documents/Data-to-share/Average-electricity-use/Ausgrid-average-electricity-consumption-by-LGA-201516.pdf.

Australian Bureau of Statistics, 2018. http://www.abs.gov.au/.

Central Coast Council, 2018. Central coast council website. https://www.centralcoast.nsw.gov.au/.

GHD Pty Ltd, April 2018. Hunter water corporation hunter water-planning technical advice adhoc thermal hydrolysis options comparison. HWC-Planning Technical Advice Adhoc, 2218668.

Government of Australia, August 2015. Intended nationally determined contribution.

S. Jain, S. Jain, I. T. Wolf, J. Lee, Y. W. Tong, 2015. A comprehensive review on operating parameters and different pretreatment methodologies for anaerobic digestion of municipal solid waste. Renewable & Sustainable Energy Reviews 52, 142–154.

Jemena, 2011. Average gas consumption. https://jemena.com.au/about/document-centre/gas/average-gas-consumption.

G. Lofrano, 2012. Green technologies for wastewater treatment: energy recovery and emerging compounds removal. Springer Science & Business Media.

MRA Consulting Group, June 2018. Market analysis of organic waste feedstocks in the lower hunter region.

NSW Government, March 2018. The NSW state infrastructure strategy 2018-2038. http://www.infrastructure.nsw.gov.au/expert-advice/state-infrastructure-strategy/.

Regional Development Australia Hunter, March 2016. Smart specialisation strategy for the hunter region: A strategy for innovation-driven growth.

Q. Sun, H. Li, J. Yan, L. Liu, Z. Yu, X. Yu, 2015. Selection of appropriate biogas upgrading technology-a review of biogas cleaning, upgrading and utilisation. Renewable & Sust. Energy Reviews 51, 521–532.

C. P. Triantafyllidis, R. H. Koppelaar, X. Wang, K. H. van Dam, N. Shah, 2018. An integrated optimisation platform for sustainable resource and infrastructure planning. Env. Mod. & Software 101, 146–168.

X. Wang, M. Guo, R. H. Koppelaar, K. H. van Dam, C. P. Triantafyllidis, N. Shah, 2018. A nexus approach for sustainable urban energy-water-waste systems planning and operation. Env. science & technology 52 (5), 3257–3266.

Anton A. Kiss, Edwin Zondervan, Richard Lakerveld, Leyla Özkan (Eds.)
Proceedings of the 29th European Symposium on Computer Aided Process Engineering
June 16th to 19th, 2019, Eindhoven, The Netherlands. © 2019 Elsevier B.V. All rights reserved.
http://dx.doi.org/10.1016/B978-0-128-18634-3.50296-4

Optimal Oversizing and Operation of the Switchable Chlor-Alkali Electrolyzer for Demand Side Management

Kosan Roh[a], Luisa C. Brée[a], Karen Perrey[b], Andreas Bulan[b] and Alexander Mitsos[a,*]

[a]*AVT.SVT, RWTH Aachen University, Forkenbeckstraße 51, Aachen 52074, Germany*
[b]*Covestro Deutschland AG, Leverkusen 51365, Germany*
alexander.mitsos@avt.rwth-aachen.de

Abstract

Flexible operation of the switchable chlor-alkali process is a novel strategy for minimizing electricity costs in markets with volatile prices (Brée et al. (2018)). This strategy allows for adjusting the process to electricity price profiles, by varying the production rate and by switching the operation between two different modes: the H_2 mode and O_2 mode. The size of the electrolyzer is a crucial factor because oversizing on the one hand leads to higher operational flexibility but on the other hand to increased capital costs. Here, we combine the optimization of the operation and electrolyzer size in the switchable chlor-alkali process for given electricity price profiles. To arrive at the optimum in reasonable computation time, we use a piecewise linear approximation and decomposition method employing the golden-section search. We verify that electrolyzer oversizing in the switchable chlor-alkali process is a suitable strategy for reducing the total production cost when the specific capital costs of the oversizing are low enough and the electricity price strongly fluctuates.

Keywords: demand side management, bifunctional cathode, chlor-alkali electrolysis, decomposition method, MILP

1. Introduction

As intermittent renewable energy penetrates the power grid, electricity prices and availability fluctuate more. A need could arise for power-intensive processes like chlor-alkali (CA) electrolysis to employ smart operation strategies. One possible strategy is demand side management (DSM), which allows for flexible operation of the process to adjust the production rate to electricity price profiles (Daryanian et al. (1989); Ghobeity and Mitsos (2010)). The production rate can be varied when an electrolyzer is oversized; and intermediates are storable or the downstream processes are also operated flexibly. Another DSM strategy for CA electrolysis is to switch its operation mode by implementing a bifunctional cathode. This novel cathode (Bulan et al. (2017)) enables to switch the operation between hydrogen (H_2) and oxygen (O_2) modes. The O_2 mode demands less electricity than the H_2 mode at the expense of not producing H_2.

Brée et al. (2018) examined the optimal operation of an oversized CA electrolyzer with the bifunctional cathode implemented. The operation was optimized by solving a mixed-

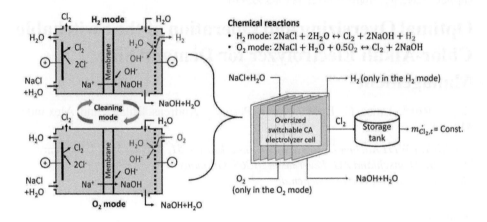

Figure 1: Description of the electrolysis and operation of the switchable CA process.

integer linear program (MILP) that minimized operating costs on the basis of a discrete-time model. The results verified that oversizing the electrolyzer can reduce operating costs when electricity prices fluctuate sharply. Moreover, additional savings are achievable by adopting the switchable operation. When electricity is expensive and H_2 and O_2 are cheap, operation in the O_2 mode is economically preferred.

Oversizing the electrolyzer, however, affects the total production cost of the CA process. It requires extra capital cost, which reduces the profitability. Therefore, the electrolyzer oversizing and reduction of operating costs are at a trade-off (Williams et al. (2012)).

Herein, we examine how electrolyzer oversizing affects the optimal operation of the switchable CA process by modifying the optimization problem formulated by Brée et al. (2018). A sizing factor is introduced as a continuous variable. The modified optimization problem minimizes the total production cost that is the sum of the operating cost and annualized additional capital cost resulting from oversizing the electrolyzer. The modification results in a large-scale mixed-integer nonlinear program (MINLP). Its solution to global optimality is computationally demanding. To reduce the computation cost, we consider the sizing factor as a complicating variable and utilize a decomposition method. At the outer, the golden-section search (GSS) technique is applied to optimize the sizing factor while the subproblems are MILPs. This strategy gives the global optimum in tractable computational time.

2. Description of Operation Strategies

Two operation strategies are applied to the switchable CA process. One is designed to vary the production rate while the other aims to switch the operation between the H_2 and O_2 modes. Figure 1 illustrates the operation scheme.

2.1. Production Rate Variation

Conventional CA electrolyzers in Germany produce chlorine (Cl_2) typically at very high utilization capacity (>95 %), so there is only little potential for flexible operation (Ausfelder et al. (2018)). When a CA electrolyzer is oversized, i.e., its production capacity exceeds the demand in the downstream process, it is possible to adjust the production rate to electricity price fluctuations by changing the voltage applied. When electricity is expensive, the production rate is reduced so that less electricity is consumed. A Cl_2 storage tank is installed as a buffer to guarantee the constant Cl_2 supply to the downstream, that is assumed to operate as is. If necessary, the process can be shut down.

2.2. Switchable Operation

In the H_2 mode, the anode compartment is filled with saturated brine to produce Cl_2 gas and sodium ions (Na^+). Na^+ ions migrate through an ion-exchange membrane, and then sodium hydroxide (NaOH) is produced along with H_2 gas at the cathode compartment. In the O_2 mode, the reactions are the same as in the H_2 mode at the anode side. At the cathode side, gaseous O_2 contacts the cathodic liquid phase and is reduced without H_2 production. The O_2 mode demands less electricity than the H_2 mode by 33 % (Jung et al. (2014)) while still producing the same amount of Cl_2 and NaOH. However, operation in the O_2 mode no longer generates the H_2 by-product while it requires additional O_2. To avoid the presence of both gases simultaneously, a cleaning process that removes either H_2 or O_2 inside the cell is required when the operation is switched.

3. Problem Formulation and Algorithm for Optimization

3.1. Problem Formulation

In this optimization, the electrolyzer size and the operation of the switchable CA process (the active operation mode and production rate) are simultaneously optimized. A sizing factor (α) is considered as a continuous variable with given bounds as $1 \leq \alpha \leq \alpha^{up}$. For example, α of 1.3 indicates the production capacity is increased by 30 %.

Mass Balance and Mode Transition Constraints to calculate the consumption and production rates ($\dot{m}_{C,t}$) of each component C and control the mode transition and level of the Cl_2 storage tank are the same as Brée et al. (2018) except for the changes described in the following. The CA process can operate only at of three modes $m \in \{H_2, O_2, \text{Shutdown}\}$. Once the process is shut down, either the H_2 or O_2 mode can be active after a certain duration for cleaning. The production rate during the operation is limited by the equation

$$\alpha \underline{\dot{m}}_{Cl2} y_{m,t} \leq \dot{m}_{Cl2,m,t} \leq \alpha \overline{\dot{m}}_{Cl2} y_{m,t}, \quad \forall m \in M, t \in T, \tag{1}$$

where $\overline{\dot{m}}_{Cl2}$ and $\underline{\dot{m}}_{Cl2}$ denote the maximum and minimum active production rates, respectively. When mode m is active at time t, a binary variable $y_{m,t}$ is 1. The production rate is kept above 50 % of the production capacity to ensure the product quality, but it cannot exceed the production capacity. Both the lower and upper bounds are proportional to α.

Power Demand Function The power demand for the reactive modes, calculated as

$$P_t = \frac{a}{\alpha} \dot{m}_{Cl2,t}^2 + b \dot{m}_{Cl2,t}, \quad t \in T, \tag{2}$$

is a quadratic function of the Cl_2 production rate. The sizing factor influences the power demand function. Here, a and b are parameters. To formulate an MILP, we perform a piecewise linear approximation by dividing the Cl_2 production rate into four ranges:

$$P'_{t,e} = a'_e \dot{m}_{Cl2,t} + b'_e \alpha + c'_e, \quad e \in \{1,2,3,4\}, \quad t \in T, \tag{3}$$

where a'_e, b'_e, and c'_e are obtained by parameter estimation from the original power demand P_t calculated.

Ramping Constraints A ramping constraint is imposed on the production rate by

$$\Delta \dot{m}_{Cl2} = \alpha \left(\overline{\dot{m}}_{Cl2} - \underline{\dot{m}}_{Cl2} \right) / \theta_{ramp}, \tag{4}$$

where θ_{ramp} denotes the number of time steps necessary to ramp the production between $\overline{\dot{m}}_{Cl2}$ and $\underline{\dot{m}}_{Cl2}$. The maximum and minimum allowable production rate of the subsequent time step $(t+1)$ depends via

$$\dot{m}_{Cl2,t+1} \leq \dot{m}_{Cl2,t} + \sum_m y_{m,t} \Delta \dot{m}_{Cl2} + \left(1 - \sum_m y_{m,t}\right) \alpha \underline{\dot{m}}_{Cl2}, \quad \forall m \in \{H_2, O_2\}, \quad t \in T, \tag{5}$$

$$\dot{m}_{Cl2,t+1} \geq \dot{m}_{Cl2,t} - \sum_m y_{m,t+1} \Delta \dot{m}_{Cl2} - \left(1 - \sum_m y_{m,t+1}\right) \alpha \underline{\dot{m}}_{Cl2}, \quad \forall m \in \{H_2, O_2\}, \quad t \in T. \tag{6}$$

Objective Function We minimize the total production cost (C^{TO}) calculated by

$$C^{TO} = C^{OP} + C^{CAP} = \sum_{t=0}^{t_n} (C_{C,t} + C_{El,t}) + (\alpha - 1) a^{CAP} f^{CAP}, \quad \forall C \in \{H_2O, O_2, H_2\}, \tag{7}$$

which is the sum of the operating cost (C^{OP}) and annualized capital cost (C^{CAP}). Inclusion of the H_2 cost indicates that H_2 should be supplemented when the O_2 mode is active. The annualized capital cost for the oversized portion of the electrolyzer is a function of α and a coefficient a^{CAP}. A 1-year depreciation factor f^{CAP} is calculated by

$$f^{CAP} = (i(1+i)^n)/((1+i)^n - 1), \tag{8}$$

where i and n denote an interest rate and electrolyzer lifetime, respectively.

3.2. Optimization Algorithm

The inclusion of the sizing factor results in an MINLP problem due to bilinear products (e.g., $\alpha y_{m,t}$) that demands heavy computation costs. Note that since the bilinear products all involve a binary variable, the MINLP can be exactly reformulated as an MILP, e.g., using the big-M method. Our computational experimentation, however, showed that these MILPs are still expensive to solve. To reduce the load, we apply a decomposition method, considering the sizing factor as a complicating variable. This leads to an MILP subproblem. The global solution is then found by the GSS on the complicating variable.

4. Simulation Results

The nominal plant capacity $(\alpha = 1)$ is 1 t_{Cl2}/h. The storage tank can store the nominal Cl_2 supply to the downstream for a maximum of 3 hours. The time horizon T for simulation

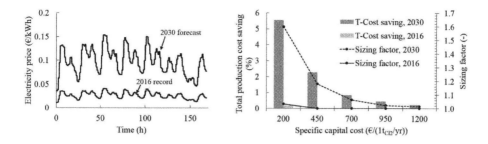

Figure 2: Electricity price profiles in 2016 and 2030 in Germany (left) and simulation results; savings in the total production cost and the optimal sizing factor (right).

is one week and discretized into 672 time steps (15 minutes for one time step). The maximum oversizing is up to 100 % ($\alpha^{up} = 2$). The specific capital cost for oversizing the electrolyzer highly depends on the individual settings of the pre-existing site. To calculate the annualized capital cost, a specific capital cost of €200 up to €1,200 per yearly production of 1 t_{Cl2} (Arnold et al. (2016)), 7 % of the interest rate, and 10 years of the electrolyzer lifetime are assumed. Two electricity price profiles recorded in first week of June in 2016 (Agora Energiewende (2018)) and assumed for 2030 by Brée et al. (2018) are considered (Figure 2, left). The latter one with the higher average price and more pronounced fluctuation reflects the prediction that employment of renewable energy sources in the German power sector will increase in 2030. H_2, O_2, and water (H_2O) prices are 2 €/kg_{H2}, 0.05 €/kg_{O2}, and 0.01 €/kg_{H2O}, respectively. The shutdown mode is active for at least 30 minutes. Once a reactive mode becomes active, such a mode lasts for at least 12 hours to prevent shortening the electrolyzer lifetime. The optimization problem implemented in GAMS 24.9.2 is solved by CPLEX 12.7.1.0 with a relative optimality tolerance of 10^{-5} and 12 threads for parallel processing.

Figure 2 (right) shows that, the specific capital cost for oversizing the electrolyzer highly influences the optimal size of the electrolyzer in the switchable CA process as well as the reduction of the total production cost compared to the reference case with no oversizing and operation switch. The lower specific capital cost leads to the larger electrolyzer and higher savings in the total production cost accordingly. The optimal size of the elec-

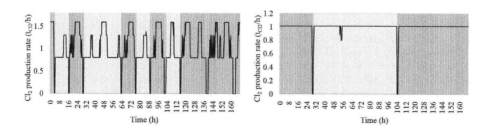

Figure 3: Optimal Cl_2 production rate profile and active mode for the specific capital cost of 200 (left, α=1.6) and 1,200 €/(1t_{Cl2}/yr) (right, α=1.02) in the 2030 scenario. Blue indicates the H_2 mode, yellow the O_2 mode, and orange the Shutdown mode.

trolyzer in the 2030 scenario is far bigger than in the 2016 scenario. When the electricity price fluctuates sharply, varying the production rate becomes advantageous. Unless the specific capital cost is very low, the oversizing is not recommended in the 2016 scenario.

Figure 3 represents the optimal Cl_2 production rate profile and active mode of the switchable CA process in the case where the specific capital cost for oversizing the electrolyzer is 200 €/($1t_{Cl2}$/yr) and 1,200 €/($1t_{Cl2}$/yr) on the basis of the 2030 scenario. More frequent shutdown of the process is observed with the lower specific capital cost: 10 times in the 200 €/($1t_{Cl2}$/yr) case and only twice in the 1,200 €/($1t_{Cl2}$/yr) case in a week. The larger electrolyzer provides more flexibility for the operation switch.

5. Conclusions

We investigated the influence of the capital cost for oversizing the electrolyzer in the switchable CA process on its optimal operation for DSM concerning electricity price profiles. We formulated an MINLP problem to minimize the sum of the operating and annualized capital costs while optimizing the electrolyzer size and operations. To find the global solution within reasonable computation time, we applied a decomposition method employing the GSS. As a result, it is noted that the specific capital cost for oversizing the electrolyzer highly affects the optimal operation. The cheaper the specific capital cost, the more frequent the plant is shut down and the higher the savings in the total production cost are. Finally, we verify that oversizing is a suitable strategy for DSM when the electricity price fluctuates strongly and moderate storage of Cl_2 (order of three hours) is allowed.

Acknowledgement

The authors gratefully acknowledge the financial support of the Kopernikus-project SynErgie by the Federal Ministry of Education and Research (BMBF) and the project supervision by the project management organization Projektträger Jülich (PtJ). We are also grateful to Pascal Schäfer, Kristina Baitalow, Hatim Djelassi, and Susanne Sass for valuable discussions.

References

Agora Energiewende, 2018. Agorameter. Accessed on 2018-10-31.
    URL https://www.agora-energiewende.de/

K. Arnold, T. Janßen, L. Echternacht, S. Höller, T. Voss, K. Perrey, 2016. Flexind - flexibilisation of industries enables sustainable energy systems.

F. Ausfelder, A. Seitz, S. v. Roon, 2018. Flexibilitätsoptionen in der grundstoffindustrie : Methodik, potenziale, hemmnisse. Tech. rep., DECHEMA, Frankfurt am Main.

L. C. Brée, K. Perrey, A. Bulan, A. Mitsos, 2018. Demand side management and operational mode switching in chlorine production. AIChE Journal, in press. DOI: 10.1002/aic.16352.

A. Bulan, R. Weber, F. Bienen, Oct 2017. Bifunktionelle elektrode und elektrolysevorrichtung für die chloralkali-elektrolyse. WO2017174563A1.

B. Daryanian, R. E. Bohn, R. D. Tabors, Aug 1989. Optimal demand-side response to electricity spot prices for storage-type customers. IEEE Transactions on Power Systems 4 (3), 897–903.

A. Ghobeity, A. Mitsos, 2010. Optimal time-dependent operation of seawater reverse osmosis. Desalination 263 (1), 76 – 88.

J. Jung, S. Postels, A. Bardow, 2014. Cleaner chlorine production using oxygen depolarized cathodes? a life cycle assessment. Journal of Cleaner Production 80, 46 – 56.

C. M. Williams, A. Ghobeity, A. J. Pak, A. Mitsos, 2012. Simultaneous optimization of size and short-term operation for an RO plant. Desalination 301, 42–52.

Anton A. Kiss, Edwin Zondervan, Richard Lakerveld, Leyla Özkan (Eds.)
Proceedings of the 29th European Symposium on Computer Aided Process Engineering
June 16th to 19th, 2019, Eindhoven, The Netherlands.
http://dx.doi.org/10.1016/B978-0-128-18634-3.50297-6

The use of optimization tools for the Hydrogen Circular Economy

M. Yáñez,[a] A. Ortiz,[a] B. Brunaud,[b] I.E. Grossmann,[b] I. Ortiz [a,*]

[a] Chemical and Biomolecular Engineering Department, University of Cantabria, Av. los Castros s/n, 39005, Santander, Spain

[b] Carnegie Mellon University, 15213, Pittsburgh, Pennsylvania, USA

ortizi@unican.es

Abstract

Hydrogen losses in industrial waste gas streams, estimated as 10 billion Nm^3 per year in Europe, constitute potential sources for hydrogen recovery. Further use of this waste hydrogen in fuelled devices promotes a world powered by hydrogen while reinforcing the paradigm of the Circular Economy. This will require the availability of effective technologies for hydrogen recovery and purification as well as the techno-economic assessment of integrating the upcycled gas into a sustainable supply chain using decision-making tools. This work using a mixed-integer programming model (MILP), and scenario analyses, develops the techno-economic modelling over the 2020-2050 period, at a regional scale comprising the north of Spain. Two main industrial waste streams are considered; one produced at integrated steel mills and coke making industries and, the second one, at chlor-alkali plants. The proposed optimization model integrates the following items: i) technology selection and operation, ii) hydrogen demand forecast, iii) geographical information, iv) capital investment models, and v) economic models. The optimal solutions that arise from the combination of all the infrastructure elements into the mathematical formulation, define the gradual infrastructure investments over time that are required for the transition towards a sustainable future energy mix, including surplus hydrogen. The results confirm that the critical factor in the configuration of the proposed energy system is the availability of industrial surplus hydrogen, contrary to conventional hydrogen energy systems that are mainly controlled by hydrogen demand. Additionally, this study confirms that the optimal levelized cost of upcycled hydrogen is in the range of 0.35 to 1.09 €/kg H_2, which is 1.5 to 2 times lower than the price of hydrogen obtained by steam conversion of natural gas.

Keywords: Hydrogen recovery, surplus hydrogen, MILP optimization model, hydrogen infrastructure

1. Introduction

Hydrogen-based energy storage systems could play in the future a key role as a bridge between intermittent electricity provided by alternative sources and the common fossil fuel-based energy system. The versatility and unique properties of hydrogen open the way to accomplish this goal. Although in recent years, the prospects of a shift to a hydrogen economy have created great interest in the scientific community and social stakeholders, the success relies on the availability of the necessary infrastructures. A number of works focused on the use of decision-support tools for the design and operation of hydrogen supply chains (HSC), have been reported addressing questions such as the design of the hydrogen fuel infrastructure applied at country, region and city levels [1–3].

Likewise, industrial waste streams with a hydrogen content higher than 50% are considered potential and promising sources for hydrogen recovery using cost-effective separation techniques [4]. Meanwhile, among the list of hydrogen-containing waste streams, some studies concentrated on the management, optimization, and utilization of steel-work off gases in integrated iron and steel plants [5,6]. However, little is focused on the optimization of various by-product gases to embed sustainability into HSC. To the best of our knowledge, reported optimization models for HSCs do not consider the competitiveness of upcycling hydrogen-rich waste gas sources for its reuse in both transportation and residential sectors. Hence, the originality of this study is to report the techno-economic feasibility of a HSC with contribution of upcycled hydrogen-rich waste gas sources to fuel both stationary and road transport applications [7].

2. Methodology

The methodology framework of this work is proposed in Figure 1. For that purpose, a mathematical formulation with the objective of maximizing the net present value (NPV) is proposed. The corresponding problem is stated as follows. Given:

- the potential sources for hydrogen recovery composition and their quality;
- a set of suppliers with their corresponding time-dependent maximum supply;
- locations of the key stakeholders in the target region: suppliers, merchants, and customers;
- a set of allowed routes between the three stakeholders, the transportation mode between them, the delivery distance between both routes; supplier-to-merchant and merchant-to-customer;
- hydrogen demand forecast by customer for both transport and residential sectors;
- raw material and product prices;
- a set of production, purification and conditioning technologies, and their yields to upgrade raw materials to hydrogen product, as well as their capacity at different scales;
- investment and operating costs of each intermediate technology, transportation mode, depreciation, and the residual values at the end of the time horizon;
- financial data (such as discount and tax rates).

The goal of the proposed model is to provide the optimum answer to the following questions: how much, where and when stakeholders shall make their investments. The outputs provided by the model are:

- optimal investment plan for all the merchants considered and related logistics;
- location (single- or multiplant), type, scale, and number of installed technologies.
- sourcing and supply routes for the raw materials and product considered;
- connections between the stakeholders, and hydrogen flows through the network.

Within the network presented, the optimization model was developed by integrating the technology selection and operation, the hydrogen demand forecast, geographical information, capital investment models, and economic models.

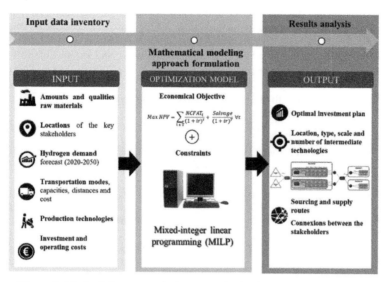

Figure 1. Methodology to optimize a sustainable hydrogen supply chain

3. Optimization model

An optimization modelling approach based on multi-scenario MILP has been developed. The mathematical model was implemented in JuMP (Julia for Mathematical Optimization) and the optimization solver used was Gurobi 7.0.2. The objective of the model is to maximize the NPV over 30 years (2020-2050). Furthermore, the operational planning model regarding plant capacity, production transportation, and mass balance relationships, is considered together with the constraints of these activities.

Because of the complexity of the proposed model, a two-stage hierarchical approach has been used in order to solve the MILP model in reasonable computational time, achieving near-optimal solutions (5% optimality gap) in less than 2 h. The first step consists of the solution of a relaxed single-period problem to determine the location of production plants at the end of the horizon. From this initial assessment, merchant companies that are not selected in the first step are eliminated. Next, in the second step, the 30-year horizon problem is solved with a reduced set of merchants. The optimality gaps have been set to 2% and 5% for the first and second step, respectively.

4. Case study: Upcycled waste gas-based HSC for the North of Spain

This work is focused on the techno-economic feasibility of the upcycling of hydrogen-containing multicomponent gas mixtures to feed stationary and portable fuel cells using optimization tools, geographically located in the north of Spain. The proposed model is focused on two main industrial waste streams.

The first hydrogen source corresponds to high purity hydrogen off gases of the chlor-alkali industry denoted as raw material R99. The second most valuable by-product considered in the optimization model, is coke oven gas (COG) which is produced at integrated steel mills and coke making industries. Hereafter, this raw material has been denoted as R50 because the average hydrogen composition is between 36-62% vol.

Moreover, the geographic distribution of the future hydrogen market comprises a number of stakeholders that correspond to the three nodes of the hydrogen network, as illustrated in Figure 2. Suppliers, which are industrial companies that produce the surplus hydrogen. Merchants, which are the major industrial gases producers and responsible of the raw material transformation into the final product. However, surplus hydrogen transformation onsite was also considered at supplier's plant sites. Finally, the potential customers that are urban areas with more than a hundred thousand inhabitants.

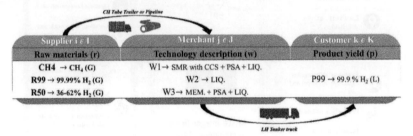

Figure 2. Waste gas streams-based HSC studied for the north of Spain

4.1. Data collection

In this study, two scenarios concerning two levels of demand for road vehicle transportation and residential/commercial sectors have been considered (see Table 1) [8]. Regarding the intermediate technologies, steam methane, CH_4, reforming (SMR) with carbon capture and storage (CCS) has been considered as benchmark technology in order to satisfy the expected demand for hydrogen [9]. With regard to the upcycling of surplus hydrogen, we have selected a combination of two of the most mature technologies for hydrogen purification: membrane technology (MEM) followed by pressure swing adsorption (PSA) [10]. The final product P99, pure liquefied hydrogen, requires a liquefaction stage (LIQ.). Each plant type incurs in fixed capital and unit production costs, as function of its capacity. Furthermore, the transportation costs depend on the selected mode and distance [11]. We considered that raw materials are transported as compressed gas hydrogen (CH2) by tube trailer or pipeline (already installed), and the final hydrogen products are shipped as liquid hydrogen (LH2) by truck. We have considered the corresponding unit transportation cost for each type of hydrogen delivery mode.

Table 1. Demand scenarios of hydrogen market penetration by end users and timeframe

	Scenario (S)	End-use (e)	2020	2030	2040	2050
Hydrogen market penetration (%)	S1	e1: Transport sector	0.7	8.4	16.6	25.3
		e2: Residential/Service sector	0.0	0.7	3.0	6.7
	S2	e1: Transport sector	1.4	16.8	33.2	50.6
		e2: Residential/Service sector	1.0	6.0	10.0	13.5
Hydrogen demand	Total S1 (tons H_2 per year)		8.9E+03	1.2E+05	2.7E+05	4.6E+05
	Total S2 (tons H_2 per year)		4.0E+04	3.4E+05	6.3E+05	9.2E+05

4.2. Results and Discussion

This section shows the main results obtained by application of the proposed multiperiod mixed-integer programming model to the optimization of the network infrastructure for the fulfilment of low hydrogen demand (S1). The solution of the model determines: i) the amount of hydrogen-rich waste streams (R50 and R99) converted into liquefied hydrogen at the supplier's plants, and ii) the optimum SMR-CCS plant site locations.

Investment Network: Case S1 leads to a solution with a NPV of 941 MM€, where the revenue derived from hydrogen sales (3370 MM€) is able to absorb the costs (2030 MM€). No single hydrogen production method will be profitable to produce enough hydrogen to fulfil the expected demand on its own. The optimal solution of the integrated surplus HSC leads to the installation of ten units of different technologies until 2050 in northern Spain: W1 (7 units), W2 (2 units) and W3 (1 unit), as shown Table 2.

Table 2. Investment plan obtained for scenario S1

j ∈ J	Location	Period	Technology	Size (ton H2/year)	Investment (MM€)
12	Asturias	2020	W3	50000	116.3
14	Huesca	2020	W2	1000	8
16	Cantabria	2020	W2	1000	8
6	Salamanca	2022	W1	200000	605.2
4	BCN (Rubí)	2028	W1	200000	605.2
7	Santander	2033	W1	200000	605.2
6	Salamanca	2037	W1	200000	605.2
5	Huesca	2041	W1	200000	605.2
6	Salamanca	2045	W1	200000	605.2
3	BCN (Parets del Vallés)	2048	W1	200000	605.2

Thus, decentralized on-site hydrogen production by the upcycling of industrial surplus hydrogen is the best choice for market uptake and for avoiding costly distribution infrastructure until the demand increases.

Surplus hydrogen flowrates: As summarized in Figure 3, in Case S1, the full amount of R99 is utilized with an inflow of 293.400 tons over the next 30 years, and R99 can meet 0.5% of the total hydrogen demand in the north of Spain for the whole time period. Whereas the amount of liquefied hydrogen produced from R50, which is 1.497.000 tons of R50, is able to cover a much larger hydrogen demand accounting for 10.1% of the total hydrogen demand. Consequently, the rest of the hydrogen produced to fulfil the total demand is obtained from CH_4 using SMR with CCS as benchmark technology while producing the least CO_2 emissions compared to the rest of the commercially available technologies.

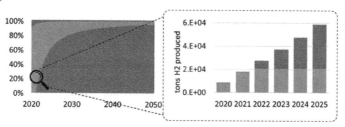

Figure 3. Share of pure H_2 produced per raw material r ∈ R; ■ CH_4; ■ R99; ■R50

However, the use of inexpensive surplus hydrogen sources may have a central role in the early phase of hydrogen infrastructure build up in the north of Spain. Therefore, industrialized hydrogen will also play an important role in initiating the transition to a hydrogen economy with localized plants of SMR with CCS; this will support the demand before expanding to less populous areas forming a more decentralized green hydrogen production. Analysing the surplus hydrogen flowrates by customer, it can be observed that though R50 is partially marketed to all final end-users, it has a pivotal contribution when the production of the final product is closer to the customers.

5. Conclusions

This research provides the methodology to assess the techno-economic feasibility of upgrading and reusing surplus hydrogen gases promoting the shift to the Circular Economy. The analysis has been performed over two scenarios of hydrogen demand (S1 and S2) and the results show that as long as both scenarios of hydrogen demand (S1 and S2) apply, all generated case studies lead to a solution with positive NPVs. The results confirm that the use of inexpensive surplus hydrogen sources such as R50 and R99 offers an economic solution to cover hydrogen demand in the very early stage of transition to the future global hydrogen-incorporated economy, especially when the industrialized hydrogen generation is closer to the demand markets.

Acknowledgments

This research was supported by the projects CTQ2015-66078-R (MINECO/FEDER, Spain) and SOE1/P1/E0293 (INTERREG SUDOE /FEDER, UE), "Energy Sustainability at the Sudoe Region: Red PEMFC-Sudoe".

References

[1]   Almansoori, A. and Shah, N. (2009) Design and operation of a future hydrogen supply chain: Multi-period model. *Int. J. Hydrogen Energy*. 34 (19),.

[2]   Ochoa Bique, A. and Zondervan, E. (2018) An outlook towards hydrogen supply chain networks in 2050 — Design of novel fuel infrastructures in Germany. *Chemical Engineering Research and Design*. 134 90–103.

[3]   Sabio, N., Gadalla, M., Jiménez, L., and Guillén-Gosálbez, G. (2010) Multi-objective optimization of a HSC for vehicle use including economic and financial risk metrics. A case study of Spain. *Computer Aided Chemical Engineering*.

[4]   Shalygin, M.G., Abramov, S.M., Netrusov, A.I., and Teplyakov, V. V. (2015) Membrane recovery of hydrogen from gaseous mixtures of biogenic and technogenic origin. *Int. J. Hydrogen Energy*. 40 (8), 3438–3451.

[5]   Lundgren, J., Ekbom, T., Hulteberg, C., Larsson, M., Grip, C.-E., Nilsson, L., et al. (2013) Methanol production from steel-work off-gases and biomass based synthesis gas. *Appl. Energy*. 112 431–439.

[6]   Chen, Q., Gu, Y., Tang, Z., Wei, W., and Sun, Y. (2018) Assessment of low-carbon iron and steel production with CO2 recycling and utilization technologies: A case study in China. *Appl. Energy*. 220 192–207.

[7]   Yáñez, M., Ortiz, A., Brunaud, B., Grossmann, I.E., and Ortiz, I. (2018) Contribution of upcycling surplus hydrogen to design a sustainable supply chain: The case study of Northern Spain. *Appl. Energy*. 231.

[8]   HyWays (2007) The European Hydrogen Energy Roadmap. *HyWays*.

[9]   Sahdai, M. (2017) Review of modelling approaches used in the HSC context for the UK. *Int. J. Hydrogen Energy*. 42 (39), 24927–24938.

[10] Alqaheem, Y., Alomair, A., Vinoba, M., and Pérez, A. (2017) Polymeric Gas-Separation Membranes for Petroleum Refining. 2017.

[11] Reuß, M., Grube, T., Robinius, M., Preuster, P., Wasserscheid, P., and Stolten, D. (2017) Seasonal storage and alternative carriers: A flexible hydrogen supply chain model. *Appl. Energy*. 200 290–302.

Anton A. Kiss, Edwin Zondervan, Richard Lakerveld, Leyla Özkan (Eds.)
Proceedings of the 29th European Symposium on Computer Aided Process Engineering
June 16th to 19th, 2019, Eindhoven, The Netherlands. © 2019 Elsevier B.V. All rights reserved.
http://dx.doi.org/10.1016/B978-0-128-18634-3.50298-8

A dynamic model for automated control of directed self-assembly of colloidal particles at low densities

Baggie W. Nyande, Yu Gao & Richard Lakerveld*

Department of Chemical and Biological Engineering, The Hong Kong University of Science and Technology, Clear Water Bay, Hong Kong S.A.R.
kelakerveld@ust.hk

Abstract

Directed self-assembly of colloidal particles is a promising route for the fabrication of advanced materials with novel properties. However, the formation of defect-free non-periodic structures is difficult to achieve without active control. Dynamic models can determine optimal input trajectories of manipulated variables during directed self-assembly. This paper presents a dynamic model for the simulation of particle trajectories in a microfluidic device in the presence of an electric field with time-varying properties, which can be used to predict the dynamic development of the particle density distribution. The model is based on first principles and has been experimentally validated at low particle densities. It has great promise to be used for model-based control of directed self-assembly of colloidal particles.

Keywords: Directed self-assembly, Model-based control, dynamic modelling, dielectrophoresis

1. Introduction

Self-assembly is defined as the spontaneous and reversible association of molecules or particles into organized structures (Whitesides and Grzybowski, 2002). The self-assembly of desired structures can be directed by external fields such as flow, magnetic or electric fields. Such directed self-assembly provides a promising bottoms-up route for the fabrication of advanced materials with micro- and nanoscale structures that may unlock novel optical, mechanical and electrical properties. The control of particle densities is necessary for the formation of defect-free non-periodic structures when specific features need to be self-assembled (Solis et al., 2010). However, controlling a local density of colloidal particles is challenging due to the stochastic nature of the particles, potential lack of local actuation and the highly nonlinear process behaviour. Controls based on free energy landscapes (Tang et al., 2016) and model-free feedback control have been implemented for the electric field assisted crystallization of colloidal particles (Jurez and Bevan, 2012) and for the defect-free alignment of colloidal particles (Gao and Lakerveld, 2018). The use of dynamic models in an open-loop or closed-loop model-based control framework offers the potential to better control the local colloidal particle density, which can ultimately minimize the occurrence of defects for non-periodic structures. However, the development and experimental validation of dynamic models that are favourable for online implementation for control of local particle density within a directed self-assembly process are in their infancy.

Directed self-assembly of micro and nano-sized particles are inherently stochastic even in the absence of measurement noise and unknown perturbations (Ulissi et al., 2013), therefore the dynamics of directed self-assembly processes are typically described with probabilistic models such as the Fokker-Planck equation (Komaee and Barton, 2016) for systems with continuous states or master equations (Lakerveld et al., 2012) for discrete-state continuous-time Markov processes. The use of master equations is limited to systems with a relatively low number of states. The full set of master equations may become computationally intractable if all states are considered due to the curse of dimensionality. In addition to the computational challenges, experimental validation of probabilistic models is difficult. Therefore, the utilization of existing dynamic models for automated control of directed self-assembly processes remains challenging.

The objective of this work is to develop and validate a dynamic model that describes the dynamic development of the particle density distribution within a microfluidic device in the presence of electric fields that can be used for control purposes. The model is deterministic and describes the prevailing dielectrophoretic and drag forces based on first principles. The model is validated with experiments. In future work, the proposed dynamic model can be extended to include noise from Brownian motion. This model can be a suitable basis for the development of novel open and closed-loop model-based control approaches for directed self-assembly of colloidal particles in microfluidic devices.

2. Approach

2.1. Microfluidic device and self-assembly system

The colloidal particles (silica, diameter of 2 μm) are suspended in DI water. The suspension is contained in a microfluidic device, which consists of two transparent electrodes (one patterned with photoresist and the other unpatterned) arranged in a parallel-plate configuration (see Fig. 1). The patterned electrode (2 cm x 2cm) is fabricated by spin-coating a 25-45 nm thick indium-tin oxide (ITO) coated glass with positive photoresist (400 nm) through standard photolithographic techniques. The two electrodes are separated by a 750 μm thick silicone rubber spacer (Grace Bio-Labs Bend, Oregon). About 23 μL suspension of silica particles is pipetted into a hole that is punched in the silicone double-adhesive rubber spacer. Electric wire leads are connected to the top and bottom electrodes. AC electric fields with a frequency between 300 kHz and 1 MHz and a peak-to-peak voltage between 2V and 10V are applied to the electrodes via a function generator (33500B, Keysight Technologies). Particles are allowed to settle in the device for about 10 minutes at the beginning of each experiment so that the effect of gravity on particle motion can be neglected. The particle assembly in the cell is continuously observed with an optical microscope (Nikon Ni-U) equipped with a digital camera (Nikon Digital Sight Qi2). Images captured by the CCD camera at a rate of 2 frame/s are analysed using image processing functions in MATLAB 2017a to obtain particle positions as a function of time in an automated fashion.

When non-uniform AC fields are applied to the microfluidic device, the particles and medium experience forces of electrical origin including dielectrophoresis (DEP) and AC electro-osmosis (ACEO). DEP is the electrically induced movement of uncharged dielectric particles in the presence of a non-uniform electric field. For two electrodes separated at a certain distance, the minimum electric field occurs at an equidistance to both electrodes which coincides with the center of the photoresist layer in our experimental setup. The maximum electric field occurs at the edges of the photoresist (Gao and Lakerveld, 2018). The direction of the particle motion is determined by the relative polarizability of the particles and the suspending medium. Particles that are more polarizable than the medium are drawn towards regions of electric field maxima (pDEP). Alternatively, particles with lower polarizability compared with the medium are directed to regions of lower electric field (nDEP). ACEO on the other hand, arises from the interaction of ions in the electrical double layer formed close to the surface of the electrodes and the tangential component of the applied electric fields. At the high AC frequencies applied in our experiments, there is insufficient time for the double layer to form, thus there is no resultant fluid flow to drive the particles (Squires and Bazant, 2004). Therefore only DEP is considered in our model development.

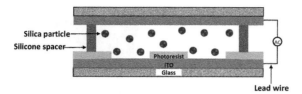

Figure 1: Schematic drawing of the microfluidic device

2.2. Model development

The dynamic trajectories of the colloidal particles in the microfluidic device in the presence of electric fields is described by a force balance over the individual particles. The translational motion of a colloidal particle in a fixed frame of reference is governed by the momentum equation:

$$m\frac{d\vec{v}}{dt} = -6\pi\eta a \vec{v} + \pi a^3 \varepsilon_m Re\left(f_{CM}\right) \nabla |E|^2. \tag{1}$$

The first term on the right-hand side of Eq. (1) is the Stokes drag force acting on the particles where $\vec{v}$ is the velocity of the particle at any given instant of time. The directed self-assembly of the colloidal particles is carried out at a frequency above the charge relaxation frequency of the ions in the electrical double layer; therefore, ACEO effects are assumed negligible. Since ACEO is not considered, the fluid velocity is assumed to be zero, a is the radius of the particle, η denotes the dynamic viscosity of the medium, m represents the mass of the particle. The effects of Brownian motion and particle-particle interactions are not considered in the model and are subject of future work.

The second term on the right-hand side of Eq. (1) is the dielectrophoretic force (Bakewell and Morgan, 2001) acting on the colloidal particles in nonuniform AC fields. The DEP force depends on the gradient of the electric field, $\nabla |E|^2$, the permittivity of the medium, ε_m and the real part of the Clausius-Mossotti function, $Re\left(f_{CM}\right)$. The Clausius-Mossotti function is defined in Eq. (2) with respect to the complex permittivity of the medium, ε_m^* and particles, ε_p^*;

$$f_{CM} = \left(\frac{\varepsilon_p^* - \varepsilon_m^*}{\varepsilon_p^* + 2\varepsilon_m^*}\right). \tag{2}$$

The complex permittivity is defined as $\varepsilon^* = \varepsilon - \left(j\sigma/\omega\right)$, where σ represents the electrical conductivity of the medium and particles, ω is the angular frequency of the electric field applied. The Clausius-Mossotti function is frequency-dependent and dictates the direction of the DEP force. For example, when the particles are more polarizable than the medium, the real part of the Clausius-Mossotti function is positive and the colloidal particles experience pDEP. Parameter values used in the simulations are obtained from literature (Yeh and Juang, 2017) and are shown in Table 1. The gradient of the electric field is obtained from the Laplace equation for electric potential, ϕ according to Eq. (3).

$$\nabla^2 \phi = 0, E = -\nabla \phi. \tag{3}$$

The Laplace equation is a mixed boundary problem which involves a boundary condition for the applied voltage on the electrode surface and a zero-flux condition in the direction normal to the electrode plane. The solution of the mixed boundary Laplace equation is computed by the finite element method using COMSOL Multiphysics to obtain the gradient of the electric field norm as function of position in the microfluidic device as shown in Fig. 2a, which completes the dynamic model.

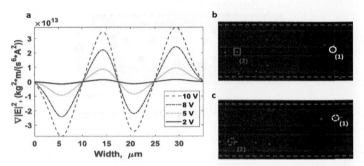

Figure 2: (a) FEM simulation of the gradient of electric field norm squared at different voltages (b) Initial positions of particles for an experiment at low particle density. (c) Particle positions after a step change from no electric field to an AC field of 1 MHz and 10 V for 32.5 s.

Table 1. Parameter values used in simulations

Parameter	ε_p	ε_m	σ_p	σ_m
Value	$3.363 \times 10^{-11} F/m$	$6.947 \times 10^{-10} F/m$	$2.56 \times 10^{-3} S/m$	$1.23 \times 10^{-2} S/m$

The model is used to simulate the dynamics of the colloidal particles in the microfluidic device with results for a step change in frequency and voltage (from 0 to 1 MHz and 0 to 10V, respectively) shown in the following section.

3. Results and Discussion

3.1. Simulation and validation of particle trajectories for a low particle density

In order to validate the predictive capabilities of the dynamic model, the model is first used to predict the trajectories of single particles in an experimental system with low particle density (see Fig. 2b), which facilitates easy particle tracking. Simulations of the single particle trajectories are performed using the initial particle positions obtained from the experiment. Under the experimental conditions, the electric-field minima for the system are located in the middle of the photoresist layer while the center of the ITO layer forms (indicated with red dashed lines in Fig. 2b and 2c) a potential barrier to particle motion. The predicted trajectories of two selected particles (indicated with circles and squares as particle (1) and (2), respectively, in Fig. 2b and 2c) obtained from simulation are shown in Fig. 3. Furthermore, the experimentally obtained particle trajectories are shown for comparison. The predicted particle trajectories from simulation are generally consistent with the experimental particle trajectories. The model slightly overestimates the response time when a step change in AC frequency and voltage is applied compared to the experimental trajectories. Nevertheless, the model predictions are satisfactory when considering that no fitting parameters have been used and the effects of Brownian motion have been neglected.

The performance of the model under pDEP conditions has also been tested, which is not shown here due to space limitations. The predicted particle trajectories when pDEP is the dominant electrokinetic phenomenon are also consistent with the experimentally observed particle trajectories. Therefore, it is concluded that the model can predict individual particles trajectories well at least in dilute systems when either nDEP or pDEP is the dominant electrokinetic phenomenon.

3.2. Simulation and validation of particle density distribution

The performance of the dynamic model at high overall particle density is investigated in this section. The particles are initially randomly distributed (see Fig. 4a) but are directed towards the

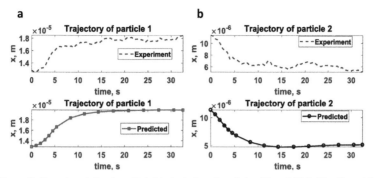

Figure 3: Experimental and predicted trajectories of particles (1) and (2) in Fig. 2b and 2c

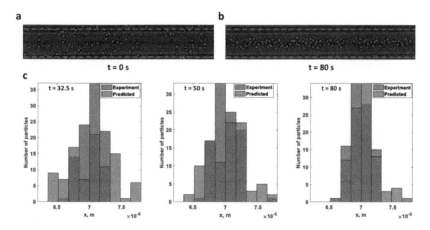

Figure 4: Particle density (a) at the start of experiment (b) when AC fields of 10 V at 1 MHz are applied for 80 s. (c) experimental and predicted particle distribution in the microfluidic device at 32.5 s, 50 s and 80 s respectively.

middle of the photoresist layer by an nDEP force acting on the particles as shown in Fig. 4b. Simulations of the particle density distribution are based on the initial particle positions of Fig. 4a. The experimental and predicted density distributions at different times are shown in Fig. 4c for comparison. It can be seen that the density distributions of the experimental system show less variation with time compared to the distributions obtained from the model. In addition, the dynamics of the model seem slower in comparison with experiments. It can also be observed that the accuracy of the model initially increases with time. However, simulations at longer times for which experimental results are not available indicate that all the particles accumulate in the center of the photoresist layer after about 400 s which is physically not possible in the experimental system because of the high overall density. The error between the model predictions and experiments may be the result of particle-particle interactions. Particle-particle interactions would have to be added to the model to predict such long-term behaviour more accurately.

4. Conclusion

A dynamic model based on the dielectrophoretic and drag forces acting on colloidal particles is presented to predict the dynamic trajectories of particles during directed self-assembly in a microfluidic device. The combined trajectories of all particles in a system can predict the dynamic development of the particle density distribution, which is of practical importance for assembling defect-free small-scale structures. The model predictions of the particle trajectories at low particle densities are in good agreement with experiments. The model is based on first principles and is computationally tractable for online implementation to develop novel control approaches for directed self-assembly. The model can be extended in future work by considering Brownian motion and particle-particle interactions from induced dipole forces (Mittal et al., 2008) and electrostatic forces that result from the particles acquiring a negative surface charge density in water. The electrostatic forces between the colloidal particles can be obtained from the DLVO theory.

Acknowledgements

The work described in this paper was supported by a grant from the Research Grants Council of the Hong Kong Special Administrative Region, People's Republic of China (Project No. 16214617).

References

D. J. Bakewell, and H. Morgan, 2001. Measuring the frequency dependent polarizability of colloidal particles from dielectrophoretic collection data. IEEE Transactions on Dielectrics and Electrical Insulation, 8(3), 566-571.

Y. Gao and R. Lakerveld, 2018. Feedback control for defect-free alignment of colloidal particles. Lab on a Chip.

J.J. Jurez and M. A. Bevan, 2012. Feedback Controlled Colloidal Self-Assembly. Advanced Functional Materials 22 (18): 3833-39.

A. Komaee and P. I. Barton, 2017. Potential canals for control of nonlinear stochastic systems in the absence of state measurements. IEEE Transactions on Control Systems Technology, 25(1), pp.161-174.

R. Lakerveld, G. Stephanopoulos, and P. I. Barton, 2012. A Master-Equation Approach to Simulate Kinetic Traps during Directed Self-Assembly. Journal of Chemical Physics 136 (18).

M. Mittal, P. P. Lele, E. W. Kaler, and E. M. Furst, 2008. Polarization and interactions of colloidal particles in ac electric fields. The Journal of chemical physics, 129(6), 064513.

E. O. P. Solis, P. I. Barton, and G. Stephanopoulos, 2010. Controlled Formation of Nanostructures with Desired Geometries. 1. Robust Static Structures. Ind. Eng. Chem. Res., 7728-45.

T. M. Squires, and M. Z. Bazant, 2004. Induced-charge electro-osmosis. Journal of Fluid Mechanics 509: 217-252.

X. Tang, B. Rupp, Y. Yang, T. D. Edwards, M. A. Grover, M. A. Bevan, 2016. Optimal feedback controlled assembly of perfect crystals. ACS Nano 10 (7): 6791-98.

Z. W. Ulissi, M. S. Strano and R. D. Braatz, 2013. Control of nano and microchemical systems. Computers & Chemical Engineering, 51, pp.149-156.

G. M. Whitesides, and B. Grzybowski, 2002. Self-assembly at all scales. Science 295 (5564): 2418-2421.

C. K. Yeh and J. Y. Juang, 2017. Dimensional analysis and prediction of dielectrophoretic crossover frequency of spherical particles. AIP Advances, 7(6), 065304.

Anton A. Kiss, Edwin Zondervan, Richard Lakerveld, Leyla Özkan (Eds.)
Proceedings of the 29th European Symposium on Computer Aided Process Engineering
June 16th to 19th, 2019, Eindhoven, The Netherlands. © 2019 Elsevier B.V. All rights reserved.
http://dx.doi.org/10.1016/B978-0-128-18634-3.50299-X

Structured millichannel multiphase reactors

J. Ruud van Ommen[*], John Nijenhuis, Johan T. Padding

Delft University of Technology, TU Delft Process Technology Institute & e-Refinery, Delft, the Netherlands

j.r.vanOmmen@tudelft.nl

Abstract

This paper examines the use of millichannels in designing multiphase reactors. We give a general analysis based on dimensionless numbers, followed by an illustration of this approach to catalytic packed-bed reactors. Finally, we discuss how such reactors can play an important role of increasing the number of electrocatalytic processes in the chemical industry. The increased degrees of freedom gives these reactors clear advantages over thermocatalytic non-structured reactors, as well as over electrocatalytic processes.

Keywords: electrocatalysis, fixed beds, degrees of freedom, reactor engineering

1. Introduction

The classical approach of randomly organized catalyst beds (either fixed or fluidized) is often outperformed by carefully designed channels or structures in which the reactions take place (Gascon et al., 2015). In recent years, microreactors have received enormous attention in academic research because they offer excellent mass and heat transfer performance, and provide a powerful tool for process intensification (Yao et al., 2015). However, their sub-mm channel diameters typically imply laminar flow, which is for many applications not optimal. Other drawbacks are high pressure drop and tendency to fouling. Moreover, a huge number of parallel channels or devices is needed to reach reasonable production volumes. Millireactors – reactors with structures and features in the mm range rather than the sub-mm range – have the potential to combine the best of both worlds: imposing a structure that enhances heat and mass transfer, while their somewhat larger feature size reduces capital and operating costs.

In this study, we explore the benefits and limitations of applying millichannel-based systems for multiphase reactions. We give a brief literature overview, and present a few cases. The interaction between reactor and catalyst will be discussed, as well as the interplay between chemistry and flow behavior. For example, we will show how filling millichannels with small particles leads to partially structured fixed beds, with attractive performance concerning mass transfer, heat transfer, and pressure drop. We will also discuss how millichannel reactors can contribute to the transition to electrochemical processes, in which the presence of sufficient electrode surface together with the ability to removed produced gases will be key. We conclude that millireactors can provide an attractive option for a range of applications in the chemical and pharmaceutical industry, giving a good compromise between physiochemical performance and cost-effectiveness.

2. Advantages of millistructured reactors

The reason why millistructured reactors offer advantages over microstructured reactors can be appreciated by estimating magnitudes of key dimensionless numbers for two

Table 1. Important dimensionless numbers for flow reactors. We use the hydraulic diameter d_h as characteristic length scale. For channels filled with particles, replace d_h by d_p.

Dimensionless number
$Re = \rho U d_h/\mu$
$Nu = \alpha d_h/\lambda = Nu(Re,Pr)$
$Sh = h d_h/D = Sh(Re,Sc)$
$Pr = c_p\mu/\lambda$
$Sc = \mu/(\rho D)$

archetypical systems: (1) An empty flow reactor of hydraulic diameter d_h and length L. (2) A flow reactor of the same dimensions, but filled with catalytic particles of diameter d_p, leading to an average porosity ε. We will consider flow of two types of fluids through these reactors, one gas-like and one liquid-like, for which we will use the physical properties of air and water at 20 ^{0}C temperature and 1 atm pressure.

A significant enhancement of mixing by convective flow is commonly experienced for sufficiently high Reynolds number Re (dimensionless numbers are defined in Table 1). Similarly, convective flows enhance the heat transfer coefficient h. A significant enhancement of heat transfer is possible for Re > 100. Similar conclusions may be drawn for mass transfer, with the roles of Nu and Pr replaced by the Sherwood and Schmidt numbers Sh and Sc (Hayhurst, 2000).

The differential pressure drop necessary to reach significant convective improvement of mixing and heat and mass transfer becomes prohibitive for small channels. For example, for an empty structured channel of hydraulic diameter d_h, the pressure drop associated with wall friction can be described by the Darcy-Weisbach equation (Romeo et al., 2002):

$$\frac{\Delta p}{L} = f\frac{\rho\, U^2}{2\, d_h} \tag{1}$$

where f is the dimensionless friction factor. For a smooth straight channel and Re < 10^3, $f = 64/Re$. For other channels and higher Re, f is larger and depends on the roughness characteristics of the channel. Fig. 1 shows $\Delta p/L$ as a function of d_h for liquid-like (black solid line) and gas-like (black dashed line) fluids at Re = 100, assuming $f = 64/Re$. At this *fixed* Re, U decreases like $1/d_h$, leading to $1/d_h^3$ scaling of the required pressure drop: for $d_h = 0.1$ mm, the pressure drop to reach Re=100 is already substantial, approximately 10 bar/m, while for $d_h = 1$ mm the necessary pressure drop is a more manageable: 0.01 bar/m.

For a channel filled with catalytic particles of diameter d_p, the differential pressure drop can be described by the Ergun equation (Ergun, 1952):

$$\frac{\Delta p}{L} = 150\frac{(1-\varepsilon)^2}{\varepsilon^3}\frac{\mu U}{d_p^2} + 1.75\frac{1-\varepsilon}{\varepsilon^3}\frac{\rho U^2}{d_p} \tag{2}$$

The first part is dominant in the laminar regime Re < 1, the second part in the turbulent regime Re > 10^3. Figure 1 shows $\Delta p/L$ as a function of d_p for liquid-like (red solid line) and gas-like (red dashed line) fluids at Re = 100 and ε=0.5. At *fixed* Re, *both* laminar and turbulent terms scale like $1/d_p^3$. Under these conditions, for $d_p = 0.1$ mm, the pressure drop necessary to reach Re=100 is enormous, approximately 1000 bar/m, while for $d_p = 1$ mm, the necessary pressure drop is much more reasonable, approximately 1 bar/m.

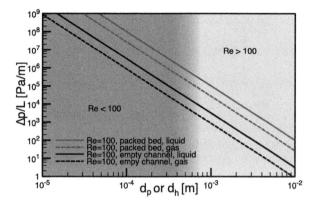

Figure 1. Pressure drop necessary to reach Re=100 as a function of hydraulic or particle diameter. Red: packed bed ($\varepsilon = 0.5$), black: empty channel. Solid lines: liquid-like fluid, dashed lines: gas-like fluid. Blue: microchannel regime, yellow: millichannel regime.

Pressure drop is not the only important consideration. Smaller reactors are also more prone to contaminants and agglomerates getting stuck in corners of the channel or pore space. Moreover, it is important to tune the residence time in the channel to optimize reaction yield or selectivity. The average residence time is $\tau = \varepsilon L/U$. At fixed Re, this leads to $\tau/L = \rho d_h/(\mu Re)$ or $\tau/L = \rho \varepsilon d_p/(\mu Re)$, for the two respective systems. In the regime where inertial flow enhancement becomes relevant (Re>100), reaching sufficiently *long* residence times requires sufficiently wide channels or sufficiently large particles.

3. Filling channels with particles

Although we have highlighted a number of clear advantages in the previous paragraph, a concern with applying structured reactors for catalytic applications is often that the catalyst holdup (i.e., the volumetric fraction of catalyst) is typically much lower than in randomly packed beds (Vervloet et al., 2013). Even if one takes into account that the catalyst effectiveness in packed beds is often below 1, still the productivity per reactor volume will be typically lower. A hybrid form of the two approaches can yield us the best of both worlds: using a system of structured channels, that are filled with particles (Calis et al., 2001, Dautzenberg and Angevine, 2004). The advantage of filling channels with particles that are of the same order of magnitude as the channel diameter is that the particle packing also becomes regular; see Fig. 2. In this way, we obtain a porosity that is importantly lower than the one for a randomly packed bed, which is 0.36 (Torquato et al., 2000). This has the advantage of a much lower pressure drop at a given particle size and fluid velocity. Alternatively, one can use higher fluid velocities and/or smaller particles to increase catalyst effectiveness and productivity of the system.

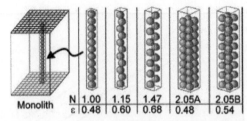

Figure 2. Packings obtained by filling small channels with particles. The aspect ratio (*N*) and porosity (ε) are given. Please note that for certain values of *N*, such as 2.05, multiple arrangements are possible (From: (Romkes et al., 2003), © Elsevier).

When the ratio of tube diameter over particle diameter becomes around 3 or larger, then the chance of getting a regular packing like in Fig. 1 strongly reduces (Calis et al., 2001). However, even in this case that are still advantages compared to regular packed beds with N typically about 104. At the wall of a packed bed, the porosity is higher than in the centre of a bed. For a channel with N around 10, the wall region covers a large part of the bed, leading to an overall porosity that is importantly lower than 0.36. Again, this has the advantage of lower pressure drop.

Vervloet et al. (2013) showed that for channels with a triangular cross-section of 7.7 mm, filled with particles of 2 mm (N=3.8), no regular packing was obtained. Still, the porosity was 0.50 (as opposed to 0.38 which was experimentally obtained for a randomly packed bed in the same study). Moreover, the heat transfer coefficient for transfer from fluid to wall (especially important for strongly exothermal reactions) was double that of a packed bed, and comparable to that of structured packings with empty channels. This shows that filling structures with particles can lead to large advantages in terms of pressure drop and heat transfer, and by analogy also mass transfer. An additional practical advantage is that one can still use the same catalyst particles as in conventional packed beds, i.e., no novel catalyst preparation procedure is required. particles.

4. Use in electrochemical processes

There is a broad consensus that the chemical industry will have to make a transition to renewable feedstocks. An important route will be to use electrochemical processes, driven by electricity from sustainable sources, to produce hydrogen from water, hydrocarbons from CO_2 and water, and ammonia from nitrogen and water. This will require novel reactor designs, with special attention to the structured design of electrodes. Although there is progress in electrochemical flow reactors, scalable macro- and micro-reaction environments need to be developed involving e.g. gas-liquid electrolyte flow (Walsh and Ponce de León, 2018).

In gas producing electrochemical cells a gas-liquid mixture is produced which impacts on energy consumption in terms of increased Ohmic resistance in the electrolyte. This is largely due to bubbles which hinder the mass transport of ionic species or block the electrode surface, and thereby reduce the e ective electrode reaction sites i.e. by sticking to the electrode and/or forming a bubble curtain (Scott, 2018).

Traditionally, parallel plate structures are used in a wide range of electrochemical synthesis applications, where gas diffusion electrodes (GDEs) and Membrane Electrode Assemblies (MEAs) are used to reduce mass transfer limitations (Merino-Garcia et al.,

2016). Gas evolution and bubble formation at electrodes plus the coalescence of these bubbles leads to the formation of a heterogeneous electrolyte system with low conductivity, higher resistance and higher cell voltage, with corresponding electrical energy losses. The temporary or permanent occupancy of the electrode surface or the electrolyte volume by gas bubbles leads to a reduced performance (Ziogas et al., 2009). Bubble disengagement is therefore key.

In electrochemical rotating cylinder reactors, a high flow rate is induced such that the electrochemical process is enhanced by convective transport. In order to obtain the mass and charge transfer from the solution to the surface of the electrodes, an optimum flow field is essential (Tomasoni et al., 2007). Nonetheless, enhancement in mass transfer to a rotating cylinder electrode depends to a large extent upon micro turbulence at the electrode surface and is at the expense of high shear losses (Scott, 2018). Other forms of flow or turbulence promotors are also used to improve gas bubble disengagement. In general, it is observed that Ohmic voltage losses and overpotentials can only be reduced to a limited extent due to practical limitations of e.g. a higher pressure drop. Gas evolution reactions occur in the three-phase (gas–liquid–solid) interface formed at the electrode when a potential is applied with a magnitude greater than the equilibrium potential. The nucleation, growth and detachment of bubbles are affected by surface tensions of gas–liquid, gas–solid and liquid–solid (Scott, 2018).

The use of fluid flow to promote greater electrochemical reaction rates by increased mass transfer to surfaces is well known and commonly practiced in flow electrolysers. However, the approach has limited success since the scalability of these microflow systems is an issue, and has not led to a compact scalable cell design, as in flat plate electrode stacks. Since bubbles are transported more easily through millichannels than through microchannels, Taylor flow in millichannel electrochemical cells would be beneficial since; a) the bubbles naturally drive the flow, b) a still narrow channel inhibits coalescence of bubbles, c) it leads to more homogeneous behaviour of the electrolyte, and d) the flow induces transportation of products away from the reaction surface. Furthermore, millichannel Taylor flow systems could lead to better scalability and a compact cell design.

The electrodes in electrochemical reactors automatically introduce a certain degree of structuring; this can be combined with other functions of the structure. Moreover, introducing structure gives more degrees freedom, and thus the possibility to solve potentially conflicting design criteria.

5. Conclusions

Millichannels have clear advantages over traditional, non-structured reactors, both in thermocatalytic and electrocatalytic processes. Analysis based on dimensionless numbers shows that the use of millichannels has important advantages over microchannels. The potential drawback of too low catalyst holdup can be overcome by filling the channels with particles. Also in electrocatalytic processes millichannels reactors can play an important role: the walls can act as electrode, while the channel configuration prevents problems with gas bubbles encountered in other types of electrocatalytic reactors.

References

Calis, H. P. A., Nijenhuis, J., Paikert, B. C., Dautzenberg, F. M. & Van Den Bleek, C. M. 2001. CFD modeling and experimental validation of pressure drop and flow profile in a novel structured catalytic reactor packing. Chemical Engineering Science, 56, 1713-1720.

Dautzenberg, F. M. & Angevine, P. J. 2004. Encouraging innovation in catalysis. Catalysis Today, 93-95, 3-16.

Ergun, S. 1952. Fluid flow through packed columns. Chem. Eng. Prog., 48, 89-94.

Gascon, J., Van Ommen, J. R., Moulijn, J. A. & Kapteijn, F. 2015. Structuring catalyst and reactor - An inviting avenue to process intensification. Catalysis Science and Technology, 5, 807-817.

Hayhurst, A. N. 2000. The mass transfer coefficient for oxygen reacting with a carbon particle in a fluidized or packed bed. Combustion and Flame, 121, 679-688.

Kang, T. G. & Anderson, P. D. 2014. The Effect of Inertia on the Flow and Mixing Characteristics of a Chaotic Serpentine Mixer. Micromachines, 5, 1270-1286.

Merino-Garcia, I., Alvarez-Guerra, E., Albo, J. & Irabien, A. 2016. Electrochemical membrane reactors for the utilisation of carbon dioxide. Chemical Engineering Journal, 305, 104-120.

Romeo, E., Royo, C. & Monzón, A. 2002. Improved explicit equations for estimation of the friction factor in rough and smooth pipes. Chemical Engineering Journal, 86, 369-374.

Romkes, S. J. P., Dautzenberg, F. M., Van Den Bleek, C. M. & Calis, H. P. A. 2003. CFD modelling and experimental validation of particle-to-fluid mass and heat transfer in a packed bed at very low channel to particle diameter ratio. Chemical Engineering Journal, 96, 3-13.

Scott, K. 2018. Process intensification: An electrochemical perspective. Renewable and Sustainable Energy Reviews, 81, 1406-1426.

Swamee, P. K. & Jain, A. K. 1976. Explicit equations for pipe-flow problems. ASCE J Hydraul Div, 102, 657-664.

Tomasoni, F., Thomas, J. F., Yildiz, D., Van Beeck, J. & Deconinck, J. Transport phenomena in an electrochemical rotating cylinder reactor. WIT Transactions on Engineering Sciences, 2007. 153-162.

Torquato, S., Truskett, T. M. & Debenedetti, P. G. 2000. Is random close packing of spheres well defined? Physical Review Letters, 84, 2064-2067.

Vervloet, D., Kapteijn, F., Nijenhuis, J. & Van Ommen, J. R. 2013. Process intensification of tubular reactors: Considerations on catalyst hold-up of structured packings. Catalysis Today, 216, 111-116.

Wakao, N., Kaguei, S. & Funazkri, T. 1979. Effect of fluid dispersion coefficients on particle-to-fluid heat transfer coefficients in packed beds. Correlation of nusselt numbers. Chemical Engineering Science, 34, 325-336.

Walsh, F. C. & Ponce De León, C. 2018. Progress in electrochemical flow reactors for laboratory and pilot scale processing. Electrochimica Acta, 280, 121-148.

Yao, X., Zhang, Y., Du, L., Liu, J. & Yao, J. 2015. Review of the applications of microreactors. Renewable and Sustainable Energy Reviews, 47, 519-539.

Ziogas, A., Kolb, G., O'connell, M., Attour, A., Lapicque, F., Matlosz, M. & Rode, S. 2009. Electrochemical microstructured reactors: Design and application in organic synthesis. Journal of Applied Electrochemistry, 39, 2297-2313.

Anton A. Kiss, Edwin Zondervan, Richard Lakerveld, Leyla Özkan (Eds.)
Proceedings of the 29th European Symposium on Computer Aided Process Engineering
June 16th to 19th, 2019, Eindhoven, The Netherlands. © 2019 Elsevier B.V. All rights reserved.
http://dx.doi.org/10.1016/B978-0-128-18634-3.50300-3

Design of Microreactor Systems with Minimization of Flow Pulsation

Osamu Tonomura,[*] Kazuki Okamoto, Satoshi Taniguchi, Shinji Hasebe

Dept. of Chemical Engineering, Kyoto University, Nishikyo, Kyoto 615-8510, Japan

tonomura@cheme.kyoto-u.ac.jp

Abstract

It is important to realize stable and accurate feed flow rates to devices in the continuous operation of micro chemical plants. Reciprocating pumps such as plunger pumps are generally used for liquid feeding. Much work of pump manufacturers has been devoted to development of reciprocating pumps with pulseless feeding, and many engineers believe that the problem of flow pulsation has been solved. However, examinations on how to design and operate a micro flow system using the pumps are not sufficient. The feed flow rates may be unstable when multiple pumps are used in the system even if the performance of the pump itself is high. This study focuses on the quantitative evaluation of flow pulsation generated by multiple pumps as well as individual pumps in micro flow systems with mixing. The causes, results, and solutions to the problem of flow pulsation are discussed from the viewpoint of system design, and a design method for micro flow systems with minimization of flow pulsation is proposed.

Keywords: Microreactor, Reciprocating pump, Flow pulsation, System design, Modeling.

1. Introduction

Micro chemical plants offer many advantages over conventional plants in terms of energy efficiency, mixing speed, reaction yield, safety, on-site/on-demand production, and so on (Hessel et al., 2005). It is important to realize uniform and accurate feed flow rates to devices in the continuous operation of micro chemical plants. Reciprocating pumps such as plunger pumps are generally used for liquid feeding. Much work of pump manufacturers has been devoted to development of reciprocating pumps with pulseless feeding, and many engineers believe that the problem of flow pulsation has been solved. However, examinations on how to design and operate a microreactor system using the pumps are not sufficient. The feed flow rates may be non-uniform when multiple pumps are used in the system even if the performance of the pump itself is high. In this research, it is intended to evaluate the magnitude of the flow pulsation in a microreactor system composed of a mixing part, a reaction part, and two fluid feeding parts based on reciprocating pumps and to propose a design method for microreactor systems with minimization of flow pulsation.

2. Experimental System Development

Figure 1 shows our developed experimental system comprising the mixing part, the reaction part, and two fluid feeding parts. In each fluid feeding part, a fluid is sent from a fluid source tank to the mixing part via a pump, a pulse damper (PD), and a resistance channel (RC). The PD is often used to reduce the flow pulsation, and in some cases it is built in the pump. The RC represents the flow resistance between the pump and the mixer. In the mixing part, a T-shaped mixer, which is one of the simplest forms of micromixer, is used. In the reaction part, a microreactor (MR) is installed and its outlet stream is discharged to a tank. The outlet pressure of the system is atmospheric.

In this study, plunger pumps are investigated, which are commonly used in the laboratory. Stainless steel round tubes are applied to RCs and MR, and their measured pressure losses per unit flow rate of water (293 K) are 1 kPa/(mL/min) and 8 kPa/(mL/min), respectively. The commercial PD (PD0235S; Uniflows) is used in the fluid feeding part. A Swagelok® union tee (SS-100-3) is used as the T-shaped mixer. The pressure loss of channels that connect pumps, PDs, RCs, mixer, MR, and tanks is negligible in the system. To quantitatively evaluate the pulsating flow, mass flow meters (LF-F; Horiba) are used before the PD and after the RC and MR, and are labeled $FI1_u$, $FI1_d$, $FI2_u$, $FI2_d$ and FI_o as shown in Figure 1. The measurement data are recorded in a data logger (NR-600; Keyence).

3. Experimental Results

In this section, the flow pulsation under two-fluid mixing is investigated. To evaluate the flow pulsation during fluid mixing, the experiments are performed by using the developed system, where water is used for the fluid in both fluid feeding parts. Plunger pump (F_{set} = 1 mL/min) is applied to the system. After the system is in a steady state, the flow rate at each flowmeter is recorded.

In Fig. 2, the flow pulsations generated by pumps are confirmed before the PDs. The attention here is the result shown in Fig. 3. It can be seen that the flow pulsation at the reaction part, namely at FI_o, is low, but the flow rates at $FI1_d$ and $FI2_d$ increase and decrease almost alternately and this observed flow pulsation is much greater than that of the non-mixing condition discussed in the previous section. This result suggests that the composition after mixing varies with time. In our view, the phenomena observed in this experiment seems to be due to two PDs combined. In the next section, modeling and simulation are carried out in order to evaluate the validity of the view and to take a solution.

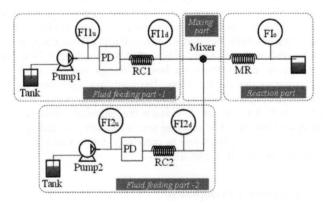

Figure 1 Experimental system for evaluation of feeding stability.

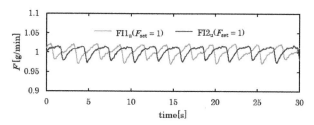

Figure 2 Flow rates measured at FI1$_\mathrm{u}$ and FI2$_\mathrm{u}$ during fluid mixing.

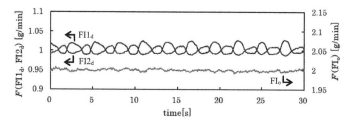

Figure 3 Flow rates measured at FI1$_\mathrm{d}$, FI2$_\mathrm{d}$ and FI$_\mathrm{o}$ during fluid mixing.

4. Design of Mixing System with Less Flow Pulsation

The modeling and numerical simulation is performed to investigate the flow pulsation during fluid mixing in a micro flow system and to propose a design guideline for constructing the system with less flow pulsation.

4.1. Model of the System

Figure 4 shows a schematic diagram of the experimental system and defines variables used in the modeling. The PD is aimed at reducing the flow pulsations. In response to the pressure increased or decreased due to flow pulsation, the PD changes its internal volume and attempts to absorb the flow pulsation. Based on the material balance, the pressure balance, and the pressure dependence of the internal volume of the PD, the following model is built:

$$F_{\mathrm{MR}} = \sum_i F_{\mathrm{d},i} \tag{1}$$

$$F_{\mathrm{d},i} = F_{\mathrm{u},i} - \frac{\mathrm{d}V_i}{\mathrm{d}t} \tag{2}$$

$$V_i = 2.366 - \frac{479.7}{P_{\mathrm{PD},i} + 97.74} + \frac{2.490 \times 10^4}{\left(P_{\mathrm{PD},i} + 97.74\right)^2} \tag{3}$$

$$P_{\mathrm{PD},i} = F_{\mathrm{d},i} R_{\mathrm{RC},i} + F_{\mathrm{MR}} R_{\mathrm{MR}} \tag{4}$$

where F is flow rate, V is internal volume of the PD, P is gauge pressure at the PD, R is channel resistance, and t is time. The subscripts i is the fluid feeding part number ($i = 1$, 2). Equations (1) and (2) are material balances at the MR and the PD, respectively. Equation (3) shows the relationship between gauge pressure and internal volume of the used PD and is derived from Fig. 5, where the closed circle points are experimental data representing such relationship and the solid line is the result of polynomial approximation. Equation (4) is the gauge pressure at each PD. In this study, MATLAB ® is used to estimate $F_{d,i}$ and F_{MR} from $F_{u,i}$ given by pumps at a recording interval.

4.2. Validation of the Constructed Model

To validate the constructed model, the calculated $F_{d,i}$ and F_{MR} are compared with the experimental results of Sec. 3. As shown in Figs. 6 and 7, it is shown that the differences between the measured values and the calculated values, which are represented by the superscripts 'exp' and 'cal', respectively, are small and that the behaviour of the flow pulsations can also be reproduced by the constructed model. Therefore, the constructed model is considered to be sufficiently practical.

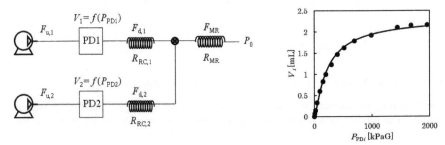

Figure 4 Schematic diagram of the system.   Figure 5 P_{PDi} vs. V_i

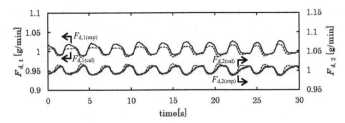

Figure 6 Comparison between the calculated and measured flow rates before and after each PD.

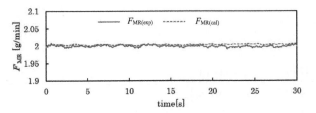

Figure 7 Comparison between the calculated and measured flow rates at MR.

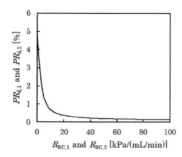

Figure 8 Relationship between the pressure loss per unit flow rate and pulsation rate.

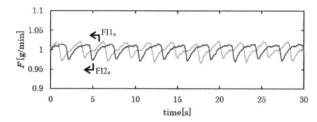

Figure 9 Flow rates measured at FI1$_u$ and FI2$_u$ during fluid mixing in the improved system.

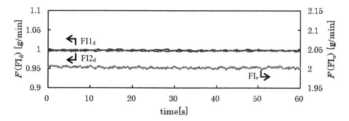

Figure 10 Flow rates measured at FI1$_d$, FI2$_d$ and FI$_o$ during fluid mixing in the improved system.

4.3. Design of Micro Flow System with Less Pulsation

To reduce the flow pulsation of the fluids fed to the mixer, observed in Sec. 3, it is proposed to increase the channel resistance of RC. However, it is not clear how much the channel resistance is needed to reduce the flow pulsation. In this study, the relationship between the channel resistance of RC and the *PR* after the PD is numerically investigated by using the constructed model. The result is shown in Fig. 8, and it is shown that 20 kPa/(mL/min) is expected to be required for $R_{RC,1}$ and $R_{RC,2}$ to make PR_d about 1/10 times of PR_u. To validate this expectation, the experiments are carried out by using the improved system where RC is changed from 1 to 20 kPa/(mL/min), which is realized by a stainless steel coiled round tube with an inner diameter of 0.25 mm and a length of 200 mm. As shown in Figs. 9 and 10, it is demonstrated that *PR* of the measured flow rates before the PD is about 1/10 times.

5. Conclusions

Our experiments with the constructed microreactor system showed that the flow pulsation at the reaction part is low but the flow rates at two fluid feeding parts increase and decrease almost alternately. This observed flow pulsation suggests that the composition after mixing varies with time. Then, in order to solve this problem efficiently, the observed phenomena were reproduced by a simplified CFD model, which is based on the material balance, the pressure balance, and the pressure dependence of the internal volume of the pulse damper inside the pump. Through the numerical study, it was proposed to increase the channel resistances between the pumps and the mixer so that the flow pulsation is reduced. The optimal value of the channel resistance was derived by simulation and its effectiveness was confirmed by experiment. The result showed that the magnitude of the flow pulsation after the optimal design was 1/10 of that before the optimization.

Acknowledgment

This work was partially supported by the Grant-in-Aid for Scientific Research (S) (no. 26220804), Scientific Research (S) (no. 25220913) and Scientific Research (B) (no. 26288049).

References

V. Hessel, S. Hardt, H. Lowe, 2005, Chemical micro process engineering: fundamentals, modelling and reactions, Wiley, VCH, Weinheim.

V. Hessel, H. Löwe, F. Schönfeld, 2005, Micromixers—a review on passive and active mixing principles, Chemical Engineering Science, 80(8-9), 2479-2501.

Anton A. Kiss, Edwin Zondervan, Richard Lakerveld, Leyla Özkan (Eds.)
Proceedings of the 29th European Symposium on Computer Aided Process Engineering
June 16th to 19th, 2019, Eindhoven, The Netherlands. © 2019 Elsevier B.V. All rights reserved.
http://dx.doi.org/10.1016/B978-0-128-18634-3.50301-5

Process Intensification and Miniaturization of Chemical and Biochemical Processes

Filip Strniša, Tomaž Urbič, Polona Žnidaršič-Plazl and Igor Plazl

Faculty of Chemistry and Chemical Technology, University of Ljubljana, Večna pot 113, SI-1001 Ljubljana, Slovenia
igor.plazl@fkkt.uni-lj.si

Abstract

In this work, the microscale (bio)process development based on scale-up/numbering-up concept in combination with model-based optimization is presented. The main features of microscale systems are reflected in fluid dynamics, therefore the understanding of fundamental mechanisms involved in fluid flow characteristics at the micro scale is essential since their behaviour affects the transport phenomena and microfluidic applications. Theoretical description of transport phenomena and the kinetics at the micro scale is discussed and illustrated on the cases of a lattice Boltzmann simulations for flow distribution in the packed bed microreactor "between two-plates" and the biocatalytic enzyme surface reaction.

Keywords: process intensification, miniaturization, model-based design

1. Introduction

Process intensification has been considered important for the future of (bio)chemical engineering for a while now. Stankiewicz and Moulijn (2000) defined process intensification as the development of novel and sustainable equipment that compared to the existing state-of-the-art, produces dramatic process improvements related to equipment sizes, waste production, and other factors. The application of microreactor technology in (bio)chemical processes meets these criteria, with the most obvious one being the equipment size reduction. However, besides spatial benefits, microreactors also provide enhanced heat and mass transfer, safety, environmental impact, and others (Pohar and Plazl, 2009). A key advantage that this technology brings is the precise process control that comes with it. This allows for repeatable conditions within the reactor system which leads to improved yields and product quality at a more consistent rate compared to the batch procedures. With the listed advantages there is no doubt that process intensification through the application of microreactor technology holds the potential to revolutionize (bio)chemical synthesis, but specific suggestions for possible replacement of established industrial processes are scarcely encountered. A number of highly innovative and systematic approaches, protocols, tools, and strategies are however currently being developed in both — industry and academia, all to minimize the gap between research and industry, and to enable a smooth transfer of lab-on-a-chip to the industrial scale. To deal with these challenges it is necessary to advance the field from (bio)catalyst discovery to (bio)catalytic microprocess design. Not only will this require a new level of understanding of the underlying reaction mechanisms and transport phenomena at the microscale, but also the development of relevant computational tools. The quest for high-performance manufacturing technology places the combination of biocatalysis and microscale technology as a key green engineering method for process development and production (Wohlgemuth et al., 2015). The aim of this work is to present the microscale (bio)process development based on scale-up/numbering-up concept in combination with model-based optimization. As microscale systems are flow-based it is essential to understand the fundamental mechanisms involved in fluid flow characteristics at the micro scale, because they affect the transport phenomena and microfluidic applications. Theoretical description of transport phenomena and the kinetics at the micro scale is discussed and illustrated on the cases of lattice Boltzmann simulations for flow distribution in packed bed microreactors "between-two-plates" and the biocatalytic enzyme surface reaction.

2. Model-based design

2.1. Macrosopic models

A theoretical description of transport phenomena and kinetics for different chemical and biochemical processes in the microfluidic devices have to be developed using the bases of continuum theory. The following set of partial differential equations regarding momentum and mass conservation (Eqs. 1, 2), heat conservation (Eq. 3), and species conservation (Eq. 4) have to be solved to describe the convection-diffusion dynamics in all three spatial directions and to depict the governing transport characteristics of processes in microfluidics:

$$\frac{\partial \vec{v}}{\partial t} + \vec{v} \cdot \nabla \vec{v} + \nabla p_k - \nu \nabla^2 \vec{v} = 0 \tag{1}$$

$$\nabla \cdot \vec{v} = 0 \tag{2}$$

$$\frac{\partial}{\partial t}(\rho h) + \nabla \cdot (\rho h \vec{v}) - \nabla \cdot [\lambda \nabla T] - S_h = 0 \tag{3}$$

$$\frac{\partial c_i}{\partial t} + \vec{v} \cdot \nabla c_i - D_i \nabla^2 c_i = 0 \tag{4}$$

where $\vec{v}$ is the velocity vector (m s^{-1}), p_k is the kinematic pressure (m^2 s^{-1}) and ν is the kinematic viscosity (m^2 s^{-1}) — Eqs. 1, 2; ρh is the thermal energy (W m^{-3}), λ is the thermal conductivity (W m^{-2} K^{-1}), T is the temperature (K) and S_h is the volumetric heat source (W m^{-3}) — Eq. 3; c_i is the concentration of the solute i, D_i is the diffusion coefficient of the species i (W m^{-2}) and $\vec{v}$ is the velocity vector obtained from Eqs. 1 and 2 — Eq. 4.

2.2. Mesoscopic models

Lattice Boltzmann methods are computational methods, that were primarily developed to solve fluid-dynamic problems, but can also be used among other to simulate heat and mass transport (Succi, 2015). The core of the method is the lattice Boltzmann equation, which is a special form of the Boltzmann equation, discretized in both — space and time. This gives the methods a base in the statistical mechanics and therefore steers them away from the "traditional" continuum-based methods, which are usually applied in chemical engineering process simulation. The lattice Boltzmann equation reads as follows:

$$f_a(\vec{x}+\vec{e}_a\Delta t, t+\Delta t) - f_a(\vec{x},t) = \Omega_a(\vec{u},\rho,\tau) \tag{5}$$

f is a discrete distribution function, which represents a packet of particles with statistically the same momentum in direction a, which is represented by a set of basic lattice velocities $\vec{e}$. $\vec{x}$ is the positional vector and t is time. Ω is the collision operator, and it can take on different forms, but they essentially all depend on the local flow velocity and fluid density $\vec{u}$, and ρ respectively. τ is the relaxation time, which is directly related to species' physical properties such as kinematic viscosity or molecular diffusivity.

2.3. Model verification and validation

Microfluidic devices have justified the high expectations and successfully demonstrated their advantages for intensification of chemical and biochemical processes in many different research areas and fields. Schematic representation of some typical microfluidic devices, known as enzymatic microreactors is given in Figure 1 (a-g). On these schemes of different microfluidic systems, one can observe the most typical enzyme (or cell) immobilization techniques as well as the most typical flow patterns of two-phase flows: the parallel and droplet flow regime, which allow for efficient performance of various processes from chemical reaction, enzyme catalyzed reaction with immobilized enzyme or cells to separation, extraction, and purification. In order to transfer these systems, successfully developed and tested on the laboratory scale, to the industrial environment and to increase the productivity from microliter scale to bigger scales at the same process efficiency and minimized costs, the optimized scale-up/numbering-up concepts have to be implemented, where the model-based design plays a key role. For this purpose, only pre-verified and validated macroscopic and lower-lever mathematical models can be used. Only verified and especially validated models bring about the understanding of basic principles and mechanisms taking place in chemical and biochemical processes at the microscale.

The model equations are usually solved numerically by writing them into computer code. Once the code is complete and error free, the model needs to be verified. In this process the model is tested whether it gives sensible results, which do not necessarily need to be correct (Plazl and Lakner, 2010). The correctness of the model's predictions is put to the test during model validation. Models can be validated either by comparing them to other already valid models or by testing them against experimental observations. Validating via experimental methods can be done in different ways, such as *e. g.* online validation, and inline validation. Common to all is that theoretical predictions and simulations matched to experimental measurements without any fitting procedures. In our previous work, we showed that macroscopic mathematical models can reliably describe various complex processes in microfluidic devices. In the case of non-homogeneous system with parallel flow pattern the steroid extraction in a microchannel system was studied theoretically and experimentally. In order to analyze experimental data and to forecast microreactor performance, a three-dimensional mathematical model with convection and diffusion terms was developed considering the velocity profile for laminar flow of two parallel phases in

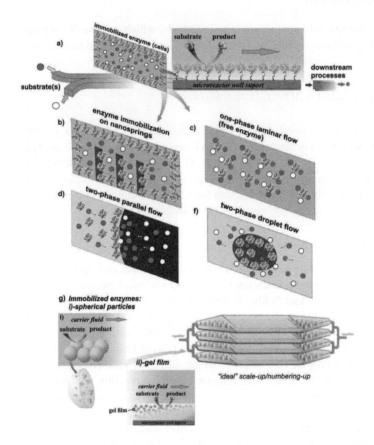

Figure 1: Schematic representation of some typical homogeneous, non-homogeneous and heterogeneous microfluidics systems: a) immobilized biocatalyst on the inner walls of microreactor with integrated down-stream processes; b) immobilized biocatalyst on the inner surface and pillars to increase surface area and biocatalyst loading; c) free enzyme in a one-phase laminar flow; d) free enzyme in a two-phase parallel flow; f) free enzyme in a droplet flow; g) packed-bed microreactor with immobilized enzymes in a spherical particles (i) and gel film (ii) with scale-up in one dimension and numbering-up of microreactor systems "between two-plates".

a microchannel at steady-state conditions. Very good agreement between model calculations and experimental data was achieved without any fitting procedure (Žnidaršič Plazl and Plazl, 2007).

Online oxygen measurements inside a microreactor with modelling of transport phenomena and enzyme catalyzed oxidation reaction was performed for online model validation of the homogeneous system with biochemical reaction. Continuum based mathematical models, with a full 3D description of transport phenomena, incorporating convection, diffusion and enzymatic reaction terms along with the parabolic velocity profile in a microchannel was developed to simulate the concentration of dissolved oxygen inside the microchannels, to assess the required model complexity for achieving precise results and to depict the governing transport characteristics at the microscale (Ungerböck et al., 2013). Droplet-based liquid-liquid extraction in a microchannel was studied, both theoretically and experimentally to online validate macroscopic models for complex two-phase droplet flow microfluidics. The finite elements method, as the most common macroscale simulation technique, was used to solve the set of partial differential equations regarding conservation of moment, mass and solute concentration in a two-domain system coupled by interfacial surface of droplet-based flow pattern (Eqs. 1, 2, 4). The model was numerically verified and validated online by following the concentrations of a solute in two phases within the microchannel by means of a thermal lens microscopic (TLM) technique coupled to a microfluidic system, which gave results of high spatial and temporal resolution. Very good agreement between model calculations and online experimental data was achieved without applying any fitting procedure to the model parameters (Lubej et al., 2015).

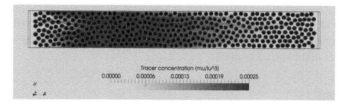

Figure 2: A snapshot of the pulse test simulation. Domain size = 2048 × 256 × 32 lattice units^3, Reynolds number = 0.95, Schmidt number = 1000. The flow is from west to east.

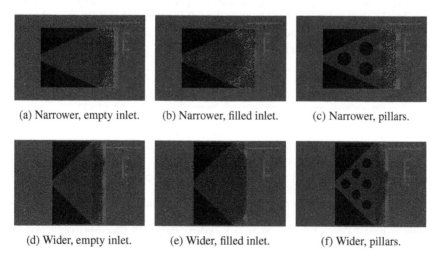

(a) Narrower, empty inlet.   (b) Narrower, filled inlet.   (c) Narrower, pillars.

(d) Wider, empty inlet.   (e) Wider, filled inlet.   (f) Wider, pillars.

Figure 3: Different scenarios for inlet flow distribution in one-dimensional scale-up. "Empty" channels do not utilize any obstacles to distribute the flow, "filled" channels use the same particles in the flow distribution zone as they do throughout the rest of the channel, and "pillars" channels have pillars in the flow distribution zone. The tracer is colored red, and flows from west to east.

The lattice Boltzmann method was applied to obtain the velocity field in a packed bed microbioreactor, and to simulate mass transport in a pulse test through it. First a computer program generated a randomly packed bed, then the flow field was calculated, and finally the mass transport of an inert tracer was simulated (Figure 2). The results were recorded and compared to experimentally collected data. In the experiments the channel was packed with enzyme-containing particles, and an inert tracer was injected through it via an HPLC-type 6-port valve. At the outlet an inline biosensor was recording the tracer concentration. The simulations and the experiments show a good agreement (Strniša et al., 2018).

A validated lattice Boltzmann model could then be further used to simulate mass transport in a "near-perfect" scaled-up microchannel, where only one dimension of the channel is increased — its width (Bajić et al., 2017). Similar simulations as above were constructed, where 3 scenarios were tested: empty space at inlet, inlet filled with biocatalytic particles, and pillars in the inlet. Unlike previously the inert tracer was supplied into the system continuously. The results (Figure 3) show that filling the inlet with particles does not create uniform flow conditions downstream, where as leaving the space empty or introducing pillars in it improves the downstream hydrodynamics. Similar simulations should be used further to determine the optimal inlet geometry, which will assure the optimal flow conditions in the microchannel.

3. Conclusions

Several studies have shown the advantages of microfluidic systems. Here it was shown how computer modelling and simulation can (or rather should) work within the research and development of such systems to optimize their performance. We now have the knowledge and technologies for building them, and the (bio)chemical industries need to recognize the potential of process intensification and miniaturization, with the consideration of new modelling approaches, such as multiscale modelling.

References

M. Bajić, I. Plazl, R. Stloukal, P. Žnidaršič Plazl, 2017. Development of a miniaturized packed bed reactor with ω-transaminase immobilized in lentikats®. Process Biochemistry 52, 63 – 72.

M. Lubej, U. Novak, M. Liu, M. Martelanc, M. Franko, I. Plazl, 2015. Microfluidic droplet-based liquid-liquid extraction: online model validation. Lab on a Chip 15, 2233—2239.

I. Plazl, M. Lakner, 2010. Modeling and finite difference numerical analysis of reaction-diffusion dynamics in a microreactor. Acta Chimica Slovenica 57 (1), 100—109.

A. Pohar, I. Plazl, 2009. Process intensification through microreactor application. Chemical and Biochemical Engineering Quarterly 23 (4), 537–544.

A. Stankiewicz, J. Moulijn, 2000. Process Intensification: Transforming Chemical Engineering. Chemical Engineering Progress 96, 22–34.

F. Strniša, M. Bajić, P. Panjan, I. Plazl, A. M. Sesay, P. Žnidaršič Plazl, 2018. Characterization of an enzymatic packed-bed microreactor: Experiments and modeling. Chemical Engineering Journal 350, 541 – 550.

S. Succi, 2015. Lattice boltzmann 2038. EPL (Europhysics Letters) 109 (5), 50001.

B. Ungerböck, A. Pohar, T. Mayr, I. Plazl, 2013. Online oxygen measurements inside a microreactor with modeling of transport phenomena. Microfluidics and Nanofluidics 14, 565—574.

P. Žnidaršič Plazl, I. Plazl, 2007. Steroid extraction in a microchannel system: mathematical modelling and experiments. Lab on a Chip 7, 883—889.

R. Wohlgemuth, I. Plazl, P. Žnidaršič Plazl, K. V. Gernaey, J. M. Woodley, 2015. Microscale technology and biocatalytic processes: opportunities and challenges for synthesis. Trends in Biotechnology 33 (5), 302 – 314.

Anton A. Kiss, Edwin Zondervan, Richard Lakerveld, Leyla Özkan (Eds.)
Proceedings of the 29th European Symposium on Computer Aided Process Engineering
June 16th to 19th, 2019, Eindhoven, The Netherlands. © 2019 Elsevier B.V. All rights reserved.
http://dx.doi.org/10.1016/B978-0-128-18634-3.50302-7

Modelling of Microfluidic Devices for Analysis of Radionuclides

Miguel Pineda[a,*], Panagiota Angeli[a], Takehiko Tsukahara[b] and Eric S. Fraga[a]

[a]*CPSE and ThAMeS Multiphase. Department of Chemical Engineering, University College London, London WC1E 6BT, UK*
[b]*Laboratory for Advanced Nuclear Energy, Tokyo Tech, Tokyo 2-12-1-N1-6 O-okayama, Meguro-ku, Japan*

m.pineda@ucl.ac.uk

Abstract

Analysis of radionuclides in radioactive waste is performed according to a number of operating protocols which are usually time-consuming and difficult to implement. The use of microdevices for liquid-liquid extraction of radionuclides is promising, since small amounts of materials are needed. In this work, a plug flow modelling approach for the liquid-liquid extraction in microdevices is presented. A multi-component chemical process based on the PUREX-TRUEX waste management treatment is analysed. We evaluate the performance of the coupled micro-units as a function of channel size, superficial mixture velocity, and initial loading. The results show that it is possible to use the model in an optimisation based design procedure for microfluidic devices.

Keywords: Microdevices, Radionuclides, Plug flow, Sensor design

1. Introduction

Liquid liquid extraction (LLE), also known as solvent extraction, is a method used to distribute one or more species (solutes) between two immiscible liquids in contact with each other due to a difference in solubility. It is among the most common separation techniques used in industrial processes. In this regard, computer simulations of LLE processes are very useful tools that can aid the design and testing of process flowsheets. They can form the basis for applications in optimisation, control, and system analysis.

The analysis of radionuclides present in high level radioactive waste (HLRW) is performed using operating protocols including several separation and purification steps which are difficult to implement in glove boxes and hot cells (Hellé, et al., 2015). Among those steps, LLE is essential to obtain a pure fraction containing the radionuclides of interest. However, one of the disadvantage of this separation method is that a large amount of organic solvent is need to dissolve the extractant and the radionuclides of interest. This entails both radiological and chemical risks. Therefore, in order to reduce the amounts of solvents used, as well as the radioactive material needed for the analysis, modifications of operating procedures are important. In the frame of studies aiming to overcome these limitations, microfluidic devices have attracted much attention, and micrototal analysis systems (μTAS) have been advocated (Ciceri, et al., 2014). Normally, in μTAS, solvent mixing, reaction, separation, and detection are integrated onto a microfluidic chip. These microfluidic technologies are suited for the efficient implementation of LLE due to the intrinsic advantages of the small dimension, i.e., high surface-to-volume ratio and laminar flow conditions. Other attractive feactures of microfluidic devices are the ability to control the contact time of the liquids involved and the use of small manipulating volumes, which decrease both chemical and radiological risk.

Micro-chemical systems that implement two or more liquid streams flowing laminarly in a micro-channel have been development for some specific radiochemical applications. These studies include the extraction of uranium U(VI) by tributylphosphate (TBP) in n-dodecane (n-DD) (Hotokezaka, et al., 2005) and the extraction of Am(III) by n-octyl(phenyl)-N,N-diisobutylcarbamoylmethylphosphine oxide (CMPO) (Ban, et. al., 2011). For industrial application of these techniques, however, one of the challenges is to have continuous and multiple separations of individual metal elements occurring in the same microtool (Hellé, et al., 2015).

Despite the interest in the development of μTAS or lab-on-chip devices, researchers typically do not have large amounts of time and money to build and test several prototypes in order to optimise performance. To reduce the number of iterations of prototyping, it is imperative to develop the best computer-aided design-forecasting methods. With the aim to improve these prediction methods, we present and validate a mathematical modelling framework to simulate continuous separations of radionuclides in a single microdevice. The long term aim is to use the mathematical model with an optimisation based design procedure. This procedure will aim to generate and evaluate alternative channel layouts and identify those that have the best performance for the online determination of radionuclides following microchip extraction by using for example the so-called thermal lens microcopy system (Hotokezaka, et al., 2005).

In the following section, we present the radiochemical system under consideration. Then, we introduce the mathematical model and present the results. We finalize with the conclusions.

2. The chemical system

In this section, we describe the radiochemical process implemented in this work.

2.1. Motivation

Reprocessing of spent nuclear fuels (SNFs) from power reactors by PUREX process is aimed at recovery of uranium and plutonium (Bascone, et al., 2017). It consists of a LLE process using 1.1 M TBP as extractant present in n-DD as diluent. Since the trivalent actinides and lanthanides (i.e. Am(III) and Eu(III)), are inextractable by TBP, they are normally rejected to HLRW. However, the trivalents can be then extracted by the TRUEX process, which is a mixture of 0.2 M CMPO - 1.2 M TBP in n-DD. In this part of the waste treatment, TBP is added as a diluent modifier in order to minimize third phase formation. It has been also observed that the extraction of trivalent actinides by CMPO decreases with the presence of uranium in the feed as it tends to saturate CMPO in the organic phase (Mathur, et al., 1993). These LLE processes are mostly carried out in pulsed columns and mixer settlers. In the following subsection, we describe a simplified version of the PUREX-TRUEX waste treatment, which will be used in the subsequent sections to validate our approach.

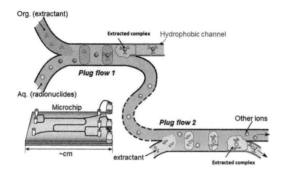

Figure 1: Conceptual diagram of a microchip involving two LLE steps, two different extractants, and three different radionuclides.

2.2. Process description

An HNO_3 aqueous phase, containing U(VI) (in the form of UO_2^{2+}) and Am(III) (in the form of Am^{+3}), is initially brought into contact with the organic phase containing 1.1 M TBP diluted in a hydrocarbon phase. This occurs in a hydrophilic small channel where dispersed plugs (elongated drops) of the organic phase are formed inside the continuous aqueous phase. It is assumed that, after the extraction of uranium, the organic stream will be separated in a side hydrophobic channel. The aqueous stream with the rest of uranium and all the americium will be mixed in a second micro-channel with another organic phase containing 0.2 M CMPO also diluted in a hydrocarbon phase for the extraction of remaining U(VI) and Am(III). We consider the plug flow pattern because it combines improved mixing with large interfacial areas. We will assume a constant HNO_3 concentration along the processes. We refer to Fig. 1 for a general schematic representation of a microchip containing two LLE steps and flow patterns as those described here. In both stages, the extraction proceeds according to the following reaction (Hellé, et al., 2015)

$$M^{n+} + nNO_3^- + nS_f \rightleftharpoons M(NO_3)_n \cdot nS, \tag{1}$$

where for the reaction along the first channel, $M=UO_2$ and S=TBP. For the reaction along the second channel $M=UO_2$ or Am and S=CMPO. The symbol S_f represents the free or unbounded extractant along the channel. For uranium $n = 2$ and for americium $n = 3$.

3. Mathematical model

As previously mentioned, the organic phase consists of dispersed plugs moving in a continuous phase containing the aqueous solution which initially contains the two radionuclides. In this work, the $i \in \{1,2\}$ channel will be modelled as an ideal plug flow reactor with the concentration of the $j \in \{U,Am\}$ radionuclide in the dispersed (organic) phase described by the so-called operational line given by

$$C_{i,j,d}(x) = C_{i,j,d}(0) - q^{-1}\left[C_{i,j,c}(x) - C_{i,j,c}(0)\right], \tag{2}$$

with the concentration of the j radionuclide in the continuous (aqueous) phase, $C_{i,j,c}(x)$, calculated from

$$v_{mix}\frac{dC_{i,j,c}(x)}{dx} = -K_L a_i \left(C_{i,j,c}(x) - C^e_{i,j,c}(x)\right), \tag{3}$$

where the equilibrium concentration of the j radionuclide in the aqueous phase is obtained using the equilibrium line, $C^e_{i,j,c}(x) = C_{i,j,d}(x)/D_{i,j}$. The parameter $q = 1$ is the ratio of the volumetric flow rates of the dispersed phase to the continuous phase. The parameter $D_{i,j}$ is the distribution coefficient defined as the ratio of the concentrations of each radionuclide in the organic and aqueous phases at equilibrium. From the complexation kinetics (see Eq. 1) we have that, $D_{1,U} = K'_{1,U}(C_{TBP_f}(x))^2$, $D_{2,U} = K'_{2,U}(C_{CMPO_f}(x))^2$, and $D_{2,Am} = K'_{2,Am}(C_{CMPO_f})^3$. We also assume that $C_{TBP_f}(x) = 1.1 - 2C_{1,U,d}(x)$ and $C_{CMPO_f}(x) = 0.2 - 2C_{2,U,d}(x) - 3C_{2,Am,d}(x)$ are the concentrations of the free extractants at position x along the channel. The parameters $K'_{1,U}$, $K'_{2,U}$, and $K'_{1,Am}$ are the conditional extraction constants. In Eqs. 2 and 3, $C_{i,j,d}(0)$ and $C_{i,j,c}(0)$ are the inlet concentrations, $K_L a_i$ is the volumetric overall mass transfer coefficient, and v_{mix} is the superficial mixture velocity. Following previous works using circular channels and plug flow conditions (Bascone, et al., 2017), we assume an average volumetric mass transfer coefficient given by

$$K_L a_i = 0.88\frac{v_{mix}}{L_i}Ca_{mix}^{-0.09}Re_{mix}^{-0.09}\left(\frac{ID_i}{L_i}\right)^{-0.1}, \tag{4}$$

where L_i and ID_i are the length and diameter of the channels. The symbols Re_{mix} and Ca_{mix} are the Reynolds and capillary numbers of the mixture, which are functions of the properties of the phases, v_{mix}, ID_i, and q. Crucial for the coupling between the two channels is the assumption that the outlet concentration of radionuclides in the continuous phase of the first channel becomes the inlet concentration of the continuous phase in the second one.

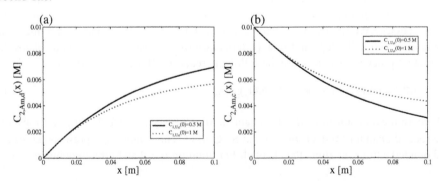

Figure 2: Evolution of the concentration of Am(III) in a) dispersed/organic phase and b) continuous/aqueous phase, for two different values of $C_{1,U,c}(0)$, with $C_{1,Am,c}(0) = 0.01$ M. The fluids along the channels have the same properties. In this case, $K'_{1,U} = K'_{2,Am} = 10^3$ and $K'_{2,U} = 10^2$. For the dimensions, $L_1 = L_2 = 0.1$ m and $ID_1 = ID_2 = 0.002$ m. For the integration of Eq. 3, we use the fourth order Runge-Kutta method with $\Delta x = 5 \times 10^{-6}$ m. Note that $C_{2,U,c}(0) = 0.127$ M and 0.468 M, for $C_{1,U,c}(0) = 0.5$ M and 1 M, respectively.

4. Results and validation of the model

It has been recognised that the extraction of trivalent actinides, like for example Am(III) decreases in the presence of large amounts of uranium in the feed as it tends to saturate CMPO in the organic phase (Mathur, et al., 1993). In the following, we show that our model reproduces, at least in a qualitative manner, this interesting finding. To this end, in Fig. 2 we plot the concentration of americium along the second channel (i.g. $C_{2,Am,d}$ and $C_{2,Am,c}$), for two large enough values of $C_{1,U,c}(0)$ (note that the concentration of Am(III) along the first channel is constant, as TBP does not extract trivalent actinides). The figure clearly shows that the extraction of americium decreases with the initial uranium loading.

Our model also predicts some interesting features that can be tested against future experiments. For example, in Fig. 3 we plot the concentration of americium as a function of the superficial velocity of the mixture v_{mix}. The figure clearly show that the extraction of americium in the microchip device should decrease with v_{mix}. This is due to the fact that an increase of the mixture velocity decreases the residence time available for mass transfer. The model also allows us to analyse the extraction as a function of the geometric

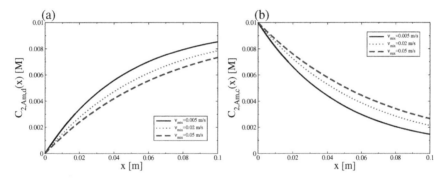

Figure 3: Evolution of the concentration of Am(III) in a) dispersed phase and b) continuous phase, versus v_{mix}. In this case, $K'_{1,U} = K'_{2,U} = K'_{2,Am} = 10^3$. For the dimensions, $L_1 = L_2 = 0.1$ m, and $ID_1 = ID_2 = 0.002$ m. The inlet concentrations are $C_{1,U,c}(0) = 0.05$ M and $C_{1,Am,c}(0) = 0.01$ M. Note that $C_{2,U,c}(0) = 0.006$ M, 0.010 M, and 0.013 M for $v_{mix} = 0.005$ m/s, 0.02 m/s, and 0.05 m/s, respectively.

properties of the microchannel. Figure 4 shows the concentration of americium along the second channel as a function of the channel diameter. The figure shows that the extraction is always higher as the channel size decreases. The interfacial area available for mass transfer decreases with the channel size, because the length of the plugs in the larger channels are longer; thus small number of plugs and reduced interfacial area are available for mass transfer.

5. Summary and conclusions

A mathematical model for the coupling of different LLE micro-unit operations in a single microchip is introduced and validated with a continuous radiochemical process, inspired by the PUREX-TRUEX waste treatment. The model reproduces, in a qualitative manner, the experimental observation that the extraction efficiency of americium decreases with

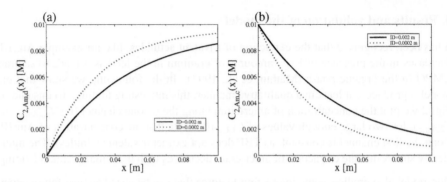

Figure 4: Similar to Fig. 3 but as a function of $ID = ID_1 = ID_2$, with $L_1 = L_2 = 0.1$ m. Note that $C_{2,U,c}(0) = 0.006$ M and 0.002 M, for $ID = 0.002$ M and 0.0002 M, respectively.

the initial loading of uranium. Our results also indicated that for fixed parameter values (i.e. channels size and flow conditions) as the mixture velocity increases the extraction efficiency of Am(III) decreases. It also shows that, for the same mixture velocity, the extraction efficiency is always higher as the diameter of the channel decreases. Our modelling approach constitutes a proof of concept and a first step towards a general modelling of analytical processes with continuous flows in microchips. The next step is to use the mathematical model with an optimisation based design procedure. For example, it could be interesting to determine the optimal channel size and layout in a constrained space for each separation step of the analytical process.

References

G. Hellé, C. Mariet, G. Cote, 2015, Liquid - liquid extraction of uranium(VI) with Aliquat 336 from HCl media in microfluidic devices:Combination of micro-unit operations and online ICP-MS determination. *Talanta*, *130* 123-131.

D. Ciceri, J. M. Perera, G. W. Stevens, 2014, The use of microfluid devices in solvent extraction. *J. Chem. Technol. Biotechnol*, *89* 771-786.

H. Hotokezaka, M. Tokeshi, M. Harada, T. Kitamori, Y. Ikeda, 2005, Development of the innovative nuclide separation system for high-level radioactive waste using microchannel chip - extraction behavior of metal ions from aqueous phase to organic phase in microchannel. *Prog. Nucl. Energy*, *47* 439-447.

Y. Ban, Y. Kikutani, M. Tokeshi, Y. Morita, 2011, Extraction of Am(III) at the interface of organic-aqueous two-layer flow in a microchannel. *J. Nucl. Sci. Technol*, *42* 1313-1318.

J. N. Mathur, M. S. Murali, P. R. Natarajan, L. P. Badheka, A. Banerji, A. Ramanujam, P. S. Dhami, V. Gopalakrishnan, R. K. Dhumwad, and M. K. Rao, 1993, Partitioning of actinides from high-level waste streams of PUREX process using mixtures of CMPO and TBP in dodecane. *Waste management*, *13* 317-325.

D. Bascone, P. Angeli E. S. Fraga, 2017, Mathematical modelling of intensified extraction for spent nuclear reprocessing. *In: Espua, A., Graells, M., Puigjaner, L., (Eds.). Proceeddings of the 27th European Symposium on Computer Aided Process Engineering - ESCAPE 27. Barcelona (Spain).* pp. 355-370.

Anton A. Kiss, Edwin Zondervan, Richard Lakerveld, Leyla Özkan (Eds.)
Proceedings of the 29th European Symposium on Computer Aided Process Engineering
June 16th to 19th, 2019, Eindhoven, The Netherlands. © 2019 Elsevier B.V. All rights reserved.
http://dx.doi.org/10.1016/B978-0-128-18634-3.50303-9

High-quality blend scheduling solution for sizing, selecting, sequencing, slotting and spotting in the processing industries

Brenno C. Menezes,[a] Jeffrey D. Kelly[b,*]

[a]*Center for Information, Automation and Mobility, Technological Research Institute, Av. Prof. Almeida Prado 532, São Paulo 05508-901, Brazil*

[b]*Industrial Algorithms Ltd., 15 Saint Andrews Road, Toronto M1P 4C3, Canada*

jdkelly@industrialgorithms.ca

Abstract

High-quality process system engineering (PSE) solutions for production scheduling evolve to a wider scope and scale when moving from simulation- to optimization-based approaches, generally by using mixed-integer programming (MIP) in discrete-time formulations. To reach as we state in this paper as the spotting level of service, into the scheduling operational details and time-step within a time-horizon of planning, the novel aspect of these types of PSE solutions is to integrate automated logistics or blend logistics decisions regarding: (1) sizing the future blend volumes for blending (or throughputs of unit-operations for processing); (2) selecting a product tank for the blends (or modes of operation for the process units); (3) sequencing the blended product grades (i.e., regular before premium, etc.) in multi-product blender (or the sequence of modes of operation for units/tanks); (4) slotting the start-times of the blends (or processing-time of units) into time-periods; and (5) spotting the future product shipments to steward to the plan. Examples of scheduling optimization effectively implemented are: (1) the production of lubes and asphalts with sequence-dependent switchovers between modes of operation; and (2) the gasoline blend scheduling operations using decomposition strategies and cuts based on nominal qualities. Faster results are obtained by tailored solutions of decomposing these industrial problems.

Keywords: Effective knowledge transfer, Industrial PSE applications, Spot markets.

1. Introduction

Integration of planning and scheduling in the processing industries can be achieved by solving a scheduling problem (in terms of time-step, level of details, etc.) within a planning time-horizon (in general for a month or longer). In such a way, the proposed planning-scheduling solution can be considered (in a product point of view) what we name in the paper as spotting, since there are potentials to increase the profit by exploring the higher gains from spot market opportunities when compared with its concurrent and complementary contract market dividends. In this case, for the producer, there is a reasonable degree of confidence in the day-by-day (optimized or prescribed) production by extending a schedule from a week to a month time-horizon. By addressing such high-quality PSE solutions into the spotting state, sellers are still in time to negotiate products in the spot market, which requires by half to one month in advance for the time-restricted contract to be made.

However, major considerations to reach the spotting stage (in a process point of view) must consider the following challenges for such integrated decision-making systems. What is the necessary time-grid of the scheduling solutions for a specific process, varying from online solutions ($\leq$ 1-hour) to 1-day discrete time-step? What planning and scheduling integration strategies exist? What degree of process details is demanded in the modelling? Which advanced manufacturing attribute can be explored and integrated in the optimization core, such as information and communication technologies (ICT) and auxiliary mechatronics (MEC) to automatically sample, analyse and communicate?

Franzoi et al. (2018a) proposed a refinery-wide scheduling approach to investigate the limits or capabilities of an enterprise-wide optimization (EWO) by solving simultaneously the supply, production, and demand chains using MILP. This solution fixes time-varying setups of unit-operations and connections that construct the flowsheet operations for further NLPs (Kelly et al., 2017a). Such EWO scope and scale may use additional detailed calculations within other edges such as in the depooling of aggregated tanks by considering their actual topology and operations (Menezes et al., 2018). Furthermore, Kelly et al. (2018a) introduce the concept of planuling that is a portmanteau of planning and scheduling where we schedule slow processes and plan fast processes together inside the slower process time-horizon, which demands an extra calculation for the fast processes, but henceforth as a scheduling problem.

To achieve the spotting solution in the matters of operational and economical standpoints may be accomplished by reducing the problem complexity when we exclude, relax and fixate part of it. These procedures must be coupled with high-performance computing (HPC) for faster calculations in parallel cores. Additionally, model and plant differences can be reduced by using closed-loop parameter feedback (Franzoi et al., 2018b) with the support of ICT and MEC capabilities. For the integrated planning and scheduling approach toward the spotting service level, we present solutions with 1-week to 2-months as time-horizon varying from 2-hours to 1-day as time-step. The effective implementation of the industrial examples uses the best technologies of modelling and solving algorithms (MSA) to optimize the problem.

2. From planning and scheduling to spotting

Benefit areas from this moving to optimization-based decisions using MILP as a prescriptive analytics application for sizing, selecting, sequencing, slotting and spotting (blend volumes and throughputs) in the processing industries are as follows:

a) optimized (vs. simulated) schedules which maximize profit and performance given pricing and discretizing on a sub-day or sub-hour basis;

b) better coordination with sales and marketing by providing optimized and achievable blend production schedules for all future product (and saleable component) shipments;

c) improved stewardship to the monthly feedstock selection and operations plan which coordinates the supply and demand of all the refinery site production;

d) increased ability to capture spot market opportunities after contract sales have been fulfilled respecting bottlenecks, in-progress blends, planned recipes, etc.;

e) blend schedules matching product quality specifications to reduce key property giveaway and to decrease the chances of off-specification blends to product tanks.

3. Effective knowledge transfer: high-quality PSE solution for spotting

Recently, state-of-the-art technology on modelling and solving algorithms (MSA), one of the ground bases of the *Industry 4.0* mandate, was implemented in an oil and gas company in Southern Asia. The company's call of technology winner has beaten the dominant petroleum scheduling technology sellers since they still deliver simulation-based scheduling solutions with limited optimization capabilities. The potentials of the cutting-edge MSA technologies are used in the examples.

Similar scope and scale to those implemented in crude-oil refineries, the examples are configured in the modelling platform named Industrial Modeling & Programming Language (IMPL) seamlessly coupled with GUROBI's superior simplex and barrier LP algorithms as one of today's most efficient solvers. These technologies optimize the production considering time-horizons for (a) 1-month with 1 day as time-step and (b) 7-days to 30-days with 2h as time-step. They are robustly and reliably solved within minutes combined for both the logistics and the quality sub-problems or any sort of decomposition. The object constructs of the examples (Figures 1 and 2) consider the unit-operation-port-state superstructure (UOPSS) formulation (Kelly, 2005) built-in with IMPL (see IMPL manual in Kelly, 2018). The unit-operation and arrow objects have binary and continuous variables (y and x, respectively) such as: a) unit-operations m for sources and sinks ($\diamond$), tanks ($\triangle$) and continuous-processes ($\boxtimes$) and b) the connectivity involving arrows ($\rightarrow$), in-port-states i ($\bigcirc$) and out-port-states j ($\otimes$).

4. Examples

4.1. Scheduling for production of lubes and asphalts

Reduced crude-oil distilled streams or atmospheric residue from crude-oil distillation units are fed to vacuum distillation units to produce fractions of lubes and asphalts (Kyungseok et al., 2017). To schedule the blocks, campaigns or conjugated modes of operation (or grades of lube base oils) in further steps of processing in solvent extraction (EU), dewaxing (DU) and hydrotreating units as in Figure 1, the MILP optimization uses the sequence-dependent switchovers introduced by Kelly and Zyngier (2007).

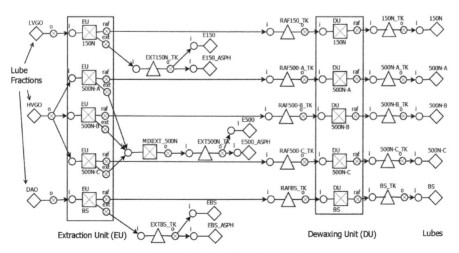

Figure 1. Production flowsheet for Lubes.

The proposed solution of the flowsheet in Figure 1 solves a 60-day horizon with 1-day time-steps, allowing opportunities of exploring the dynamics of the contract and spot markets. The MILP for a 60-day time-horizon with 1-day time-steps (60 time-periods) is solved in 30.7 seconds (with GUROBI 8.1.0) and 40.9 seconds (with CPLEX 12.8.0) at 0.0% of MILP relaxation gap using an Intel Core i7 machine at 3.4 GHz (in 8 threads) with 64 GB of RAM. There are 10,247 constraints (2,942 equality) for 5,102 continuous variables and 3,120 binary variables with 5,280 degrees-of-freedom. Figure 2 shows the Gantt chart for the dewaxing (DU) operations in time-slot positions.

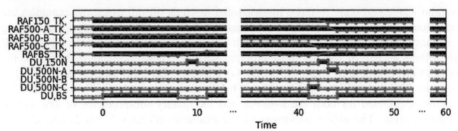

Figure 2. Dewaxing (DU) modes of operation and the holdup in their feed tanks (TK).

4.2. Gasoline blend scheduling

The gasoline blend scheduling is solved for the logistics (MILP) problem including nominal quality balances (Kelly et al., 2018b) in the MILP solution instead of neglecting the quality relationships completely. In the problem, without preferred recipes initialization as in Mendez et al. (2006), slack or surplus variables are calculated to balance the amounts of quality around the blender unit-operations and a successive MILP can be iterated to meet the quality specifications by using the slack or surplus variables and their distances from their lower and upper respective specifications. In such a way, cuts based on nominal qualities may avoid impending quality NLP infeasibilities if the NLP is needed after the MILP setup variables fixation. In this case, successive linear programming (SLP) technology can be used for these industrially-sized nonlinear dynamic optimizations since any infeasibilities or inconsistencies due to poor opening quantities and qualities, as example, are quickly detected and identified by the user in terms of the offending variable and/or constraint.

The proposed solution solves a 7- to 30-day horizon (time-step of 2 hours) for the problem in Figure 3. Table 1 shows the MILP solution (using nominal quality cuts) at 2% of MILP relaxation gap in an Intel Core i7 machine at 3.4 GHz (in 8 threads) with 64 GB of RAM. Although all cases converged, to reduce the CPU time in the full space MILP when > 60 minutes, we use a temporal decomposition heuristic known as rolling-horizon with crossover (RHC) in the time windows of the neighbouring time-horizon splits (Kelly, 2002) as shown in the additional CPU (in minutes).

Table 1. Statistics for 7, 14, 21 and 30 days (2 h time-step).

CPU (min) GUROBI (8.1.0)				CPU (min) CPLEX (12.8.0)				Equations (for 14d)	Continuous/ Binary Variables (for 14d)
7d	14d	21d	30d	7d	14d	21d	30d		
2.4	6.1	> 60	> 60	25.9	> 60	> 60	> 60	99,300	47,091/ 13,992
RHC		+ 11.1	+8.5	RHC	+ 37.3	> 60	> 60		

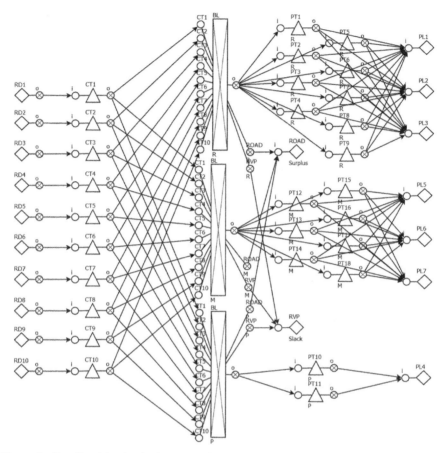

Figure 3. Gasoline blend scheduling with regular, medium and premium grade modes.

As an example of a result in the gasoline blend scheduling problem, Figure 4 shows the Gantt chart of the blender unit-operations (R, M and P) for 14 days as time-horizon solved using MILP with cuts on nominal qualities or factors for qualities (Kelly et al., 2018) in octane number (Road) and Reid vapor pressure (RVP) as shown in the flowsheet in Figure 3. The sequence of modes of operation (R, M and P in the blender BL unit) attend the supply of detailed and specific demands of the different grades of gasoline. The gasoline premium grade (P) has a reduced demand and production with punctual deliveries as seen in the holdups of the PT10 and PT11 tanks of the P grade.

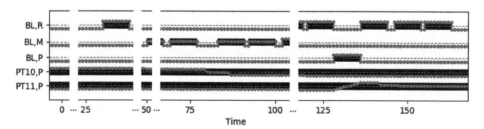

Figure 4. Blender (BL) modes of operation and tanks (PT10 and PT11) holdups (14 d).

5. Conclusions

We have exposed the knowledge around PSE solutions transferred to the processing industries. They are configured into a process industry specific domain programming language (IMPL) and solved using one of the best commercial solvers GUROBI. By solving the problems in a planning horizon with a scheduling level of detail (i.e., in time-step and operational relationships), any company's profit and performance may be improved by exploring their spot markets in a timely fashion. It may reduce management gaps between production and market segments inside the plant and teams in headquarters. The dynamics of both contracted and spot market require decisions within quarters and months in advance. For that, in the examples of PSE solutions in this paper, the planning and scheduling merge in the necessary spotting level of service.

Acknowledgements

The first author gratefully acknowledges the financial support from the São Paulo Research Foundation (FAPESP) under grant #2018/26252-0.

References

R.E. Franzoi, B.C. Menezes, J.D. Kelly, J.W. Gut, 2018a, Refinery-Wide Scheduling for Optimization of Multiple Unit-Operations in the Supply, Production, and Demand Chains in Fuels, Lubes, Asphalts and Petrochemicals Industries. In: AICHE Annual Meeting, Pittsburgh, United States.

R.E. Franzoi, B.C. Menezes, J.D. Kelly, J.W. Gut, 2018b, Effective Scheduling of Complex Process-Shops Using Online Parameter Feedback in Crude-Oil Refineries. Computer Aided Chemical Engineering, 44, 1279-1284.

J.D. Kelly, 2018, IMPL Manual, www.industrialgorithms.ca.

J.D. Kelly, 2002, Chronological Decomposition Heuristic for Scheduling: Divide and Conquer Method. AIChE Journal, 48, 2995-2999.

J.D. Kelly, 2005. The Unit-Operation-Stock Superstructure (UOSS) and the Quantity-Logic-Quality Paradigm (QLQP) for Production Scheduling in The Process Industries. In Multidisciplinary International Scheduling Conference Proceedings: New York, United States, 327-333.

J.D. Kelly, B.C. Menezes, F. Engineer, I.E. Grossmann, 2017, Crude-Oil Blend Scheduling Optimization of an Industrial-Sized Refinery: a Discrete-Time Benchmark. In: Foundations of Computer Aided Process Operations, Tucson, United States.

J.D. Kelly, R.E. Franzoi, B.C. Menezes, J.W. Gut, 2018a, Planuling: A Hybrid Planning and Scheduling Optimization to Schedule Slow and Plan Fast Processes. In: AICHE Annual Meeting, Pittsburgh, United States.

J.D. Kelly, B.C. Menezes, I.E. Grossmann, 2018b, Successive LP Approximation for Nonconvex Blending in MILP Scheduling Optimization Using Factors for Qualities in the Process Industry. Industrial and Engineering Chemistry Research, 57, 11076-11093.

N. Kyungseok, S. Joohyun, J.H. Lee, 2017, An Optimization Based Strategy for Crude Selection in a Refinery with Lube Hydro-Processing. Computer and Chemical Engineering, 116, 91-111.

C.A. Mendez, I.E. Grossmann, I. Harjunkoski, P.A. Kabore, 2006, A Simultaneous Optimization Approach for Off-Line Blending and Scheduling of Oil-Refinery Operations. Computer and Chemical Engineering, 30, 614-634.

B.C. Menezes, J.D. Kelly, I.E. Grossmann, 2018, Logistics Optimization for Dispositions and Depooling of Distillates in Oil-Refineries: Closing the Production Scheduling and Distribution Gap. Computer Aided Chemical Engineering, 40, 1135-1140.

Anton A. Kiss, Edwin Zondervan, Richard Lakerveld, Leyla Özkan (Eds.)
Proceedings of the 29th European Symposium on Computer Aided Process Engineering
June 16th to 19th, 2019, Eindhoven, The Netherlands. © 2019 Elsevier B.V. All rights reserved.
http://dx.doi.org/10.1016/B978-0-128-18634-3.50304-0

Flexible and efficient solution for control problems of chemical laboratories

Tibor Nagy[a*], Florian Enyedi[a], Eniko Haaz[a], Daniel Fozer[a], Andras Jozsef Toth[a] and Peter Mizsey[a,b]

[a]*Dept. of Chemical and Environmental Process Engineering, Budapest University of Technology and Economics, Budafoki út 8. F II. ,Budapest H-1111, Hungary*
[b]*Department of Fine Chemicals and Environmental Technologies, University of Miskolc, H-3515, Miskolc, Egyetemváros, Hungary*
tibor.nagy@mail.bme.hu

Abstract

In chemical laboratories there are always a permanent need to apply flexible, target oriented and efficient control solutions and data acquisition. Traditional equipment and methods utilized are however either time consuming or relatively expensive causing chemical laboratories often to be either not equipped with proper data acquisition and control systems or the installed setup is not easily reconfigured etc... Our aim is to show a simple and promising example that provides flexible and efficient control solution in chemical laboratories.

Keywords: distillation column control, fixed-, variable reflux

1. Introduction

Automation and process control is carried in increasing depth since the 50's. With such solution, operation of plants has become more precise, cost efficient and more safe. In smaller scales, such as chemical laboratories where various processes are investigated, reconfigured and developed, processes are regulated either manually or with high relative investment. Our choice for control as a demonstration is a distillation column that is particularly important since it is such a separation process that includes the common control problems and challenges that may occur in operation Luyben (1989). These common problems are found as a cause of high difference of time constants of process variables (PV), delay, process stiffness, interacting loops etc. studied by Morari et al. Grosdidier et al. (1985). Some of the mentioned problems and challenges occur only in MIMO (multiple input multiple output) control while others are also appear also in SISO systems. In our case we investigate a SISO (single input single output) case of column bottom and head temperature control that is in direct function of the product purity. With various control solutions and setup, cost-efficient and flexible operation can be carried out in laboratory sizes as well.

2. Methods and experimental Setup

In our work we define two cases:

- case A: purifying chlorobenzene from impurities

• case B: separating n-heptane and hexane

The column (see figure 1) for case A glass wall distillation column with the height of 30 cm height of Mellapak structured packing from Sulzer Chemtech and for case B it is built of two packed stages filled with random glass raschig ring type packings. In both cases the top of the column includes a gas vapor separator below the condenser. The condensed liquid is driven out of the column to a glass valve that is able vary the reflux ratio based on its on/off time ratio. The glass valve is driven by a 24 V solenoid valve and mosfet structure driven by the controller digital output. Preliminary tests showed, that the minimal time of stay/state should set to be at least 2 seconds. For normal operation we set the periodic time to be 10 seconds.

The reboiler of the distillation column is a 750 ml triple neck type round-bottom flask filled to with 500 ml of the initial liquid mixture.The heat is applied at the bottom by a heating basket by Bovimex MBO that is powered by 50 Hz 230 V AC and actuated with a solid state relay (SSR) that is regulated by the digital output of the Arduino board with the periodic time of 2 seconds. As the solid state relay is only capable of cutting power at 0 V crossings of the power source, 2 seconds is chosen as time period so the minimal resolution of the power adjustment is 1%.

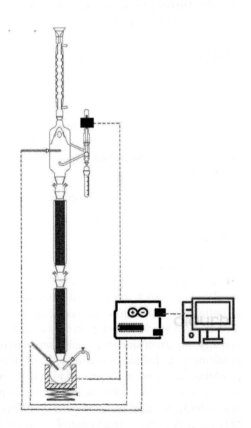

In case of distillation, the temperature measurement should be particularly precise as in such case it is the process variable (PV) that gives the basis of the quality of control. In the investigated scenario, the temperature range is set to 0-200 °C with a precision range of +/- 0.1 °C. In order to satisfy these, RTD PT100 temperature sensors are applied with two approaches. The measured signal is either applied to a MAX31865 temperature sensor amplifier DAC from Adafruit or is sent to a signal transmitter to form 4-20 mA analog signals. In case of the utilization of MAX31865, in order to receive the most accurate signals the measured resistance is

Figure 1: Distillation column and control scheme

continually compared to a lookup table including resistance and temperature value pairs of DIN 43760/IEC 751 standard. The thermometer verification is carried out by the boiling of penthane at atmospheric pressure (36.1 °C) and as the steady state is achieved, the thermometer oscillates within +/- 0.1 °C range. In case of the analog signal processing two point temperature verification is carried out, by adjusting the range minimum and span of the transmitter.

The microcontrollers utilized are Arduino UNO and Due that are particularly popular for process control studies Rubio-Gomez et al. (2019). In case of analog signal input the transmitted signal is converted to 0-3 V to match the controller input range. These analog input reads are then refined with running average that takes 200 samples and gives two digital signals/second. Such frequency of data acquisition in such system is sufficient and will not cause stability problems. In case of digital input, SPI communication is utilized. Output signals are connected to the digital output pins with PWN. The Arduino is programmable by integrated development environment (IDE) with a PC by defining the variables of the different input and output pin-s, setpoints and generating the control algorithm etc. In other cases it is also possible to retransmit the Arduino's I/O and perform the data acquisition and control at a higher level.

As for controller tuning, different methods are available. In order to test the built system and closed loop a number of tuning methods and setups are investigated, these are:

- PID applying step response (open Z-N tuning) tuning
- PID applying Matlab PID tuner
- Cohen Coon method

2.1. Open Loop Ziegler-Nichols Method

The Ziegler-Nichols Method is one of the best known tuning methods, that is based an open loop step-response function of time (Ziegler (1942)). The produced response function is approached by a first order transfer function with dead time. Applying the extracted parameters (K, τ and D_s) heuristic tables are utilized for tuning.

2.2. Cohen Coon method

The Cohen Coon tuning method also approximates the process with a first order element with dead time, but it is calculated differently (Cohen and Coon (1953)). Some literature favor the Cohen Coon method as it is more flexible with dead time compared to the Z-N method. To determine the tuning parameters just like with Z-N method, tuning tables are used.

2.3. Matlab PID tuner

Matlab PID tuner is an application of Matlab software that allows the instant tuning of a processes. The tuner application includes process identification in which the One pole + delay model is used. The application automatically calculates the linear model of the process and it provides interactive tuning where performance and robustness can be balanced. In our case the control loop is tuned for both performance and robustness to have only a moderate overshoot and after damping to settle on the new setpoint.

2.4. Model based control

A model based composition control is carried out with the model mixture of n-heptane and hexane. The internal model is divided to a reflux plan and the reflux control part. In order to achieve the planed product purity the number of theoretical plate (NTP) for the column is measured. In order to keep the composition of the distillate constant, the slope of the distillation working line must change in time and thus the reflux ratio changes as well. The reflux control part executes the process including startup, setting the correct reflux ratio to the corresponding reboiled temperature and the shut down procedure.

3. Results

In figure 2 the temperature response of the step change of 50% can be observed that is applyied to demonstrate a startup. The initial temperature is room temperature and the final value settled at ~120 °C. The tendency of the function shows dead time of ~5 minutes, while the rising time has an inflection at 26 minutes and the final value is reached at ~32 minutes.

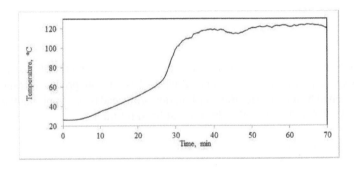

Figure 2: Response of the open loop distillation column

Based on the step response, the different tuning method show different control qualities for the column bottom temperature control (see figure 3).

Table 1: Evaluation of different tuning methods

rating parameter	Open Z-N Method	Cohen-Coon method	MATLAB PID Tuner
rise time [min]	82	23	17,5
maximal overshoot [°C]	2.65	18.2	10,6
settling time [min]	183	61.5	90
steady state error [°C]	$(-0.1) - (+0.4)$	$(+0.2) - (+0.6)$	± 0.5
ISE	5041300	2956470	2187620

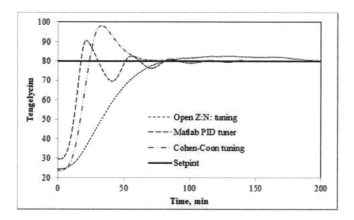

Figure 3: Column bottom temperature control with different tuning methods

The bang-bang control of the top product is shown on figure 4 for purifying chloroben-zene from impurities. In The figure the top temperature of the column to and the reboiler is shown in the function of time. A results show a typical startup and operation proce-dure, where the dead time of the column top temperature is roughly over 15 min. As the column reaches the steady state within the target temperature range the reflux valve starts it operation with a reflux ratio of 3. It can be observed, that as product is taken away from the system, the bottom temperature starts rising as the composition changes in the reboiler.

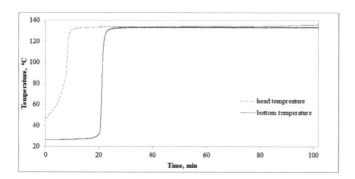

Figure 4: onoff control

The case of varying reflux composition control is shown in figure 5. The figure shows the case of separating n-hapten from hexane. The top and the bottom temperature and the reflux ratio in the function of time. After the steady state is established the reflux ratio is adjusted according to the calculated reflux plan to maintain the desired product purity. At the reflux ratio of 14 the product rate becomes so small, that the system shuts the reflux valve and the heating off, shutting the process down.

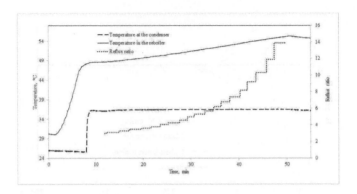

Figure 5: Variable reflux ratio control

4. Conclusions

There is permanent demand for target oriented and flexibly operated apparatus in chemical laboratories for efficient research activities. This challenge can be overcome by the utilization of microcontrollers. In many cases the additional knowledge for implementing such devices make such solution time consuming and effort intensive, while professional out of the box devices prices are high. Increasing popularity of microcontrollers these barriers are constantly stretched and by now user-friendly solutions are available. Applying such solution data acquisition and process control at laboratory scale is made practical and easy and cost efficient. As for demonstration in a chemical laboratory, batch distillation process is investigated with various cases of SISO control. In these cases The column bottom and head temperature is controlled either by the reboiler duty or the reflux ratio. Including on-line in-situ data acquisition, process analysis and controller tuning is carried out. The control structure is also investigated applying on-off, PID, and model based control strategies.

5. Acknowledgement

This publication was supported by the János Bolyai Research Scholarship of the Hungarian Academy of Sciences, ÚNKP-18-4-BME-209 New National Excellence Program of the Ministry of Human Capacities, NTP-NFTÁŰ-18-B-0154, OTKA 112699 and 128543. This research was supported by the European Union and the Hungarian State, co-financed by the European Regional Development Fund in the framework of the GINOP-2.3.4-15-2016-00004 project, aimed to promote the cooperation between the higher education and the industry.

References

G. Cohen, G. Coon, 1953. Theoretical consideration of retarded control. Trans. ASME 75, 827–834.

P. Grosdidier, M. Morari, B. R. Holt, 1985. Closed-loop properties from steady-state gain information. Industrial & Engineering Chemistry Fundamentals 24 (2), 221–235.

W. L. Luyben, 1989. Process Modeling,Simulation and Control for Chemical Engineers, 2nd Edition. McGraw-Hill Higher Education.

G. Rubio-Gomez, L. Corral-Gomez, J. A. Soriano, A. Gomez, F. J. Castillo-Garcia, 2019. Vision based algorithm for automated determination of smoke point of diesel blends. Fuel 235, 595 – 602.

N. B. Ziegler, J.G & Nichols, 1942. Optimum settings for automatic controllers. Transactions of the ASME 64:, 759âĂŞ768.

Anton A. Kiss, Edwin Zondervan, Richard Lakerveld, Leyla Özkan (Eds.)
Proceedings of the 29th European Symposium on Computer Aided Process Engineering
June 16th to 19th, 2019, Eindhoven, The Netherlands. © 2019 Elsevier B.V. All rights reserved.
http://dx.doi.org/10.1016/B978-0-128-18634-3.50305-2

Facilitating learning by failure through a pedagogical model-based tool for bioprocesses

Simoneta Caño de Las Heras[a], Björn Gutschmann[b], Krist V.Gernaey[a], Ulrich Krühne[a] and Seyed Soheil Mansouri[a,*]

[a]*Process and Systems Engineering Centre (PROSYS), Department of Chemical and Biochemical Engineering, Technical University of Denmark, Søltofts Plads, Building 229, DK-2800 Kgs. Lyngby, Denmark*
[b]*Chair of Bioprocess Engineering, Institute of Biotechnology, Technische Universität Berlin, Ackerstraße 76, ACK24, 13355 Berlin, Germany*
seso@kt.dtu.dk

Abstract

Facing uncertain situations and deciding a course of action is an opportunity to help students understand the practical implications of the theoretical knowledge and prepares them for the future challenges as engineers. However, allowing "failures and making wrong decisions" is becoming more and more challenging in industry due to safety and economic reasons and must be essentially allowed in academia. However, in academia it is also very challenging due to limitations in resources and safety concerns. Meanwhile, simulators are generally valuable tools for education that allow an intuitive learning based on action and offer the students an autonomy in their decision-making process. In spite of these advantages, the simulators commonly used in engineering education are meant for commercial purposes and incorporate a rigorous description of the process. Therefore, they are not inherently ready to be deployed in an educational program. This work presents a model-based pedagogical simulation tool, *FermProc*, having a clear objective for use in bio-manufacturing education. The in *FermProc* implemented computer-aided framework devises an active learning design based on the creation and analysis of different realistic and unrealistic scenarios such that students can safely confront common operational problems. Therefore, this model-based tool is intended to provide the users with the experience and the confidence to face and resolve uncomfortable situations encountered while operating a bioprocess.

Keywords: Learning by failure, Educational Simulator, Bioprocess simulator, Decision-making.

1. Introduction

A survey conducted by Nguyen (1998), stated that industry "requires an engineering graduate with an equivalent of a person with 10 years of experience" and "a person with motivation, who is creative, a good team member, risk-taking and decisive". Up-to-date, the current changing technological landscape makes it even more necessary to provide engineering graduates with these qualities. Most of these attributes are strongly connected to training. However, providing the opportunity for such training can be difficult considering the increasing number of students, the time restrictions, and the available resources. Simulators, with learning based on action and enhancing exploration, can help in fulfilling this need (Balamuralithara and Woods 2009). However, the question to answer is if "simulators can solve the problems we are interested in?" Even so, the majority of the commercial simulators used in engineering education lack a pedagogical aim in the design; which entails limitations in the learning process. For example, in the case of infeasible operational conditions, commercial simulators are not prepared to provide information about the corresponding inappropriateness. On the other hand, the acquisition of knowledge and skills can be facilitated through an enjoyable experience and considering the new habits and interests of the students (Kiili 2005).

In this work, different approaches inside a model-based software that allows experiencing failure with the support of learning content for bioprocesses are presented. Furthermore, game-based elements are implemented inside the computational tool to encourage an enjoyable experience and the conceptualisation of complex knowledge. Therefore, the current version of the virtual software, *FermProc*, is explained, with its design requirements and advantages for allowing the learning by failure.

2. *FermProc*: A pedagogical model-based tool for bioprocesses

FermProc is a software designed with prime pedagogical objectives based on a previously developed methodology (Caño et al. 2018). Hence, it includes the simulation of mathematical models, as well as the possibility to modify them, learning-hints, questionnaires, mini-games, and the possibility to confront realistic technical problems. Furthermore, the process simulation is developed in an open-source platform, thus offering the possibility for students to access the mathematical models and gain a profound understanding about their complexity.

2.1 Design requirements of FermProc

This model-based tool is based on three pillars: 1) a careful learning design, 2) a motivational approach based on the use of gamification and, 3) the use of template-based models that can be displayed, reused and modified (Fedorova, Sin, and Gani 2015). Template-based modelling is an approach that requires the development of models on divisible pieces that compose a template. Through this approach, a model is decomposed into three sets of equations; i) balance equations, ii) constitutive equations and iii) connection and conditional equations (Cameron and Gani 2011). Each existing concept can be treated as a building block which can be combined with other suitable building blocks to create a new model. Meanwhile, the rational combination of equations to create a new model requires previous knowledge of the user or it can evolve to the creation of an additional process model. In both scenarios, the users can explore and learn about the construction of mathematical models. Furthermore, due to the educational aims, this software provides an organized and versatile learning experience in a user-friendly environment that engages the user in the learning process. This is promoted by the presence of rules and explanation across the tool.

In order to fulfil the design requirements, a relational database is developed in SQLite. This database scheme with the corresponding data types can be seen in Figure 1. The database collects and reuses interrelated rules for a comprehensive modification of the kinetic models (shown in Figure 2). Moreover, it uses the process conditions to correlate an "expert system" with problem and solutions. The problem-solution feature in *FermProc* will be further explained in Section 2.2.

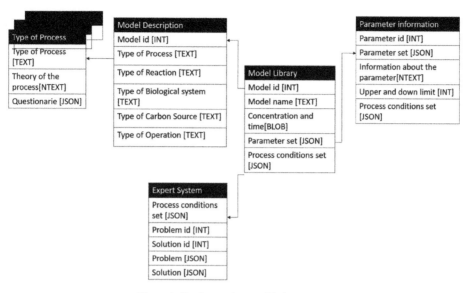

Figure 1. Database scheme with data types.

2.2 *Advantages of FermProc*

FermProc is a platform developed "*by* students *for* students" with the support of highly skilled teachers. This is done through involving the students in all the steps of the process as co-designer and co-developers of the computational tool. Therefore, in contrast with other commercial simulators, *FermProc* has clear learning goals. At its current stage, they are the acquisition of i) knowledge for the understanding of the mathematical model associated with a bioprocess, ii) skills in the evaluation of fictional and non-fictional scenarios in a bioconversion and iii) competences in meaningful decision-making across the design of the bioprocess. Moreover, this software has the objective to promote creativity and critical-thinking as it allows to make and take control of mistakes. This is done by the support of theory in every decision taken by the user, the need for applying their knowledge in different scenarios and a continuous assessment of the choices of the user. In the current stage of development, this is put into practice with the modification of the simulated model and other activities implemented as mini-games or multimedia content.

The feature for the modification of a chosen model can be seen in Figure 2. Initially, the model is selected based on a series of questions related to the bioprocess to be simulated. These choices are supported by theoretical content and the structure of the selected model is explained with a visual help or a video. Once the model is selected and solved inside the tool, the user can select a parameter or operational condition from the mathematical model and modify different values in a range depending on the parameter or selected conditions. If the modified value is out-of-range due to, for example, a thermodynamic restriction, the software will display a new window with the theoretical explanation and the corresponding literature reference. This allows enhanced awareness of the feasible range of values and provides a source of information that the users can rely on to intensify knowledge about the applied model.

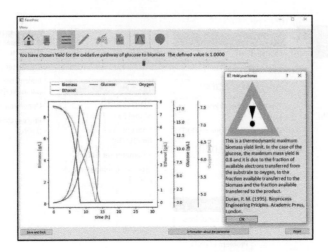

Figure 2. Example of the screen for the modification of the model in FermProc with a popup window that will appear when an out-of-range value is selected.

In addition, other activities are implemented to engage the creativity of the user and to make complex knowledge more approachable with the inclusion of game elements. Some examples of the implemented mini-games inside *FermProc* are:

- A *memory game* for matching combinations of definitions, nomenclature and international units of the parameters involved in a complex fermentation model. With different levels of difficulty, the user starts connecting nomenclature and definition and advances to connect nomenclature and units.

- A *"what is what?"* for the analysis of graphs. The software solves the model, chosen by the user, with different initial model parameter values (1%, 25%, 50% and 100%) and generates a graph for each simulation. Then, the user is required to choose the graph with the highest value of the parameter. In the current state, the parameters are associated with the biomass growth and the choice is made based on the plot of the evolution of the biomass concentration against the time. This system can provide the user with the training for the critical thinking needed to associate the variation of parameters that are not directly represented in the graphs but have an effect in the process. In addition, the understanding of mechanistic models is strongly intensified using this approach.

- An *operational problem*. In this game, the user needs to recognize an operational issue and select a solution from a problem and solution data set. This can be seen in Figure 3 and Figure 4. The operational problem is randomly picked inside a database (in Figure 1) considering the process modelled. For example, in Figure 3, the user is facing a very low *Saccharomyces cerevisiae* growth rate and he/she needs to use the available information about glucose, ethanol, oxygen and the set value temperature and the measured temperature to identified possible problems in the operation process. In this specific case, the low efficiency of the microorganisms is due to a low fixed temperature value (of 15°C) when yeast has an optimal growth rate at 30°C (Sonnleitner and Käppeli 1986).

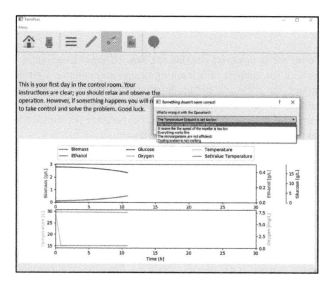

Figure 3. The screen of the first step of the operational problem in which the problem needs to be categorized.

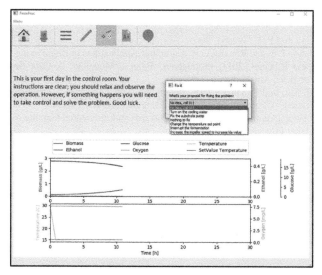

Figure 4. The second step of the operational process mini-game. A solution for the issue needs to be proposed based on the available solutions. Due to lack of space, the screen for the modification of the parameter (such as, like in this case, the change of the temperature setpoint and the choices based on the type of microorganism you are using) hasn't been included in this manuscript

The problem-solution database, that is part of the Figure 1, is created with the experiences gained by industry and academia, and therefore the users can confront real-life situations. It is expected that as the users are solving those operational issues, they are building confidence in its decision-making abilities and developing a skilled mind-set with awareness of the most common failures and the capability of solving hand-on problems.

Therefore, *FermProc* is designed considering how to allow meaningful failures and to make choices with the support of a learning design and content. Consequently, as the user goes through the software, he/she develops the abilities to analyse and confront problematic situations and make decisions about the process. It is important to highlight that *FermProc* is still in development and a user experience is planned for summer 2019 to corroborate the design and features of the software.

3. Conclusions

In order to allow students to gain experience without exposing themselves or put the installations at risk and with a limited consumption of time and resources, new and innovative model-based tools need to be developed. In the case of bioprocesses, several technical problems can happen during the operation that will require a fast analysis and critical response. Considering those issues, *FermProc,* which is a model-based tool for the simulation of bioprocesses, is developed. It integrates learning design and game elements. In its current state of development, *FermProc* is prepared to facilitate comprehension of kinetic parameters and their limits. *FermProc* also has implemented other activities such as problem-solution database that will help students to face technical problems while gaining theoretical knowledge in a safe environment.

References

B. Balamuralithara and P. C. Woods, 2009, "Virtual Laboratories in Engineering Education: The Simulation Lab and Remote Lab" Computer Applications in Engineering Education.

I. T. Cameron and R. Gani, 2011, Product and Process Modeling: A Case Study Approach. Elsevier B.V., Amsterdam.

M. Fedorova, G. Sin, and R. Gani, 2015, "Computer-Aided Modelling Template: Concept and Application." Computers and Chemical Engineering, 83: 232-247

K. Kiili, 2005. "Digital Game-Based Learning: Towards an Experiential Gaming Model." Internet and Higher Education 8(1):13–24.

S. Caño, et al. , 2018, "A Methodology for Development of a Pedagogical Simulation Tool Used in Fermentation Applications." Computer Aided Chemical Engineering 44:1621–26.

D. Q. Nguyen, 1998, "The Essential Skills and Attributes of an Engineer: A Comparative Study of Academics, Industry Personnel and Engineering Students." Business 2(1):65–76.

B. Sonnleitner, and O. Käppeli, 1986, "Growth of Saccharomyces Cerevisiae Is Controlled by Its Limited Respiratory Capacity: Formulation and Verification of a Hypothesis." Biotechnology and Bioengineering 28(6):927–37.

Author Index

Printed and bound by CPI Group (UK) Ltd, Croydon, CR0 4YY

03/10/2024

01040329-0005